ASO-S Mission: Inflight Performance and First Results

Weiqun Gan • John Leibacher
Cristina H. Mandrini • Lidia van Driel-Gesztelyi
Editors

ASO-S Mission: Inflight Performance and First Results

Previously published in *Solar Physics*
Volume 298–300, Issue 1–12, 2023–2025

Springer

Editors
Weiqun Gan
Purple Mountain Observatory
Chinese Academy of Sciences
Nanjing, China

Cristina H. Mandrini
Institute of Astronomy and Space Physics
Ciudad Autónoma Buenos Aires, Argentina

John Leibacher
National Solar Observatory
Tucson, AZ, USA

Lidia van Driel-Gesztelyi
Mullard Space Science Laboratory
Univ College London, Holmbury House
Dorking, UK

Spin-off from journal: "Solar Physics" Volume 298–300, Issue 1–12, 2023–2025

ISBN 978-94-024-2331-0

Cover illustration: The Lyman-alpha solar disk image captured by SDI/LST/ASO-S on 26 October 2022 at 17:02:36 UT

This Springer imprint is published by the registered company Springer Nature B.V.
The registered company address is: Van Godewijckstraat 30, 3311 GX Dordrecht, The Netherlands

Contents

Part 2: FMG Related Research

Solar Physics (2025) 300:59
https://doi.org/10.1007/s11207-025-02473-5

EDITORIAL

ASO-S Mission: Inflight Performance and First Results

W.Q. Gan[1] · John Leibacher[2,3,4] · Cristina H. Mandrini[5] · Lidia van Driel-Gesztelyi[6,7]

Accepted: 23 April 2025 / Published online: 6 May 2025

Abstract
China's first comprehensive solar physics mission, the Advanced Space-based Solar Observatory (ASO-S), was successfully launched on 9 October 2022. The mission focuses on exploring the connections between the solar magnetic field, solar flares, and coronal mass ejections (CMEs). ASO-S carries three instruments: the Full-disk solar vector MagnetoGraph (FMG), the solar Hard X-ray Imager (HXI), and the Lyman-alpha Solar Telescope (LST). The unique combination of these payloads allows simultaneous observations of the vector magnetic field of the full Sun, high-energy imaging spectroscopy of solar flares, and formation and evolution of CMEs on the disk and in the inner corona. This topical collection, along with other publications, provides an overview of ASO-S's achievements over the first one and a half years, including mission validation, payload calibration, data flow description, analysis software demonstration, and numerous investigations based on ASO-S observed data. We hope this collection will be valuable for researchers seeking to understand the mission or utilize the data.

Keywords ASO-S mission · Solar magnetic field · Solar flare · Coronal mass ejection

1. Topical Collection Summary

The general descriptions of ASO-S up to Phase-B were published in an early special issue of *Research in Astronomy and Astrophysics* (**19**, No. 11, 2019). The subsequent progress, particularly the performance prior to launch and in orbit, along with data quality and investigations based on the first nearly one and a half years of observations, are the primary focus

✉ W.Q. Gan

1 Purple Mountain Observatory, Chinese Academy of Sciences, Nanjing 210023, China

2 Université Paris-Saclay, CNRS, Institut d'Astrophysique Spatiale, Orsay, France

3 Lunar and Planetary Laboratory, University of Arizona, Tucson, AZ, USA

4 National Solar Observatory, Boulder, CO, USA

5 Instituto de Astronomía y Física del Espacio, IAFE, Consejo Nacional de Investigaciones Científicas y Técnicas - Universidad de Buenos Aires, Buenos Aires, Argentina

6 University College London, Mullard Space Science Laboratory, Dorking, UK

7 Konkoly Observatory, Budapest, Research Centre for Astronomy and Earth Sciences, Budapest, Hungary

of this topical collection. Four natural categories have been established: an overview of the mission and instruments, research related to FMG, research related to HXI, and research related to LST.

For the overview of the mission and instruments, there are five articles in total. Gan et al. (2023) presented the final technical status of the entire mission just before the launch, along with the first-light results from the three payloads. Bai et al. (2024), Su et al. (2024), and Chen et al. (2024) provided detailed descriptions of the in-flight performance and calibration of FMG, HXI, and LST, respectively. Huang et al. (2024) showcased the Science Operation and Data Analysis Center of ASO-S, detailing data products, analysis software, and user-friendly interfaces.

For the research related to FMG, nine articles are included. Among them, five focus on detailed calibrations, such as correcting the influence of orbital velocity (Liu et al., 2023), validating the most magnetic-sensitivity spectral position (Liu et al., 2024), improving the weakening magnetic field within sunspot umbrae (Xu et al., 2024b), and comparing FMG observations with other instruments (Xu et al., 2024a; Chen et al., 2025). Idrees et al. (2024) used FMG data to study the detailed decay of a beta-configuration sunspot group. Additionally, magnetic helicity derived from FMG was shown to be consistent with that obtained from the Helioseismic and Magnetic Imager (HMI: Yang et al., 2025). Furthermore, two studies utilized multi-wavelength observations, including FMG, to investigate the initiation process of a large-scale filament eruption (Sun et al., 2024) and the formation of multiple magnetic flux ropes leading to an eruption (Song et al., 2024).

Nine papers related to HXI are included in this collection. The temporal behavior of solar flares, including the Neupert effect and quasi-periodic pulsations (QPP), was investigated by Li et al. (2024) and Shi et al. (2024), respectively. Mrozek et al. (2024) and Ryan et al. (2024) performed joint diagnostics using HXI and the Spectrometer/Telescope for Imaging X-rays (STIX) to determine the three-dimensional geometry of solar flares. Shamsutdinova et al. (2024) combined observations from the Solar Radioheliograph (SRH) with HXI data to diagnose the source of a limb flare. Gou et al. (2024) utilized HXI data in their study of coronal jets. In terms of technical advancements, Yu et al. (2024) provided detailed information about the design and implementation of the solar aspect system for HXI. Xia et al. (2024) developed a novel HXR image reconstruction algorithm based on machine learning for HXI. Additionally, Li et al. (2025) completed a fundamental work on spectral cross-calibration between HXI and other X-ray missions.

There are seven articles related to LST. A notable work is about the white-light flare phenomenon (WLF). Jing et al. (2024) conducted a statistical analysis of WLFs observed by the White-light Solar Telescope (WST) of LST, demonstrating that WLFs are more common than previously thought. An individual WLF was also studied by Li et al. (2024). A series of events observed by LST was studied together with other datasets, such as a failed prominence eruption (Xue et al., 2024), the interaction of two prominences (Wei et al., 2024), and the asymmetry in flare loop footpoints (Shi et al., 2024). Ying et al. (2024) modeled a CME and its driven shock via a 3D MHD simulation and synthesized Lyman-alpha and white light images of the CME and shock within a few solar radii. Lu et al. (2024) outlined the principles for automatically detecting solar flares based on observations from the Solar Disk Imager (SDI) of LST.

By February 2025, in addition to the 30 papers in this topical collection, over 100 other publications directly related to ASO-S have been published elsewhere, such as in the *Astrophysical Journal*, *Astronomy & Astrophysics*, and *Monthly Notices of the Royal Astronomical Society*, and so on. Together, these articles provide a comprehensive overview of ASO-S's achievements during its early operational phase.

2. Topical Collection Articles

Bai, X.Y., Deng, Y.Y., Zhang, H.Y., Yang, J.F., Li, F., Su, J.T., et al.: 2024, Calibration and performance of the Full-Disk Vector MagnetoGraph (FMG) on board the Advanced Space-Based Solar Observatory (ASO-S), *Sol. Phys.* **299**, 157. DOI.

Chen, B., Feng, L., Zhang, G., Li, H., He, L.P., Song, K.F., et al.: 2024, Inflight performance and calibrations of the Lyman-Alpha Solar Telescope on board the Advanced Space-Based Solar Observatory, *Sol. Phys.* **299**, 118. DOI.

Chen, J., Su, J.T., Xie, W.B., Deng, Y.Y., Bai, X.Y., Hu, Z.Y., et al.: 2025, Direct measurement of the longitudinal magnetic field in the solar photosphere with Zeeman effect, *Sol. Phys.* **300**, 49. DOI.

Gan, W.Q., Zhu, C., Deng, Y.Y., Zhang, Z., Chen, B., Huang, Y., et al.: 2023, The Advanced Space-Based Solar Observatory (ASO-S), *Sol. Phys.* **298**, 68. DOI.

Gou, T.Y., Liu, R., Su, Y., Veronig, A.M., Pan, H.Y., Luo, R.B., et al.: 2024, High-resolution observation of blowout jets regulated by sunspot rotation, *Sol. Phys.* **299**, 99. DOI.

Huang, Y., Li, Y.P., Liu, S., Su, J.T., Feng, L., Li, Y., et al.: 2024, Science Operation and Data Analysis Center of the Advanced Space-Based Solar Observatory (ASO-S) Mission, *Sol. Phys.* **299**, 135. DOI.

Idrees, S., Su, J.T., Chen, J., Deng, Y.Y.: 2024, Investigation of decaying β-configuration sunspot in active region NOAA 13229, *Sol. Phys.* **299**, 91. DOI.

Jing, Z.C., Li, Y., Feng, L., Li, H., Huang, Y., Li, Y.P., et al.: 2024, A statistical study of solar white-light flares observed by the White-Light Solar Telescope of the Lyman-Alpha Solar Telescope on the Advanced Space-Based Solar Observatory (ASO-S/LST/WST) at 360 nm, *Sol. Phys.* **299**, 11. DOI.

Li, D., Dong, H.Y., Chen, W., Su, Y., Huang, Y., Ning, Z.J.: 2024, A statistical investigation of the Neupert effect in solar flares observed with ASO-S/HXI, *Sol. Phys.* **299**, 57. DOI.

Li, Q., Li, Y., Su, Y., Song, D.C., Li, H., Feng, L., et al.: 2024, Spectral and imaging observations of a C2.3 white-light flare from the Advanced Space-Based Solar Observatory (ASO-S) and the Chinese Hα Solar Explorer (CHASE), *Sol. Phys.* **299**, 73. DOI.

Li, Z.T., Su, Y., Liu, W., Chen, W., Yu, F., Gan, W.Q.: 2025, Spectral cross-calibrations of ASO-S/HXI with other X-ray missions, *Sol. Phys.* **300**, 56. DOI.

Liu, S., Su, J.T., Bai, X.Y., Deng, Y.Y., Chen, J., Song, Y.L., et al.: 2023, The magnetic-field calibration of the Full-Disk Magnetograph onboard the Advanced Space Based Solar Observatory (ASO-S/FMG), *Sol. Phys.* **298**, 141. DOI.

Liu, S., Su, J.T., Bai, X.Y., Deng, Y.Y., Chen, J., Song, Y.L., et al.: 2024, A Study on magnetic-sensitivity wavelength position of the working line used by the Full-Disk Magnetograph onboard the Advanced Space Based Solar Observatory (ASO-S/FMG), *Sol. Phys.* **299**, 70. DOI.

Lu, L., Tian, Z.Y., Feng, L., Shan, J.H., Li, H., Su, Y., et al.: 2024, Automatic solar flare detection using the Solar Disk Imager onboard the ASO-S mission, *Sol. Phys.* **299**, 72. DOI.

Mrozek, T., Li, Z.T., Karlický, M., Chrysaphi, N., Su, Y., Chen, W., et al.: 2024, Kink-and-Disconnection Failed Eruption in 3D, *Sol. Phys.* **299**, 81. DOI.

Ryan, D.F., Massa, P., Battaglia, A.F., Dickson, E.C.M., Su, Y., Chen, W., et al.: 2024, Triangulation of hard X-ray sources in an X-Class solar flare with ASO-S/HXI and Solar Orbiter/STIX. *Sol. Phys.* **299**, 114. DOI.

Shamsutdinova, J.N., Kashapova, L.K., Li, Z.T., Su, Y., 2024, Spatial and spectral evolution of microwave and X-ray sources during the solar limb flare on February 5, 2023, *Sol. Phys.* **299**, 80. DOI.

Shi, F.P., Li, D., Ning, Z.J., Warmuth, A., Chen, W., Su, Y., et al.: 2024, Multiwavelength observations of quasiperiodic pulsations in the impulsive phase of an eruptive flare with the Hard X-Ray Imager on board ASO-S and other instruments, *Sol. Phys.* **299**, 30. DOI.

Shi, G.L., Feng, L., Chen, J., Ying, B.L., Li, S.T., Li, Q., et al.: 2024, Asymmetric hard X-ray radiation of two ribbons in a thermal-dominated C-class flare, *Sol. Phys.* **299**, 104. DOI.

Song, Y.L., Su, J.T., Zhang, Q M., Zhang, M., Deng, Y.Y., Bai, X.Y., et al.: 2024, Observation of a large-scale filament eruption initiated by two small-scale erupting filaments pushing out from below, *Sol. Phys.* **299**, 85. DOI.

Su, Y., Zhang, Z., Chen, W., Chen, D.Y., Yu, F., Hu, Y., et al.: 2024, The tests and calibrations of the Hard X-ray Imager aboard ASO-S, *Sol. Phys.* **299**, 153. DOI.

Sun, Z., Li, T., Hou, Y.J., Tian, H., Wu, Z.Q., Li, K., et al.: 2024, The solar origin of an intense geomagnetic storm on 1 December 2023: successive slipping and eruption of multiple magnetic flux ropes, *Sol. Phys.* **299**, 93. DOI.

Wei, H.Y., Huang, Z.H., Zhang, Q.M., Li, Y., Fu, H., Xiong, M., et al.: 2024, Rising of two crossing prominences and the resulting oscillations, *Sol. Phys.* **299**, 64. DOI.

Xia, Y.H., Su, Y., Liu, H., Yu, W H., Li, Z.T., Chen, W., et al.: 2024, A new solar hard X-ray image reconstruction algorithm for ASO-S/HXI based on deep learning, *Sol. Phys.* **299**, 158. DOI.

Xu, H.Q., Su, J.T., Liu, S., Deng, Y.Y., Yang, S.B., Bai, X.Y., et al.: 2024a, Comparison of line-of-sight magnetic field observed by ASO-S/FMG, SDO/HMI and HSOS/SMAT, *Sol. Phys.* **299**, 17. DOI.

Xu, H.Q., Su, J.T., Liu, S., Deng, Y.Y., Bai, X.Y., Chen, J., et al.: 2024b, Correction for the weakening magnetic field within the sunspot umbra observed by ASO-S/FMG, *Sol. Phys.* **299**, 82. DOI.

Xue, J.C., Feng, L., Li, H., Zhang, P., Chen, J., Shi, G.L., et al.: 2024, Association between a failed prominence eruption and the drainage of mass from another prominence, *Sol. Phys.* **299**, 89. DOI.

Yang, S.B., Liu, S., Su, J.T., Deng, Y.Y.: 2025, Comparison of relative magnetic helicity flux calculation results based on the line-of-sight magnetograms of ASO-S/FMG and SDO/HMI, *Sol. Phys.* **300**, 26. DOI.

Ying, B.L., Shi, G.L., Feng, L., Lu, L., Xue, J.C., Li, S.T., et al.: 2024, Parameter effects on the total intensity of H I Lyα line for a modeled coronal mass ejection and its driven shock, *Sol. Phys.* **299**, 92, DOI.

Yu, J.R., He, Y.H., Tao, J.Y., Su, Y., Zhang, Z., Yang, J.F., et al.: 2024, The solar aspect system of the Hard X-ray Imager onboard ASO-S, *Sol. Phys.* **299**, 122. DOI.

Declarations

Competing Interests The authors declare no competing interests.

Part 1: Overview and Inflight Performance

Solar Physics (2023) 298:68
https://doi.org/10.1007/s11207-023-02166-x

RESEARCH

The *Advanced Space-Based Solar Observatory* (ASO-S)

Weiqun Gan[1] · Cheng Zhu[2] · Yuanyong Deng[3] · Zhe Zhang[1] · Bo Chen[4] · Yu Huang[1] · Lei Deng[2] · Haiyan Wu[5] · Haiying Zhang[6] · Hui Li[1] · Yang Su[1] · Jiangtao Su[3] · Li Feng[1] · Jian Wu[1] · Jijun Cui[5] · Chi Wang[5] · Jin Chang[1] · Zengshan Yin[2] · Weiming Xiong[5] · Bin Chen[5] · Jianfeng Yang[7] · Fu Li[7] · Jiaben Lin[3] · Junfeng Hou[3] · Xianyong Bai[3] · Dengyi Chen[1] · Yan Zhang[1] · Yiming Hu[1] · Yaoming Liang[5] · Jianping Wang[2] · Kefei Song[4] · Quanfeng Guo[4] · Lingping He[4] · Guang Zhang[4] · Peng Wang[4] · Haicao Bao[2] · Caixia Cao[2] · Yanping Bai[2] · Binglong Chen[2] · Tao He[2] · Xinyu Li[2] · Ye Zhang[2] · Xing Liao[2] · Hu Jiang[2] · Youping Li[1] · Yingna Su[1] · Shijun Lei[1] · Wei Chen[1] · Ying Li[1] · Jie Zhao[1] · Jingwei Li[1] · Yunyi Ge[1] · Ziming Zou[5] · Tai Hu[5] · Miao Su[5] · Haidong Ji[8] · Mei Gu[9] · Yonghuang Zheng[10] · Dezhen Xu[11] · Xing Wang[12]

Received: 27 January 2023 / Accepted: 13 May 2023 / Published online: 24 May 2023

Abstract
The *Advanced Space-based Solar Observatory* (ASO-S) was successfully launched at 23:43 UT on 8 October 2022. Here we describe the final technical status of the whole mission right before the launch, including the spacecraft platform and the three onboard payloads. The mission's science goals, organizations, preliminary performance in the commissioning phase, and the first-light results of the three payloads are briefly presented.

1. Introduction

The *Advanced Space-based Solar Observatory* (ASO-S, Chinese nickname *Kuafu*-1), the first comprehensive Chinese dedicated solar observatory in space (Gan et al., 2019a,c; Gan, Feng, and Su, 2022; Gan et al., 2019b), was successfully launched by a LM-2D rocket at 23:43 UT on 8 October 2022 from the Jiuquan Satellite Launch Center, Gansu Province, China. ASO-S was supported within the framework of the newly established Strategic Priority Program on Space Science of the Chinese Academy of Sciences (CAS), and underwent a conceptual study (2011 – 2013), background study (2014 – 2016), and comprehensive evaluation (2016 – 2017). At the end of 2017, ASO-S was formally approved by the CAS. Since then, ASO-S entered into the engineering phases via the official Phase-B (January 2018 – April 2019), Phase-C (May 2019 – August 2021), and Phase-D (September 2021 – September 2022).

The primary scientific objectives of the ASO-S can be simply expressed by "1M2B", where M stands for solar magnetic field and 2B for the two most violent eruptions on the Sun – solar flares and coronal mass ejections (CMEs), that is to observe and study their formation, evolution, interaction, and mutual relevance. Specifically, the three relationships between solar magnetic field and solar flares, solar flares and CMEs, and CMEs and solar magnetic field are the focus of attention for ASO-S. To realize these scientific goals, ASO-S

Extended author information available on the last page of the article

Table 1 Organization of ASO-S mission.

Mission level	
Chief Scientist	Weiqun Gan (Purple Mountain Observatory: PMO)
Chief Designer	Jijun Cui (National Space Science Center: NSSC)
Chief Administrator	Chi Wang (NSSC)
Principal investigators	
Spacecraft	Cheng Zhu, Lei Deng, Haicao Bao, Caixia Cao (Innovation Academy for Microsatellites: Microsat)
FMG	Yuanyong Deng (National Astronomical Observatories: NAOC), Haiying Zhang (Nanjing Institute of Astronomical Optics & Technology: NIAOT), Jianfeng Yang (Xi'an Institute of Optics and Precision Mechanics: XIOPM), Jiangtao Su (NAOC)
HXI	Zhe Zhang, Yang Su, Jian Wu (PMO)
LST	Bo Chen (Changchun Institute of Optics, Fine Mechanics and Physics: CIOMP), Hui Li, Li Feng (PMO)
SMOC	Ziming Zou, Haiyan Wu, Tai Hu (NSSC)
SODC	Hui Li, Yu Huang, Youping Li (PMO)

employs three payloads: the *Full disc vector MagnetoGraph* (FMG), for measuring the full-disk solar vector magnetic field; the *Hard X-ray Imager* (HXI), for observing non-thermal signals from ≈ 30 to 200 keV emitted by energetic electrons; and the *Lyman-α Solar Telescope* (LST), for observing both the solar disk and inner corona up to 2.5 solar radii in Lyman-α and visible light. The characteristics of the mission, beside the exclusive scientific objectives of "1M2B" and therefore the unique deployment of the payloads, also include a single near-Earth-orbit platform to perform simultaneous observations of full-disc solar vector magnetic field, nonthermal hard X-ray spectral imaging, and the formation as well as the propagation of CMEs from the disc to the nearby corona in dual wavebands. Each payload has its own special features. For instance, FMG balances both high sensitivity and high temporal resolution; HXI obtains accurate flux and more spatial-modulation Fourier components than a similar instrument before such as *Hard X-ray Telescope* onboard *Yohkoh* (Kosugi et al., 1991), and LST will take seamless observations in Lyman-α waveband from the disk center to 2.5 solar radii for the first time. Some comparisons between the ASO-S payload and other missions' payloads were made by Gan et al. (2015). The organization of the mission is briefly listed in Table 1.

The early introductory publications related to ASO-S until the end of Phase-B were collected by Gan et al. (2019b). Here we provide the status of the mission since then and the final parameters just before the launch, briefly mention the in-flight performance, present roughly a part of the first light of the payloads, and discuss briefly the scientific return and plans in the near future.

2. Spacecraft

The architecture of the ASO-S is composed of two modules: platform module and payload module. The platform module contains various of instruments belonging to seven different subsystems of the spacecraft (SC). The payload module contains the three payloads. The two modules are designed such that multiple instruments from different payloads can work most tranquilly and stably. That is to say, the state-of-the-art fabrication of the SC makes it

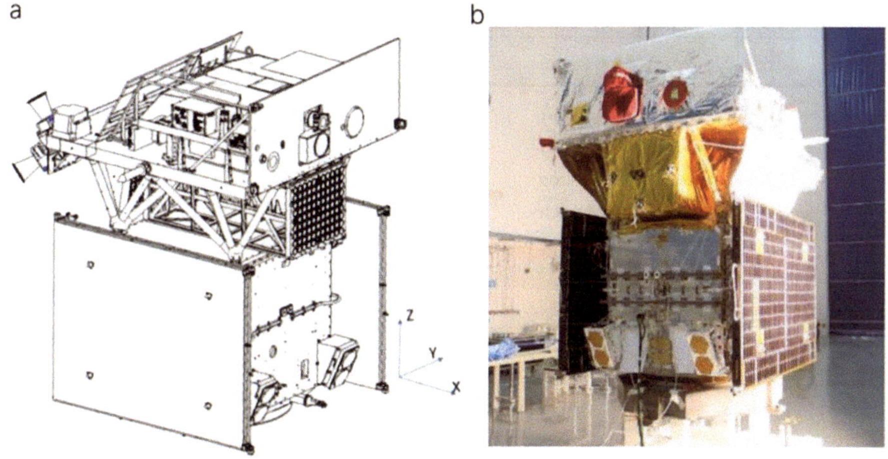

Figure 1 ASO-S designed (**a**) and actual configuration before the launch (**b**). The envelope is 1953 mm × 9184 mm × 2450 mm in the X-, Y-, and Z-dimensions.

possible for little vibration influence on the payload functioning due to oscillation sources from solar arrays and attitude controlling, for example.

The attitude–orbit control subsystem (AOCS) includes attitude-sensing devices such as the star tracker, gyroscope, Sun sensors, and attitude actuators such as momentum wheels, magnetic-torque device, and magnetometer. The effective combination of attitude devices and innovative algorithms of the AOCS contributes to a good performance: the pointing accuracy is better than $36''$ (3σ), and the pointing stability is better than $1.8''\ \mathrm{s}^{-1}$ (3σ). Such attitude specifics can meet the requirements of efficient payload observations, together with the self-carried stability subsystem of the payloads. Another feature of the SC is that phased arrays of antennas are introduced for data transmission, which make the data transmission both efficient and the state of the art. The SC has two solar arrays deployed along the +Y-axis and −Y-axis in satellite-body fixed coordinates (cf. Figure 1). The envelope of the SC is 1953 mm × 9184 mm × 2450 mm in X-, Y-, Z-dimensions when it flies freely in space. The SC mass is about 857 kg. Its main working attitude is Sun pointing. The X-axis of the spacecraft body fixed coordinate system is along the Sun direction, and the Z-axis parallel to the normal of Earth's ecliptic plane.

The SC is now in a Sun-synchronous orbit with an altitude of about 720 km above the Earth's surface, and the descending node time is at 0600 local time. Such an orbit makes it more efficient to observe the Sun, although it causes some eclipse in each orbit from May to August, and the maximum eclipse time is 18 minutes. The atmosphere at an altitude of 720 km is so sparse that it will not shorten the lifespan of SC. The design lifespan is not less than four years. The final specifications for the ASO-S are listed in Table 2. The designed and actual configuration before the launch are shown in Figure 1.

3. Payloads

The design and test results of the payload engineering model (EM or phase-B study) have been introduced in detail (Deng et al., 2019; Li et al., 2019; Chen et al., 2019; Zhang et al., 2019a,b). Here, we mainly present the updated information after EM, including the technical

Table 2 Specifications of the ASO-S spacecraft.

Items		Specifications
Orbit	Orbit type	Sun-synchronous
	Altitude	720 ± 5 km
	Inclination	98.27°
	Eccentricity	0
	Local time for SC passing the descending node	0600 local
Attitude	Pointing	Sun pointing three-axis stabilized
	Accuracy	better than $36''$ (3σ)
	Stability	better than $1.8''$ s^{-1} (3σ)
SC size	Envelope before launch	$\leq \varphi$ 2900 mm × 2760 mm
Timing of SC	Timing accuracy	≤ 5 ms
Data transmission and storage	Data transmission	X band, transmission rate ≥ 1000 Mbps
	Allowed data storage	≥ 4 Tbits
Energy	Solar array	GaAs cell, solar array area ≥ 6 m^2
	Battery	Li battery pack $\geq$90 Ah
TT&C (Tracking, Telemetry and Control)	Data rate	S band upload rate: 2000 bps, download data rate: 8192 bps
	Orbit determination accuracy	better than 50 m (1σ)
Power	Average power consumption	≤ 980 W
	Peak power consumption	≤ 1417 W
Mass	Total mass	857 kg
Lifespan	Design lifespan	≥ 4 years
Reliability	Reliability	≥ 0.6 (at the end of nominal lifespan)

status change in the later qualification model (QM) and the flight model (FM), as well as some test results.

3.1. FMG

FMG is a birefringent-filter-based magnetograph, aiming to measure the photospheric magnetic fields of the entire solar disk with high spatial and temporal resolution, and high magnetic sensitivity. Most of the technical status of QM and FM are consistent with the EM (Deng et al., 2019; Su et al., 2019a), except the *Image Detecting System* (IDS) in the *Data Acquisition, Processing and Control System* (DAPCS) panel. We use another kind of CMOS sensor, named KP1010, in the QM and FM. The total array of KP1010 is 10k × 10k, the size of a single pixel is 6.4 μm. In our application, we use just 8k × 8k pixels in its central part. After exposure, we first carry out 2 × 2 binning onboard and then output 4k × 4k image. The performance of the KP1010 (after 2 × 2 binning) is listed in Table 3.

Table 3 Main performance specifics for the FM of FMG.

Items	Measurements
FOV	33.67′
Spatial resolution	1.04″
FWHM of Birefringent (BF) filter (@ 532.42 nm)	0.01014 nm
Stability of the BF passband	0.0016 nm
Out of band suppression of BF	13.0%
Sensitivity of polarization measurement	2650:1 @ 1080 s (deep integration mode)
	1020:1 @ 120 s (routine observation mode)
Max frame rate of KP1010 (after 2 × 2 binning)	10 fps
Full well capacity of KP1010	42,310 e^-
Photo response non-uniformity of KP1010	0.64%
Readout noise of KP1010	25.38 e^- $pixel^{-1}$
Onboard data-processing capability of the DAPCS	16.13 fps
Response time for polarization analyzer	$\leq$ 50 ms
Adjusting range of pointing	Y: $\pm 2.5'$, Z: $\pm 1.75'$
Accuracy of image stabilization	$\leq 0.25''$ (1σ) per 30 s

We conducted the test observations with the QM from 25 July to 14 August 2020 at the Huairou Solar Observing Station (HSOS), NAOC. The main purpose of this test observation was to evaluate the performance of the magnetic-field measurement, as we have changed the most important unit, the image sensor in QM and FM, and to evaluate the other performances. We carried out the polarization measurement at −0.008 nm from the center of the Fe I 532.19 nm line. In the "routine observation" mode, we collected left- and right-polarization images alternately, which was repeated 128 times to deduce the Stokes-*V* signal, and further got the longitudinal magnetogram after calibration. The "deep integration" mode will be realized by repeating 9 times the "routine observation", and it takes about 18 minutes. Figure 2 shows a filtergram (Panel a) and corresponding longitudinal magnetogram (Panel b) of the "deep integration" mode. The magnetic sensitivity is estimated by a "quiet" region near the disk center. We use the standard deviation of this "quiet" region as the sensitivity for magnetic field, which is about 5.2 Gauss and 2.5 Gauss, in "routine observation" mode and "deep integration" mode, respectively.

The FM was only tested in the laboratory. There are two reasons why we did not carry out test observations. On the one hand, it is difficult to ensure that the instrument would not be contaminated during the trial observation; on the other hand, the focal plane of FM is designed according to the vacuum wavelength, and there would be no clear image on the camera in the air. Figure 3 shows the FM of FMG, and Table 3 lists the main performance measured for the final FMG, which satisfy well the designed requirements.

3.2. HXI

HXI is a Fourier-transform-type X-ray indirect-imaging telescope aiming to explore the nonthermal radiation process of solar flares. It is designed to image the full solar disk in the energy range of 30 – 200 keV with a spatial resolution of better than 3.2″ at 30 keV. The design and verification of the EM and QM of HXI can be retrieved in a series of articles (Zhang et al., 2019a,b; Su et al., 2019b; Chen et al., 2020, 2021). In short, HXI includes three main

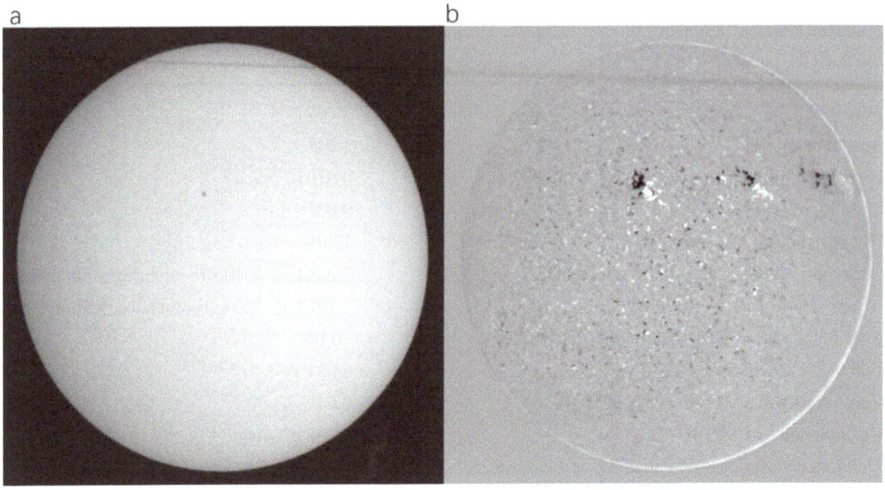

Figure 2 FMG test observations with the QM: a filtergram of Fe I 5324.185 Å (**a**) and corresponding longitudinal magnetogram (**b**).

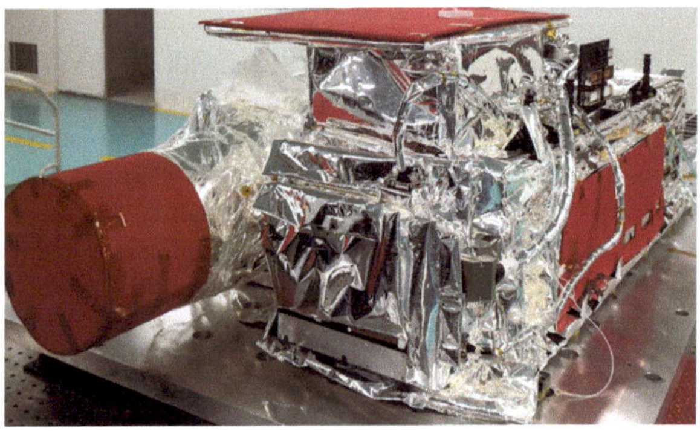

Figure 3 The flight model of FMG (dimension: 1510 mm × 636 mm × 350 mm).

parts: the *collimator* (HXI-C), the *spectrometer* (HXI-S), and the *electronics control box* (HXI-E), which are mainly used for X-ray flux modulation, modulated X-ray photon detection, and operation management, respectively. Ninety-one pairs of tungsten grids, namely sub-collimators, modulate the incident X-ray flux simultaneously with a series of spatial frequencies. Moreover, a *solar aspect system* (SAS) attached to the HXI-C is employed to monitor the deformation of HXI-C and resolve the solar aspect information.

The FM of HXI, as shown in Figure 4, had been produced and tested during 2020 and 2021 (Wan et al., 2022; Liu et al., 2022). Its design is almost the same as that of the EM and QM on the basis of good and reliable preliminary work. At the very beginning of 2022, the FM of HXI was delivered to Microsat in Shanghai for assembly, integration, and testing with the spacecraft before launch. Some major performance specifications for the FM of HXI had been finally measured during these tests, which are listed in Table 4.

For the FM of HXI, we only mention briefly the work on the deformation monitoring of HXI-C and the X-ray beam test. The deformation of HXI-C, including relative shift and rotation between front and rear grids, has a serious impact on the imaging performance of HXI. Therefore, the deformation is carefully monitored during each test or transportation

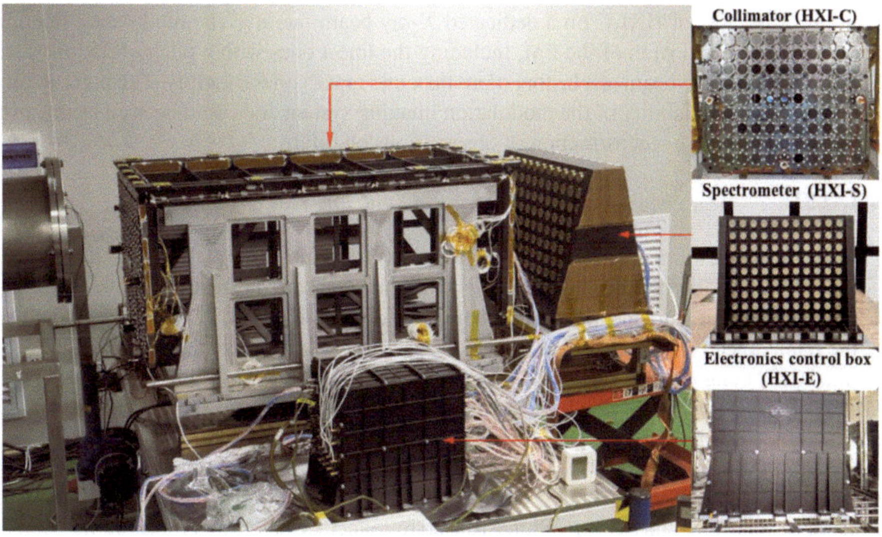

Figure 4 The flight model of HXI, running in the beam test. The dimensions for HXI-C and HXI-S are 708 mm × 700 mm × 1618 mm and for HXI-E 406 mm × 296 mm × 339 mm.

Table 4 Main performance specifications for the FM of HXI.

Items	Measurements
Energy range	30 keV to 200 keV for imaging, better than 30 keV to 400 keV for spectrum
Energy resolution (FWHM)	< 23.7% @ 32 keV
Spatial resolution	3.2″ @ 30 keV
Temporal resolution	0.125 s, 0.25 s, 0.5 s, and 1 s for eruption periods, 4 s for quiet periods
FOV	Full Sun, 40.3′
Pointing precision of SAS	0.63″ for the Sun center
Deformation of the collimator	relative rotation < 5.7″ and relative shift < 18.4 μm before launch
Deadtime for detectors	≈ 1.8 μs
Gain difference between detectors	< 16.2%
Updating frequency for SAS dataset	4 Hz

accomplished in the whole period from assembly to launch. The SAS was used as the only measurement method to monitor the deformation since the HXI-C was integrated into the spacecraft. The other methods, such as the coordinate-measuring machine, are no longer suitable due to interference with the spacecraft. The FM of HXI-C maintained good stability during the whole period according to the monitoring results that the relative shift varies less than 18.4 μm and relative rotation less than 5.7″. These values are much better than the design specifications, which were required to keep less than 30 μm for shift and 10″ for rotation before launch. On the other hand, knowledge about the X-ray flux modulation relation of each sub-collimator is considered as the basis of modulation imaging of HXI. For this reason, an X-ray beam test is employed to directly verify and calibrate the modulation relation of the FM of HXI-C on a dedicated X-ray beam facility. The modulation relation curves of all 91 grid pairs of the FM, including the finest ones with a pitch of 36 μm, have been obtained by the beam test lasting more than one month on the facility. The high-quality results prove the reliability of the modulation imaging system and could be used as imaging parameters for image reconstruction of observations on orbit.

3.3. LST

LST is the first Chinese space telescope to observe the Sun from disk center to the inner corona up to 2.5 $R_\odot$ in both the Lyman-α and white-light wavebands (cf. Table 5) for diverse solar activities and the quiet Sun. The scientific objectives of LST and its contributions to the main objectives of the ASO-S have been extensively described in previous articles (Li et al., 2019; Feng et al., 2019). LST is a suite of imaging instruments, including the *Solar Corona Imager* (SCI), which is a Lyot-type coronagraph working in both the Lyman-α and 700 nm wavebands (called SCIUV and SCIWL, respectively), the *Solar Disk Imager* (SDI), and the *White-light Solar Telescope* (WST) working in the Lyman-α and 360 nm waveband, respectively (see Table 5 for specific information). SCIUV and SCIWL share the same optical design, and a beam-splitter is used to reflect Lyman-α and transmit the 700 nm waveband. Each of the four instruments of LST has its default observation modes/cadences, which have been described in detail in previous articles (Li et al., 2019; Feng et al., 2019). The EM's optical, mechanical, electrical, and thermal designs have been presented in detail (Chen et al., 2019). The stabilization capability of the spacecraft does not satisfy the requirement of SDI and SCI images. To meet the stringent requirement, the SCI primary mirror is an active mirror driven by a piezoelectric table, which tracks the Sun according to the pointing signal from the *guide telescope* (GT) of the LST and has an angular compensation of $\pm 214''$. The secondary mirror of SDI is also an active mirror and has an angular compensation of $\pm 69''$. The SCI, the SDI, and the WST share a common mechanical bench with sufficient stability and stiffness to retain their optical axes consistency within $\pm 10''$. After the EM of the LST, some changes and updates have been made to optimize the parameters and improve the characteristics and performance. Here we briefly report on several major ones.

In order to achieve the stringent requirement for stray-light suppression of the SCI, a gradient coating is deposited on the SCI's secondary-mirror inner edge, which smoothly decreases reflectivity from 1.3 $R_\odot$ to 1.1 $R_\odot$ region for the white-light waveband while that for the Lyman-α waveband is almost unchanged. Meanwhile, the Lyot stop also plays an important role in stray-light suppression by reducing the effective aperture to about 48 mm, which is an appropriate size instead of the 60 mm clear aperture of the main mirror. Compared with the EM, another Lyman-α filter was added near the entrance of the SDI to improve the spectral purity. Therefore, in the QM and FM, the SDI has two pre-filters at the entrance and one filter on the filter wheel. Calculations show that this combination reduces the out-of-band radiation down to around 2%. In order to improve the in-flight flat-field calibration accuracy by the KLL algorithm (Kuhn, Lin, and Loranz, 1991), a defocusing lens is added in the front of SDI's CMOS detector (on the filter wheel) to blur the apparent structures of the solar disk in the observing waveband so that the acquired images are more suitable for the KLL algorithm (Li et al., 2021).

Before launch, all on-ground tests, measurements, and calibrations of LST have been done at CIOMP. Figure 5 shows an FM photo of LST. The main test and calibration results are shown in Table 5. The center wavelengths and bandwidths (FWHM) of the four instruments in Table 5 were derived from the spectral response curves obtained on-ground before launch.

Table 5 Main test and measured parameters for the FM of LST.

Items	Instruments	Measurements
Center wavelength	SCI	701.0 nm and 122.6 nm
	SDI	121.6 nm
	WST	360.0 nm
Bandwidth (FWHM)	SCI	62.7 nm @ 700.0 nm, 6.0 nm @ 121.6 nm
	SDI	9.5 nm
	WST	3.7 nm
FOV	SCI	Inner $33.7'$, Outer $80.2'$ @ 700.0 nm
		Inner $33.7'$, Outer $80.6'$ @ 121.6 nm
	SDI	$38.5'$
	WST	$38.7'$
Spatial resolution	SCI	$4.4''$
	SDI	$1.2''$
	WST	$1.1''$
Image stabilization (1σ)	SCI	$0.2''$ per 120 s
	SDI	$0.1''$ per 60 s
Stray-light suppression (700 nm)	SCI	1.5×10^{-7} $B_\odot$ @ 1.1 $R_\odot$, 3.5×10^{-8} $B_\odot$ @ 2.5 $R_\odot$
Scope for image stabilization	SCI	$\pm 214''$
	SDI	$\pm 69''$

Figure 5 The flight model of LST (dimensions: 1380 mm × 818 mm × 474 mm).

4. Operation

For most of the commissioning time and later normal working time, both the *Space Science Mission Operation Center* (SMOC) at NSSC, CAS and the ASO-S *Science Operation and Data Center* (SODC) (Huang et al., 2019) at PMO, CAS are together responsible for managing the spacecraft in orbit. The original data are transferred to SMOC from three ground stations located at Sanya, Kashi, and Miyun. These three ground stations can provide the download of all of the data observed daily. Every day, there are six to nine contacts between the spacecraft and ground stations. The maximum data latency to the ground station is about ten hours. After decoding and pre-processing, the data are transferred from SMOC to the SODC, where they are processed further into high-level data, which are stored and will be open, together with analysis software, to users all over the world afterwards.

After the commissioning phase, the mission will work automatically for most of the time, but there are indeed different modes, which can be commanded from the ground or adjusted autonomously through detection software running onboard, for the former, for example, from routine (two-minute cadence) to deep integration mode for FMG; for the latter, for example, the eruption mode or quiet mode for HXI. In order to satisfy specific campaigns or specific goals, SODC can upload one day ahead a series of commands.

At the time of writing (three months since launch), the mission works well and various commissioning-phase tests are being done step-by-step. We plan to finish all commissioning tests and in-flight calibrations within six months, and then we will open all the data to the community in real time via the SODC data access at the mission home page at aso-s.pmo.ac.cn/en_index.jsp. Before that time, the ASO-S science team plans to hold a training course (most probably via online meetings) to teach users how to access and analyze the data.[1] The data policy of the ASO-S was made by the science committee over two years ago, in which it emphasizes the equal right for users or the team members to access the scientific data of the ASO-S mission, encouraging collaboration with team members for the initial two years, and beyond. The full version of the data policy can be referred to at the home page of the ASO-S mission.

The ASO-S science team has the duty to organize and promote the research based on the ASO-S data. Beside focused projects supporting analyzing the ASO-S data in China, the ASO-S science team also plans to issue an international cooperation program in due time, in which over ten guest investigators are annually invited to visit SODC for a couple of months to perform joint or independent research related to ASO-S data. The regular ASO-S international workshop or a joint workshop with other missions are being considered, depending on the performance of the mission. We hope indeed that with a rich scientific return, the ASO-S could make a remarkable contribution to the solar studies in Solar Cycle 25.

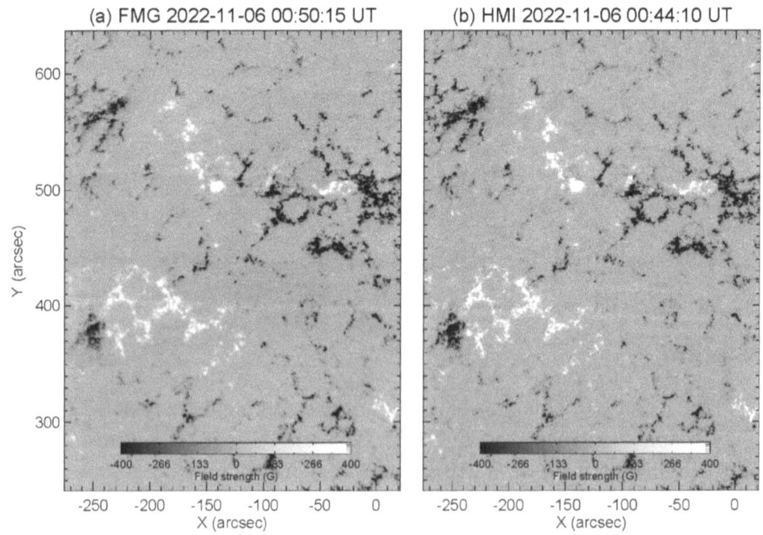

Figure 6 Line-of-sight magnetogram in quiet Sun acquired by FMG on 06 November 2022 00:50:15 UT (**a**) and the corresponding HMI magnetogram (**b**) is shown for comparison.

[1] At the time that this article was accepted for publication, the tutorial on ASO-S data had already been held from 11 to 12 April 2023 online. All the materials of the tutorial have been put on the website of the mission.

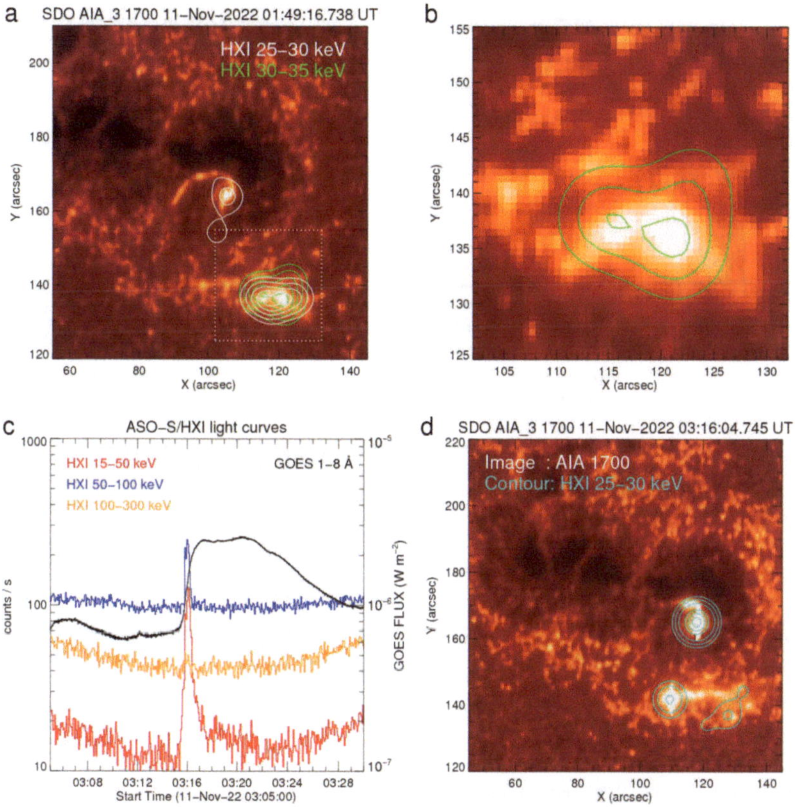

Figure 7 HXI/ASO-S X-ray light curves and images of two flares observed on 11 November 2022. (**a**) The HXI/ASO-S images reconstructed using the HXI Clean algorithm (Su et al., 2019) in the energy ranges of 25 – 30 (*white*) and 30 – 35 keV (*green*) with contour levels of 15, 30, 50, 70, and 90 percent of the peak intensity for the flare that occurred at 01:48 UT. The background 1700 Å image was taken by SDO/AIA at 01:49:16 UT. (**b**) Enlarged view of the southern ribbon with HXI 30 – 35 keV contours (30, 60, and 90 percent contours). (**c**) ASO-S/HXI light curves in the energy ranges of 15 – 50, 50 – 100, and 100 – 300 keV and GOES 1 – 8 Å light curve for the GOES C3 flare with a start time at $\approx$ 03:13 UT. (**d**) The ASO-S/HXI Clean image for the energy range of 25 – 30 keV (*cyan* contours at levels of 5, 10, 20, 50, 70 and 90 percent of the peak intensity). The background 1700 Å image was taken by SDO/AIA at 03:16:04 UT.

5. First Light

During the first three months since launch, the first light of the three payloads (except for SCI) have been obtained, although more elaborate calibrations are still needed.

Figure 6a shows a cutout longitudinal magnetogram of FMG in quiet Sun taken at 00:50 UT on 06 November 2022, in comparison with that observed by the *Helioseismic and Magnetic Imager* (HMI: Scherrer et al., 2012) onboard the *Solar Dynamics Observatory* (SDO: Pesnell, Thompson, and Chamberlin, 2012), as shown in Figure 6b. The FMG transverse fields as well as the full-disc magnetograms are still under test.[2]

[2] At the time that this article was accepted for publication, the FMG transverse fields have not yet been obtained. The possible reason is attributed to the performance of one of the two liquid crystal variable retarders.

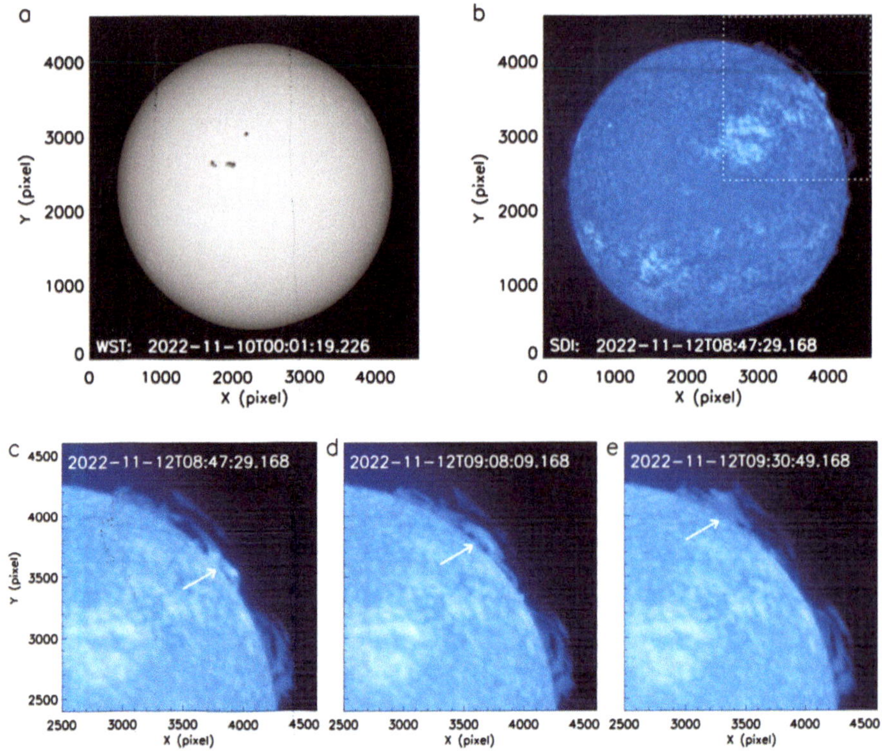

Figure 8 Panels **a** and **b**: WST and SDI full-disk images at 360 nm and 121.6 nm. The images are calibrated with in-flight dark-current and flat-field data. Panels **c** – **e**: enlargement of a subregion marked by the *dotted lines* in Panel **b** to show the fine structures of prominences and the northward plasma flows indicated by *three arrows*.

Figure 7 shows HXI CLEAN images reconstructed for a GOES C4 flare that started at 01:45 UT and the GOES C3 flare that started at 03:10 UT on 11 November 2022, overlaid on the image at 1700 Å taken by the *Atmospheric Imaging Assembly* (AIA: Lemen et al., 2012). The HXI images have been rotated so that the solar North is pointing up and corrected for the pointing shifts during the image duration. A manual shift of $\approx 91''$ (the shift will not be necessary after the full imaging calibration in the future, which requires more observations of flares) was then applied to the images to match the HXR sources with the bright ribbons seen by AIA. It is revealed that the two hard X-ray sources in the energy range of 25 – 30 keV are well consistent with the bright ribbons in the 1700 Å image (Figure 7a, d). At higher energies of 30 – 35 keV (Figure 7b), the fine structures of the two separate bright features in the southern ribbon are resolved as two sources, which suggests a correct reconstruction of HXI images at its finest resolution of $3.2''$, even though the sub-collimators are not calibrated yet.

Figure 8 demonstrates the imaging capability of WST and SDI of the LST. Panels a – b show the full-disk images of WST and SDI, respectively. In the WST image, sunspot umbra, penumbra, and light bridges can be distinguished. The SDI captures various structures on

Besides, there is a darker region in the whole-disc image. The reason and remedy are still under investigation. The details will be described in an upcoming article specifically for the instrument.

the Sun, from omnipresent chromospheric network, active regions on the disk, to the off-limb prominences. To show the fine structures of the prominences, the dotted subregion in Panel b is enlarged. The northward plasma flows indicated by three arrows can be observed, as shown in Panels c – e. At the time of this article, the SCI is still under test.[3]

The details of each instrument's performance in orbit and quality of the images will be discussed later in separate articles by each of payloads when the commissioning phase is finished. At this moment, three months since launch, the spacecraft platform is normal, and the HXI is close to finishing its tests, while FMG and LST are still under test.

Acknowledgments The ASO-S mission is a large scientific project and has lasted over ten years from the early concept study. Hundreds of people have been involved in the project for different periods of the time. All the participants are heartfeltly acknowledged! SDO is the first mission under NASA's Living With a Star (LWS) program. The SDO/AIA and SDO/HMI data used in this article were downloaded from the JSOC data center and they are courtesy of NASA and the science teams.

Author contributions Beside the contributors listed in Table 1, the contributors on the mission-level coordination include also W. Xiong, Bin Chen, and M. Su; on the spacecraft platform Z. Yin, H. Bao, C. Cao., Y. Bai, Binglong Chen, T. He, X. Li, Ye Zhang, X. Liao, and H. Jiang; on the FMG F. Li, J. Lin, J. Hou, and X. Bai; on the HXI D. Chen, Yan Zhang, Y. Hu, Y. Liang, and J. Wang; on the LST K. Song, Q. Guo, L. He, G. Zhang, and P. Wang; on the SODC Yingna Su, S. Lei, W. Chen, Ying Li, J. Zhao, J. Li, and Y. Ge; on the missions' other systems H. Ji, M. Gu, Y. Zheng, D. Xu, and X. Wang. The manuscript was organized by W. Gan with the help from all authors.

Funding The ASO-S is supported by the Strategic Priority Research Program on Space Science, Chinese Academy of Sciences (Grant No. XDA15320000, XDA15320100, XDA15320102, XDA15320103, XDA15320104, XDA15320300, XDA15052200). Part of members' own work is also supported by the Ministry of Science and Technology of China (Grant No. 2022YFF0503000), the National Natural Science Foundation of China (Grant No. 12233012, 12022302, 12173100, 11921003, U1731241, 11973097), and the Youth Innovation Promotion Association CAS.

Declarations

Competing interests The authors declare no competing interests.

References

Chen, B., Li, H., Song, K.F., Guo, Q.F., Zhang, P.J., He, L.P., et al.: 2019, The Lyman-alpha Solar Telescope (LST) for the ASO-S mission—II. Design of LST. *Res. Astron. Astrophys.* **19**, 159. DOI. ADS.

Chen, D.Y., Hu, Y.M., Zhang, Z., Zhang, Y., Zhang, Y.Q., Huang, Y., Ma, T., Chang, J.: 2020, Environmental tests of the HXI spectrometer for the ASO-S mission. *J. Instrum.* **15**, T10008. DOI. ADS.

Chen, D.Y., Hu, Y.M., Ma, T., Su, Y., Yang, J.F., Wang, J.P., et al.: 2021, Design and verification of the HXI collimator on the ASO-S mission. *Res. Astron. Astrophys.* **21**, 136. DOI. ADS.

Deng, Y.Y., Zhang, H.Y., Yang, J.F., Li, F., Lin, J.B., Hou, J.F., et al.: 2019, Design of the full-disk MagnetoGraph (FMG) onboard the ASO-S. *Res. Astron. Astrophys.* **19**, 157. DOI. ADS.

Feng, L., Li, H., Chen, B., Li, Y., Susino, R., Huang, Y., et al.: 2019, The Lyman-alpha Solar Telescope (LST) for the ASO-S mission—III. Data and potential diagnostics. *Res. Astron. Astrophys.* **19**, 162. DOI. ADS.

Gan, W.Q., Feng, L., Su, Y.: 2022, A Chinese solar observatory in space. *Nat. Astron.* **6**, 165. DOI. ADS.

Gan, W.Q., Deng, Y.Y., Li, H., Wu, J., Zhang, H.Y., Chang, J., et al.: 2015, ASO-S: Advanced Space-based Solar Observatory. *Proc. SPIE* **9604**, 96040T. DOI.

[3] At the time that this article was accepted for publication, the SCI was found to have a defect in technological design. So far, coronal structures are almost unrecognizable in 700 nm images from SCIWL, while the Lyman-α images from SCIUV look better except for some obscuration. We are still optimizing the observation parameters and modes to improve the image quality. The details will be described in an upcoming article specifically for the instrument.

et al. 2007; Dominguez-Tagle et al. 2022; Elmore et al. 1992; Gosain, Venkatakrishnan, and Venugopalan 2006; Hagyard et al. 1982; Jones et al. 1992; Livingston, Harvey, and Slaughter 1971; Mickey et al. 1996; van Noort and Rouppe van der Voort 2008; Qu 2011; Martinez Pillet et al. 2023; Raja Bayanna et al. 2014). The Michelson Doppler Imager (MDI) on board the Solar and Heliospheric Observatory (SOHO), launched in 1995, marked a new era of spaceborne solar magnetic field observations (Scherrer et al. 1995). Subsequently, space missions, such as, Hinode, the Solar Dynamics Observatory (SDO), and Solar Orbiter, have also carried instruments to measure solar magnetic fields (Lites and Ichimoto 2013; Lites et al. 2013; Scherrer et al. 2012; Solanki et al. 2020).

The Full-disk vector MagnetoGraph (FMG), a payload on board the Advanced Space-based Solar Observatory (ASO-S), represents China's first spaceborne magnetograph. ASO-S was successfully launched on 8 October 2022 to a Sun-synchronous orbit with an altitude of about 720 km above the Earth's surface. The descending node time of the orbit is at 06:00 local time. Its primary objective is to investigate the relationship between solar eruptions and the dynamic evolution of magnetic fields. Deng et al. (2019) comprehensively described the design of FMG and shared the test observation results for the engineering model. Su et al. (2019) discussed the planned data processing methods for FMG within the ASO-S data center, as outlined by Huang et al. (2019). Gan et al. (2023) presented preliminary on-orbit results from ASO-S, including data captured by FMG. Following an extensive six-month on-orbit testing and calibration phase, the observation and calibration modes for in-flight operations, along with parts of the data processing pipeline, have been refined and consolidated. On 1 April 2023, FMG began to release Level 2 filtergram and longitudinal magnetic field data products for solar active regions (http://aso-s.pmo.ac.cn/sodc/dataArchive.jsp). Figure 1a and 1b illustrates the full-disk Stokes I and V/I maps acquired by FMG on 7 March 2023. The maps of Stokes V/I clearly display the large-scale distribution and polarities of the longitudinal magnetic field associated with diverse solar features. Furthermore, the Stokes Q/I and U/I images of individual sunspots, shown in Figure 1d and 1e, reveal four-petal structures, thus validating the accuracy of FMG data.

In the article, we present the ground calibration and in-flight performance of FMG. We also provide a brief overview of the current status of the released high-level science data and the corresponding data processing pipeline. The organization of this article is as follows. Section 2 presents the results of imaging and spectral performance, signal-to-noise ratio (SNR), and polarization sensitivity. Section 3 outlines the data processing procedures used to generate the high-level data products from FMG raw observations, following with a conclusion section.

2. Ground Calibration and Inflight Performance

FMG is a sophisticated instrument that consists of an optical package, an electronics box, and a thermal control system (Deng et al. 2019). FMG optical package (shown in Figure 2) has:

- an optical system that images the Sun with a field of view (FOV) of 37' × 37' and an aperture of 140 mm,
- a polarimeter that combines with two liquid crystal variable retarders (LCVR) and a polarized beam splitter to measure the full Stokes I, Q, U, and V parameters (Hou et al. 2020),

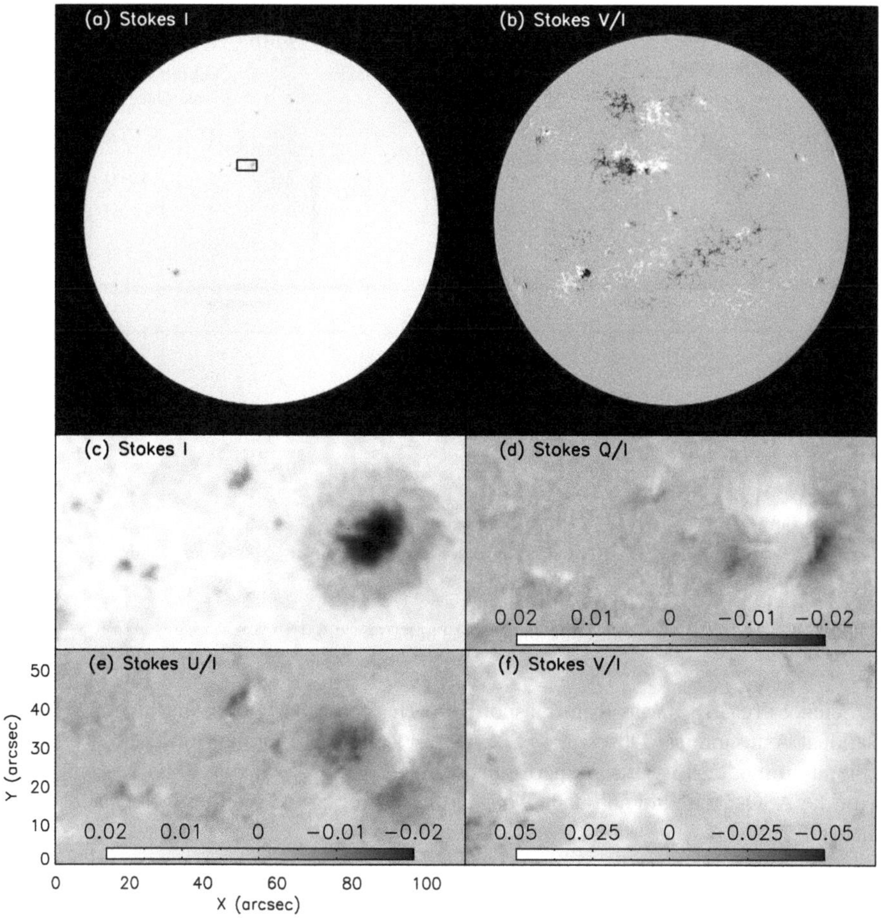

Figure 1 Panels a and b: Full-disk Stokes I (limb darkening corrected) and V/I maps, taken by FMG on 7 March 2023. Panels c to f: Zoomed-in Stokes I, Q/I, U/I, and V/I maps of the black rectangular region in panel a.

- a wide-field and tunable seven-stage Lyot filter with an aperture of 37 mm and a width about 0.01 nm that acquires narrow band filtergrams at specific wavelength (at −0.008 nm from the Fe I 532.42 nm line center),
- an Image-Stabilization System (ISS) that compensates for satellite jitters and pointing consistency errors between the FMG and the Lyman-alpha Solar Telescope (LST), which is another payload on the ASO-S (Li et al. 2019; Chen et al. 2019),
- a data acquisition and processing system that includes a 4k × 4k CMOS camera with a pixel size of 12.8 μm × 12.8 μm and a pixel resolution of 0.55 arcsec, a temperature control module for the LVCR polarimeter (within 35 ±0.1 °C), a voltage driver for the LCVR, and a data processing unit that integrates series of raw images with various polarized modulation states.

The FMG has completed its integration and alignment phase for the flight model in March 2022. The thermal, electronic tests, and the spatial resolution calibration of the instrument were carried out at Xi'an Institute of Optics and Precision Mechanics, Chinese Academy

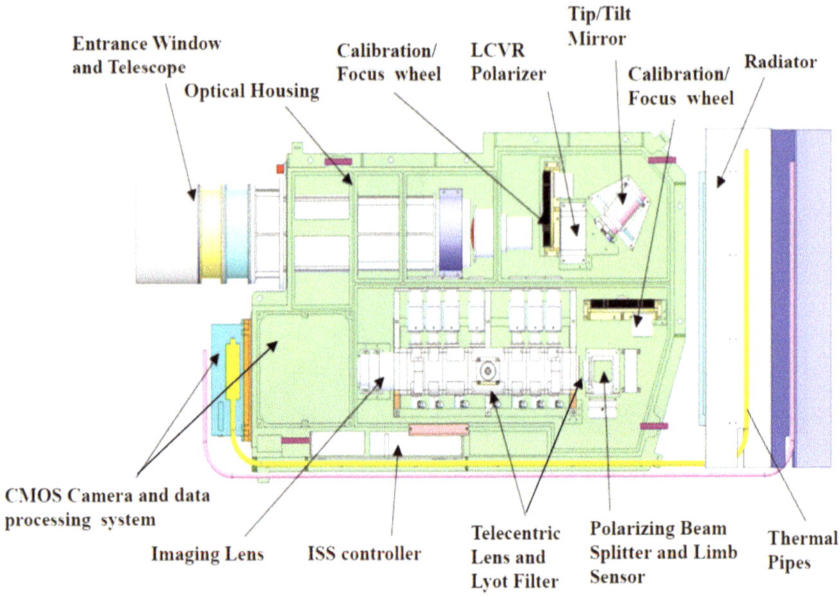

Figure 2 Opto-mechnical drawing of the flight model of FMG.

of Sciences (CAS). It was then transported to Huairou Solar Observing Station (HSOS), National Astronomical Observatories, CAS, for detailed calibration of its observing and calibration mode as well as its performance in April 2022. Finally, FMG was integrated with the ASO-S satellite and underwent further testing at the satellite factory and launch site.

2.1. Imaging Performance

On the day of ASO-S launch, the FMG instrument started to power up and its thermal control system was activated. After several orbits, the optical box stabilized at a temperature of 22 ±0.1 °C. On 16 October 2022, FMG captured its first light image, as shown in Figure 3a. This was a single-exposure filtergram with an exposure time of 20 ms. The Lyot filter was tuned at approximately − 0.1 nm from the Fe I 532.42 nm line center. Two LVCRs were powered off during the image capture. Figure 3a shows a large-scale non-uniformity resulting from the optical system the reason of which is still unknown. As there are no corresponding dark and flat field calibration files available, these non-uniformities cannot be corrected. However, this does not affect the excellent image quality of FMG. To demonstrate the excellent image quality, we take a zoom-in image of the rectangular region in Figure 3a with a FOV of $129'' \times 68''$, and present it in Figure 3b. The granulations and intergranular lanes are clearly resolved, indicating that the best focus position calibrated during ground calibration remains the same after launch.

In Figure 3c, we present the intensity of the slice along a granule (line A in Figure 3b) and perform a Gaussian fit. The fitted full width at half maximum (FWHM) is 1.24 arcseconds. During ground calibration, a resolution test target was used at the focus position of a collimator with a focal length of 10 m. The calibrated spatial resolution is 1.04 arcseconds. The inflight spatial resolution for a single exposure filtergram is consistent with that from ground calibration.

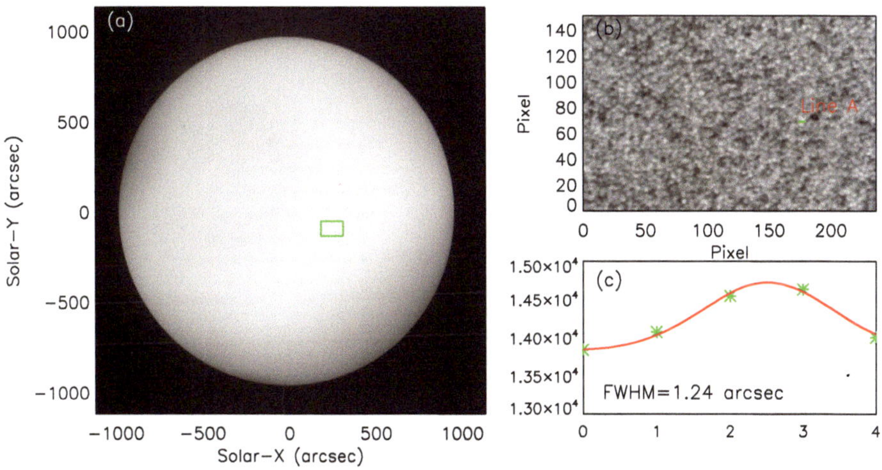

Figure 3 (a): First light filtergram taken by FMG on 16 October 2022. (b): Zoomed-in view of the rectangular region in (a). (c): Intensity along the line A in (b). The red line marks the Gaussian fitting result.

FMG employs a multi-frame integration method to increase the SNR of the polarization modulated images taken in the routine observation mode. This involves sequentially acquiring six modulated images, (I+Q)/2, (I-Q)/2, (I+U)/2, (I-U)/2, (I+V)/2, and (I-V)/2, within the cadence of 2 minutes. Each image is automatically integrated 128 times using the on board data acquisition and processing system. This integration process may result in a different spatial resolution compared to single exposure filtergrams, primarily due to the performance characteristics of the ISS, which are further tested. Series of single exposure filtergrams are taken with the ISS both on and off. The solar disk center is determined by fitting the solar limb, allowing for a quantitative evaluation of the performance of the ISS.

In Figure 4a and Figure 4b, the relative drift of the solar disk center along the x and y axes of the CMOS detector over a period of 10 minutes is presented. Without ISS stabilization, the RMS values are 1.5 and 2.1 arcsec for the x- and y-axis, respectively. However, after ISS activation, the RMS values decreased to 0.03 and 0.07 arcseconds, indicating a significant reduction in satellite jitters. We also calculated the drift of solar disk center within one day. The RMS values are 0.23 and 0.39 arcsec for the x- and y-axis, respectively. This small drift further highlights the excellent performance of the ISS. The pointing and tracking of FMG with ISS activation is less than half pixel, ensuring high-quality of the polarization modulated filtergrams. Therefore, FMG data taken within one or several days do not require further alignment for the high-level data production. This simplifies post-processing of the data. Xu et al. (2024a) conducted an analysis of the spatial resolution of the longitudinal magnetograms of FMG and found that it is about 1.5 arcseconds.

As the FMG optical box is maintained at a stable temperature of 22 ±0.1 °C, there is no significant variation for the spatial resolution of FMG due to thermal fluctuations. Throughout the period from approximately June to August each year, the FMG passes the Earth's shadow at each orbit, resulting in alterations in the external heat flux, particularly affecting the entrance window. With the solar limb profiles, we estimated the relative variation in spatial resolution as a function of orbital velocity, and the result is shown in Figure 5. The spatial resolution undergoes changes of ±0.5 arcsec at the maximum spacecraft Sun-FMG orbital velocity, which is approximately ±4 km/s. For the minimum Sun-FMG orbital velocity, around ±0.5 km/s, the spatial resolution varies by approximately ±0.2 arcsec. Given

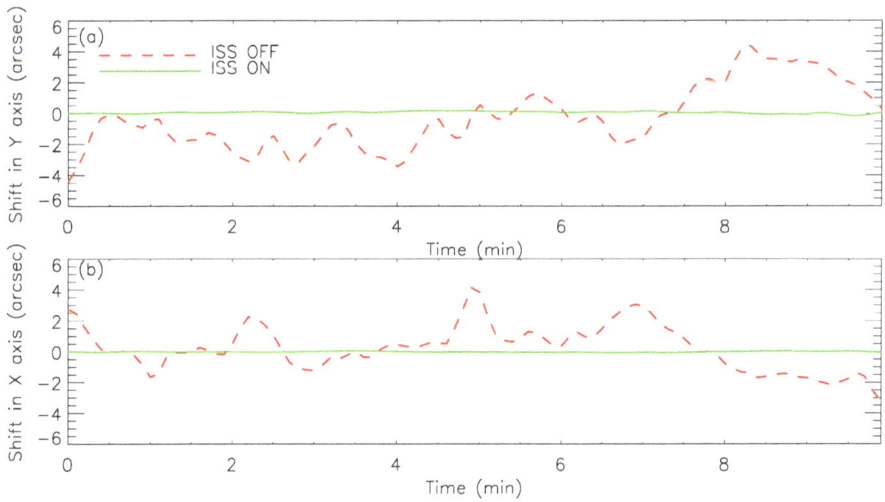

Figure 4 Relative drift of the solar disk center within 10 minutes along y- (panel a) and x-axis (panel b) of the CMOS detector. Red and green lines indicate that the ISS is off and on, respectively.

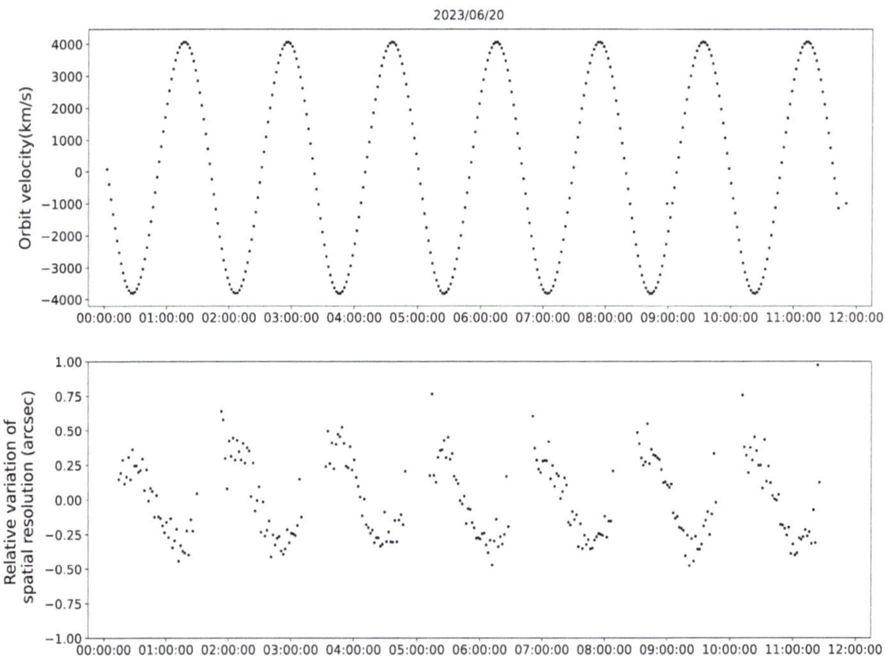

Figure 5 Variation of Sun-FMG orbital velocity on 20 June 2023 (upper panel). Relative variation of the spatial resolution of FMG with orbital velocity (lower panel).

that the orbital period is about 100 minutes, we currently did not adjust the focus wheels each orbit to compensate for these variations in spatial resolution.

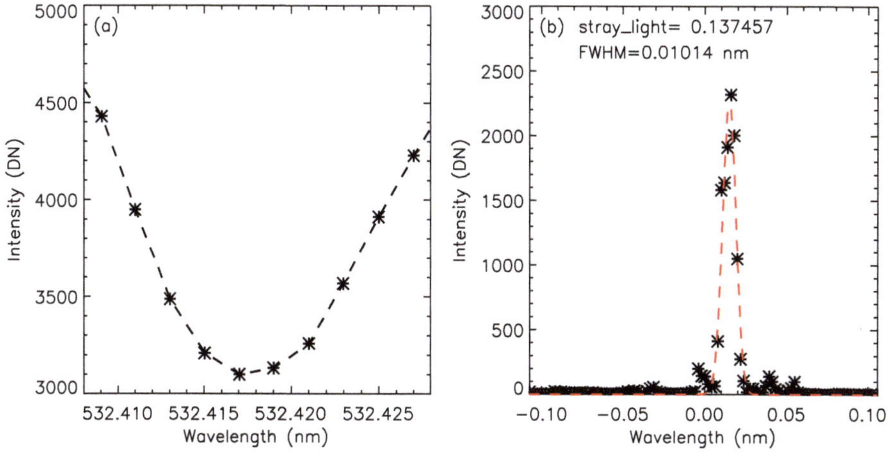

Figure 6 In-flight scanned spectral profile of the Fe I 532.42 nm line (panel a) and transmission profile of the Lyot filter (panel b) from the ground calibration. The red-dashed line in panel b is the Gaussian fitting result.

2.2. Spectral Performance

We adjusted the half waveplate at each stage of the Lyot filter elements to scan the Fe I 532.42 nm spectral line profiles from − 0.01 nm to 0.01 nm relative to the line center on 19 October 2022. The step size is 0.002 nm and the result is shown in Figure 6a. When determining the in-flight line center, we also take into account the impact of the radial velocity of the satellite orbit, which is about 0.9 km/s for the observation period. The asterisk points in Figure 6b indicate the measured transmission profiles of the Lyot filter after integrating it with the FMG optical system during ground calibration. Similar to the method employed by HMI, we used a dye laser and a collimator to illuminate the FMG entrance pupil (Couvidat et al. 2012). The spectral resolution of the dye laser was firstly calibrated with a gas cell according to the method in Bai et al. (2023). The calibrated line width of the dye laser was about 0.002 nm, much lower than the designed width of the Lyot filter. Subsequently, FMG performed a wavelength scan from − 0.1 nm to 0.1 nm relative to the Fe I 532.42 nm line center. The FWHM of the Lyot filter is 0.01014 nm, as determined by the Gaussian fitting with an out-of-band suppression of 13.7%.

FMG has a FOV of 37’ × 37’. To minimize the wavelength drift, i.e. peak value of the transmitted filter profile within the large FOV, the Lyot filter was arranged in a telecentric optical system. However, there was still a residual wavelength drift due to the alignment errors in the optical elements of FMG. During ground calibration of the flight model of FMG, we also measured this residual wavelength drift by inserting two calibration lenses in two calibration wheels into the optical path of FMG. Similar calibration lenses are also employed in HMI (Schou et al. 2012). The sunlight was fed into the laboratory using a coelostat at HSOS. The entrance pupil was imaged on the CMOS detector, and the incident angle in the Lyot filter was almost the same as that for the routine solar observations, ensuring uniform illumination and wavelength of the input light. We tuned the Lyot filter ranging from − 0.03 to 0.03 nm relative to the Fe I 532.4 nm line center. The line center position was obtained for each pixel, and the relative wavelength drift was calculated across the whole FOV, as shown in Figure 7c. The maximum value is about 0.003 nm, which corresponds to a velocity of 1.7 km/s. The precise in-flight wavelength position of FMG for each pixel is

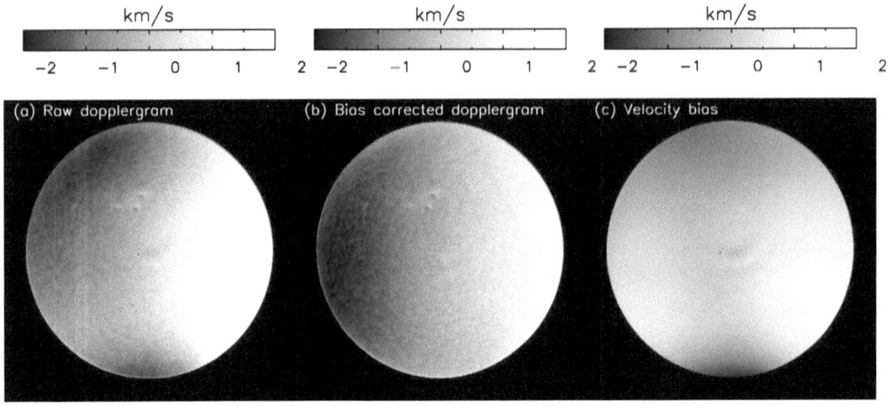

Figure 7 (a): Raw Dopplergram. (b): Bias corrected Dopplergram. (c): Doppler/velocity bias caused by non-uniformity of the transmission profiles of the Lyot filter over the entire FOV. The reason for the fringe near disk center is unknown so far.

the combination of the instrumental wavelength drift with that caused by the radial velocity from the satellite orbit.

The wavelength drift can also be interpreted as an instrumental velocity bias in the solar Dopplergram measurements. Figure 7a and Figure 7b show the raw Dopplergram taken during ground calibration, as well as the one after velocity bias correction. In the raw Dopplergram, the Sun's south polar region has a negative velocity, which is also present in the velocity bias image (Figure 7c), indicating that it is caused by the instrument rather than by the velocity from the Sun. After correction, the velocity near the solar south pole in the velocity bias corrected Dopplergram in Figure 7b is almost zero. The symmetric distribution of the Dopplergram in the east-west direction, caused by the solar rotation, becomes more apparent, confirming the accuracy of our wavelength calibration. As FMG is in vacuum focus and the test is conducted at HSOS in the air for the ground calibration, the spatial resolution of the Dopplergram is not the best. The Dopplergram cannot be obtained in the routine observation mode of FMG since it takes observations at a single wavelength point. The calibrated wavelength drift is mainly used for accurate calibration of the magnetic field. Moreover, the method presented here can also be employed for future magnetographs that measure magnetic fields and Dopplergrams at multiple wavelength points of the Fe I 532.42 nm line (Deng et al. 2023).

2.3. Signal to Noise Ratio and Polarized Sensitivity

The CMOS camera of the flight model of FMG underwent a series of tests in the laboratory, including linearity, gain, readout noise, etc., using an integrating sphere. We also measured its quantum efficiency using a monochromator, and the result was 56.2%. The gain used for the in-flight routine observations is 3.15 e^-/DN, corresponding to a full-well charge of 42.3 ke^-. The reason for selecting this gain is that it provides the best SNR for single exposure images. The corresponding readout noise and bias value are approximately 25.3 e^- and 170 DN, respectively.

During the ground calibration at HSOS, we also test the SNR of FMG. By rotating two calibration lenses into the optical path and feeding the sunlight into FMG, we successfully achieved uniform illumination of the entrance pupil again. With the aforementioned gain,

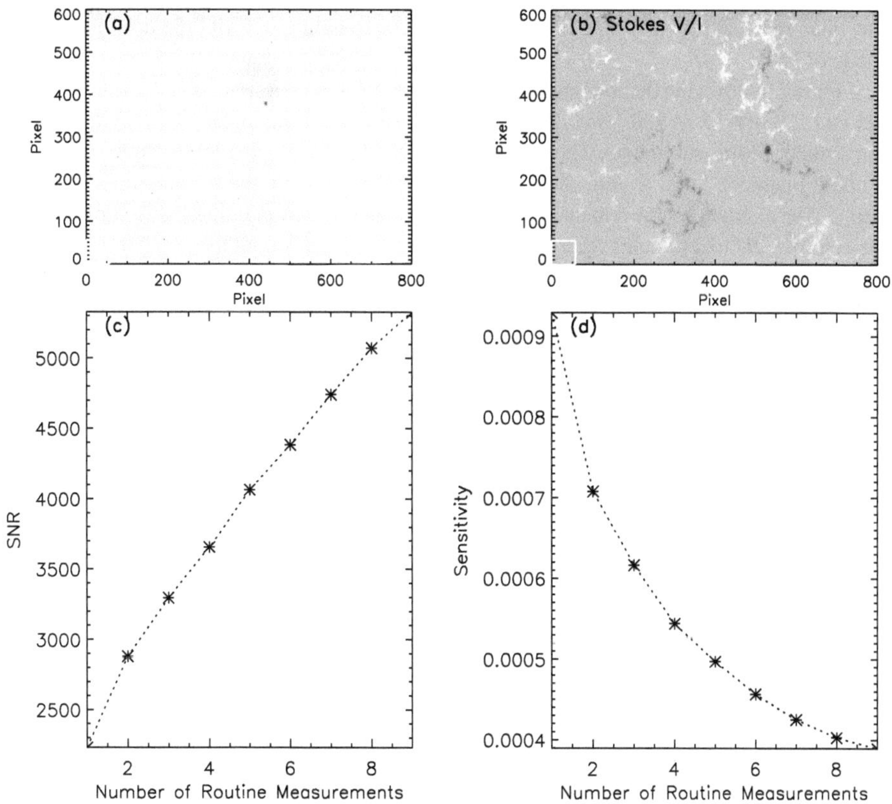

Figure 8 (a): Local region from FMG for the SNR test in the ground calibration. (b): Local in-flight Stokes V/I of FMG. (c): Relationship between the SNR and the number of routine mode measurements for the box region in (a). (d): Relationship between polarized sensitivity and the number of routine mode measurements for the box region in (b).

two polarization-modulated images, i.e. (I+V)/2 and (I-V)/2 were obtained. Each has an integration of 128 frames in routine mode. The measurement was repeated nine times to simulate the deep integration mode of FMG. Figure 8a is an example of the (I+V)/2 images taken in the ground calibration. We utilized a 50×50 pixel region to calculate the SNR in the demodulated Stokes V/I image and achieved a SNR of approximately 2230 in routine mode. The theoretical SNR can be derived according to:

$$\mathrm{SNR} = \frac{I \times N}{\sqrt{I \times N + \mathrm{dark} \times N + N \times \mathrm{noise}^2}}, \tag{1}$$

where I is the bias-corrected intensity in a single exposure image, measured in units of electrons and N is the number of integration frames. Dark and noise represent the dark current and readout noise values. I is 25,515 e^- in the SNR test. The exposure time is 35 ms, making the dark current negligible. As both (I+V)/2 and (I-V)/2 images have 128-frames integration, N is 256. The theoretical SNR is 2500, based on the above equation. The observed SNR is slightly less than the theoretical value due to the presence of additional unknown noises during the measurement. We also plot the variation of the SNR with the measurement times, as illustrated in Figure 8c. The SNR increases with more measurements and the value reaches 5300 for nine measurements, corresponding to FMG the deep integration mode of FMG.

Figure 8b shows the demodulated inflight Stokes V/I map observed on 8 November 2022, with a FOV of 440" × 330". We also tested the sensitivity of the Stokes V/I map by varying the number of measurements. The sensitivity is calculated using the RMS value of a region without magnetic field, as indicated in the box in Figure 6b. In the routine mode, the sensitivity for a single measurement with 128-frame intergration is 0.0009 with a cadence of 40 s (equivalent to 2 minutes for a complete Stokes observation). As the number of measurements increases, the sensitivity decreases. The sensitivity is 0.00039 for nine measurements (corresponding to a deep integration mode with 1152-frame integration). From the ground SNR test, the polarized sensitivity is about 0.00044 according to the SNR value of 2230 in routine mode, which is the multiplicative inverse of the SNR. The difference with the inflight value 0.0009 is mainly due to the influence of the satellite orbit and the evolution of solar granulations, since the (I+V)/2 and (I-V)/2 images are not taken simultaneously. As FMG takes observation at a single wavelength point of the Fe I 532.4 nm line, the actual wavelength position varies with the orbit, as shown in Su et al. (2019) and Liu et al. (2023). The photons or electrons received in the raw polarized modulated images change with the orbit, meaning that the sensitivity of the originally observed Stokes V/I maps is slightly different for various orbit radial velocities.

3. Data Processing and Products

All of the imaging and telemetry data recorded by the ASO-S satellite are initially received by the Space Science Mission Operation Network (SMOC) from the National Space Science Center, CAS. The FMG data undergo further decoding, renaming according to its file naming rules, and are transfered to the ASO-S Science Operation and Data Center (SODC), operated by Purple Mountain Observatory, CAS. High-level calibration and scientific data are generated using a data processing pipeline developed by NAOC and PMO. Quick-look and high-level science data can be downloaded through the ASO-S website (http://aso-s.pmo.ac.cn/sodc/dataArchive.jsp).

The FMG data are classified into Level 0, Level 0.5, Level 1.0, Level 1.5 and Level 2. Level 2 data have been released since 1 April 2023. The data received by the ASO-S SODC are in the FITS format produced by the SMOC. These FITS format data have a size of 4 k × 4 k in 16 bits, and the FITS header contains the information regarding orbit velocity, observing mode, exposure time, gain value, and filter position, among others. Level 0 data are further classified into calibration data and scientific data based on the observing plan and modes and stored in the corresponding directories.

Level 0.5 data consist of calibration files derived from a series of observed files in the calibration mode, including the dark field, flat field, velocity bias, and polarized calibration matrix, etc. Level 1.0 data are the Stokes I, Q/I, U/I, and V/I maps that have been corrected for various instrument effects, including dark, flat field, and polarization bias, etc. Based on Stokes Q/I, U/I and V/I data without polarization bias correction for one week, the polarization bias images, i.e. cross talk from Stokes I to Stokes Q, U, and V, are obtained using statistical methods (Bai et al. 2018; Xu et al. 2019). Level 1.5 data are the Stokes I, Q/I, U/I, and V/I maps after crosstalk from V to Q/U and geometry correction. Geometry correction is a process to convert the coordinates from the detector coordinate system to the helioprojective-Cartesian coordinate system. The pixel value for Level 0, Level 0.5, Level

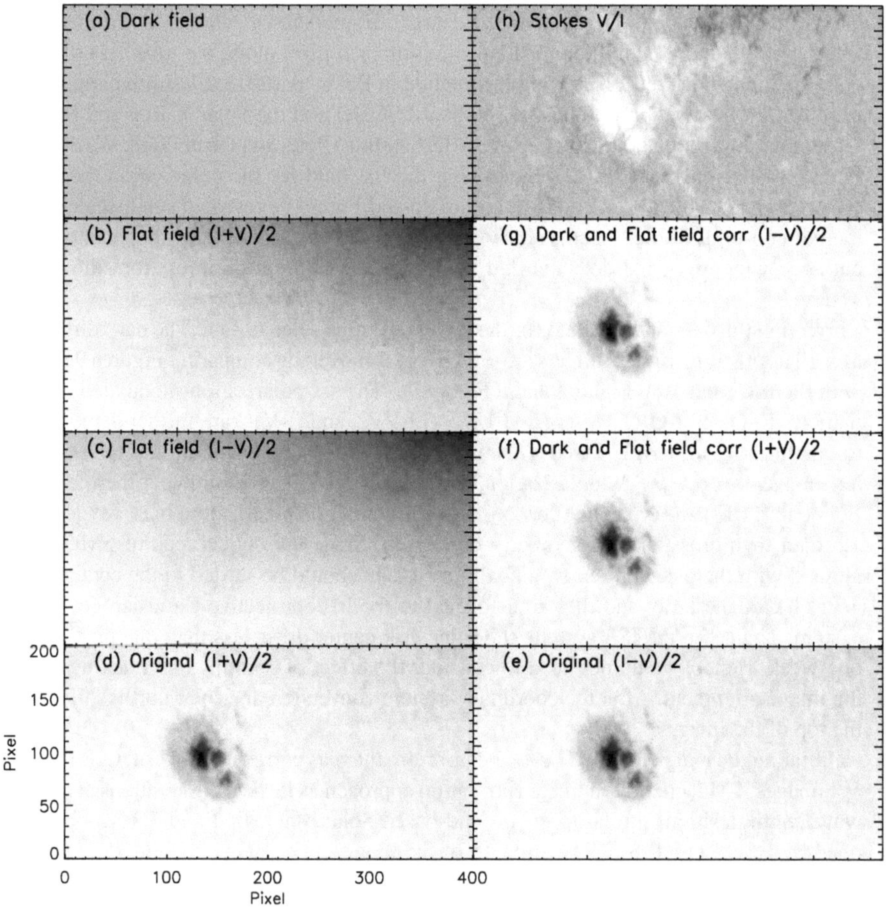

Figure 9 An example of the data processing of FMG from Level 0 to Level 1.0.

1.0, and Level 1.5 data is the unit DN. Level 2 data are the calibrated magnetograms using the linear calibration method under the weak-field approximation assumption (Liu et al. 2023). Data at levels higher than 1.5 use the helioprojective-Cartesian coordinate system instead of the detector coordinate system used in Levels 0, 0.5, and 1.0. The data processing pipeline also generates quick-look images in PNG format.

Most of the algorithms used in the FMG data processing pipeline are described in detail by Su et al. (2019). In the following, we briefly summarize the processing steps of routine observation data from Level 0 to Level 2. Based on the in-flight performance of FMG, some of the processing algorithms have been optimized.

- Generating the FMG dark field (Level 0.5 data) involves capturing one hundred images taken by rotating the baffle in the calibration wheel into the optical path. The Level 0 data are firstly corrected for the dark field, which is updated every two weeks. Operating at approximately −10 °C and with an exposure time of 35 ms per frame, the CMOS sensor experiences negligible dark current. The dark field is dominated by the bias, as seen in Figure 9a.

- Generating the FMG flat field (Level 0.5 data) with small and medium scale from the routine science data by taking the median value for each pixel along the time axis of a data cube spanning one week, following the method in Bai et al. (2018). A similar approach is used for flat field calibration of MDI on board SOHO and the Solar X-Ray and Extreme Ultraviolet Imager on board the FengYun-3E Satellite (Potts and Diver 2008; Wachter and Schou 2009; Song et al. 2022). Examining the flat field for the polarization-modulated (I+V)/2 and (I-V)/2 observations in Figure 9b and Figure 9c revealed small-scale instrumental non-uniformities, such as dirty spots from the dust of the optical elements or the entrance window of the detector, and medium-scale non-uniformities from the optical system.
- Correcting the dark and flat field for the Level 0 routine science data. The dark dirty spots as well as other instrument effects are effectively removed by comparing Figures 9g and 9f with the raw images in Figure 9d and Figure 9e. The six polarization-modulated images, namely (I+Q)/2, (I-Q)/2, (I+U)/2, (I-U)/2, (I+V)/2, and (I-V)/2 are further demodulated to obtain Stokes I, Q/I, U/I, and V/I images (Level 1.0 data). Using Stokes Q/I, U/I, and V/I images for one week, polarization bias images are derived following the method in Bai et al. (2018) and Xu et al. (2019). After correcting the polarization bias, we get Level 1.5 data with crosstalk and geometry correction. Since the optical axis of FMG is not aligned with the detector center, all calibrated data should be shifted to the center of the 4 k× 4 k image array and then transformed to the helioprojective-Cartesian coordinate system. Thanks to the ISS system, the solar disk center drifts less than one pixel within one week. Hence, we do not need to consider the effect of pointing and tracking during the image alignment. After the coordinate system conversion, the solar north is flipped to the top of the images.
- Calibrating the magnetogram (Level 2 data) from the bias corrected Stokes I, Q/I, U/I, and V/I images. FMG employs a linear calibration approach as it takes observations at a single wavelength position, similar to the ground-based Solar Magnetic Field Telescope developed by HSOS (Ai 1987). The strength of the longitudinal magnetic field B_l is derived according to:

$$\mathrm{B_l} = \mathrm{C_l} \times \frac{\mathrm{V}}{\mathrm{I}}. \tag{2}$$

  Currently, the calibration coefficient C_l for the longitudinal magnetic field is set to 21,913, as determined through comparison with the HMI data (Liu et al. 2023; Xu et al. 2024a). In Section 2.3, the on-orbit polarization sensitivity is approximately 0.00039 and 0.0009 for the deep integration and routine modes, corresponding to a cadence of 18 and 2 minutes, respectively. The corresponding sensitivity of the longitudinal magnetic field is 8.5 G and 20 G with the current linear calibration coefficient. So far, only local longitudinal magnetograms for active regions have been released, rather than full-disk data.

Xu et al. (2024a) compared four active regions and a pore region taken from FMG, HMI, and the Solar Magnetism and Activity Telescope (SMAT) at HSOS. They found that the spatial distribution of the magnetic-field structure is basically consistent for the three instruments. As FMG works at a fixed wavelength and is not tunable, a magnetic saturation problem occurs, especially at sunspot umbrae (e.g. the Stokes V/I shown in Figure 1f). The corresponding magnetic field strength in the umbra observed by FMG is weaker than the one of HMI. Xu et al. (2024b) also developed a method to correct the magnetic saturation. After correction, the unsigned magnetic flux in sunspot umbrae measured by FMG aligns more closely with that of HMI.

As the Lyot filter of FMG is not tuned, the actual working wavelength of FMG fluctuates with orbital positions, resulting the variation in $\frac{V}{I}$ signals, even if the magnetic field strength remains constant. The main factors influencing this working wavelength include the varying Sun–FMG radial velocities across orbits, ranging from ±0.5 km/s to ±3.95 km/s annually, the solar rotation itself causing a ± 2 km/s Doppler shift at the east-west limb, and the Doppler shifts due to solar p-mode oscillations contributing about ± 0.5 km/s. When utilizing the magnetic-field data from FMG with a 48-minute cadence, the linear calibration coefficient remains consistent as the orbit velocity is the same for a given solar disk position. Using HMI data as a reference, the linear calibration coefficient varies from 19,000 to 23,000 for the spacecraft Sun-FMG orbital velocity of ±0.5 km/s. Accounting for p-mode oscillations, the variation in the linear calibration coefficient is similar. Considering the maximum orbital velocity range of ± 4 km/s, the linear calibration coefficient roughly spans from 17,000 to 30,000. Liu et al. (2023) provide an illustrative example of the magnetic flux evolution from FMG for the active region NOAA 13230. The periodic variation in magnetic flux attributed to the orbital velocity can be identified using a linear calibration coefficient and this periodic variation can be effectively mitigated by applying a varying calibration coefficient with the orbital velocity derived from the comparison with HMI data. Recently, we are exploring the possibility of non-linear magnetic field calibration employing machine learning techniques. Hu et al. (2022) trained a neural network with Stokes polarization images from HMI at a single wavelength, along with the magnetic field strength from Stokes inversions considering satellite orbital velocity and solar rotation. The preliminary results are encouraging. With the FMG data taken in 2023, we are training the model now. In the future, we are planning to update the non-linear magnetic field calibration to remove the influence of magnetic saturation and Doppler velocity from the above-mentioned effects.

As one of the LVCRs in the polarimeter of FMG does not seem to work as expected, efforts are still ongoing to derive the horizontal magnetic field component from routine observations. Figures 1d and 1e are derived in the satellite rotation mode by rotating the satellite from 0 to 360 degrees with a step of 45 degrees. This approach enables FMG to obtain full Stokes images. In other words, the full disk vector magnetogram can be only observed based on user application. Moreover, the Lyot filter on board the spacecraft has bubbles, which are also present in the Narrowband Filter Imager on board Hinode (Welsch et al. 2012), resulting a darker region at the upper left corner of the full-disk filtergram. The image quality is degraded in the darker region. Currently, this issue limits data releases to sunspot or local regions.

Despite the presence of bubbles in the full-disk filtergram and longitudinal magnetic field data (Level 1 and 1.5), the 24-hour uninterrupted observation makes it highly valuable for monitoring and forecasting solar activity. Currently, we are cooperating with colleagues from the Chinese space weather prediction department to utilize FMG data for solar activity prediction. Recently, we also generated the Carrington synoptic magnetograms as shown in Figure 10 using the full-disk longitudinal magnetograms covering a solar Carrington rotation. The Carrington synoptic magnetograms were released to the public since 2024.

4. Conclusion

In the article, we present the ground calibration and on-orbit performance of the FMG on board ASO-S. As China's first spaceborne solar magnetograph, FMG has the ability to capture full-disk vector magnetic fields. The spatial resolution of the filtergrams and longitudinal magnetograms are about 1.5 arcseconds. The FWHM of the Lyot filter band is 0.01 nm.

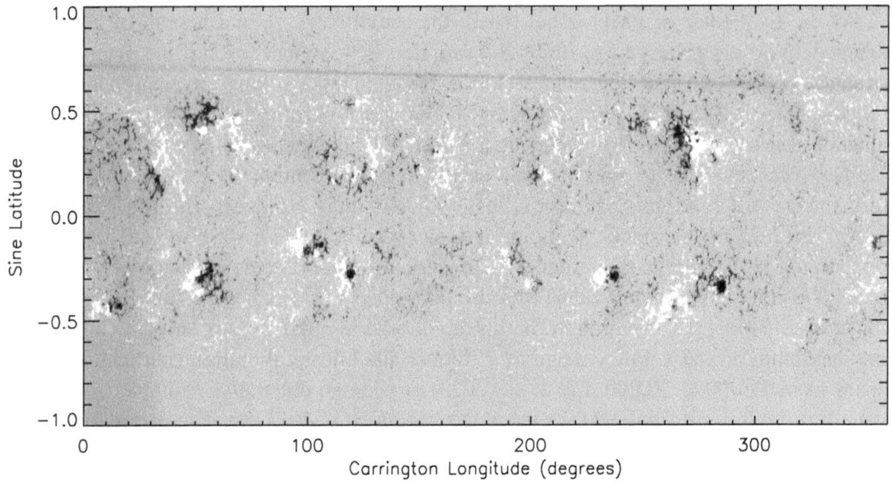

Figure 10 An example of LOS Carrington synoptic magnetogram from FMG for CR 2271.

Utilizing two calibration wheels, we have measured the wavelength drift across each pixel within the fields of view of FMG. During the ground calibration, the SNR achieved 5300 in the deep integration mode. In orbit, the polarization sensitivity of Stokes V/I maps from FMG is approximately 0.00039 in deep integration mode with a cadence of 18 min and 0.0009 in routine mode with a cadence of 2 min. The corresponding sensitivity of the longitudinal magnetic field is 20 G and 8.5 G for the routine and deep intergration mode with the current linear calibration coefficient of 21,913. Since 1 April 2023, FMG has been consistently releasing Level 2 products, specifically filtergrams and longitudinal magnetic field data for active regions. Line-of-sight Carrington synoptic magnetograms collected over a 27-day solar rotation can also be produced and are released since February 2024. In the future, we will strive to minimize the impact of the unexpected inflight peformance and to improve the data quality (such as correcting the effect of orbital velocity).

Acknowledgments FMG was developed by the National Astronomical Observatories of China of CAS, Nanjing Institute of Astronomical Optics and Technology of CAS, and Xi'an Institute of Optics and Precision Mechanics of CAS. We also thank the help and support during the ground development and on-orbit test of FMG from the following institutes: Purple Mountain Observatory of CAS, National Space Science Center of CAS, the Innovation Academy for Microsatellites of CAS, Xi'an Microelectronics Technology Institute of China Aerospace Science and Technology Corporation, Institute of Fluid Physics, China Academy of Engineering Physics, Changchun Institute of Optics, Fine Mechanics and Physics of CAS.

Author contributions The contributors for FMG's ground calibration include: Y. Deng, H. Zhang, J. Yang, F. Li, D. Wang, J. Lin, J. Hou, Y. Sun, W. Duan, Q. Song, Z. Zheng, H. Ni, Y. Zeng, Z. Wu, J. Wang, W. Ge, J. Lv, L. Shen, N. Wang, J. He, C. Wang. The contributors for FMG's in-flight performance and data processing pipeline are: J. Su, S. Liu, Y. Song, K. Ji, Y. Huang, X. Yang, Y. Bai, X. Wang, H. Xu, J. Chen, Z. Hu. The manuscript was organized by X. Bai and S. Liu. All authors reviewed the manuscript.

Funding The authors are supported by the Strategic Priority Research Program of the Chinese Academy of Sciences with the Grant No. XDB0560000, the National Science Foundation of China (NSFC) under Grants 12273059, 12073077, 12373115, 12173049, and the Youth Innovation Promotion Association of CAS (2023061).

Data Availability No datasets were generated or analyzed during the current study.

Declarations

Competing interests The authors declare no competing interests.

References

Ai, G.-X.: 1987, *Publ. Beijing Astron. Obs.* **9**, 27.

Babcock, H.W.: 1953, *Astrophys. J.* **118**, 387. DOI.

Bai, X.Y., Deng, Y.Y., Wang, Y.R., Liu, S.Q., Mao, X.J.: 2018, *Chin. Sci. Bull.* **63**, 301. DOI.

Bai, X., Feng, Z., Zhang, Z., Deng, Y., Zhu, X., Ji, K., et al.: 2023, *Res. Astron. Astrophys.* **23**, 125010. DOI.

Beck, C., Bellot Rubio, L.R., Kentischer, T.J., Tritschler, A., Del Toro Iniesta, J.C.: 2010, *Astron. Astrophys.* **520**, A115. DOI.

Cao, W., Gorceix, N., Coulter, R., Ahn, K., Rimmele, T.R., Goode, P.R.: 2010, *Astron. Nachr.* **331**, 636. DOI.

Chen, B., Li, H., Song, K.-F., Guo, Q.-F., Zhang, P.-J., He, L.-P., et al.: 2019, *Res. Astron. Astrophys.* **19**, 159. DOI.

Collados, M., Lagg, A., Díaz García, J.J., Hernández Suárez, E., López López, R., Páez Mañá, E., et al.: 2007, *The Physics of Chromospheric Plasmas* **368**, 611.

Couvidat, S., Schou, J., Shine, R.A., Bush, R.I., Miles, J.W., Scherrer, P.H., et al.: 2012, *Solar Phys.* **275**, 285. DOI.

Deng, Y.-Y., Zhang, H.-Y., Yang, J.-F., Li, F., Lin, J.-B., Hou, J.-F., et al.: 2019, *Res. Astron. Astrophys.* **19**, 157. DOI.

Deng, Y.-Y., Zhou, G.P., Dai, S.W., Wang, Y., Feng, F.X.S., et al.: 2023, *Chin. Sci. Bull.* **68**, 298. DOI.

Dominguez-Tagle, C., Collados, M., Lopez, R., Cedillo, J.J.V., Esteves, M.A., Grassin, O., et al.: 2022, *J. Astron. Instrum.* **11**, 2250014. DOI.

Elmore, D.F., Lites, B.W., Tomczyk, S., Skumanich, A.P., Dunn, R.B., Schuenke, J.A., et al.: 1992, *Polarization Analysis and Measurement* **1746**, 22. DOI.

Gan, W., Zhu, C., Deng, Y., Zhang, Z., Chen, B., Huang, Y., et al.: 2023, *Solar Phys.* **298**, 68. DOI.

Gosain, S., Venkatakrishnan, P., Venugopalan, K.: 2006, *J. Astrophys. Astron.* **27**, 285. DOI.

Hagyard, M.J., Cumings, N.P., West, E.A., Smith, J.E.: 1982, *Solar Phys.* **80**, 33. DOI.

Hale, G.E.: 1908, *Astrophys. J.* **28**, 315. DOI.

Hou, J.-F., Wang, H.-F., Wang, G., Luo, Y.-Q., Li, H.-W., Zhang, Z.-L., et al.: 2020, *Chin. Phys. B* **29**, 074208. DOI.

Hu, Z., Ji, K., Chen, J., Deng, Y., Su, J., Bai, X., et al.: 2022, *Astron. Astrophys.* **666**, A93. DOI.

Huang, Y., Li, H., Gan, W.-Q., Li, Y.-P., Su, J.-T., Deng, Y.-Y., et al.: 2019, *Res. Astron. Astrophys.* **19**, 164. DOI.

Jones, H.P., Duvall, T.L., Harvey, J.W., Mahaffey, C.T., Schwitters, J.D., Simmons, J.E.: 1992, *Solar Phys.* **139**, 211. DOI.

Li, H., Chen, B., Feng, L., Li, Y., Huang, Y., Li, J.-W., et al.: 2019, *Res. Astron. Astrophys.* **19**, 158. DOI.

Lites, B.W., Ichimoto, K.: 2013, *Solar Phys.* **283**, 601. DOI.

Lites, B.W., Akin, D.L., Card, G., Cruz, T., Duncan, D.W., Edwards, C.G., et al.: 2013, *Solar Phys.* **283**, 579. DOI.

Liu, S., Su, J.T., Bai, X.Y., Deng, Y.Y., Chen, J., Song, Y.L., et al.: 2023, *Solar Phys.* **298**, 141. DOI.

Livingston, W., Harvey, J., Slaughter, C.: 1971, *Publ. - R. Obs., Edinb.* **8**, 52.

Martinez Pillet, V., Cauzzi, G., Tritschler, A., Harra, L., Andretta, V., Vourlidas, A., et al.: 2023, *IAU Symp.* **372**, 3. DOI.

Mickey, D.L., Canfield, R.C., LaBonte, B.J., Leka, K.D., Waterson, M.F., Weber, H.M.: 1996, *Solar Phys.* **168**, 229. DOI.

Potts, H.E., Diver, D.A.: 2008, *Astron. Astrophys.* **492**, 863. DOI.

Qu, Z.Q.: 2011, *Solar Polarization 6* **437**, 423.

Raja Bayanna, A., Mathew, S.K., Venkatakrishnan, P., Srivastava, N.: 2014, *Solar Phys.* **289**, 4007. DOI.

Scherrer, P.H., Bogart, R.S., Bush, R.I., Hoeksema, J.T., Kosovichev, A.G., Schou, J., et al.: 1995, *Solar Phys.* **162**, 129. DOI.

Scherrer, P.H., Schou, J., Bush, R.I., Kosovichev, A.G., Bogart, R.S., Hoeksema, J.T., et al.: 2012, *Solar Phys.* **275**, 207. DOI.

Schou, J., Scherrer, P.H., Bush, R.I., Wachter, R., Couvidat, S., Rabello-Soares, M.C., et al.: 2012, *Solar Phys.* **275**, 229. DOI.

Solanki, S.K., del Toro Iniesta, J.C., Woch, J., Gandorfer, A., Hirzberger, J., Alvarez-Herrero, A., et al.: 2020, *Astron. Astrophys.* **642**, A11. DOI.

Song, Q., Bai, X., Chen, B., Hu, X., Chen, Y., Hou, Z., et al.: 2022, *Res. Astron. Astrophys.* **22**, 105001. DOI.

Su, J.-T., Bai, X.-Y., Chen, J., Guo, J.-J., Liu, S., Wang, X.-F., et al.: 2019, *Res. Astron. Astrophys.* **19**, 161. DOI.
van Noort, M.J., Rouppe van der Voort, L.H.M.: 2008, *Astron. Astrophys.* **489**, 429. DOI.
Wachter, R., Schou, J.: 2009, *Solar Phys.* **258**, 331. DOI.
Welsch, B.T., Kusano, K., Yamamoto, T.T., Muglach, K.: 2012, *Astrophys. J.* **747**, 130. DOI.
Xu, H., Su, J., Bai, X., Chen, J., Gao, Y., Liu, S., et al.: 2019, *Solar Phys.* **294**, 129. DOI.
Xu, H., Su, J., Liu, S., Deng, Y., Bai, X., Chen, J., et al.: 2024b, *Solar Phys.* **299**, 82. DOI.
Xu, H., Su, J., Liu, S., Deng, Y., Yang, S., Bai, X., et al.: 2024a, *Solar Phys.* **299**, 17. DOI.

Publisher's Note Springer Nature remains neutral with regard to jurisdictional claims in published maps and institutional affiliations.

Authors and Affiliations

Xianyong Bai[1,2,3] · Yuanyong Deng[1,2,3] · Haiying Zhang[4,2] · Jianfeng Yang[5] · Fu Li[5] · Jiangtao Su[1,2,3] · Suo Liu[1,3] · Yongliang Song[1,3] · Kaifan Ji[6] · Yu Huang[7] · Xiao Yang[1,3] · Dongguang Wang[1,3] · Jiaben Lin[1,3] · Junfeng Hou[1,2,3] · Yingzi Sun[1,3] · Wei Duan[1,2] · Qian Song[1] · Yang Bai[1,3] · Xiaofan Wang[1,3] · Haiqing Xu[1,3] · Jie Chen[1,3] · Ziyao Hu[1,2,3] · Zhaoying Zheng[4] · Houkun Ni[4] · Yizhong Zeng[4] · Zhen Wu[4] · Jianing Wang[4] · Wei Ge[5] · Juan Lv[5] · Lun Shen[5] · Nange Wang[5] · Jiawei He[5] · Chenjie Wang[5]

✉ X. Bai
xybai@bao.ac.cn

✉ S. Liu
lius@nao.cas.cn

1 National Astronomical Observatories, Chinese Academy of Sciences, Beijing, 100101, China

2 School of Astronomy and Space Science, University of Chinese Academy of Sciences, Beijing 101408, China

3 Key Laboratory of Solar Activity and Space Weather, National Space Science Center, Chinese Academy of Sciences, Beijing 100190, China

4 Nanjing Institute of Astronomical Optics Technology, Chinese Academy of Sciences, Nanjing 210042, China

5 Xi'an Institute of Optics and Precision Mechanics, Chinese Academy of Sciences, Xi'an 710119, China

6 Yunnan Observatories, Chinese Academy of Sciences, Kunming 650011, China

7 Purple Mountain Observatory, Chinese Academy of Sciences, Nanjing 210013, China

Solar Physics (2024) 299:153
https://doi.org/10.1007/s11207-024-02392-x

RESEARCH

The Tests and Calibrations of the Hard X-ray Imager Aboard ASO-S

Yang Su[1,2] · Zhe Zhang[1,2] · Wei Chen[1,2] · Dengyi Chen[1] · Fu Yu[1] · Yiming Hu[1,2] · Yan Zhang[1] · Fanxiaoyu Xia[1] · Changxue Chen[1,2] · Zhentong Li[1] · Xiankai Jiang[1,2] · Yu Huang[1,2] · Yongqiang Zhang[1] · Wei Liu[1] · Tao Ma[1,2] · Dong Li[1] · Wenhui Yu[1,2] · Youping Li[1] · Mingsheng Cai[1,2] · Jianhua Guo[1,2] · Yongyi Huang[1] · Haoxiang Wang[1,2] · Yaoming Liang[3] · Miao Ma[3] · Jianping Wang[4] · Shanshan Zhu[4] · Jinyou Tao[5] · Jirui Yu[5] · Jianfeng Yang[5] · Jian Wu[1] · Weiqun Gan[1]

Received: 15 March 2024 / Accepted: 9 October 2024 / Published online: 31 October 2024

Abstract
The Hard X-ray Imager (HXI) aboard the Advanced Space-based Solar Observatory (ASO-S) is an instrument designed to observe hard X-ray (HXR) spectra and images of solar flares. Having 91 subcollimators to modulate incident X-rays, HXI can obtain 91 modulation data and 45 visibilities to reconstruct images with a spatial resolution as high as ~ 3.1 arcsec. HXI was launched on 9 October 2022 and powered up on 17 October 2022. After the on-orbit testing phase lasting for three months, HXI was ready for regular observations on 18 January 2023. With fine-tuning of the detectors and electronics, we were able to expand the energy range from $\sim 30-200$ keV to ~ 10 keV – 300 keV, which significantly raised the scientific values of the data and the number of detected flare events. This paper presents the changes and improvements of HXI instrument since 2019, the important ground tests, on-orbit tests, and calibration works. We also present the light curves, spectra, and reconstructed images of one flare observed by HXI on 6 January 2023.

Keywords Spectrum, X-ray · X-ray bursts, spectrum · Flares, spectrum · Flares, energetic particles · Instrumental effects

1. Introduction

The Hard X-ray Imager (HXI) (Zhang et al. 2019b; Su et al. 2019) is one of the three payloads aboard the ASO-S mission (Gan et al. 2019, 2023) launched on 9 October 2022. Its scientific objectives are to investigate how energy is released in solar flares, how electrons are accelerated by magnetic reconnection, and the relationship between flares and eruptions, including jets, failed eruptions, and coronal mass ejections (CMEs).

HXI observes the fluxes, spectra, and images of hard X-ray (HXR) emissions of solar flares in the energy range from ~ 10 to 300 keV. It also provides imaging spectroscopy for the analysis of individual HXR sources. HXI provides unique HXR observations of solar flares from the viewpoint of Earth during the upcoming solar maximum of the 25th solar

Extended author information available on the last page of the article

Table 1 The characteristics of HXI: design (Zhang et al. 2019b; Su et al. 2019) and measurements.

	Design	Measurements
Energy range	30 – 200 keV	~ 15 – 300 keV (imaging) ~ 10 – 300 keV (spectra)
Energy resolution (FWHM)	≤ 27% @ 32 keV	~ 16.6% @ 32 keV
Spatial resolution	≤ 3.5 arcsec	~ 3.1 arcsec
Field of view	≥ 40 arcmins	> 40.3 arcmins
Temporal resolution	≤ 0.5 s	regular mode: 4 s burst mode: configurable from 0.125 to 2 s
Pointing accuracy (SAS)	< 2 arcsecs	< 0.2 arcsecs
Mass		153.95 kg
Power		111.75 W
Size		collimator: 1190 mm in length the front grids: 36 mm in diameter the rear grids: 22 mm in diameter $LaBr_3$ scintillator: 1 inch in diameter grid pitch: 36 – 1224 μm

cycle. More importantly, by combining HXI data and the data observed from a different viewpoint by Solar Orbiter/STIX (Krucker et al. 2020), we can obtain, for the first time, stereoscopic X-ray images of solar flares, a milestone in research of flare magnetic reconnection and process of particle acceleration.

The design and simulation works for HXI were published in 2019 (Zhang et al. 2019b; Su et al. 2019). Since then, a number of modifications and improvements have been made to HXI, and a series of successful ground tests have been carried out. The simulated data, data production software, and early beta versions of analysis software (HXI GUI) were also developed before the launch of ASO-S. During the commissioning phase, we carried out a series of on-orbit performance tests and adjustments. On 18 January 2023, HXI entered the conventional observation mode.

In this paper, we mainly describe the changes and improvements of HXI instrument, the important ground tests, the on-orbit tests, and the calibration works, aiming to provide a useful summary for all HXI-related tests and calibrations. The ground tests mainly include beam test of HXI grids, deformation tests of HXI collimator, environmental tests, detector calibration, and SAS calibration. The on-oribt tests mainly include HXI performance tests, deformation measurements of HXI collimator, environmental tests, SAS calibration and optimization, detector calibration and optimization, imaging calibration, and cross-calibration. For the tests published in separate papers, we briefly introduce the related work and present the corresponding reference.

2. HXI Design

Aiming to image the hard X-ray emission generated from solar flare activities, HXI is designed as a Fourier-transform-type X-ray telescope as its wildly successful predecessors, Yohkoh/HXT (Kosugi et al. 1991), RHESSI (Lin et al. 2002), and Solar Orbiter/STIX (Krucker et al. 2020). Its characteristics have been summarized in Table 1.

As described in detail by Zhang et al. (2019b) and Su et al. (2019), HXI consists of three main parts: The collimator (HXI-C) is used for modulating incident X-rays and locating the

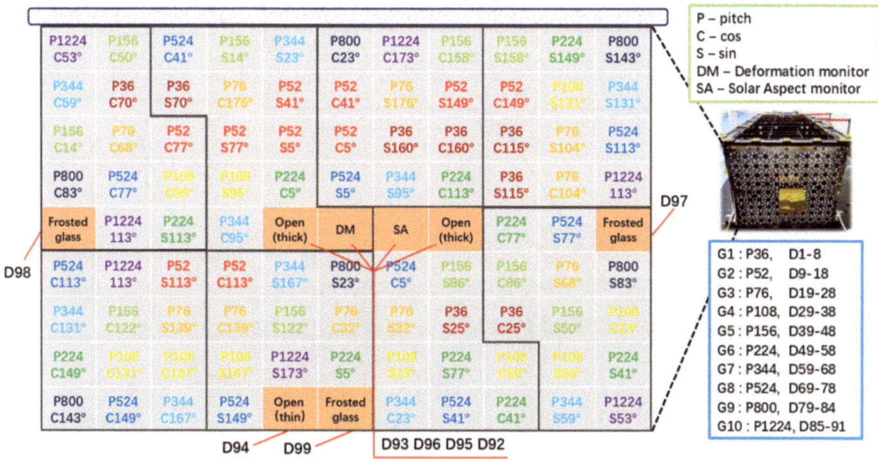

Figure 1 The layouts of all HXI detectors and grids (toward the Sun). Here G stands for grid Group, and P stands for Pitch in μm. The positioning angles are displayed in the location of the corresponding grid mount. D92, D93, D94 are total flux monitors, and D95–D99 are background monitors.

solar disk center (Chen et al. 2021). 91 pairs of tungsten bigrids, namely subcollimators, with 10 different pitches from 36 μm to 1224 μm, and 10 different orientations between $0-180$ degrees, are mounted in an 11×9 array on the HXI-C frame (Figure 1). A solar aspect system (SAS) and three open units occupy the remaining eight positions in the array. The SAS includes two cameras, the solar aspect (SA) unit for pointing information and the deformation monitor (DM) unit for deformation information. The lens of the SA unit is mounted on the front plate of the frame, whereas the CMOS detector shares the PCB and mounting structure with the DMs on the rear plate. In addition to the lens system integrated with the detector, the DM also includes three sets of frosted glass and an entrance window located on the front plate. The open units without grids are used to determine the total X-ray flux and spectra. Three positions on the rear plate corresponding to the frosted glasses are closed by 2-mm-thick tungsten to stop the X-ray signal from the Sun. These units are used to detect the detector background in orbit.

The spectrometer (HXI-S) is used for the number and energy of X-ray photon detection. 99 detector modules, which comprise a $LaBr_3$ scintillator coupled with a photomultiplier (PMT), are arranged as an 11×9 array corresponding to that of HXI-C. A ^{133}Ba and ^{241}Am mixed source is mounted in each module to provide on-orbit energy calibration signals for the detector. The acquisition signals from detector modules are digitized and organized by eight front–end electronics units in HXI-S (Zhang et al. 2019a). It is worth mentioning that 91 detectors record the modulated flux; five monitor the background, and three monitor the total flux, one of which has thin front window and therefore is more sensitive at low energies. The detailed layouts of all detectors and grids are shown in Figure 1.

The electronics control box (HXI-E) is used for power supply and data management. The housekeeping and science data from HXI-C and HXI-S are organized and transferred to the spacecraft. Also, it receives remote commands from spacecraft and controls the observation process of HXI in orbit.

It should be mentioned that the HXI-C and HXI-S are mounted upside down on the optical reference plate outside the spacecraft, whereas the HXI-E is mounted inside the spacecraft. As a conventional space thermal control method, the HXI-C and HXI-S are covered

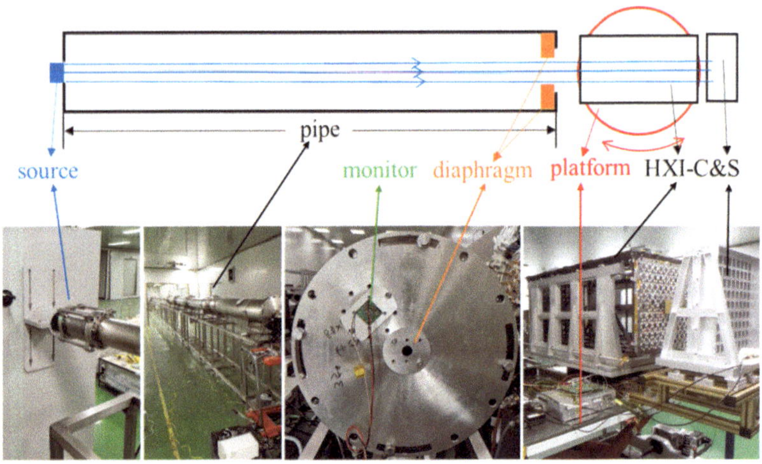

Figure 2 Schematic (top) and photos (bottom) of the beam facility (Jiang et al. 2023b).

with several layers of multilayer insulation material. Moreover, an active thermal control system is employed to keep the temperature differences between the front and rear grids to be less than 1 °C.

3. Beam Test

The observation performance of X-ray telescopes like HXI cannot be verified through direct observation of celestial objects on the ground due to the strong absorption of X-rays by the atmosphere. Fortunately, the X-ray modulation relation of each subcollimator is the basis of modulation imaging. Therefore we can test the modulation relation of HXI-C to confirm the imaging performance of this instrument. The X-ray beam test is first employed by our team to directly verify and calibrate the modulation relation of the full size of HXI-C. The details of this test have been summarized by Jiang et al. (2023b), so only a brief introduction is given here.

Based on a analytical mathematical model (Jiang et al. 2023a) describing the X-ray modulation process in the beam facility and the collimator system, an X-ray beam facility has been built and improved during the last several years (Chen et al. 2023). The facility includes an X-ray tube with a stable high-voltage power supply up to 225 keV, a 26-meter beam pipe with adjustable exit diaphragms, a high-precision experiment platform that can translate and rotate, and a set of auxiliary detectors and electronics systems (see Figure 2). The facility characteristics have been listed in Table 2.

The modulation relation curves of all 91 grid pairs of HXI-C, even including the bigrid with pitch as small as 36 μm, have been finally obtained by the beam test lasting more than one month in 2021 on the facility. The results, as partly shown in Figure 3, have strongly proved the reliability of the modulation imaging system and also could be used as imaging parameters for image reconstruction of observation on orbit.

Table 2 The characteristic of the facility in beam test.

Characteristic	Parameter
High voltage of the source	10 – 225 keV
Diameter of the source	0.4 mm
Beam pipe length	26 m
Diameter of diaphragms	3.6 – 15 mm
Translation accuracy	< 0.02 mm
Rotation accuracy	∼ 0.0002°
Stability of the platform	< 0.01 mm/12 h

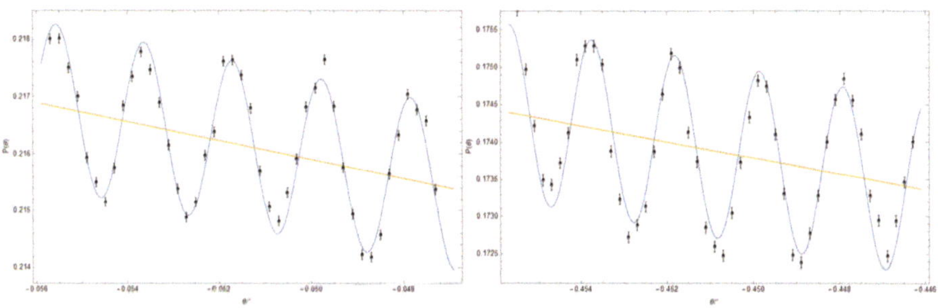

Figure 3 Two measured modulation relation curves of bigrids with a pitch of 36 μm (Jiang et al. 2023b).

4. Deformation Tests Before and After Launch

It is obvious that the relative deformation between the front and rear grids of HXI-C should be kept well to acquire a better X-ray image. The relative displacement of the front and rear panels would cause shift between the front and rear grids, resulting in changes in grid phase, which, if uncorrected, affect source locations and image quality. The relative rotation should be kept small (mainly depending on the finest grid pitch) for the imaging principle to work properly. Therefore the relative displacement and rotation were monitored from assembly to on-orbit operation and must be maintained under 36 μm and 10 arcsecs (estimated from the design parameters of HXI-C during the early development stage), respectively.

The SAS was developed to measure the relative deformation on the ground and orbit, whereas the absolute value could be measured by a coordinate measuring machine soon after the integration and the tests of HXI-C (details of the SAS system, the inverse calculation and regularization process, and the deformation test can be found in Zhang et al. 2019b; Yu et al. 2024a and Yu et al. 2024b). After HXI turned into a thermal balanced status on orbit, deformations including relative displacement and distortion were acquired by the SAS test. Figure 4 presents the comprehensive measuring results. It shows a quite stable state for the mechanical and thermal design of HXI. It will be beneficial for imaging calibration and scientific findings.

5. Environmental Tests and on-Orbit Performance

Like most other space-borne instruments, HXI has taken a series of environmental tests before its launch. The primary purpose of these tests conducted on flight models is to pre-

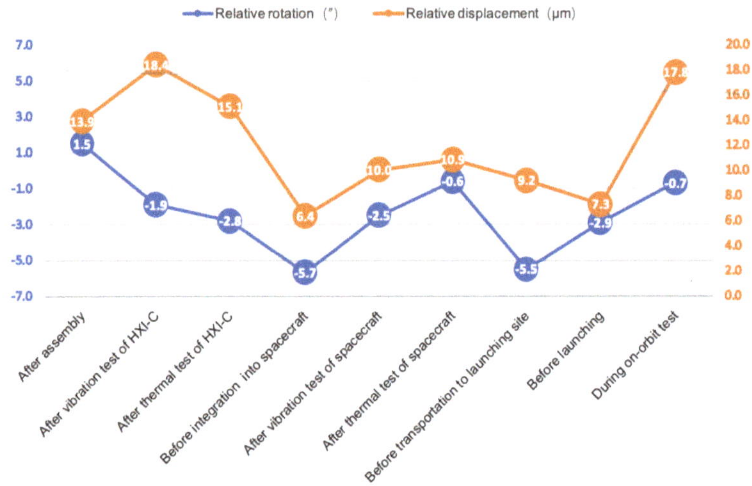

Figure 4 The summary of the HXI-C's deformation monitoring results during ground tests, integration, transportation, and on-orbit test. These values (except the last one) are the maximum displacements measured on ground by the CMM machine. The uncertainties of these values were accounted for the precision of the 3D CMM machine, which is 0.005 mm. For the twist angle measurement, the error is less than $0.2''$according to the calculation. Theses measurements suggest that the relative twist and shift between the front and rear panels are well kept within the tolerances, $10''$and 36 μm, respectively.

cipitate incipient failures from latent defects and to verify conformance to specification requirements. Typical tests include mechanical vibration and thermal tests (thermal cycle and thermal vacuum). Dynamic vibration usually occurs during transportation and launch; as a consequence, we do not discuss them in this paper since HXI was successfully sent into space. The following will mainly focus on the thermal ones including tests on the ground, performance on orbit, and relative deformation conducted by temperature differences.

Deformation of the equipment was mainly influenced by vibration during launch and temperature differences in orbit. Thus the stability of the mechanical and thermal environment is of great significance to such a high-precision astronomy payload. According to the specifications of the satellite, the HXI has gone through all environmental tests with limited deformation. On ground test, we also considered the realistic thermal environment and set the corresponding simulation conditions outside thermal flux. For example, the front grids would be directly facing the Sun, whereas the rear ones as well as the HXI-S detectors were not. More details may be found in Chen et al. (2020, 2021).

Figure 5 presents the photos of HXI with the integrated ASO-S satellite during the thermal balance and vibration tests from 8 to 16 June 2022. The test results (Figure 6) show that the temperature differences are less than 0.7 °C between the front and rear grids during the thermal tests for HXI-C. Temperature uniformity is less than 3 °C for the 99 detector modules on HXI-S. These results suggest that HXI may be surrounded in a better thermal environment compared to premium design.

Before and after these tests, the SAS was powered on to monitor the relative deformation of HXI-C. The HXI-S was continuously working during the thermal test as well. Variation of the energy resolution of detector modules was used to check whether there were any property deductions of the instrument. The instrument was surely operated nominally in the thermal vacuum test. The temperatures reported here were obtained after the instrument

Figure 5 The flight model of HXI in (a) the vibration test of the spacecraft and (b) the thermal balance test of the spacecraft.

reached thermal balance. According to the thermal vacuum test and the on-orbit temperature results, it takes about 12 hours to be stable.

HXI was sent to space on 9 October and first turned on a week later. After a period of on-orbit optimization and debugging, HXI operated regularly since January 2023. During the past year, the temperature distribution of HXI-C and HXI-S is reported as Figure 7. From these graphs, temperature uniformities for grids of HXI-C and detector modules of HXI-S are better than 0.7 °C and 2 °C, respectively. Thermal performances are much better than on-ground tests, which demonstrates the design is effective.

Note that the temperatures of two front monitors (Figure 7(a)) rise slowly after September 2023 and get stable after January 2024. We believe that the temperature variation is affected by the heat leak around the lens of SAS, which was anticipated before launch. That is the reason we arranged coarse grids around the center of the front panel (Figure 1) in case the heat may build up slowly and affect the surrounding grids. It may be caused by two reasons. One is the possible decreasing reflection capability of the lens, and the other one is related to the varying Sun–Earth distance. The good news is that long-time temperature monitoring shows that the temperature has reached a stable level and even decreased slightly after March 2024.

6. Detector Calibration

6.1. Ground Calibration of Detectors

The performance of detector units of HXI-S also is important for image reconstruction on orbit. Therefore several kinds of calibrations and tests have been performed on FM of HXI before launch. Liu et al. (2022) have introduced and summarized the results of these tests, such as detection efficiency, energy resolution, and energy to ADC channel (Energy-Channel) relationship. Detection efficiency characterization was calibrated on a Monoenergetic X-ray beam facility, the hard X-ray Calibration Facility (HXCF) (Guo, Wu, and Hou 2021) at the National Institute of Metrology (NIM) of China. HXCF employs the Bragg diffraction principle to generate a single-energy X-ray beam from the continuum produced by the X-ray

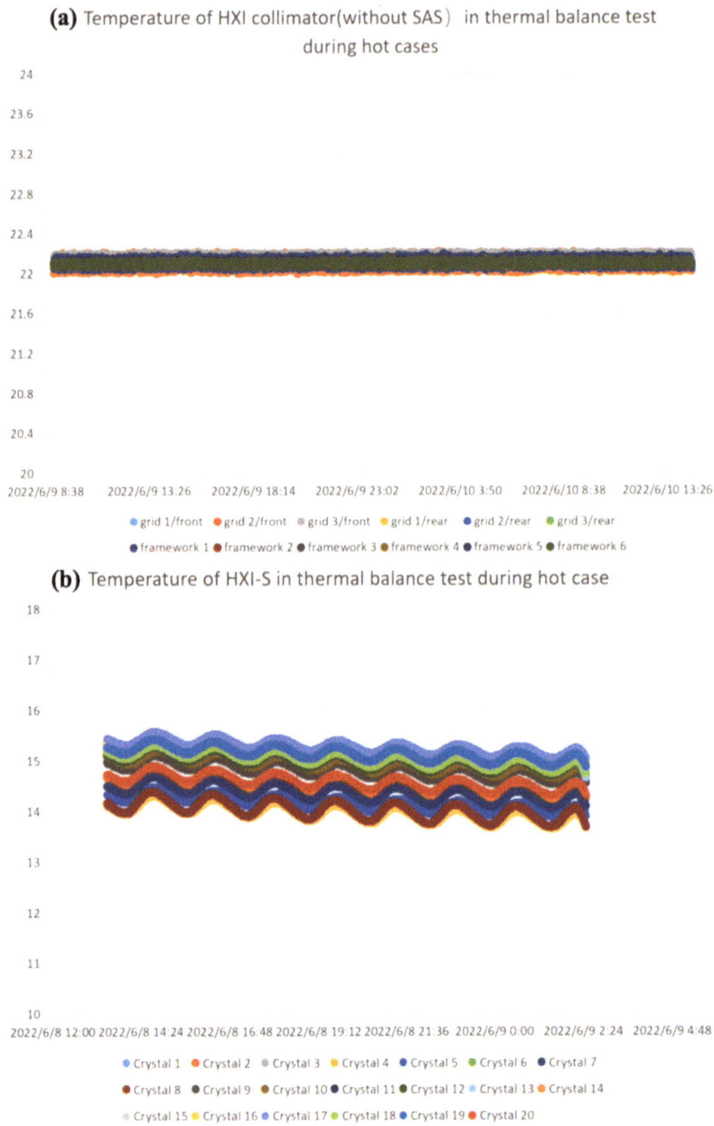

Figure 6 Temperature during the thermal test: (a) the grids of HXI-C, (b) the detector modules of HXI-S (Chen et al. 2020, 2021).

tube, with better than 0.1% monochrome performance. The detection efficiency of all 100 detector units, including a backup unit, was calibrated between 30 keV and 80 keV in several energy steps, which is set at 2 keV between 30 and 38 keV, 0.1 keV between 38 and 39 keV, and 5 keV between 40 and 80 keV. Meanwhile, Monte Carlo simulations were also carried out for the detection efficiency of HXI-S under real conditions. It has been concluded by Liu et al. (2022) that the experimental results are consistent with the simulations and that the difference in detection efficiency between detector modules is well controlled within ±3%. Radioactive sources ^{133}Ba and ^{137}Cs are chosen to calibrate the energy resolution and

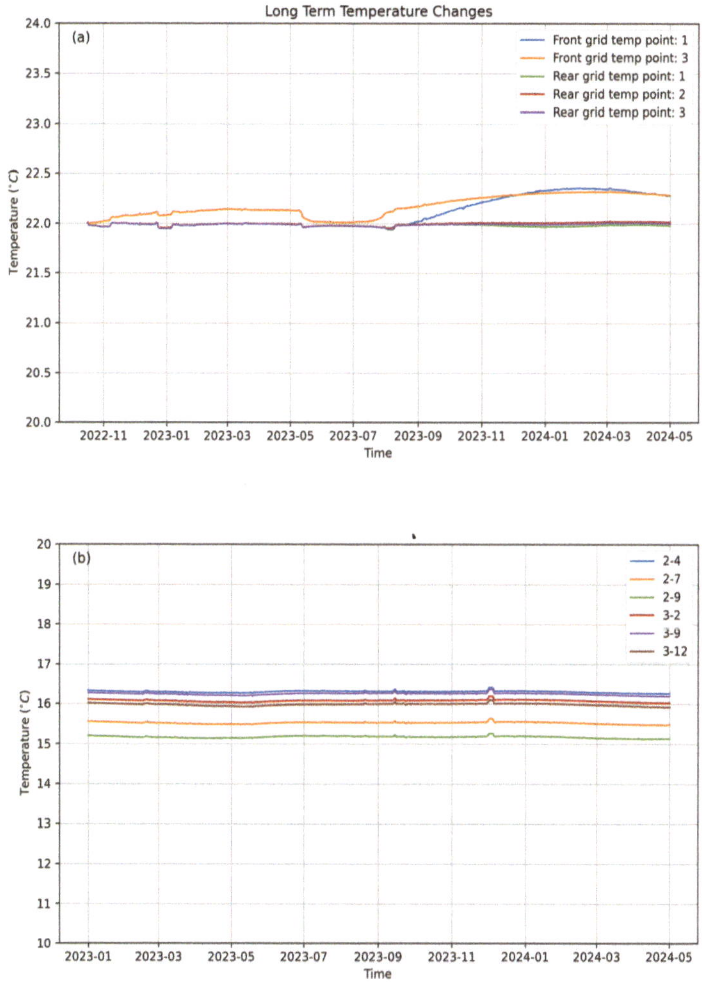

Figure 7 On-orbit thermal control performance of HXI during 2023. (a) HXI-C: five monitoring points on the front and rear plate, respectively. (b) HXI-S: six monitoring points on the detector modules.

E-C relationship. The calibration results indicate that the energy resolutions of all detector units of FM are better than 26% at 32 keV, whereas the difference between 99 units is less than 8%. The obtained E-C relationship with voltage variation demonstrates that the detector gains decreasing can be well aided on orbit by the high voltage adjustment.

6.2. Differential Nonlinearity (DNL) Effect Correction

Periodic bumps or dips in certain bins of the spectrum are a common distortion of spectrometer data. These bumps or dips increase errors in the Gaussian fit, which in turn affect the accuracy of energy calibration and resolution measurements. The primary cause of this problem is the DNL error of the ADC chips in the front–end electronics, which appears as a nonuniform width of ADC channels. Its effect on energy calibration can be significantly

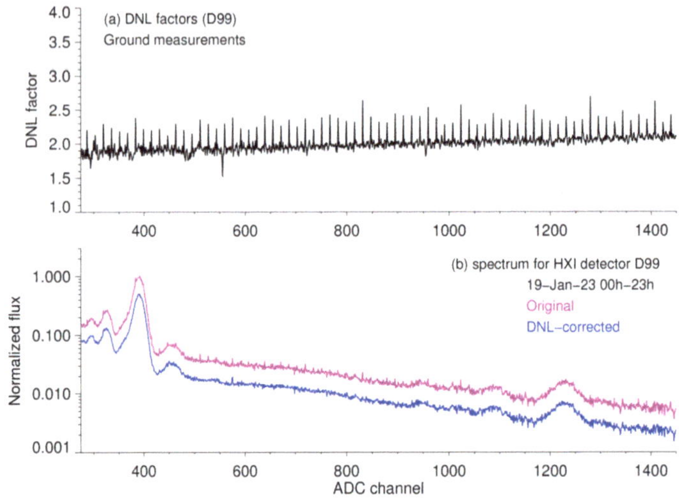

Figure 8 (a) The DNL factors for detector D99 measured from the ground tests. They represent the width of each ADC channel. (b) The calibration spectrum of D99 taken in the time range from 19 January 2023 00h to 23h. The purple curve is the original spectrum, and the blue one is the DNL-corrected.

mitigated or eliminated by DNL correction data measured on ground, i.e., the bin sizes of ADC channels.

During the ground test stage of the HXI-S flight model, the DNL correction data for all the 99 detectors was obtained by independently scanning in each channel using a dedicated charge injection circuit (Figure 8a). Preliminary results in the ground test stage show that the bumps or dips have been eliminated and the nonlinearity of the energy spectrum in the whole energy range could be optimized from 0.8% to 0.3% after DNL correction (Wan et al. 2022). An example is shown in Figure 8b.

DNL correction data was also utilized to produce Level 1.0 science data, resulting in a smoother energy spectrum and more accurate energy calibration compared to Level 0.5 science data.

6.3. Energy Calibration

The energy calibration is important work to calibrate the corresponding energy (keV) for each of the ADC channels of each detector. Mixed calibration sources of ^{133}Ba and ^{241}Am are housed inside every detector module (Zhang et al. 2019b), and their characteristic lines (26.34, 59.54 keV from ^{241}Am; 81, 356.02 keV from ^{133}Ba), marked in Figure 9, are used for the calibration.

It was done by the following steps. First, we apply the DNL correction to get the ADC spectrum with 2048 channels, which are background spectra accumulated automatically during data production excluding flare time, particle events, and SAA passages. Second, we fit the spectrum in two ADC ranges to determine the channel locations of the 26.34, 59.54, and 356.02 keV lines of the build-in calibration sources (^{241}Am and ^{133}Ba, Zhang et al. 2019b). The fitting model contains a continuum background model plus a number of Gaussian functions (an example is given in Figure 9b). Then we get two slopes for the linear relationship between energy and ADC channels below and above a break energy (assumed at 59.5 keV), respectively. One is determined from $(59.54 - 26.34)/(Chan_{59} - Chan_{26})$, and the other one from $(356 - 59.54)/(Chan_{356} - Chan_{59})$. After this step, we obtain the E-C relation function and the corresponding energy of each ADC channel. This calibration task is done automatically during the production process of Lev1.0 data.

The reason we use two slopes is because we found that a single linear E-C function determined from 59.5 and 356 keV lines cannot produce accurate spectrum at low energies for some detectors, indicating that nonlinearity effect still exists in some of the detector systems after DNL correction. Two slopes can solve this problem. However, it should be pointed out that the actual break energy is in fact unknown. Here we use a fixed value at 59.5 keV as the break energy, which is probably the best choice. The 81 keV line is relatively weak and wide, resulting in larger uncertainty in locations of the line center and therefore in resulted slopes. Besides, the relatively small difference between the two slopes minimized the effect of a fixed break energy at 59.5 keV.

In the spectra observed during the early stage of HXI, the 26.34 keV line is not clearly visible for many detectors due to the high starting energy. Instead, the 511 keV line in the background is well observed and used for energy calibration (Figure 9a), which is the electron–positron annihilation line caused by particles hitting the instrument, a signal commonly seen in background spectra of high-energy detectors in space. This line was out of the effective ADC channel range after the commissioning phase (see Section 6.4) and therefore no longer used for calibration.

Since the energy calibration was done with a number of calibration lines fitted at the same time with Gaussians plus a background model, the fitting results best describe the location of line centers and the Gaussian width σ. Therefore we used the fitting results of the strongest line at 59.5 keV to get the spectral resolutions for the dates of 19 January 2023 and 19 January 2024. First, we determine the Gaussian width σ of the 59.5 keV line through calibration fitting and obtain the energy resolution $\eta_{59\mathrm{keV}} = \Delta E_{59}/E_{59\mathrm{keV}} = 2.35\sigma/E_{59\mathrm{keV}} \approx 12.2\%$, which is defined as the ratio of the full width at half maximum (FWHM) to the energy at line center (Knoll 2000). Subsequently, this value is further scaled to the 32 keV energy level (the 32 keV line from the internal background from ^{138}La was used for on-ground resolution test) for all 99 detectors, using the following specified formula:

$$\eta_{32\mathrm{keV}} = \eta_{59\mathrm{keV}} \times \sqrt{\frac{E_{59\mathrm{keV}}}{E_{32\mathrm{keV}}}}.$$

The results in Figure 10 show an average resolution of 16.6% and 16.3% @32 keV for 19 January 2023 and 19 January 2024, respectively, which suggests that the detectors are stable in terms of spectral resolution.

6.4. Optimization of Detector Observational Capability

The objective energy range of HXI observation 30 – 200 keV was proposed during the early stage of the ASO-S mission. Therefore the design of HXI, including the detectors, bigrids, electronic readout system, etc., was carried out and optimized accordingly. However, the missing range at low energies is important for flare detection and for science, especially for spectral analysis where low-energy spectra are needed for accurate determination of thermal and nonthermal electron parameters and imaging of flares where images in 10 – 30 keV complete the picture of high-energy processes. Therefore the HXI team made great efforts to improve HXI's observation capacity by optimizing the observation mode and conditions of the detector system during the on-orbit testing phase and managed to expand the energy range to $\sim 10 - 300$ keV.

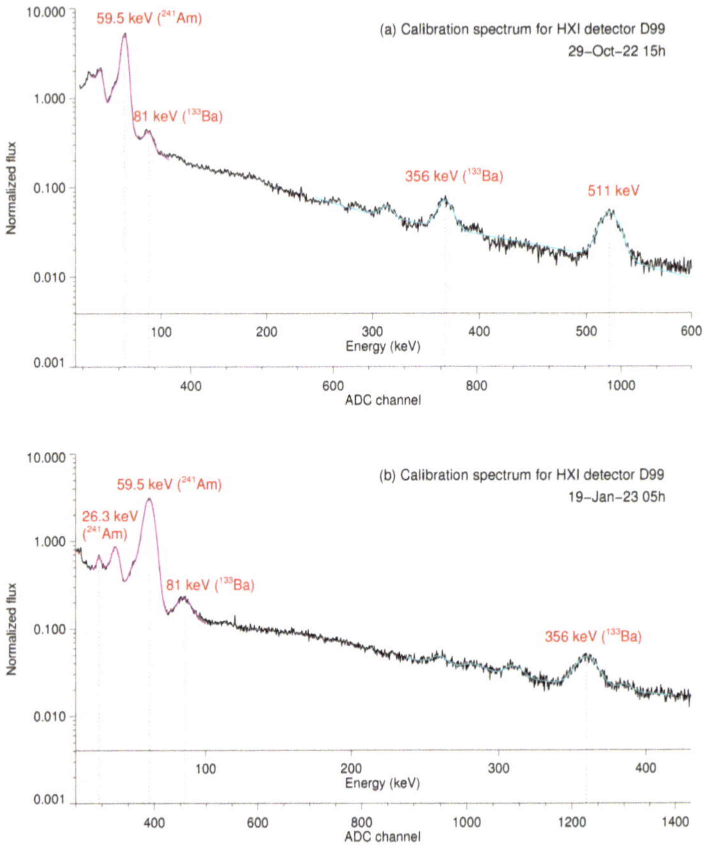

Figure 9 (a) The ADC spectrum and the calibration result for detector D99 on 29 October 2022. The three calibration lines at 59.5, 81, and 356 keV of the built-in calibration sources are marked in red. The electron–positron annihilation line in the background at 511 keV is also identified. The spectral fitting used to determine the corresponding ADC channels of these lines was done in two separate ADC ranges, each with a fitting model consisting of a smooth decreasing background model and three Gaussians. The fitting results are shown in purple and cyan curves. (b) Same as (a) but for the spectrum obtained on 19 January 2023. Here the 511 keV line is not visible. The 26.34 keV line is used for the correction of the ADC bin width at low energies.

The fine adjustments made to each of the 99 detector systems include threshold, high voltage, amplification factor, gain factor, etc. Figure 11 compares the calibrated energy ranges of all the detectors for 21 October 2022, shortly after launch, and 19 January 2023, after the series of adjustments to the detector system. It is clear that the observational energy range extended below 15 keV, which greatly raised the scientific values of the data. In addition, these adjustments are also beneficial for extending the lifetime of the detectors.

6.5. Detector Response Matrix

Detector response matrix describes how an HXR detector records count flux from incident photon flux. It is the key to convert count spectra and images back to photon spectra and images of the X-ray source. The first version of HXI detector response matrix was obtained from Geant 4 physical simulations (Su et al. 2019) based on the HXI mass model (from

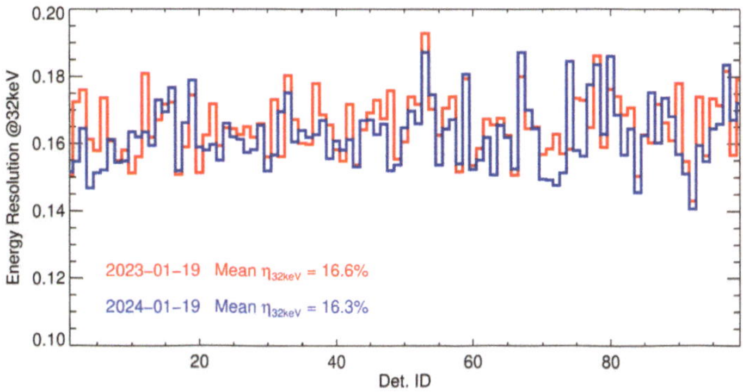

Figure 10 Energy resolution for each of the 99 detectors obtained on 19 January 2023 (red) and 19 January 2024 (blue), respectively. The unit of the Y-axis is percentage @ 32 keV.

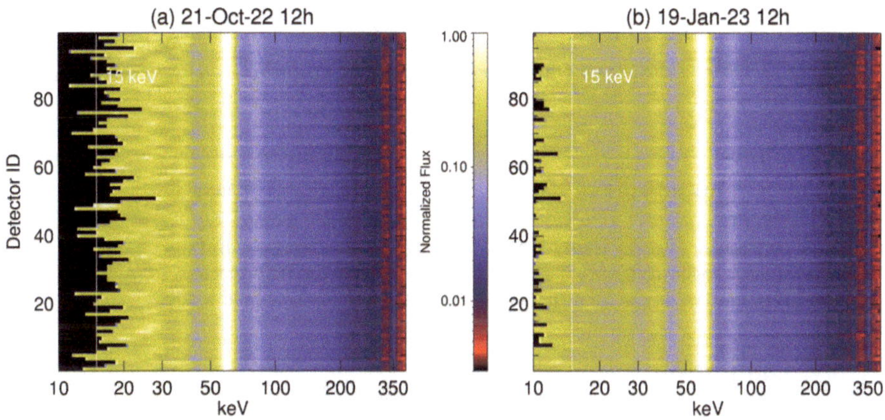

Figure 11 (a) The effective energy ranges for all HXI detectors on 21 October 2022 before the fine adjustments. Each row here is the calibrated spectrum for each detector, independently normalized by the maximum count flux. (b) The effective energy ranges for all HXI detectors on 19 January 2023 after the series of adjustments made to the detector system. Note that these ranges may vary slowly with time or change due to energetic particle events.

the early design) plus an aluminum shell to simulate the effect of the platform and other instruments. Since then, a number of improvements have been included in the simulation, such as the aluminium front window of the detectors and a more accurate mass model of HXI and the whole satellite. The latest version of the response matrix (V10, to be released with HXI GUI V1.50 beta, Figure 12) also takes into account the effect of multilayers in front of the collimator and detectors. The version V11 will be released after cross calibration between HXI and other HXR instruments (see Li et al. 2024).

Figure 13(b). These two types of extracted key information are transmitted to the ground and solved by two aspect algorithms (see Yu et al. 2020), i.e., the four-quadrant method (FQM) and least square method (LSM), respectively. As we will see in Section 7.2, both algorithms produce comparable results and are operated simultaneously as a form of confirmation and redundancy. FQM performs by solving a system of overdetermined equations with four areas as constraints, whereas LSM uses the least square method to fit the solar disk based on the edge points extracted from 14 rows of pixels. Both algorithms can derive the center (i.e., aspect solution) and radius parameters (x, y, r) of the solar disk.

Figure 13(c) shows a description of Figure 13(a) in the HPC system, which is constructed based on the aspect solution and rotation angle (provided by the platform), whereas Figure 13(d) also displays an observation from SDO/HMI continuum for comparison. Although the SAS is not specifically designed for scientific observation, and the CMOS pixel array is only 2048 × 2048, the obtained solar images have good quality, which is conducive to exporting high-quality information on orbit for aspect solutions.

7.2. Testing Considerations

The testing of SAS includes simulation and on-orbit calibration and optimization. Before the launch of ASO-S, two algorithms (LSM and FQM) had been developed and tested based on simulated data generated from the visible-light images observed by the Atmospheric Imaging Assembly (AIA, Lemen et al. 2012) onboard the Solar Dynamics Observatory (SDO, Pesnell, Thompson, and Chamberlin 2012), taking into account instability factors as much as possible. Their accuracy could reach $0.25''$ and $0.14''$, respectively (see Yu et al. (2020) for details).

For the on-orbit testing (Yu 2023), the reference values (x_{i0}, y_{i0}, r_{i0}) are obtained by fitting all the edge points available of the full solar disk in each tested image recorded in calibration mode. Meanwhile, the measurement values (x_i, y_i, r_i) for each tested image are derived respectively with FQM based on the four corner areas $(S1, S2, S3, S4)$ and with LSM based on 14 rows of data. The error of the solution is evaluated by the difference between the measured and reference values.

Initially, there are 130 calibration images for testing, which come from observations of different exposure times, Sun–Earth distances, and pointing states. In addition to the stability of the satellite platform toward the Sun and the variation of the Sun–Earth distance, which modulate the various offsets and sizes of the solar image on CMOS, the exposure time for imaging and the threshold for image binarization also affect the size and integrity of the solar image. These factors are important aspects that affect the algorithm.

Different threshold settings for image binarization, in a sense, are also equivalent to considering more observation states of the CMOS exposure time and the Sun–Earth distance effect. To increase the stability and credibility of the evaluation, the dataset was expanded to more than 50,000 tested images by considering different thresholds for binarization within a certain range for each calibration image.

The parameters of the tested images are displayed in Figure 14(a), including the distance from the solar disk center to the CMOS center, CMOS exposure time, and threshold width (i.e., the number of binarization thresholds for each calibration image).

In addition, a scheme has been considered for the selection of binarization thresholds and was used in the present testing. It is based on the sharp decrease feature in the brightness of the solar limb. First, we sort all pixels of a calibration image as an ascending sequence (labeled SQ1) by intensity from low to high. The steep changes in the SQ1 indicate a sharp decrease in brightness at the edge of the solar disk, and the intensity values at these positions

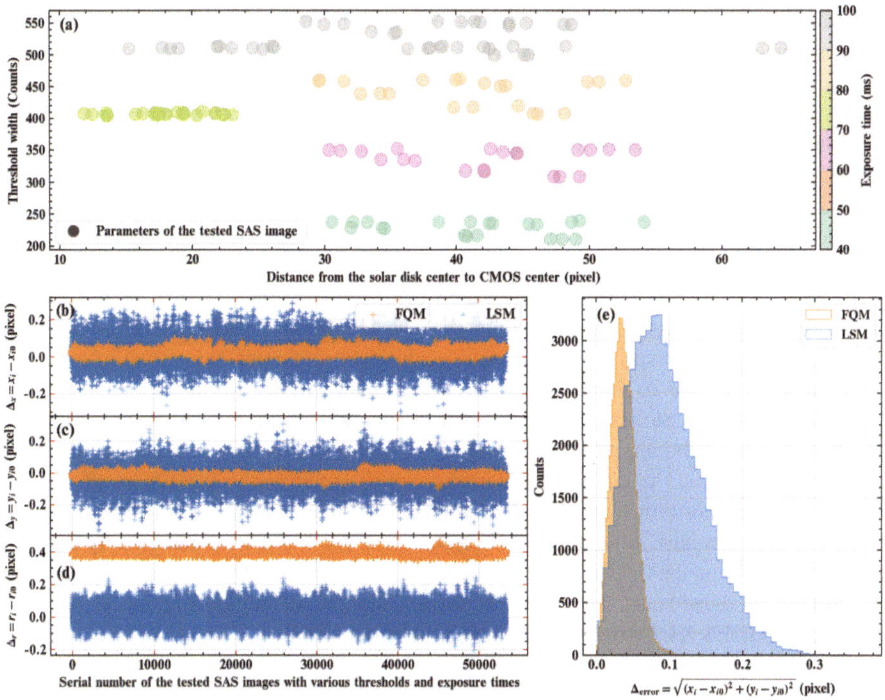

Figure 14 (a) Image parameters for testing the SAS algorithm. (b)–(d) The errors of x, y, and r in the aspect solutions, and different colors mark the FQM and LSM algorithms. (e) The statistical distribution of the composite error of x and y. (b)–(e) is an update from Yu (2023) with new calibration images considered.

are suggested as thresholds for binarization. Therefore the derivative of the SQ1 (labeled SQ2) is calculated to find these sharp edges. Gaussian fitting is performed on the SQ2 to determine the $\pm 1\sigma$ positions, from which we get the best intensity range of the threshold.

The test results on threshold and exposure time helped in guiding the selection of threshold and exposure time in routine observations and in improving the pointing accuracy of the SAS system.

7.3. Reliability and Long-Term Status of Aspect Solution

According to the testing details mentioned in Section 7.2, the residuals between the measured and reference values of the aspect solutions (x, y) and radius r for all test samples are displayed respectively in Figures 14(b)–14(d), whereas the statistical distribution of the total residuals which combined in the x- and y-directions are also shown in Figure 14(e). Obviously, the pointing measurement errors of FQM and LSM are mainly dominated in the low-value range of $< 0.1''$ and $< 0.2''$, respectively, which also meets the expectations of simulation testing based on SDO/AIA data on the ground.

The accuracy of the aspect solution depends on whether CMOS can collect appropriate data, and the key lies in the integrity, size, and position of the binarized solar disk. Specifically, it is reflected in factors such as the stability of the satellite platform, exposure time, image threshold, Sun–Earth distance, as well as solar eclipses and Earth shadows. Based on the observations of more than one year, under the normal observation mode, the offset of the satellite platform can well meet the requirements of maintaining a reasonable relative position between the solar disk and CMOS. In Figure 15(a) the area trend of four corners on the CMOS during the time period from 1 December 2022 to 1 February 2024 is shown with over 130 million sampling points, partially excluding data with a significant offset of the platform due to the on-orbit testing.

Figure 15(b) shows the aspect solutions (x, y) obtained using FQM based on the areas $(S1, S2, S3, S4)$ from Figure 15(a) as inputs. Meanwhile, the tendency of radius r is also displayed which is smoothly modulated by the Sun–Earth distance, indirectly reflecting the stability and rationality of the aspect solutions derived by FQM. Note, the satellite may have experienced the Earth shadow regularly between May and August 2023, so the solution during the corresponding time period may be affected. Figures 15(c)–15(d) correspond to Figures 15(a)–15(b), respectively, but focus on the case within one day.

In addition, Figure 16 shows the statistical distribution of the aspect solutions from Figure 15(b), including the individual and joint distributions in the x- and y-directions. According to the distribution on the xy-plane, the aspect solutions are concentrated in a small region near the center of the CMOS detector, which is within the maximum offset range allowed by the algorithm. Therefore the satellite platform is relatively stable, and the aspect solutions is also robust. Note that the effectiveness of the FQM is constrained by the areas $(S1, S2, S3, S4)$ on the CMOS. In the case of only two effective areas available, the machine learning method can be used to derive the aspect solution, which has been successfully applied to the GOES X1.1 flare on 20 June 2023 (Li et al. 2023).

8. Imaging Calibration

Imaging calibration includes two major tasks, the solar aspect solution and calibration of subcollimators.

8.1. Solar Aspect Solution

Since the solar limb is invisible in HXRs, solar aspect solution is needed for locating flare HXR sources on the solar disk and determining their orientations. This task is highly dependent on the design of the HXR imager itself and the solar aspect system (SAS).

For HXI, there are three major coordinate systems (Figure 17), the Helioprojecive (HPC) coordinates system, the HXI coordinate system (HXI imaging FOV), and the SA coordinate system. The HXI imaging FOV and the SA FOV are related by the observed solar disk (although invisible in HXI X-ray FOV), whose center location changes with the pointing of the ASO-S platform. Since the three-axis stabilized spacecraft points at the Sun (the x-axis in the platform coordinate) with platform xy-plane kept in the ecliptic plane (Gan et al. 2019), HXI x-axis (Figure 17) is parallel to HPC x-axis (with an uncertainty less than 36 arcsec).

HXI imaging process is done in the following steps. First, imaging parameters (such as image center, pixel size, image size, etc.) are received from HXI analysis software GUI. Second, some parameters and data, such as image center in HPC $[xc, yc]$, pointing data in the SAS coordinates, etc., are converted to the coordinates in HXI FOV (of the collimator). Third, images are reconstructed in HXI FOV based on imaging parameters and instrument parameters, which are stored in the HXI software and database. Two types of imaging methods can be used here: one is based on patterns, which are the grid responses at each pixel in the corresponding HXI FOV (Su et al. 2019), and the other one is based on visibilities.

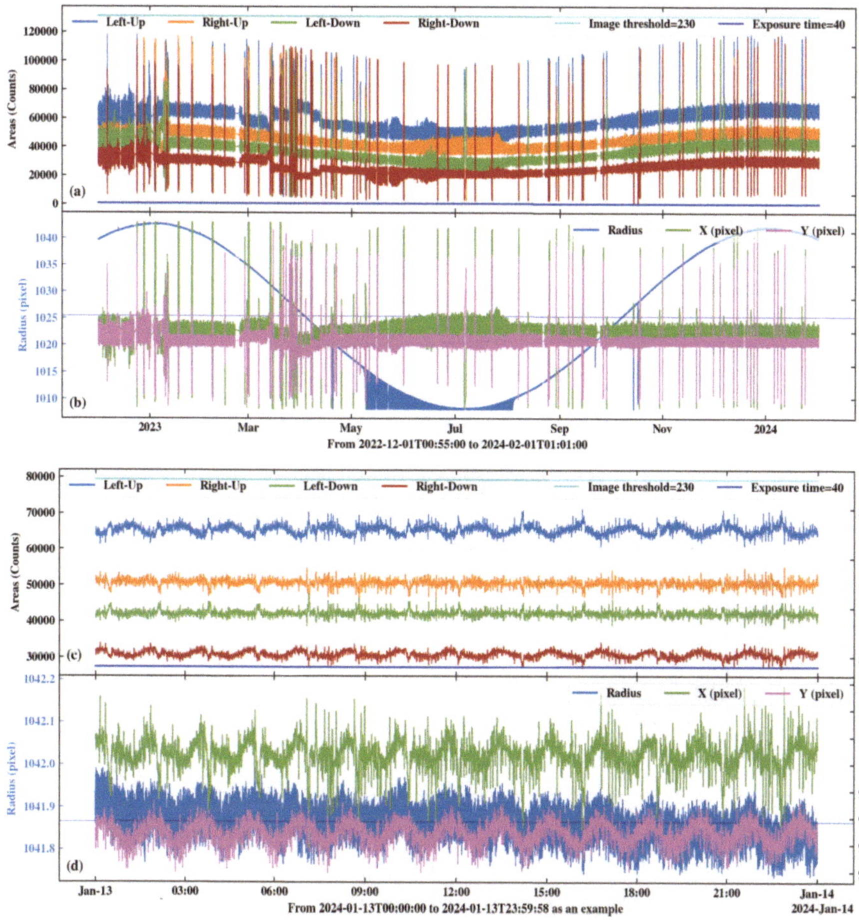

Figure 15 (a) The evolution of exposure time, binarization threshold, and the area of the four corners for the HXI/SAS images with more than 130 million samples. (b) The aspect solutions obtained using the FQM algorithm based on the areas in panel (a). The significant changes in area and radius (i.e., the prominent spikes) are mainly due to the calibrations of other payloads when the satellite pointed away from the disk center, and the satellite night time in Earth's shadow from May to August 2023. (c)–(d) Similar to (a)–(b), but as the zoom in on a short time interval in (a)–(b). This figure is an update from Yu (2023).

Fourth, images are converted and rotated to the HPC coordinates. Here a solar aspect solution is needed in the second and fourth steps.

To get a solar aspect solution, we need at least a few well observed flares that have high HXR flux for better statistics, clean background for confident background subtraction, compact source(s) for clear flare locations, and clear EUV/UV sources for coalignment. Here we take the flare occurred on 9 May 2023 as an example. The HXI image reconstructed by HXI reveals double footpoint sources, consistent with those observed in SDO/AIA EUV and UV channels. Using their source centroids (SC) $[SC_{HPCx}, SC_{HPCy}]$ and $[SC_{HXIx}, SC_{HXIy}]$ in two different coordinate systems, we obtained the coordinates of the solar disk center (SDC) in HXI FOV, $[SDC_{HXIx}, SDC_{HXIy}]$, which varies with time. Combing it with the coordinates of solar disk center in SA CMOS detector $[SDC_{SAx}, SDC_{SAy}]$ (relative to the CMOS center), we further obtained the coordinate transformation between the SA center (SAC) and

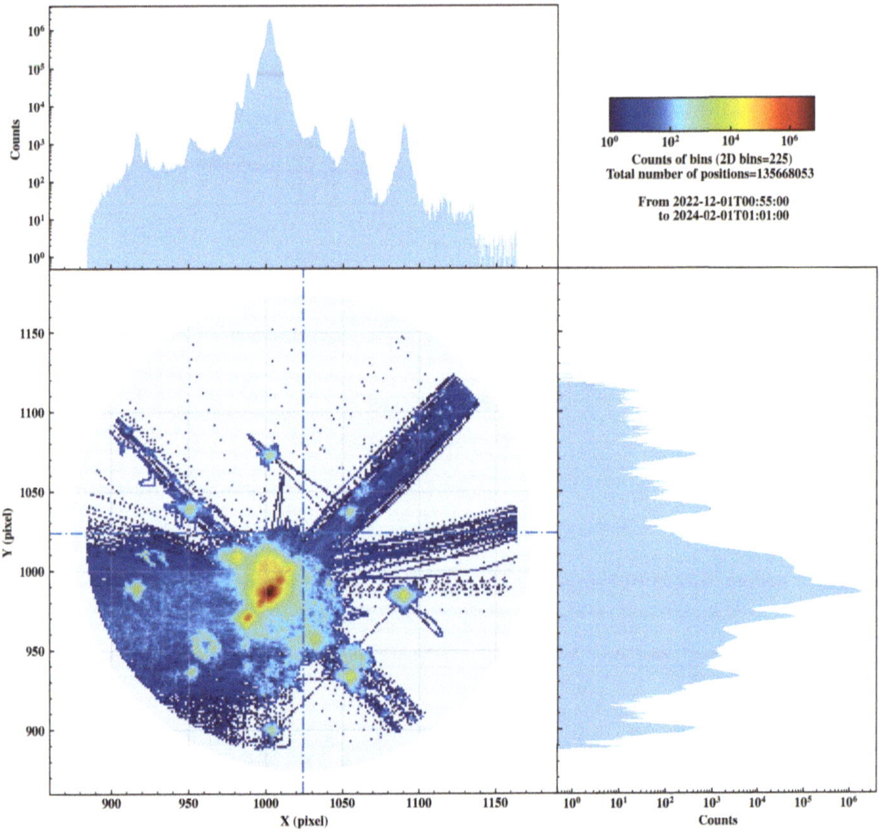

Figure 16 The statistical distribution of the aspect solutions in observations from Figure 15. The 2D histogram in the xy-plane is a joint distribution for x and y. The shaded blue circle has a radius of about 150 pixels, which is the reliable region for aspect solutions in simulation testing. The 1D histograms are for x and y, respectively. This figure is an update from Yu (2023).

HXI collimator [SAC_{HXIx}, SAC_{HXIy}] [15.2, −3.1] arcsec based on the double sources in the flare on 9 May 2023, which should be stable due to the stable thermal environment. We expect a change smaller than 2 arcsec. However, more flares will be used to determine more accurate values and whether they change with orbit and time.

In the coordinate transform, two angles are considered. One is the solar rotational obliquity angle to the ecliptic in Earth view α. The other one is the orientation angle of the SA CMOS detector relative to HXI X β. We used the "auto_align_images" provided in the Solar SoftWare (SSW) to align the white light image taken by the SA camera in calibration mode and SDO/HMI image and found β of 44.54 degree (Figure 18), only slightly different from the design, 45 degree.

8.2. Calibration of Subcollimators

Accurate image reconstruction requires a good knowledge of the bigrid subcollimators and their on-orbit status. These include (effective) grid slit width, tilt angles (the angle between grid plane and the panel plane), grid phases (or relative shifts between the front and rear

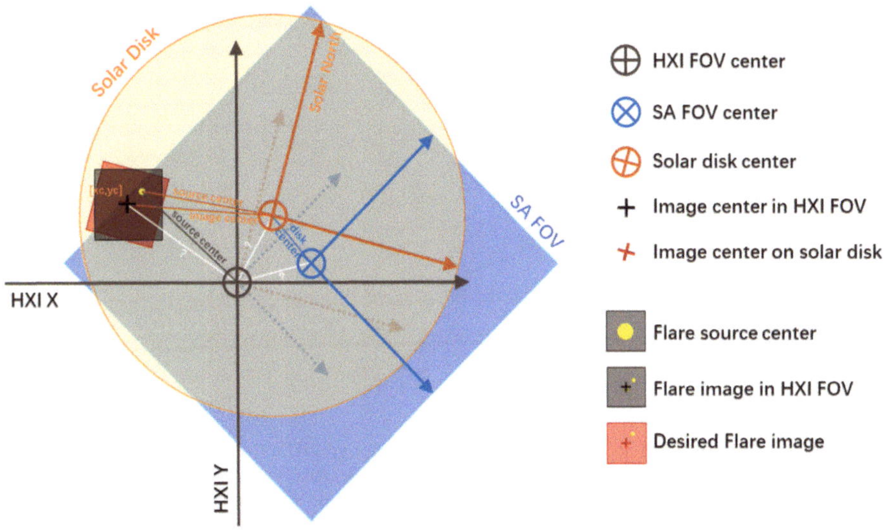

Figure 17 Diagram of HXI solar aspect solution. There are three major coordinate systems, the HXI FOV, SA FOV, and HPC. The image center $[xc, yc]$ is in HPC coordinates.

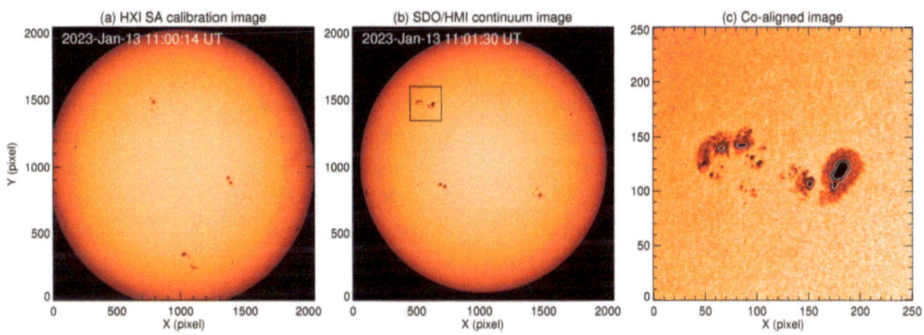

Figure 18 (a) The white light image taken on 13 January 2023 by the SA camera in calibration mode; (b) the SDO/HMI continuum image taken on 13 January 2023. (c) Coaligned SA image with the HMI image.

grids), positioning angles (the angle between grid slits and HXI X-axis), the alignment of grid layers, relative twist of the front and rear grids, etc. Although most of the parameters can be measured on ground, some must be measured or calibrated on-orbit again due to possible changes after launch.

The bigrid calibration method based on phase corrections with point sources or compact sources has been described by Su et al. (2019). Using this method, we have obtained preliminary grid phases of grid groups G3–G10 based on the flare observations at 13h on 1 May 2023, 10h on 3 May 2023, and 12 h on 23 May 2023 and released these parameters in the HXI GUI (database).

More accurate calibration requires flare observations that satisfy the following conditions: (1) the flux (count rate) must be high enough for good statistics; (2) the background should be low and easy to remove for all the imaging detectors; (3) the flare source must be point source or compact; (4) double sources are fine as long as both of them are compact;

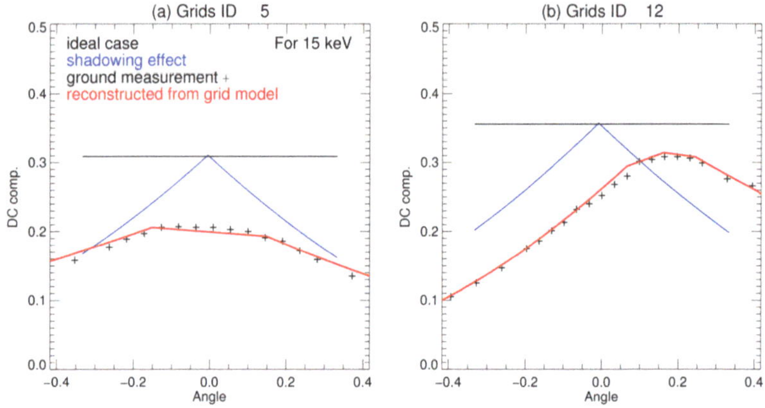

Figure 19 (a) The calculated and measured DC distribution for grid ID5; (b) same for grid ID12. The DC component at 15 keV is displayed as a function of angles in degrees for four cases: ideal case (the grid has no thickness and it has infinitely high density, black curve), case with the grid shadowing effect taken into account (grid has finite thickness and density, blue curve), ground measurement from beam test (black crosses), and the reconstructed curve from preliminary grid models (red).

(5) we need as many compact flare cases as possible for the calibration; (6) all the above conditions should be met for low energies in 15 – 30 keV, as this energy range is most sensitive to the nonideal factors in the imaging system. Photons with higher energies can easily penetrate any sticking out single grid layer.

In practice, the imaging calibration is much more complicated process as more nonideal factors need to be taken into account. However, their effects are mixed together, making it difficult to calibrate. Here we show the preliminary result of two subcollimators in Figure 19. By building the grid models we reconstructed the DC components (red curves) that match well with the ground measurements (crosses). More details and results of grid calibration will be presented in a separate paper (Jiang et al., in preparation).

9. Flare Images and Spectra

Here we took the first GOES X-class flare observed by HXI on 6 January 2023 as an example. The GOES and HXI light curves of the three total flux detectors (D92, D93, D94) are shown in Figure 20. Note that the decay phase is affected by the particle events in the radiation belt.

We fit the spectrum taken in the time range from 6 January 2023 00:55:40.899 to 00:55:50.899 UT to get the best fit parameters. The fitting model consists of a thermal component (f_vth) and a nonthermal component (f_thick2_vnorm), plus albedo correction. The fitting results reveal a temperature of 19.6 MK for the thermal component, energetic electron distribution power-law index δ_e of 4.3, and low energy cutoff of 21.8 keV (Figure 21). According to the fitting, most of the emissions below 15 keV are thermal in origin. The HXI images are reconstructed for the same time interval of the spectrum by HXI_CLEAN with subcollimator groups G2–G10, resulting in a spatial resolution of ~ 4.5 arcsec. At low energies, the thermal sources are found mostly in the corona (14 – 18 keV image, green contours in Figure 22). The nonthermal sources above 40 keV are located in the flare ribbons, matching well with AIA observations (cyan and purple contours in Figure 22), but we do

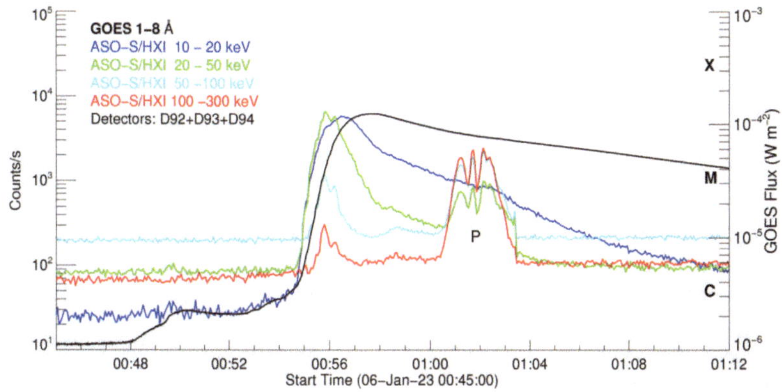

Figure 20 HXI and GOES light curves for the X1.2 flare on 6 January 2023. The total flux monitors D92, D93, and D94 are used to produce the light curves. Note that the multiple high-energy peaks in the decay phase (marked by "P") are caused by the particles during the transit through the Van Allen radiation belts.

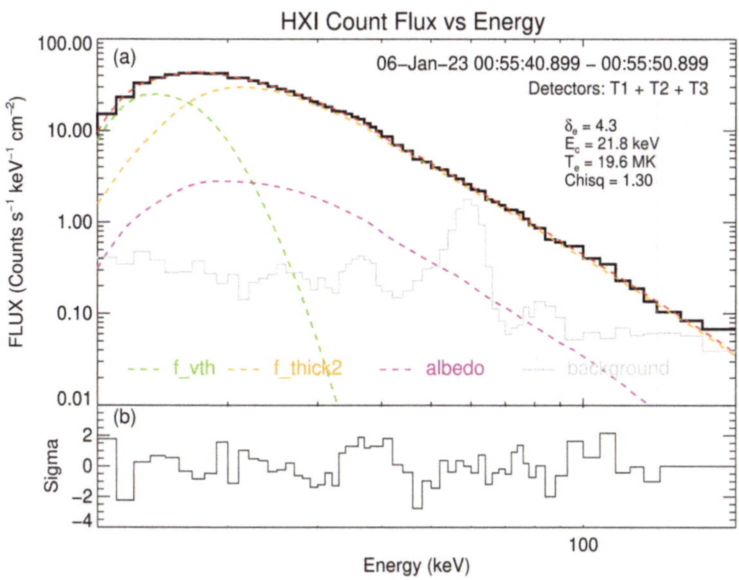

Figure 21 (a) Spectral fitting results for the HXI observation taken in the time range from 00:55:40.899 to 00:55:50.899 UT on 6 January 2023. The spectrum (black) was made from the three total flux detectors. The fitting model includes a thermal component (green) and a thick target 2 (orange). (b) Normalized residuals.

not see a clear corresponding source in this energy range for the southern ribbon visible in the AIA 1700 Å image.

10. Summary

After on-orbit testing phase, HXI's observational capability has been pushed to its limits, but the instrument is kept in even better conditions. Its performance is significantly improved

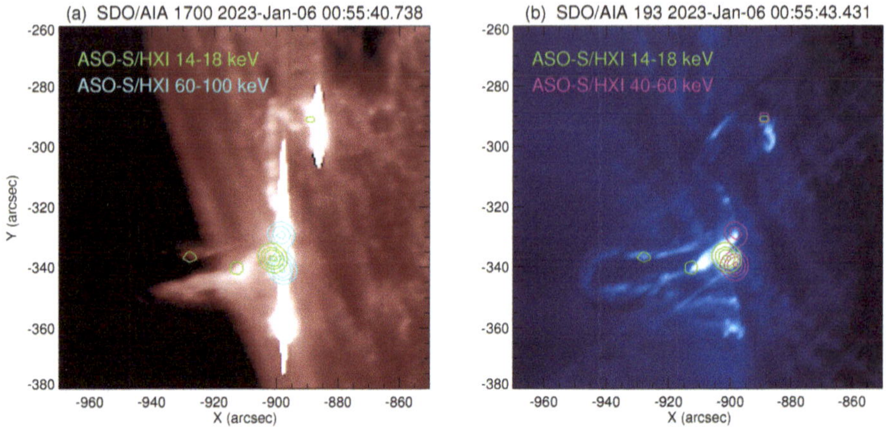

Figure 22 Reconstructed HXI images overplotted on SDO/AIA images for the flare on 6 January 2023. (a) HXI images in 14 – 18 keV and 60 – 100 keV overplotted on an AIA 1700 image. (b) HXI images in 14 – 18 keV and 40 – 60 keV overplotted on an AIA 193 image.

with most indicators better than the design indexes. The energy calibration used four calibration lines, which should provide accurate energies of the ADC channels. The imaging calibration take into account both ground measurements and on-orbit observations, as well as detailed calculations of the effects of nonideal factors.

However, there are still more tasks to be done in the near future. So far we do not find enough flare observations suitable for imaging calibrations. More compact events or double compact sources are required for more accurate calibrations. With more events in the future, we expect a better understanding of the state of the bigrid system and therefore better image reconstruction. Another important ongoing work is the cross calibration between HXI and other X-ray instruments (Li et al. 2024) like STIX and Fermi. It is particularly important as HXI and STIX provide the best opportunity for studies of X-ray directivity and three-dimensional property of HXR sources. The calibration result will help us further review the response matrix and improve both spectroscopy and imaging.

Acknowledgements The work is supported by the National Key R&D Program of China 2022YFF0503002, the Strategic Priority Research Program of the Chinese Academy of Sciences (Grant No. XDB0560000), and the NSFC 12333010, 11820101002, 12233012. ASO-S mission is supported by the Strategic Priority Research Program on Space Science, the Chinese Academy of Sciences, Grant No. XDA15320000. SDO/AIA data are obtained from the JSOC Data Archive.

Author contributions Z.Z. and J.W. led the design, development, and tests of HXI. Y.S. designed the imaging system, developed the imaging software and carried out energy calibration and imaging calibration. W.C. developed most of the code for HXI spectral study and contributed in energy calibration and design of the imager. Y.S., W.C., F.Y., Y.H., Y.L. developed HXI data processing and production software, studied on-orbit testing data, designed some of the on-orbit tests. F.Y. developed two algorithms for the SAS. D.C., Y.H., Y.Z., Y.Z., and J.G. developed the HXI-C and HXI-S and carried out the ground test of them. T.M., W.L., X.J., Y.H., and H.W. organized and carried out the detector calibration on the ground. M.C. developed the high-voltage power supply of HXI. Y.L. and M.M. developed most parts of HXI-E and carried out the ground test of it. J.W. and S.Z. developed the active thermal control system of HXI and carried out the engineering implementation. J.T., J.Y., and J.Y. built the HXI-C and developed the SAS. They take charge of the test of HXI-C and its deformation monitoring. F.X., F.Y., C.C. developed the main frame of HXI analysis software, and Z.L., Y.S., W.C., W.Y. helped improve it. W.G. is the ASO-S PI and also contributed during the design phase and on-orbit testing phase of HXI. Y.S., Z.Z., D.Y., F.Y., X.J., Y.Z., W.C., W.G. wrote the main manuscript.

Data Availability No datasets were generated or analysed during the current study.

Declarations

Competing interests The authors declare no competing interests.

References

Chen, D., Hu, Y., Zhang, Z., Zhang, Y., Zhang, Y., Guo, J., Huang, Y., Ma, T., Chang, J.: 2020, Environmental tests of the HXI spectrometer for the ASO-S mission. *J. Instrum.* **15**, T10008. DOI. ADS.

Chen, D.-Y., Hu, Y.-M., Ma, T., Su, Y., Yang, J.-F., Wang, J.-P., Xu, G.-Z., Jiang, X.-K., Guo, J.-H., Zhang, Y.-Q., Zhang, Y., Chen, W., Chang, J., Zhang, Z.: 2021, Design and verification of the HXI collimator on the ASO-S mission. *Res. Astron. Astrophys.* **21**, 136. DOI. ADS.

Chen, D., Zhang, Z., Jiang, X., Hu, Y.: 2023, Development of an X-ray modulation characterization system for HXI payload onboard ASO-S mission. *Chin. J. Space Sci.* **43**, 747. DOI. ADS.

Gan, W.-Q., Zhu, C., Deng, Y.-Y., Li, H., Su, Y., Zhang, H.-Y., Chen, B., Zhang, Z., Wu, J., Deng, L., Huang, Y., Yang, J.-F., Cui, J.-J., Chang, J., Wang, C., Wu, J., Yin, Z.-S., Chen, W., Fang, C., Yan, Y.-H., Lin, J., Xiong, W.-M., Chen, B., Bao, H.-C., Cao, C.-X., Bai, Y.-P., Wang, T., Chen, B.-L., Li, X.-Y., Zhang, Y., Feng, L., Su, J.-T., Li, Y., Chen, W., Li, Y.-P., Su, Y.-N., Wu, H.-Y., Gu, M., Huang, L., Tang, X.-J.: 2019, Advanced Space-based Solar Observatory (ASO-S): An overview. *Res. Astron. Astrophys.* **19**, 156. DOI. ADS.

Gan, W., Zhu, C., Deng, Y., Zhang, Z., Chen, B., Huang, Y., Deng, L., Wu, H., Zhang, H., Li, H., Su, Y., Su, J., Feng, L., Wu, J., Cui, J., Wang, C., Chang, J., Yin, Z., Xiong, W., Chen, B., Yang, J., Li, F., Lin, J., Hou, J., Bai, X., Chen, D., Zhang, Y., Hu, Y., Liang, Y., Wang, J., Song, K., Guo, Q., He, L., Zhang, G., Wang, P., Bao, H., Cao, C., Bai, Y., Chen, B., He, T., Li, X., Zhang, Y., Liao, X., Jiang, H., Li, Y., Su, Y., Lei, S., Chen, W., Li, Y., Zhao, J., Li, J., Ge, Y., Zou, Z., Hu, T., Su, M., Ji, H., Gu, M., Zheng, Y., Xu, D., Wang, X.: 2023, The Advanced Space-based Solar Observatory (ASO-S). *Solar Phys.* **298**, 68. DOI. ADS.

Guo, S., Wu, J.-J., Hou, D.: 2021, The development, performances and applications of the monochromatic X-rays facilities in (0.218–301) keV at NIM, China. *Nucl. Sci. Techniq.* **32**, 65. https://api.semanticscholar.org/CorpusID:237817055.

Jiang, X.-K., Wu, J., Chen, D.-Y., Hu, Y.-M., Wang, H.-X., Liu, W., Zhang, Z.: 2023a, Construction and validation of a geometry-based mathematical model for the Hard X-ray Imager. *Res. Astron. Astrophys.* **23**, 025004. DOI. ADS.

Jiang, X.-K., Zhang, Z., Hu, Y.-M., Wang, H.-X., Liu, W., Zhang, Y.-Q., Zhang, Y., Wu, J., Chen, D.-Y.: 2023b, X-ray beam experiment for the Hard X-ray Imager onboard ASO-S. *Nucl. Instrum. Methods Phys. Res., Sect. A, Accel. Spectrom. Detect. Assoc. Equip.* **1053**, 168379. DOI. ADS.

Knoll, G.F.: 2000, Radiation detection and measurement. ADS.

Kosugi, T., Makishima, K., Murakami, T., Sakao, T., Dotani, T., Inda, M., Kai, K., Masuda, S., Nakajima, H., Ogawara, Y., Sawa, M., Shibasaki, K.: 1991, The Hard X-ray Telescope (HXT) for the SOLAR-A mission. *Solar Phys.* **136**, 17. DOI. ADS.

Krucker, S., Hurford, G.J., Grimm, O., Kögl, S., Gröbelbauer, H.-P., Etesi, L., Casadei, D., Csillaghy, A., Benz, A.O., Arnold, N.G., Molendini, F., Orleanski, P., Schori, D., Xiao, H., Kuhar, M., Hochmuth, N., Felix, S., Schramka, F., Marcin, S., Kobler, S., Iseli, L., Dreier, M., Wiehl, H.J., Kleint, L., Battaglia, M., Lastufka, E., Sathiapal, H., Lapadula, K., Bednarzik, M., Birrer, G., Stutz, S., Wild, C., Marone, F., Skup, K.R., Cichocki, A., Ber, K., Rutkowski, K., Bujwan, W., Juchnikowski, G., Winkler, M., Darmetko, M., Michalska, M., Seweryn, K., Białek, A., Osica, P., Sylwester, J., Kowalinski, M., Ścisłowski, D., Siarkowski, M., Stęślicki, M., Mrozek, T., Podgórski, P., Meuris, A., Limousin, O., Gevin, O., Le Mer, I., Brun, S., Strugarek, A., Vilmer, N., Musset, S., Maksimović, M., Fárník, F., Kozáček, Z., Kašparová,

J., Mann, G., Önel, H., Warmuth, A., Rendtel, J., Anderson, J., Bauer, S., Dionies, F., Paschke, J., Plüschke, D., Woche, M., Schuller, F., Veronig, A.M., Dickson, E.C.M., Gallagher, P.T., Maloney, S.A., Bloomfield, D.S., Piana, M., Massone, A.M., Benvenuto, F., Massa, P., Schwartz, R.A., Dennis, B.R., van Beek, H.F., Rodríguez-Pacheco, J., Lin, R.P.: 2020, The Spectrometer/Telescope for Imaging X-rays (STIX). *Astron. Astrophys.* **642**, A15. DOI. ADS.

Lemen, J.R., Title, A.M., Akin, D.J., Boerner, P.F., Chou, C., Drake, J.F., Duncan, D.W., Edwards, C.G., Friedlaender, F.M., Heyman, G.F., Hurlburt, N.E., Katz, N.L., Kushner, G.D., Levay, M., Lindgren, R.W., Mathur, D.P., McFeaters, E.L., Mitchell, S., Rehse, R.A., Schrijver, C.J., Springer, L.A., Stern, R.A., Tarbell, T.D., Wuelser, J.-P., Wolfson, C.J., Yanari, C., Bookbinder, J.A., Cheimets, P.N., Caldwell, D., Deluca, E.E., Gates, R., Golub, L., Park, S., Podgorski, W.A., Bush, R.I., Scherrer, P.H., Gummin, M.A., Smith, P., Auker, G., Jerram, P., Pool, P., Soufli, R., Windt, D.L., Beardsley, S., Clapp, M., Lang, J., Waltham, N.: 2012, The Atmospheric Imaging Assembly (AIA) on the Solar Dynamics Observatory (SDO). *Solar Phys.* **275**, 17. DOI.

Li, D., Li, Z., Shi, F., Su, Y., Chen, W., Yu, F., Li, C., Qiu, Y., Huang, Y., Ning, Z.: 2023, Observational signature of continuously operating drivers of decayless kink oscillation. *Astron. Astrophys.* **680**, L15. DOI. ADS.

Lin, R.P., Dennis, B.R., Hurford, G.J., Smith, D.M., Zehnder, A., Harvey, P.R., Curtis, D.W., Pankow, D., Turin, P., Bester, M., Csillaghy, A., Lewis, M., Madden, N., van Beek, H.F., Appleby, M., Raudorf, T., McTiernan, J., Ramaty, R., Schmahl, E., Schwartz, R., Krucker, S., Abiad, R., Quinn, T., Berg, P., Hashii, M., Sterling, R., Jackson, R., Pratt, R., Campbell, R.D., Malone, D., Landis, D., Barrington-Leigh, C.P., Slassi-Sennou, S., Cork, C., Clark, D., Amato, D., Orwig, L., Boyle, R., Banks, I.S., Shirey, K., Tolbert, A.K., Zarro, D., Snow, F., Thomsen, K., Henneck, R., Mchedlishvili, A., Ming, P., Fivian, M., Jordan, J., Wanner, R., Crubb, J., Preble, J., Matranga, M., Benz, A., Hudson, H., Canfield, R.C., Holman, G.D., Crannell, C., Kosugi, T., Emslie, A.G., Vilmer, N., Brown, J.C., Johns-Krull, C., Aschwanden, M., Metcalf, T., Conway, A.: 2002, The Reuven Ramaty High-Energy Solar Spectroscopic Imager (RHESSI). *Solar Phys.* **210**, 3. DOI. ADS.

Li, Z., Su, Y., Liu, W., Chen, W., Yu, F., Gan, W.: 2024, Spectral cross-calibrations of ASO-S/HXI with other X-ray missions. *Solar Phys.* Submitted.

Liu, W., Zhang, Z., Wu, J., Ma, T., Zhang, Y., Hu, Y.M., Zhang, Y.Q., Huang, Y.Y., Wang, H.X., Jiang, X.K., Chen, D.Y.: 2022, Pre-launch characterization of the spectrometer of Hard X-ray Imager (HXI) onboard the ASO-S mission. *J. Instrum.* **17**, P05045. DOI. ADS.

Pesnell, W.D., Thompson, B.J., Chamberlin, P.C.: 2012, The Solar Dynamics Observatory (SDO). *Solar Phys.* **275**, 3. DOI.

Su, Y., Liu, W., Li, Y.-P., Zhang, Z., Hurford, G.J., Chen, W., Huang, Y., Li, Z.-T., Jiang, X.-K., Wang, H.-X., Xia, F.-X.-Y., Chen, C.-X., Yu, W.-H., Yu, F., Wu, J., Gan, W.-Q.: 2019, Simulations and software development for the Hard X-ray Imager onboard ASO-S. *Res. Astron. Astrophys.* **19**, 163. DOI. ADS.

Wan, Q., Zhang, Y., Guo, J.H., Zhang, Y.Q., Xu, X.: 2022, Design of front-end electronics for HXI spectrometer flight model on-board ASO-S satellite. *J. Instrum.* **17**, P02020. DOI. ADS.

Warmuth, A., Önel, H., Mann, G., Rendtel, J., Strassmeier, K.G., Denker, C., Hurford, G.J., Krucker, S., Anderson, J., Bauer, S.-M., Bittner, W., Dionies, F., Paschke, J., Plüschke, D., Sablowski, D.P., Schuller, F., Senthamizh Pavai, V., Woche, M., Casadei, D., Kögl, S., Arnold, N.G., Gröbelbauer, H.-P., Schori, D., Wiehl, H.J., Csillaghy, A., Grimm, O., Orleanski, P., Skup, K.R., Bujwan, W., Rutkowski, K., Ber, K.: 2020, The STIX aspect system (SAS): the Optical Aspect System of the Spectrometer/Telescope for Imaging X-rays (STIX) on Solar Orbiter. *Solar Phys.* **295**, 90. DOI. ADS.

Yu, F.: 2023, Research on flare magnetic process and ASO-S/HXI Solar Aspect System. Doctoral thesis, University of Science and Technology of China.

Yu, F., Su, Y., Zhang, Z., Huang, Y.: 2020, Research on the Solar Aspect System algorithm of ASO-S/HXI. *Acta Astron. Sin.* **61**, 40. ADS.

Yu, J.-R., Ruan, P., Su, Y., He, Y.-H., Tao, J.-Y., Zhang, Z., Guo, S., Xue, B., Yang, J.-F.: 2024a, Inverse calculation and regularization process for the Solar Aspect System (SAS) of HXI payload on ASO-S spacecraft. *Res. Astron. Astrophys.* **24**, 045003. DOI. ADS.

Yu, J., He, Y., Tao, J., Yang, S. Zhe, Z., Yang, J., Wang, N., Guo, S., Lv, B., Chen, X., Xue, B., Hu, Y., Chen, D., Yu, .F, Ding, M., Ruan, P.: 2024b, The solar aspect system of the hard X-ray imager onboard ASO-S . *Solar Phys.* **299**, 122. DOI. ADS.

Zhang, Z., Chen, D.-Y., Wu, J., Chang, J., Hu, Y.-M., Su, Y., Zhang, Y., Wang, J.-P., Liang, Y.-M., Ma, T., Guo, J.-H., Cai, M.-S., Zhang, Y.-Q., Huang, Y.-Y., Peng, X.-Y., Tang, Z.-B., Zhao, X., Zhou, H.-H., Wang, L.-G., Song, J.-X., Ma, M., Xu, G.-Z., Yang, J.-F., Lu, D., He, Y.-H., Tao, J.-Y., Ma, X.-L., Lv, B.-G., Bai, Y.-P., Cao, C.-X., Huang, Y., Gan, W.-Q.: 2019b, Hard X-ray Imager (HXI) onboard the ASO-S mission. *Res. Astron. Astrophys.* **19**, 160. DOI. ADS.

Zhang, Y., Guo, J.-H., Zhang, Z., Chen, D.-Y., Hu, Y.-M., Zhang, Y.-Q., Ma, T., Huang, Y.-Y.: 2019a, Spectrometer of hard X-ray imager payload onboard the ASO-S mission. *Nucl. Sci. Technol.* **30**, 128. DOI.

Publisher's Note Springer Nature remains neutral with regard to jurisdictional claims in published maps and institutional affiliations.

Authors and Affiliations

Yang Su[1,2] · Zhe Zhang[1,2] · Wei Chen[1,2] · Dengyi Chen[1] · Fu Yu[1] · Yiming Hu[1,2] · Yan Zhang[1] · Fanxiaoyu Xia[1] · Changxue Chen[1,2] · Zhentong Li[1] · Xiankai Jiang[1,2] · Yu Huang[1,2] · Yongqiang Zhang[1] · Wei Liu[1] · Tao Ma[1,2] · Dong Li[1] · Wenhui Yu[1,2] · Youping Li[1] · Mingsheng Cai[1,2] · Jianhua Guo[1,2] · Yongyi Huang[1] · Haoxiang Wang[1,2] · Yaoming Liang[3] · Miao Ma[3] · Jianping Wang[4] · Shanshan Zhu[4] · Jinyou Tao[5] · Jirui Yu[5] · Jianfeng Yang[5] · Jian Wu[1] · Weiqun Gan[1]

✉ Y. Su
yang.su@pmo.ac.cn

✉ Z. Zhang
zhangzhe@pmo.ac.cn

W. Chen
w.chen@pmo.ac.cn

D. Chen
dychen@pmo.ac.cn

F. Yu
fuyu@pmo.ac.cn

Y. Hu
huyiming@pmo.ac.cn

Y. Zhang
yzhang@pmo.ac.cn

F. Xia
xiafxy@pmo.ac.cn

C. Chen
cxchen@pmo.ac.cn

Z. Li
ztli@pmo.ac.cn

X. Jiang
jiangxk@pmo.ac.cn

Y. Huang
huangyu@pmo.ac.cn

Y. Zhang
yqzhang@pmo.ac.cn

W. Liu
liuwei1@pmo.ac.cn

T. Ma
matao@pmo.ac.cn

D. Li
lidong@pmo.ac.cn

W. Yu
yuwh@pmo.ac.cn

Y. Li
yplee@pmo.ac.cn

M. Cai
caims@pmo.ac.cn

J. Guo
jhguo@pmo.ac.cn

Y. Huang
huangyy@pmo.ac.cn

H. Wang
wanghx@pmo.ac.cn

Y. Liang
liangym1208@nssc.ac.cn

M. Ma
mamiao@nssc.ac.cn

J. Wang
wangjp@microsate.com

S. Zhu
zhuss@microsate.com

J. Tao
tjy@opt.ac.cn

J. Yu
yujirui@opt.ac.cn

J. Yang
yangjf@opt.ac.cn

J. Wu
wujian@pmo.ac.cn

W. Gan
wqgan@pmo.ac.cn

[1] Key Laboratory of Dark Matter and Space Astronomy, Purple Mountain Observatory, Chinese Academy of Sciences, Nanjing, 210023, China

[2] School of Astronomy and Space Science, University of Science and Technology of China, Hefei, 230026, China

[3] National Space Science Center, Chinese Academy of Sciences, Beijing, 100190, China

[4] Innovation Academy for Microsatellites, Chinese Academy of Sciences, Shanghai, 201203, China

[5] Xi'an Institute of Optics and Precision Mechanics, Chinese Academy of Sciences, Xi'an, 710119, China

Solar Physics (2024) 299:118
https://doi.org/10.1007/s11207-024-02354-3

RESEARCH

Inflight Performance and Calibrations of the Lyman-Alpha Solar Telescope on Board the Advanced Space-Based Solar Observatory

Bo Chen[1] · Li Feng[2] · Guang Zhang[1] · Hui Li[2] · Lingping He[1] · Kefei Song[1] · Quanfeng Guo[1] · Ying Li[2] · Yu Huang[2] · Jingwei Li[2] · Jie Zhao[2] · Jianchao Xue[2] · Gen Li[2] · Guanglu Shi[2] · Dechao Song[2] · Lei Lu[2] · Beili Ying[2] · Haifeng Wang[1] · Shuang Dai[1] · Xiaodong Wang[1] · Shilei Mao[1] · Peng Wang[1] · Kun Wu[1] · Shuai Ren[1] · Liang Sun[1] · Xianwei Yang[1] · Mingyi Xia[1] · Xiaoxue Zhang[1] · Peng Zhou[1] · Chen Tao[1] · Yang Liu[1] · Sibo Yu[1] · Xinkai Li[1] · Shuting Li[2] · Ping Zhang[2] · Qiao Li[2] · Zhengyuan Tian[2] · Yue Zhou[2] · Jun Tian[2] · Jiahui Shan[2] · Xiaofeng Liu[2] · Zhichen Jing[2] · Weiqun Gan[2]

Received: 28 February 2024 / Accepted: 27 July 2024 / Published online: 29 August 2024

Abstract
The Lyα Solar Telescope (LST) is the first instrument to achieve imaging of the full solar disk and the coronal region in both white light (WL) and ultraviolet (UV) H I Lyα, extending up to 2.5 solar radii (Rs), contributing to solar physics research and space weather forecasting. Since its launch on 9 October 2022, LST has captured various significant solar activity phenomena, including flares, filaments, prominences, and coronal mass ejections (CMEs). On-orbit observation and test results show that LST covers a continuous spatial range and the wavelengths of 121.6, 360, and 700 nm. The Lyα Solar Disk Imager (SDI) has a field of view (FOV) of 38.4′ and a spatial resolution of around 9.5″, while the White-Light Solar Telescope (WST) has an FOV of 38.43′ and a spatial resolution of around 3.0″. The FOV of the Lyα Solar Corona Imager (SCI) reaches 81.1′ and its spatial resolution is 4.3″. The stray-light level in the 700 nm waveband is about 7.8×10^{-6} MSB at 1.1 Rs and 7.6×10^{-7} MSB at 2.5 Rs, and in Lyα waveband it is around 4.3×10^{-3} MSB at 1.1 Rs and 4.1×10^{-4} MSB at 2.5 Rs (MSB: mean solar brightness). This article will detail the results from on-orbit tests and calibrations.

Keywords Lyα waveband · Solar telescope · Coronagraph · Calibration

✉ B. Chen
chenb@ciomp.ac.cn

✉ L. Feng
lfeng@pmo.ac.cn

1 Changchun Institute of Optics, Fine Mechanics and Physics, Chinese Academy of Sciences, Changchun, China

2 Purple Mountain Observatory, Chinese Academy of Sciences, Nanjing, China

1. Introduction

Space-based solar observations have been conducted for decades. However, imaging observations of the hydrogen Lyman α (Lyα) spectral line, which is crucial for studying the hydrogen-dominated solar transition zone, remain rare. This rarity is largely due to the complex radiation mechanisms in this wavelength, compounded by extensive absorption and scattering effects that obscure changes in radiation brightness and distributions. Additionally, significant challenges in developing detectors, optical components, and radiation calibration arise due to severe absorption and technological complexities in this waveband. As a result, there are few imaging instruments specifically designed for this band, leading to a scarcity of observational data.

Over the past few decades, numerous solar missions have provided high-quality imaging and spectroscopic data. Space-based coronagraphs, dating back to the 1970s, began with the Orbiting Solar Observatory 7 (OSO-7) coronagraph, a pioneer in discovering coronal mass ejections (CMEs). Following OSO-7, several other coronagraphs were deployed before the Solar and Heliospheric Observatory (SOHO). The first coronagraphic observations in the ultraviolet (UV) including the Lyα line were performed by the UV Coronal Spectrometer (UVCS) on board of Spartan 201 (Kohl et al. 1994), which was a prototype of the UV Coronal Spectrometer (UVCS) on board the Solar and Heliospheric Observatory (SOHO: Kohl et al. 1995). The Large Angle Spectroscopic Coronagraph (LASCO) C1, C2, and C3 coronagraphs on board SOHO (Brueckner et al. 1995) have significantly contributed to CME and solar wind research. In 2020, the Solar Orbiter (SolO) was launched, featuring the Metis coronagraph (Antonucci et al. 2020), capable of observing the solar corona in both visible light and hydrogen Lyα wavelengths, with a field of view (FOV) ranging from 1.6° to 3.4°. Few instruments have been developed for imaging the solar disk in the ultraviolet (UV) spectrum, especially in Lyα. The first solar disk Lyα observations date back to the Skylab Apollo Telescope Mount (ATM: Reeves 1976). With rocket experiments in the 1990s, the Multi-Spectral Solar Telescope Array (MSSTA, Hoover et al. 1991) and the Multiple XUV Imager (MXUVI: Auchere et al. 1999) captured full disk Lyα images. Subsequently, the Very High Angular Resolution Ultraviolet Telescope (VAULT: Korendyke et al. 2001), the Transition Region and Coronal Explorer (TRACE: Handy et al. 1999), and the Extreme-Ultraviolet Imager (EUI: Rochus et al. 2020) on SolO conducted high spatial resolution imaging of Lyα in specific solar regions.

Since 2009, the Changchun Institute of Optics, Fine Mechanics and Physics (CIOMP) has conducted extensive research on space X-ray and EUV instruments, which have been applied to the Change'3 (CE-3) EUV camera (Chen et al. 2014), Fengyun-3 D satellite (FY-3 D) Wide-field Auroral Imager (WAI: Zhang et al. 2019), and Fenyun-3 E satellite (FY-3 E) Solar X-EUV Imager (Chen et al. 2022). Based on this foundation, several key technologies have been developed, including stray-light suppression, detection technology, and Lyα radiometric calibration. LST is capable of stably observing the full solar disk and corona in the Lyα band. Additional details are available in previous articles (Chen et al. 2019). LST has successfully captured numerous full solar disk images in the Lyα and 360 nm wavebands, as well as dual-waveband coronal images in the Lyα and 700 nm bands. Data from the Solar Disk Imager (SDI) and the White-Light Solar Telescope (WST) are now accessible at http://aso-s.pmo.ac.cn/sodc/asos.jsp. Based on LST's observing capabilities, its scientific objective and data products have been documented in previous papers (Li et al. 2019; Feng et al. 2019a).

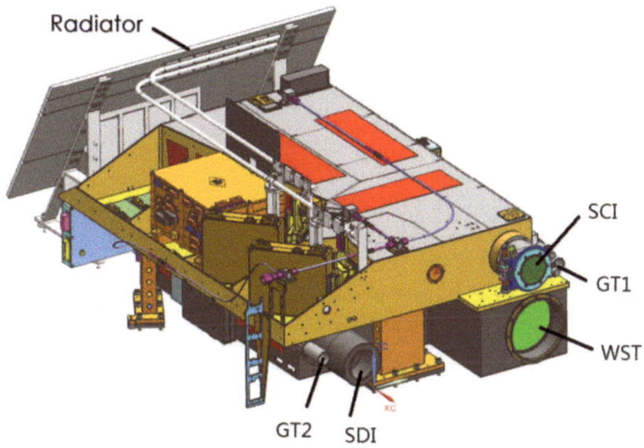

Figure 1 Schematic diagram of LST structure.

2. Instrument Overview

LST, an innovative solar telescope in orbit, spans observations from the solar disk to 2.5 Rs and is equipped with four solar imaging channels. It comprises two coronal imaging channels within the Solar Corona Imager (SCI), named SCIUV and SCIWL, which are capable of imaging and observing the corona in the Lyα and 700 nm wavebands, respectively. Additionally, LST features two full solar disk imaging channels that are SDI for the Lyα waveband and WST for the 360 nm waveband. To optimize space utilization and minimize weight, SCI, SDI, and WST are mounted on a common optical bench (Chen et al. 2019). The two SCI channels are positioned on the top of the bench, while the SDI and WST are situated at the bottom, as depicted in Figure 1. Comprehensive specifications of LST are detailed in previous publications (Gan et al. 2023; Chen et al. 2019).

SCI features a unique design that combines far ultraviolet and visible light in a dual waveband, utilizing an internally occulted reflecting imaging scheme. The SCIUV and SCIWL channels employ the same three-mirror off-axis optical system. A beam splitter in the system reflects the radiation at 121.6 nm while transmitting the visible light at 700 nm, enabling simultaneous imaging of the Lyα and visible light corona. This optical layout is detailed in a previous publication (Chen et al. 2019). The Lyα beam splitter in SCI serves dual functions. First, it reflects Lyα radiation while simultaneously transmitting visible light. Second, it acts as an Lyα reflective filter, characterized by a narrow bandpass and high out-of-band suppression capability. The reflectance of the splitter at the Lyα wavelength is 67.0%, and it has an in-band to out-of-band reflectance suppression ratio of 11.6:1. Its transmittance at the 700 nm visible light wavelength is 93.1%. Constructed from six periodic layers of LaF_3/MgF_2 multilayer on a fused silica substrate, the splitter features LaF_3 and MgF_2 layers with thicknesses of 18.0 nm and 13.5 nm, respectively. LaF_3 forms the top layer film, while MgF_2 is in direct contact with the substrate, forming the bottom layer film.

Besides the Lyα beam splitter, SCI employs a Lyα filter, which, in conjunction with a narrowband transmission filter and mirror reflectivity distributions, achieves suppression ratios of 93:1 from 115.5 to 900 nm for 121.6 nm and 120:1 from 550 to 850 nm for 700 nm. The optical system of SCI exhibits transmittance values of 1.4×10^{-2} at 121.6 nm and 1.6×10^{-1} at 700 nm, as illustrated in Figure 2.

Both SDI and WST adopt an off-axis two-mirror reflective optical layout, focusing the solar disk on the detectors with an FOV of up to 38′. SDI is dedicated to work in the Lyα line, while WST captures images in the 360 nm waveband. To enhance out-of-band suppression, SDI incorporates three Lyα filters from Acton. Two of these filters are positioned

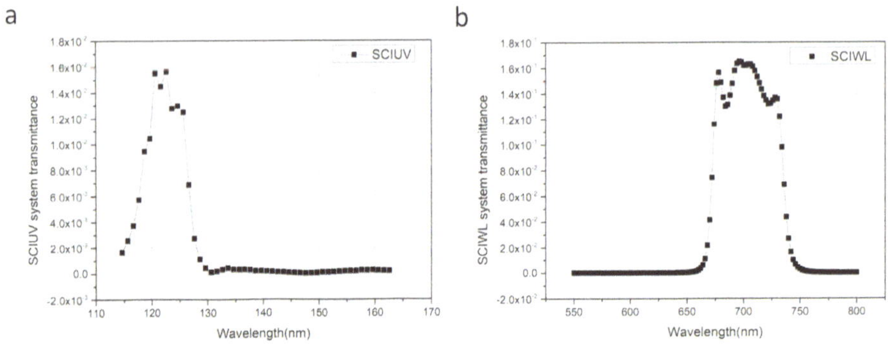

Figure 2 Spectral response distribution of SCI in 121.6 nm (**a**) and 700 nm (**b**).

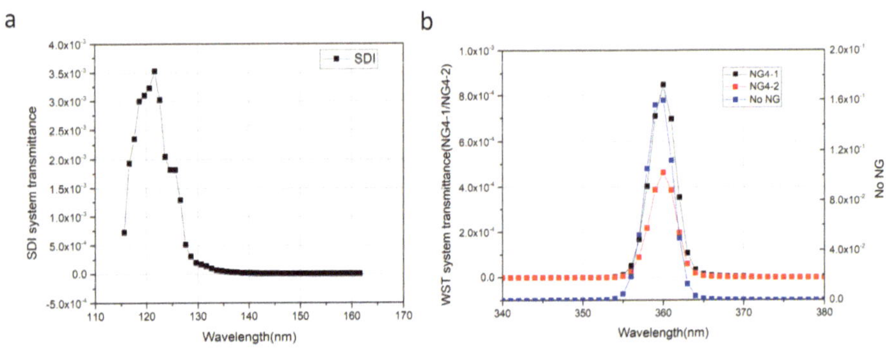

Figure 3 The system transmission of SDI (**a**) and WST (**b**) for different attenuator windows such as NG4-1, NG4-2 (left coordinate axis), and No NG (right coordinate axis).

at the entrance of SDI to block out-of-band solar radiation, and the third narrow-band filter is located in the front of the detector. This configuration effectively reduces out-of-band radiation, allowing SDI to capture Lyα radiation with a high spectral purity.

Figure 3a presents the system transmission of SDI. It is characterized by a spectral bandwidth of 9.0 nm and an out-of-band suppression ratio of 50:1. The dual entrance filters of SDI play a crucial role in blocking over 90% of the Sun's intense radiation, significantly contributing to the thermal management of the instrument. This effective thermal control is essential for maintaining the stability and performance of SDI during solar observations.

WST, operating at the wavelength of 360 nm, is designed to match SDI in both FOV and spatial resolution. At its entrance, WST employs a filter that blocks over 80% of out-of-band radiation, crucial for reducing hot radiation and selectively filtering specific wavelengths. Positioned in front of the CMOS detector, a filter wheel in WST comprises one empty hole and two attenuator windows, offering attenuation levels of 5.3×10^{-3} and 2.9×10^{-3}. These are designed to cover the range of solar irradiation encountered in the 360 nm waveband. The transmittance of the system at 360 nm is 1.6×10^{-1}, 8.5×10^{-4}, 4.6×10^{-4} for No NG, NG4-1 and NG4-2, respectively, with its spectral response distribution illustrated in Figure 3b.

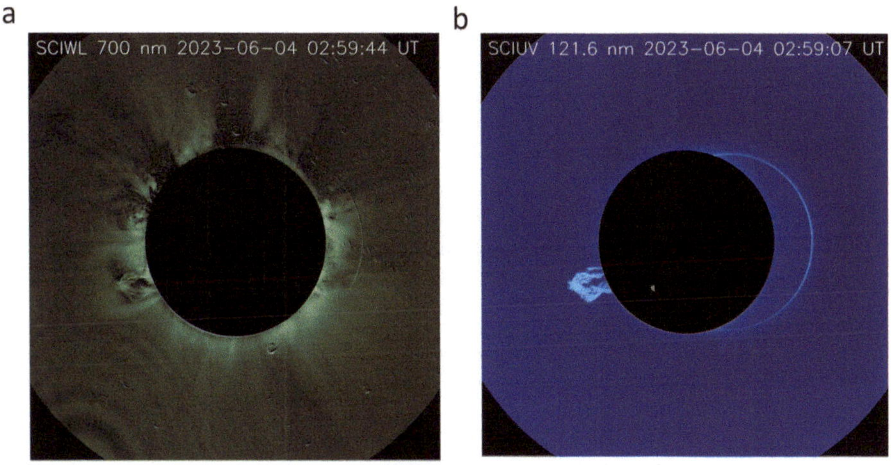

Figure 4 Images captured by SCI in the dual wavebands of 700 nm (**a**) and 121.6 nm (**b**) simultaneously on 4 June 2023.

3. On-Orbit Tests

Since the launch of LST, it has captured a substantial number of solar disk and coronal images. These observations have provided insights into various performances of LST. Additionally, specific on-orbit observation modes have been designed to verify other aspects of the performance of the telescope. The on-orbit performance test results have been derived by analyzing the observation and test data. The detailed work on these on-orbit tests and the analysis of the observation results are outlined as follows.

3.1. Observations and on-Orbit Tests for SCI

After the self-cleaning of SCI on orbit, it has successfully captured some CMEs and numerous Lyα prominences. Figure 4 showcases images of a prominence and an associated CME. The left panel of the figure displays the prominence in white light, while the right panel shows it in the Lyα waveband. These images have been obtained by subtracting the respective daily minimum images as background. In the white-light image, features such as the prominence, cavity, and the CME front are distinctly visible. Conversely, in the Lyα waveband, only the prominence is evident. However, by integrating data from both SDI and SCIUV, it has become possible to trace the evolution of the prominence from the solar disk to the coronal region. Ongoing calibration and data processing efforts are expected to refine these observations further. On-orbit observation results indicate that SCI can achieve dual waveband imaging. In addition, a large prominence was observed simultaneously in Lyα and white-light band on 28 July 2023 by SCI; see the supplementary movie for details. In the white-light prominence movie, the bright ring of the original image is removed using a machine learning process, while the Lyα prominence movie is obtained by removing the quiet solar image and a machine learning process to perceptively show the prominence more clearly. The machine learning process is as follows: learn the bright ring information with an on-orbit image sequence, extract the pattern of bright ring appearance, and then remove the bright ring in the new image with the learned pattern.

In Figure 4 and the movies, the black disks represent the occulted sunlight within 1.1 Rs. The images have been processed by subtracting the daily-minimum background. For the

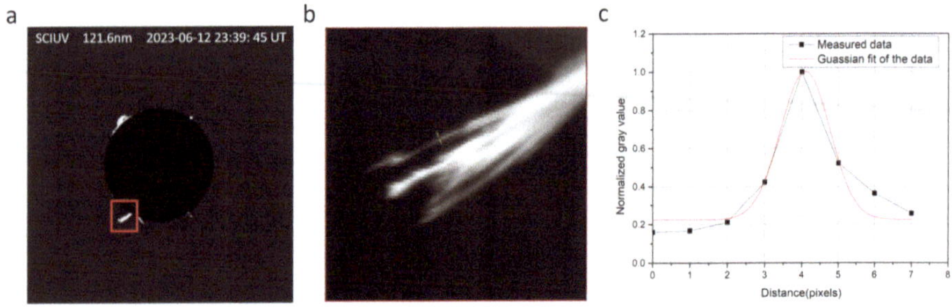

Figure 5 Lyα prominence image observed by SCIUV on 12 June 2023 (**a**), enlarged image inside the red line (**b**), and normalized gray distribution of the fine prominence structure (**c**).

SCIWL channel at 700 nm, the final image is a sum of three images taken at three different polarizer positions. In the polarization wheel of SCI, we installed three polarizers orientated at −60, 0, and 60 degrees, respectively (Feng et al. 2019b). Subsequently, three such images are summed to create Figure 4a and provide the total brightness of the corona. A different combination of the images would provide also the polarized brightness. The bright ring in the figure results from the illumination of the inner edge of the secondary mirror by the solar edge, and a ghost image is generated after passing through the beam splitter. The position of the bright ring is fixed relative to the center of the secondary mirror. To assess the spatial resolution of SCI, the images of prominences with fine structures captured on 12 June 2023, were selected for analysis, as shown in Figure 5. By examining the minimum prominence structures and measuring the brightness distribution along their radial direction in Figure 5b, the resolution was evaluated based on the full width at half maximum (FWHM) of these structures. The FWHM of the Lyα prominence structure was determined to be approximately 1.8 pixels from Figure 5c. Given the pixel size of 2.4″, this equals to a spatial resolution of 4.3″ and SCIWL gives a similar result. Thus, the spatial resolution of SCI is determined to be better than its designated resolution of 4.8″.

To evaluate the stray-light suppression level of SCI, the coronal distribution was measured and compared with the solar disk brightness at the same time, leading to the determination of the stray-light suppression distribution of SCI. For the SCIWL channel, the distribution of stray-light suppression level was ascertained using both the visible corona image and the mean brightness of the solar disk (mean solar brightness, MSB), as depicted in Figure 6. The results reveal that the average stray-light level at 1.1 Rs and 2.5 Rs are approximately 6.1 to 7.8 $\times$ 10^{-6} MSB and 6.2 to 7.6 $\times$ 10^{-7} MSB, respectively, with two different approaches. However, it was observed that the stray-light suppression capability of SCI is worse than the on-ground measurement results in previous articles (Gan et al. 2023). This can be primarily attributed to the illumination of the inner edge of the secondary mirror by the solar edge, resulting in a bright ring at the inner edge of the FOV. This phenomenon adversely affects the stray-light suppression performance of SCI. Refer to the Appendix A for the detailed calculation procedure.

In addition, the stray-light level of SCIUV is evaluated by the Lyα corona image captured on 8 July 2023. It was calculated using the captured Lyα image combined with the average brightness data of the solar disk, as depicted in Figure 7. In this analysis, the solar disk brightness in the Lyα band is referenced from SDI at the same time. The resulting stray-light distribution along the yellow line is presented in Figure 7. The average stray-light levels measured at 1.1 Rs and 2.5 Rs are approximately 5.5 $\times$ 10^{-3} MSB and 5.3 $\times$ 10^{-4}

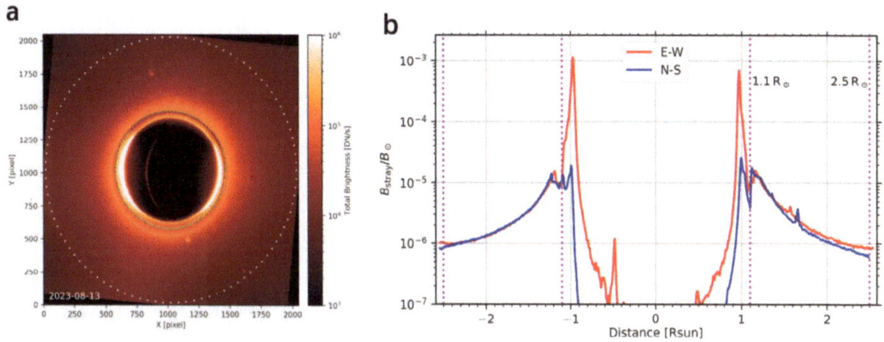

Figure 6 (**a**) The minimum image of the eight measurements rotated every 45 degrees with respect to the Sun-satellite line by SCIWL. (**b**) SCIWL stray-light distribution along the east–west and north–south directions crossing the FOV center.

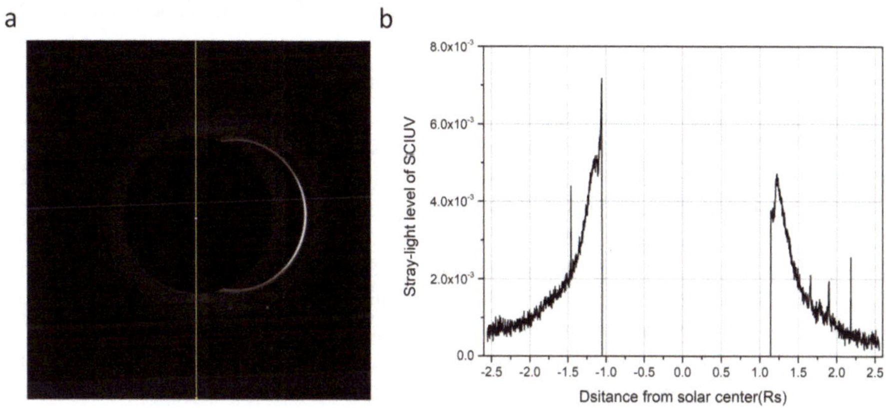

Figure 7 Image of SCI WL channel (**a**) and stray-light distribution along the yellow line (**b**).

MSB, respectively. The standard deviations are 3.7×10^{-3} MSB at 1.1 Rs and 3.1×10^{-4} MSB at 2.5 Rs. Refer to the Appendix B for the detailed calculation procedure.

3.2. Observations and on-Orbit Tests for SDI

SDI has captured a number of solar active phenomena in the H I Lyα line, including flares and eruptive prominences. In Figure 8, an eruptive prominence observed on the south–east limb on 4 June 2023 is shown. The image has undergone several processing steps, including dark current and flat field calibration, rotation to align with solar north, and a despiking procedure to remove hot pixels and cosmic ray interference. Additionally, the image reveals bright active regions and dark filaments across the solar disk. It is observed that filaments in the Lyα waveband appear more diffuse than those typically seen in the Hα waveband. SDI has the capability to shift between routine mode and burst mode. When a local enhancement is detected, for instance, a flare, SDI changes its cadence from 1 minute to 6 seconds, and data in a cutout window of 1024×1024 pixels are transmitted to the ground. More details of the triggering and termination scheme for LST instruments can be found in Lu et al. (2020).

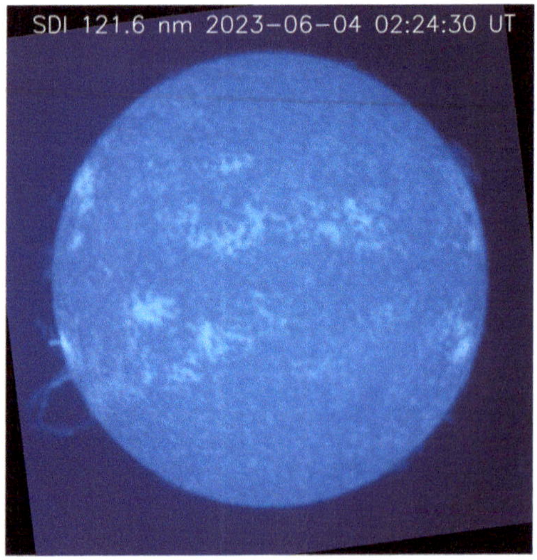

Figure 8 A Lyα full-disk image with an eruptive prominence at the south–east limb observed by SDI on 4 June 2023.

The results from on-orbit observations indicate that the FOV of SDI is consistent with the design specification. The imaging range of a solar radius in SDI is calibrated by the size of the Sun on that day, and then the FOV is calculated according to the effective pixel diameter of SDI imaging. The FOV of SDI is calculated by the Sun size on 25 June 2023. The corresponding effective FOV is 38.4'.

However, the observed spatial resolution of SDI appears to be lower than designed. Owing to the lack of sufficiently fine features in the Sun's Lyα waveband, the fast Fourier transform (FFT) method has been employed for evaluating the resolution of SDI. According to this evaluation, the spatial resolution is approximately 9.5" for the first-light images on 26 October 2022, with more detailed information available in Appendix C. The primary cause of this deviation may be identified as the wavefront errors of the entrance Lyα filter, leading to defocusing and image blurring.

3.3. Observations and on-Orbit Tests for WST

WST has acquired a substantial number of images at 360 nm and successfully captured more than 50 white-light flares since the launch of the ASO-S satellite (Jing et al. 2024). These flares are distinguished by an enhancement in the Balmer continuum of the hydrogen atom. The results from on-orbit observations indicate that the FOV of WST is also consistent with the design specification. The WST FOV test method is the same as that of SDI and is calculated by the Sun size on 16 March 2023. The corresponding effective FOV for WST is 38.4'.

Figure 9 shows a calibrated full-disk image captured by WST, featuring a white-light flare indicated by intensity brightening, marked by the arrow in the enlarged square. For an in-depth analysis of the physical processes of these white-light flares as observed by WST, refer to Jing et al. (2024) in this topical collection. The on-orbit observation results demonstrate that the FOV of WST is consistent with the design specifications, and it is capable of capturing full solar disk images. WST also has a burst mode which obtains the triggering signal from that of SDI. In the burst mode, WST has a time cadence of one second for the first five minutes and two seconds for the second five minutes and has a cutout window of 1024 × 1024 pixels as well.

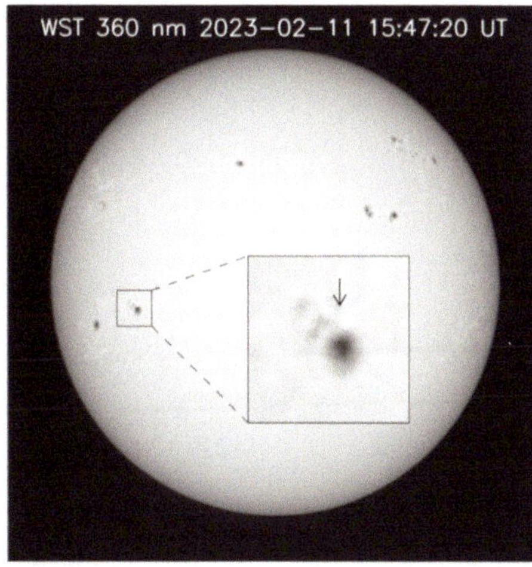

Figure 9 A 360 nm full-disk image with an enlarged white-light flare observed by WST on 11 February 2023.

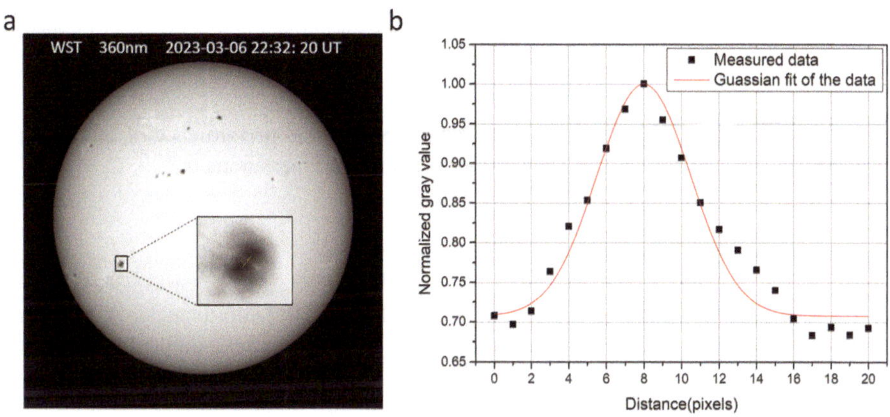

Figure 10 The sunspot light bridge observed by WST on 06 March 2023 (**a**) and the distribution map of the structure along the yellow line (**b**).

To assess the resolution, an image of a tiny sunspot light bridge structure captured by WST was analyzed. This structure, observed within a sunspot on 06 March 2023, and shown in Figure 10, was used to determine the spatial resolution. By extracting the smallest bright bridge feature and analyzing the brightness distribution along its vertical direction, the spatial resolution was calculated based on the FWHM. The FWHM of this bright feature is approximately 6 pixels. Given that each pixel corresponds to a resolution of 0.5", the measured structure corresponds to an angular size of approximately 3.0". The image in Figure 10 illustrates that the spatial resolution of WST is lower than the expected 1.2". A significant factor contributing to this reduced resolution is a temperature gradient of approximately 2 °C between the center and edge of the entrance filter when subjected to intense sunlight irradiation. This temperature differential causes a wavefront change within the entrance filter, maybe resulting in defocusing and image blurring.

In summary, we have conducted on-orbit tests for over a year and completed most of the planned tests. While some measured performance has met the design requirements, three other performances, including the spatial resolution of SDI, the spatial resolution of WST

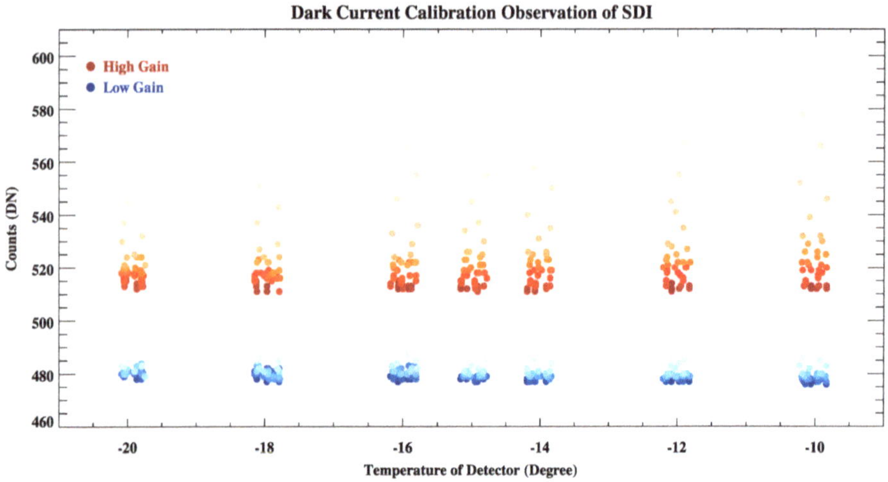

Figure 11 Measured dark current counts as a function of detector temperature and exposure time for SDI. Red and blue symbols are used for the CMOS high- and low-gain modes, respectively. The counts for different exposure times from low to high values are indicated by dots from deep to light colors.

and the stray-light level of SCIWL (that influences also the performance of SCIUV) are lower than expected and require further improvements and measurements.

4. On-Orbit Calibrations

We describe the LST on-orbit calibrations for WST and SDI. Unfortunately, SCI does not work as expected, we are still working on the dark current, flat field, and radiometric calibrations for SCI.

4.1. Dark-Field Calibrations of WST and SDI

To derive the dark fields of WST and SDI for their working temperature and exposure time, the dark current models are devised for WST and SDI. These models are functions of detector temperature and exposure time. For WST, the CMOS detector works at -25 ± 0.2 °C. The temperature sequence has seven values and is centered at -25 °C, i.e., -30 °C, -28 °C, -26 °C, -25 °C, -24 °C, -22 °C, and -20 °C. The exposure time sequence has 30 values from 0.003 to 1.0 s.

In order to decrease the contamination rate and extend the time interval of the CMOS clearing, the CMOS detector of SDI mostly works at -15 ± 0.2 °C. The temperature sequence contains -20 °C, -18 °C, -16 °C, -15 °C, -14 °C, -12 °C, and -10 °C, and the exposure time covers from 0.0019 to 120 s with a total of 30 values as well. Figure 11 shows an example of the SDI dark current measurements as a function of detector temperature and exposure time.

Based on these measurements, the dark current models are established. With several tests, we find that the dark field is best fitted with a function such as $I_{\mathrm{dc}} = a_0\ y + a_1 + (a_2\ y^2 + a_3\ y + a_4)\ x$, where y is the temperature of the detector, x is the exposure time minus the minimum exposure that is used for obtaining the bias of the detector, and I_{dc} is the measured counts of the dark current. Subsequently, the dark field for the detector temperature and

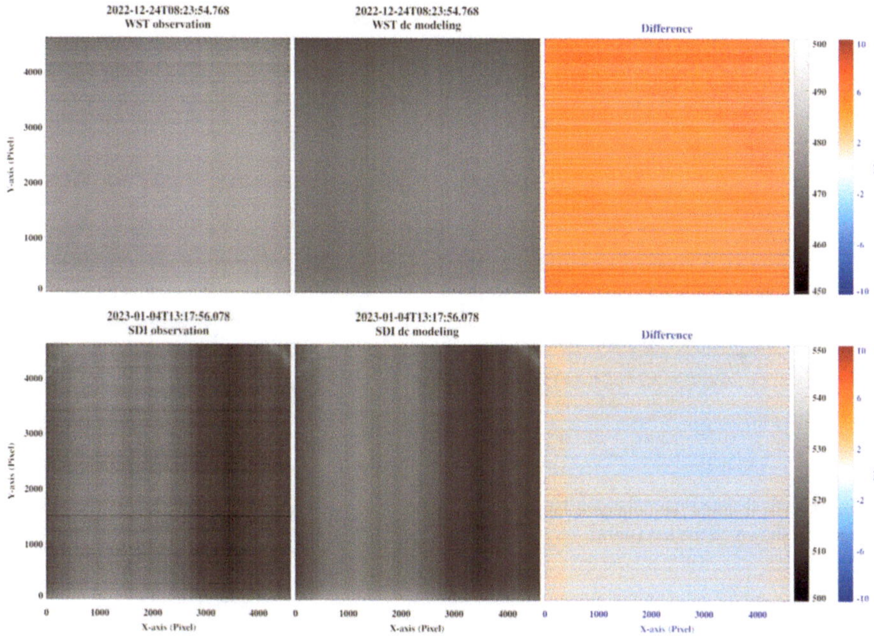

Figure 12 Comparison of the observed dark field around the working detector temperature and exposure time and the corresponding dark field model for WST (upper panel) and SDI (lower panel), respectively. The difference is the observed dark field minus the modeled dark field.

exposure time for each image acquisition can be derived. In Figure 12, the dark field data are obtained which are around the working detector temperature and exposure time and the corresponding dark field models. The upper and lower panels show the results for WST and SDI, respectively. We find that the model can recover the observed dark field with median difference values of 5.3 DN and -0.37 DN for WST and SDI, respectively.

4.2. Flat-Field Calibrations of WST and SDI

WST and SDI flat fields have been measured in flight by using some different methods. SDI has three flat-field calibration methods which include two KLL (Kuhn, Lin and Loranz, for short KLL, Kuhn, Lin, and Loranz 1991) methods and one light-emitting diode (LED) monitoring method. One KLL method off-points the instrument from the Sun–Earth line by the spacecraft with relatively large displacements (SAT-KLL method, for short), and the other KLL method with small displacements using the piezoelectric table (PZT; PZT-KLL method, for short). The LED-monitoring method uses an LED in front of the filter to illuminate the detector for monitoring its attenuation. WST has two in-flight flat-field calibration methods, one is the normal SAT-KLL method and the other is the LED method (Li et al. 2021).

We adopt the proposed 21 positions by Li et al. (2020a, 2020b) so that the in-flight flat fields of SDI and WST can be done simultaneously. A set of data takes about 80 minutes, which is the time required for the attitude adjustment and stabilization of the satellite as well as image acquisition. Three images with the same exposure time were collected from each position, we used all these images for the 21 positions and then used the 21 median images to produce a flat field.

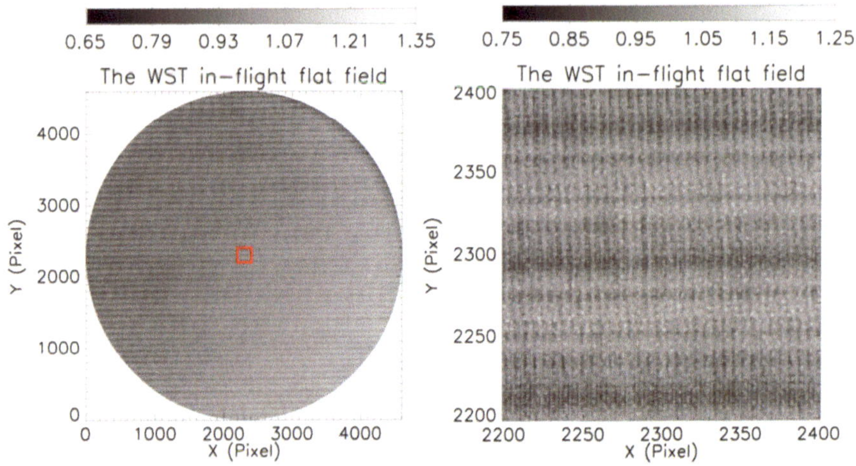

Figure 13 The in-flight flat field on 4 January 2023, in the nominal FOV of WST (left) and partial image in the red box shown on the left (right).

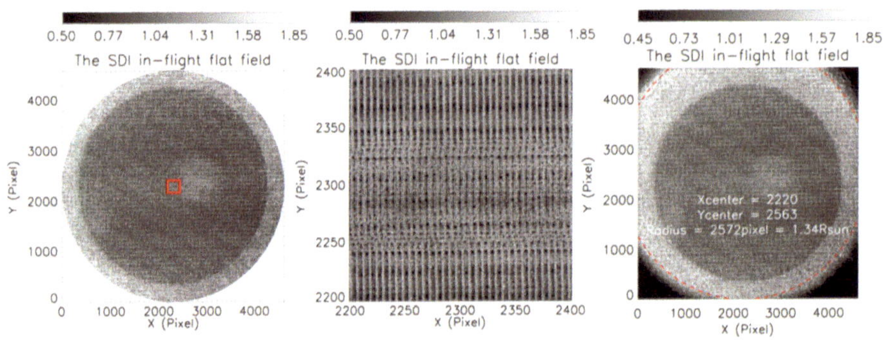

Figure 14 The high gain in-flight flat field on 4 January 2023 in the nominal FOV of SDI (left) and partial image in the red box shown on the left (middle). The red dashed line in the right panel circles the area with reliable flat field measurements.

For WST, the Sun is an ideal source for the in-flight flat-field calibration with the SAT-KLL method. The full-disk photosphere images in general show little temporal evolution during the calibration. A sample in-flight KLL flat-field for WST is shown in Figure 13. The WST flat field has many stripe structures due to laser annealing of the CMOS. The precision of the flat field in each pixel is defined as the ratio of the standard deviation to the mean value for a few repetitive measurements. Subsequently, the precision of the flat field in the designed FOV within 1.2 Rs is the average of the precision of all pixels in the FOV. It is approximately 0.5% for WST, which is better than the results from the simulated data in Li et al. (2021).

The SDI in-flight flat field shows more the evolutions of the solar structures than the flat field of WST. A sample in-flight flat field for SDI is shown in Figure 14, the SDI flat field also has many stripe structures due to laser annealing of the CMOS. The red dashed line in the right panel circles the area with reliable flat field measurements and defines the effective FOV of SDI. The in-flight flat field precision is around 2.5%.

Table 1 ASTM G173-03 reference spectra around 360 nm.

Wavelength (nm)	Irradiance (erg s^{-1} cm^{-2} nm^{-1})
358.0	7.8916E+2
358.5	7.3100E+2
359.0	8.5805E+2
359.5	1.0321E+3
360.0	1.0890E+3
360.5	1.0265E+3
361.0	9.4150E+2
361.5	9.1800E+2
362.0	9.5800E+2

4.3. Radiometric Calibrations of WST and SDI

Different methods are used to calibrate the observed fluxes of WST at 360 nm and SDI in Lyα. For WST, we use a standard reference solar spectral irradiance at 360 nm from the American Society of Testing and Materials (ASTM G173-03) to calibrate the flux. For SDI, we use the calibrated solar flux in Lyα observed by the SOlar Stellar Irradiance Comparison Experiment (SOLSTICE) on the Solar Radiation and Climate Experiment (SORCE) or by the Extreme Ultraviolet Sensor on board the Geostationary Operational Environmental Satellite (GOES/EUVS) that has a similar waveband as SDI to calibrate the SDI flux.

ASTM G173-03 tables provide a standard reference solar spectral irradiance at 280 – 4000 nm. Table 1 lists the reference spectra from 358 to 362 nm (360 ± 2 nm, i.e., the waveband of WST), with a constant interval of 0.5 nm. Integrating these irradiances over the wavelength, we then get a standard flux of 3.7349×10^3 erg cm^{-2} s^{-1} at 360 nm. By using the observed flux of $(3.6043 \pm 0.0045) \times 10^{12}$ (for filter 1 with a transmittance of ≈ 0.0053) and $(1.8342 \pm 0.0029) \times 10^{12}$ DN s^{-1} (for filter 2 with a transmittance of ≈ 0.0029) during November 2022, we can obtain the radiometric calibration factors of $(1.0362 \pm 0.0013) \times 10^{-9}$ and $(2.0362 \pm 0.0032) \times 10^{-9}$ erg cm^{-2} DN^{-1} for the two filters of WST, respectively. As shown in the upper panel of Figure 15, the calibration factor increases with time from August 2023 to January 2024, mainly due to the degradation of WST. The precision is defined as the standard deviation of the calibration factors in a few weeks which is about 2%.

The radiometric calibration of SDI is done with an indirect method by using the GOES/EUVS irradiance in the Lyα passband. The lower panel of Figure 15 presents the calibration factor as a function of time. The long-term variation of the degradation of SDI has been monitored and then the radiometric calibration factors are corrected during the lifetime of ASO-S. Note that there is a periodic variation with a period of about 25 days in the time series of the radiometric calibration factors of SDI. This could be explained by two reasons: (1) there exists a periodic variation of solar rotation in the transition region represented by the Lyα irradiance (Zhang et al. 2023); (2) the wavelength ranges plus response functions of SDI and GOES/EUVS are not exactly the same, which could not remove the periodic effect of the solar rotation in the transition region.

5. Discussion

LST has been working in orbit for over a year and has successfully undergone tests. However, three specifications have not met expectations yet, necessitating further on-orbit testing

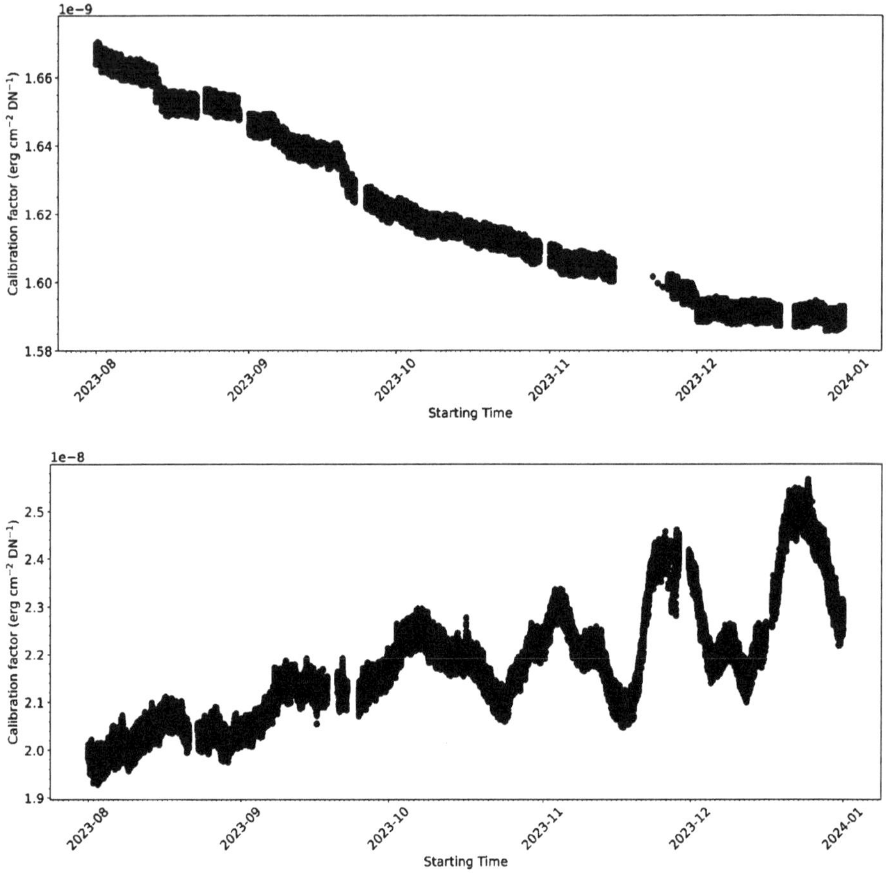

Figure 15 WST (upper panel) and SDI (lower panel) radiometric calibration factor as a function of time.

and adjustments to enhance image resolution, minimize stray-light influence, and achieve higher quality images.

For SDI and WST, the primary factors affecting resolution could be wavefront changes in the entrance filter, leading to system defocus and the deterioration of the point spread function (PSF). Therefore, the image deconvolution can be used to improve the quality of degraded images. At present, the image quality can be significantly improved by extracting the PSF of SDI from the on-orbit images. To refine the accuracy of deconvolution, the next step involves conducting further on-orbit testing to obtain a high-accuracy PSF for the entire image plane, thus improving the overall image quality. In addition, adjustments to LST's orientation are planned to capture images of UV stars with known brightness for extended exposure radiometric brightness calibration, aiming to enhance the precision of radiometric calibration.

Regarding SCI, additional processing is required to minimize the effects of the bright rings, reducing their impact on coronal imaging. Moreover, SCI and SDI share an overlapping FOV between 1.1 Rs and 1.2 Rs. By comparing images of solar prominences captured in this region by the two instruments, it is theoretically possible to obtain SCIUV images with an absolute brightness. Adjustments in satellite orientation will also be made to image

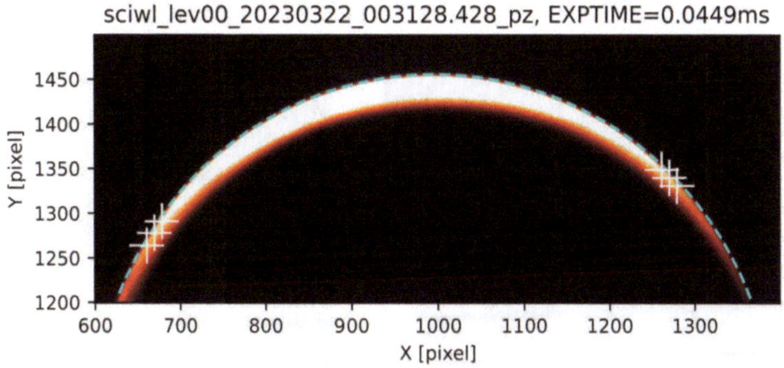

Figure 16 SCI off-pointing image in which six positions are marked to measure the brightness close to the limb on the solar disk and subsequently calculate the MSB in units of DN.

UV stars with known brightness, possibly enabling radiometric brightness calibration for SCI.

Appendix A: Estimate of the SCIWL Stray-Light Level

The stray-light suppression level is the ratio of the stray-light brightness to the mean solar brightness (MSB). The stray-light brightness is approximated by the minimum image of the measurements rotated every 45 degrees with respect to the Sun-satellite line. The MSB is calculated by using the solar disk image of SCIWL on 22 March 2023, when part of the solar disk entered the field of view (FOV) of SCIWL, and the two sharp corners in the solar image were not saturated (see Figure 16). The secondary mirror is made of gradient reflective film near the internal FOV, and its reflectivity is about 10% of that of the nongradient region. By fitting the edge of the solar disk (green dotted line), using the brightness at the plus signs, the limb darkening model at 700 nm, and the reflectivity of the second mirror, it is estimated that the MSB is about 1.1×10^{10} DN/s with a standard deviation of 6.9×10^{8} DN/s. The stray-light levels of SCIWL can be obtained as the ratio of the brightness of the stray-light image to the MSB. The average stray-light levels at 1.1 Rs and 2.5 Rs are approximately 7.8×10^{-6} MSB and 7.6×10^{-7} MSB, respectively, with the standard deviation of 5.1×10^{-7} MSB at 1.1 Rs and 5.0×10^{-8} MSB at 2.5 Rs. Alternatively, we evaluate the stray-light levels with a second method by converting the stray image from DN to MSB using the radiometric calibration on the ground. It turns out that the stray-light levels at 1.1 Rs and 2.5 Rs are about 6.1×10^{-6} MSB and 6.2×10^{-7} MSB, respectively, which are similar but a bit lower than the first method.

Appendix B: Estimate of the SCIUV Stray-Light Level

The imaging wavebands of SCIUV and SDI are both 121.6 nm, and there are overlapping FOV between 1.1 Rs and 1.2 Rs. The brightness of the same prominence is seen by SDI and SCI, so the response relationship between SCIUV and SDI images is established.

Firstly, the same Lyα prominence observed by SCIUV channel and SDI channel at 22:45 UT on 27 February 2023 have been compared (see Figure 17). The grayscale level of the prominence from SCIUV is 985.6 DN/s after subtracting the neighboring background. The

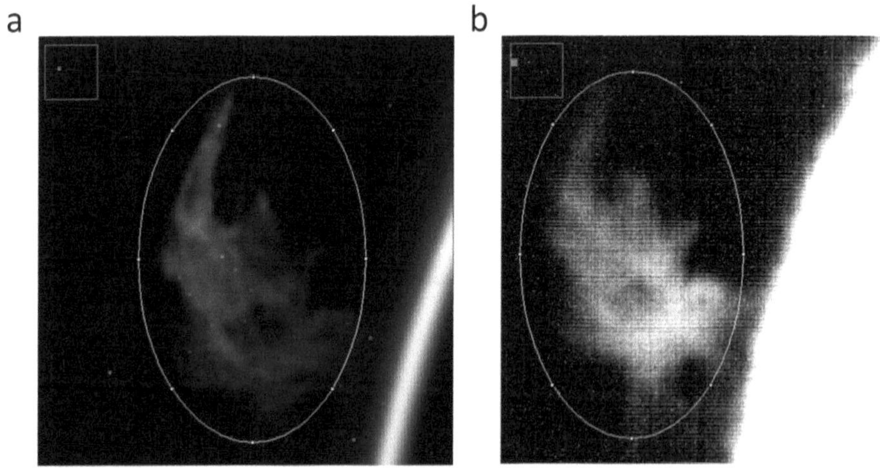

Figure 17 Prominence observed by SCIUV channel (**a**) and the same prominence observed by SDI channel (**b**).

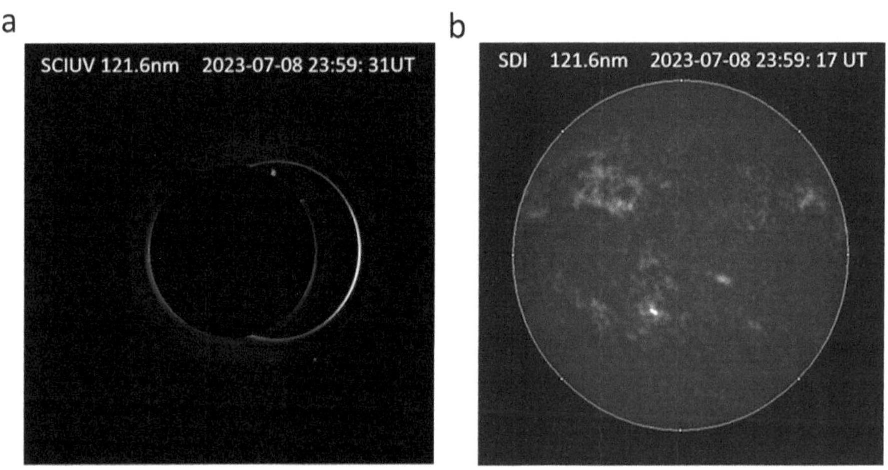

Figure 18 Images for calculating SCIUV stray-light level (**a**) and the solar disk image by SDI (**b**).

grayscale of the prominence from SDI is about 6.96 DN/s after subtracting the neighboring background. Therefore, the grayscale of SCIUV low-gain image is about 141.6 times that of SDI high-gain image after normalizing the exposure time.

In Figure 18a, the grayscale of the coronal image during the solar quiet period of SCIUV on 8 July 2023 is selected to evaluate the stray-light. The grayscale of the image after background subtraction is 20.4 and 1.98 DN/s, respectively, at 1.1 Rs and 2.5 Rs. In Figure 18b, the average grayscale of the normalized time of SDI solar disk image after background removal is 33.8 DN/s. In strict terms, the stray-light level may be evaluated by the minimum grayscale of each pixel corresponding to multiple images when the satellite rotates 45 degrees, and the result calculated by the 45 degrees rotation method is better than that in this paper.

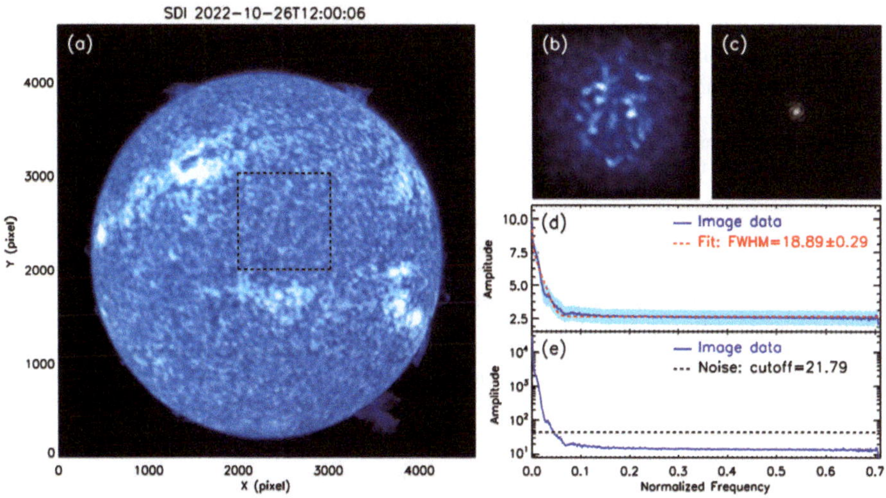

Figure 19 (**a**) Full-disk image of the SDI observed on 26 October 2022 with a dashed box indicating the selected quiet region with 1024 × 1024 pixels. (**b**) Zoomed image of SDI filtered using a Hanning window. (**c**) Transformed image in the frequency domain after applying the FFT. (**d**) Results of the fitting of the PSF function: its abscissa and ordinate indicate the normalized frequency (in unit length of the reciprocal of image size) and amplitude, respectively. The blue line represents the amplitude spectrum, and the red dashed line represents the fitting curve. The Gaussian FWHM is about 19 pixels. (e) Results of the SDI signal cutoff. The black line indicates the cutoff level of three times the background noise. The normalized cutoff frequency is around 0.046, corresponding to a spatial period of 21.79 pixels.

Therefore, the stray-light grayscale of normalized exposure time divided by the average grayscale of the solar disk is the stray-light level. Finally, the stray-light level derived is approximately 4.3×10^{-3} MSB and 4.1×10^{-4} MSB at 1.1 Rs and 2.5 Rs, respectively.

Appendix C: Estimate of the SDI Spatial Resolution

We evaluate the spatial resolution of the SDI instrument using the fast Fourier transform (FFT) method. We perform two approaches to estimate the resolution, one via the fitting of a Gaussian point spread function (PSF) (Brostrøm and Mølhave 2022), and the other via the estimate of the signal cutoff based on the amplitude spectrum in the frequency domain. Panel a of Figure 19 presents a full-disk image observed by SDI on 26 October 2022, where a dashed box represents a randomly selected quiet region with a size of 1024 × 1024 pixels. The zoomed image of this region and its transformed image after FFT in the frequency domain are shown in panels b and c, respectively. We assume the quiet-Sun region in the SDI image can be approximated by the convolution of a step function and a Gaussian PSF. The amplitude spectrum of the SDI image after FFT can be fitted by a convolution function. In panel d, the blue line represents the amplitude of the SDI image in the frequency domain as a function of the normalized frequency (length in units of reciprocal of image size), derived from panel c. The red dashed line represents the optimal fitting curve with a Gaussian full width at half maximum (FWHM) of 18.89 ± 0.29 pixels. For the second method, we calculate the average value $\bar{A}$ of the amplitude spectrum within the frequency range of 0.2 to 0.5, which is used to estimate the noise level ($3 \times \bar{A}$) indicated by the black dashed line. The normalized cutoff frequency of the SDI signal is 0.046, corresponding to a spatial period of

21.79 pixels, which is slightly higher than that derived from the fitting method, as it treats some high-frequency signals as the background noise. Therefore, the spatial resolution we obtained for SDI is around 9.5″.

Supplementary Information The online version contains supplementary material available at https://doi.org/10.1007/s11207-024-02354-3.

Acknowledgments The ASO-S mission is supported by the Strategic Priority Research Program on Space Science, the Chinese Academy of Sciences (CAS).

Author contributions Bo Chen and Li Feng contributed equally to this work; they are co-first authors and co-corresponding authors listed in no particular order. LST team of CIOMP carried out the ground spectral response calibrations, some in-flight performance measurements and in-flight calibrations. Bo Chen proposed the relative in-flight measurement and calibration scheme. Some persons proposed in-flight operation models, carried out in-flight performance measurements, calibrations and the data processing algorithm, including Kefei Song, Lingping He, Guang Zhang, Quanfeng Guo, Haifeng Wang, Shuang Dai, Xiaodong Wang, Shilei Mao, Peng Wang, Kun Wu, Shuai Ren, Liang Sun, Xianwei Yang, Mingyi Xia, Xiaoxue Zhang, Pen Zhou, Shen Tao, Yang Liu, Sibo Yu. Some persons carried out the ground spectral response calibrations, including Lingping He, Haifeng Wang and Xiaodong Wang. Guang Zhang wrote the relevant part of the manuscript. LST team at Purple Mountain Observatory (PMO, led by Li Feng and Hui Li) planned and carried out different in-flight calibration modes and corresponding analyses for in-flight dark current (Jie Zhao, Shuting Li), flat field (Jingwei Li), radiometric calibrations (Ying Li, Gen Li), etc. Yu Huang submitted all the in-flight test commands to the ground segment, which subsequently were transmitted to the spacecraft. The PMO team also developed the data processing pipe line to produce higher-level scientific data (Li Feng, Ying Li, Lei Lu, Beili Ying and all the others) for the evaluation of the in-flight performance of WST, SDI, and SCI, including SCI stray-light level estimates (Jianchao Xue), spatial resolution (WST: Dechao Song, SDI: Guanglu Shi, SCI: Beili Ying), field of view (WST: Qiao Li, SDI: Beili Ying, SCI: Lei Lu), etc. Weiqun Gan as PI of the ASO-S mission supervised all the activities.

Funding This work is supported by the Strategic Priority Research Program of the CAS (Grant No. XDB0560000); National Natural Science Foundation of China (Grant No. 12233012, 11921003, 12273040); National Key R&D Program of China (Grant No. 2022YFF0503003, 2022YFF0503004, 2022YFF0503000); the Joint Research Fund in Astronomy (U2031122) under cooperative agreement between the National Natural Science Foundation of China (NSFC) and Chinese Academy of Science (CAS).

Data Availability No datasets were generated or analysed during the current study.

Declarations

Competing interests The authors declare no competing interests.

References

Antonucci, E., Romoli, M., Andretta, V., Fineschi, S., Heinzel, P., Moses, J.D., et al.: 2020, Metis: the Solar Orbiter visible light and ultraviolet coronal imager. *Res. Astron. Astrophys.* **642**, A10. DOI.

Auchere, F., Hassler, D.M., Slater, D.C., Woods, T.N.: 1999, SwRI/LASP sounding rocket intercalibration with the EIT instrument on board SoHO. In: Siegmund, O.H., Flanagan, K.A. (eds.) *EUV, X-ray, and Gamma-Ray Instrumentation for Astronomy X, Society of Photo-Optical Instrumentation Engineers (SPIE) Conference Series* **3765**, 351. DOI.

Brostrøm, A., Mølhave, K.: 2022, Spatial Image Resolution Assessment by Fourier Analysis (SIRAF). *Microsc. Microanal.* **28**, 469. DOI.

Brueckner, G.E., Howard, R.A., Koomen, M.J., Korendyke, C.M., Michels, D.J., Moses, J.D., et al.: 1995, The large angle spectroscopic coronagraph (LASCO) visible light coronal imaging and spectroscopy. *Solar Phys.* **162**, 357. DOI.

Chen, B., Song, K., Li, Z., Wu, Q., Ning, Q.L., Wang, X.D., et al.: 2014, Development and calibration of the Moon-based EUV camera for Changer-3's. *Res. Astron. Astrophys.* **14**, 1654. DOI.

Chen, B., Li, H., Song, K.F., Guo, Q.F., Zhang, P.J., He, L.P., et al.: 2019, The Lyman-alpha Solar Telescope (LST) for the ASO-S mission–II. Design of LST. *Res. Astron. Astrophys.* **19**, 159. DOI.
Chen, B., Zhang, X.X., He, L.P., Song, K.F., Liu, S.J., Ding, G.X., et al.: 2022, Solar X-ray and EUV imager on board the FY-3E satellite. *Light: Sci. Appl.* **11**, 329. DOI.
Feng, L., Li, H., Chen, B., Li, Y., Susino, R., Huang, Y., et al.: 2019, The Lyman-alpha Solar Telescope (LST) for the ASO-S mission–III. Data and potential diagnostics. *Res. Astron. Astrophys.* **19**, 162. DOI.
Feng, L., Li, H., Inhester, B., Chen, B., Ying, B.-L., Lu, L., et al.: 2019, On the error analyses of polarization measurements of the white-light coronagraph aboard ASO-S. *Res. Astron. Astrophys.* **19**, 059. DOI.
Gan, W.Q., Zhu, C., Deng, Y.Y., et al.: 2023, The Advanced Space-Based Solar Observatory (ASO-S). *Solar Phys.* **298**, 68. DOI.
Handy, B.N., Acton, L.W., Kankelborg, C.C., Wolfson, C.J., Akin, D.J., Bruner, M.E., et al.: 1999, The transition region and coronal explorer. *Solar Phys.* **187**, 229. DOI.
Hoover, R.B., Walker, A.B.C. Jr., Lindblom, J., Barbee, T.W. Jr.: 1991, The Multi-Spectral Solar Telescope Array. *Bull. Am. Astron. Soc.* **23**, 1038.
Jing, Z.C., Li, Y., Feng, L., Li, H., Huang, Y., Li, Y.P., et al.: 2024, A statistical study of Solar White-Light flares observed by the White-Light Solar Telescope of the Lyman-Alpha Solar Telescope on the Advanced Space-Based Solar Observatory (ASO-S/LST/WST) at 360 nm. *Solar Phys.* **299**, 11. DOI.
Kohl, J.L., Gardner, L.D., Strachan, L., Hassler, D.M.: 1994, Ultraviolet spectroscopy of the extended solar corona during the SPARTAN 201 mission. *Space Sci. Rev.* **70**, 253. DOI.
Kohl, J.L., Esser, R., Gardner, L.D., Habbal, S., Daigneau, P.S., Dennis, E.F., et al.: 1995, The ultraviolet coronagraph spectrometer for the solar and heliospheric observatory. *SOHO Mission.* **162**, 313. DOI.
Korendyke, C.M., Vourlidas, A., Cook, J.W., Dere, K.P., Howard, R.A., Morrill, J.S., et al.: 2001, High-resolution imaging of the upper solar chromosphere: first light performance of the Very-high-resolution Advanced Ultraviolet Telescope. *Solar Phys.* **200**, 63. DOI.
Kuhn, J.R., Lin, H., Loranz, D.: 1991, Gain calibrating nonuniform image-array data using only the image data. *Publ. Astron. Soc. Pac.* **103**(668), 1097.
Li, H., Chen, B., Feng, L., Li, Y., Huang, Y., Li, J.W., et al.: 2019, The Lyman-alpha Solar Telescope (LST) for the ASO-S mission—I. Scientific objectives and overview. *Res. Astron. Astrophys.* **19**, 158. DOI.
Li, J.W., Li, H., Feng, L., Li, Y., Huang, Y., Zhao, J., et al.: 2020a, Influence of the flat field data acquisition interval on the Lyman-alpha solar telescope flat field accuracy. *Acta Astron. Sin.* **61**, 39.
Li, J.W., Li, H., Li, Y., Huang, Y., Zhao, J., Lu, L., et al.: 2020b, Selection of in-flight flat-field calibration dither pattern for the Solar Disk Imager of the Lyman-alpha Solar Telescope. *Proc. SPIE* **11443**, 1146. DOI.
Li, J.W., Li, H., Li, Y., Feng, L., Huang, Y., Zhao, J., et al.: 2021, Methodology for in-flight flat-field calibration of the Lyman-alpha Solar Telescope (LST). *Res. Astron. Astrophys.* **21**, 121. DOI.
Lu, L., Li, H., Huang, Y., Feng, L., et al.: 2020, The trigger and termination scheme for the event mode of the Lyman-alpha Solar Telescope (LST) onboard the ASO-S mission. *Chin. Astron. Astrophys.* **44**, 490. DOI.
Reeves, E.M.: 1976, The EUV chromospheric network in the quiet Sun. *Solar Phys.* **46**, 53. DOI.
Rochus, P., Auchere, F., Berghmans, D., Harra, L., Schmutz, W., Schuehle, U., et al.: 2020, The solar orbiter EUI instrument: the extreme ultraviolet imager. *Astron. Astrophys.* **642**, A8. DOI.
Zhang, X.X., Chen, B., He, F., Song, K.F., He, L.P., Liu, S.J., et al.: 2019, Wide-field auroral imager onboard the Fengyun satellite. *Light: Sci. Appl.* **8**, 47. DOI.
Zhang, X., Deng, L., Fei, Y., Li, C., Tian, X.: 2023, Temporal Variation of the Rotation in the Solar Transition Region. *Astrophys. J. Lett.* **951**, L3. DOI.

Publisher's Note Springer Nature remains neutral with regard to jurisdictional claims in published maps and institutional affiliations.

Solar Physics (2024) 299:135
https://doi.org/10.1007/s11207-024-02368-x

RESEARCH

Science Operation and Data Analysis Center of the Advanced Space-Based Solar Observatory (ASO-S) Mission

Yu Huang[1,2] · Youping Li[1] · Suo Liu[3] · Jiangtao Su[3] · Li Feng[1,2] · Ying Li[1,2] · Yang Su[1,2] · Wei Chen[1,2] · Dong Li[1] · Shijun Lei[1] · Yingna Su[1,2] · Yunyi Ge[1] · Lei Lu[1] · Jingwei Li[1] · Hui Li[1] · Weiqun Gan[1]

Received: 1 February 2024 / Accepted: 24 August 2024 / Published online: 27 September 2024

Abstract
A reliable data analysis center plays a crucial role in the successful execution of a space mission. The Science Operation and Data analysis Center (SODC) of Advanced Space-based Solar Observatory (ASO-S) is a bridge between the science team of ASO-S and data users (Huang et al. 2019). ASO-S plays a crucial role in understanding the solar eruptions (such as flares and coronal mass ejections) and the magnetism behind them. In this article, we outline the current status of ASO-S, as well as its data products and analysis software. These resources aid in the enhanced understanding of solar magnetism and its associated energetic eruptions.

Keywords ASO-S · Science operation · Data analysis · Data product

1. Introduction

The Advanced Space-based Solar Observatory (ASO-S) is the first Chinese comprehensive solar space mission (Gan et al. 2019, 2023). The mission aims to provide a greater understanding of the relationships between the magnetic field, solar flare and coronal mass ejection (CME).

ASO-S was successfully launched aboard a LM-2D rocket at 07:43 AM Beijing time on October 9, 2022 from the Jiuquan Satellite Launch Center, China. According to the plan, it was launched into a Sun-synchronous orbit at an altitude of about 720 kilometers. The functionality and performance of the satellite platform were first tested after launch.

The ASO-S mission encompasses three payloads, specifically:

• The Full-disk Vector MagnetoGraph (FMG) performs precise, high-cadence measurements of the solar photospheric magnetic field across the entire disk, achieving a resolution

✉ Y. Huang
huangyu@pmo.ac.cn

1 Purple Mountain Observatory, Chinese Academy of Sciences, Nanjing 210023, China

2 School of Astronomy and Space Science, University of Science and Technology of China, Hefei 230026, China

3 National Astronomical Observatories, Chinese Academy of Sciences, Beijing 100101, China

of $\sim$1.5 arcseconds through the use of 4096 $\times$ 4096 pixel images (Deng et al. 2019; Su et al. 2019a).

• The Lyman-alpha Solar Telescope (LST) is composed of three telescopes, each serving a distinct purpose (Chen et al. 2019; Li et al. 2019; Feng et al. 2019). Two of these telescopes, the Solar Disk Imager (SDI) and the White-light Solar Telescope (WST), capture full-disk images in the Lyman-alpha (121.6 nm) line and white-light waveband (360 nm), respectively. These telescopes offer a design resolution of 1.2 arcseconds, achieved through the use of 4608 $\times$ 4608 pixel images. The third telescope, known as the Solar Corona Imager (SCI), specializes in imaging the solar corona, capturing its structure from 1.1 to 2.5 solar radii in both the Lyman-alpha line and the white light (700 nm) waveband. The SCI operates with a resolution of 4.8 arcseconds, utilizing 2048 $\times$ 2048 pixel images.

• The Hard X-ray Imager (HXI) takes imaging spectroscopy of solar non-thermal X-ray emission from $\sim$10 to 300 keV (Zhang et al. 2019; Su et al. 2019b). The HXI is based on a Fourier-transform imaging technique identical to that of the Hard X-ray Telescope (HXT) on the Yohkoh mission. The HXI implements a set of 91 subcollimators to record the energy spectra in 1024 channels.

The ASO-S satellite operates in a Sun-synchronous orbit, such an orbit enables more efficient observation of the Sun, although it causes some eclipse in each orbit from May to August. During routine observations, all telescopes need to point at the center of the solar disk simultaneously. In order to maintain the consistency of the telescope optical axes as much as possible, FMG, SDI and SCI are equipped with image stabilization systems, which can perform small adjustments in direction and improve the accuracy of image stabilization. HXI and WST can ensure full-disk observations through their large field of view.

This paper focuses on the work of ASO-S's data processing, and provides an overview of the payloads' operational status and the data products. Section 2 briefly introduces the data pipeline and current status of SODC. Section 3 introduces the operational status and data processing procedures of the three payloads. Section 4 discusses the application of observational plans. Section 5 concludes with a simple summary.

2. The Status of SODC

The ASO-S satellite operates in a Sun-synchronous orbit, meaning data can only be transmitted when it passes over China. As a result, there are about eight orbits per day where data can be downloaded. This results in a delay of several hours when the ground station receives the data. Once the telemetry data is received by the ground station, it is transmitted to the National Space Science Center (NSSC) for preliminary processing. Following the production of level 0 data products by NSSC, they are distributed to SODC at the Purple Mountain Observatory. The major functions of SODC include planning scientific observations for the ASO-S mission, receiving level 0 data from the NSSC, and generating high-level data products and images for scientific users (Huang et al. 2019). Every day, approximately 0.5 TB of level 0 data and 2.0 TB of high-level data products need to be stored. We have established a storage system with an effective capacity of 6 petabytes (PB) for storing scientific data. To ensure data safety, we have also implemented a backup system for the raw data. To enhance the timeliness of data releases, we have developed a fully automated data production process. As the data production softwares for the three payloads were all developed using IDL, to ensure compatibility, the production control software was also developed using IDL. We have established a production control database to store information on production data. This database serves as a centralized repository for managing and accessing critical data throughout the production process. Once new data is transmitted to SODC, the data production process automatically begins. Once the high-level data products are generated, the information is entered into the database. By automating this process, SODC is able to streamline its operations, reduce human error, and improve the overall quality and timeliness of its data products. Scientific data and data analysis software for ASO-S satellite can be downloaded from the website: 'http://aso-s.pmo.ac.cn/sodc/dataArchive.jsp'. Users can access the ASO-S data analysis softwares via the SolarSoftWare(SSW) by updating sswidl using the command of ssw_upgrade. Currently, the data analysis software is written in IDL and has been tested on the Linux, Mac, and Windows operating systems. Additionally, data analysis guides and demos can also be downloaded from the website.

3. Data Products and Analysis

After the successful launch of the ASO-S satellite, all three payloads have begun their on-orbit testing. After approximately nine months, the payloads have completed their testing and most of the on-orbit calibration. However, some of the specifications and observation results from the FMG and LST payloads have not met the expected performance. Due to this, significant adjustments have been made to the production process, as well as the data production softwares. The team is working closely to address the issues and refine the operations to optimize the performance of the payloads.

3.1. FMG Status and Data Products

Magnetic field permeating the solar atmosphere provides the energy for the activity of flares and CMEs. The FMG payload onboard ASO-S is designed to measure solar magnetic fields, which contribute to studies of solar activity. During routine observation, FMG obtains a set of full-Sun Stokes Q, U and V images at the wavelength of the Fe I 5324.179 Å absorption line (e.g., -0.08 Å from the center) within 2 minute. After the commissioning phase, FMG is currently operational in routine observation mode. However, two significant issues remain unresolved. Firstly, bubbles have emerged in the Lyot filter, leading to a significant dark area in the full-disk images. Therefore, we only provide images of solar active regions (ARs) that are not affected by black spot area. Additionally, the LCVR polarizer is unable to capture Stokes U images. Without the U image, we cannot obtain the vector magnetic field, and we can only provide the longitudinal magnetic field. Consequently, for FMG, only the longitudinal magnetic field of ARs outside the affected black spot area can be provided.

According to the requirements, ASO-S can conduct observation in rotation calibration mode, and Q, U, V images can be observed through satellite rotation at 45° intervals. Due to the slow rotation speed of the satellite, Q and U images are observed at different times, which affects the accuracy of the vector magnetic field. Here, the definitions of FMG data levels and the production process are provided under the assumption that Stokes U images are available. There are six levels of data designed for FMG magnetic field observation: level 0, 0.5, 1.0, 1.5, and 2.0, and quick look product level Q. Level 0 data are raw images constructed from the telemetry stream, including Stokes Q, U and V maps and header keywords. For level 0.5 data, a group of raw data that can be used to deduce polarization images and produce a vector magnetogram is reorganized into a datacube with corresponding dark and flat field corrections. Based on level 0.5 data, intensity and polarization images are created by demodulating filter images with dark and flat field corrections. Then, the median of polarization data (Stokes Q, U, V) over a week is obtained as the polarization bias. For level 1.0 data, intensity and polarization images are produced with dark field, flat field, and polarization bias corrections. Level 1.5 data includes intensity and polarization images with geometry correction based on observation information such as satellite position and angle. The solar images in level 1.5 are centered and the north pole of the Sun is located directly above the image. For level 2.0 data, magnetic fields are calibrated under weak-field approximation from level 1.5, and intensity images are included. Additionally, the active region magnetic field is also defined as level 2.0 by cutting out from level 2.0 full-disk images using NOAA information as a reference. Level Q data products are quick-look images of ARs based on the level 2.0 data. However, due to the dark region in the upper left corner of the full-disk filtergram, only the longitudinal magnetogram for ARs of level 2.0 has been released to date.

The magnetic-field calibration method of FMG is a linear calibration approach based on the weak-field approximation (Jefferies, Lites, and Skumanich 1989; Jefferies and Mickey 1991).

$$B_{\mathrm{L}} = C_{\mathrm{L}} V, \tag{1}$$

$$B_{\mathrm{T}} = C_{\mathrm{T}} (Q^2 + U^2)^{1/4}, \tag{2}$$

$$\theta = \arctan(B_{\mathrm{L}}/B_{\perp}), \tag{3}$$

$$\phi = 1/2 \arctan(U/Q), \tag{4}$$

here B_{L} and B_{T} are the longitudinal and transverse photosphere magnetic field, respectively. θ and ϕ are the magnetic-field inclination and azimuth, respectively. C_{L} and C_{T} are the calibration coefficients for the longitudinal and transverse magnetic field, respectively.

The spatial resolution of magnetic field observed by FMG is approximately 1.5 arcseconds. The temporal resolution is under two minutes in routine observation mode. The sensitivity of the longitudinal magnetic field is approximately 15 Gauss. These specifications ensure that FMG observations provide high-quality data with good spatial and temporal resolution.

There are three methods to obtain the dark current of the FMG on orbit: using a shutter, observing during the Earth shadow period, and observing when the satellite is pointed away from the Sun. Currently, the shutter dark current method is used, and the difference in dark field at different times is only about 1%. The flat field of FMG cannot be directly obtained by imaging a uniform source. Instead, the flat field is acquired through a statistical method. The median of all filter images observed in a week is regarded as the flat field for the corresponding observation.

3.2. Analysis for FMG

Two analysis programs of FMG were provided, Data reading and polarity inversion line (PIL) calculations, respectively. The routine 'read_fmg.pro' can be used to retrieve both header information and data into IDL variables. PILs in ARs are the location with large magnetic gradient and are closely associated with solar eruptions (Falconer, Moore, and Gary 2003; Schrijver 2007; Guo, Zhang, and Chumak 2007).

The computation of PILs solely relies on the longitudinal magnetic field. The algorithm compares the sign of the magnetic field at a specific position (pixel) with those of its surrounding values. If a change in magnetic field sign is detected, it indicates the presence of a magnetic-field strength reversal point at that corresponding position. The magnitude of this change is recorded as the gradient at that location. PILs are then defined as the lines that can be traced by connecting these magnetic reversal points.

The routine 'fmg_arpil.pro' is used to calculate the location of PILs, and also returns the position of the maximum value of the magnetic field gradient in the data field of view, as well as the longitude and latitude of the position of the maximum value for an input of cutout magnetic fields.

3.3. HXI Status and Data Products

The scientific objective of HXI is to examine magnetic reconnection, particle acceleration, and plasma heating processes that occur during solar flares (Su et al. 2019b). The HXI collimator comprises 91 sub-collimators, serving to modulate hard X-rays. Additionally, it has a Solar Aspect System (SAS) that monitors the position of the HXI, along with any relative shifts and twists between the front and rear panels (Zhang et al. 2019). During the commissioning phase, we extended the energy range of HXI from the design objective $30-200$ keV to $\sim 10-300$ keV by optimizing the detector system with fine adjustments of threshold, high voltage, amplification factor, gain factor, etc. This improvement significantly raised the scientific values of the data and made it possible to detect more flare events. The images are reconstructed on ground using Fourier transform imaging technology, based on spatially modulated signals. The highest spatial resolution is $\sim$3.1 arcseconds. HXI can automatically switch the observing cadence between routine mode and burst mode according to the variation of X-ray flux on-orbit. The temporal resolution of HXI working at routine mode is 4 s, and the burst mode with fast cadence can be pre-selected from one of the four settings, 0.125, 0.25, 0.5 or 1 s. There are four levels of data products, namely level 0, 0.5, 1.0 and level Q1. Level 0 data are count spectra constructed from the raw telemetry stream, specifically divided into detector data FITS and SAS data FITS. The binary tables in level 0 data are read out, decoded, and subsequently saved as level 0.5 FITS files. Background spectra, identified during non-flare and low particle rate time windows, are accumulated and stored in corresponding .SAV files for each hour (FITS). The level 0.5 detector data mainly contains ADC spectra data of the 99 detectors, each has 1024 uncalibrated ADC channels. These data will be released on demand. Level 1.0 detector data comprise calibrated spectra, which have been rebinned into 80 energy bins in the range of 10 to 300 keV. These spectra are energy-calibrated and corrected for differential non-linearity (DNL). Additionally, Level 1.0 SAS data encompass the centroids of three frosted glasses, the centroid of solar disk (DM), and the locations of solar disk center derived from both the least square method (LSM, circular limb fitting) and the four-quadrant method (FQM). These measurements are acquired four times per second. Level Q1 data provides quicklook data and images for light curves, monitor data, pointing, etc. Both fits and png files are provided and the data are also stored by hour. Users can browse the quicklook products in the 'Image Browser' of HXI analysis software to gain a comprehensive overview of the flare and determine the time range of the data. The flare list is under development, which will present the list of automatically detected HXI flare events along with pertinent parameters.

3.4. HXI Analysis Software

The objective of HXI analysis software is to offer a user-friendly interface for X-ray imaging and analysis with HXI data. Like most of the X-ray imagers, HXI is also an indirect imager. Since the data processing and analysis is usually much more complicated than that of optical images, a useful analysis tool is crucial to draw more users' attention and interest to use the

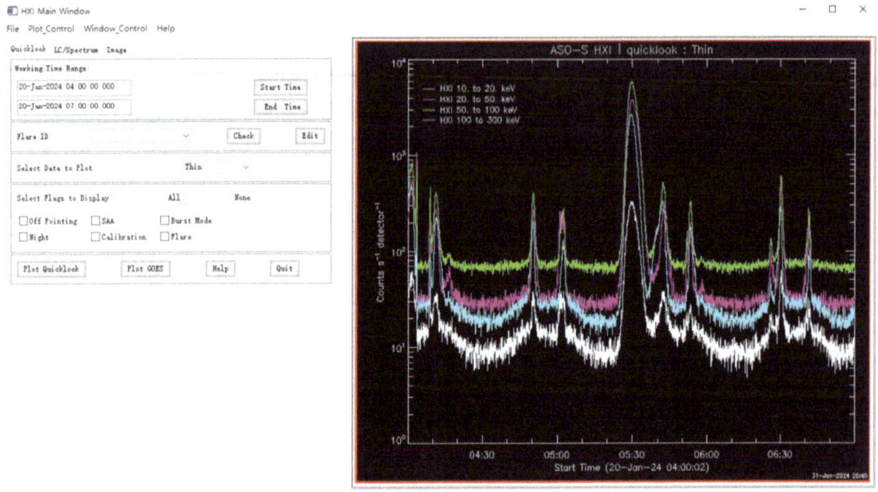

Figure 1 HXI Graphical User Interface.

data and quickly get science out of the data. The Graphical User Interface (GUI) to the RHESSI data analysis software (RHESSI GUI) is known as a powerful tool set for analysis of RHESSI images, spectra, lightcurve, and summary data (Lin et al. 2002). This tool was developed almost 20 years ago and is still being maintained and updated. To allow users to quickly get used to HXI software and to minimize users' adaptation time, we adopted some components of the RHESSI software and developed a new interface. In this new interface, the basic data processing flow is consistent with the RHESSI GUI. The HXI GUI is also object-oriented. It includes a GUI and over 30 procedures/functions, which provide three major functions with three tabs: quicklook, light curve and spectrum, imaging.

The interface of the HXI analysis software is shown in Figure 1. The Quicklook tab offers a functional zone dedicated to generating quicklook plots of flare X-ray light curves (including total and background fluxes), monitor data (such as high voltage, temperatures, etc.), and pointing data. This feature enables users to either load a specific Level Q1 FITS file through the 'Open' option in the main menu or automatically search for the corresponding Level Q1 FITS files based on a defined time range. Subsequently, the software reads the data, combines it, and generates time profiles of the selected parameters for visualization. The Light curve/spectra tab facilitates users in plotting light curves and spectra for customized energy ranges. It enables users to save the data in either SAV or FITS file formats. This functional zone is compatible for HXI Level 1.0 FITS files. Users have the flexibility to choose different detectors for generating light curves and spectra. It is worth noting that D92 and D93 represent total flux monitors with a thick front window, while D94 is the total flux monitor with a thin front window. D95 to D99 are background detectors. The result can be exported as a FITS file, compatible with OPSEX for spectral analysis. The Imaging tab offers an intuitive interface for creating images with HXI data. Users have the flexibility to modify or add multiple time intervals and energy ranges. The detector range determines which detectors are utilized in the imaging process. Users can specify the detector ID range directly or choose from grid groups 1 through 10, which represent varying pitches or spatial resolutions. The available imaging methods encompass HXI_BP, HXI_Clean, HXI_FF, VIS_BP, VIS_Clean, VIS_FF, etc. X-center and Y-center designate the coordinates of the map's center, typically defined as the flare location. 'Nxy' determines the number of pixels along the

X or Y axis, while 'psize' represents the size of each individual pixel in arcseconds. Please note that HXI software will be updated from time to time.

3.5. LST Status and Data Products

LST is comprised of several key components, including a Solar Disk Imager (SDI), a Solar Corona Imager (SCI), a White-light Solar Telescope (WST), and two Guide Telescopes (Chen et al. 2019). The WST and SDI are designed to image the Sun from the disk center to 1.2 solar radii in the violet narrow-band continuum (360 $\pm$ 2 nm) and Lyman-alpha (Lyα) waveband (121.6 $\pm$ 4.5 nm), respectively. SDI has three observation modes: routine, burst and user-defined modes. The cadence of SDI working at routine mode is 60 s, and it automatically enters the burst mode (Lu et al. 2020). The burst mode observation of SDI works with a much higher cadence of 6 seconds, using a 1024 $\times$ 1024 windowed size. In this mode, the full-disk imaging still runs about every minute. Similar, WST has three observation modes: routine, burst and user-defined modes. The routine mode of WST has a cadence of 120 s, and enters the burst mode synchronized with SDI, with the same position window. The burst mode of WST works with a much higher cadence of 1 seconds for the first five minutes. Then, WST continues to operate in burst mode for another 5 minutes at a cadence of 2 seconds. After that, WST returns to the routine mode. WST and SDI are both designed to have a spatial resolution of 1.2 arcseconds. Unfortunately, the spatial resolutions of both WST and SDI are below expectations, with values approximately 5 and 10 arcseconds, respectively. On the other hand, there are a few known instrumental artifacts present in WST and SDI images that need to be noted. A brightening spot on the Sun appears as multiple unwanted extra spots in the SDI image. For strong white-light flares, there are extra horns in the WST image due to the imperfection of some optical elements. These artifacts may be the main reason for the poor spatial resolution of the WST and SDI images.

SDI and WST generate six levels of data products, each designed to progressively enhance the quality and usability of the images. Level 0 data are the raw images constructed directly from the telemetry stream, providing the most basic representation of the observed data. Level 1 data are processed images that include flat-fielding, dark current correction, spike map generation, and bad pixel indexing. For SDI level 1 data, high-gain and low-gain images captured by the CMOS sensor are merged to optimize the dynamic range and image quality. Level 1.5 data are derived from Level 1 images and undergo additional processing steps. This includes plate scaling, rotation and shift correction and repairing bad pixels and spike pixels. Level 2 data represent images of physical quantities, that have been radiometrically calibrated from the level 1 data. Level 2.5 data are created from level 2 data, with processing of plate scaling, rotation and shifting, and repairing bad pixels and spike pixels. Currently, level 1 data for both SDI and WST have been released.

The dark fields of WST and SDI were obtained through baffle observation with different CMOS temperatures and exposure times. Details on how to obtain the flat field of WST and SDI using the KLL method can be found in Li et al. (2021).

SCI is designed to capture images of the inner solar corona, ranging from 1.1 to 2.5 solar radii, in both Lyα (122.6 $\pm$ 3 nm) and white-light (700 $\pm$ 32 nm) wavelengths. A beam-splitter is employed to divide the incoming coronal light into two separate beams: the reflected beam feeds the Lyα channel, while the transmitted beam feeds the white-light channel. The secondary mirror of SCI serves as the focal plane with a cone-shaped hole, acting as the internal occulter to block solar disk light. However, due to the small size of the hole in the secondary mirror, light from the solar disk edge is reflected into the optical path, resulting in strong scattered light and unwanted ghost images. This makes it difficult for SCI to observe CMEs in both channels. Additionally, saturated solar edges, ghost images, and excessive scattered light make it challenging to process the few prominence eruptions observed by SCI in Lyα. As a result, no SCI data products have been released yet. We hope to release processed CME event data in the near future.

3.6. LST Analysis Software

Currently, only WST and SDI data products have been released, and the introduction of LST analysis software is mainly aimed at these two instruments. The analysis tools for LST are similar to those of Solar Dynamics Observatory (SDO) (Pesnell, Thompson, and Chamberlin 2012). Basically, users need to invoke the 'read_lst.pro' command to access the WST or SDI level 1 data. Subsequently, they can employ the 'lst_prep.pro' routine to preprocess the level 1 data to higher levels. However, due to the contamination control and efficiency issues of SDI, the working temperature of the CMOS and exposure time are higher and longer than their optimal values. The increased thermal noise in the image can be attributed to this factor. To mitigate this issue, we strongly advise users to include the '/despike_on' option when utilizing 'lst_prep.pro'. Additionally, if users are only interested in a subset of the full field of view (FOV) data, we recommend utilizing cutout data to enhance processing efficiency and accuracy.

4. Observation Proposal

All three payloads onboard ASO-S, including FMG, LST, and HXI, have routine observation mode and fast/burst mode. For more information on these modes, please refer to Huang et al. (2019). In the routine mode, HXI takes full-size observations with a lower cadence. If a flare is detected, the fast mode with a higher cadence is activated. SDI and WST record full-size images with a moderate cadence in the routine mode. Once a flare is detected by SDI, they switch to burst mode (Feng et al. 2019).

In order to expand the scientific research scope of the mission, LST can conduct observations based on user requirements. For instance, coordinated observations with ground-based facilities and space-borne instruments are encouraged. For LST, various requirements, including changing time cadence, exposure time, image size and instrument combination, can be fulfilled in the user observation mode. For example, if users want to study active region dynamics, LST can track the active region with a higher cadence and a small field of view (e.g., 512×512, 1024×512, etc). If one wants to observe features on the quiet Sun, such as coronal holes and quiescent filaments, the exposure time of LST can be increased.

Users are encouraged to submit the observation applications at least one month ahead, so the science schedule coordinators have sufficient time to discuss and refine the proposals to fit the current ASO-S situation. This initial request should include preferred timing and target of observations, as well as the science rationale, such as abstract, short summary of the request, results from previous ASO-S observations, and a list of relevant publications. Once the observation request is approved, the users will be notified via email. Three to five days before the observations, the users should visit the designated link to confirm the exact timing and pointing information with the ASO-S planner.

5. Summary

As of this writing, SODC of ASO-S has completed the development of most basic functions and is supporting satellite operation on orbit. The satellite has finished the testing stage and entered the routine operating stage. The observation plan is uploaded every two days. The data products of FMG, HXI, WST and SDI have been released. The event data products of SCI will be released in the future. Data analysis software has also been released and is constantly being updated, especially for the analysis software of HXI, where functions have been continuously improved. In the future, SODC will also provide lists of flares and CMEs. The ASO-S Solar Eruptive Events Database (A-SEED) is under development, such a database is important for marking eruptive events out of the large amount of ASO-S data. We hope to provide users with high-quality data products and easy-to-use data download services.

Author contributions Huang Yu wrote the main manuscript text, Li Youping provided the content of data analysis, Liu Suo and Su Jiangtao provided the data product information and analysis software introduction of FMG, Feng Li and Li Ying provided the data product information and analysis software introduction of LST, Su Yang and Chen Wei provided the data product information and analysis software introduction of HXI, Li Dong participated in the writing of the article, Lei Shijun and Ge Yunyi provided the database information, Su Yingna introduced the observation plan, Lu Lei provided the relevant content of LST burst mode, Li Jingwei provided the flat field information of LST, Li Hui and Gan Weiqun discussed the composition and structure of the article.

Funding This work is supported by the Strategic Priority Research Program of the Chinese Academy of Sciences, Grant No. XDB0560000. This work is also funded by the National Key R&D Program of China 2022YFF0503002 (2022YFF0503000).

Data Availability No datasets were generated or analysed during the current study.

Declarations

Competing interests The authors declare no competing interests.

References

Chen, B., Li, H., Song, K.F., Guo, Q.F., Zhang, P.J., He, L.P., et al.: 2019, The Lyman-alpha Solar Telescope (LST) for the ASO-S mission - II. Design of LST. *Res. Astron. Astrophys.* **19**, 159.

Deng, Y.Y., Zhang, H.Y., Yang, J.F., Li, F., Lin, J.B., Hou, J.F., et al.: 2019, Design of the Full-disk MagnetoGraph (FMG) onboard the ASO-S. *Res. Astron. Astrophys.* **19**, 157.

Falconer, D.A., Moore, R.L., Gary, G.A.: 2003, A measure from line-of-sight magnetograms for prediction of coronal mass ejections. *J. Geophys. Res.* **108**, SSH11.

Feng, L., Li, H., Chen, B., Li, Y., Susino, R., Huang, Y., et al.: 2019, The Lyman-alpha Solar Telescope (LST) for the ASO-S mission - III. Data and potential diagnostics. *Res. Astron. Astrophys.* **19**, 162.

Gan, W.Q., Zhu, C., Deng, Y.Y., Li, H., Su, Y., Zhang, H.Y., et al.: 2019, Advanced Space-based Solar Observatory (ASO-S): An overview. *Res. Astron. Astrophys.* **19**, 156.

Gan, W.Q., Zhu, C., Deng, Y.Y., Zhang, Z., Chen, B., Huang, Y., et al.: 2023, The Advanced Space-based Solar Observatory (ASO-S). *Solar Phys.* **298**, 68.

Guo, J., Zhang, H.Q., Chumak, O.V.: 2007 Magnetic properties of flare-CME productive active regions and CME speed. *Astron. Astrophys. Suppl. Ser.* **462**, 1121.

Huang, Y., Li, H., Gan, W.Q., Li, Y.P., Su, J.T., Deng, Y.Y., et al.: 2019, The Science Operations and Data Center (SODC) of the ASO-S mission. *Res. Astron. Astrophys.* **19**, 164.

Jefferies, J.T., Lites, B.W., Skumanich, A.: 1989, Transfer of line radiation in a magnetic field. *Astrophys. J.* **343**, 920.

Jefferies, J.T., Mickey, D.L.: 1991, On the inference of magnetic field vectors from Stokes profiles. *Astrophys. J.* **372**, 694.

Li, H., Chen, B., Feng, L., Li, Y., Huang, Y., Li, J.W., et al.: 2019, The Lyman-alpha Solar Telescope (LST) for the ASO-S mission — I. Scientific objectives and overview. *Res. Astron. Astrophys.* **19**, 158.

Li, J.W., Li, H., Li, Y., Feng, L., Huang, Y., Zhao, J., et al.: 2021, Methodology for in-flight flat-field calibration of the Lyman-alpha solar telescope (LST). *Res. Astron. Astrophys.* **21**, 121.

Lin, R.P., Dennis, B.R., Hurford, G.J., Smith, D.M., Zehnder, A., Harvey, P.R., et al.: 2002, The Reuven Ramaty High-Energy Solar Spectroscopic Imager (RHESSI). *Solar Phys.* **210**, 3.

Lu, L., Li, H., Huang, Y., Feng, L., Zhu, B., Whang, P., et al.: 2020, The trigger and termination scheme for the event mode of the Lyman-alpha Solar Telescope (LST) onboard the ASO-S Mission. *Chin. Astron. Astrophys.* **44**, 4.

Pesnell, W.D., Thompson, B.J., Chamberlin, P.C.: 2012, The Solar Dynamics Observatory (SDO). *Solar Phys.* **275**, 3.

Schrijver, C.J.: 2007, A characteristic magnetic field pattern associated with all major solar flares and its use in flare forecasting. *Astrophys. J. Lett.* **655**, L117.

Su, J.T., Bai, X.Y., Chen, J., Guo, J.J., Liu, S., Wang, S.F., et al.: 2019a, Data reduction and calibration of the FMG onboard ASO-S. *Res. Astron. Astrophys.* **19**, 161.

Su, Y., Liu, W., Li, Y.P., Zhang, Z., Hurford, G.J., Chen, W., et al.: 2019b, Simulations and software development for the Hard X-ray Imager onboard ASO-S. *Res. Astron. Astrophys.* **19**, 163.

Zhang, Z., Chen, D.Y., Wu, J., Chang, J., Hu, Y.M., Su, Y., et al.: 2019, Hard X-ray Imager (HXI) onboard the ASO-S mission. *Res. Astron. Astrophys.* **19**, 160.

Part 2: FMG Related Research

Solar Physics (2023) 298:141
https://doi.org/10.1007/s11207-023-02236-0

RESEARCH

The Magnetic-Field Calibration of the *Full-Disk Magnetograph* Onboard the *Advanced Space Based Solar Observatory* (ASO-S/FMG)

S. Liu[1,2,3] · J.T. Su[1,2,3] · X.Y. Bai[1,2,3] · Y.Y. Deng[1,2,3] · J. Chen[1,2] · Y.L. Song[1,2] · X.F. Wang[1,2] · H.Q. Xu[1,2] · X. Yang[1,2]

Received: 7 August 2023 / Accepted: 23 November 2023 / Published online: 6 December 2023

Abstract
The full-disk magnetograph is a main scientific payload onboard the *Advanced Space based Solar Observatory* (ASO-S/FMG) that through Stokes-parameters observation measures the vector magnetic field. The accuracy of magnetic-field values is an important aspect of checking the quality of the FMG magnetic-field measurement. According to the design of the FMG, the linear calibration method under the weak-field approximation is the preferred scheme for magnetic-field calibration. However, the spacecraft orbital velocity can affect the position of observed spectral lines, which then results in a change of the polarization-signal strength. Thus, the magnetic field is modulated by the orbit velocity of the spacecraft. In this article, through cross-calibration between FMG and HMI (*Helioseismic and Magnetic Imager* onboard the *Solar Dynamic Observatory*), the effects of spacecraft orbital velocity on the coefficient of magnetic-field calibration are investigated. By comparing the magnetic field of FMG and HMI with spacecraft orbital velocity as an auxiliary reference, revised linear-calibration coefficients that depend on spacecraft orbital velocity are obtained. The magnetic field of FMG corrected by the revised calibration coefficients removing the effect of spacecraft orbital velocity will be more accurate and suitable for scientific research.

Keywords Solar magnetic field · Polarimetric measurements · Radiative transfer

1. Introduction

The solar magnetic field provides an essential force that dominates solar activity. The energy and mass released from solar activity, such as solar flares, filament eruptions, corona

✉ S. Liu
lius@nao.cas.cn

J.T. Su
sjt@nao.cas.cn

1 National Astronomical Observatories, Chinese Academy of Science, Beijing, 100101, China

2 Key Laboratory of Solar Activity and Space Weather, National Space Science Center, Chinese Academy of Science, Beijing, 100190, China

3 School of Astronomy and Space Sciences, University of Chinese Academy of Sciences, Beijing, 100101, China

mass eruptions, and small-scale activity mostly originate from stored magnetic energy (Krall et al., 1982; Shibata et al., 1995; Wang et al., 1994; Tsuneta, 1996; Forbes, 2000; Bond et al., 2001; Priest and Forbes, 2002; Nindos, Patsourakos, and Wiegelmann, 2012). The full-disk magnetograph onboard the *Advanced Space based Solar Observatory* (ASO-S/FMG: Gan et al., 2019; Deng et al., 2019) measures magnetic fields on the solar photosphere, which contribute to studies of solar activity (Gan et al., 2023). Fe I λ 5234.19 Å is chosen as a magnetic-sensitive line by FMG to measure the photosphere magnetic field. FMG uses a liquid-crystal variable retarder (LCVR) to observe polarized Stokes images (I, Q, U, V), which are calibrated to yield the magnetic field on the photosphere (Deng et al., 2019). During FMG operation, one or a few limited positions in the range of the spectral line are fixed to observe the Stokes parameters.

The magnetic-field calibration is a necessary and important process for producing reliable magnetic fields, and magnetic-field calibration methods vary for different types of instruments. A Stokes polarimeter has the ability to measure the full spectra of Stokes parameters I, Q, U, V of a spectral line, then an inversion method is applied to derive the magnetic field under reasonable modes with other necessary thermal parameters. Since the full information of a spectral line is used by the Stokes polarimeter, it can measure the magnetic fields with higher accuracy, but lower temporal resolution than magnetograph-type instruments (Tsuneta et al., 2008; Ichimoto et al., 2008). FMGs obtain magnetograms at higher temporal resolutions than a spectragraph-based magnetograph, hence magnetic fields with rapid evolution can be studied by FMG observations. For FMG, very limited Stokes information (one spectral position is fixed in routine observation) can be obtained, so the calibration of the magnetic field needs to be improved and more carefully implemented. The magnetic-field calibration method applied by FMG is linear calibration under the weak-field approximation (Jefferies, Lites, and Skumanich, 1989; Jefferies and Mickey, 1991):

$$B_{\mathrm{L}} = C_{\mathrm{L}} V, \tag{1}$$

$$B_{\mathrm{T}} = C_{\mathrm{T}} (Q^2 + U^2)^{1/4}, \tag{2}$$

$$\theta = \arctan(B_{\mathrm{L}}/B_{\perp}), \tag{3}$$

$$\phi = 1/2 \arctan(U/Q), \tag{4}$$

where B_{L} and B_{T} are the line-of-sight (LOS) and transverse photosphere magnetic field, respectively. θ is the inclination between the vector magnetic field and the line of sight, and ϕ is the magnetic-field azimuth, measured from solar north. C_{L} and C_{T} are the calibration coefficients for the LOS and transverse magnetic fields, respectively.

The calibration coefficients C_{L} and C_{T} are decisive parameters for the determination of the photosphere magnetic field, and C_{L} and C_{T} are spectral-position-related parameters (Unno, 1956; Rachkovsky, 1967; Landolfi and Landi Degl'Innocenti, 1982). The ASO-S spacecraft runs in a Sun-synchronous orbit with an altitude of about 720 km, an orbital inclination angle of 98.27°, and an orbital period of about 99.2 minutes, which result in Sun–FMG radial velocity changes between -3.9 to $+3.9$ km s^{-1}. The maximum value of the radial velocity corresponds to a wavelength shift from -0.07 Å to $+0.07$ Å, which results in the changes of the observed spectral position. As a result, calibration coefficients C_{L} and C_T should be corrected at every moment in time based on the velocity of the ASO-S spacecraft. Hence, to obtain and apply the calibration coefficients C_{L} and C_{T} depending on the velocity of ASO-S spacecraft would improve the FMG magnetic-field observation.

In this article, the longitudinal magnetic field of FMG and HMI are investigated and used to calculate the calibration coefficients that related to spacecraft orbital velocity, with the purpose of improving the accuracy and reliability of the FMG magnetic-field measurement. We present the observations in Section 2; the methods and results analysis are given in Section 3; the discussions and conclusions are included in Section 4.

2. Observations

The FMG consists of one birefringent Lyot filter (FWHM = 0.1 Å, wavelength stabilization < 0.02 Å) and two LCVR polarimeters to measure the magnetic field using the spectral line Fe I λ5234.19 Å. The aperture of FMG is 14 cm and its effective focal length is 5355.76 mm. The spatial resolution of the magnetic field observed by FMG is about 1.5 arcsec, for full-disk images (4k × 4k) the spatial sampling is about 0.55 arcsec per pixel. During routine observation its temporal resolution is about two minutes and the longitudinal sensitivity is 15 G. Stokes Q, U, V are observed by various combinations of azimuth and retardation of two LCVRs (cf. Table 1, Deng et al., 2019; Hou et al., 2020). Data processing includes dark-field, flat-field, polarization bias, and geometry correction, then the polarization signals are calibrated to magnetic field under the weak-field approximation (Su et al., 2019; Xu et al., 2021; Liu et al., 2022). The active-region data are cut out from full-disk images based on external auxiliary time and position information obtained from NOAA (www.solarmonitor.org/data/2023/04/01/meta/).

The *Helioseismic and Magnetic Imager* is one of the instruments onboard *Solar Dynamic Observatory* (SDO/HMI: Scherrer et al., 2012; Schou et al., 2012). HMI has an aperture of 14 cm and a 4k × 4k CCD observes full-disk magnetograms in the photospheric absorption line Fe I centered at 6173.3 Å. The spatial resolution of HMI is about 0.9 arcsec, while the pixel size is about 0.5 arcsec. The temporal resolution is 45 seconds for the longitudinal magnetic field, while for other observations there are different temporal resolutions (jsoc.stanford.edu/ajax/lookdata.html). Stokes Q, U, V are observed at six fixed wavelength positions by HMI, then inversion algorithms are applied to the Stokes observations to deduce the photosphere magnetic field. In this article, the longitudinal magnetic fields of FMG and HMI with temporal resolution of 45 seconds are employed to investigate the calibration of FMG, with the aim to correct the coefficient of FMG modulated by and related to the spacecraft velocity. We note that HMI also has an orbital velocity variation, which has been removed in its calibration of the HMI data shown here. Figure 1 shows an example of an original FMG and HMI longitudinal magnetic field observed at 04 April 2023 03:24 UT quasi-simultaneously. After geometric correction using position information during the observation, these two longitudinal magnetic-field signals have very good consistency, however, there are still differences in the size and spatial resolution for these two images.

3. Methods and Results

In order to compare calibration coefficients of FMG and HMI, first these two magnetic-field images are coaligned by adjusting the size and pixel resolution. Based on observational information, especially coordinate information, the two magnetic-field images are basically aligned when they are converted to a heliospheric coordinate system. However, due to the existence of various system errors, there may still be slight position deviations between the two magnetic-field images of FMG and HMI. Therefore, an image-alignment method of a scale-invariant feature transform (SIFT: Lowe, 1999) based on image features was adopted to coalign FMG and HMI magnetic-field images. The three main steps of SIFT for image

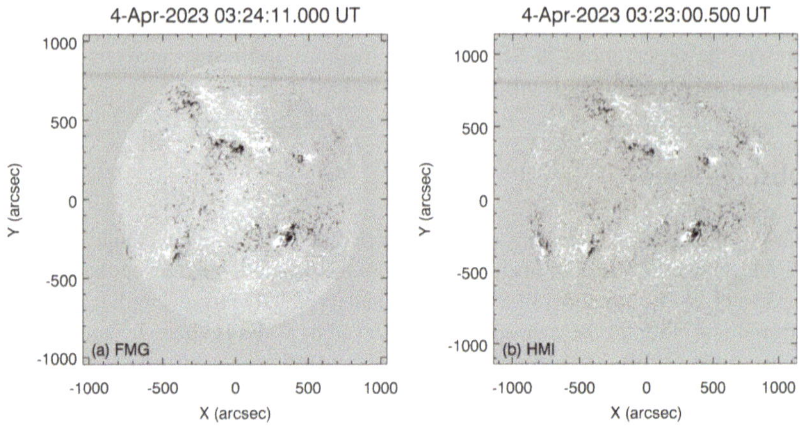

Figure 1 Original FMG (*left*) and HMI (*right*) full-disk longitudinal magnetic field observed 04 April 2023 at 04:24 UT quasi-simultaneously.

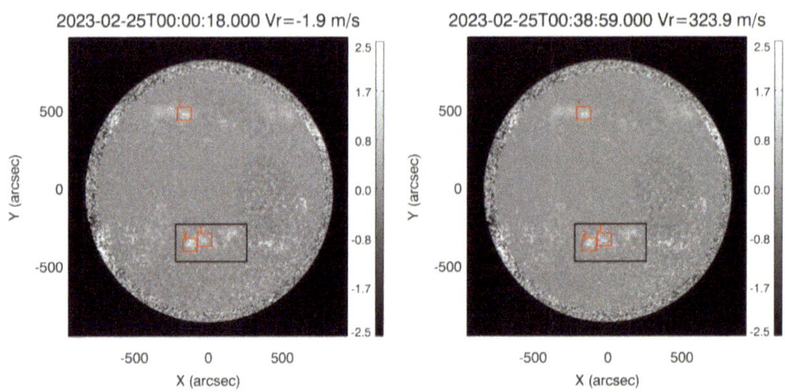

Figure 2 Distribution of the ratios of the line-of-sight magnetic field between the FGM and HMI for $v_r = -1.9\ \mathrm{m\,s^{-1}}$ (*left*) and $v_r = 323.9\ \mathrm{m\,s^{-1}}$ (*right*) labeled, respectively. The region indicated by a *black rectangle* is the sub-region shown in Figure 3. The subregions 1, 2, and 3 indicated by an *orange rectangle* are the active regions for calculating pixel-to-pixel calibration coefficients in Figure 4.

matching include: i) the feature-point identification; ii) the feature-point matching; iii) determination of image-registration parameters. Feature-point identification searches for rotation, scaling, and translation at different image spatial scales of invariant feature points. Hence, SIFT is suitable to solve the image rotation, affine transformations, intensity, and viewpoint changes in matching features. The SIFT method has been successfully applied in solar image processing (Yang et al., 2018; Ji et al., 2019). After coalignment of the FMG and HMI magnetic field, the calibration coefficients of FMG corresponding to HMI can be obtained pixel-to-pixel on the solar surface directly. Figure 2 shows the full-disk distribution of the ratio of line-of-sight magnetic field between the FGM and HMI for line-of-sight velocity $v_r = -1.9\ \mathrm{m\,s^{-1}}$ and $v_r = 323.9\ \mathrm{m\,s^{-1}}$, respectively. While Panels a and b in Figure 3 show the enlarged images cut from Figure 2 indicated by a black rectangle to investigate the distribution of coefficients of FMG comprising to HMI in detail, Panels c and d in Figure 3 give the errors of these ratios corresponding to Panels a and b, respectively. From Figures 2

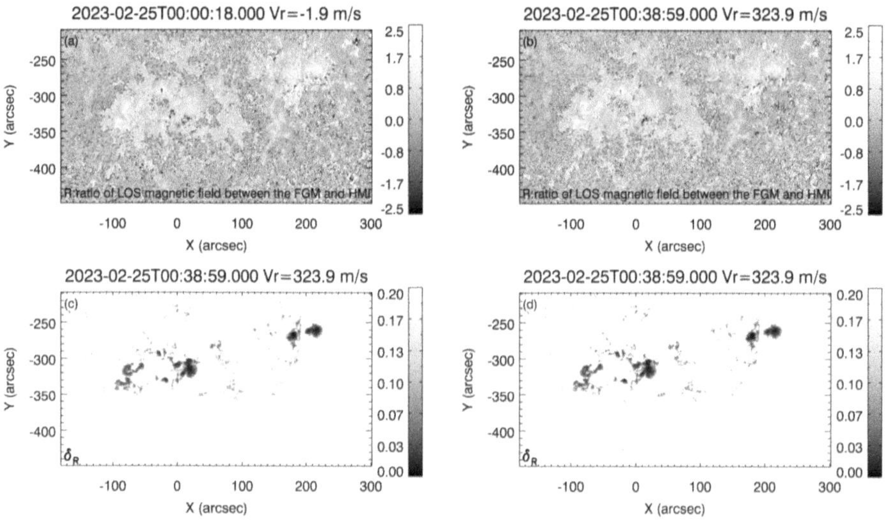

Figure 3 Panels **a** and **b** are the ratio of the line-of-sight magnetic field between the FGM and HMI, which are the same as Figure 2, but for a sub-region cut from full-disk image in Figure 2 indicated by a *black rectangle*. Panels **c** and **d** are the errors of the ratio of the line-of-sight magnetic field between the FGM and HMI for Panels **a** and **b**, respectively.

and 3, it can be found that the distribution of the ratio of line-of-sight magnetic field between the FGM and HMI within the activity region is relatively flat and consistent, which can also be seen from the errors of these ratios shown in Panels c and d in Figure 3, where the relative errors are small in active regions. The fluctuation of these ratios in the quiet region is relatively large, as some magnetic fields with small value in the quiet region may be close to the noise level. Hence, the comparison of calibration coefficients between FMG and HMI is focused on the active region, to obtain the reliable results. In Figure 2, three small subregions with strong magnetic fields labeled 1, 2, and 3 are chosen to investigate the ratio of the line-of-sight magnetic field between FGM and HMI. A series of full-disk images coaligned by SIFT, the same as Figure 2, are rotated to a reference time, and the subregions are cut out from full-disk images by their pixel positions, then the median values of the ratio of the line-of-sight magnetic field between the FGM and HMI for these series of subregions, which can indicate the effective values of these ratios on the whole, are calculated. Figure 4a shows the ratio of the line-of-sight magnetic field between the FGM and HMI of these time series for subregions 1, 2, and 3, and v_r with units of m s^{-1} (line-of-sight component of spacecraft velocity) is plotted as an additional left y-axis to investigate the relations between calibration coefficients and the spacecraft velocity. Here, the mean of errors of these ratios for subregions are calculated and labeled δ_R, with the values of $\delta_R = 0.118$, 0.158, and 0.064 for subregions of 1, 2, and 3, respectively. Figures 4b, c, and d show the scatter plots of the ratio of the line-of-sight magnetic field between the FGM and HMI vs. v_r for subregions 1, 2, and 3 labeled. It can be seen that the ratios of the line-of-sight magnetic field between the FGM and HMI have a consistent trend with the longitudinal velocity of the spacecraft, and the monotonicity of the ratio and spacecraft longitudinal velocity can also be clearly seen from Figures 4b, c, and d. This means that the spacecraft longitudinal velocity regularly modulates the magnetic-field calibration coefficient due to the shift of position of the observed spectral line by the Doppler effect resulting from spacecraft velocity.

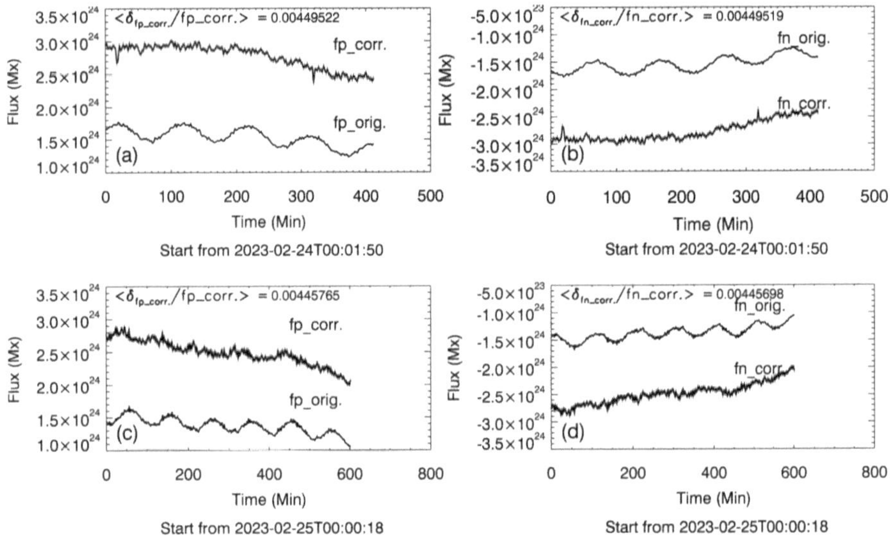

Figure 7 Examples of the evolution of the magnetic flux for active regions before and after cross-calibration coefficients correction. Here, the cross-calibration coefficients are calculated by fitting. Panels **a** and **c** are active region NOAA 13230 observed on 24 February 2023, Panels **b** and **d** are NOAA 13236 observed on 25 February 2023, positive and negative magnetic flux before and after correction are indicated by fp_{orig}, fp_{corr}, fn_{orig}, and fn_{corr}, respectively. For plots of fp_{corr} and fn_{corr}, the error bars are added.

these active regions before correction are modulated by spacecraft velocity, evidently with the same periodicity, while after the correction the periodicities of evolution of the magnetic flux become significantly weaker. Therefore, the coefficients corrected FMG magnetic field will be more suitable for quantitative research on the solar magnetic field.

4. Discussions and Conclusions

The FMG onboard the ASO-S spacecraft operates in a Sun-synchronous orbit with an orbital period of about 99.2 minutes. Subsequently Sun–FMG radial velocities also have periodicity with longitudinal velocity [v_r] between Sun and FMG changing throughout the year from -3.9 to $+3.9$ km s^{-1}, as well as throughout each orbit. The longitudinal velocity of FMG affects the calibration of magnetic field obviously for routine FMG observations, when one spectral position is fixed during an observation. The method of correcting the magnetic field affected by spacecraft velocity can be investigated by comparing the calibration coefficients with other instruments.

In this article, the HMI longitudinal magnetic field with a 45-second cadence working as reference to calibrate the FMG magnetic field is tested. From the test results in Figure 7, it can be seen that this type of calibration method can largely eliminate the influence of spacecraft periodic velocity on the magnetic field of FMG. The corrected magnetic field is closer to the actual conditions, which would definitely improve the reliability of the FMG magnetic field. The consistency with the HMI magnetic field is not the goal of FMG calibration, the real goal is to eliminate the impact of the spacecraft orbital velocity period on the FMG magnetic field, and to have a precise measurement of magnetic field without biases or systematic errors. For example, in Figure 7, the effect of magnetic-flux evolution being consistent with spacecraft velocity period should be corrected, as it is a false signal. Since the linear calibration method is adopted by FMG, the consistency of FMG magnetic-field values itself without the influence of spacecraft orbital period is very important.

FMG observes magnetic fields over the full solar disk, and the calibration coefficient is very sensitive to the position on the solar surface. There exists a large-scale background Doppler velocity field related to the solar position, which inevitably participates in the influence of spacecraft velocity on the FMG magnetic field. Therefore, a complete data sample, in which the active-region observations exist on positions on the solar surface, is a necessity in order to fully and effectively correct the FMG magnetic field influenced by the periodicity of spacecraft velocity. However, in this article, the focus of attention has been placed on the impact of spacecraft velocity.

Another aspect that needs to be mentioned is that calibration coefficients are spectral-position related, since the relation between Stokes signal strength and magnetic field varies for different observation positions of a spectral line when the periodic velocity affects the shape of spectral lines. In the routine observations of FMG at present, the observed spectral line is fixed at -0.08 Å away from the center of the line. Hence, the dataset of correcting calibration coefficients in this article is suitable to correct the FMG magnetic that is observed at the line wing of -0.08 Å. Additionally, the dataset of calibration coefficients for the observation at line wing of -0.08 Å is not complete, and more active regions trying to cover the solar disk are need.

An alternative method to correct FMG magnetic field is a post-processing method based on the periodicity of spacecraft velocity with high certainty. For example, the evolution of magnetic flux in Figure 7, filtering technology (such as the FFT) can be applied to these curves to obtain the component with the period of 99.2 minutes, then a reliable magnetic field can be deduced by eliminating the 99.2-minute periodic component. Additionally, machine-learning methods of artificial intelligence may also correct the FMG magnetic field after the acquisition of a large observation sample with enough information of observation spectral position, solar surface position, and a full range of spacecraft velocity.

From the results in the article, it can be seen that through cross-comparison with HMI, it is largely possible to correct the periodic influence of spacecraft orbital velocity on the FMG magnetic field, and obtain a more realistic magnetic field observed by FMG. However, it should be noted that HMI is not the only sample used as a reference standard, since the self-consistency of the FMG magnetic field avoiding spacecraft-velocity impact is the main goal of correction. Another observation that can be chosen as reference is data obtained by SMAT (*Solar Magnetism* and *Activity Telescope* at the Huairou Solar Observing Station); the advantage of FMG and SMAT is that they have common design parameters. To correct FMG magnetic fields more comprehensively, the main aspect is to establish a more comprehensive calibration database with adequate information of observation spectral position, solar-surface position, and full range of spacecraft velocity, all of these need continuous long-term stable observations of FMG.

Acknowledgments This work is supported by the National Key R&D Program of China 2022YFF0503000 (2022YFF0503001), the Strategic Priority Research Program on Space Science, the Chinese Academy of Sciences (Grant Nos. XDA15320301, XDA15320302, and XDA15052200), the Natural Science Foundation of China (Grant Nos. 11203036, 11703042, U1731241, 11427901, 11473039, 11427901, and 11178016), the Young Researcher Grant of National Astronomical Observations, Chinese Academy of Sciences, and the Key Laboratory of Solar Activity National Astronomical Observations, Chinese Academy.

Author contributions S. Liu and J.T. Su carried out this work and were responsible for the data analysis. X.Y. Bai and Y.Y. Deng provided suggestions on data processing and data analysis. J. Chen, Y.L. Song, X.F. Wang, H.Q. Xu, and X. Yang gave overall suggestions for discussing and publishing. All the authors were supported by National Key R&D Program of China 2022YFF0503000 (2022YFF0503001).

Declarations

Competing interests The authors declare no competing interests.

References

Bond, H.E., Mullan, D.J., O'Brien, M.S., Sion, E.M.: 2001, Detection of coronal mass ejections in V471 tauri with the Hubble space telescope. *Astrophys. J.* **560**, 919. DOI. ADS.

Deng, Y.Y., Zhang, H.Y., Yang, J.F., Li, F., Lin, J.B., Hou, J.F., Wu, Z., Song, Q., Duan, W., Bai, X.Y., Wang, D.G., Lv, J., Ge, W., Wang, J.N., Zheng, Z.Y., Wang, C.J., Wang, N.G., Ni, H.K., Zeng, Y.Z., Zhang, Y., Yang, X., Sun, Y.Z., Zhang, Z.Y., Wang, X.: 2019, *Res. Astron. Astrophys.* DOI. ADS.

Forbes, T.G.: 2000, A review on the genesis of coronal mass ejections. *J. Geophys. Res.* **105**, 23153. DOI. ADS.

Gan, W.-Q., Zhu, C., Deng, Y.-Y., Li, H., Su, Y., Zhang, H.-Y., Chen, B., Zhang, Z., Wu, J., Deng, L., Huang, Y., Yang, J.-F., Cui, J.-J., Chang, J., Wang, C., Wu, J., Yin, Z.-S., Chen, W., Fang, C., Yan, Y.-H., Lin, J., Xiong, W.-M., Chen, B., Bao, H.-C., Cao, C.-X., Bai, Y.-P., Wang, T., Chen, B.-L., Li, X.-Y., Zhang, Y., Feng, L., Su, J.-T., Li, Y., Chen, W., Li, Y.-P., Su, Y.-N., Wu, H.-Y., Gu, M., Huang, L., Tang, X.-J.: 2019, Advanced space-based solar observatory (ASO-S): an overview. *Res. Astron. Astrophys.* **19**, 156. DOI. ADS.

Gan, W., Zhu, C., Deng, Y., Zhang, Z., Chen, B., Huang, Y., Deng, L., Wu, H., Zhang, H., Li, H., Su, Y., Su, J., Feng, L., Wu, J., Cui, J., Wang, C., Chang, J., Yin, Z., Xiong, W., Chen, B., Yang, J., Li, F., Lin, J., Hou, J., Bai, X., Chen, D., Zhang, Y., Hu, Y., Liang, Y., Wang, J., Song, K., Guo, Q., He, L., Zhang, G., Wang, P., Bao, H., Cao, C., Bai, Y., Chen, B., He, T., Li, X., Zhang, Y., Liao, X., Jiang, H., Li, Y., Su, Y., Lei, S., Chen, W., Li, Y., Zhao, J., Li, J., Ge, Y., Zou, Z., Hu, T., Su, M., Ji, H., Gu, M., Zheng, Y., Xu, D., Wang, X.: 2023, The advanced space-based solar observatory (ASO-S). *Solar Phys.* **298**, 68. DOI. ADS.

Hou, J.-F., Wang, D.-G., Deng, Y.-Y., Zhang, Z.-Y., Sun, Y.-Z.: 2020, Interference effect on the liquid-crystal-based Stokes polarimeter. *Chin. Phys. B* **29**, 124211. DOI. ADS.

Ichimoto, K., Lites, B., Elmore, D., Suematsu, Y., Tsuneta, S., Katsukawa, Y., Shimizu, T., Shine, R., Tarbell, T., Title, A., Kiyohara, J., Shinoda, K., Card, G., Lecinski, A., Streander, K., Nakagiri, M., Miyashita, M., Noguchi, M., Hoffmann, C., Cruz, T.: 2008, Polarization calibration of the solar optical telescope onboard hinode. *Solar Phys.* **249**, 233. DOI. ADS.

Jefferies, J., Lites, B.W., Skumanich, A.: 1989, Transfer of line radiation in a magnetic field. *Astrophys. J.* **343**, 920. DOI. ADS.

Jefferies, J.T., Mickey, D.L.: 1991, On the inference of magnetic field vectors from Stokes profiles. *Astrophys. J.* **372**, 694. DOI. ADS.

Ji, K.F., Liu, H., Jin, Z.Y., Shang, Z.H., Qiang, Z.P.: 2019, An automatic approach of mapping the solar high-resolution image to helioprojective-Cartesian coordinates system. *Chin. Sci. Bull.* **64**, 1738. DOI. ADS.

Krall, K.R., Smith, J.J.B., Hagyard, M.J., West, E.A., Cummings, N.P.: 1982, Vector magnetic field evolution, energy storage, and associated photospheric velocity shear within a flare-productive active region. *Solar Phys.* **79**, 59. DOI. ADS.

Landolfi, M., Landi Degl'Innocenti, E.: 1982, Magneto-optical effects and the determination of vector magnetic fields from Stokes profiles. *Solar Phys.* **78**, 355. DOI. ADS.

Liu, S., Su, J.T., Bai, X.Y., Deng, Y.Y., Chen, J., Song, Y.L., Wang, X.F., Xu, H.Q., Yang, X.: 2022, A study on correcting the effect of polarization crosstalk in full-disk solar photospheric magnetic field observations. *Solar Phys.* **297**, 6. DOI. ADS.

Lowe, D.G.: 1999, Object recognition from local scale-invariant features. In: *Proc. Seventh IEEE Internat. Conf. Comp. Vis.* **2**, 1150. DOI.

Nindos, A., Patsourakos, S., Wiegelmann, T.: 2012, On the role of the background overlying magnetic field in solar eruptions. *Astrophys. J. Lett.* **748**, L6. DOI. ADS.

Priest, E.R., Forbes, T.G.: 2002, The magnetic nature of solar flares. *Astron. Astrophys. Rev.* **10**, 313. DOI. ADS.

Rachkovsky, D.N.: 1967, Magnetic rotation effects in spectral lines. *Izv. Krym. Astrofiz. Obs.* 37. ADS.

Scherrer, P.H., Schou, J., Bush, R.I., Kosovichev, A.G., Bogart, R.S., Hoeksema, J.T., Liu, Y., Duvall, T.L., Zhao, J., Title, A.M., Schrijver, C.J., Tarbell, T.D., Tomczyk, S.: 2012, The helioseismic and magnetic imager (HMI) investigation for the solar dynamics observatory (SDO). *Solar Phys.* **275**, 207. DOI. ADS.

Schou, J., Scherrer, P.H., Bush, R.I., Wachter, R., Couvidat, S., Rabello-Soares, M.C., Bogart, R.S., Hoeksema, J.T., Liu, Y., Duvall, T.L., Akin, D.J., Allard, B.A., Miles, J.W., Rairden, R., Shine, R.A., Tarbell, T.D., Title, A.M., Wolfson, C.J., Elmore, D.F., Norton, A.A., Tomczyk, S.: 2012, Design and ground calibration of the helioseismic and magnetic imager (HMI) instrument on the solar dynamics observatory (SDO). *Solar Phys.* **275**, 229. DOI. ADS.

Shibata, K., Masuda, S., Shimojo, M., Hara, H., Yokoyama, T., Tsuneta, S., Kosugi, T., Ogawara, Y.: 1995, Hot-plasma ejections associated with compact-loop solar flares. *Astrophys. J. Lett.* **451**, L83. DOI. ADS.

Su, J.-T., Bai, X.-Y., Chen, J., Guo, J.-J., Liu, S., Wang, X.-F., Xu, H.-Q., Yang, X., Song, Y.-L., Deng, Y.-Y., Ji, K.-F., Deng, L., Huang, Y., Li, H., Gan, W.-Q.: 2019, Data reduction and calibration of the FMG onboard ASO-S. *Res. Astron. Astrophys.* **19**, 161. DOI. ADS.

Tsuneta, S.: 1996, Structure and dynamics of magnetic reconnection in a solar flare. *Astrophys. J.* **456**, 840. DOI. ADS.

Tsuneta, S., Ichimoto, K., Katsukawa, Y., Nagata, S., Otsubo, M., Shimizu, T., Suematsu, Y., Nakagiri, M., Noguchi, M., Tarbell, T., Title, A., Shine, R., Rosenberg, W., Hoffmann, C., Jurcevich, B., Kushner, G., Levay, M., Lites, B., Elmore, D., Matsushita, T., Kawaguchi, N., Saito, H., Mikami, I., Hill, L.D., Owens, J.K.: 2008, The solar optical telescope for the hinode mission: an overview. *Solar Phys.* **249**, 167. DOI. ADS.

Unno, W.: 1956, Line formation of a normal Zeeman triplet. *Publ. Astron. Soc. Japan* **8**, 108. ADS.

Wang, H., Ewell, J.M.W., Zirin, H., Ai, G.: 1994, Vector magnetic field changes associated with X-class flares. *Astrophys. J.* **424**, 436. DOI. ADS.

Xu, H.-Q., Liu, S., Su, J.-T., Deng, Y.-Y., Plotnikov, A., Bai, X.-Y., Chen, J., Yang, X., Guo, J.-J., Wang, X.-F., Song, Y.-L.: 2021, Automatic detection and correction algorithms for magnetic saturation in the SMFT/HSOS longitudinal magnetograms. *Res. Astron. Astrophys.* **21**, 067. DOI. ADS.

Yang, P., Zeng, S.G., Liu, S., Zheng, S., Lin, G.H., He, H.L.: 2018, Registration and location method of solar magnetic field images based on scale-invariant feature point matching. *Astron. Res. Tech.* **15**, 59.

Solar Physics (2024) 299:17
https://doi.org/10.1007/s11207-024-02260-8

RESEARCH

Comparison of Line-of-Sight Magnetic Field Observed by ASO-S/FMG, SDO/HMI and HSOS/SMAT

Haiqing Xu[1,2] · Jiangtao Su[1,2,3] · Suo Liu[1,2,3] · Yuanyong Deng[1,2,3] · Shangbin Yang[1,2,3] · Xianyong Bai[1,2,3] · Jie Chen[1,2] · Xiaofan Wang[1,2] · Xiao Yang[1,2] · Yongliang Song[1,2] · Shahid Idrees[1,2,3]

Received: 27 October 2023 / Accepted: 19 January 2024 / Published online: 9 February 2024

Abstract
The Full-disk MagnetoGraph (FMG) onboard the Advanced Space based Solar Observatory has obtained a series of line-of-sight magnetic-field measurements since its launch in October 2022. It is important to compare its observational data with other existing solar telescopes. In this paper, we make a detailed comparison of four active regions and a pore region simultaneously observed by FMG, the Helioseismic and Magnetic Imager (HMI) onboard the Solar Dynamic Observatory, and the Solar Magnetism and Activity Telescope (SMAT) at Huairou Solar Observing Station. We find that the magnetic-field structure and spatial distribution are basically consistent among these three instruments. The initial correlation coefficient of magnetic field is approximately 0.90. The coefficient slightly increases when sunspot umbra regions are excluded, and it increases significantly up to 0.98 for the magnetic field in penumbra regions. The magnetic field observed by FMG tends to be weaker than the HMI in strength in sunspot umbra due to saturation effect, whereas larger outside sunspot. The differences are probably due to different noise levels, seeing conditions (SMAT is affected by the Earth's atmosphere) and observational and calibration methods.

Keywords Solar magnetic field · FMG magnetogram

1. Introduction

Observation of full-disk solar vector magnetic fields can provide an overview of the global variations of the magnetic field, which is very important for the study of large-scale magnetic activity associated with flares and coronal mass ejections (CMEs). It also provides

✉ H. Xu
xhq@bao.ac.cn

✉ J. Su
sjt@bao.ac.cn

1 National Astronomical Observatories, Chinese Academy of Sciences, Beijing, 100101, China

2 Key Laboratory of Solar Activity and Space Weather, National Space Science Center, Chinese Academy of Sciences, Beijing 100190, China

3 School of Astronomy and Space Sciences, University of Chinese Academy of Sciences, Beijing 100049, China

data for numerical simulations and space weather models to forecast solar and geoeffective events. The scientific objectives of the Advanced Space-based Solar Observatory (ASO-S) are to study the relationships between the solar magnetic field, solar flares and coronal mass ejections (Gan et al., 2019). The Full-disk MagnetoGraph (FMG) is one payload onboard the ASO-S for measuring solar photospheric magnetic fields. The aperture of FMG is 14 cm and a CMOS camera with 4096 × 4096 pixels is used. FMG observes polarized Stokes images using a Liquid Crystal Variable Retarder (Hou et al., 2020). The routine observations of FMG are taken at a single wavelength position of the Fe I λ 5324.19 Å and a linear calibration method is adopted. The sensitivity for longitudinal magnetic field is 15 G in normal mode. The temporal resolution is 30 seconds for single component and 2 minutes for vector magnetogram in normal mode. The spatial resolution is approximately 1.5″ with a 0.5″ pixel size (Deng et al., 2019, Su et al., 2019). The spatial resolution for the flight model (FM) of FMG is 1.04″ (Gan et al., 2023).

For a new instrument, it is necessary to compare its observational data with those of other existing solar telescopes. In this paper, we compare the line-of-sight (LOS) magnetic field observed by FMG with the Helioseismic and Magnetic Imager onboard the Solar Dynamic Observatory (SDO/HMI: Scherrer et al., 2012, Schou et al., 2012) and the Solar Magnetism and Activity Telescope (SMAT: Zhang et al., 2007) at Huairou Solar Observing Station (HSOS). These two instruments have been continuously observing magnetic field for over 10 years. HMI is currently the most commonly used space-based full-disk magnetograph in the world. The aperture of HMI is 14 cm and two CCDs with 4096 × 4096 pixels are used. The routine observations of HMI are taken at six wavelength positions of the Fe I λ 6173.3 Å line. The Stokes parameters are then inverted to produce the vector magnetic field using an inversion algorithm. The spatial resolution of HMI is 1″ with a 0.5″ pixel size. SMAT is a ground-based instrument which consists of an aperture of 10 cm telescope performing full-disk vector magnetic-field observations. SMAT uses the same working spectral line of Fe I λ 5324.19 Å as FMG. The spatial resolution of the full-disk vector magnetograms is better than 5″ and the temporal resolution for observing a full-disk vector magnetogram is about (or less than) 10 minutes. The LOS magnetic field of SMAT is calibrated through the symmetry measurements of Stokes parameter V, normalized by the intensity I, to remove the large-scale background of V signal and increase the signal-to-noise ratio (Su and Zhang, 2007). The calibration coefficient is obtained using the weak-approximation method (Ai, Li, and Zhang, 1982). A 2k × 2k CCD is currently used. The output data after calibration is 992 × 992 pixels and the pixel size is approximately 2″.

There are several studies comparing the magnetic field obtained by HMI and SMAT with those of other magnetographs. For example, Liu et al. (2012) performed a detailed comparison of LOS magnetic field between HMI and the Michelson Doppler Imager (MDI) onboard the Solar and Heliospheric Observatory (SOHO: Scherrer et al., 1995). Hayashi, Yang, and Deng (2016) used the synoptic maps of solar photospheric magnetic field from four observatories, including HMI and SMAT, to compare the potential field solutions for Carrington Rotation 2144. Demidov et al. (2018) compared the LOS magnetograms obtained by SMAT, HMI and the Solar Telescope for Operative Predictions (STOP). Sinjan et al. (2023) compared the magnetic field maps from the High Resolution Telescope (HRT) of the Polarimetric and Helioseismic Imager (PHI) on the Solar Orbiter mission (SO: Solanki et al., 2020) and HMI. These studies found a general consistency among different instruments, but there were also some differences. The cross-comparison can help us to search for the reasons for these differences and investigate the method to improve the quality of the data.

This paper is organized as follows: In Section 2, we describe the data reduction and the results of the comparison. In Section 3, we discuss the factors influencing the correlation of

Table 1 Data sample.

NOAA	Location	Date	Time (UT)		
			FMG	HMI	SMAT
Pore	N37W02	2022–12–06	00:50:15	00:44:10	–
13173	N24W15	2022–12–29	18:58:49	18:58:21	–
13176	N19E12	2022–12–30	22:32:28	22:32:06	–
13229	N25W05	2023–02–22	05:24:23	05:24:39	05:24:21
13282	N11E10	2023–04–17	03:18:52	03:18:46	03:19:09

magnetic field obtained by FMG and HMI. In Section 4, we summarize the main conclusions.

2. Data Reduction and Results

We selected four active regions and a pore region simultaneously observed by FMG and HMI. Of these, only the active regions NOAA 13229 and 13282 were simultaneously observed by FMG, HMI and SMAT. The basic information about these data samples is listed in Table 1, which includes the NOAA number, location, observing date and time. The data of four active regions (pore region) were obtained in normal mode (deep integration mode). The FMG data in normal mode has been accessible on the website http://aso-s.pmo.ac.cn/sodc/dataArchive.jsp since April 1, 2023. We used the HMI 45-second magnetograms for our comparison. To achieve a detailed comparison between any two magnetograms observed by different instruments, we first linearly interpolated the full-disk magnetograms onto the same pixel size (approximately 0.6″). Next, we employed the cross-correlation method to align the magnetograms and extracted the same region from each full-disk magnetogram. For FMG and SMAT data, we adjusted their magnetic-field strength to match that of HMI by multiplying a constant factor determined from the slope of the linear fit to the magnetic field scatter plot (e.g., slope = 1 in Figure 1(d)). This adjustment of strength does not affect the correlation coefficient of magnetic field between different instruments. The constant factor varies for different active regions (time) due to the effect of satellite orbit velocity on FMG LOS magnetic field, as discussed by Liu et al. (2023). The calibration coefficient of SMAT LOS magnetic field was obtained by using the weak-approximation method and the magnetic field strength may differ slightly from that of HMI. We will give a detailed analysis for each selected region in the following subsections.

2.1. NOAA 13229

Active region (AR) NOAA 13229 comprises multiple sunspots with a complex magnetic field. Figures 1(a) – (c) demonstrate the remarkable similarity of the magnetic structures observed by FMG, HMI and SMAT. In FMG data, we observe a decrease in magnetic field strength within the umbra, a phenomenon commonly referred to as “saturation”. Some fine structures cannot be recognized due to lower spatial resolution of SMAT data. Figure 1(d) illustrates the relationship between the LOS magnetic fields measured by FMG and HMI with a correlation coefficient of 0.90. Figure 1(e) shows the relationship between FMG and SMAT LOS magnetic fields, yielding a correlation coefficient of 0.87. Figure 1(f) depicts

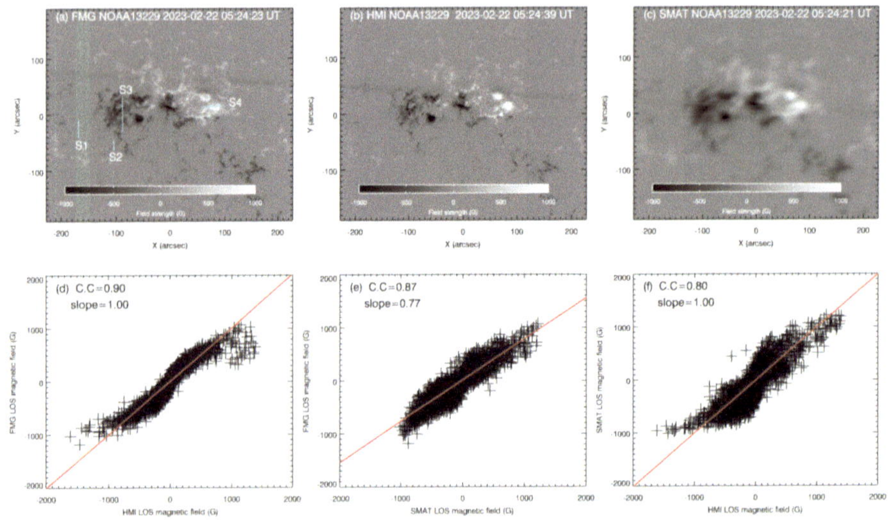

Figure 1 (a)–(c): LOS magnetograms of NOAA 13229 on Feb. 22, 2023, observed by FMG, HMI and SMAT, respectively. (d)–(f): Scatter plots of LOS magnetic fields observed by FMG and HMI, FMG and SMAT, HMI and SMAT, respectively. The red solid lines are linear fit to the data and the slopes are marked on the corresponding panel. C.C is the abbreviation for the correlation coefficient of magnetic field.

the relationship between HMI and SMAT LOS magnetic fields, with a correlation coefficient of 0.80.

We created four slices representing distinct magnetic-field structure. Figure 2 shows the magnetic-field variation along slices S1 – S4, as indicated in Figure 1(a). In the quiet region (Figure 2(a)), the magnetic field observed by FMG (blue line) and HMI (black line) primarily fluctuates within ± 20 G, exhibiting low consistency, while the magnetic field observed by SMAT exhibits smoother variations. We created slice S2 passing through a small magnetic-field structure, and the magnetic-field distribution along this slice is shown in Figure 2(b). For FMG and HMI, the magnetic-field strength exhibits a significant increase from quiet region to the small magnetic-field structure. In contrast, the magnetic-field strength observed by SMAT remains relatively constant. Gaussian fit was applied to the observed data from FMG (blue dotted line) and HMI (black dotted line). The full width at half maximum (FWHM) of Gaussian fit is $1.40''$ for FMG and $1.37''$ for HMI. When the magnetic structure is sufficiently small, the FWHM can represent the spatial resolution of an instrument. As evident from Figure 2(c), the magnetic-field variation along slice S3 observed by FMG and HMI exhibits high consistency; however, the magnetic-field strength of FMG exceeds that of HMI at most positions. The magnetic-field variation trend observed by SMAT aligns with that of FMG and HMI, although the fine structure has been lost. Slice S4 passes through the sunspot. Figure 2(d) reveals that the magnetic-field strength gradually increases toward the center of the slice, while the magnetic-field strength observed by FMG decreases at positions 15 – 28 arcsec, consistent with the saturation phenomena observed in magnetograms.

2.2. NOAA 13282

NOAA 13282 is a bipolar active region. Figures 3(a) – (c) reveal that the magnetic structures observed by FMG, HMI and SMAT are remarkably similar. Besides FMG and SMAT, the saturation phenomenon can also be found in HMI magnetogram, but over the smaller area.

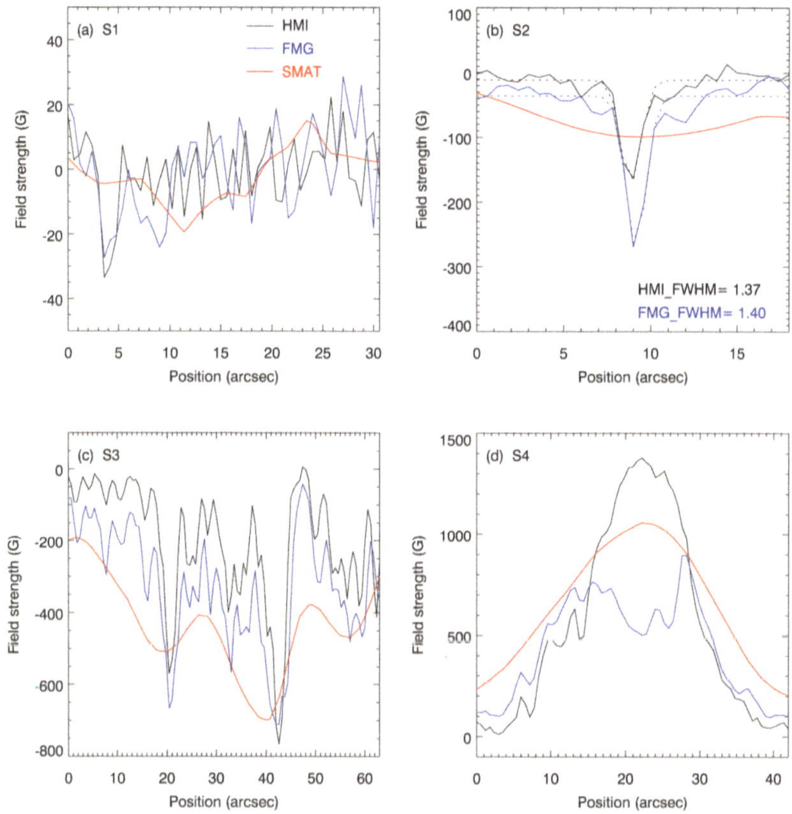

Figure 2 Distribution of magnetic field along slices (S1 – S4) in Figure 1(a). Each slice starts from the left or lower end. The black, blue and red solid lines in each panel represent HMI, SMAT and FMG data, respectively. The black and blue dotted lines in panel (b) represent Gaussian fits to the data of HMI and FMG. The full width at half maximum (FWHM) is marked.

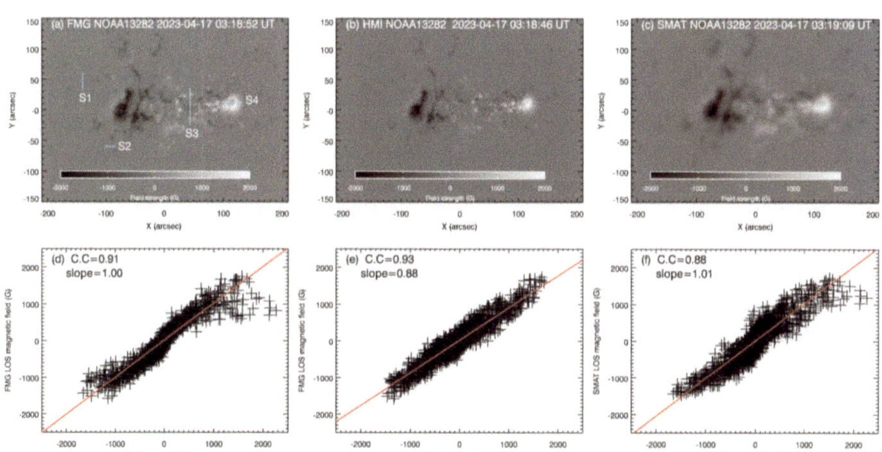

Figure 3 As in Figure 1, but for NOAA 13282.

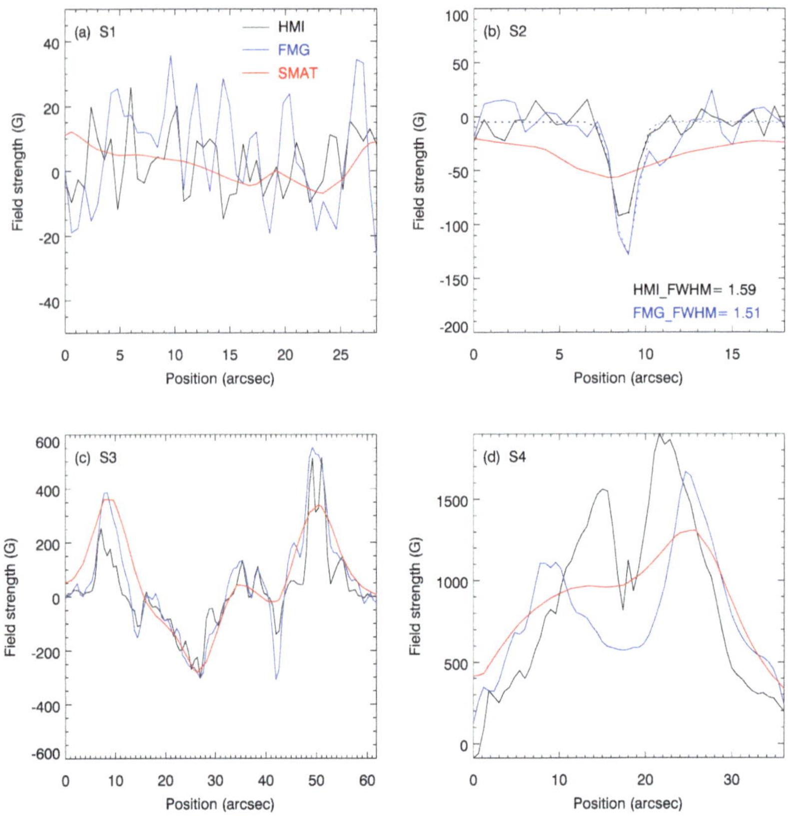

Figure 4 As in Figure 2, but for NOAA 13282.

Figure 3(d) shows the relationship between the LOS magnetic fields observed by FMG and HMI with a correlation coefficient of 0.91. Figure 3(e) shows the relationship between the LOS magnetic fields observed by FMG and SMAT with a correlation coefficient of 0.93. Figure 3(f) shows the relationship between the LOS magnetic fields observed by SMAT and HMI with a correlation coefficient of 0.88.

Figure 4 illustrates the magnetic-field variation along slices S1 – S4, as indicated in Figure 3(a). It is evident that the magnetic field observed by SMAT exhibits a smooth variation with position, while a finer structure is found in FMG and HMI observations. The magnetic-field variation trend is very consistent, except for the quiet region. As shown in Figure 4(b), the FWHM of Gaussian fit is 1.51″ for FMG and 1.59″ for HMI. Figure 4(d) reveals that the magnetic field observed by FMG, HMI, and SMAT decreases near the center of slice (sunspot umbra).

2.3. NOAA 13173

AR NOAA 13173 comprises a small sunspot. The magnetograms of FMG and HMI are shown in Figure 5(a) and (b), respectively. These images reveal a remarkably similar magnetic structure. The saturation phenomenon is also found in the FMG magnetogram. Figure 5(c) shows the relationship between the LOS magnetic fields observed by FMG and HMI with a correlation coefficient of 0.89.

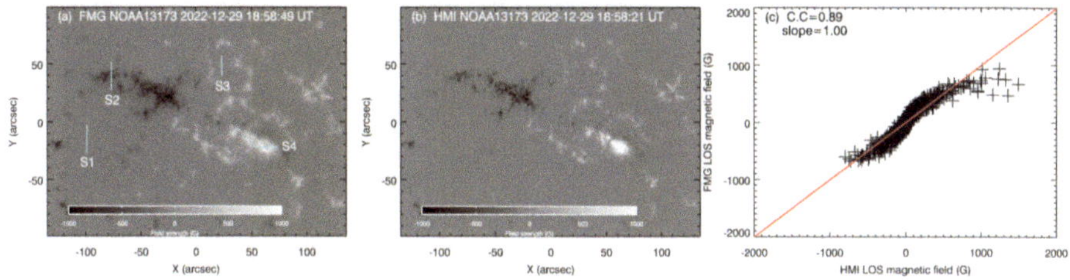

Figure 5 As in Figure 1, but for NOAA 13173.

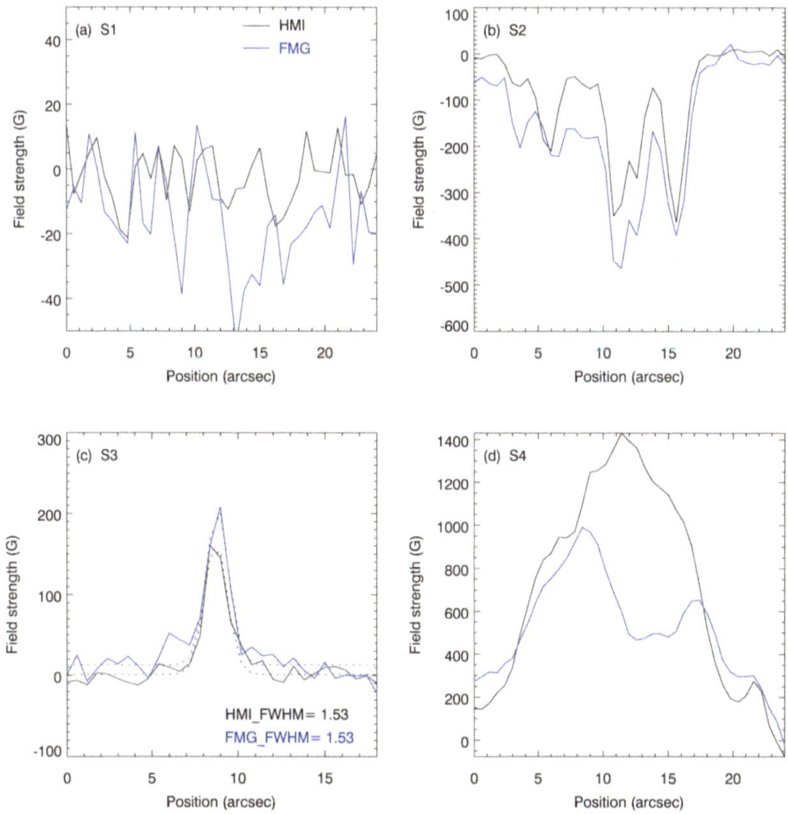

Figure 6 As in Figure 2, but for NOAA 13173.

We selected distinct magnetic-field structures to make slices for detailed analysis. Figure 6 shows the magnetic-field variation along slices S1 – S4, as indicated in Figure 5(a). Similar results to those mentioned earlier were observed. Figure 6(b) reveals that the FWHM of Gaussian fit is 1.53″ for both FMG and HMI.

2.4. NOAA 13176

AR NOAA 13176 comprises multiple sunspots. The magnetograms of FMG and HMI are shown in Figure 7(a) and (b), respectively, revealing a highly similar magnetic structure.

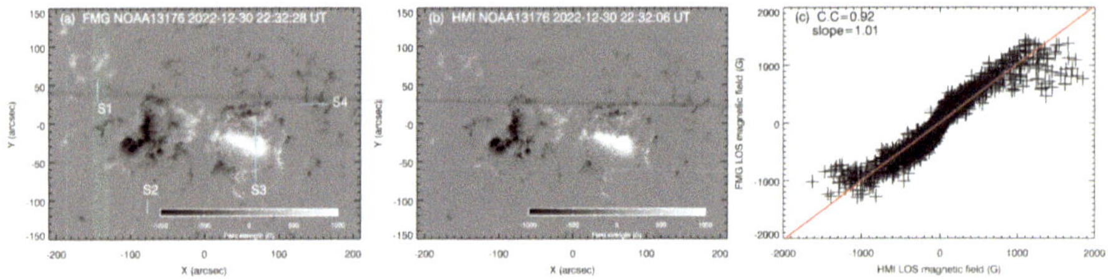

Figure 7 As in Figure 1, but for NOAA 13176.

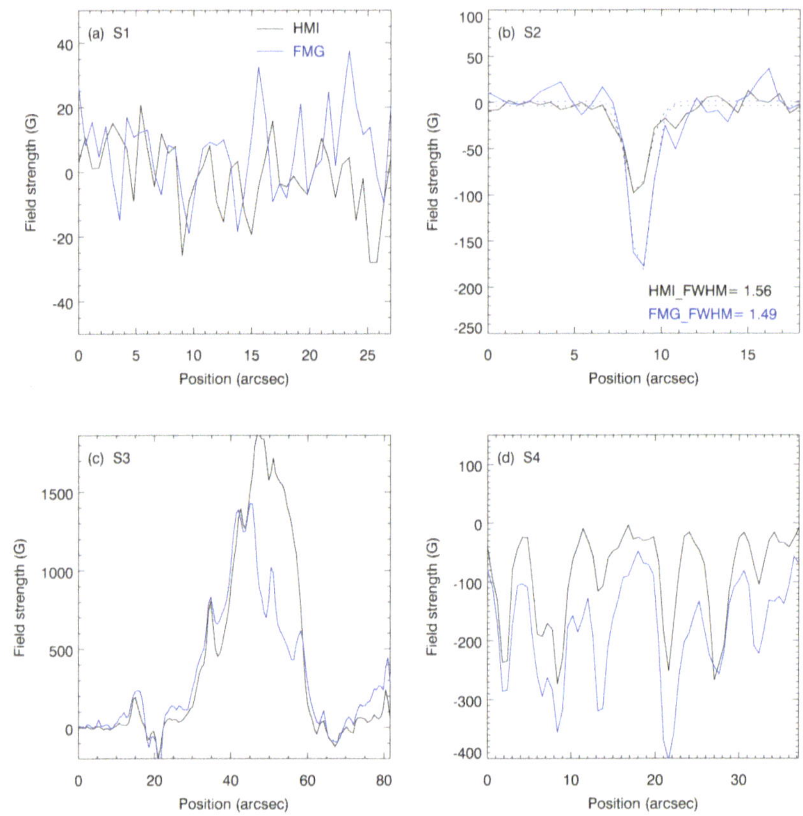

Figure 8 As in Figure 2, but for NOAA 13176.

The saturation phenomenon is evident in the FMG magnetogram. Figure 7(c) shows the relationship between the LOS magnetic fields observed by FMG and HMI with a correlation coefficient of 0.92. Figure 8 shows the magnetic-field variation along slices S1 – S4, as indicated in Figure 7(a). Similar results to those mentioned earlier were observed. Figure 8(b) shows that the FWHM of Gaussian fit is 1.49″ for FMG and 1.56″ for HMI.

We also performed a similar analysis for a pore region. The magnetograms of FMG and HMI are shown in Figure 9(a) and (b), respectively, demonstrating remarkably similar magnetic structures. Figure 9(c) shows the relationship between the LOS magnetic fields

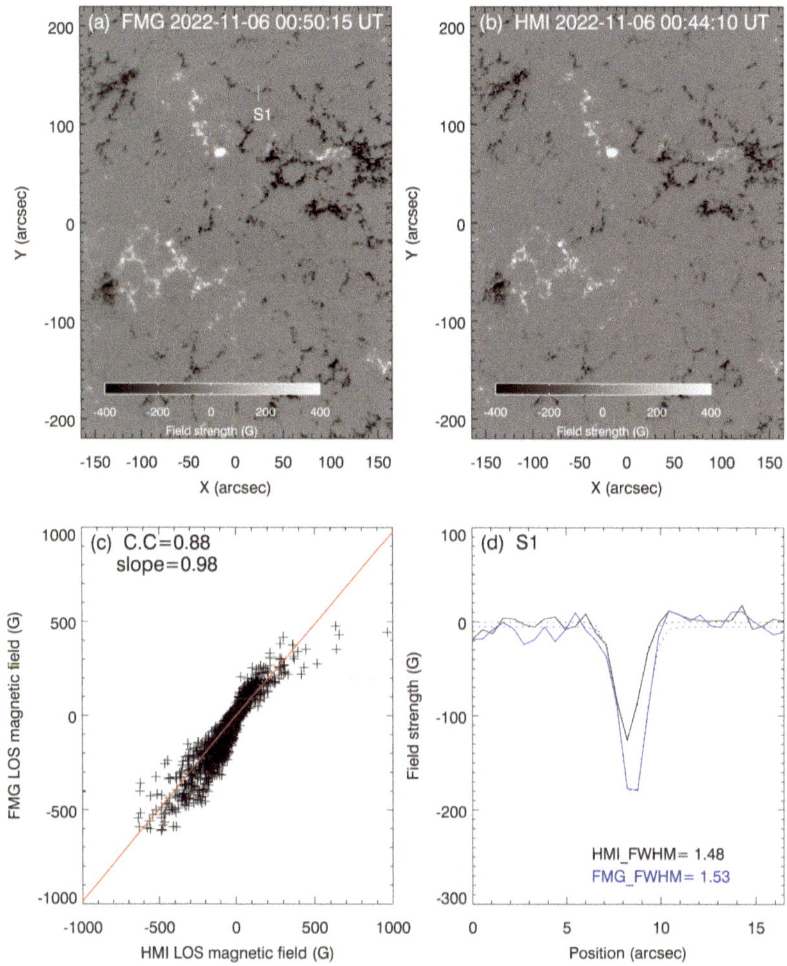

Figure 9 (a) and (b): LOS magnetograms of a pore region, observed by FMG and HMI on Nov. 6, 2022. (c): The scatter plot of magnetic field. The red solid line is linear fit to the data and the slope is marked. (d) Distribution of magnetic field along slice S1, which starts from the lower end. The black and blue solid lines represent HMI and FMG data, respectively. The black and blue dotted lines represent Gaussian fits to the observational data.

observed by FMG and HMI, with a correlation coefficient of 0.88. Figure 9(d) shows that the FWHM of Gaussian fit is 1.53″ for FMG and 1.48″ for HMI.

3. Discussion

The saturation can be found in sunspot umbra for all four active regions observed by FMG due to its routine observation at a single wavelength position and the adoption of linear calibration. Xu et al. (2021) developed a method to correct for saturation in FMG magnetic field based on the data obtained by the Solar Magnetic Field Telescope at HSOS. We will discuss how to remove the saturation effect from FMG data in a seperate work. To avoid

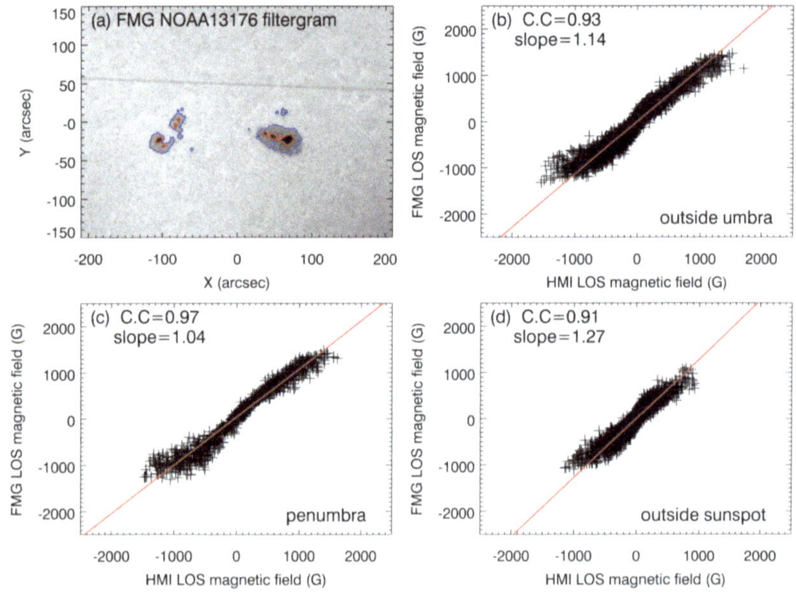

Figure 10 (a) Filtergram of NOAA 13176, observed by FMG. The red and blue contours show umbral and penumbral boundaries, respectively; (b)–(d) scatter plots of magnetic field (FMG vs. HMI) outside umbra, inside penumbra and outside sunspot, respectively. The red solid lines are linear fit to the data and the slopes are marked on the corresponding panel.

the influence of the saturation effect on the correlation of magnetic fields, the umbra region can be removed when analyzing the correlation of magnetic fields between FMG and HMI. Figure 10 illustrates this method in detail. First, the normalized light intensity is used to divide the active region into different parts, and then the correlation of magnetic fields in different parts between FMG and HMI is analyzed. In Figure 10(a), the red and blue contours represent the umbra and penumbra boundaries, respectively, dividing the active region into three parts: umbra (inside the red contour), penumbra (between the red and blue contours) and outside sunspots (outside the blue contour). Figure 10(b) shows the scatter plot of magnetic fields outside the umbra with a correlation coefficient of 0.93 and a linear fit slope of 1.14. The correlation coefficient increases to 0.97 in the penumbra and the slope is 1.04 (Figure 10(c)). Outside the sunspot, the correlation coefficient decreases to 0.91 and the slope increases to 1.27 (Figure 10(d)). The same analysis was done for other three active regions and the results are summarized in Table 2. The upper part of Table 2 shows the linear correlation coefficients of magnetic fields observed by FMG and HMI for cases: full active region (second column), without umbra (third column), penumbra region (fourth column), and outside sunspot region (last column). The bottom part of Table 2 shows the slopes of linear fit to the scatter plots of magnetic fields. As evident from Table 2, one can see that the correlation coefficients and slopes (except for NOAA 13173) all exhibit a slight increase when the umbra region is excluded. The correlation coefficients all show a significant increase in the penumbra, with the highest value reaching 0.98. On the other hand, the slopes decrease in two active regions and increase in other two active regions. Compared to the entire active region, the correlation coefficients outside the sunspot show a slight increase for two active regions and a decrease for other two active regions, but the slopes all significantly increase.

Table 2 Relationship between magnetic fields observed by FMG and HMI.

NOAA	AR FOV	Outside umbra	Penumbra	Outside sunspot
C.C				
13173	0.89	0.91	0.95	0.91
13176	0.92	0.93	0.97	0.91
13229	0.90	0.92	0.98	0.92
13282	0.91	0.93	0.97	0.89
Slope				
13173	1.00	1.00	0.66	1.18
13176	1.01	1.14	1.04	1.27
13229	1.00	1.09	0.86	1.19
13282	1.00	1.20	1.16	1.25

As evident from Table 2, one can also find that the correlation of magnetic fields between FMG and HMI is primarily affected by the outside sunspot region, rather than the umbra and penumbra. This may be attributed to the relatively small area occupied by the umbra and penumbra, which ranges from 0.37% to 2.01% of the entire active region. The outside sunspot region includes numerous small-scale and weak magnetic fields, rendering it more susceptible to noise-induced reduction in correlation compared to other regions. When considering the correlation coefficients exceeding 3σ (the standard deviation of quiet region), a slight increase to 0.93 – 0.94 is observed. This is particularly evident for NOAA 13282, where the correlation coefficient increases from 0.89 to 0.93.

4. Summary

A detailed pixel-by-pixel comparison of the LOS magnetic field observed by FMG, HMI, and SMAT revealed that the magnetic structure and distribution are essentially consistent among the three instruments. The space-based instruments FMG and HMI show finer magnetic structures with high correlation. The observed differences likely arise from the influence of noise, seeing condition (SMAT is affected by the Earth's atmosphere), observational and calibration methods. The main conclusions are as follows:

(1) Within the entire field of view of the active region, the correlation coefficient of magnetic field between any two instruments is approximately 0.90. The correlation coefficient of magnetic fields between FMG and HMI is the highest in the penumbra, reaching up to 0.98. However, in the quiet region, the correlation is relatively low due to the noise interference.

(2) The magnetic-field strength outside sunspots observed by FMG is stronger than that observed by HMI. The magnetic field in sunspot umbra observed by FMG is seriously affected by saturation effect, resulting in a lower strength compared to that observed by HMI.

(3) The spatial resolution of FMG and HMI is nearly identical, as estimated from the LOS magnetogram.

Acknowledgments We are sincerely grateful to the anonymous reviewer whose comments helped us to significantly improve the article. We acknowledge the free data usage policy of HSOS/SMAT and SDO/HMI.

Author contributions H.Q. Xu and J.T. Su carried out this work and were responsible for the data analysis. S. Liu provided suggestions on FMG data processing and conducted the data alignment. Y.Y. Deng proposed the overall idea. S.B. Yang provided the SMAT data processing. X.Y. Bai, J. Chen, X.F. Wang, X. Yang, and Y.L. Song gave suggestions for discussion and paper revision. S. Idrees corrected the language. All authors reviewed the manuscript.

Funding This work is supported by National Key R&D Program of China (Grant No. 2022YFF0503001, 2021YFA1600500, 2022YFF0503800), National Natural Science Foundation of China (Grant No. 12273059, 11427901, 12173049, 12073040, 12003051, 12373057), Beijing Natural Science Foundation (Grant No. 1222029), and the Strategic Priority Research Program of the Chinese Academy of Sciences (Grant No. XDB0560000). ASO-S mission is supported by the Strategic Priority Research Program on Space Science, the Chinese Academy of Sciences (Grant No. XDA15320000).

Data Availability The HMI data used in this paper is taken from http://jsoc.stanford.edu.

Declarations

Competing interests The authors declare no competing interests.

References

Ai, G.X., Li, W., Zhang, H.Q.: 1982, *Acta Astron. Sin.* **23**, 39. ADS.
Demidov, M.L., Wang, X.F., Wang, D.G., Deng, Y.Y.: 2018, *Solar Phys.* **293**, 146. DOI. ADS.
Deng, Y.Y., Zhang, H.Y., Yang, J.F., Li, F., Lin, J.B., Hou, J.F., et al.: 2019, *Res. Astron. Astrophys.* **19**, 157. DOI. ADS.
Gan, W.Q., Zhu, C., Deng, Y.Y., Li, H., Su, Y., Zhang, H.Y., et al.: 2019, *Res. Astron. Astrophys.* **19**, 156. DOI. ADS.
Gan, W.Q., Zhu, C., Deng, Y.Y., Zhang, Z., Chen, B., Huang, Y., et al.: 2023, *Solar Phys.* **298**, 68. DOI. ADS.
Hayashi, K., Yang, S.B., Deng, Y.Y.: 2016, *J. Geophys. Res.* **121**, 1046. DOI. ADS.
Hou, J.F., Wang, D.G., Deng, Y.Y., Zhang, Z.Y., Sun, Y.Z.: 2020, *Chin. Phys. B* **29**, 124211. DOI. ADS.
Liu, Y., Hoeksema, J.T., Scherrer, P.H., Couvidat, S., Bush, R.I., Duvall, T.L., et al.: 2012, *Solar Phys.* **279**, 295. DOI. ADS.
Liu, S., Su, J.T., Bai, X.Y., Deng, Y.Y., Chen, J., Song, Y.L., et al.: 2023, *Solar Phys.* **298**, 141. DOI. ADS.
Scherrer, P.H., Bogart, R.S., Bush, R., Hoeksema, J.T., Kosovichev, A.G., Schou, J., et al.: 1995, *Solar Phys.* **162**, 129. DOI. ADS.
Scherrer, P.H., Schou, J., Bush, R.I., Kosovichev, A.G., Bogart, R.S., Hoeksema, J.T., et al.: 2012, *Solar Phys.* **275**, 207. DOI. ADS.
Schou, J., Scherrer, P.H., Bush, R.I., Wachter, R., Couvidat, S., Rabello-Soares, M.C., et al.: 2012, *Solar Phys.* **275**, 229. DOI. ADS.
Sinjan, J., Calchetti, D., Hirzberger, J., Kahil, F., Valori, G., Solanki, S.K., et al.: 2023, *Astron. Astrophys.* **673**, A31. DOI. ADS.
Solanki, S.K., del Toro Iniesta, J.C., Woch, J., Gandorfer, A., Hirzberger, J., Alvarez-Herrero, A., et al.: 2020, *Astron. Astrophys.* **642**, A11. DOI. ADS.
Su, J.T., Zhang, H.Q.: 2007, *Astrophys. J.* **666**, 559. DOI. ADS.
Su, J.T., Bai, X.Y., Chen, J., Guo, J.J., Liu, S., Wang, X.F., et al.: 2019, *Res. Astron. Astrophys.* **19**, 161. DOI. ADS.
Xu, H.Q., Liu, S., Su, J.T., Deng, Y.Y., Plotnikov, A., Bai, X.Y., et al.: 2021, *Res. Astron. Astrophys.* **21**, 67. DOI. ADS.
Zhang, H.Q., Wang, D.G., Deng, Y.Y., Hu, K.L., Su, J.T., Lin, J.B., et al.: 2007, *Chin. J. Astron. Astrophys.* **7**, 281. DOI. ADS.

Solar Physics (2024) 299:70
https://doi.org/10.1007/s11207-024-02311-0

RESEARCH

A Study on Magnetic-Sensitivity Wavelength Position of the Working Line Used by the *Full-Disk Magnetograph* Onboard the *Advanced Space Based Solar Observatory* (ASO-S/FMG)

S. Liu[1,2,3] · J.T. Su[1,2,3] · X.Y. Bai[1,2,3] · Y.Y. Deng[1,2,3] · J. Chen[1,2,3] · Y.L. Song[1,2,3] · X.F. Wang[1,2,3] · H.Q. Xu[1,2,3] · X. Yang[1,2,3] · Shahid Idrees[3,1,2]

Received: 17 January 2024 / Accepted: 27 April 2024 / Published online: 24 May 2024

Abstract
Utilizing data from the *Solar Magnetism and Activity Telescope* (SMAT), analytical solutions of polarized radiative transfer equations, and in-orbit test data for the *Full-disk Magnetograph* (FMG) onboard the *Advanced Space based Solar Observatory* (ASO-S), this study reveals the magnetic-sensitivity spectral positions for the Fe I λ 5234.19 Å working line used by FMG. From the experimental data of SMAT, it is found that the most sensitive position is located at the line center for linear polarization (Stokes Q/U), while it is about −0.07 Å away from the line center for circular polarization (Stokes V). Moreover, both the theoretical analysis and the in-orbit test data analysis of FMG confirm the above results. Additionally, the theoretical analysis suggests the presence of distinct spectral pockets (centered at 0.08 – 0.15 Å) from the line, harboring intense magnetic sensitivity across all three Stokes parameters. Striking a balance between high sensitivity for both linear and circular polarization, while capturing additional valuable information, a spectral position of −0.08 Å emerges as the champion for routine FMG magnetic field observations.

Keywords Solar magnetic fields · Polarimetric measurements · Radiative transfer theories

1. Introduction

Measuring the solar magnetic field on the photosphere typically involves analyzing the inversion of polarized spectra based on the Zeeman effect. Two main instruments serve this

✉ S. Liu
lius@nao.cas.cn

J.T. Su
sjt@nao.cas.cn

1 National Astronomical Observatories, Chinese Academy of Science, Beijing, 100101, China

2 Key Laboratory of Solar Activity and Space Weather, National Space Science Center, Chinese Academy of Science, Beijing, 100190, China

3 School of Astronomy and Space Sciences, University of Chinese Academy of Sciences, Beijing, 100101, China

purpose: one is known as a polarization spectrograph, used to measure the Stokes parameter profiles of magnetically sensitive lines. Subsequently, a specific atmospheric model is used to invert these profiles and derive the magnetic field. Examples include the *Advanced Stokes Polarimeter* at the *Vacuum Tower Telescope* (VTT/ASP: Elmore et al., 1992) of the National Solar Observatory, and the *Spectro-Polarimeter* of the *Solar Optical Telescope* (SOT/SP: Tsuneta et al. 2008) onboard *Hinode*. The other, called a vector video magnetograph (filter instrument), employs limited spectral sampling to capture Stokes parameter images. Calibration or inversion techniques then translate these images into magnetic field information. Examples include the *Solar Magnetic Field Telescope* at Huairou Solar Observing Station (HSOS/SMFT: Ai and Hu, 1986), the *Michelson Doppler Imager* onboard the *Solar and Heliospheric Observatory* (SOHO/MDI: Scherrer et al., 1995), and the *Helioseismic and Magnetic Imager* onboard the *Solar Dynamical Observatory* (SDO/HMI: Scherrer et al., 2012; Schou et al., 2012). While high spectral resolution grants polarization spectrographs superior magnetic field measurement accuracy compared to vector magnetographs, they sacrifice temporal resolution. However, increasing the spectral sampling of vector magnetographs can enhance the agreement between the two instrument types (Lites and Skumanich, 1985; LaBonte, Mickey, and Leka, 1999).

For filter-type instruments utilizing single spectral sampling, choosing a magnetically sensitive position within the working line is crucial. The increased sensitivity translates directly to more reliable magnetic field measurements. Hence, investigations into spectral line positions with optimal magnetic sensitivity are of paramount importance. The full-disk vector magnetograph onboard the *Advanced Space based Solar Observatory* (ASO-S/FMG) is dedicated to full-disk photospheric magnetic field observations (Deng et al., 2019; Gan et al., 2019; Su et al., 2019). In its routine mode, it acquires polarization images at a single spectral position. Therefore, determining the magnetically sensitive position of the working line is critical for FMG. Its chosen line, Fe I λ 5234.19 Å, exhibits strong magnetic sensitivity due to its Landé factor of 1.5 ($g = 1.5$) and total half-width of 0.334 Å (Deng et al., 2019; Su et al., 2019; Hou et al., 2020; Gan et al., 2023).

This study embarks on a comprehensive exploration of the magnetic sensitivity within the working line utilized by the FMG. We leverage three distinct vantage points. (i) As our experimental data, we draw upon photospheric magnetic field observations acquired by the *Solar Magnetism and Activity Telescope* (SMAT) at the Huairou Solar Observing Station (Zhang et al., 2007). (ii) We delve into the analytical solutions of radiative transfer equations, investigating the resulting response functions and their relationship to magnetic field strength (Landolfi and Landi Degl'Innocenti, 1982; Ruiz Cobo and del Toro Iniesta, 1994; Orozco Suárez and Del Toro Iniesta, 2007). (iii) Finally, we analyze in-orbit test data from FMG itself, offering an additional perspective on our experimental and theoretical findings. This valuable insight also provides the optimal operation of FMG.

The structure of this article is arranged as follows. Section 2 details the descriptions and experimental results obtained with SMAT. Section 3 delves into the theoretical analysis of radiative transfer equations and their corresponding response functions to varying magnetic field strengths. Section 4 dissects the analysis of FMG in-orbit test data. Finally, Section 5 provides a comprehensive discussion and conclusive remarks stemming from the combined knowledge gleaned from each data source.

2. Analysis of the Experimental Data of SMAT

In 2006, the *Solar Magnetism and Activity Telescope* (SMAT), with its 10 cm aperture and effective field of view of $32' \times 32'$, embarked on a mission to map the full-disk magnetic

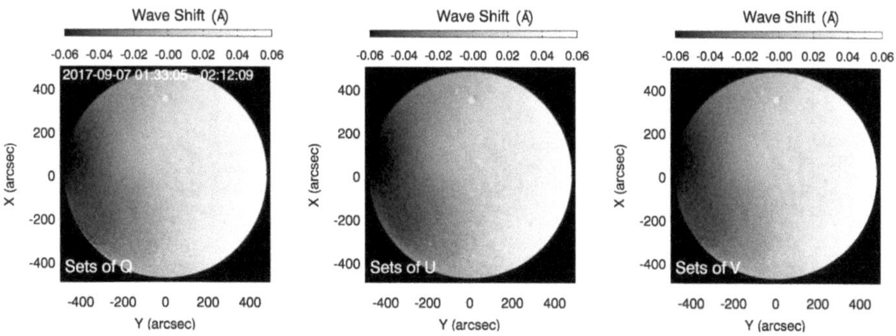

Figure 1 Wavelength shifts in unit of the Å relative to that of the top heliocentric center obtained for three sets of Q, U, and V observations (I ± Q, I ± U, and I ± V).

field of the solar photosphere. Equipped with a 0.125 Å bandpass birefringent filter, SMAT operates in two distinct modes. The first is known as a routine observation mode, in which the magnetograph focuses on a single predetermined wavelength within the working line, capturing Stokes Q, U, and V images. The second, called scanning mode, unlocks a wider spectral range, scanning from −1 Å in the blue wing to +1 Å in the red wing relative to the working line center. This broader perspective allows for the construction of full-disk magnetograms using a 992 × 992 pixel CCD camera. Our experiment leveraged SMAT's scanning mode on 7 September 2017, acquiring 93 datasets (encompassing 31 individual sets for each of Q, U, and V). The scan meticulously covered a range from −0.3 to 0.3 Å around the line center with a step size of 0.02 Å, taking approximately 30 minutes to complete.

Figure 1 visually portrays the full-disk distributions of the measured wavelength shifts relative to the heliocentric center. These shifts were calculated by: (i) fitting the line centers of intensity profiles at various positions across the solar disk; (ii) subtracting the line center values at each position from the value at the disk center. Encouragingly, the wavelength shifts obtained from the Q, U, and V profiles display remarkable consistency, further highlighting the telltale signature of solar rotation in the data.

Figure 2 showcases the Q, U, and V profiles extracted from an active region exhibiting strong polarization. The horizontal axis denotes the relative position of the spectral line to its center, labeled as $\lambda - \lambda_0$ in unit of Å. The top panels a – c depict the nonpolarization intensity (I) profiles for each polarization state (Q, U, and V). Notably, the line centers are marked within each panel. Below, the bottom panels d – f showcase the profiles of Q/I, U/I, and V/I, respectively. These normalized profiles reveal fundamental characteristics: V exhibits an antisymmetric profile, while Q and U show symmetric profiles. Figure 2 serves as a reliable reference for our analysis due to its representation of typical spectral line profiles.

Moving on to Figure 3, the top row displays full-disk Stokes Q images captured at different wavelengths. The image at the line center occupies the central position, with the relative wavelength deviations $(\lambda - \lambda_0)$ labeled above each panel. Examining these images, it appears that the line center exhibits the highest magnetic sensitivity. However, for a more precise evaluation, we calculate the variance (δ) of the Stokes images within each image. We define δ as $\delta_x = 1/(N-1)\sum_{i=0}^{N-1}\sqrt{(\bar{x}-x_i)}^2$ for a series of values x_i as signals, where $\bar{x}$ is the average of x_i, and N is the number of pixels. Generally, the larger the δ_x, the stronger the signal. To quantify this, we calculate the δ_x values for Q, U, and V within a quiet region at the disk center (e.g., indicated by a white square in the middle column of top row), marking them on each panel. Interestingly, for Stokes Q measurements, the peak sensitivity

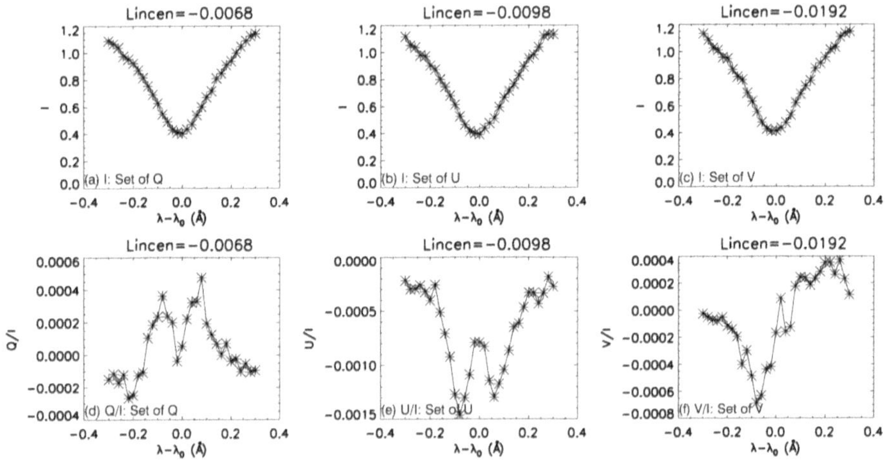

Figure 2 Stokes I, Q, U and V profiles in an active region, the x-axis indicates the wavelength offset relative to line center (range: −0.3 Å to 0.3 Å). **(a)**, **(b)**, and **(c)**: Non-polarization intensity (I) profiles obtained from three separate sets of Q, U, and V observations. The respective line centers are calculated and highlighted within each panel. **(d)**, **(e)** and **(f)**: Normalized profiles of Q/I, U/I, and V/I, respectively. These ratios reveal fundamental characteristics of the spectral line, with V exhibiting an antisymmetric profile and Q and U showing symmetric profiles.

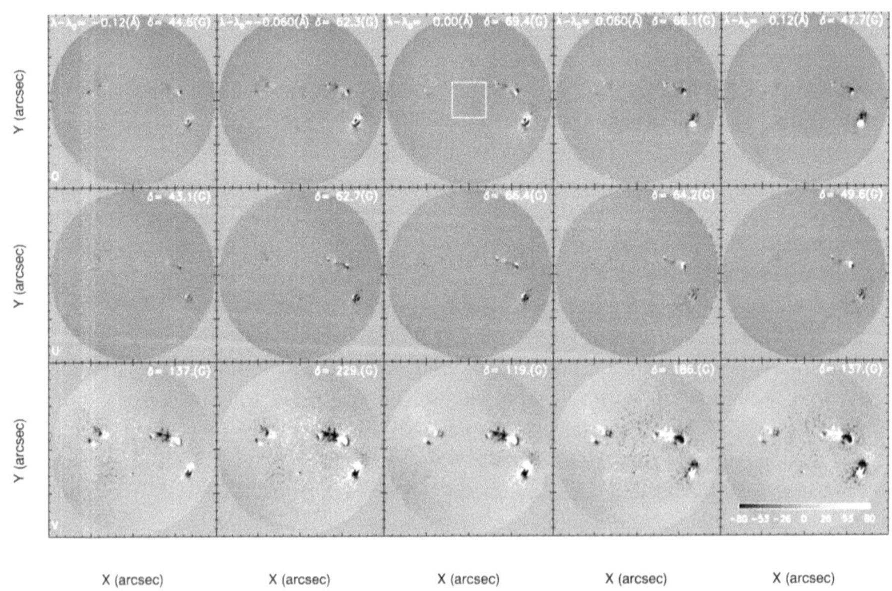

Figure 3 Top row: Stokes Q images captured on 07 September 2017 at different wavelength positions relative to the line center ($\lambda - \lambda_0$ in Å). Each panel displays the variance (δ) of the magnetic field within a quiet region (outlined by a white square) at its corresponding wavelength. Similar images for Stokes U and V are presented in the middle and bottom rows, respectively.

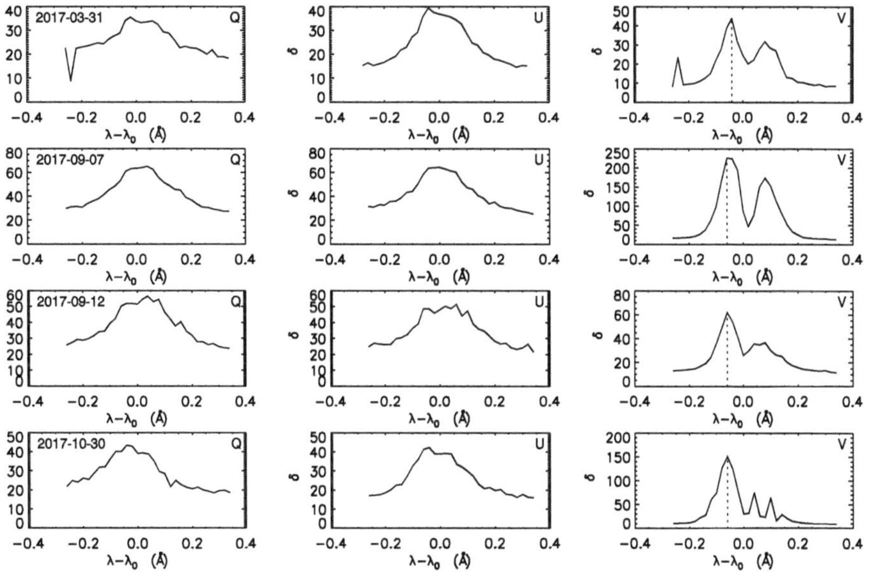

Figure 4 Variations in magnetic sensitivity for Stokes Q, U, and V within a quiet region (marked by a white square in Figure 3) over four days. The horizontal axis represents the wavelength offset from the line center ($\lambda - \lambda_0$) in Å), while the vertical dotted lines in the third column pinpoint the peak variance (δs) values for each parameter.

occurs around the line center with a δ_x value of 69.4 G. Similarly, the middle and bottom rows reveal analogous results for Stokes U and V. While U also exhibits peak sensitivity at the line center, V demonstrates its highest sensitivity slightly blueward of the line center, at approximately −0.06 Å.

Figure 4 delves deeper into the nuances of magnetic sensitivity, showcasing the variations in variance (δ) for Stokes Q, U, and V within a quiet region at the disk center (as seen in Figure 3) over four days. The horizontal axis represents the observed wavelength positions relative to the line center, spanning from −0.3 to 0.3 Å. Notably, the vertical dotted lines in the third column pinpoint the peak δ values for Stokes V, ranging from −0.06 and −0.08 Å with a median of −0.07 Å. These results echo the findings of Figure 3: Stokes Q and U exhibit peak sensitivity at the line center, while Stokes V reaches its maximum sensitivity slightly blueward of the center.

3. Theoretical Analysis

Building upon the Milne–Eddington (M–E) atmosphere model, we leverage analytical solutions for the radiative transfer equations governing polarized light (Unno, 1956; Rachkovsky, 1967; Landolfi and Landi Degl'Innocenti, 1982). In this study, we fix specific parameters associated with these solutions: the spectral line damping parameter ($a = 0.95$), line center wavelength ($\lambda_0 = 5324.186$ Å), line center opacity ratio ($\eta_0 = 10.47$), Doppler width ($\Delta\lambda_D = 40.50$), and source function ($\mu B_1 = -1.16$). However, the magnetic field parameters – total field strength [B], inclination [γ], and azimuth [ϕ] – remain variable.

Response functions quantify the sensitivity of Stokes parameter profiles to changes in atmospheric properties. They are formulated as partial derivatives of the Stokes vector with

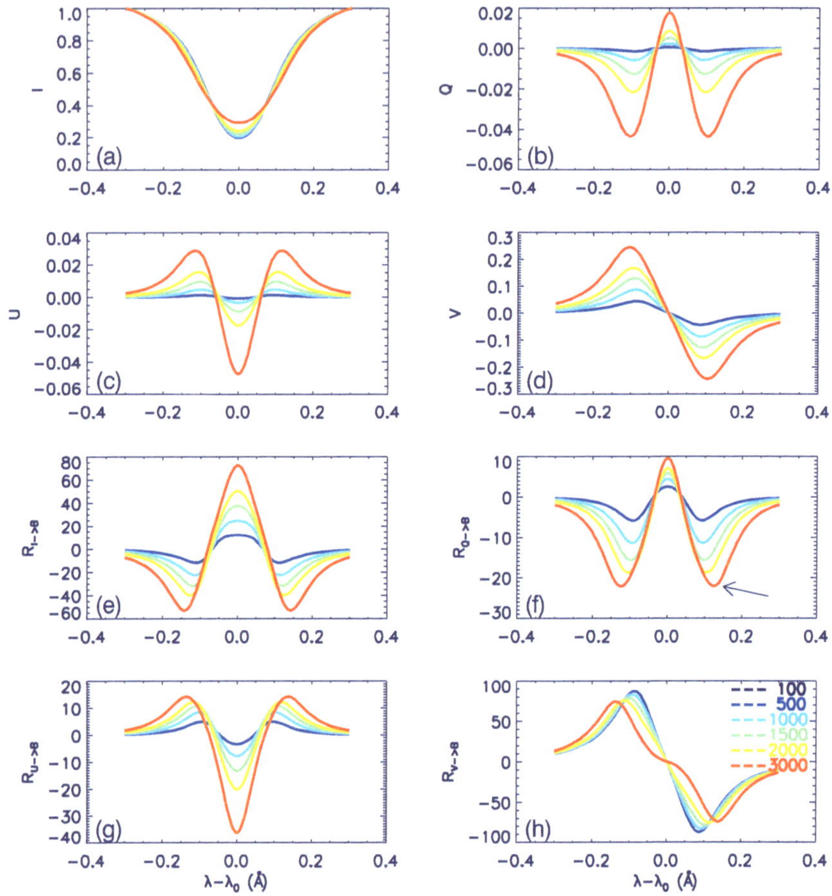

Figure 5 **(a)**, **(b)**, **(c)**, and **(d)** Individually show I, Q, U, and V profiles and **(e)**, **(f)**, **(g)**, and **(h)** Give their response to field strength, where the thermodynamic parameters are fixed with $a = 0.95$, $\lambda_0 = 5324.186$ Å, $\eta_0 = 10.47$, $\Delta\lambda_D = 0.04050$ Å, $\mu B_1 = -1.16$, and inclination (γ) and azimuth (ϕ) angles are assigned 45.0° and 67.5°, while the magnetic field strength (B) changes from 100 to 3000 G as indicated by the colored curves. Where an arrow in **(f)** indicates the local peak position of the spectral line in the response fuction.

respect to the modeled parameters (Ruiz Cobo and del Toro Iniesta, 1994; Orozco Suárez and Del Toro Iniesta, 2007). Mathematically, they take the form

$$R_x(\lambda) = \frac{\partial I(\lambda)}{\partial x}, \tag{1}$$

where $R_x(\lambda)$ is a function of wavelength and x represents any parameter within the analytical solutions. Here, we focus on response functions relating to magnetic field strength and utilize these results as valuable guidance for FMG to select an optimal wavelength position for routine observations.

Figure 5 shows the responses of I, Q, U, and V to B with the thermodynamic parameters, inclination (γ) and azimuth (ϕ) angles fixed, and the magnetic field strength (B) assigned to the values of 100, 500, 1000, 1500, 2000, and 3000 G. In this figure, the panels a–d show the profiles of I, Q, U, and V under different magnetic field strengths, while the panels

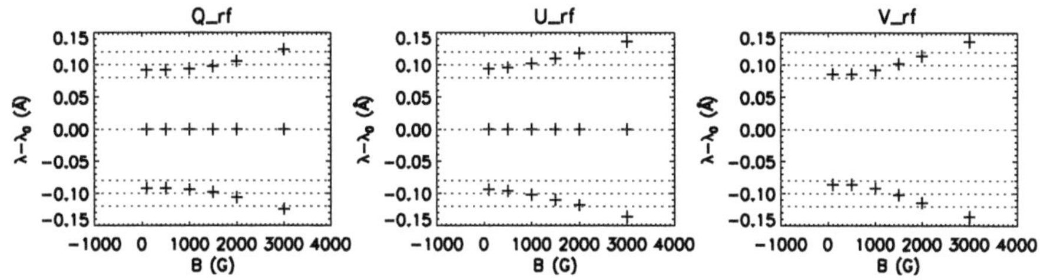

Figure 6 Local peak of the response functions vs. field strength extracted from Figure 5. The horizontal lines label $\lambda - \lambda_0 = 0, \pm 0.08, \pm 0.1, \pm 0.12$ (Å).

e – h show the corresponding response functions, respectively. For the curves of the response function in this figure, the spectral positions with local peak mean the better sensitivities. It can be found that the more sensitive positions for Q and U are located at line center, while those for V are located at the wings. Additionally, with the field strength increasing, the sensitive positions of V shift toward line wings. For Q and U, there are local peak values at line wings as indicated by an arrow in panel f, which also shift to line wings with field strength increasing. To show the results more intuitively, the local peak values among these response functions in Figure 5 are extracted out and their variations with field strength are shown in Figure 6. In the figure, the horizontal lines with values of $\lambda - \lambda_0 = 0, \pm 0.08, \pm 0.1, \pm 0.12$ Å are plotted. It can be found that below 3000 G, for Q and U there are two spectral bands from 0.08 to 0.15 Å, which basically match with that for V.

4. Analysis of the in-Orbit Test Data of FMG

ASO-S satellite was launched with the CZ-2D rocket at 15:43:55 UT on 9 October 2022. FMG started its operation on 16 October 2022. It was an scanning mode for spectral calibration, however, for safety this mode is not available at present. An alternative scheme based on the velocity of the satellite can be used to adjust the spectral position approximately. The ASO-S satellite operates in a Sun synchronous orbit (SSO) with an altitude of about 720 km, an orbital inclination angle of 98.27 deg, and an orbital period of about 99.2 minutes. All these together result in the Sun–FMG radial velocity [v_r] changing from -3.9 to $+3.9\ \mathrm{km\,s^{-1}}$ (the positive direction is the normal direction of the Sun – the Sun points towards the satellite). Correspondingly, a wavelength shift ranging from -0.07 to $+0.07$ Å will superimpose on the selected spectral position. Hence, the magnetic field can be obtained at various spectral positions depending on variations of the Sun–FMG radial velocity. In FMG routine observation, if the filter position is fixed at -0.08 Å away from the line center, then the actual spectral position can be deduced with v_r. In Figure 7, panel a shows a full-disk line-of-sight (LOS) magnetogram observed at 01:24:11 UT on 25 August 2023, where the white square region is selected for investigating the signals of the LOS magnetic field obtained at different v_r. Panels b – d show the LOS magnetogram in the quiet region when $v_r = -4.4\ \mathrm{m\,s^{-1}}$, $v_r = 2063.0\ \mathrm{m\,s^{-1}}$, and $v_r = -2309.4\ \mathrm{m\,s^{-1}}$ observed on 25 February 2023, respectively. We assume that the changes of the LOS magnetic field during this time interval mainly result from the variations of spectral position caused by v_r. It can be found that the signal of the LOS magnetic field observed at $v_r = -4.4\ \mathrm{m\,s^{-1}}$ ($\delta = 28.3$) is better than those at $v_r = 2063.0\ \mathrm{m\,s^{-1}}$ ($\delta = 18.7$) and $v_r = -2309.4\ \mathrm{m\,s^{-1}}$ ($\delta = 27.2$). Figure 8 shows the variance (δ) in the quiet-Sun region as indicated by the white square in

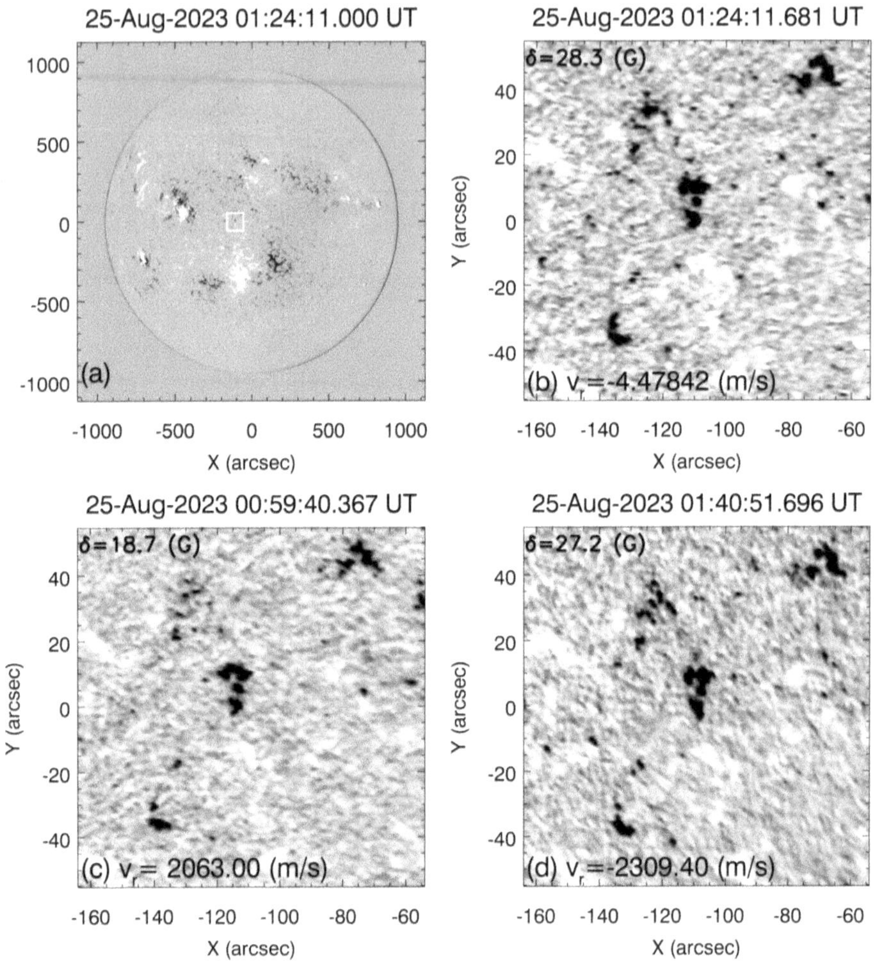

Figure 7 **(a)** Shows a full-disk LOS magnetic field observed by FMG on 25 August 2023, the *white square* identifies a quiet region selected for further investigation. **(b)** – **(d)** Show the LOS magnetic field within this region at three distinct Sun–FMG radial velocities $v_r = -4.4$ m/s ($\delta = 28.3$ G), 2063.0 m/s ($\delta = 18.7$ G), and -2309.4 m/s ($\delta = 27.2$ G), respectively.

Figure 7, varying with the corresponding $\delta\lambda$s observed on 15 August 2023, which are deduced from the values of v_r. It is found that the local peaks of the variance are located at $\delta\lambda$ around -0.065 Å (in a range from -0.08 to -0.05 Å), i.e., those indicated by three red vertical lines. This suggests that these results are basically consistent with those of the SMAT experimental data, as well as the theoretical analysis of the radiative transfer equations.

5. Discussion and Conclusions

This study delved into the magnetic sensitivity of the Fe I λ 5234.19 Å line to varying field strengths, drawing upon a multifaceted approach: SMAT observations at HSOS, in-orbit

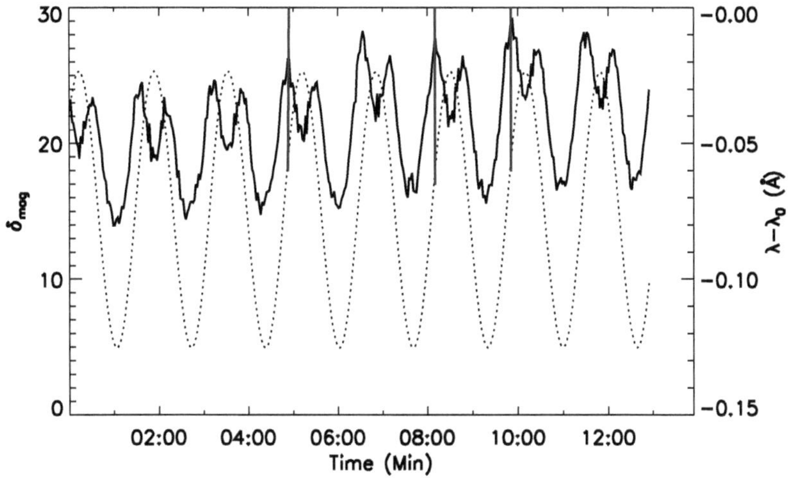

Figure 8 Variance values in the quiet-Sun region outlined by a *white square* in Figure 7 vs. the corresponding $\delta\lambda = \lambda - \lambda_0$, observed on 15 August 2023. Three vertical red lines indicate the local peaks of the variance.

tests of the FMG onboard ASO-S, and theoretical analysis of radiative transfer response functions. The insights gained provide valuable guidance for selecting the optimal spectral position for routine FMG observations.

Both SMAT and FMG were specifically designed for full-disk magnetic field measurements using the same Fe I λ 5234.19 Å line, making SMAT an ideal candidate for this investigation. Our analysis revealed that the most sensitive positions differ for different Stokes parameters: Q and U exhibit peak sensitivity at the line center, while V reaches its maximum about −0.07 Å away from the center. It is important to note, however, that our study focused on a quiet-Sun region at the disk center with relatively low field strengths.

The theoretical investigation, employing analytical solutions and their corresponding response functions, corroborated the findings from the SMAT data. Both approaches identified the line center as the most sensitive position for Q and U, while V demonstrated peak sensitivity within the line wing region. Interestingly, for field strengths below 3000 G, both Q/U and V exhibit shared zones (bands) of increased sensitivity around 0.08 – 0.15 Å. This highlights the advantage of choosing a common spectral position for FMG observations for all three Stokes parameters. Additionally, within these bands, the peak sensitivity for both Q/U and V tends to shift towards the line wings as the magnetic field strength increases. Thus, it can be inferred that different spectral line positions should be used to measure different magnetic field intensities. Therefore, when paying attention to the magnetic field of different features of the structure of an active region, the corresponding spectral line positions should be selected for specific observation targets. Consequently, it is a difficult task to choose a spectral position to observe the magnetic field when the whole active region is regarded as an observation target. It is difficult to simultaneously balance the weak magnetic field (e.g. < 100 G) in the quiet area and the strong magnetic field (e.g. > 2000 G) in the umbral area for a fixed wavelength position. In practical observations, it is crucial to meticulously and comprehensively consider the determination of observation spectral line positions in order to measure the magnetic field and generate high-quality data with minimal errors. For instance, when an active region is considered as an object with an expected magnetic field strength ranging from 500 to 1500 G, a magnetic field intensity of 1000 G can be utilized as a benchmark for selecting the appropriate spectral position, which indicates that the entire magnetic field is measured using the optimal spectral position, resulting in the minimum error in magnetic field measurement.

Optimizing the selection of magnetically sensitive spectral line positions is paramount for precise observations of magnetic fields. However, calibrating the magnetic field is equally crucial to ensure accurate measurements. Under the weak field approximation and using a linear calibration, the varying strength of the magnetic field across distinct solar features leads to varying error amplitudes (Zhang, 2019). Zhang (2019) studied the relationship between Stokes V/I and magnetic field strength across various spectral positions. The results showed that the curve shapes of the V/I-magnetic field strength relationship vary depending on the spectral position. Therefore, even for a magnetic field with a fixed strength, the errors in determining the magnetic field under the weak field approximation would differ when different spectral positions are used. For instance, if we consider a fixed magnetic field of 1500 G, the errors in the magnetic field determination using a linear calibration under the weak field approximation would be 84 and 72 G for the spectral lines selected at −0.035 and −0.075 Å, respectively. It should be noted that the magnetic field error also depends on the linearity degree of the magnetic field and V/I at the fixed spectral position when a linear calibration is used.

From the relations between variances (δs) of FMG LOS magnetic field versus the radial velocity of ASO-S relative to Sun (v_r), it can be found that the most sensitive position of Stokes V is located at the line wing of about −0.065 Å. It should be noted that a response function is symmetric in theory, which means that the magnetic-sensitivities are the same at symmetric positions of red and blue shifts. However, for SMAT and FMG data, the most magnetic-sensitive positions tend to be located at the blue wing of the spectral line, which can be seen from the peak of the curves in the third column in Figure 4, where variances (δs) in the red-shift position are lower than those in the corresponding blue-shift symmetric position. The results of the FMG test confirm the conclusions of SMAT and theoretical analysis for Stokes V (LOS magnetic field related); and the spectral line position that is set at −0.08 Å away from the line center should be a suitable selection for the magnetic field measurement with FMG.

Author contributions S. Liu, J.T. Su and X.Y. bai carried out this work and S. Liu was responsible for the data analysis. X.Y. Bai provided suggestions on data processing and data analysis. J. Chen, Y.L. Song, X.F. Wang, H.Q. Xu, and X. Yang gave overall suggestions for discussion and publication.

Funding This work is supported by National Key R&D Program of China (Nos. 2022YFF0503001, 2021YFA1600500, and 2021YFA1600503), the Strategic Priority Research Program on Space Science of Chinese Academy of Sciences (Grant No. XDB0560000), and the Natural Science Foundation of China (Grant Nos. 12273059, 11203036, 11703042, U1731241, 11427901, 11473039, 11427901, and 11178016), and the Youth Innovation Promotion Association of CAS(2023061). ASO-S mission is supported by the Strategic Priority Research Program on Space Science, the Chinese Academy of Sciences, Grant No. XDA15320000.

Data Availability No datasets were generated or analysed during the current study.

Declarations

Competing interests The authors declare no competing interests.

References

Ai, G.-X., Hu, Y.-F.: 1986, Principles of a solar magnetic field telescope. *Acta Astron. Sin.* **27**, 173. ADS

Deng, Y.Y., Zhang, H.Y., Yang, J.F., Li, F., Lin, J.B., Hou, J.F., Wu, Z., Song, Q., Duan, W., Bai, X.Y., Wang, D.G., Lv, J., Ge, W., Wang, J.N., Zheng, Z.Y., Wang, C.J., Wang, N.G., Ni, H.K., Zeng, Y.Z., Zhang, Y., Yang, X., Sun, Y.Z., Zhang, Z.Y., Wang, X.: 2019, *Res. Astron. Astrophys.* DOI. ADS

Elmore, D.F., Lites, B.W., Tomczyk, S., Skumanich, A.P., Dunn, R.B., Schuenke, J.A., Streander, K.V., Leach, T.W., Chambellan, C.W., Hull, H.K.: 1992, The advanced Stokes polarimeter – a new instrument for solar magnetic field research. In: Goldstein, D.H., Chipman, R.A. (eds.) *Polarization Analysis and Measurement* **1746**, 22. DOI. ADS

Gan, W.-Q., Zhu, C., Deng, Y.-Y., Li, H., Su, Y., Zhang, H.-Y., Chen, B., Zhang, Z., Wu, J., Deng, L., Huang, Y., Yang, J.-F., Cui, J.-J., Chang, J., Wang, C., Wu, J., Yin, Z.-S., Chen, W., Fang, C., Yan, Y.-H., Lin, J., Xiong, W.-M., Chen, B., Bao, H.-C., Cao, C.-X., Bai, Y.-P., Wang, T., Chen, B.-L., Li, X.-Y., Zhang, Y., Feng, L., Su, J.-T., Li, Y., Chen, W., Li, Y.-P., Su, Y.-N., Wu, H.-Y., Gu, M., Huang, L., Tang, X.-J.: 2019, Advanced Space-based Solar Observatory (ASO-S): an overview. *Res. Astron. Astrophys.* **19**, 156. DOI. ADS

Gan, W., Zhu, C., Deng, Y., Zhang, Z., Chen, B., Huang, Y., Deng, L., Wu, H., Zhang, H., Li, H., Su, Y., Su, J., Feng, L., Wu, J., Cui, J., Wang, C., Chang, J., Yin, Z., Xiong, W., Chen, B., Yang, J., Li, F., Lin, J., Hou, J., Bai, X., Chen, D., Zhang, Y., Hu, Y., Liang, Y., Wang, J., Song, K., Guo, Q., He, L., Zhang, G., Wang, P., Bao, H., Cao, C., Bai, Y., Chen, B., He, T., Li, X., Zhang, Y., Liao, X., Jiang, H., Li, Y., Su, Y., Lei, S., Chen, W., Li, Y., Zhao, J., Li, J., Ge, Y., Zou, Z., Hu, T., Su, M., Ji, H., Gu, M., Zheng, Y., Xu, D., Wang, X.: 2023, The Advanced Space-Based Solar Observatory (ASO-S). *Solar Phys.* **298**, 68. DOI. ADS

Hou, J.-F., Wang, D.-G., Deng, Y.-Y., Zhang, Z.-Y., Sun, Y.-Z.: 2020, Interference effect on the liquid-crystal-based Stokes polarimeter. *Chin. Phys. B* **29**, 124211. DOI. ADS

LaBonte, B.J., Mickey, D.L., Leka, K.D.: 1999, The imaging vector magnetograph at Haleakalā – II. Reconstruction of Stokes spectra. *Solar Phys.* **189**, 1. DOI. ADS

Landolfi, M., Landi Degl'Innocenti, E.: 1982, Magneto-optical effects and the determination of vector magnetic fields from Stokes profiles. *Solar Phys.* **78**, 355. DOI. ADS

Lites, B.W., Skumanich, A.: 1985, The inference of vector magnetic fields from polarization measurements with limited spectral resolution. In: Hagyard, M.J. (ed.) *Measurements of Solar Vector Magnetic Fields.* ADS

Orozco Suárez, D., Del Toro Iniesta, J.C.: 2007, The usefulness of analytic response functions. *Astron. Astrophys.* **462**, 1137. DOI. ADS

Rachkovsky, D.N.: 1967, Magnetic rotation effects in spectral lines. *Izv. Krym. Astrofiz. Obs.* **37**, 259. ADS

Ruiz Cobo, B., del Toro Iniesta, J.C.: 1994, On the sensitivity of Stokes profiles to physical quantities. *Astron. Astrophys.* **283**, 129. ADS

Scherrer, P.H., Bogart, R.S., Bush, R.I., Hoeksema, J.T., Kosovichev, A.G., Schou, J., Rosenberg, W., Springer, L., Tarbell, T.D., Title, A., Wolfson, C.J., Zayer, I., MDI Engineering Team: 1995, The solar oscillations investigation – Michelson Doppler imager. *Solar Phys.* **162**, 129. DOI. ADS

Scherrer, P.H., Schou, J., Bush, R.I., Kosovichev, A.G., Bogart, R.S., Hoeksema, J.T., Liu, Y., Duvall, T.L., Zhao, J., Title, A.M., Schrijver, C.J., Tarbell, T.D., Tomczyk, S.: 2012, The Helioseismic and Magnetic Imager (HMI) investigation for the Solar Dynamics Observatory (SDO). *Solar Phys.* **275**, 207. DOI. ADS

Schou, J., Scherrer, P.H., Bush, R.I., Wachter, R., Couvidat, S., Rabello-Soares, M.C., Bogart, R.S., Hoeksema, J.T., Liu, Y., Duvall, T.L., Akin, D.J., Allard, B.A., Miles, J.W., Rairden, R., Shine, R.A., Tarbell, T.D., Title, A.M., Wolfson, C.J., Elmore, D.F., Norton, A.A., Tomczyk, S.: 2012, Design and ground calibration of the Helioseismic and Magnetic Imager (HMI) instrument on the Solar Dynamics Observatory (SDO). *Solar Phys.* **275**, 229. DOI. ADS

Su, J.-T., Bai, X.-Y., Chen, J., Guo, J.-J., Liu, S., Wang, X.-F., Xu, H.-Q., Yang, X., Song, Y.-L., Deng, Y.-Y., Ji, K.-F., Deng, L., Huang, Y., Li, H., Gan, W.-Q.: 2019, Data reduction and calibration of the FMG onboard ASO-S. *Res. Astron. Astrophys.* **19**, 161. DOI. ADS

Unno, W.: 1956, Line formation of a normal Zeeman triplet. *Publ. Astron. Soc. Japan* **8**, 108. ADS

Zhang, H.: 2019, From polarimetry to helicity: studies of solar magnetic fields at the Huairou Solar Observing Station. *Sci. China Ser. G, Phys. Mech. Astron.* **62**, 999601. DOI. ADS

Zhang, H.-Q., Wang, D.-G., Deng, Y.-Y., Hu, K.-L., Su, J.-T., Lin, J.-B., Lin, G.-H., Yang, S.-M., Mao, W.-J., Wang, Y.-N., Hu, Q.-Q., Xue, J.-S., Lu, H.-T., Ni, H.-K., Chen, H.-L., Zhou, X.-J., Zhu, Q.-S., Yuan, L.-J., Zhu, Y.: 2007, Solar magnetism and the activity telescope at HSOS. *Chin. J. Astron. Astrophys.* **7**, 281. DOI. ADS

Publisher's Note Springer Nature remains neutral with regard to jurisdictional claims in published maps and institutional affiliations.

Solar Physics (2024) 299:82
https://doi.org/10.1007/s11207-024-02326-7

RESEARCH

Correction for the Weakening Magnetic Field Within the Sunspot Umbra Observed by ASO-S/FMG

Haiqing Xu[1,2] · Jiangtao Su[1,2,3] · Suo Liu[1,2,3] · Yuanyong Deng[1,2,3] · Xianyong Bai[1,2,3] · Jie Chen[1,2] · Xiaofan Wang[1,2] · Xiao Yang[1,2] · Yongliang Song[1,2]

Received: 18 January 2024 / Accepted: 15 May 2024 / Published online: 14 June 2024

Abstract
The magnetic field inside the sunspot umbra, as observed by the Full-disk MagnetoGraph (FMG) onboard the Advanced Space based Solar Observatory (ASO-S), was found to be experiencing a weakening. To address this issue, we employed a method developed by Xu et al. (2021) to correct the weakening in the data of 20 active regions observed by FMG during the period from 29 December 2022 to 23 July 2023. Research has revealed that the onset of magnetic field weakening occurs at a minimum magnetic field strength of 705 G, with the peak strength reaching up to 1931 G. We computed the change ratio R_1 of the unsigned magnetic flux within the sunspot umbra, considering measurements both before and after correction. The change ratio R_1 spans from 26% to 124%, indicating a significant increase in the unsigned magnetic flux within sunspot umbrae observed by FMG after correction. To illustrate this, we selected four active regions for comparison with data from the Helioseismic and Magnetic Imager (HMI). After correction, we found that the unsigned magnetic flux in sunspot umbrae measured by FMG aligns more closely with that of HMI. This supports the effectiveness of the corrective method for FMG, despite imperfections, particularly at the umbra–penumbra boundary.

Keywords Magnetic fields · Photosphere

1. Introduction

The strong magnetic field of sunspots has been the focus of exploration for over a century, dating back to its initial discovery by Hale (1908). A local relationship between the con-

✉ H. Xu
xhq@bao.ac.cn

✉ J. Su
sjt@bao.ac.cn

1 National Astronomical Observatories, Chinese Academy of Sciences, Beijing, 100101, China

2 Key Laboratory of Solar Activity and Space Weather, National Space Science Center, Chinese Academy of Sciences, Beijing, 100190, China

3 School of Astronomy and Space Sciences, University of Chinese Academy of Sciences, Beijing, 100049, China

We selected 20 active regions observed by FMG from 29 December 2022 to 23 July 2023. Four of these active regions have been utilized by Xu et al. (2024), and the same data reduction method was employed. The remaining 16 active regions were chosen when they were closest to the solar disk center and had the smallest satellite orbit velocity.

3. Results

3.1. Test of the Method

We take active region NOAA 13310 observed by FMG as an example to test the method. The relationship between the values of I/I_m and $|V/I|$ is given in Figure 1(a), which is similar to Figure 2 in Xu et al. (2021). It is found that the second-order polynomial (red line) gives a good fit when $|V/I| > 0.01$ and $I/I_m \leq 0.75$. The I/I_m corresponding to the apex (green asterisk) is 0.428. I/I_m and $|V/I|$ have a negative correlation for the area $I/I_m > 0.428$ and a positive correlation for the area $I/I_m < 0.428$, which is inside the red contour in Figure 1(c). The weakening magnetic field is found for the area inside the red contour in Figure 1(d), owing to the linear calibration adopted by FMG. The I/I_m corresponding to the apex can be as the threshold (hereafter denoted I_0) for occurrence of magnetic weakening. We reconstruct Stokes V/I for pixels where $I/I_m < I_0$ using Equation 4 (Equation 3 in Xu et al., 2021).

$$\frac{V^s}{I} = \left|\frac{a_2}{a_1}\left(\left|\frac{V}{I}\right| - c_1\right) + c_2\right| \cdot sign\left(\frac{V}{I}\right), \tag{4}$$

where (a_1, c_1) and (a_2, c_2) are linear fitting coefficients corresponding to the areas $I/I_m < I_0$ and $I/I_m \geq I_0$, respectively, a_1 and a_2 are slopes, and c_1 and c_2 are constants. After recalculating, the relationship between the values of I/I_m and $|V/I|$ is given in Figure 1(b), and a negative correlation is found.

The FMG LOS magnetic field B_L can be recalibrated from Equations 5 and 6:

$$B_L = C_L \frac{V^s}{I}, \quad \frac{I}{I_m} < I_0, \tag{5}$$

and

$$B_L = C_L \frac{V}{I}, \quad \frac{I}{I_m} \geq I_0, \tag{6}$$

where C_L is the calibration coefficient. FMG adopts 30,000 G for routine calibration. The recalibrated B_L map is shown in Figure 1(e). We can see that the weakening magnetic field in sunspot umbra has been corrected, but a discontinuity (yellow arrow) arises at the umbra–penumbra boundary. We applied a Gaussian smoothing function to smooth the corrected data by 4 pixels. The smoothed B_L map is displayed in Figure 1(f). We can see that the discontinuity has been eliminated; however, the spatial resolution has also been reduced.

3.2. Correction for FMG Data

Figures 2(a1)–(a4) present the original magnetograms of four active regions observed by FMG. Within the umbrae, a decrease in magnetic field strength is observed. Figures 2(b1)–(b4) display the magnetograms after correcting for magnetic field weakening

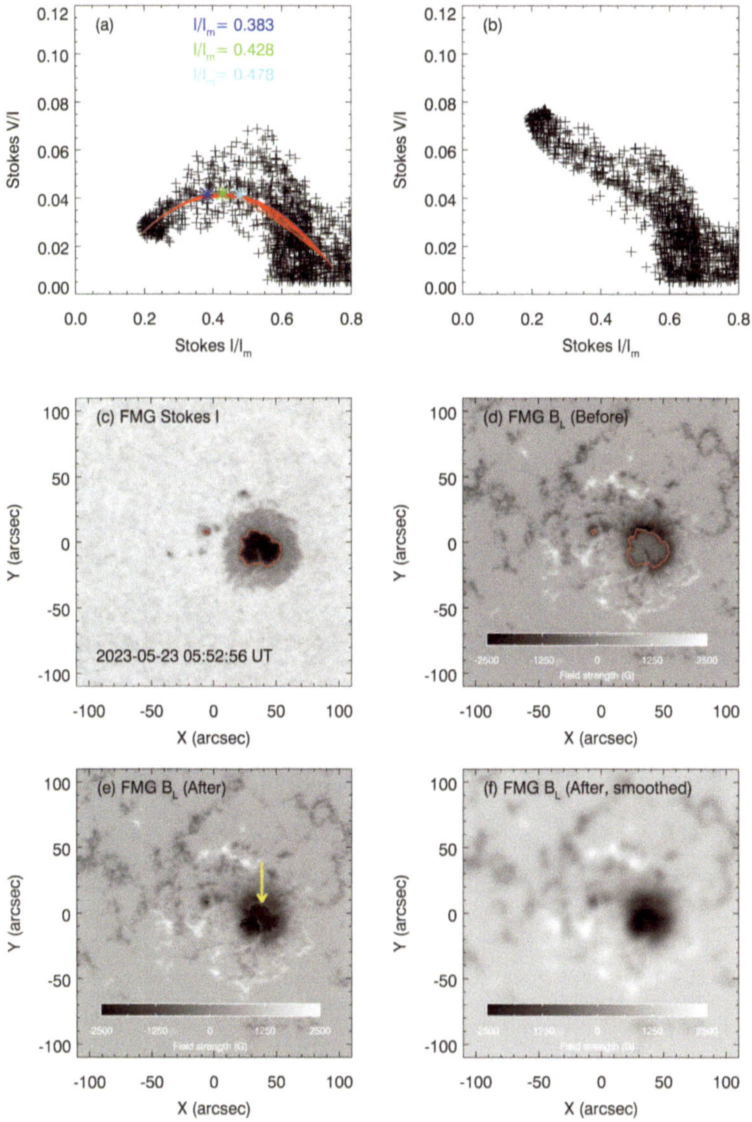

Figure 1 Panel (a) displays the $I/I_m - |V/I|$ relationship of NOAA 13310 observed by FMG. The green asterisk presents the apex of the second-order polynomial fit (red line). The cyan and blue asterisks present the $\pm 1\sigma$ uncertainty of apex. The corresponding I/I_m values are marked using the same color as asterisks. Panel (b) displays the $I/I_m - |V/I|$ relationship after correcting for the weakening magnetic field in sunspot umbra. Panel (c) is the map of Stokes I and panel (d) shows the LOS magnetogram before correction. The red contour represent $I/I_m = 0.428$. Panels (e) and (f) show the corrected LOS magnetograms without and with smoothing the data, respectively.

in sunspot umbrae. It is evident that the magnetic field strength increases within sunspot umbrae, aligning more closely with the HMI 45-s magnetograms in Figures 2(c1)–(c4). However, a discontinuity indicated by yellow arrows at the boundary of umbrae and penumbrae is visible. To conduct a detailed analysis, four slices passing through the sunspots were

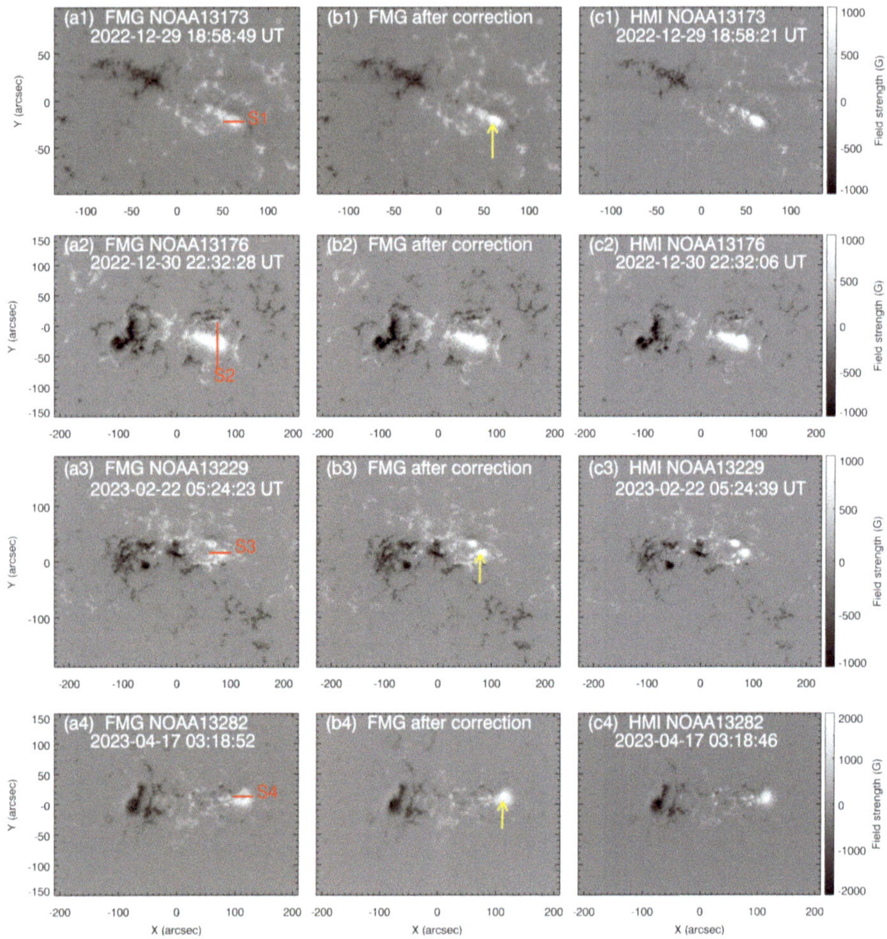

Figure 2 Panels (a1) – (a4) display the original LOS magnetograms of four active regions observed by FMG. Panels (b1) – (b4) represent the LOS magnetograms after correcting for magnetic field weakening in sunspot umbrae observed by FMG. Panels (c1) – (c4) show the LOS magnetograms observed by HMI.

created, as indicated by red lines in Figures 2(a1) – (a4). The distributions of magnetic field along slices S1 – S4 are depicted in Figures 3(a) – (d). It is observed that the magnetic field strength in sunspot umbrae significantly increases after correction, as indicated by comparing the red and blue lines. Specifically for slices S2 and S3, the magnetic fine structures of FMG align well with those of HMI (black lines). However, we also identify discontinuities or errors at the positions where the magnetic field starts to decrease. To eliminate this discontinuity, we applied a Gaussian smoothing function to the corrected data. The cyan and green lines in Figures 3(a) – (d) indicate that the corrected magnetic field has been smoothed by 2 and 4 pixels, respectively. It is observed that the discontinuity has been eliminated, but the magnetic field outside the umbrae was also altered. It is evident that the more data are smoothed, the more effective the elimination of discontinuities. The scatter plots of magnetic field (FMG vs. HMI) after correcting for magnetic field weakening in sunspot umbra observed by FMG are presented in Figures 3(e) – (h). The relationship is approximately lin-

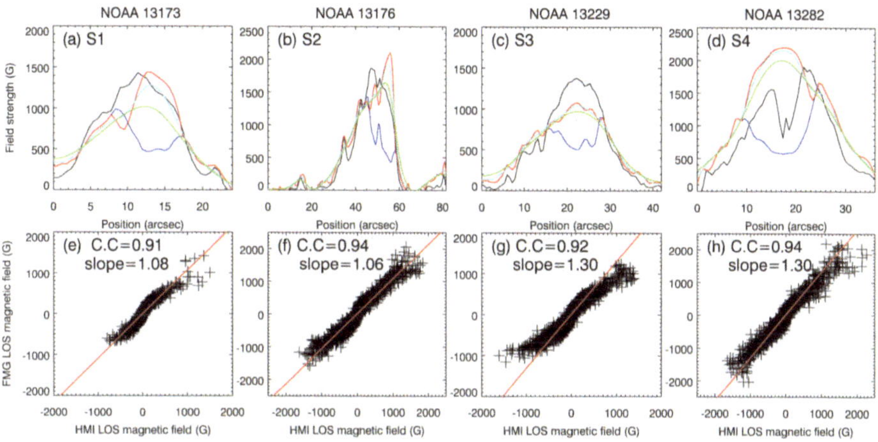

Figure 3 Panels (a) – (d) present the magnetic field distribution along slices S1 – S4 as indicated in Figure 2. The black and blue solid lines in each panel represent HMI and FMG, respectively. The red, cyan, and green solid lines represent the FMG magnetic field after correcting for weakening of the magnetic field in sunspot umbra. The red line indicates that the corrected magnetic field is not smoothed, whereas the cyan and green lines indicate that the corrected magnetic field has been smoothed by 2 and 4 pixels. Panels (e) – (h) feature the scatter plots of magnetic field (FMG vs. HMI) after correcting for the weakening of magnetic field in sunspot umbra observed by FMG, no smoothing applied. C.C is the correlation coefficient of the magnetic field.

Table 1 The information regarding four active regions simultaneously observed by FMG and HMI.

NOAA	Location	Date	Time (UT)	B_s (G)[1]	R_1 (%)[2]	R_2 (%)[3]	R_3 (%)[4]
13173	N24W15	2022-12-29	18:58:49	958	60	5	41
13176	N19E12	2022-12-30	22:32:28	1015	57	5	33
13229	N25W05	2023-02-22	05:24:23	705	44	19	43
13282	N11E10	2023-04-17	03:18:52	1071	69	16	30

[1]The magnetic field strength at the point where it starts to decrease.

[2]The change ratio of unsigned magnetic flux after and before correcting the weakening of magnetic field in sunspot umbrae observed by FMG.

[3]The ratio of unsigned magnetic flux differences between HMI and FMG after correction.

[4]The ratio of unsigned magnetic flux differences between HMI and FMG before correction.

ear. We employed a linear fitting to the data with errors included. The error is 10.2 G for HMI 45-s magnetograms (Liu et al., 2012) and 15 G for FMG in normal mode (Deng et al., 2019). The correlation coefficients and slopes are all slightly higher compared to those reported in Xu et al. (2024).

Additionally, we calculated the magnetic field strength B_s at the point where it starts to decrease, as well as R_1, R_2, and R_3. The results are summarized in Table 1. The lowest B_s value is 705 G, whereas the highest is 1071 G, and R_1 ranged from 44% to 69%, indicating that the unsigned magnetic flux in sunspot umbrae observed by FMG increased by 44% – 69% after correction. Furthermore, R_2 was significantly smaller than R_3, indicating that the magnetic flux in sunspot umbrae observed by FMG aligns more closely with that of HMI after correction.

Table 2 The information regarding 16 active regions observed by FMG.

NOAA	Location	Date	Time (UT)	B_s (G)	R_1 (%)
13273	N10W11	2022-12-06	02:36:55	1532	35
13285	S17E04	2023-04-26	17:36:15	1626	58
13288	S22E10	2023-04-26	17:36:15	1135	26
13293	N13W07	2023-05-06	12:36:39	1769	36
13294	N08E03	2023-05-07	20:48:44	1931	66
13296	N08E03	2023-05-07	03:30:10	1696	27
13297	N08W04	2023-05-08	01:46:34	1616	32
13310	S20E02	2023-05-23	05:52:56	1257	96
13314	N15W04	2023-05-23	05:52:56	1817	124
13315	S16E05	2023-05-26	19:56:44	1259	34
13321	S15W01	2023-06-04	00:53:01	1347	99
13339	S18E01	2023-06-22	09:06:27	991	76
13340	S21E05	2023-06-22	09:06:27	1855	27
13354	S15W03	2023-06-28	02:28:02	1557	48
13373	N07W01	2023-07-19	01:34:25	1533	68
13377	S08W07	2023-07-23	03:13:18	1486	80

For the other 16 active regions, we make the same analysis and find a similar result. Table 2 gives the information and results of the 16 active regions. The lowest B_s is 991 G, and the highest is 1931 G. The lowest R_1 is 26%, and the highest is 124%; B_s and R_1 are different for different active regions, but we did not find any relationship between B_s and R_1.

To investigate the effect of location on the correction algorithm, we selected a relatively stable active region NOAA 13310 and traced it from east to the solar disk center (from 18 to 22 May 2023). We used the FMG data when the satellite orbit velocity was near zero to diminish its influence on magnetic field strength. From Figures 4(a1) – (b5) we can see that the weakening of magnetic field persists for all magnetograms. The red contours represent I/I_m values calculated by the algorithm, indicating the occurrence of magnetic weakening. It is found that the algorithm is independent of the solar disk location when detecting the weakening magnetic field. Figures 4(c1) – (d5) show the magnetograms after correcting for the weakening magnetic field. Some abnormal points (yellow arrows) are found, and the magnetic field in certain areas has not been fully corrected (pink arrows) when the active region is far from the disk center. The correction is more accurate when the active region is close to the disk center. We calculated the unsigned magnetic flux from the corrected magnetograms. From Figure 5 we can see that the unsigned flux gradually increases from east toward the solar disk center (black line). We employed a linear fitting to the data and then subtracted the fitting result from the unsigned flux to remove the general trend. There is no systematic variation in the residual flux (red line).

4. Conclusions and Discussions

We conducted a study on the weakening magnetic field within the sunspot umbrae of 20 active regions observed by FMG. It has been demonstrated that the method developed by

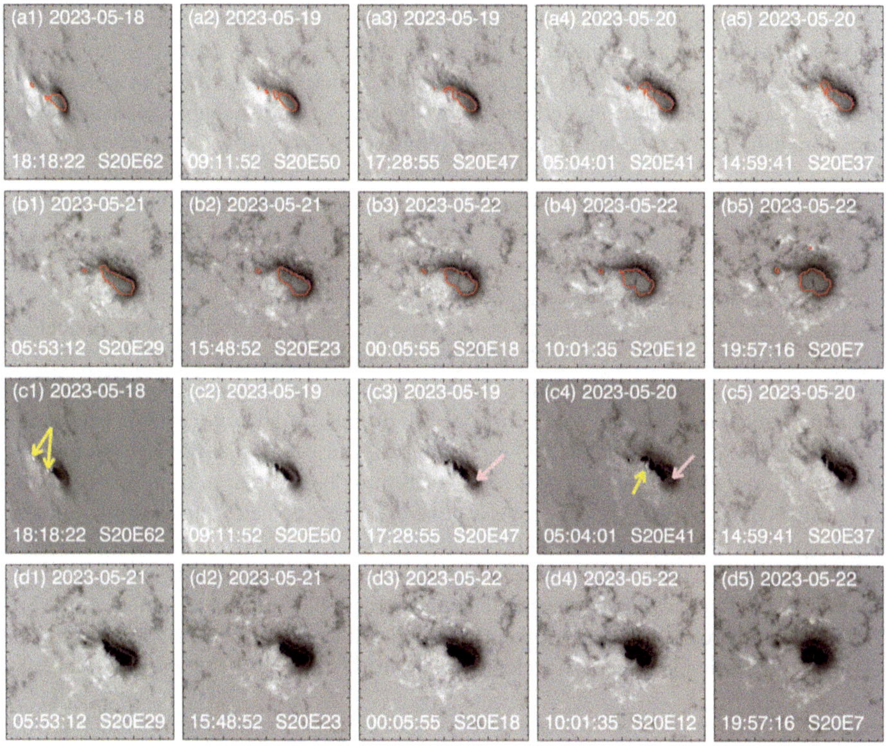

Figure 4 Panels (a1) – (b5) display the original LOS magnetograms of NOAA 13310 observed by FMG from 18 to 22 May, 2023. The weakening magnetic field persists inside the red contours. Panels (c1) – (d5) exhibit the corrected LOS magnetograms of the same active region. The yellow arrows mark the abnormal points. The pink arrows point out the uncorrected area. The size of magnetograms is $192.5'' \times 192.5''$.

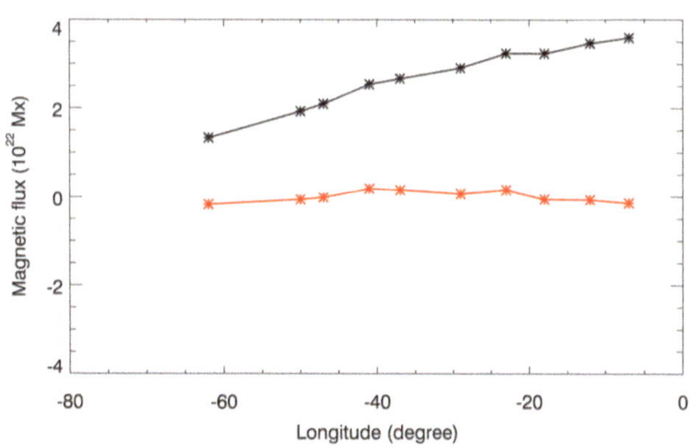

Figure 5 The variation of magnetic flux with longitude of NOAA 13310 from 18 to 22 May 2023. The black and red lines represent the unsigned magnetic flux and residual flux, respectively.

Xu et al. (2021) can be directly applied to FMG data, with only minor modifications required for the range of Stokes V/I and I/I_m when performing polynomial fits. The thresholds for magnetic field weakening to occur range from 705 G to 1931 G. After correcting for the magnetic field weakening, the unsigned magnetic flux within sunspot umbrae observed by

FMG increased by 26% – 124%, and the difference ratio of unsigned magnetic flux between HMI and FMG decreased from 30% – 43% to 5% – 19%.

Despite its effectiveness, this method possesses certain limitations. Although smoothing the data eliminates the discontinuity at the boundary between umbra and penumbra, it also diminishes spatial resolution, thus compromising the recognition of fine structures. The method relies on the relationship between V/I and I/I_m, which is an extension of the $I - B$ relationship that varies across different umbrae (Norton and Gilman, 2004). Notably, the correction accuracy diminishes when active regions are situated far from the disk center. Therefore this method has not yet been integrated into the data production process, and its use should be determined based on the specific research context.

The Fe I 5324.19 Å spectral line is the most commonly used to observe magnetic fields at HSOS. A numerical calculation by Zhang (2019) showed that the Stokes parameters of the spectral line have different sensitivities in the quiet Sun and sunspot atmosphere. The linear approximation for Stokes V can only be used for relatively weak magnetic fields below 2000 G for the quiet Sun and 1000 G for the umbra. Our observations reveal that the magnetic field strength at which magnetic field weakening begins ranges from 705 G to 1931 G (average value $\sim$ 1400 G). This suggests that the magnetic field range suitable for using linear approximation varies depending on the sunspot. The calibration process from polarized light to magnetic field is complex, and both numerical calculations and observations suggest that different calibration coefficients should be used for different regions of the Sun. Additional research is required for the calibration of filter-type magnetographs measuring circular polarization at a single-wavelength position.

Acknowledgments We are sincerely grateful to the anonymous reviewer whose comments helped us to significantly improve the paper. We acknowledge the free data usage policy of the SDO/HMI. The HMI data used in this paper are taken from http://jsoc.stanford.edu.

Author contributions H.Q. Xu and J.T. Su carried out this work and were responsible for the data analysis. S. Liu provided suggestions on data processing. Y.Y. Deng, X.Y. Bai, J. Chen, X.F. Wang, X. Yang, and Y.L. Song gave suggestions for discussing and paper revision. All authors reviewed the manuscript.

Funding This work is supported by National Key R&D Program of China (Grant Nos. 2022YFF0503800, 2022YFF0503001, 2021YFA1600500), National Natural Science Foundation of China (Grant Nos. 12273059, 11427901, 12173049, 12073040, 12003051, 12373057), Beijing Natural Science Foundation (Grant No. 1222029), and the Strategic Priority Research Program of the Chinese Academy of Sciences (Grant No. XDB0560000). ASO-S mission is supported by the Strategic Priority Research Program on Space Science, the Chinese Academy of Sciences, Grant No. XDA15320000.

Data Availability The data presented in this study are available on request from the corresponding author. The ASO-S/FMG original data is available on the website: http://aso-s.pmo.ac.cn/sodc/dataArchive.jsp. The SDO/HMI original data is available on the website: http://jsoc.stanford.edu/.

Declarations

Competing interests The authors declare no competing interests.

References

Ai, G.X., Hu, Y.F.: 1986, *Acta Astron. Sin.* **27**, 173 (in Chinese). ADS.

Chae, J., Moon, Y.-J., Park, Y.-D., Ichimoto, K., Sakurai, T., Suematsu, Y., et al.: 2007, *Publ. Astron. Soc. Japan* **59**, S619. DOI. ADS.

Deng, Y.Y., Zhang, H.Y., Yang, J.F., Li, F., Lin, J.B., Hou, J.F., et al.: 2019, *Res. Astron. Astrophys.* **19**, 157. DOI. ADS.
Gan, W.Q., Zhu, C., Deng, Y.Y., Li, H., Su, Y., Zhang, H.Y., et al.: 2019, *Res. Astron. Astrophys.* **19**, 156. DOI. ADS.
Hale, G.E.: 1908, *Astrophys. J.* **28**, 315. DOI. ADS.
Kosugi, T., Matsuzaki, K., Sakao, T., Shimizu, T., Sone, Y., Tachikawa, S., et al.: 2007, *Solar Phys.* **243**, 3. DOI. ADS.
Leonard, T., Choudhary, D.P.: 2008, *Solar Phys.* **252**, 33. DOI. ADS.
Liu, Y., Hoeksema, J.T., Scherrer, P.H., Couvidat, S., Bush, R.I., Duvall, T.L., et al.: 2012, *Solar Phys.* **279**, 295. DOI. ADS.
Liu, S., Su, J.T., Bai, X.Y., Deng, Y.Y., Chen, J., Song, Y.L., et al.: 2023, *Solar Phys.* **298**, 141. DOI. ADS.
Martínez, V., Vázquez, M.: 1993, *Astron. Astrophys.* **270**, 494. ADS.
Moon, Y.-J., Kim, Y.-H., Park, Y.-D., Ichimoto, K., Sakurai, T., Chae, J., et al.: 2007, *Publ. Astron. Soc. Japan* **59**, S625. DOI. ADS.
Norton, A.A., Gilman, P.A.: 2004, *Astrophys. J.* **603**, 348. DOI. ADS.
Plotnikov, A., Kutsenko, A., Yang, S.B., Xu, H.Q., Bai, X.Y., Zhang, H.Q., Kuzanyan, K.: 2021, *Solar Phys.* **296**, 165. DOI. ADS.
Sakurai, T., Ichimoto, K., Nishino, Y., Shinoda, K., Noguchi, M., et al.: 1995, *Publ. Astron. Soc. Japan* **47**, 81. ADS.
Schou, J., Borrero, J.M., Norton, A.A., Tomczyk, S., Elmore, D., Card, G.L.: 2012, *Solar Phys.* **275**, 229. DOI. ADS.
Su, J.T., Bai, X.Y., Chen, J., Guo, J.J., Liu, S., Wang, X.F., et al.: 2019, *Res. Astron. Astrophys.* **19**, 161. DOI. ADS.
Ulrich, R.K., Evans, S., Boyden, J.E., Webster, L.: 2002, *Astrophys. J. Suppl.* **139**, 259. DOI. ADS.
Xu, H.Q., Liu, S., Su, J.T., Deng, Y.Y., Plotnikov, A., Bai, X.Y., et al.: 2021, *Res. Astron. Astrophys.* **21**, 67. DOI. ADS.
Xu, H.Q., Su, J.T., Liu, S., Deng, Y., Yang, S., Bai, X., Chen, J., Song, Y.L., et al.: 2024, *Solar Phys.* **299**, 17. DOI. ADS.
Zhang, H.Q.: 2019, *Sci. China Ser. G, Phys. Mech. Astron.* **62**, 999601. DOI. ADS.

Solar Physics (2025) 300:49
https://doi.org/10.1007/s11207-025-02455-7

REVIEW

Direct Measurement of the Longitudinal Magnetic Field in the Solar Photosphere with the Zeeman Effect

Jie Chen[1,2] · Jiangtao Su[1,2,3] · Wenbin Xie[4] · Yuanyong Deng[1,2,3] · Xianyong Bai[1,2,3] · Ziyao Hu[1,2,3] · Haiqing Xu[1,2] · Suo Liu[1,2,3] · Yin Zhang[1,2] · Yingzi Sun[1,2] · Yongliang Song[1,2] · Xiaofan Wang[1,2]

Received: 31 January 2024 / Accepted: 16 March 2025 / Published online: 4 April 2025

Abstract
The primary objective of our research is to validate direct measurements of the solar magnetic field through calculating the splitting of polarized spectral lines. The data are collected by the Solar Magnetism and Activity Telescope (SMAT) located at the Huairou Solar Observing Station (HSOS). The number of sampling points of the spectral line profile was varied from 5 to 31. By fitting the profiles of left- and right-circularly polarized light intensities, we determined the spectral line splitting, and the error between the magnetic-field strength determined from this and the true value was less than 2%. We found that the correlation between the magnetic-field strength measured in the active regions by SMAT and that observed by Helioseismic and Magmetic Imager (HMI) on board the Solar Dynamics Observatory (SDO) reached about 80%, but in the quiet region, this correlation was very low.

Keywords Magnetic field · Velocity field · Stokes parameters · Photosphere

1. Introduction

The Sun's magnetic field is extremely significant, as it is the driving force behind various solar activities like solar jets (Shibata et al. 1992, 1994; Sterling et al. 2015; Zhang and Ji

✉ J. Chen
chenjie@bao.ac.cn

J. Su
sjt@bao.ac.cn

W. Xie
wb_xie@bao.ac.cn

1 National Astronomical Observatories, Chinese Academy of Sciences, Beijing, 100101, China

2 Key Laboratory of Solar Activity and Space Weather, National Space Science Center, Chinese Academy of Sciences, Beijing, 100190, China

3 School of Astronomy and Space Sciences, University of Chinese Academy of Sciences, Beijing 100049, China

4 College of Mathematics, Physics and Electronic Information Engineering, Guangxi Minzu Normal University, Chongzuo 532200, China

2014; Zhang, Ji, and Su 2016, Chen et al., 2015, 2017, 2022), flares (Sturrock 1966; Toriumi and Wang 2019; Li et al. 2021; Zhang 2024), and coronal mass ejections (Gopalswamy and Thompson 2000; Gopalswamy et al. 2012). Accurately measuring the solar magnetic field has become crucial.

The Solar Magnetic Field Telescope (SMFT), designed by Ai, Li, and Zhang (1982), has been a crucial instrument at the Huairou Solar Observing Station (HSOS). This telescope offers high-resolution photospheric magnetograms, but it only captures local magnetic fields. With a 35-cm aperture, SMFT has a field of view (FOV) spanning 4 arcmin $\times$ 3.5 arcmin and a pixel resolution of 0.242 arcsec. One of the key features of SMFT is its ability to acquire numerous solar photospheric vector magnetograms at Fe I λ 5324.19 Å. This spectral line, a normal triplet in the magnetic field, has a Landé factor g of 1.5. The excitation potential of the low energy level of this line is 3.197 eV, and the equivalent width of the line is 0.334 Å. The residual intensity at the core of the line is approximately 0.17, as reported by Kurucz et al. (1984). In 2005, the Solar Magnetism and Activity Telescope (SMAT), which is a full-disk solar telescope, was developed and installed at HSOS (Zhang et al. 2007). There are two instruments, one is a 10-cm telescope performing vector magnetic-field observations, the other is a 20-cm telescope performing Hα observations. The effective field of view is $32' \times 32'$, the original temporal and spatial resolutions are about 5 min and $5''$, respectively. The magnetograph uses the same spectral line as SMFT.

The Advanced Space-based Solar Observatory (ASO-S), launched in 2022, aims to study the connection between the solar magnetic field, solar flares, and coronal mass ejections (CMEs) (Gan et al. 2019). One of its three payloads is the Full-disk vector MagnetoGraph (FMG), which measures the vector magnetic field in the photosphere. FMG employs a CMOS camera with 4096 $\times$ 4096 pixels (Deng et al. 2019). Designed to capture the full-disk solar vector magnetic field at a high frequency of 2 min, FMG has a spatial resolution of approximately 1.5 arcsec. There are two observation modes for FMG, one is the routine mode, and the other is the user-defined mode. In routine observations, Stokes Q, U, and V observations are taken at a wavelength position 0.08 Å away from the line center in the blue wing of the Fe I λ 5324.19 Å absorption line (Su et al. 2019; Liu et al. 2022).

All the telescopes mentioned above employ filter-based magnetographs that measure Stokes I, Q, U, and V maps at a single wavelength point. The vector magnetic fields are derived using a linear calibration method under the weak-field approximation (Jefferies et al. 1989). To calculate the calibration coefficients, several methods are employed: theoretical calibration (Ai, Li, and Zhang 1982) using solar-atmosphere models provided by Allen (1973), empirical calibration with a standard line profile, velocity calibration (Wang, Ai, and Deng 1996) based on the relationship between Dopplergrams and magnetograms, nonlinear least-squares fitting calibration (Su and Zhang 2004), and neural-network calibration using existing data (Guo et al. 2021; Hu et al. 2022).

There are two kinds of instruments measuring solar photospheric magnetic fields. One is the filter-based magnetograph, the advantage of this instrument is imaging observation with high temporal resolutions. The other is a spectral polarimeter, differently from a filter-based magnetograph, which is based on the measurements of Stokes parameters distributions across the full spectral line profile (or even several lines simultaneously), providing a major advantage for a deep diagnostics of magnetic fields. An excellent review of a detailed analysis of different spectroscopic methods of the solar magnetic field is given by Solanki (1993).

The FMG payload offers a user-defined observation mode, which allows users to observe magnetic fields using multiple sampling points within the working spectral line. In this mode, our objective is to directly assess the solar line-of-sight magnetic field by calculating the splitting of polarized spectral lines. A key aspect of our investigation focuses on determining the optimal wavelength positions for magnetic-field measurements within the spectral line and assessing the associated measurement uncertainties. The method and data selection will be described in Section 2. The analysis and results will be given in Section 3 and the conclusion and discussion will be presented in Section 4.

2. Method

The magnetic field of the Sun can be obtained through the study of solar polarized light. The radiative-transfer equation for Stokes parameters can be written as

$$\frac{d\mathbf{I}}{d\tau_c} = \mathbf{KI} - j, \tag{1}$$

where **I** denotes the four Stokes parameters I, Q, U, and V, τ_c is the optical depth, **K** is the absorption matrix, and j is the emission vector.

In the process of studying the Stokes spectral radiation of the solar magnetic atmosphere, the successful formulation of a theory for the transfer of polarized radiation was developed originally by Unno (1956). More universal analytical solutions of the radiative-transfer equations for polarized radiation were obtained by Rachkovsky (1967) and Landolfi and Landi Degl'Innocenti (1982). Several inversion codes have been developed for detection of magnetohydrodynamic parameters with depth in the solar atmosphere and/or across the field of view. For example, Very Fast Inversion of the Stokes Vector (VFISV) routinely analyses pipeline data from the Helioseismic and Magnetic Imager (HMI) on board the Solar Dynamics Observatory (SDO) (Borrero et al. 2011). However, there is no need for real inversion and a detection of the effective magnetic-field strength is enough in many applications.

By measuring the Zeeman splitting, the effective longitudinal magnetic-field strength can be obtained. The relationship between the longitudinal magnetic field B_l and Zeeman splitting $\delta\lambda$ can be written as

$$B_l = \frac{\delta\lambda}{2} \times \frac{1}{4.67 \times 10^{-13} g_{eff} \lambda_0^2}, \tag{2}$$

where λ_0 is the wavelength used here, which is Fe I λ 5324.19 Å. g_{eff} is the Landé factor with a value of 1.5.

There are several methods to measure the Zeeman splitting. One method is the center-of-gravity (COG) method. The COG wavelength λ_{COG} of a line profile I is defined as the centroid of its residual intensity profile, the detailed consideration of the different aspects of the COG method was described by Uitenbroek (2003). Another method of measuring the Zeeman splitting in each point is a bisector method, a detailed description and application of this method can be found in Berlicki, Mein, and Schmieder (2006).

Based on the analytical solution obtained by Landolfi and Landi Degl'Innocenti (1982), the polarized-light intensity follows a Gaussian distribution or normal distribution. Different from the previous method, we obtain the center of the wavelength λ through Gaussian fitting, and the parameters of the Gaussian distribution function are obtained by Levenberg–Marquardt least-squares fitting (mpfitexpr.pro in Solar Software). Thus, we obtain the line center of the polarized light. Then, the Zeeman splitting is obtained by subtracting the intensity of the center of the left-circularly polarized light $I - V$ from that of the right-circularly polarized light $I + V$. Figure 1 clearly demonstrates the method to obtain the Zeeman splitting using a schematic diagram. The left-circular polarization is represented by the blue

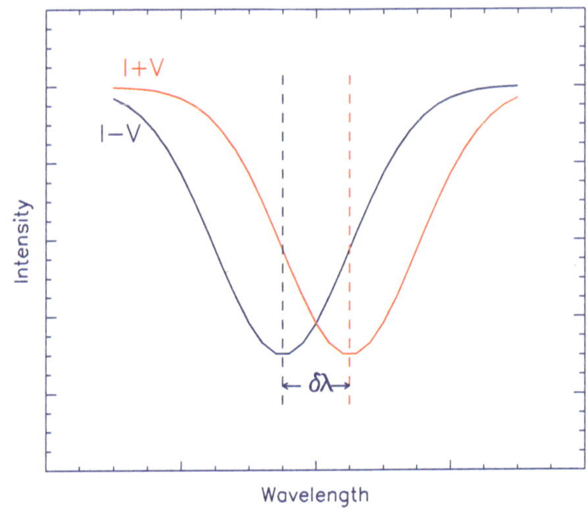

Figure 1 Schematic diagram for obtaining the Zeeman splitting, with the blue line representing left-circular polarization and the red line representing right-circular polarization. The two dashed lines represent the centers of the lines, respectively.

curve, and the right-circular polarization is represented by the red curve. The line centers are represented by the blue dashed line and the red dashed line, respectively. The distance between the blue and red dashed lines, the Zeeman splitting $\delta\lambda$, is the parameter we want to obtain.

Using 5324.19 Å as the line center, with a total magnetic-field strength of 500 G, and an inclination angle of 53°, we obtain the Stokes I and V profiles, the left-circularly polarized is obtained by $I - V$ and the right-circularly polarized light is obtained by $I + V$, which are shown in Figure 2. With a step size of 0.02 Å, taking 31 points, we will obtain the Stokes I and V values for these 31 points, with a wavelength range spanning −0.3 Å to 0.3 Å. By calculating $I - V/I + V$, we obtain the intensity of left-circularly/right-circularly polarized light at each wavelength point. The intensity of left-circularly/right-circularly polarized light is represented by black/red dots, as displayed in Figure 3. For the left-circularly/right-circularly polarized light intensity obtained at the 31 wavelength points, we fitted their Gaussian distribution functions based on the least-squares fitting method. These are represented by black/red curves, respectively, in Figure 3a. In order to compare with the analytical solution, we normalize the observed intensities. The observed normalized intensities of left-circularly/right-circularly polarized light that were obtained by SMAT on September 7, 2017 is shown in Figure 3b.

Based on Equation 2, we calculated the longitudinal magnetic-field strength and error for different wavelength points measured at different wavelength intervals. The wavelength points corresponding to different wavelength intervals are as follows: 0.02 Å with 31 points, 0.04 Å with 15 points, 0.06 Å with 11 points, 0.06 Å with 7 points, and 0.06 Å with 5 points.

The Sun rotates, which produces the Doppler effect. Plasma motion can also produce the Doppler effect. For the Doppler effect, the relationship between observed frequency f' and emitted frequency f is given by:

$$f' = \frac{v + v_0}{v - v_s} f, \tag{3}$$

where v is the propagation speed of waves, v_0 is the speed of the observer's movement, and v_s is the speed of the wave source. Here, v is the speed of light, c, the value of v_0 is 0, and v_s is the speed produced by solar rotation and plasma motion, which is also the Doppler

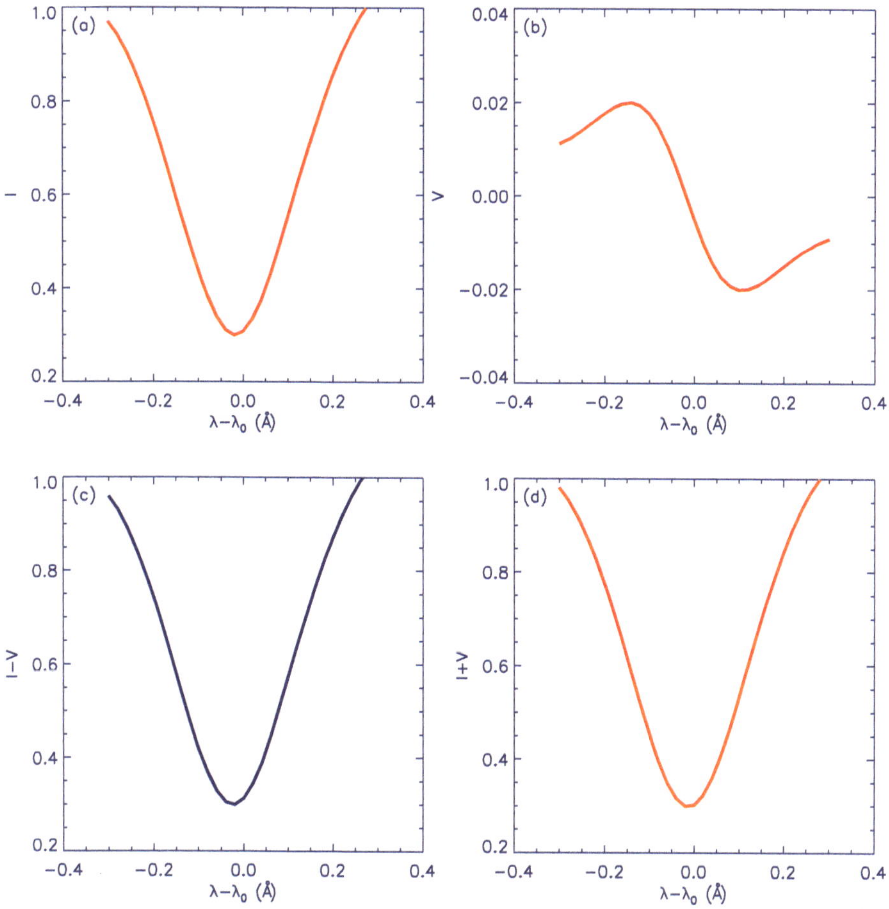

Figure 2 Synthetic profiles of I, V, and $I-V$, $I+V$, where B_{total} is 500 G and the inclination angle is 53°.

velocity v_d. After substituting the above terms, v_d can be obtained:

$$v_d = v_s = c\frac{\Delta\lambda}{\lambda}. \tag{4}$$

$\Delta\lambda$ is the spectral drift due to the Doppler velocity, which is obtained by subtracting the theoretical line center from the line center fitted from observed different wavelength points. Based on Equation 4, the Doppler velocities induced by solar rotation and plasma motion for the aforementioned wavelength point numbers and corresponding wavelength intervals were also calculated. The results are presented in Table 1.

3. Results

3.1. Inversion of the Longitudinal Magnetic Field Based on Observations

A full-disk vector magnetograph is one of the instruments in SMAT. This magnetograph measures Stokes I and V signals using the wavelength Fe I λ 5324.19 Å. On May 18, 2020,

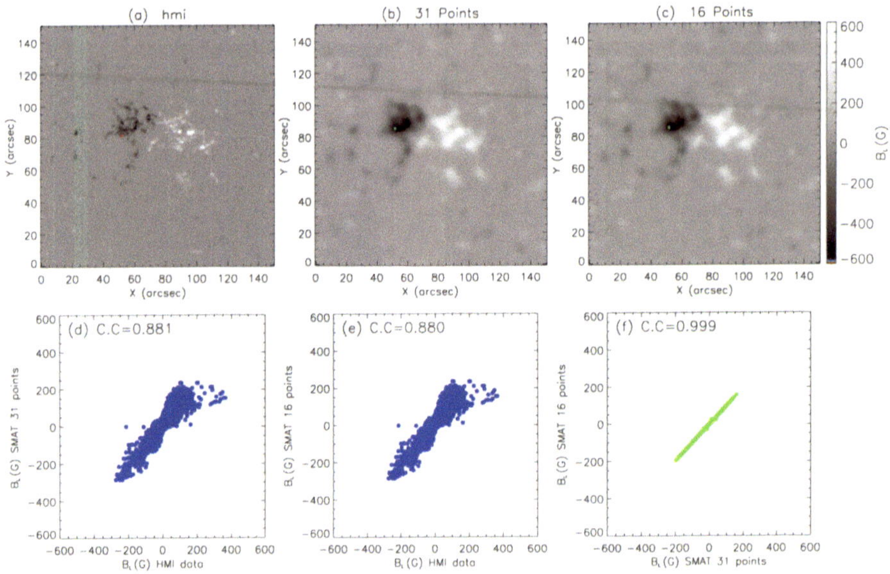

Figure 6 Longitudinal magnetograms of NOAA AR 12760 obtained by HMI and SMAT from different wavelength points, and the relationship of the longitudinal magnetic field between HMI and SMAT observations made on April 28, 2020 around 3:34 UT. (a) A longitudinal magnetogram observed by HMI. (b) and (c) Longitudinal magnetograms inversed by SMAT with 31 and 16 wavelength points, respectively. (d) and (e) The relationship of the longitudinal magnetic field between 31 points and 16 points obtained from SMAT and HMI. (f) The relationship of longitudinal magnetic field between 31 points and 16 points obtained from SMAT.

Table 2 Correlation between HMI and SMAT for NOAA AR 12760 observed on April 28, 2020 and NOAA AR 12759 observed on April 4, 2020.

Date	C. C. HMI and SMAT				
	31	16	11	7	5
28-Apr-20	0.88	0.88	0.85	0.89	0.89
4-Apr-20	0.81	0.74	0.76	0.79	0.77

on April 28, 2020, was compared with HMI data. NOAA AR 12760 is a simple α-type region that was located at S07E10 on the beginning of April 28, 2020. After coalignment, the HMI and SMAT longitudinal magnetograms of NOAA AR 12760 are displayed in Figure 6. Figure 6a exhibits the magnetogram observed by HMI, Figures 6b and c exhibit the magnetograms observed by SMAT with 31 wavelength points and 16 wavelength points, respectively.

After smoothing the HMI data by 4 points, we performed a linear Pearson correlation analysis between the HMI and SMAT data. The linear Pearson correlation coefficients between HMI and SMAT at different wavelength points are displayed in Table 2.

The correlation between HMI and SMAT at different wavelength points is shown in Figures 6d–f. Figures 6d and e show the correlation between HMI and SMAT at different wavelength points. The correlation coefficients are greater than 0.85, indicating a good correlation. Figure 6f shows the correlation of SMAT between 31 wavelength points and 16 wavelength points. The correlation coefficient is close to 0.99. This extremely high correlation shows that there are no essential differences in the results when different number of

Table 3 Fitted ratio value between SMAT (various wavelength points at different wavelength intervals) and HMI.

Date 28-Apr-2020					Date 4-Apr-2020				
HMI NOAA 12760					HMI NOAA 12759				
31	15	11	7	5	31	15	11	7	5
0.68	0.68	0.69	0.72	0.76	0.52	0.52	0.53	0.51	0.51

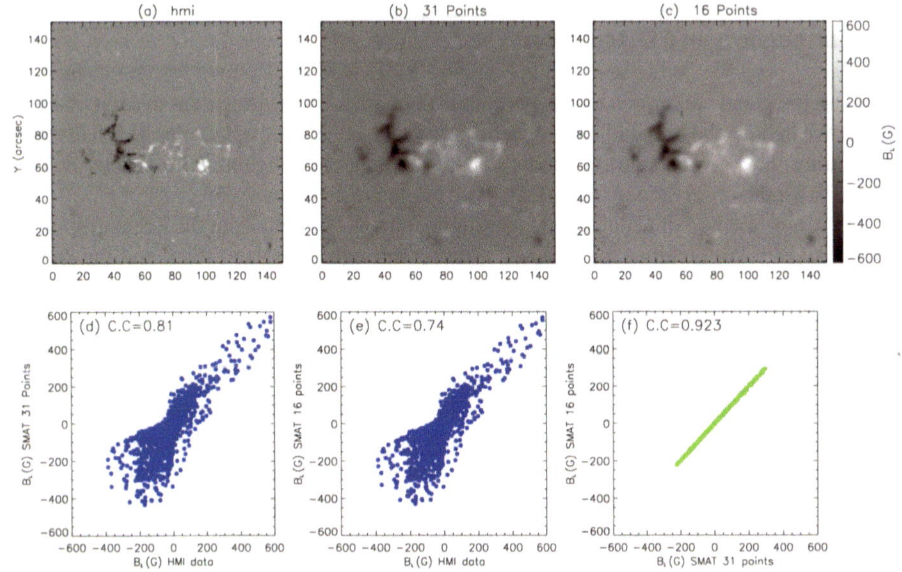

Figure 7 The same as in Figure 6, but for another active region (NOAA AR 12759).

points over the spectral line profile are used. The ratios of magnetic-field strengths (regression coefficient) between HMI and SMAT at different wavelength points were calculated, and the detailed ratios are listed in Table 3. For NOAA AR 12760, the regression coefficient of magnetic-field strengths between SMAT and HMI is approximately 0.7. To make the slope of the data in Figure 6 to be 45°, we multiplied a coefficient by the y-axis for each data point.

To further validate our results, we selected another active region, NOAA AR 12759. NOAA AR 12759 is a simple α-type region that was located at N27E10 at the beginning of April 4, 2020. Figure 7 shows the magnetogram of HMI and SMAT at different wavelength points. After smoothing the HMI data with 4 points, the correlation between HMI and SMAT was analyzed. The correlation between HMI and SMAT at different wavelength points is shown in Figure 7. The linear Pearson correlation coefficient between HMI and SMAT at different wavelength points are around 0.8, and the detailed correlation coefficients are listed in Table 2. As can be seen from Table 3, for NOAA AR 12759, the regression coefficient of magnetic-field strengths between SMAT at different wavelength points and HMI is approximately 0.5. To make the slope of the data in Figure 7 to be 45°, we multiplied a coefficient by the y-axis for each data point.

1. Introduction

Filaments are observed as dark elongated structures on the solar disk in the Hα and extreme ultraviolet (EUV) images. Over the limb, they are observed as bright structures and are called prominences. They are usually located along the magnetic polarity inversion lines (PILs), and are observed as cooler and denser plasma, suspending in the dips of coronal magnetic structures. There are two typical models that describe the magnetic topology of filaments: the sheared arcade model and the flux rope model (e.g. Aulanier, DeVore, and Antiochos, 2002; Mackay et al., 2010; Guo et al., 2010; Xia, Keppens, and Guo, 2014; Jiang et al., 2014a). The typical plasma temperature and density of filaments are about 6000 K and 10^{11} cm^{-3}, respectively; they are about 100 times cooler and denser than their surrounding corona (e.g. Mackay et al., 2010; Priest, 2014).

Filaments can be classified into three types, according to their locations, sizes, and activities. The first type of filaments are located in active regions, with each of their two legs rooted in or near sunspots of different magnetic polarities. Such filaments are called "active-region filaments". They are usually very active and easy to erupt. However, the scale of this type of filaments is usually not large, with their lengths confined within their active regions. The second type of filaments are called "quiescent filaments". They are often located above the PILs of enhanced magnetic networks. Such filaments are relatively stable and not easy to erupt. But they are usually much thicker and longer than active-region filaments. Their lengths can reach up to 1000 Mm (e.g. Priest, 2014; Chen, Xu, and Ding, 2020). The third type of filaments are called "intermediate filaments". They are those between the former two types and usually are located outside active region cores, with their two legs rooted at the boundary of the active region, or one leg at the active region boundary and the other in the quiet Sun (Priest, 2014).

Filaments play a significant role in solar activity. Most of their eruptions lead to solar flares and coronal mass ejections (CMEs). There are mainly two types of physical mechanisms to explain filament eruptions. One is magnetic reconnection. This includes flux emergence, tether-cutting, breakout, and catastrophic models (e.g. Antiochos, DeVore, and Klimchuk, 1999; Chen and Shibata, 2000; Lin and Forbes, 2000; Moore et al., 2001; Zhang et al., 2020). The other is an ideal magnetohydrodynamic (MHD) instability. This mechanism includes kink instability, torus instability, and disturbances on filaments by other small-scale type of activity (e.g. Hood and Priest, 1981; Ji et al., 2003; Fan, 2005; Török and Kliem, 2005; Démoulin and Aulanier, 2010; Chen et al., 2018).

For some active regions, the magnetic topology could be very complicated (e.g. Yan et al., 2015; Xu et al., 2020; Zheng et al., 2021; Li et al., 2022). Several filaments can coexist simultaneously or one filament can have several branches (e.g. Chandra et al., 2011; Liu et al., 2012; Cheng et al., 2014; Zhu et al., 2015; Hou et al., 2018; Su et al., 2018; Qiu et al., 2020). One well-known model is the so-called "double-decker" model (Liu et al., 2012; Kliem et al., 2014). It refers to the filament that consists of two vertically-distributed segments, of two magnetic flux ropes (MFRs) or of one MFR sitting over a sheared arcade magnetic structure. It is often used to explain the partial eruption of filaments (e.g. Zhu and Alexander, 2014; Zhang et al., 2015; Awasthi, Liu, and Wang, 2019; Hou et al., 2023; Sun et al., 2023). Observations also show that two different filaments can make up a "double-decker" configuration (e.g. Zhu et al., 2015; Hu et al., 2022).

When an active region possesses two or more filaments simultaneously, the magnetic topology is usually very complicated. Collisions and interactions usually occur between these filaments, particularly when one of them begins to erupt (e.g. Linton, Dahlburg, and Antiochos, 2001; Su et al., 2007; Kumar, Manoharan, and Uddin, 2010; Török et al., 2011;

Jiang et al., 2014b; Zhu et al., 2015). However, such cases are very rare in observations, especially for magnetic systems with more than two filaments.

In this study, we present a very special case where two erupting active-region filaments push the above large-scale filament (the third) out to eruption. This event gives us a unique insight into the dynamics of the filaments during their eruptions. The paper is organized as follows. Section 2 describes the data. Section 3 presents the analysis and results. Section 4 gives the summary and discussion.

2. Data

The multifilament eruption we study occurred on 25 February 2023, in the northwest of the solar disk and was very close to the solar limb (see Figure 1). We carry out a comprehensive analysis using the data from both space- and ground-based telescopes. The *Atmospheric Imaging Assembly* (AIA: Lemen et al., 2012) on board the *Solar Dynamics Observatory* (SDO) can provide multiwavelength images of the solar atmosphere. The time cadences for the extreme ultraviolet (EUV) passbands and the ultraviolet (UV) passbands are 12 and 24 s, respectively. The spatial sampling is about $0.6''$ per pixel. The *Extreme-Ultraviolet Imager* (EUVI: Wuelser et al., 2004) on board the *Solar Terrestrial Relations Observatory Ahead* spacecraft (STEREO-A: Kaiser et al., 2008) also has good observational data for this event. EUVI provides four-passband observations, including 171, 195, 284, and 304 Å, with a spatial pixel size of about $1.6''$. We also use the observations of the Ne VII 46.5 nm spectral line, obtained by the *Solar Upper Transition Region Imager* (SUTRI: Bai et al., 2023) on board the *Space Advanced Technology demonstration satellite series* (SATech-01). SUTRI can image the upper transition region with a temperature of about 0.5 MK (Tian, 2017). The spatial pixel size and time resolution of SUTRI are about $1.2''$ and 30 s, respectively.

The *Advanced Space-based Solar Observatory* (ASO-S: Gan et al., 2019), launched on 9 October 2022, is the first Chinese space-borne comprehensive solar observatory. Its scientific objective is to study the relationship between the solar magnetic field and solar eruptions. Three instruments are deployed on ASO-S: the *Full-disk vector MagnetoGraph* (FMG: Deng et al., 2019), the *Lyα Solar Telescope* (LST: Chen et al., 2019; Li et al., 2019) and the *Hard X-ray Imager* (HXI: Zhang et al., 2019). In this study, we mainly use the magnetograms taken by FMG. It observes the Sun at the wavelength of Fe I 5324.19 Å with a $4\,\mathrm{K}\times4\,\mathrm{K}$ CMOS camera. The spatial pixel size is about $0.5''$. Normally, it takes line-of-sight (LOS) magnetograms with a cadence of 30 s, for the vector magnetograms the cadence is 120 s.

The Hα images taken by the *Chinese Hα Solar Explorer* (CHASE: Li et al., 2022) and the *Global Oscillation Network Group* (GONG: Harvey et al., 2011) are also used. CHASE provides high-quality raster scanning observations of the full Sun with the spectral ranges of 6559.7 – 6565.9 Å (Hα) and 6567.8 – 6570.6 Å (Fe I). The spatial, temporal, and spectral resolutions are about $1''$, 1 minute, and 0.072 Å, respectively. In order to show the CME, the data from the *Large Angle Spectroscopic Coronagraph* (LASCO: Brueckner et al., 1995) on board the *Solar and Heliospheric Observatory* (SOHO: Domingo, Fleck, and Poland, 1995) and COR2 coronagraph on board STEREO-A are also used.

3. Analysis and Results

Figure 1 is the overview of the multifilament eruptions with the observations of the EUVI 304 Å, AIA 211 Å, SUTRI 465 Å, CHASE and GONG Hα. Before the eruption, two sets

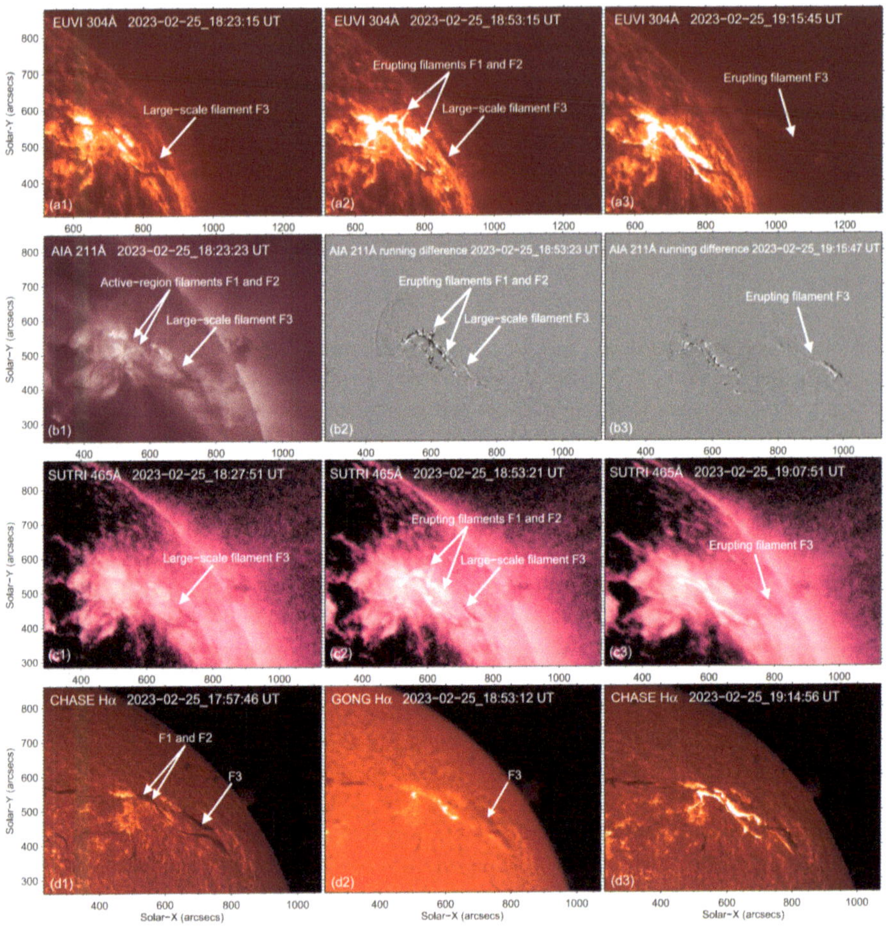

Figure 1 Overview of the filament eruptions in active region NOAA 13229. (a1)–(a3), (b1)–(b3), (c1)–(c3), and (d1)–(d3) Show the STEREO-A/EUVI 304 Å, SDO/AIA 211 Å, and the running differences, SUTRI 465 Å, CHASE and GONG Hα images, respectively. From left to right, we show the preeruption, the eruptions of active-region filaments F1 and F2, and large-scale filament F3 eruption being pushed out by the erupting F1 and F2. Animations for this multifilament eruptions observed by EUVI 304 Å and AIA 211 Å running difference are available online.

of filaments can be seen clearly: the low-lying filaments F1 and F2, and the large-scale filament F3, with the last one higher up in the solar atmosphere. These two sets of filaments can be easily distinguished in the Hα observation (Figure 1d1). However, it is a challenge to observe the filaments F1 and F2 only through Hα images. Employing the AIA 211 Å image, we can more clearly see these two low-height filaments F1 and F2 (Figure 1b1). Filaments F1 and F2 appear to intertwine with each other.

From the central panels of Figure 1, we can see that when F1 and F2 had already erupted and even when they had already risen to a somewhat higher height, F3 was still and remained stable. There was a significant height difference between the two sets of filaments before their eruptions. During the eruptions of F1 and F2, it is evident that their left legs are intertwined (Figure 1b2). The flare ribbons are just located below F1 and F2. Finally, the

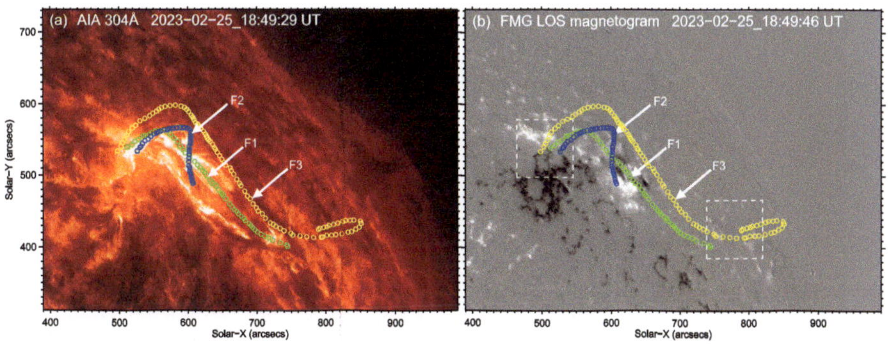

Figure 2 The configuration of the three filaments in NOAA 13229. (a) and (b): Show the locations of the filaments, overlaid on the AIA 304 Å image and the ASO-S/FMG line-of-sight (LOS) magnetogram, resepctively, at 18:49 UT, when F1 and F2 are already erupting and F3 is still stable. Here all three filaments are outlined by circles but with different colors. Two white dashed boxes in (b) mark the foot areas of F3, which will be shown in Figure 8.

large-scale filament F3 is pushed out to eruption by the erupting F1 and F2, and then we see two much longer flare ribbons (Figure 1d3).

Figure 2 shows the configuration of the three filaments, during the eruptions of F1 and F2 while F3 is still stable. We can see that F1 and F2 are two active-region filaments in NOAA 13229. F1 is located along the PIL of the active region, and F2 just crosses the PIL and lies above F1. Their left legs are very close and seem to be intertwined. F3 is a large-scale filament that crosses the active region with a length of about 360 Mm. Its left leg is rooted in the network of negative magnetic polarity, at the northern edge of the active region. Its right leg is rooted in a quiet-Sun region, far away from the active region. Most of this filament is nearly parallel to F1, above the PIL of the active region.

Based on these observations, we regard filaments F1 and F2 as one magnetic system that coincides with the "double-decker" model (Liu et al., 2012). We regard filament F3 as another independent magnetic system, due to its larger scale and higher altitude, relative to F1 and F2. We see that in this event the energy is first released by the lower magnetic system, that is, in the F1 and F2 eruptions, then by the upper magnetic system, that is, in the F3 eruption.

Figure 3 shows the interactions between the active-region filaments F1, F2, and the large-scale filament F3 from 18:36 UT to 19:10 UT. The associated flare starts at 18:40 UT. We see that the two filaments F1 and F2 are activated and begin to rise. At the early stage, both filaments are seen as very bright all along (see Figure 3b). It should come from the interchange magnetic reconnection between two filaments due to their special configuration. When F1 rises, it is accompanied by a slight expansion and rotating motion in the early stage about from 18:38 UT to 18:45 UT. At about 18:50 UT, F1 first collides with F3 and begins to push it rising. We see a bright emission below F3, as marked by the green elliptical circle in Figure 3c. At the same time, F2 is still rising and F3 begins to rise slowly. About four minutes later, at 18:54 UT, F2 collides with the left "shoulder" of F3 and begins to push F3 (see the left green elliptical circle in Figure 3d). Then, F3 rises much faster and the flare ribbons elongate.

The interaction between F2 and F3 is more noticeable in AIA 131 Å and 94 Å observations (Figure 4). It may be due to the high-temperature but low-density plasma in the left shoulder of F3 where F2 collides with it. We can see a bright structure sliding up-

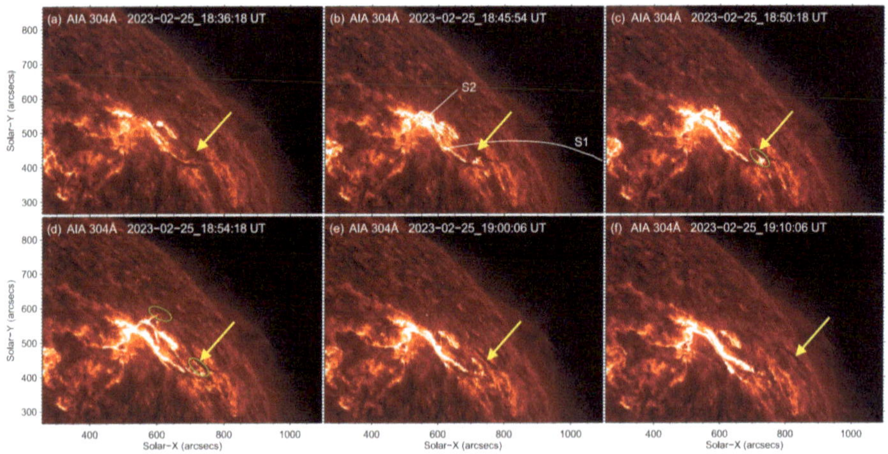

Figure 3 Interactions between active-region filaments F1, F2, and large-scale filament F3. Yellow arrows mark the F3 at different times. Green elliptical circles indicate the positions where F1 and F2 collide with F3 and subsequently begin to push it. Slits S1 and S2 correspond to the eruptions of F1 and F2 and their interactions with F3, respectively, time-distance maps of which will be shown in Figure 5. An animation for this figure is available online.

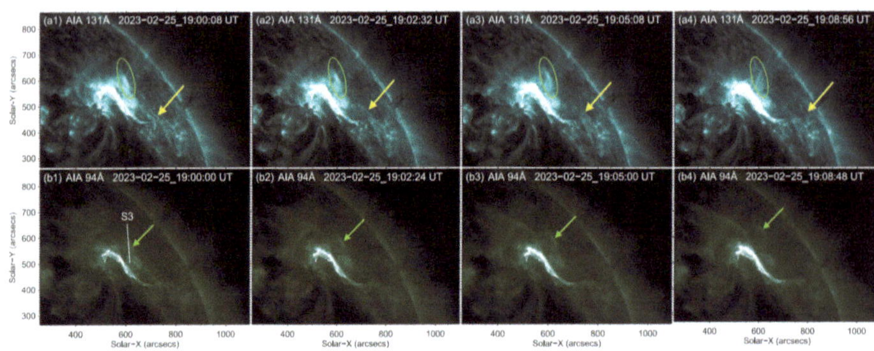

Figure 4 Interaction between filaments F2 and F3, from 19:00 UT to 19:10 UT. Upper (a1)–(a4) and bottom (b1)–(b4) images show the AIA high-temperature passbands 131 Å and 94 Å images, respectively. In the upper panels, yellow arrows mark the erupting F3. Green elliptical circles indicate the positions where F2 pushes F3 with a bright structure moving upwards. Green arrows in the bottom panels point to the moving bright structure. Slit S3 is used to show the motion of the bright structure, time-distance maps of which will also be shown in Figure 5, together with S1 and S2. An animation for this figure is available online.

wards smoothly (see the green elliptical circles in Figure 4a1–a4 and green arrows in Figure 4b1–b4). This sliding motion persists for more than 10 minutes, indicating that F2 is pushing F3 to erupt. Simultaneously, we see that the large-scale filament F3 erupts more rapidly with the velocity increasing from about 39.0 to about 90.6 km/s (see Figure 6b).

Figure 5 is the time-distance maps for slits S1, S2, and S3, where the filament eruptions can be seen clearly. The flare starts at 18:40 UT. At the early stage of the flare, F1 and F2 erupt and rapidly rise but F3 remains stable (Figures 5a and 5b). At around 18:50 UT, F1 first encounters F3 and begins to push it upwards. Then, we see that F3 begins to rise slowly. There is a slight brightening at the lower boundary of F3. In this process, we do not see

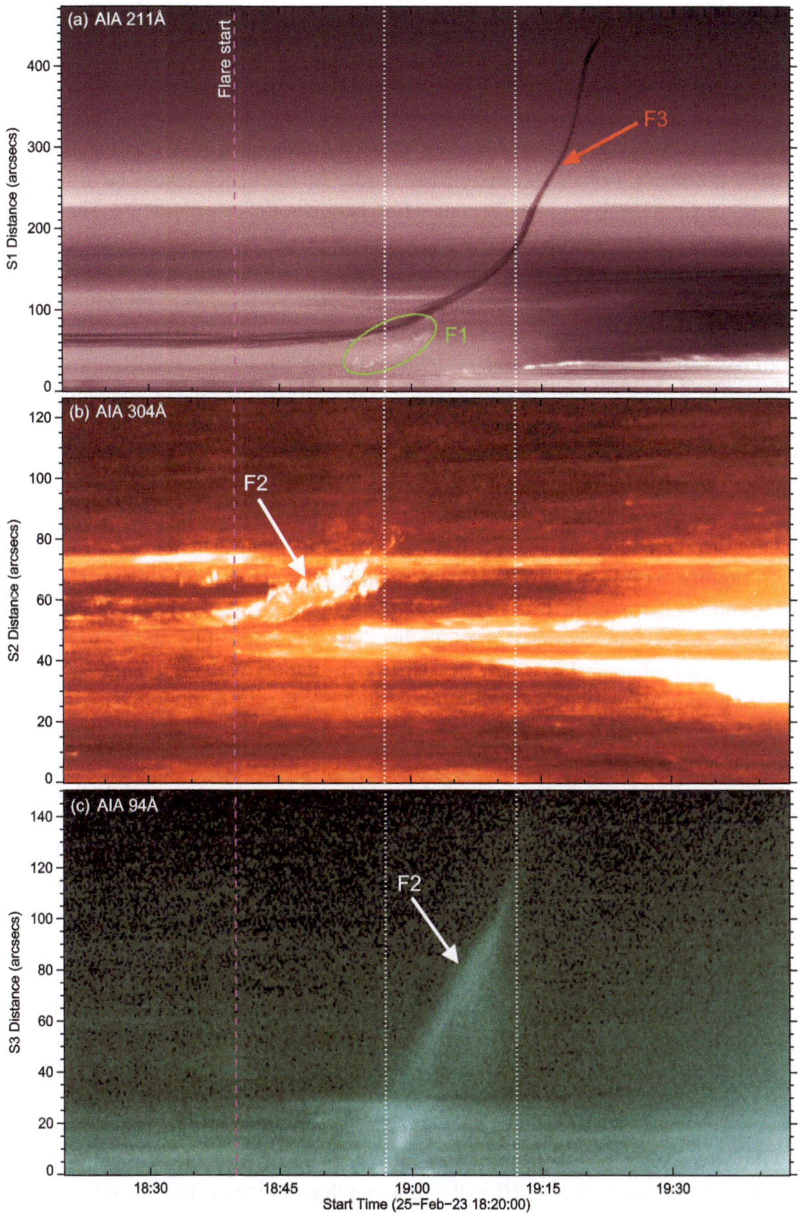

Figure 5 Time-distance maps of the three slits. From top to bottom, they are of S1, S2, and S3, using the AIA 211 Å, 304 Å, and 94 Å data, respectively. The positions of slits S1, S2, and S3 can be seen in Figures 3 and 4. The vertical magenta dashed lines outline the flare start time. The green elliptical circle in panel a indicates the eruption of F1, associated with expansion, rising and rotating motions, and then pushing F3 to eruption. White arrows in panel b point to the eruption of F2 and its interaction with F3 in panel c. Red arrow marks the evolution of F3. The two vertical white dotted lines correspond to the period when F2 is pushing F3.

very evident or strong reconnection between F1 and F3. This may be due to the fact that the two filaments run nearly parallel to each other along the PIL of the active region and with the same magnetic direction (Figure 2). As pointed out by Linton, Dahlburg, and Antiochos (2001), magnetic reconnection between two flux tubes significantly depends on their relative orientation. However, small-scale and weak reconnections could still occur between the magnetic field lines in the outer shells of two filaments, since the magnetic field could still make an angle because of a helical or sheared structure. This slight brightening may come from the heating of the plasma at the border of filaments, resulting from the collision, squeezing and weak magnetic reconnection between the two filaments. The interaction between F1 and F3 lasts about 15 minutes.

F2 collides with F3 at around 18:55 UT (Figures 5b and 5c). The interaction between F2 and F3 also lasts about 15 minutes. When only F1 pushes, F3 rises very slowly. However, when F1 and F2 simultaneously push F3 from about 18:56 UT to 19:03 UT, we see that F3 begins to rise rapidly. Finally, after the interactions with F1 and F2, the large-scale filament F3 is ejected at a higher speed, possibly with a conversion of its own nonpotential magnetic energy into kinetic energy together with that gained from the interactions with F1 and F2 in the form of pushing.

Figure 6 shows the corresponding relationship between the GOES 1–8 Å X-ray flux and the eruption of filament F3. Obviously, the development of the event can be divided into three stages. In the first stage from 18:40 UT to 18:52 UT, we see that the GOES X-ray flux increases rapidly. At this stage, the filaments F1 and F2 erupt, but F3 remains stable.

The second stage is from 18:52 UT to 19:09 UT. During this period, we see a gradual deceleration in the increase of GOES X-ray flux. The eruptions of F1 and F2 are interrupted by the large-scale filament F3. The interactions with the above large-scale filament F3 decrease the ejection velocities of the erupting F1 and F2. Part of the kinetic energy of F1 and F2 is transferred to F3. At this stage, it is evident that F3 undergoes two distinct acceleration phases. In the first phase, that is, during its interaction with F1, its rising velocity increases from 0 to about 39.0 km/s. In the second phase, when it is interacting with both F1 and F2, its rising velocity increases from about 39.0 to about 90.6 km/s.

The third stage is from 19:09 UT to the end of the flare. After interaction with F1 and F2, the large-scale filament F3 escapes the Sun at a significantly higher speed, reaching about 684.7 km/s. The GOES X-ray flux increases rapidly again and finally reaches the level of M6.4.

Figure 7 is a cartoon that shows the interactions between the filaments F1 and F2 with F3. F1 and F2 are two active-region filaments located above the PIL with a "double-decker" configuration (Liu et al., 2012). F1 is longer and makes a relatively smaller angle with the PIL. F2 rides over F1 and has a relatively larger angle with the PIL. F3 is a large-scale filament crossing over the active region with its axis nearly parallel to the PIL. F1 and F2 erupt almost simultaneously. During the eruption, interchange reconnections occur between F1 and the two legs of F2. Both filaments rapidly rise and F1 is also accompanied by an expansion and rotational motion. F1 is the first one that collides with F3 and begins to push F3, since the lower part of F3 is situated just above F1. It is about five minutes later that F2 collides with the left shoulder of F3 and begins to push F3 upwards along the shoulder. At last F3 is pushed out to eruption causing a large CME. F1 probably erupts successfully. But F2 is a failed eruption (see Figures 1 and 3, and the animations for these figures). Such different fates of F1 and F2 are probably due to the different angles between F1, F2, and F3.

Figure 8 shows the temporal evolution of the LOS magnetic field in the two regions where filament F3 rooted before the flare. The left and right foot areas correspond to the network region and quiet Sun, respectively (Figures 8a1–a4 and 8b1–b4). During a period of about

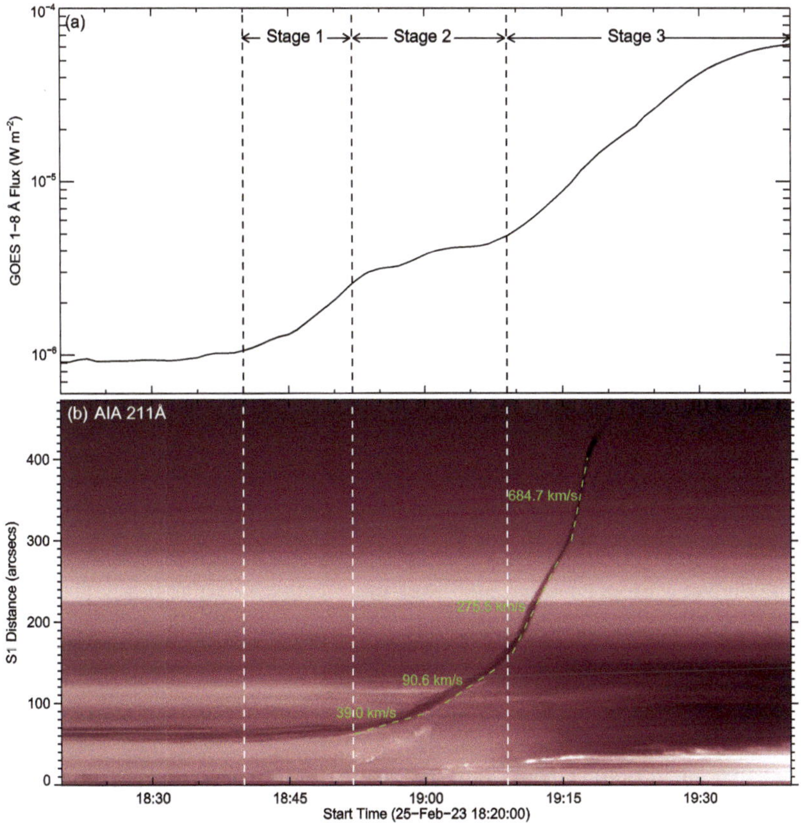

Figure 6 The three stages of the eruption. The upper panel shows GOES X-ray 1–8 Å flux. The bottom is the time-distance plot of slit S1 using AIA 211 Å data. Three stages can be seen that correspond to the eruptions of the two active-region filaments F1 and F2 (Stage 1), the interactions between both F1 and F2 with the large-scale filament F3 (Stage 2), and the eruption of F3 (Stage 3).

2 hours before the flare, we see there is no obvious flux emergence or cancellation in both areas. There is no approaching or separation between the positive and negative magnetic fields in the left foot area (Figures 8c and 8d). The right foot area is mainly associated with the positive magnetic field; we see no obvious converging or diffusing motions of the field. In summary, there are no significant photospheric changes in the two foot areas of F3 before the flare. That means the large-scale and high-lying filament F3 is unlikely to erupt because of photospheric changes in this period of time. Therefore, the pushing of F1 and F2 is an important process for the eruption of F3.

Figure 9 shows the CME resulting from the multifilament eruptions. Interestingly, the eruptions occur at the northwest of the solar disk close to the limb, yet they produce a halo CME with a velocity that reaches 1170.0 km/s (https://cdaw.gsfc.nasa.gov/CME_list/UNIVERSAL_ver1/2023_02/univ2023_02.html). The CME results from the eruption of the large-scale filament F3 and the combined pushing by filaments F1 and F2, which makes F3 to gain a significant amount of initial kinetic energy in a very short period of time. This CME also drives a shock wave, which can effectively accelerate electrons and other particles. It is

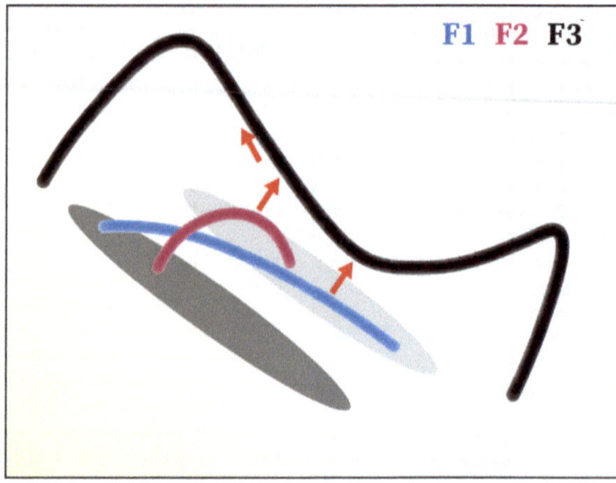

Figure 7 A cartoon that shows the interactions between the two active-region filaments F1 and F2 with the large-scale filament F3. Red arrows indicate the directions in which F1 and F2, respectively, collide with F3 and subsequently push F3 to eruption. The grey and dark ellipses outline the locations of positive and negative magnetic fields in the active region.

noteworthy that this halo CME triggered a geomagnetic storm (https://www.swpc.noaa.gov/news/strong-g3-geomagnetic-storms-possible-february-27).

Figure 10 is a schematic diagram that illustrates the basic idea of a large-scale filament, crossing over an active region, forced to erupt by a small-scale active-region filament pushing it out. In this picture, two types of filaments are involved: a large-scale intermediate or quiet filament carrying a large amount of material, which is usually stable and difficult to erupt, and an active-region filament, which is usually small but very active. The directions of the two filaments are almost coincident on the horizontal plane and almost parallel on the vertical plane. Their magnetic field directions are also similar. When the small-scale active-region filament erupts and begins to rise, it will then push the large-scale filament above to erupt. In this process, the collision, squeezing, and reconnection of the magnetic field lines between the surface layers of two filaments may occur, as in our observations in Figures 3 and 4. This figure suggests that we should pay special attention to large-scale filaments suspended above an active region, as they could be triggered into eruption by underlying activity and eject a significant amount of material into the interplanetary space, posing significant hazards to our near-Earth environment.

4. Summary and Discussion

We present observations of a multifilament eruption which is associated with an M6.4 flare and a halo CME. This event gives us a unique insight into the dynamics of multifilament eruptions and their corresponding effects on the interplanetary space.

Three filaments are involved in this eruption: two active-region filaments F1, F2, and a large-scale filament F3. They present a unique configuration: the two active-region filaments are located along the PIL, with F2 riding over F1, which coincides with the "double-decker" model (Liu et al., 2012); the higher-altitude large-scale filament F3 extends crossing over the active region, lying along the direction of the PIL (Figures 2 and 7). Considering GOES X-ray flux and multiwavelength observations, this event can be divided into three stages: 1) the eruptions of F1 and F2, 2) the interactions between the erupting F1 and F2 with F3, and 3) the eruption of F3 (Figure 6).

We observe that the eruption of F3 is a completely forced process, being pushed out by the erupting F1 and F2. This process has not been described and included in the conventional mechanisms of filament eruptions, namely the MHD instabilities (e.g. Hood and

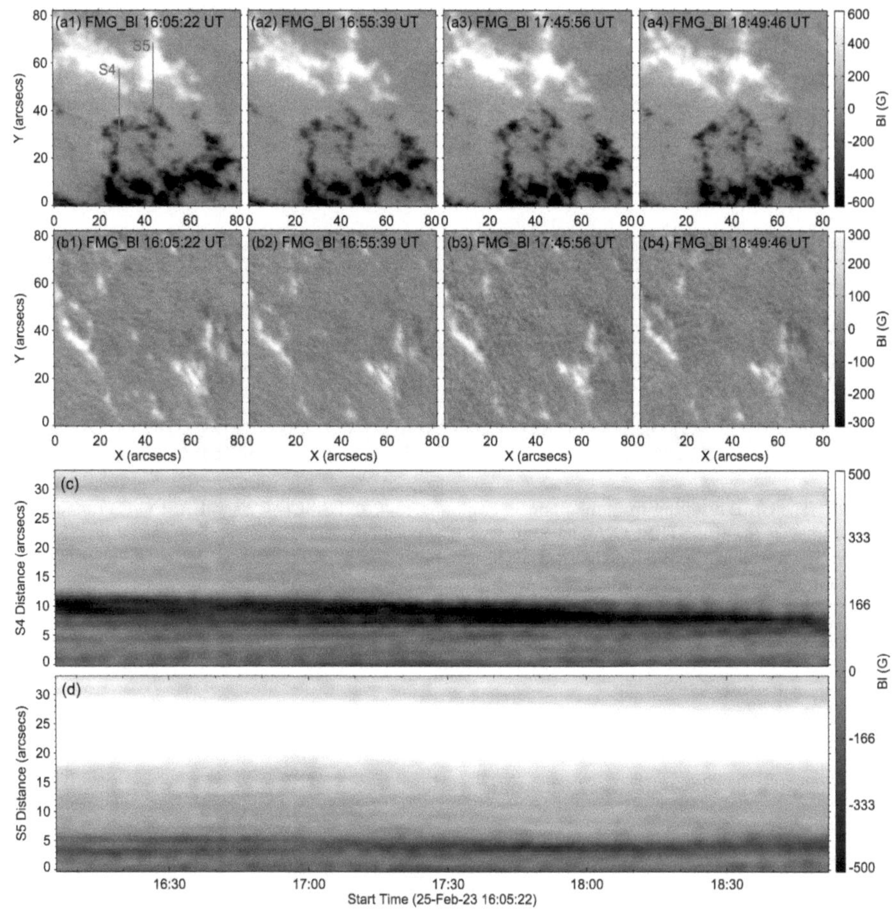

Figure 8 Temporal evolution of the LOS magnetic field in the two regions where filament F3 is rooted as observed by ASO-S/FMG. The two regions are marked by white dashed boxes in Figure 2b. (a1)–(a4) and (b1)–(b4): Show the left and right foot areas, respectively. Slits S4 and S5 are used to show the motions of different magnetic field polarities, time-distance maps of which are shown in panels c and d, respectively.

Priest, 1981; Ji et al., 2003; Fan, 2005; Török and Kliem, 2005; Démoulin and Aulanier, 2010) and magnetic reconnection related processes (e.g. Antiochos, DeVore, and Klimchuk, 1999; Chen and Shibata, 2000; Lin and Forbes, 2000; Moore et al., 2001). In the traditional picture, the filament eruption is caused by either the instability or triggered by magnetic reconnection occurring between the filament and the surrounding magnetic field. The initial kinetic energy of the filament eruption comes from their own magnetic system. However, in our case study, the initial kinetic energy of F3 comes from the erupting F1 and F2 and the pushing of F1 and F2 on F3, which lasts for more than 15 minutes.

There are also models on the interaction of two (or multi) flux ropes, such as the tilt–kink instability (e.g. Keppens, Porth, and Xia, 2014) and coalescence instability (e.g. Tajima et al., 1985; Murtas, Hillier, and Snow, 2021). When two adjacent parallel flux loops develop opposite currents, the tilt–kink instability will occur. This instability has been suggested to be involved in the eruption that triggers flux ropes with a "double-decker" configuration (Keppens, Porth, and Xia, 2014). The coalescence instability could result in impulsive en-

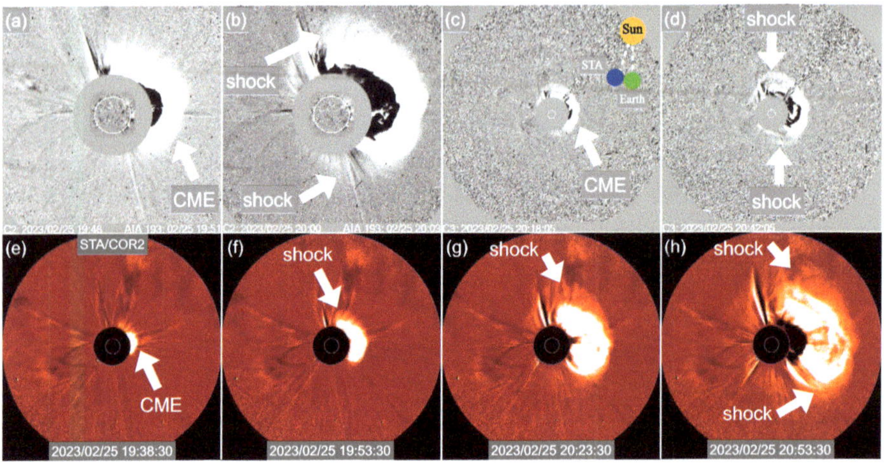

Figure 9 The CME resulting from the filament eruptions. (a)–(d): LASCO/C2 and LASCO/C3 running difference images. (e)–(h): STEREO-A/COR2 images. The positions of STEREO-A (STA) satellite, the Earth, and the Sun are shown in the top right corner of panel c. The separation angle of STA with respect to the Sun–Earth line is 12.85°. The CME and CME-driven shock are pointed by white arrows.

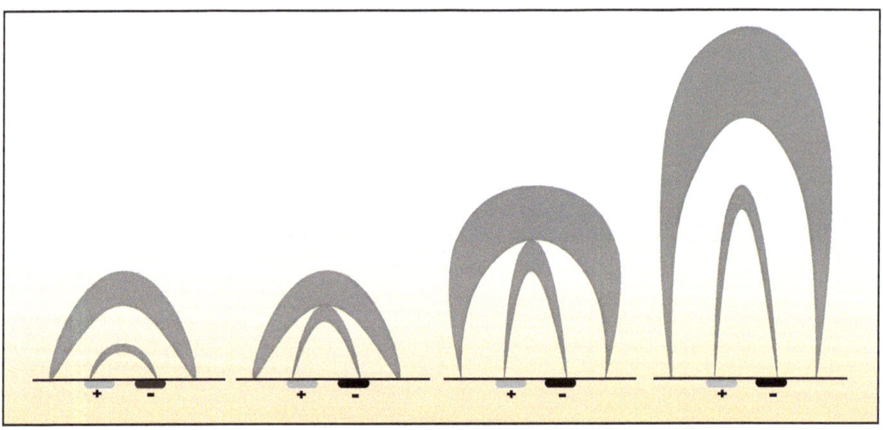

Figure 10 A schematic diagram that shows the large-scale filament eruption initiated by small-scale active-region filaments pushing out from below; "+" and "-" denote the signs of magnetic polarities in the active region.

ergy release with fast magnetic reconnection associated with the merging of two current carrying flux loops (Tajima et al., 1985; Murtas, Hillier, and Snow, 2021). In this study, we cannot conclude on the currents carried by the filaments and we do not observe clear signatures of strong reconnection between filaments F1 and F2 with filament F3. So, we are not sure whether the tilt–kink or coalescence instability would occur during the interactions between filaments F1 and F2 with filament F3. Nevertheless, any of them could still be possible.

Zhu et al. (2015) reported a "double-decker" filament eruption with a contact angle of about 20° between the two filaments. Their observations showed that the two filaments merged as the lower one erupted and collided with the upper one. In this process, a large amount of energy was released due to the reconnection between the magnetic fields of the two filaments. Their "double-decker" filament eruption ultimately resulted in an M2.9 flare and a CME. Here, our study presents a different story. No significant or violent magnetic reconnection is observed, between the lower two active-region filaments F1, F2, and the upper large-scale filament F3, possibly due to their unique magnetic configurations. F1 and F3 are almost parallel to each other and are nearly on the same vertical plane with the same magnetic field direction. Therefore, strong magnetic reconnection is unlikely to occur between them, although there is expansion and untwisting motions during the rise of F1 eruption.

When F2 is pushing F3, a bright structure moving upwards along the left shoulder of F3 is evident in the high-temperature AIA wavelengths (Figure 4). This bright structure should come from the heating of the plasma by squeezing and magnetic reconnection between the top of F2 and F3. Since the uplift process of F2 is also accompanied by untwisting, it is difficult to determine how large the contact angle is between the magnetic fields at the top of F2 and the left shoulder of F3. We cannot judge the specific contribution of magnetic reconnection to the plasma heating. However, even if partial magnetic reconnection occurs between them, it is not enough to destroy the overall magnetic structure of F3. As we see, F3 is eventually pushed out to eruption nearly without being restructured. This is evidently different from the observations of Zhu et al. (2015).

An interesting question is whether F3 could erupt spontaneously without the pushing from F1 and F2. F3 is a large-scale filament with a length of about 360 Mm. We see a lot of cold and dense plasma gathered in the central dip of the filament (Figures 1, 2, and 7). It may take a long time to heat and evaporate the plasma in the filament to reduce gravity. It is also difficult for the plasma within the filament to be spontaneously transported down to the lower atmosphere through the two legs of the filament by some kind of disturbance. As we see, even though the eruptions of F1 and F2 had already started for several minutes, F3 still remained stable, while the pushing by F1 and F2 on F3 persisted for more than 15 minutes to make the F3 finally erupt.

It is generally believed that the magnetic topology of a filament plays a key role in determining whether it will erupt or not (Chen, Xu, and Ding, 2020). However, in our case it is difficult to derive the magnetic structure of F3 through extrapolation. F3 is very large and rooted in a region of weak magnetic field. Being close to the limb also brings in a significant projection effect. For the photospheric magnetic field in the regions where F3 is rooted, we do not see any shearing, rotating, converging, or diffusing motions (Figure 8). This means that the magnetic topology of F3 should remain unchanged without external disturbances. This makes F3 to be a relatively stable filament. We speculate that F3 would not erupt, at least in a short time period, without the pushing from F1 and F2.

Our study provides an important indication on the impact of combined eruptions on space weather. Active-region filaments are very active and easy to erupt. However, their lengths are limited to the size of the active region, thus further restricting the scale of their eruptions. Quiet and intermediate filaments are large in size and, so, their mass is significant, but they are stable and difficult to erupt. If a large-scale filament is just suspended above an active region with a small-scale filament below it, as in our study, when the lower active-region filament erupts, it will then push the upper one to erupt (Figures 7 and 10). This means that a large amount of plasma will be ejected into the interplanetary space at a high initial velocity, resulting in a large space weather influence. So our study suggests that we should pay particular attention to configurations where a large-scale filament suspends above an active region with one or more underneath active-region filaments. This type of topology has a great potential to result in severe space weather events.

Acknowledgments We are grateful to Dr. Hechao Chen at Yunnan University, Dr. Yijun Hou at National Astronomical Observatories of Chinese Academy of Sciences (CAS) and Dr. Zhenyong Hou at Peking University for their valuable discussions. The authors thank the ASO-S, SUTRI, CHASE, SDO, STERO, and GOES teams for providing the data. ASO-S mission is supported by the Strategic Priority Research Program on Space Science of CAS (Grant No. XDA15320000). The CHASE mission is supported by the China National Space Administration (CNSA). SDO is a space mission in the Living With a Star Program of NASA. SOHO is a project of international cooperation between ESA and NASA. SUTRI is a collaborative project conducted by the National Astronomical Observatories of CAS, Peking University, Tongji University, Xi'an Institute of Optics and Precision Mechanics of CAS, and the Innovation Academy for Microsatellites of CAS. The CME catalog used is generated and maintained at the CDAW Data Center by NASA and The Catholic University of America in cooperation with the Naval Research Laboratory.

Author contributions Y.L. Song and J.T. Su carried out this study. Q.M. Zhang analyzed the associated CME. M. Zhang and Y.Y. Deng provided suggestions on the design of study. X.Y. Bai, S. Liu, X. Yang, J. Chen, H.Q. Xu, K.F. Ji and Z.Y. Hu provided suggestions on the data analysis. Y.L. Song wrote the manuscript. J.T. Su and M. Zhang revised the manuscript. All the authors helped to improve the manuscript.

Funding This work is supported by National Key R&D Program of China (Grant No. 2022YFF0503001, 2021YFA0718601), National Natural Science Foundation of China (Grant No. 12173049, 11973056, 11803002) and Beijing Natural Science Foundation (Grant No. 1222029).

Data Availability ASO-S data is available upon reasonable request. SDO, SUTRI data used in this work are publicly accessible on their websites (http://jsoc.stanford.edu/ajax/lookdata.html and https://sun10.bao.ac.cn/SUTRI/). The data of STEREO-A and SOHO can be obtained through the Virtual Solar Observatory (https://sdac.virtualsolar.org/cgi/search). The Hα data from CHASE and GONG can be accessed from their official websites (https://ssdc.nju.edu.cn/home and https://gong2.nso.edu/archive/patch.pl?menutype=s).

Declarations

Competing interests The authors declare no competing interests.

References

Antiochos, S.K., DeVore, C.R., Klimchuk, J.A.: 1999, A model for solar coronal mass ejections. *Astrophys. J.* **510**, 485. DOI. ADS.

Aulanier, G., DeVore, C.R., Antiochos, S.K.: 2002, Prominence magnetic dips in three-dimensional sheared arcades. *Astrophys. J. Lett.* **567**, L97. DOI. ADS.

Awasthi, A.K., Liu, R., Wang, Y.: 2019, Double-decker filament configuration revealed by mass motions. *Astrophys. J.* **872**, 109. DOI. ADS.

Bai, X., Tian, H., Deng, Y., Wang, Z., Yang, J., Zhang, X., Zhang, Y., Qi, R., Wang, N., Gao, Y., Yu, J., He, C., Shen, Z., Shen, L., Guo, S., Hou, Z., Ji, K., Bi, X., Duan, W., Yang, X., Lin, J., Hu, Z., Song, Q., Yang, Z., Chen, Y., Qiao, W., Ge, W., Li, F., Jin, L., He, J., Chen, X., Zhu, X., He, J., Shi, Q., Liu, L., Li, J., Xu, D., Liu, R., Li, T., Feng, Z., Wang, Y., Fan, C., Liu, S., Guo, S., Sun, Z., Wu, Y., Li, H., Yang, Q., Ye, Y., Gu, W., Wu, J., Zhang, Z., Yu, Y., Ye, Z., Sheng, P., Wang, Y., Li, W., Huang, Q., Zhang, Z.: 2023, The Solar Upper Transition Region Imager (SUTRI) onboard the SATech-01 satellite. *Res. Astron. Astrophys.* **23**, 065014. DOI. ADS.

Brueckner, G.E., Howard, R.A., Koomen, M.J., Korendyke, C.M., Michels, D.J., Moses, J.D., Socker, D.G., Dere, K.P., Lamy, P.L., Llebaria, A., Bout, M.V., Schwenn, R., Simnett, G.M., Bedford, D.K., Eyles, C.J.: 1995, The Large Angle Spectroscopic Coronagraph (LASCO). *Solar Phys.* **162**, 357. DOI. ADS.

Chandra, R., Schmieder, B., Mandrini, C.H., Démoulin, P., Pariat, E., Török, T., Uddin, W.: 2011, Homologous flares and magnetic field topology in active region NOAA 10501 on 20 November 2003. *Solar Phys.* **269**, 83. DOI. ADS.

Chen, P.F., Shibata, K.: 2000, An emerging flux trigger mechanism for coronal mass ejections. *Astrophys. J.* **545**, 524. DOI. ADS.

Chen, P.-F., Xu, A.-A., Ding, M.-D.: 2020, Some interesting topics provoked by the solar filament research in the past decade. *Res. Astron. Astrophys.* **20**, 166. DOI. ADS.

Chen, H., Duan, Y., Yang, J., Yang, B., Dai, J.: 2018, Witnessing tether-cutting reconnection at the onset of a partial eruption. *Astrophys. J.* **869**, 78. DOI. ADS.

Chen, B., Li, H., Song, K.-F., Guo, Q.-F., Zhang, P.-J., He, L.-P., Dai, S., Wang, X.-D., Wang, H.-F., Liu, C.-L., Zhang, H.-J., Zhang, G., Wang, Y., Liu, S.-J., Zhang, H.-X., Liu, L., Mao, S.-L., Liu, Y., Peng, J.-H., Wang, P., Sun, L., Liu, Y., Han, Z.-W., Wang, Y.-L., Wu, K., Ding, G.-X., Zhou, P., Zheng, X., Xia, M.-Y., Wu, Q.-W., Xie, J.-J., Chen, Y., Song, S.-M., Wang, H., Zhu, B., Chu, C.-B., Yang, W.-G., Feng, L., Huang, Y., Gan, W.-Q., Li, Y., Li, J.-W., Lu, L., Xue, J.-C., Ying, B.-L., Sun, M.-Z., Zhu, C., Bao, W.-M., Deng, L., Yin, Z.-S.: 2019, The Lyman-alpha Solar Telescope (LST) for the ASO-S mission – II. Design of LST. *Res. Astron. Astrophys.* **19**, 159. DOI. ADS.

Cheng, X., Ding, M.D., Zhang, J., Sun, X.D., Guo, Y., Wang, Y.M., Kliem, B., Deng, Y.Y.: 2014, Formation of a double-decker magnetic flux rope in the sigmoidal solar active region 11520. *Astrophys. J.* **789**, 93. DOI. ADS.

Démoulin, P., Aulanier, G.: 2010, Criteria for flux rope eruption: non-equilibrium versus torus instability. *Astrophys. J.* **718**, 1388. DOI. ADS.

Deng, Y.-Y., Zhang, H.-Y., Yang, J.-F., Li, F., Lin, J.-B., Hou, J.-F., Wu, Z., Song, Q., Duan, W., Bai, X.-Y., Wang, D.-G., Lv, J., Ge, W., Wang, J.-N., Zheng, Z.-Y., Wang, C.-J., Wang, N.-G., Ni, H.-K., Zeng, Y.-Z., Zhang, Y., Yang, X., Sun, Y.-Z., Zhang, Z.-Y., Wang, X.-F.: 2019, Design of the full-disk MagnetoGraph (FMG) onboard the ASO-S. *Res. Astron. Astrophys.* **19**, 157. DOI. ADS.

Domingo, V., Fleck, B., Poland, A.I.: 1995, The SOHO mission: an overview. *Solar Phys.* **162**, 1. DOI. ADS.

Fan, Y.: 2005, Coronal mass ejections as loss of confinement of kinked magnetic flux ropes. *Astrophys. J.* **630**, 543. DOI. ADS.

Gan, W.-Q., Zhu, C., Deng, Y.-Y., Li, H., Su, Y., Zhang, H.-Y., Chen, B., Zhang, Z., Wu, J., Deng, L., Huang, Y., Yang, J.-F., Cui, J.-J., Chang, J., Wang, C., Wu, J., Yin, Z.-S., Chen, W., Fang, C., Yan, Y.-H., Lin, J., Xiong, W.-M., Chen, B., Bao, H.-C., Cao, C.-X., Bai, Y.-P., Wang, T., Chen, B.-L., Li, X.-Y., Zhang, Y., Feng, L., Su, J.-T., Li, Y., Chen, W., Li, Y.-P., Su, Y.-N., Wu, H.-Y., Gu, M., Huang, L., Tang, X.-J.: 2019, Advanced space-based solar observatory (ASO-S): an overview. *Res. Astron. Astrophys.* **19**, 156. DOI. ADS.

Guo, Y., Schmieder, B., Démoulin, P., Wiegelmann, T., Aulanier, G., Török, T., Bommier, V.: 2010, Coexisting flux rope and dipped arcade sections along one solar filament. *Astrophys. J.* **714**, 343. DOI. ADS.

Harvey, J.W., Bolding, J., Clark, R., Hauth, D., Hill, F., Kroll, R., Luis, G., Mills, N., Purdy, T., Henney, C., Holland, D., Winter, J.: 2011, Full-disk solar H-alpha images from GONG. In: *AAS/Solar Physics Division Abstracts #42, AAS/Solar Physics Division Meeting* **42**, 17.45. ADS.

Hood, A.W., Priest, E.R.: 1981, Critical conditions for magnetic instabilities in force-free coronal loops. *Geophys. Astrophys. Fluid Dyn.* **17**, 297. DOI. ADS.

Hou, Y.J., Zhang, J., Li, T., Yang, S.H., Li, X.H.: 2018, Eruption of a multi-flux-rope system in solar active region 12673 leading to the two largest flares in Solar Cycle 24. *Astron. Astrophys.* **619**, A100. DOI. ADS.

Hou, Y., Li, C., Li, T., Su, J., Qiu, Y., Yang, S., Yang, L., Li, L., Guo, Y., Hou, Z., Song, Q., Bai, X., Zhou, G., Ding, M., Gan, W., Deng, Y.: 2023, Partial eruption of solar filaments. I. Configuration and formation of double-decker filaments. *Astrophys. J.* **959**, 69. DOI. ADS.

Hu, H., Liu, Y.D., Chitta, L.P., Peter, H., Ding, M.: 2022, Spectroscopic and imaging observations of spatially extended magnetic reconnection in the splitting of a solar filament structure. *Astrophys. J. Lett.* **940**, L12. DOI. ADS.

Ji, H., Wang, H., Schmahl, E.J., Moon, Y.-J., Jiang, Y.: 2003, Observations of the failed eruption of a filament. *Astrophys. J. Lett.* **595**, L135. DOI. ADS.

Jiang, C., Wu, S.T., Feng, X., Hu, Q.: 2014a, Nonlinear force-free field extrapolation of a coronal magnetic flux rope supporting a large-scale solar filament from a photospheric vector magnetogram. *Astrophys. J. Lett.* **786**, L16. DOI. ADS.

Jiang, Y., Yang, J., Wang, H., Ji, H., Liu, Y., Li, H., Li, J.: 2014b, Interaction and merging of two sinistral filaments. *Astrophys. J.* **793**, 14. DOI. ADS.

Kaiser, M.L., Kucera, T.A., Davila, J.M., St. Cyr, O.C., Guhathakurta, M., Christian, E.: 2008, The STEREO mission: an introduction. *Space Sci. Rev.* **136**, 5. DOI. ADS.

Keppens, R., Porth, O., Xia, C.: 2014, Interacting tilt and kink instabilities in repelling current channels. *Astrophys. J.* **795**, 77. DOI. ADS.

Kliem, B., Török, T., Titov, V.S., Lionello, R., Linker, J.A., Liu, R., Liu, C., Wang, H.: 2014, Slow rise and partial eruption of a double-decker filament. II. A double flux rope model. *Astrophys. J.* **792**, 107. DOI. ADS.

Kumar, P., Manoharan, P.K., Uddin, W.: 2010, Evolution of solar magnetic field and associated multiwavelength phenomena: flare events on 2003 November 20. *Astrophys. J.* **710**, 1195. DOI. ADS.

Lemen, J.R., Title, A.M., Akin, D.J., Boerner, P.F., Chou, C., Drake, J.F., Duncan, D.W., Edwards, C.G., Friedlaender, F.M., Heyman, G.F., Hurlburt, N.E., Katz, N.L., Kushner, G.D., Levay, M., Lindgren, R.W., Mathur, D.P., McFeaters, E.L., Mitchell, S., Rehse, R.A., Schrijver, C.J., Springer, L.A., Stern, R.A., Tarbell, T.D., Wuelser, J.-P., Wolfson, C.J., Yanari, C., Bookbinder, J.A., Cheimets, P.N., Caldwell,

D., Deluca, E.E., Gates, R., Golub, L., Park, S., Podgorski, W.A., Bush, R.I., Scherrer, P.H., Gummin, M.A., Smith, P., Auker, G., Jerram, P., Pool, P., Soufli, R., Windt, D.L., Beardsley, S., Clapp, M., Lang, J., Waltham, N.: 2012, The Atmospheric Imaging Assembly (AIA) on the Solar Dynamics Observatory (SDO). *Solar Phys.* **275**, 17. DOI. ADS.

Li, H., Chen, B., Feng, L., Li, Y., Huang, Y., Li, J.-W., Lu, L., Xue, J.-C., Ying, B.-L., Zhao, J., Yang, Y.-T., Gan, W.-Q., Fang, C., Song, K.-F., Wang, H., Guo, Q.-F., He, L.-P., Zhu, B., Zhu, C., Deng, L., Bao, H.-C., Cao, C.-X., Yang, Z.-G.: 2019, The Lyman-alpha Solar Telescope (LST) for the ASO-S mission — I. Scientific objectives and overview. *Res. Astron. Astrophys.* **19**, 158. DOI. ADS.

Li, C., Fang, C., Li, Z., Ding, M., Chen, P., Qiu, Y., You, W., Yuan, Y., An, M., Tao, H., Li, X., Chen, Z., Liu, Q., Mei, G., Yang, L., Zhang, W., Cheng, W., Chen, J., Chen, C., Gu, Q., Huang, Q., Liu, M., Han, C., Xin, H., Chen, C., Ni, Y., Wang, W., Rao, S., Li, H., Lu, X., Wang, W., Lin, J., Jiang, Y., Meng, L., Zhao, J.: 2022, The Chinese Hα Solar Explorer (CHASE) mission: an overview. *Sci. China Phys. Chem.* **65**, 289602. DOI. ADS.

Li, L., Song, H., Peter, H., Chitta, L.P.: 2022, Failed solar eruption of a multithermal flux rope. *Astrophys. J. Lett.* **941**, L1. DOI. ADS.

Lin, J., Forbes, T.G.: 2000, Effects of reconnection on the coronal mass ejection process. *J. Geophys. Res.* **105**, 2375. DOI. ADS.

Linton, M.G., Dahlburg, R.B., Antiochos, S.K.: 2001, Reconnection of twisted flux tubes as a function of contact angle. *Astrophys. J.* **553**, 905. DOI. ADS.

Liu, R., Kliem, B., Török, T., Liu, C., Titov, V.S., Lionello, R., Linker, J.A., Wang, H.: 2012, Slow rise and partial eruption of a double-decker filament. I. Observations and interpretation. *Astrophys. J.* **756**, 59. DOI. ADS.

Mackay, D.H., Karpen, J.T., Ballester, J.L., Schmieder, B., Aulanier, G.: 2010, Physics of solar prominences: II—magnetic structure and dynamics. *Space Sci. Rev.*, **151**, 333. DOI. ADS.

Moore, R.L., Sterling, A.C., Hudson, H.S., Lemen, J.R.: 2001, Onset of the magnetic explosion in solar flares and coronal mass ejections. *Astrophys. J.* **552**, 833. DOI. ADS.

Murtas, G., Hillier, A., Snow, B.: 2021, Coalescence instability in chromospheric partially ionized plasmas. *Phys. Plasmas* **28**, 032901. DOI. ADS.

Priest, E.: 2014, Magnetohydrodynamics of the Sun. DOI. ADS.

Qiu, Y., Guo, Y., Ding, M., Zhong, Z.: 2020, The magnetic topology and eruption mechanism of a multiple-ribbon flare. *Astrophys. J.* **901**, 13. DOI. ADS.

Su, J., Liu, Y., Kurokawa, H., Mao, X., Yang, S., Zhang, H., Wang, H.: 2007, Observation of interactions and eruptions of two filaments. *Solar Phys.* **242**, 53. DOI. ADS.

Su, Y., Liu, R., Li, S., Cao, W., Ahn, K., Ji, H.: 2018, High-resolution observations of flares in an arch filament system. *Astrophys. J.* **855**, 77. DOI. ADS.

Sun, Z., Li, T., Tian, H., Hou, Y., Hou, Z., Chen, H., Bai, X., Deng, Y.: 2023, Observation of two splitting processes in a partial filament eruption on the sun: the role of breakout reconnection. *Astrophys. J.* **953**, 148. DOI. ADS.

Tajima, T., Brunel, F., Sakai, J.-I., Vlahos, L., Kundu, M.R.: 1985, The coalescence instability in solar flares. In: Kundu, M.R., Holman, G.D. (eds.) *Unstable Current Systems and Plasma, Instabilities in Astrophysics* **107**, 197. ADS.

Tian, H.: 2017, Probing the solar transition region: current status and future perspectives. *Res. Astron. Astrophys.* **17**, 110. DOI. ADS.

Török, T., Kliem, B.: 2005, Confined and ejective eruptions of kink-unstable flux ropes. *Astrophys. J. Lett.* **630**, L97. DOI. ADS.

Török, T., Chandra, R., Pariat, E., Démoulin, P., Schmieder, B., Aulanier, G., Linton, M.G., Mandrini, C.H.: 2011, Filament interaction modeled by flux rope reconnection. *Astrophys. J.* **728**, 65. DOI. ADS.

Wuelser, J.-P., Lemen, J.R., Tarbell, T.D., Wolfson, C.J., Cannon, J.C., Carpenter, B.A., Duncan, D.W., Gradwohl, G.S., Meyer, S.B., Moore, A.S., Navarro, R.L., Pearson, J.D., Rossi, G.R., Springer, L.A., Howard, R.A., Moses, J.D., Newmark, J.S., Delaboudiniere, J.-P., Artzner, G.E., Auchere, F., Bougnet, M., Bouyries, P., Bridou, F., Clotaire, J.-Y., Colas, G., Delmotte, F., Jerome, A., Lamare, M., Mercier, R., Mullot, M., Ravet, M.-F., Song, X., Bothmer, V., Deutsch, W.: 2004, EUVI: the STEREO-SECCHI extreme ultraviolet imager. In: Fineschi, S., Gummin, M.A. (eds.) *Telescopes and Instrumentation for Solar Astrophysics, SPIE Conf. Ser.* **5171**, 111. DOI. ADS.

Xia, C., Keppens, R., Guo, Y.: 2014, Three-dimensional prominence-hosting magnetic configurations: creating a helical magnetic flux rope. *Astrophys. J.* **780**, 130. DOI. ADS.

Xu, H., Su, J., Chen, J., Ruan, G., Awasthi, A.K., Zhang, H., Zhang, M., Ji, K., Zhang, Y., Liu, J.: 2020, Multiwavelength observation of a failed eruption from a helical kink-unstable prominence. *Astrophys. J.* **901**, 121. DOI. ADS.

Yan, X.L., Xue, Z.K., Pan, G.M., Wang, J.C., Xiang, Y.Y., Kong, D.F., Yang, L.H.: 2015, The formation and magnetic structures of active-region filaments observed by NVST, SDO, and Hinode. *Astron. Astrophys. Suppl. Ser.* **219**, 17. DOI. ADS.

Zhang, Q.M., Ning, Z.J., Guo, Y., Zhou, T.H., Cheng, X., Ji, H.S., Feng, L., Wiegelmann, T.: 2015, Multiwavelength observations of a partially eruptive filament on 2011 September 8. *Astrophys. J.* **805**, 4. DOI. ADS.

Zhang, Z., Chen, D.-Y., Wu, J., Chang, J., Hu, Y.-M., Su, Y., Zhang, Y., Wang, J.-P., Liang, Y.-M., Ma, T., Guo, J.-H., Cai, M.-S., Zhang, Y.-Q., Huang, Y.-Y., Peng, X.-Y., Tang, Z.-B., Zhao, X., Zhou, H.-H., Wang, L.-G., Song, J.-X., Ma, M., Xu, G.-Z., Yang, J.-F., Lu, D., He, Y.-H., Tao, J.-Y., Ma, X.-L., Lv, B.-G., Bai, Y.-P., Cao, C.-X., Huang, Y., Gan, W.-Q.: 2019, Hard X-ray Imager (HXI) onboard the ASO-S mission. *Res. Astron. Astrophys.* **19**, 160. DOI. ADS.

Zhang, Q., Wang, Y., Liu, R., Zhang, J., Hu, Y., Wang, W., Zhuang, B., Li, X.: 2020, Eruption of solar magnetic flux ropes caused by flux feeding. *Astrophys. J. Lett.* **898**, L12. DOI. ADS.

Zheng, R., Zhang, L., Wang, B., Kong, X., Song, H., Wu, Z., Feng, S., Chen, H., Chen, Y.: 2021, Compound eruptions of twin flux ropes in a solar active region. *Astrophys. J. Lett.* **921**, L39. DOI. ADS.

Zhu, C., Alexander, D.: 2014, Eruption of a bifurcated solar filament. *Solar Phys.* **289**, 279. DOI. ADS.

Zhu, C., Liu, R., Alexander, D., Sun, X., McAteer, R.T.J.: 2015, Complex flare dynamics initiated by a filament–filament interaction. *Astrophys. J.* **813**, 60. DOI. ADS.

Publisher's Note Springer Nature remains neutral with regard to jurisdictional claims in published maps and institutional affiliations.

Solar Physics (2024) 299:91
https://doi.org/10.1007/s11207-024-02322-x

RESEARCH

Investigation of Decaying β-Configuration Sunspot in Active Region NOAA 13229

Shahid Idrees[1,2] · Jiangtao Su[1,2] · Jie Chen[1,2] · Yuanyong Deng[1,2]

Received: 5 December 2023 / Accepted: 13 May 2024 / Published online: 24 June 2024

Abstract
In this study, we investigate the decay of sunspots in the active region NOAA 13229 using data from the ASO-S/FMG and SDO/HMI. We closely examine the decay patterns of sunspots S1 and S2, which reveal different decay rates and features due to the mechanisms of magnetic cancellation, dispersion, and the role of horizontal flows. Our analysis highlights the significant impact of magnetic flux changes, including the decrease of both the sunspot area and magnetic flux over time, which adheres to distinct decay laws. This study elucidates the complex interplay between magnetic submergence, cancellation, and dispersion in the sunspot decay process, contributing to our understanding of the underlying mechanisms driving these phenomena. Our results emphasize the importance of horizontal flow dynamics in shaping the decay characteristics of sunspots, providing insights for the role played by the magnetic and plasma processes in solar active regions.

Keywords Sunspots · Solar activity · Solar flares · Active regions

1. Introduction

Sunspot study is an important and hot topic of research in solar physics due to some main reasons: (i) sunspots have a significant impact on the Sun's magnetic field, (ii) sunspots are the major source of solar activity such as CMEs, solar flares, etc., (iii) sunspots may be utilized to investigate the underlying structure and dynamics of the Sun. Sunspots are the most essential feature of the photosphere, consisting of the umbra and penumbra. The difference between a sunspot and a pore is the presence of penumbra. A lot of research

✉ S. Idrees
shahidbsp1341@gmail.com

✉ J. Su
sjt@bao.ac.cn

J. Chen
chenjie@bao.ac.cn

Y. Deng
dyy@nao.cas.cn

1 University of Chinese Academy of Sciences, Beijing 100049, China

2 National Astronomical Observatories, Chinese Academy of Sciences, Beijing 100012, China

has been conducted from a simulation and observation perspective towards the generation of sunspot penumbra (Leka and Skumanich, 1998; Schlichenmaier et al., 2010; Romano et al., 2020; Li et al., 2018; Murabito et al., 2016; Romano et al., 2014; Lim et al., 2013; Shimizu, Ichimoto, and Suematsu, 2012; Rempel, 2011, 2012; MacTaggart, Guglielmino, and Zuccarello, 2016; Thomas et al., 2002). On the other hand, the process behind sunspot decay is still poorly understood.

Sunspot decay can be studied from two perspectives. One of them is the study of the area decay of sunspots, termed photometric decay. The other is the study of the process of diminished magnetic flux from the photosphere, known as magnetic flux diffusion. It was studied by Bumba (1963) who observed that the area of the sunspot declines linearly with respect to time. He also made significant contributions by studying the decay rates of leader and follower sunspots. Different decay laws have been proposed by previous studies related to sunspot decay, like linear decay law (Martinez Pillet, Moreno-Insertis, and Vazquez, 1993; Rüdiger and Kitchatinov, 2000; Chapman et al., 2003; Gafeira et al., 2014; Muraközy, 2020), parabolic decay law (Litvinenko and Wheatland, 2015; Petrovay and van Driel-Gesztelyi, 1997; Petrovay, Martínez Pillet, and van Driel-Gesztelyi, 1999), quadratic decay law, exponential decay law, etc. Solanki (2003) proposed that 95% of sunspot decay follows the linear decay law.

The decay mechanisms responsible for the linear and parabolic decay laws of the sunspot area are likely to be distinct. Meyer et al. (1974) proposed a model to explain the linear decay law, attributing it to turbulent magnetic diffusion within a sunspot. This turbulent diffusion model suggests that the loss of magnetic flux within the sunspot occurs uniformly throughout its entire area, regardless of the sunspot size or perimeter length (Krause and Ruediger, 1975; Martínez Pillet, 2002). In contrast, the parabolic decay law is more consistent with the turbulent erosion model. According to this model, the decay of a sunspot is primarily driven by the erosion of its outer boundary. In other words, the decay rate is expected to be directly proportional to the length of the sunspot perimeter (Petrovay and van Driel-Gesztelyi, 1997; Petrovay and Moreno-Insertis, 1997; Martínez Pillet, 2002; Solanki, 2003; Litvinenko and Wheatland, 2015).

In addition to the decay of sunspot area, the decay of magnetic flux is a significant process during the decay of sunspots. Skumanich and Lites (1994) observed a linear decrease in the total magnetic flux of symmetric sunspots over time, with a decay rate of 0.9×10^{20} Mx day^{-1}. Verma and Denker (2012) conducted a study on active region NOAA 11126 using the Helioseismic and Magnetic Imager (HMI) and reported decay rates of 2.3 and 4.7×10^{20} Mx day^{-1} for the positive and negative magnetic fluxes, respectively. Similarly, Sheeley et al. (2017) analyzed 36 sunspots using HMI magnetograms and continuum images, finding a nearly linear decreasing trend in the magnetic fluxes of certain sunspots, with an average decay rate of $2-4 \times 10^{20}$ Mx day^{-1}. Rempel (2015) performed simulations of a sunspot and observed a linear decay of magnetic flux within its umbra. The decay rate increases linearly with group size until a certain point, beyond which it stabilizes. The decay rate per individual sunspot indicates a diffusion process. Variations in decay rates are noted across latitudes, with limited evidence for differences between solar cycles (Hathaway and Choudhary, 2008).

Among the most significant observational evidence of sunspot decay are moving magnetic features (MMFs). It has been always considered that MMFs are the cause of magnetic flux transfer from a sunspot during sunspot decay (Chen et al., 2015). As a result of an interplay between the Evershed flow, moat flows, and penumbra magnetic field, MMFs are produced during the sunspot decay (Martínez Pillet, 2002). There is a strong relation between the Evershed flows, moat flows, and MMFs (Cabrera Solana et al., 2006; Sainz Dalda and Martínez Pillet, 2005). Deng et al. (2007) suggested that the magnetic flux removal is proceeded by three steps: breaking up of the umbra, flux cancellation of MMFs, and dispersion of flux via MMFs. Verma and Denker (2012) studied MMFs as the main reason for magnetic flux removal and they also noticed the horizontal flows in the vicinity of sunspots during sunspot decay. Afterwards, horizontal flow fields have been observed by using HMI Doppler maps (Strecker and Bello González, 2018). Moat flow has been found around the decaying sunspots whose velocity decreased with the decay of sunspots. In addition, various studies have suggested that in the phase of sunspot decay the horizontal magnetic fields within the penumbra undergo a transformation, becoming vertical (Bellot Rubio, Tritschler, and Martínez Pillet, 2008; Watanabe, Kitai, and Otsuji, 2014; Verma et al., 2018)). As these penumbral magnetic fields rise towards the chromosphere due to buoyancy, the penumbra of the sunspot gradually diminishes in the photosphere. Similarly, the rapid fading of the penumbra after a solar flare is attributed to the rearrangement of the magnetic field lines (Wang et al., 2004; Deng et al., 2005)). In simulations conducted by Rempel (2015), it was observed that the submergence of a magnetic field played a significant role in sunspot decay. However, due to challenges linked to observing flow motions beneath the photosphere, there is currently no evidence to support this perspective.

Previous research into the decay of sunspots has primarily focused on configurations of sunspots with attention given to the decay process of individual configurations. To highlight this gap, we conducted a study specifically examining the decay phase of β-configuration sunspots. We closely analyzed two sunspots (S1 and S2) during their decay stage. For an account of our observations and methods used in this study, please refer to Section 2. The implications of our gathered data and subsequent discussions are presented in Section 3. Finally, we outline the conclusions drawn from our findings in Section 4.

2. Method and Observation

The focus of our observational study is the decay of sunspots in the active region (AR) NOAA 13229 which occurred from 00:01 UT on 2023 February 22 to 17:06 UT on 2023 February 25. The decay of a specified sunspot is interesting because it relates to four flares occurring during its decay period (Shimizu, Ichimoto, and Suematsu, 2012). Data has been taken from the ASO-S (Advanced Space-based Solar Observatory) and SDO (Solar Dynamics Observatory). The ASO-S (Chinese nickname Kuafu-1) is the first Chinese comprehensive dedicated solar observatory in space (Gan et al., 2019, 2023). Its primary scientific objective is to improve our understanding of a variety of solar phenomena such as solar flares, CMEs, solar magnetic field, and their relationships. In addition, it has consequences for understanding the dynamics of the Sun and its impact on space weather. The Full-disk vector MagnetoGraph (Deng et al., 2019) on board the ASO-S is designed to measure the photospheric magnetic fields over the entire solar disk through Fe I 532.42 nm line with high spatial and temporal resolutions, and high magnetic sensitivity. The Full-disk MagnetoGraph (FMG) boasts a 14-cm aperture and a high-resolution 4096 × 4096 pixel CMOS camera. FMG observes polarized Stokes images using a Liquid Crystal Variable Retarder (Hou et al., 2020). FMG measures longitudinal magnetic fields with a sensitivity of 15 G in the normal mode. Temporal resolution varies, offering 30-second updates for single-component magnetograms and 2-minute intervals for vector magnetogram acquisition in normal mode. Spatially, the instrument achieves roughly 1.5 arcseconds resolution with a 0.55 arcsecond pixel size (Deng et al., 2019; Su et al., 2019), further enhanced to 1.04 arcseconds for the

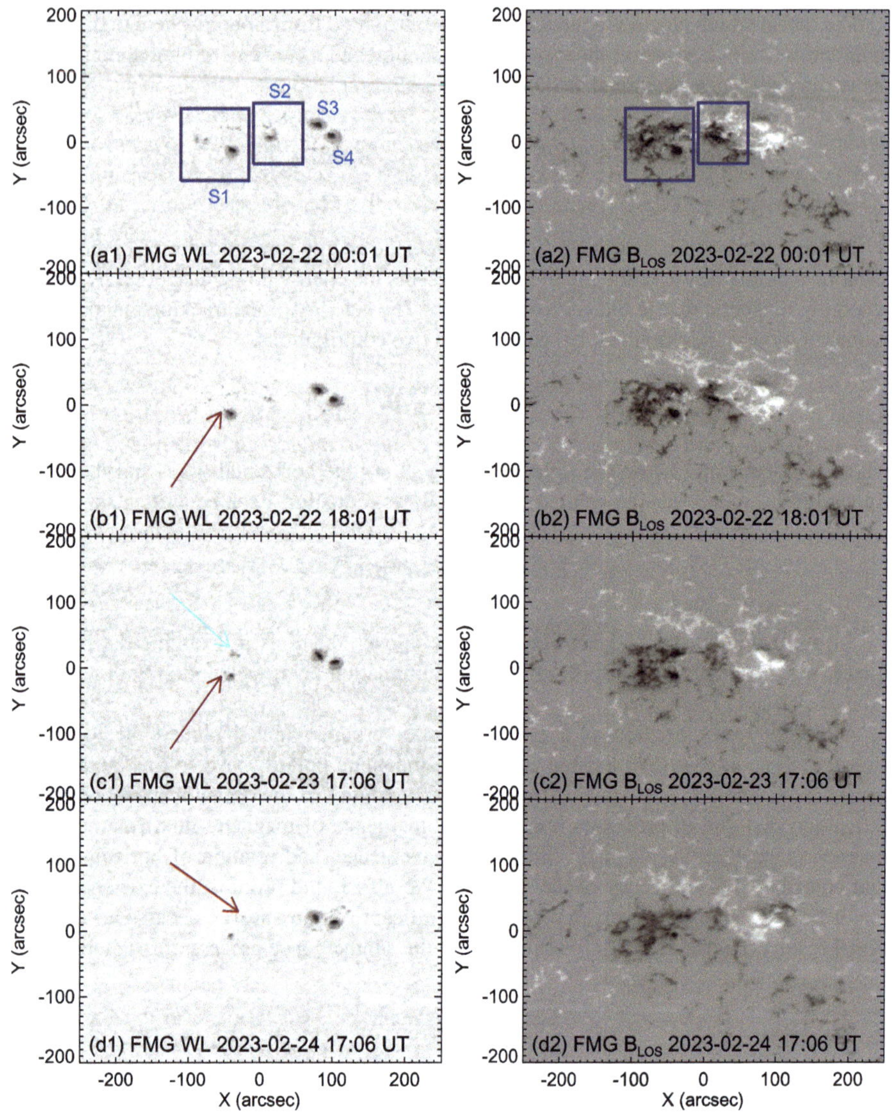

Figure 1 Left panels (a1, b1, c1, d1) show the area decay pattern of sunspots (focused on S1 and S2) with respect to time. The red arrows indicate the rotation of a separated part and the light blue arrow represents disintegration of a sunspot. Right panels (a2, b2, c2, d2) represent the magnetic flux decay with respect to time. Both rectangles in the panels show the sunspots we focused on (S1 and S2). The map has a field of view of 500 arcsec × 400 arcsec.

The sunspot S2 decayed more rapidly from 00:01 UT on 2023 February 22 to 17:06 UT on 2023 February 24. It has completely disappeared at last and nothing remained. The negative flux in the sunspot S2 was surrounded by positive flux on its north and west sides. Both fluxes decreased significantly, which can be observed in the last panel (d2). So, it can be clearly seen that magnetic flux cancellation is dominant in the sunspot S2 compared to the sunspot S1 because the positive flux quantity is comparable to the negative flux so they can

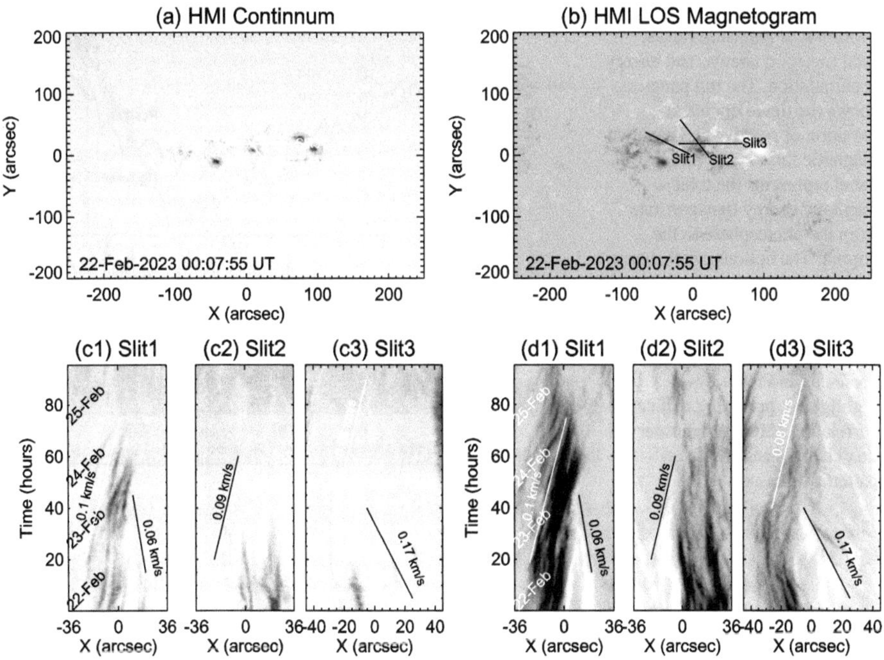

Figure 2 Top: (a) Three slit positions (slit1, slit2, slit3) drawn on the HMI continuum (b) HMI LOS magnetogram. Bottom: panels c1–c3 and d1–d3 represent the space–time diagrams of magnetic cancellation. Where oblique straight lines are used to calculate the positive and negative flux speeds and these panels are used to verify the magnetic cancellation time up to 50 hours (similar to Figure 3). The map has the field of view of 500 arcsec × 400 arcsec.

cancel or annihilate each other. This may be the reason why the sunspot S2 decayed more rapidly than the sunspot S1 and the sunspot S2 disappeared completely at the end. Meanwhile, the leading sunspots, S3 and S4, seemed to decay the least, and there were many tiny pores which emerged on their north-west at 17:06 UT on 2023 February 24 (panel d1). To verify the magnetic cancellation, we take three slits marked on the HMI longitudinal magnetogram and continuum as shown in Figure 2 (a and b). They mainly pass through the magnetic cancellation regions (encountering area of positive and negative flux) of the sunspots S1 and S2. In the sunspot S1, the magnetic cancellation (i.e., when opposite magnetic polarity fluxes approach each other, merge, and, as the result of the encounter, magnetic flux decreases due to cancellation of a positive flux concentration by a negative one) occurred in the north-west corner as shown in Figure 1. In sunspot S2, the positive flux surrounded and cancelled the negative flux from its north-western areas. The bottom panels, c1–c3 and d1–d3, show space–time diagrams of magnetic cancellation, which are stacked with slits 1 and 2 taken at different times. The oblique straight line is used to calculate the velocities of positive and negative fluxes as shown in Figure 2 (panels c1, c2, c3, d1, d2, d3).

We analyzed the time evolution of positive and negative magnetic flux, total energy transport from the photosphere to the corona, and energy accumulation in the corona as shown in Figure 3, and found proof for the existence of submergence. The top panel of the figure shows a decrease of positive and negative fluxes which verify the process of cancellation of magnetic fluxes. The middle panel of the figure shows the negative transport rate of magnetic energy from the photosphere to the corona, i.e., there is a net loss of magnetic energy

Table 1 Information related to diffusion and decay rate of the sunspot S1.

No.	Time (UT)	Diffusion features	Negative flux decay rate (Mx s^{-1})	Area decay rate or magnetic diffusivity ($m^2 s^{-1}$)
1	2023-02-22(00:10) to 02-23(23:06:11)	Fast decay without magnetic flux change	-2.0×10^{14}	5.6×10^8
2	2023-02-23(06:11) to 02-23(16:16)	No decay and no magnetic flux change	-2.0×10^{14}	0
3	2023-02-23(16:16) to 02-24(22:29)	Fast decay with flux decrease	-2.3×10^{16}	1.8×10^9
4	2023-02-24(22:29) to 02-25(23:50)	Slow decay with flux decrease	-2.3×10^{16}	2.8×10^6

Table 2 Information related to diffusion and decay rate of the sunspot S2.

No.	Time (UT)	Diffusion features	Positive flux decay rate (Mx s^{-1})	Negative flux decay rate (Mx s^{-1})	Area decay rate or magnetic diffusivity ($m^2 s^{-1}$)
1	2023-02-22(05:07) to 02-23(23:16:07)	Fast decay with decreasing flux	-5.0×10^{15}	-8.5×10^{15}	1.9×10^{15}
2	2023-02-23(16:07) to 02-24(02:07)	Flux decrease	-5.0×10^{15}	-8.5×10^{15}	–
3	2023-02-24(02:07) to 02-24(22:07)	Negative flux decrease	$+5.4 \times 10^{14}$	-2.6×10^{15}	–
4	2023-02-24(22:07) to 02-25(13:07)	Slow Change in flux	-9.6×10^{14}	$+2.4 \times 10^{14}$	–

Table 3 Variation of average horizontal velocities in different stages and in different regions R1, R2, R3, R4.

Region	Stage-I $\mathbf{v}_x$ ($m s^{-1}$)	Stage-II $\mathbf{v}_x$ ($m s^{-1}$)	Stage-III $\mathbf{v}_x$ ($m s^{-1}$)	Stage-IV $\mathbf{v}_x$ ($m s^{-1}$)
R1	10 ± 9	10 ± 14	7 ± 12	0 ± 18
R2	-39 ± 27	-24 ± 20	-15 ± 11	-8 ± 11
R3	11 ± 27	23 ± 27	24 ± 22	13 ± 16
R4	-34 ± 29	-16 ± 21	-12 ± 32	-9 ± 28

magnetogram, which was taken at 00:10 UT on 2023 February 22. The horizontal velocity averaged over 00:10 UT on 2023 February 22 and 05:58 UT on 2023 February 23 is superimposed on them, which is obtained by the LCT method with respect to HMI longitudinal magnetograms. Usually, the characteristics of flows in and around the sunspot were revealed by Denker and Verma (2011). We focused on the four regions marked with white rectangles in Figure 5. The averaged velocity in the x-direction in four stages is listed in Table 3.

We have seen in stage-I that the horizontal flows are more divergent and less turbulent (i.e., less spread or irregular due to stronger flow initially) in region R2 as shown in Figure 5.

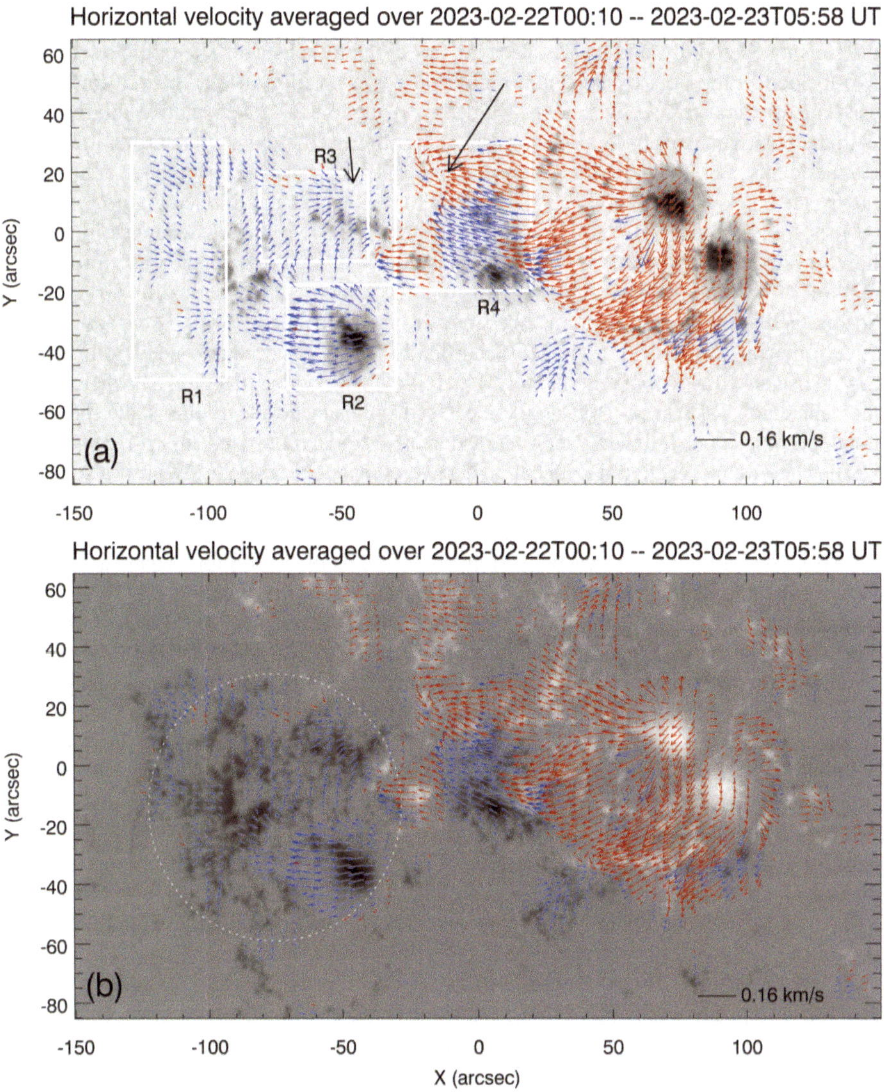

Figure 5 Evolution of horizontal flows over time in stage I. Panel (a) shows the HMI continuum on which horizontal velocity flow patterns (averaged over 00:10 UT on 2023 February 22 and 05:58 UT on 2023 February 23) are superimposed. Panel (b) represents the HMI radial magnetogram on which horizontal velocity flow patterns (averaged over 00:10 UT on 2023 February 22 and 05:58 UT on 2023 February 23) are superimposed. The map has the field of view of 300 arcsec × 150 arcsec.

The bulk flow in region R2 was relatively stronger towards the east, but it met a reverse weak flow from that in region R1. This flow also moved up towards region R3 in the y-direction, but the flow of region R3 met a strong flow from R4. It seems that the flow in region R2 was tightly constricted by the flows from regions R1, R3, and R4. The sunspot S1 in region R2 shows a fast decay in the area at first due to more divergent flows, but the dominant negative flux decreased very slowly due to less turbulence in horizontal flows, likely as the diffused flux cannot escape freely in this region and only the magnetic cancellation in this region annihilated the negative flux. Hence, initially, the flow in region R2 is fast and erodes the sunspot S1 rapidly, and its area decay is fast at the initial stage. The bulk flow in region R4, located in the polarity inversion region (also the position of S2), moved towards the east and was constricted by the westward flow of region R3. The negative flux cancelled with the positive fluxes from its east, west, and northward directions. Therefore, its area and magnetic fluxes of both polarity decreased rapidly. There is also a convergence of horizontal flows (see black arrows) within and very close to regions R3 and R4 (see Figure 5).

It is found in stage II (see Figure 6) that there were two obvious changes in the bulk flow in regions R2–R4. One is that the flow velocity in region R2 decreased from -39 to $-24\ \mathrm{m\,s^{-1}}$ shown in Table 3. This suggests that the velocity of magnetic diffusion declined. Second is that the flow in region R3 became strong (from 11 to 23 $\mathrm{m\,s^{-1}}$) because it faced fewer opposing flows from region R4, and that in region R4 became weak (from 34 to 16 $\mathrm{m\,s^{-1}}$) because flows from regions (R1, R2, and R4) were added in region R3 whereas some flow in region R4 was left out and some faced opposite flows. At this stage, horizontal flows became more turbulent (more spreader or irregular flow) and less divergent. So, it verified the more negative flux decay and fewer sunspot area decay. We can also see the convergence of horizontal flows (see black arrows) within and very near to regions R3 and R4 (see Figure 6).

We plotted a circle that nearly encircles the regions of R1–R3 and superimposed it in Figure 6(b). We plotted the same circle and superimposed it in Figure 5(b). Obviously, by comparison, we noticed that the whole flow field shape of regions R1–R3 is converted into a nearly round shape in this stage, but that stage shows an oval shape and the long axis was roughly in the 135° direction (0° in the x-direction). It may be caused by a large compression from the flows of regions R2 and R4 (i.e., flows became less divergent and more turbulent). But now, their flow velocity has decreased. If there was a columnar magnetic flux located in the region of R1–R3, it was in a stable or balanced state. The magnetic diffusion temporally stopped and there was weak magnetic cancellation. In other words, magnetic pressure was balanced from all sides due to a lack of magnetic diffusion. There was no mass interchange between regions R3 and R4 due to the compressed balance. There was nearly no magnetic diffusion in this stage. This may reflect the fact that the magnetic diffusion of the sunspot S1 became saturated temporally. The magnetic cancellation continued and both fluxes in the sunspot S2 continued to decrease.

It is seen in stage III (see Figure 7) that the flow velocities decrease in regions R1, R2, and R4, except for region R3. Now, horizontal flows can be classified as more turbulent and chaotic (i.e., irregular or more spread flow because of decreased magnetic flux). The average velocity in region R3 was 24 $\mathrm{m\,s^{-1}}$, which is larger than that in region R4. There was a flow in region R3 moving to region R4. We can say that the mass flow of continuity is followed in region R3 in this stage, i.e., the outflow is nearly equal to the inflow (23 to 24 $\mathrm{m\,s^{-1}}$) that is why its velocity remains almost same. There was also the field diffused out of the circle mainly on its east side (see Figure 7(b)). As the field moved outside of region R1, the field compression may have decreased due to the least divergence of flows (i.e., the magnetic pressure exerted by flows from region R1 on region R2 is decreased because some flows leaked from the south and east side. So, magnetic pressure is not balanced from all sides due to the lack of magnetic pressure exerted by region R1). The sunspot S2 decayed with a larger amplitude ($1.8 \times 10^{9}\ \mathrm{m^2\,s^{-1}}$) than that ($5.6 \times 10^{8}\ \mathrm{m^2\,s^{-1}}$) in the first stage (see Table 1). The magnetic flux began to decrease rapidly due to both the magnetic cancellation and the field escaping from the region R1 (i.e., less addition of flows in region R3 and lack of convergence of flows seen in Figure 7). As a result, flows were escaping (from the west side) rapidly from region R3 because it did not face any opposition and convergence. This can be a reason for the faster decay rate of sunspot S2 in this stage compared with stage I.

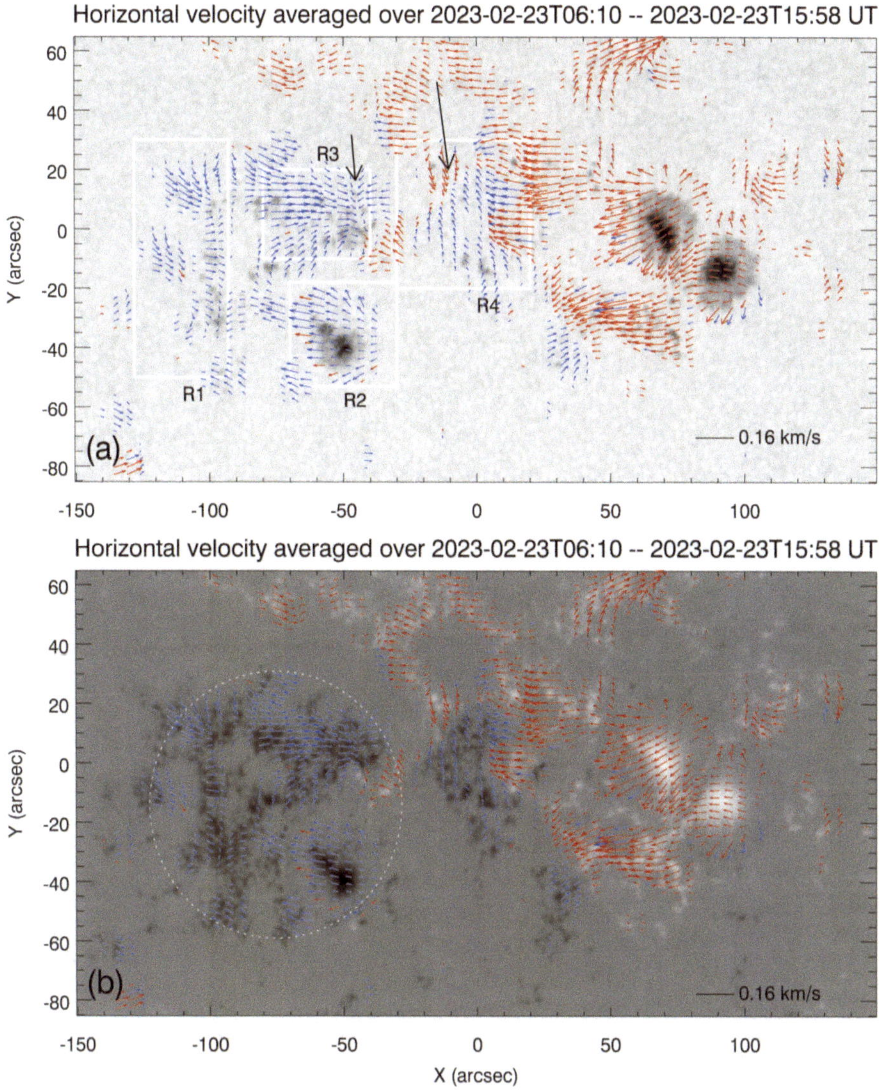

Figure 6 Evolution of horizontal flows over time in stage II. Panel (a) shows the HMI continuum on which horizontal velocity flow patterns (averaged over 06:10 UT on 2023 February 23 and 15:58 UT on 2023 February 23) are superimposed. Panel (b) represents the HMI radial magnetogram on which horizontal velocity flow patterns (averaged over 06:10 UT of 2023 February 23 and 15:58 UT on 2023 February 23) are superimposed. The map has the field of view of 300 arcsec × 150 arcsec.

It is noticed in stage IV (see Figure 8) that the flow fields in regions R1 to R4 all continued to decrease with large amplitude and now flows were very weak and disorganized. The average field velocity in region R1 was zero. The flow field divided into two parts (in region R1), one moved towards the east and the other towards the west. There were more fluxes moved out of the circle from its east side (see Figure 8(b)), and a continuous flux flow was formed located on its south side. The positive flux completely disappeared in this stage. The decay of the sunspot S1 decreased to 2.8×10^6 m^2 s^{-1}, but the rate of flux decrease

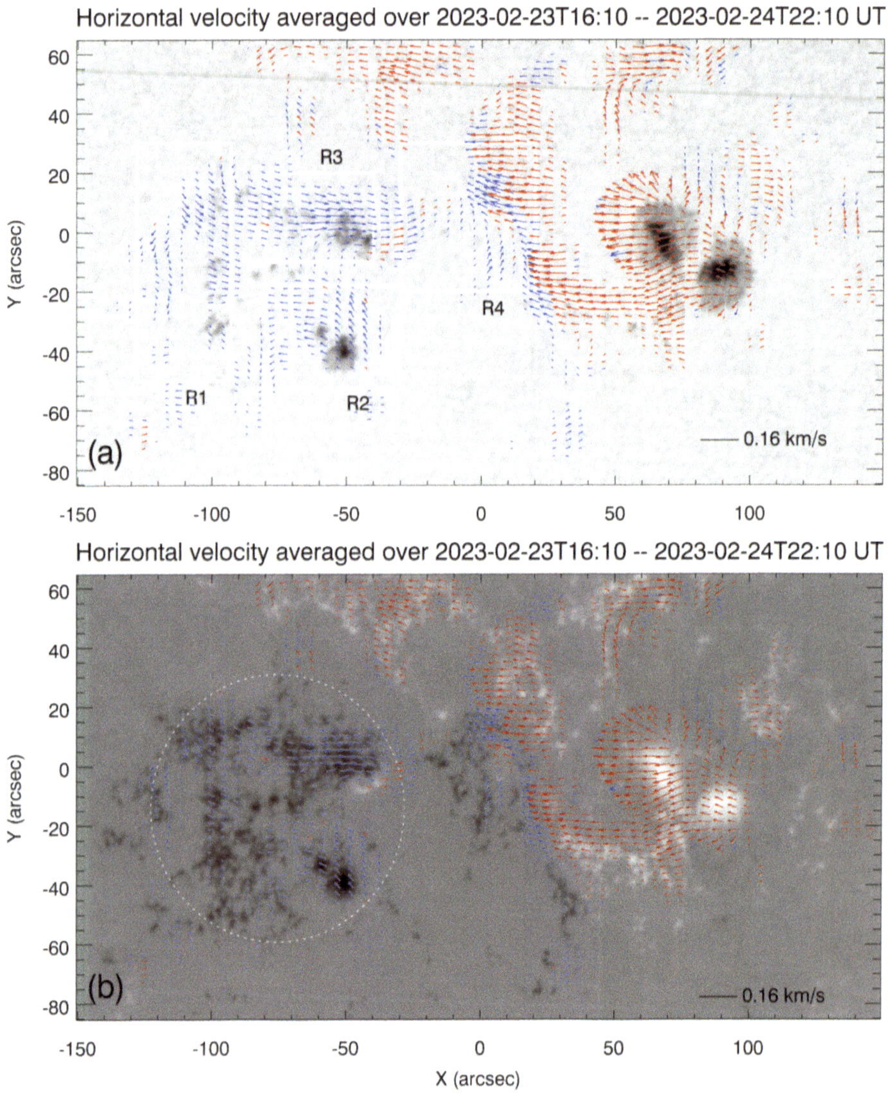

Figure 7 Evolution of horizontal flows over time in stage III. Panel (a) shows the HMI continuum on which horizontal velocity flow patterns (averaged over 16:10 UT of 2023 February 23 and 22:10 UT of 2023 February 24) are superimposed. Panel (b) represents the HMI radial magnetogram on which horizontal velocity flow patterns (averaged over 16:10 UT of 2023 February 23 and 22:10 UT of 2023 February 24) are superimposed. The map has the field of view of 300 arcsec × 150 arcsec.

was similar to the last stage. At the last stage, the flow speeds became weaker so the sunspot decay rates decreased because the erosion of sunspots became smaller. In addition, the net horizontal flows in region R4 (sunspot S2 location) was zero because sunspot S2 had completely disappeared. The sunspot decay can also be verified by horizontal flows. It has been found that horizontal flows play a key role in the decay of sunspots via erosion, magnetic diffusion, and fragmentation.

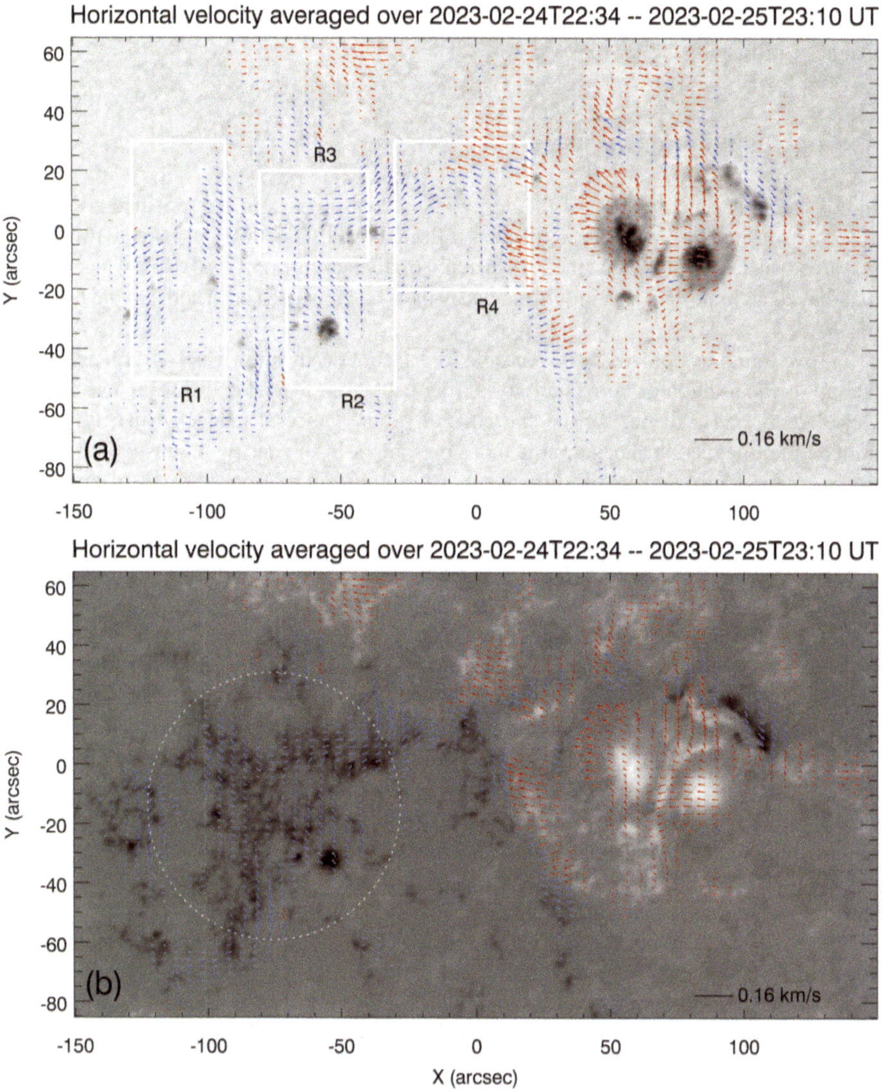

Figure 8 Evolution of horizontal flows over time in stage IV. Panel (a) shows the HMI continuum on which horizontal velocity flow patterns (averaged over 22:34 UT of 2023 February 24 and 23:10 UT of 2023 February 25) are superimposed. Panel (b) represents the HMI radial magnetogram on which horizontal velocity flow patterns (averaged over 22:34 UT of 2023 February 24 and 23:10 UT of 2023 February 25) are superimposed. The map has the field of view of 300 arcsec × 150 arcsec.

As a result of these flows, the reconnection of the magnetic fields near the magnetic polarity inversion line may lead to the submergence of the surrounding magnetic field to the magnetic polarity inversion line which is the direct cause of the rapid attenuation of sunspot S1. Moreover, the convergence of horizontal flows may also have a relation to the sunspot decay because the convergence phenomenon is dominant initially and sunspot decay (area and especially magnetic flux) is also fast at the start. The maximum magnetic diffusivity

(η_m) of sunspots was calculated as $\eta_m = 1.8 \times 10^9$ m^2 s^{-1}. It was higher than the normal magnetic diffusivity ($\eta_m = 10^4$ m^2 s^{-1}).

4. Conclusion and Discussion

In this study, an investigation of the decay of β-type sunspots in the specified AR NOAA 13229 has been done. We utilized the data collected by ASO-S/FMG and SDO/HMI. In this research paper, area decay, magnetic dispersion of distinct sunspots, and the key mechanisms that played an important role in sunspot decay have been focused on. There are the following results:

(1) We found five processes responsible for sunspot decay known as convergence of horizontal flows, rotation of horizontal flows, fragmentation, magnetic diffusion and magnetic cancellation, and evidence for the existence of submergence of magnetic flux. It has been noticed that magnetic cancellation is more prominent in the decay of sunspot S2 whereas the fragmentation and magnetic diffusion played significant roles in the decay of sunspot S1. The area decay rate of the sunspot S1 varied from 5.6×10^8 to 2.8×10^6 m^2 s^{-1}. Meanwhile, the decay rate of sunspot S2 remained 1.9×10^{15} m^2 s^{-1}. The calculated value of magnetic diffusivity was 1.8×10^9 m^2 s^{-1} which is an anomalous magnetic diffusion (normal $\sim 10^4$ m^2 s^{-1}).

(2) The sunspots S1 and S2 had different decay rates and magnetic features. The decay rate of sunspot S2 was faster than that of sunspot S1 and they followed the exponential and parabolic decay laws, respectively. In parabolic decay, the initial decay rate is faster than in later stages and it obeys descending order. The sunspot S2 disappeared completely in 35 hours from 05:07 UT on 2023 February 22 to 11:07 UT on 2023 February 23.

(3) The positive flux of sunspot S1 was negligible and its negative flux decreased very slowly (slightly change) at the start for up to 50 hours and then declined linearly. On the other hand, the negative magnetic flux (2.4×10^{21} Mx) of sunspot S2 was greater than its positive magnetic flux (1.3×10^{21} Mx). Both declined with the same pattern (same rate), showing the magnetic cancellation process.

(4) The flow maps (horizontal velocities flows) obtained using the LCT technique revealed the key role of the horizontal flows in the area decay and magnetic flux diminished characteristics of sunspots. It has been found that the horizontal velocities are decreased with the decay of sunspots. The decreased trend of horizontal velocities was observed from -39 to -8 m s^{-1} and from -34 to -9 m s^{-1} in regions R2 (sunspot S1) and R4 (sunspot S2), respectively. We also found the convergence of horizontal flows initially in the first two stages where the decay process and horizontal velocities flows were fast.

(5) At the start, more divergent (i.e., carrying material away from the center of the sunspot) and less turbulent flows (i.e., breaking up the sunspots) have been observed. In the middle stage, flows became more turbulent and less divergent. In the last stage, flows converted into being more turbulent and chaotic. It revealed the fact that divergence of flows plays a significant role in the area decay, and the turbulence of flows plays a prominent role in magnetic diffusion of sunspots.

(6) The evidence for the existence of an important mechanism has been found which is known as the submergence of magnetic fluxes, which is also a key reason for the sunspot decay.

Previously, a simulation-based study has revealed that the process of submergence of magnetic fluxes plays an important role in the decay of sunspots (Rempel, 2015) but it has not been verified by observational studies. In this study, we observed the submergence of magnetic flux in the middle of regions R3 and R4 which played a significant role in the sunspot decay. Our detailed case study of a specific active region contributes to the overall understanding of sunspot dynamics. We observed processes like magnetic cancellation, fragmentation, and, crucially, submergence, which support findings by Howard (1992), Norton et al. (2017), Muraközy (2021, 2022). By integrating these mechanisms with established models, we highlight the complexity of sunspot evolution, where spatial positioning and magnetic field dynamics within active regions play a critical role.

Previous studies suggested that the area decay of sunspots is either a linear (Solanki, 2003; Li et al., 2021) or quadratic (Litvinenko and Wheatland, 2015) function of time. Most of the research showed the linear area decay of different sunspots. The area growth rates of 399 sunspot groups have been investigated by using the solar and Heliospheric Observatory/Michelson Doppler Imager–Derecen Data (SDD) sunspot catalogue and discovered that an asymmetric Gaussian function well suited the sunspot groups' growth and decay processes (Muraközy, Baranyi, and Ludmány, 2014). Our results expressed different decay rates and magnetic features of two sunspots S1 and S2. One of our interesting findings is the exponential decay of sunspot S2 and its complete disappearance within a short time. On the other hand, the sunspot S1 decay pattern showed different rates during the four stages of its decay, and it remained as a pore at the end.

According to some study results, the magnetic flux diminishes linearly as the sunspot decays (Verma et al., 2016; Deng et al., 2007; Li et al., 2021). This is not justified by our results. In our case, magnetic flux decreased in different stages. The positive flux of sunspot S1 remained almost constant but its negative flux varied in two steps. Initially, it decreased slightly and, in the second stage, it declined linearly. On the other hand, the diminishing trend of the negative and positive fluxes of sunspot S2 is nearly the same, but their declining patterns were also not purely linear. They decreased in descending rate steps. Moreover, the flux decay of the sunspot not only happened on its edges but throughout the sunspot. The turbulent diffusion model follows these findings (Solanki, 2003; Meyer et al., 1974; Martínez Pillet, 2002). The negative magnetic flux of sunspot S1 decline rate varied from -2.0×10^{14} to -2.3×10^{16} Mx. On the other hand, positive and negative magnetic fluxes of sunspot S2 changed from -5.0×10^{15} to -9.6×10^{14} Mx and from -8.5×10^{15} to 2.4×10^{14} Mx, respectively. There will be a lot of MMFs pouring out of the sunspot throughout the destruction phase. MMFs may be responsible for the sunspot's magnetic flux decline. A previous study suggests that during sunspot decline, the magnetic flux in the sunspot is transmitted via MMFs to the surrounding network (Verma and Denker, 2012; Deng et al., 2007). It is also suggested by some researchers that an extension of Evershed flows is considered MMFs (Cabrera Solana et al., 2006).

Acknowledgments Thanks to the referee's constructive comments and suggestions, the manuscript has undergone significant improvements. This research was supported by the National Key R & D Program of China (Nos. 2022YFF0503001, 2021YFA1600500, and 2022YFF0503800), the Strategic Priority Research Program of the Chinese Academy of Sciences (Grant NO. XDB0560301), the National Natural Science Foundation of China (grant Nos. 12273059 and 12373057) and the Beijing Natural Science Foundation (Grant No. 1222029). We would like to thank the ASO-S and SDO/HMI teams for providing the data on the NOAA AR 13229, and the ANSO funding organization for supporting author's doctoral study.

Author contributions All authors contributed and reviewed the manuscript.

Data Availability The datasets used in this study were obtained from publicly available sources: SDO/HMI data from JSOC Lookdata (JSOC Lookdata, Stanford University) and ASOS/FMG data from the ASOS Data Center (ASOS Data Archive).

Declarations

Competing interests The authors declare no competing interests.

References

Bellot Rubio, L.R., Tritschler, A., Martínez Pillet, V.: 2008, Spectropolarimetry of a decaying sunspot penumbra. *Astrophys. J.* **676**, 698. DOI. ADS.
Bumba, V.: 1963, Development of SPOT group areas in dependence on the local magnetic field. *Bull. Astron. Inst. Czechoslov.* **14**, 91. ADS.
Cabrera Solana, D., Bellot Rubio, L.R., Beck, C., del Toro Iniesta, J.C.: 2006, Evershed clouds as precursors of moving magnetic features around sunspots. *Astrophys. J. Lett.* **649**, L41. DOI. ADS.
Chae, J.: 2001, Observational determination of the rate of magnetic helicity transport through the solar surface via the horizontal motion of field line footpoints. *Astrophys. J. Lett.* **560**, L95. DOI. ADS.
Chapman, G.A., Dobias, J.J., Preminger, D.G., Walton, S.R.: 2003, On the decay rate of sunspots. *Geophys. Res. Lett.* **30**, 1178. DOI. ADS.
Chen, J., Su, J., Yin, Z., Priya, T.G., Zhang, H., Liu, J., Xu, H., Yu, S.: 2015, Recurrent solar jets induced by a satellite spot and moving magnetic features. *Astrophys. J.* **815**, 71. DOI. ADS.
Démoulin, P., Berger, M.A.: 2003, Magnetic energy and helicity fluxes at the photospheric level. *Solar Phys.* **215**, 203. DOI. ADS.
Deng, N., Liu, C., Yang, G., Wang, H., Denker, C.: 2005, Rapid penumbral decay associated with an X2.3 flare in NOAA active region 9026. *Astrophys. J.* **623**, 1195. DOI. ADS.
Deng, N., Choudhary, D.P., Tritschler, A., Denker, C., Liu, C., Wang, H.: 2007, Flow field evolution of a decaying sunspot. *Astrophys. J.* **671**, 1013. DOI. ADS.
Deng, Y.-Y., Zhang, H.-Y., Yang, J.-F., Li, F., Lin, J.-B., Hou, J.-F., Wu, Z., Song, Q., Duan, W., Bai, X.-Y., Wang, D.-G., Lv, J., Ge, W., Wang, J.-N., Zheng, Z.-Y., Wang, C.-J., Wang, N.-G., Ni, H.-K., Zeng, Y.-Z., Zhang, Y., Yang, X., Sun, Y.-Z., Zhang, Z.-Y., Wang, X.-F.: 2019, Design of the Full-disk MagnetoGraph (FMG) onboard the ASO-S. *Res. Astron. Astrophys.* **19**, 157. DOI. ADS.
Denker, C., Verma, M.: 2011, Velocity fields in and around sunspots at the highest resolution. In: Prasad Choudhary, D., Strassmeier, K.G. (eds.) *Physics of Sun and Star Spots* **273**, 204. DOI. ADS.
Gafeira, R., Fonte, C.C., Pais, M.A., Fernandes, J.: 2014, Temporal evolution of sunspot areas and estimation of related plasma flows. *Solar Phys.* **289**, 1531. DOI. ADS.
Gan, W.-Q., Zhu, C., Deng, Y.-Y., Li, H., Su, Y., Zhang, H.-Y., Chen, B., Zhang, Z., Wu, J., Deng, L., Huang, Y., Yang, J.-F., Cui, J.-J., Chang, J., Wang, C., Wu, J., Yin, Z.-S., Chen, W., Fang, C., Yan, Y.-H., Lin, J., Xiong, W.-M., Chen, B., Bao, H.-C., Cao, C.-X., Bai, Y.-P., Wang, T., Chen, B.-L., Li, X.-Y., Zhang, Y., Feng, L., Su, J.-T., Li, Y., Chen, W., Li, Y.-P., Su, Y.-N., Wu, H.-Y., Gu, M., Huang, L., Tang, X.-J.: 2019, Advanced Space-based Solar Observatory (ASO-S): An overview. *Res. Astron. Astrophys.* **19**, 156. DOI. ADS.
Gan, W., Zhu, C., Deng, Y., Zhang, Z., Chen, B., Huang, Y., Deng, L., Wu, H., Zhang, H., Li, H., Su, Y., Su, J., Feng, L., Wu, J., Cui, J., Wang, C., Chang, J., Yin, Z., Xiong, W., Chen, B., Yang, J., Li, F., Lin, J., Hou, J., Bai, X., Chen, D., Zhang, Y., Hu, Y., Liang, Y., Wang, J., Song, K., Guo, Q., He, L., Zhang, G., Wang, P., Bao, H., Cao, C., Bai, Y., Chen, B., He, T., Li, X., Zhang, Y., Liao, X., Jiang, H., Li, Y., Su, Y., Lei, S., Chen, W., Li, Y., Zhao, J., Li, J., Ge, Y., Zou, Z., Hu, T., Su, M., Ji, H., Gu, M., Zheng, Y., Xu, D., Wang, X.: 2023, The Advanced Space-based Solar Observatory (ASO-S). *Solar Phys.* **298**, 68. DOI. ADS.
Hathaway, D.H., Choudhary, D.P.: 2008, Sunspot group decay. *Solar Phys.* **250**, 269. DOI. ADS.
Hoeksema, J.T., HMI Project Team: 2007, *The Helioseismic & Magnetic Imager on the Solar Dynamics Observatory*. In: American Astronomical Society Meeting Abstracts 210, 24.15. ADS.
Hou, J.-F., Wang, D.-G., Deng, Y.-Y., Zhang, Z.-Y., Sun, Y.-Z.: 2020, Interference effect on the liquid-crystal-based Stokes polarimeter. *Chin. Phys. B* **29**, 124211.
Howard, R.F.: 1992, Some characteristics of the development and decay of active region magnetic flux. *Solar Phys.* **142**, 47. DOI. ADS.
Jiang, C., Feng, X.: 2013, Extrapolation of the solar coronal magnetic field from SDO/HMI magnetogram by a CESE-MHD-NLFFF code. *Astrophys. J.* **769**, 144. DOI. ADS.
Krause, F., Ruediger, G.: 1975, On the turbulent decay of strong magnetic fields and the development of sunspot areas. *Solar Phys.* **42**, 107. DOI. ADS.
Leka, K.D., Skumanich, A.: 1998, The evolution of pores and the development of penumbrae. *Astrophys. J.* **507**, 454. DOI. ADS.

Li, Q., Yan, X., Wang, J., Kong, D., Xue, Z., Yang, L., Cao, W.: 2018, The formation of a sunspot penumbra sector in active region NOAA 12574. *Astrophys. J.* **857**, 21. DOI. ADS.

Li, Q., Zhang, L., Yan, X., Wang, J., Kong, D., Yang, L., Xue, Z.: 2021, The decay of α-configuration sunspots. *Astrophys. J.* **913**, 147. DOI. ADS.

Lim, E.-K., Yurchyshyn, V., Goode, P., Cho, K.-S.: 2013, Observation of a non-radial penumbra in a flux emerging region under chromospheric canopy fields. *Astrophys. J. Lett.* **769**, L18. DOI. ADS.

Litvinenko, Y.E., Wheatland, M.S.: 2015, Modeling sunspot and starspot decay by turbulent erosion. *Astrophys. J.* **800**, 130. DOI. ADS.

Liu, Y., Scherrer, P.H., Hoeksema, J.T., Schou, J., Bai, T., Beck, J.G., Bobra, M., Bogart, R.S., Bush, R.I., Couvidat, S., Hayashi, K., Kosovichev, A.G., Larson, T.P., Rabello-Soares, C., Sun, X., Wachter, R., Zhao, J., Zhao, X.P., Duvall, J.T.L., DeRosa, M.L., Schrijver, C.J., Title, A.M., Centeno, R., Tomczyk, S., Borrero, J.M., Norton, A.A., Barnes, G., Crouch, A.D., Leka, K.D., Abbett, W.P., Fisher, G.H., Welsch, B.T., Muglach, K., Schuck, P.W., Wiegelmann, T., Turmon, M., Linker, J.A., Mikić, Z., Riley, P., Wu, S.T.: 2012, A first look at magnetic field data products from SDO/HMI. In: Bellot Rubio, L., Reale, F., Carlsson, M. (eds.) *4th Hinode Science Meeting: Unsolved Problems and Recent Insights, Astronomical Society of the Pacific Conference Series* **455**, 337. ADS.

MacTaggart, D., Guglielmino, S.L., Zuccarello, F.: 2016, The pre-penumbral magnetic canopy in the solar atmosphere. *Astrophys. J. Lett.* **831**, L4. DOI. ADS.

Martínez Pillet, V.: 2002, Decay of sunspots. *Astron. Nachr.* **323**, 342. https://doi.org/10.1002/1521-3994(200208)323:3/4<342::AID-ASNA342>3.0.CO;2-5. ADS.

Martinez Pillet, V., Moreno-Insertis, F., Vazquez, M.: 1993, The distribution of sunspot decay rates. *Astron. Astrophys.* **274**, 521. ADS.

Meyer, F., Schmidt, H.U., Weiss, N.O., Wilson, P.R.: 1974, The growth and decay of sunspots. *Mon. Not. Roy. Astron. Soc.* **169**, 35. DOI. ADS.

Murabito, M., Romano, P., Guglielmino, S.L., Zuccarello, F., Solanki, S.K.: 2016, Formation of the penumbra and start of the Evershed flow. *Astrophys. J.* **825**, 75. DOI. ADS.

Muraközy, J.: 2020, Study of the decay rates of the umbral area of sunspot groups using a high-resolution database. *Astrophys. J.* **892**, 107. DOI. ADS.

Muraközy, J.: 2021, On the decay of sunspot groups and their internal parts in detail. *Astrophys. J.* **908**, 133. DOI. ADS.

Muraközy, J.: 2022, Variations of the internal asymmetries of sunspot groups during their decay. *Astrophys. J.* **925**, 87. DOI. ADS.

Muraközy, J., Baranyi, T., Ludmány, A.: 2014, Sunspot group development in high temporal resolution. *Solar Phys.* **289**, 563. DOI. ADS.

Norton, A.A., Jones, E.H., Linton, M.G., Leake, J.E.: 2017, Magnetic flux emergence and decay rates for preceder and follower sunspots observed with HMI. *Astrophys. J.* **842**, 3. DOI. ADS.

November, L.J., Simon, G.W.: 1988, Precise proper-motion measurement of solar granulation. *Astrophys. J.* **333**, 427. DOI. ADS.

Petrovay, K., Martínez Pillet, V., van Driel-Gesztelyi, L.: 1999, Making sense of sunspot decay – II. Deviations from the mean law and plage effects. *Solar Phys.* **188**, 315. DOI. ADS.

Petrovay, K., Moreno-Insertis, F.: 1997, Turbulent erosion of magnetic flux tubes. *Astrophys. J.* **485**, 398. DOI. ADS.

Petrovay, K., van Driel-Gesztelyi, L.: 1997, Making sense of sunspot decay. I. Parabolic decay law and Gnevyshev-Waldmeier relation. *Solar Phys.* **176**, 249. DOI. ADS.

Rempel, M.: 2011, Penumbral fine structure and driving mechanisms of large-scale flows in simulated sunspots. *Astrophys. J.* **729**, 5. DOI. ADS.

Rempel, M.: 2012, Numerical sunspot models: robustness of photospheric velocity and magnetic field structure. *Astrophys. J.* **750**, 62. DOI. ADS.

Rempel, M.: 2015, Numerical simulations of sunspot decay: on the penumbra-Evershed flow-moat flow connection. *Astrophys. J.* **814**, 125. DOI. ADS.

Romano, P., Guglielmino, S.L., Cristaldi, A., Ermolli, I., Falco, M., Zuccarello, F.: 2014, Evolution of the magnetic field inclination in a forming penumbra. *Astrophys. J.* **784**, 10. DOI. ADS.

Romano, P., Murabito, M., Guglielmino, S.L., Zuccarello, F., Falco, M.: 2020, Restoring process of sunspot penumbra. *Astrophys. J.* **899**, 129. DOI. ADS.

Rüdiger, G., Kitchatinov, L.L.: 2000, Sunspot decay as a test of the eta-quenching concept. *Astron. Nachr.* **321**, 75. https://doi.org/10.1002/(SICI)1521-3994(200003)321:1<75::AID-ASNA75>3.0.CO;2-B. ADS.

Sainz Dalda, A., Martínez Pillet, V.: 2005, Moving magnetic features as prolongation of penumbral filaments. *Astrophys. J.* **632**, 1176. DOI. ADS.

Scherrer, P.H., Schou, J., Bush, R.I., Kosovichev, A.G., Bogart, R.S., Hoeksema, J.T., Liu, Y., Duvall, T.L., Zhao, J., Title, A.M., Schrijver, C.J., Tarbell, T.D., Tomczyk, S.: 2012, The Helioseismic and Magnetic

Imager (HMI) investigation for the Solar Dynamics Observatory (SDO). *Solar Phys.* **275**, 207. DOI. ADS.

Schlichenmaier, R., Rezaei, R., Bello González, N., Waldmann, T.A.: 2010, The formation of a sunspot penumbra. *Astron. Astrophys.* **512**, L1. DOI. ADS.

Schou, J., Scherrer, P.H., Bush, R.I., Wachter, R., Couvidat, S., Rabello-Soares, M.C., Bogart, R.S., Hoeksema, J.T., Liu, Y., Duvall, T.L., Akin, D.J., Allard, B.A., Miles, J.W., Rairden, R., Shine, R.A., Tarbell, T.D., Title, A.M., Wolfson, C.J., Elmore, D.F., Norton, A.A., Tomczyk, S.: 2012, Design and ground calibration of the Helioseismic and Magnetic Imager (HMI) instrument on the Solar Dynamics Observatory (SDO). *Solar Phys.* **275**, 229. DOI. ADS.

Sheeley, J.N.R., Stauffer, J.R., Thomassie, J.C., Warren, H.P.: 2017, Tracking the magnetic flux in and around sunspots. *Astrophys. J.* **836**, 144. DOI. ADS.

Shimizu, T., Ichimoto, K., Suematsu, Y.: 2012, Precursor of sunspot penumbral formation discovered with hinode solar optical telescope observations. *Astrophys. J. Lett.* **747**, L18. DOI. ADS.

Skumanich, A., Lites, B.W.: 1994, Vector field structure of a small sunspot. In: Schüssler, M., Schmidt, W. (eds.) *Solar Magnetic Fields*, 200. ADS.

Solanki, S.K.: 2003, Sunspots: An overview. *Astron. Astrophys. Rev.* **11**, 153. DOI. ADS.

Strecker, H., Bello González, N.: 2018, Evolution of the flow field in decaying active regions. Transition from a moat flow to a supergranular flow. *Astron. Astrophys.* **620**, A122. DOI. ADS.

Su, J.-T., Bai, X.-Y., Chen, J., Guo, J.-J., Liu, S., Wang, X.-F., Xu, H.-Q., Yang, X., Song, Y.-L., Deng, Y.-Y., Ji, K.-F., Deng, L., Huang, Y., Li, H., Gan, W.-Q.: 2019, Data reduction and calibration of the FMG onboard ASO-S. *Res. Astron. Astrophys.* **19**, 161. DOI. ADS.

Thomas, J.H., Weiss, N.O., Tobias, S.M., Brummell, N.H.: 2002, Downward pumping of magnetic flux as the cause of filamentary structures in sunspot penumbrae. *Nature* **420**, 390. DOI. ADS.

Verma, M., Denker, C.: 2012, Horizontal flow fields observed in Hinode G-band images. III. The decay of a satellite sunspot and the role of magnetic flux removal in flaring. *Astron. Astrophys.* **545**, A92. DOI. ADS.

Verma, M., Denker, C., Balthasar, H., Kuckein, C., González Manrique, S.J., Sobotka, M., Bello González, N., Hoch, S., Diercke, A., Kummerow, P., Berkefeld, T., Collados, M., Feller, A., Hofmann, A., Kneer, F., Lagg, A., Löhner-Böttcher, J., Nicklas, H., Pastor Yabar, A., Schlichenmaier, R., Schmidt, D., Schmidt, W., Schubert, M., Sigwarth, M., Solanki, S.K., Soltau, D., Staude, J., Strassmeier, K.G., Volkmer, R., von der Lühe, O., Waldmann, T.: 2016, Horizontal flow fields in and around a small active region. The transition period between flux emergence and decay. *Astron. Astrophys.* **596**, A3. DOI. ADS.

Verma, M., Denker, C., Balthasar, H., Kuckein, C., Rezaei, R., Sobotka, M., Deng, N., Wang, H., Tritschler, A., Collados, M., Diercke, A., González Manrique, S.J.: 2018, High-resolution imaging and near-infrared spectroscopy of penumbral decay. *Astron. Astrophys.* **614**, A2. DOI. ADS.

Wang, H., Liu, C., Qiu, J., Deng, N., Goode, P.R., Denker, C.: 2004, Rapid penumbral decay following three X-class solar flares. *Astrophys. J. Lett.* **601**, L195. DOI. ADS.

Watanabe, H., Kitai, R., Otsuji, K.: 2014, Formation and decay of rudimentary penumbra around a pore. *Astrophys. J.* **796**, 77. DOI. ADS.

Publisher's Note Springer Nature remains neutral with regard to jurisdictional claims in published maps and institutional affiliations.

Solar Physics (2024) 299:93
https://doi.org/10.1007/s11207-024-02329-4

RESEARCH

The Solar Origin of an Intense Geomagnetic Storm on 1 December 2023: Successive Slipping and Eruption of Multiple Magnetic Flux Ropes

Zheng Sun[1] · Ting Li[2,3,4] · Yijun Hou[2,3,4] · Hui Tian[1,4] · Ziqi Wu[1] · Ke Li[1] · Yining Zhang[2] · Zhentong Li[5] · Xianyong Bai[2] · Li Feng[5,6] · Chuan Li[7,8] · Zhenyong Hou[1] · Qiao Song[9,10,11] · Jingsong Wang[9,10,11] · Guiping Zhou[2,3,4]

Received: 27 February 2024 / Accepted: 23 May 2024 / Published online: 27 June 2024

Abstract
The solar eruption that occurred on 28 November 2023 (SOL2023-11-28) triggered an intense geomagnetic storm on 1 December 2023. The associated terrestrial auroras manifested at the most southern latitudes in the northern hemisphere observed in the past two decades. In order to explore the profound geoeffectiveness of this event, we conducted a comprehensive analysis of its solar origin to offer potential factors contributing to its impact. Magnetic flux ropes (MFRs) are twisted magnetic structures recognized as significant contributors to coronal mass ejections (CMEs), thereby impacting space weather greatly. In this event, we identified multiple MFRs in the solar active region and observed distinct slipping processes of the three MFRs: MFR1, MFR2, and MFR3. All three MFRs exhibit slipping motions at a speed of 40 – 137 $\mathrm{km\,s^{-1}}$, extending beyond their original locations. Notably, the slipping of MFR2 extends to ~ 30 Mm and initiates the eruption of MFR3. Ultimately, MFR1's eruption results in an M3.4-class flare and a CME, while MFR2 and MFR3 collectively produce an M9.8-class flare and another halo CME. This study shows the slipping process in a multi-MFR system, showing how one MFR's slipping can trigger the eruption of another MFR. We propose that the CME–CME interactions caused by multiple MFR eruptions may contribute to the significant geoeffectiveness.

Keywords Coronal mass ejections · Initiation and propagation · Magnetic reconnection · Observational signatures · Magnetosphere · Geomagnetic disturbances

1. Introduction

Magnetic flux ropes (MFRs) are constituted by a collection of helical magnetic field lines winding around a common axis (Titov and Démoulin, 1999; Krall et al., 2001; Liu, 2020). Extensive observations and simulations highlight the significant role of MFRs in driving solar flares and coronal mass ejections (CMEs), which have important impacts on the Earth and planetary space environments (Inoue and Kusano, 2006; Cheng et al., 2011; Song et al., 2015; Jiang et al., 2019). Statistical analysis reveals that approximately half of the CMEs are accompanied by the presence of MFRs (Nindos et al., 2015). Therefore, investigating the

Extended author information available on the last page of the article

detailed mechanisms of the MFR's formation, initiation, and evolution processes is essential for a better understanding of their effects on space weather.

Two prevailing assumptions regarding the formation of an MFR exist. Firstly, MFRs can be preexisting before an eruption, originating from the convection zone through flux emergence (Leka et al., 1996; Manchester et al., 2004; Fan, 2009; Török et al., 2014) or from magnetic reconnection preceding the eruption (Patsourakos, Vourlidas, and Stenborg, 2013; Chintzoglou, Patsourakos, and Vourlidas, 2015; Cheng et al., 2023). Secondly, MFRs may be formed during the eruption through magnetic reconnection (Cheng et al., 2011; Wang et al., 2017; Liu et al., 2018). An additional scenario involves a preexisting seed MFR, which continues its buildup as the eruption progresses (Cheng, Guo, and Ding, 2017; Veronig et al., 2018; Chen et al., 2019).

Coronal sigmoids, S-shaped magnetic structures above polarity-inversion lines (PILs) observed in soft X-rays and EUV, are often considered MFR progenitors (Chen, 2011; Jiang et al., 2013; Archontis, Hood, and Tsinganos, 2014). They could either be the MFR itself or sheared arcades providing magnetic conditions for MFR buildup (Gibson et al., 2006; Liu et al., 2010; Savcheva et al., 2012). Solar filaments, dense and cold plasma suspended in the tenuous and hot corona, can also be regarded as MFRs or sheared arcades (Canou and Amari, 2010; Li et al., 2018b; Sun et al., 2023). Observations indicate that they can act as seed MFRs and increase poloidal magnetic flux during eruptions (Liu et al., 2019; Xie et al., 2023). There can be simultaneous multiple MFRs in one eruption, which may cause a complicated eruption process (Ding, Hu, and Wang, 2006; Hou et al., 2018, 2023; Awasthi et al., 2018).

Upon formation, the MFR exhibits distinct magnetic connectivities compared to the surrounding field lines due to their twisted nature. Consequently, the twisted MFR is wrapped by quasiseparatrix layers (QSLs), serving as a boundary between the MFR and the surrounding magnetic field (Demoulin et al., 1996, 1997; Titov and Démoulin, 1999; Titov, Hornig, and Démoulin, 2002). The cross-section between the QSLs and the photosphere, known as the QSL footprints, often manifests a double J shape, with the hook part encircling the MFR footpoints. This characteristic has been validated through numerous simulations and observations (Aulanier, Janvier, and Schmieder, 2012; Janvier et al., 2013; Zhao et al., 2016). Notably, several studies indicate a correspondence between the flare ribbons and QSL footprints (Savcheva et al., 2015; Zhao et al., 2016; Jiang et al., 2018).

Three-dimensional (3D) magnetic reconnection occurring in the QSLs is often associated with apparent slipping motion of field lines, known as the "slipping reconnection" (Priest and Forbes, 1992; Priest and Démoulin, 1995; Li, Priest, and Guo, 2021). In the 3D MFR eruption models, overlying sheared arcades can configure into hyperbolic flux tube (HFT) structures beneath the MFR during or before the eruption (Aulanier, Janvier, and Schmieder, 2012). The slipping reconnection within the HFT typically results in continuous brightenings observed at the footprints, often at sub-Alfvénic speeds (Aulanier et al., 2006; Janvier et al., 2013; Dudík et al., 2014). This reconnection process enhances the poloidal flux of the MFR, thereby facilitating its buildup. Previous observations and simulations have shown the slipping reconnection in the process of the buildup of an MFR (Janvier et al., 2013; Dudík et al., 2014, 2016; Li and Zhang, 2014; Li et al., 2016, 2018a).

The eruption of MFRs forms the CMEs, which propagate into interplanetary space and are typically referred to as interplanetary CMEs (ICMEs). Some ICMEs directed towards the Earth can be geoeffective, leading to geomagnetic storms and aurora on the Earth. The disturbance storm time (Dst) and Kp indices are used to quantify the geomagnetic effect. In this study, our focus centers on a solar eruption characterized by two successive solar flares, first an M3.4-class flare followed by an M9.8-class flare, occurring on 28 November

2023. This particular eruption leads to an intense geomagnetic storm on 1 December 2023, with a Dst index of -105 nT. This event stands out as one of the six intense geomagnetic storms ($Dst \leq -100$ nT) recorded since the start of Solar Cycle 25. Particularly notable is that the auroras associated with this event manifest at the most southern latitudes in the northern hemisphere for the past two decades. Given the exceptional geoeffectiveness of this event, it is necessary to carry out a thorough analysis of the solar source of the eruption. Our objective is to unravel the possible reasons contributing to its profound geomagnetic effects from the solar observations. The remainder of this paper is organized as follows. Section 2 describes the observations and data analysis used in our study. In Section 3, we present our observational results in detail. The major findings are thoroughly discussed and summarized in Section 4. Finally, we draw our conclusions in Section 5.

2. Observations and Data Analysis

The Atmospheric Imaging Assembly (AIA) onboard the Solar Dynamics Observatory (SDO) can provide both extreme ultraviolet (EUV) and ultraviolet (UV) images of the Sun since 2010, capturing atmospheric dynamics in a wide range of temperatures from $\log(T) \approx 4.0$ K to $\log(T) \approx 7.0$ K (Lemen et al., 2012). We analyzed 94, 131, 171, 193, and 304 Å data with a spatial resolution of 0.6″/pixel and a temporal resolution of 12 seconds in our study. UV 1600 Å observations with a resolution of 0.6″ pixel^{-1} and 24 seconds were also used. The Advanced Space-based Solar Observatory (ASO-S; Gan et al., 2019) is the first comprehensive Chinese dedicated solar observatory in space, proposed for the 25th solar maximum. Three payloads are deployed: the Hard X-ray Imager (HXI; Zhang et al., 2019), the Full-disk vector MagnetoGraph (FMG; Deng et al., 2019), and the Lyman-α Solar Telescope (LST; Li et al., 2019). The time cadence varies from 4 seconds in regular observation mode to approximately 0.125 seconds in burst mode. Our study utilizes the full-disk light curve from three total flux monitors (D92, D93, D94). The data is subtracted by the summed count rates of the corresponding three background monitors (D95, D96, D99), focusing on the energy range of 10 – 50 keV. The White-light Solar Telescope (WST) onboard LST works in the 360 ± 2 nm waveband and has a field of view of 1.2 solar radii. The images taken by WST have a size of 4608×4608 pixels and the pixel size is $\sim 0.5''$. FMG can measure the solar photospheric magnetic fields through the Fe I 532.42 nm line with high spatial and temporal resolutions, providing line-of-sight (LOS) magnetograms. Solar X-ray Extreme Ultraviolet Imager (X-EUVI; Chen et al., 2022) onboard Fengyun-3E meteorological satellite (FY-3E) can provide EUV images in 195.5 Å, with a cadence of ~ 14 s and a pixel size of 2.5″. The Solar Upper Transition Region Imager (SUTRI; Bai et al., 2023) contributes to our study by providing EUV images at the wavelength of Ne VII 465 Å, corresponding to a temperature of approximately 0.5 MK (Tian, 2017). SUTRI's capabilities include a spatial resolution of approximately 8″ and a time resolution of about 30 seconds. Moreover, Hα images from the New Vacuum Solar Telescope (NVST; Liu et al., 2014b) and the Chinese Hα Solar Explorer (CHASE; Li et al., 2022) are used to show the filaments involved in the event. NVST offers Hα images with a field of view (FOV) of $150'' \times 150''$, a cadence of 45 seconds, and a spatial resolution of 0.165″ per pixel. Calibration of the data follows the methodology outlined by Cai et al. (2022). CHASE's Hα Imaging Spectrograph (HIS) has a pixel spectral resolution of 0.024 Å and a pixel spatial resolution of 0.52″. The Geostationary Operational Environmental Satellite (GOES) can provide soft X-ray (SXR) flux data in 1 – 8 Å and 0.5 – 4 Å ranges, with a cadence of 1 second. The white-light observations of CMEs are acquired from the Large Angle Spectrometric Coronagraph (LASCO;

Brueckner et al., 1995) instrument aboard the Solar and Heliospheric Observatory (SOHO) spacecraft (Brueckner et al., 1995). We utilize LASCO-C3, which offers an FOV spanning from 3 to 32 solar radii, enabling us to observe the CME within the near-Sun corona. In addition, the Wind satellite (Harten and Clark, 1995) provides magnetic and plasma observations from the L1 point, allowing for detailed monitoring of solar wind properties and magnetic field dynamics in the vicinity of the Earth.

To derive the 3D coronal magnetic field, we utilized the photospheric vector magnetic field data provided by SDO/HMI with a cadence of 720 seconds. A nonlinear force-free field (NLFFF) extrapolation was performed using an optimization approach method (Wheatland, Sturrock, and Roumeliotis, 2000; Wiegelmann, 2004; Wiegelmann et al., 2012). To ensure the force-free assumption is satisfied in our input data, we applied the preprocessing method introduced by Wiegelmann, Inhester, and Sakurai (2006). The extrapolation box was defined as $740 \times 480 \times 256$, with a grid resolution of $0.5''$. Utilizing the extrapolated field, we computed the twist number T_w employing the procedure developed by Liu et al. (2016). The twist number T_w serves as an indicator of the degree of twisting in magnetic field lines, which can help us identify the MFRs (Berger and Prior, 2006).

3. Results

On 28 November 2023, NOAA active region (AR) 13500 was positioned at S20°W04°, slightly south of the center of the solar disk (Figure 1(b)). Figure 1 presents the overview of this AR. The AR exhibited a significant solar eruption, first an M3.4-class flare (SOL2023-11-28T19:07) starting at $\sim$ 19:07 UT, reaching its peak at $\sim$ 19:32 UT. Following this event, a subsequent M9.8-class flare (SOL2023-11-28T19:35) occurred, beginning at $\sim$ 19:35 UT and peaking at $\sim$ 19:50 UT. Figure 1(a) illustrates the SXR and HXR light curves corresponding to these associated flares. In panel (c), zoomed-in images of the EUV 94 Å wavelength reveal the presence of a sigmoid located around the PIL, suggesting the existence of an MFR or sheared magnetic arcade.

Figure 2 presents the NLFFF extrapolation results of the AR 13500 at 18:36 UT, which is approximately half an hour before the eruption. This AR exhibits three distinct MFRs: MFR1 and MFR2 are near each other, with their western ends anchored in the northwest negative-polarity sunspot. The magnetic twist of MFR1 is calculated in the range of 1.0 – 2.1, and the twist of MFR2 is smaller than MFR1 with its maximum twist of about 1.25. MFR3 is approximately 30 Mm away from MFR2, anchored in the southern negative-polarity sunspot. The twist of the MFR3 is in the range of 1.0 – 1.63. Preeruptive observations of the AR are shown in the first row of Figure 3. Hα observations in panels (a) and (b) reveal multiple filaments, with particular emphasis on two (outlined by yellow and red dashed lines) located in the northwest. Combined with the magnetogram in panel (c), it is found that the locations of the two filaments are close to those of MFR1 and MFR2 extrapolated in Figure 2. The east end of the extrapolated MFR2 is somewhat north to the observed filament (red dashed lines in Figure 3(a)–(c)), which may imply that the observed filament is composed of both twisted magnetic fields and sheared arcades (Guo et al., 2010). These filaments can be considered as the counterparts of the extrapolated MFRs (Canou and Amari, 2010; Li et al., 2018b). Though there is no associated filament for the extrapolated MFR3, the EUV observations show the presence of high-temperature sheared or twisted loop bundles at the location of MFR3 (Figure 3(d1)–(d3)), which probably correspond to the twisted loop bundles of MFR3 (Cheng et al., 2011; Liu, 2020). Panel (c) reveals that the AR possesses a $\beta\gamma\delta$ magnetic configuration, indicative of a complicated magnetic field configuration that

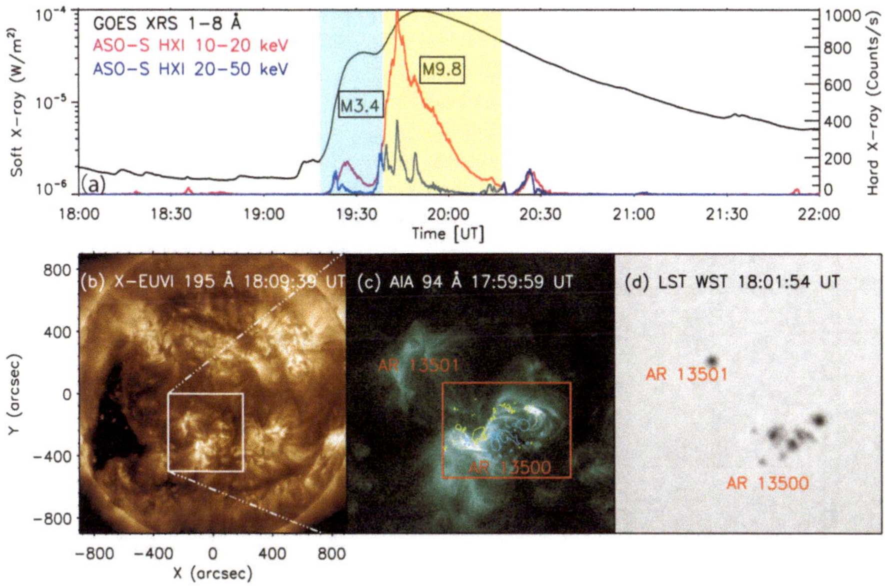

Figure 1 Overview of the eruption. Panel (a): temporal evolution of the GOES XRS 1 – 8 Å (black), ASO-S HXI 10 – 20 keV (red), and 20 – 50 keV (blue). The blue and yellow rectangles indicate the time range of the M3.4-class flare and the M9.8-class flare, respectively. Panel (b): overview of the AR 13501 and AR 13500 in FY-3E/X-EUVI 195 Å. The white rectangle denotes the field of view (FOV) of panels (c)–(d). Panels (c)–(d): zoomed-in area of AR 13500 and AR 13501 in AIA 94 Å (c) and in LST/ WST 3600 Å (d). The orange rectangle shows the FOV of Figure 2. Panel (c) is superimposed by contours of the magnetic field at ± 400 Gauss levels, with yellow indicating positive values and blue representing negative values.

tends to erupt as obtained through statistical works (Hale and Nicholson, 1938; Zhongxian and Jingxiu, 1994; Toriumi and Wang, 2019). Beneath F1 and near the western footpoint of F2, the emergence of numerous discrete magnetic poles of opposite polarity is observed. Motions in the photosphere may lead to magnetic cancellation, making the system unstable and initiating the eruption.

The second to fourth rows of Figure 3 illustrate the initial eruption of MFR1. Brightenings near the location of F1 occur at ∼ 19:13 UT as shown in panels (d1) and (e1). Then the eastern end of F1 exhibited the apparent slipping motion towards the southeast (yellow arrows in panels (d2)–(f2)). Anchored at the newly-drifted eastern footpoint, a loop bundle (LB1) marked as yellow dots can be observed. This observed feature in 131 Å has been investigated by previous studies and often considered as part of the hot channel, which is one of the indicators for MFRs (Cheng et al., 2011; Chintzoglou, Patsourakos, and Vourlidas, 2015; Gou et al., 2019; Zheng et al., 2021). At ∼ 19:19 UT, the slipping motion of the eastern end of LB1 continued with the extension of about 10 Mm (panels (d3)–(f3)). At the same time, another loop bundle LB2 near LB1 occurred, marked with the red dotted line. After that, LB1 and F1 showed an ascending and then erupted (seen in Figure 3(f3) and the online animation), however, LB2 and F2 remained. It can be speculated that the LB1 are associated with F1, which represents an indicator of twisted magnetic field lines of MFR1. On the other hand, LB2 and F2 are probably both associated with MFR2. This slipping motion of MFR during eruptions has been reported by many previous studies (Dudík et al., 2014; Li and Zhang, 2014; Li et al., 2016; Wang et al., 2017), indicating the buildup of MFRs.

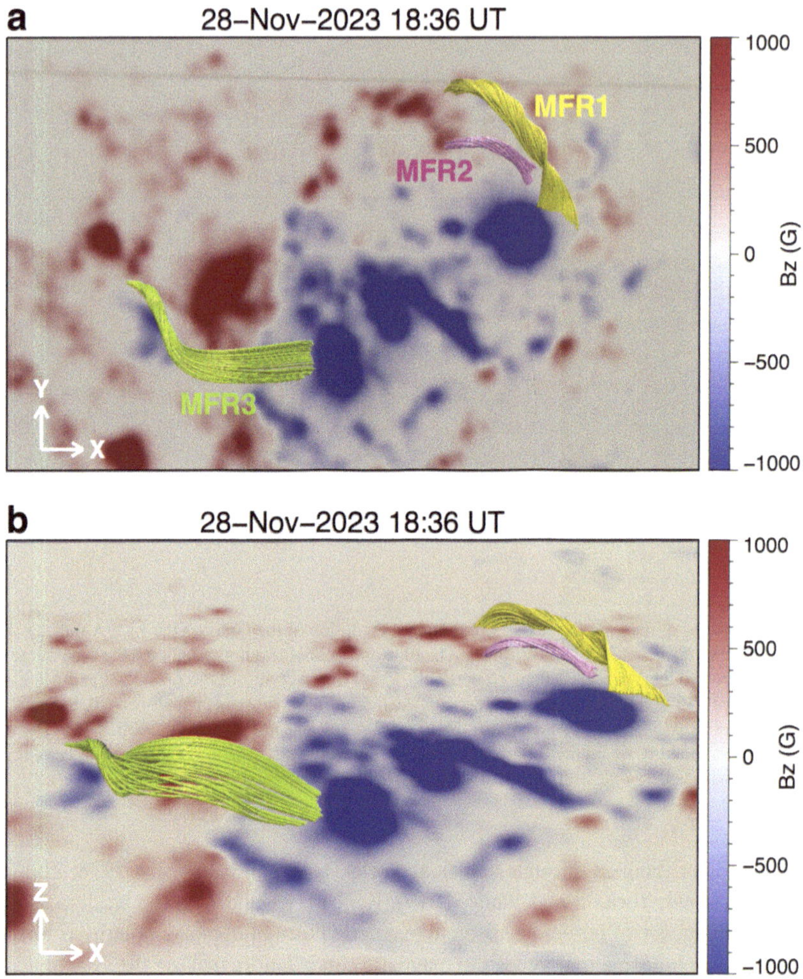

Figure 2 Multi-MFR configuration of the AR 13500 revealed by NLFFF extrapolation at 18:36 UT. Panels (a) and (b) show the top view and side view of the extrapolated results, respectively. Three identified MFRs are marked as MFR1, MFR2 and MFR3. The extrapolation is performed within the area with a twist number above 1.0.

After the eruption of F1, continual eastward slipping motion started from about 19:21 UT (red arrows in Figure 4). The eastern end of the slippage extended towards the southeast and passed through the positive-polarity sunspot at the east of the AR (panels (c1)–(c3)). The slippage lasted for about 10 min and showed a sequence of brightening of approximately 30 Mm from the initial eastern end of seed MFR2. Moreover, the AIA observations showed the presence of dense jet-like structures at the base of the east end of the slippage. Combining the observations, we thought that these jet-like structures are generated from the loop bundles of MFR2. In this way, the slipping process is likely to be associated with buildup of MFR2. Finally, the slippage halted near the location of LB3 (marked as green dotted line in panel (b3)), followed by LB3's ascending (Figure 5), indicating that the slippage might initiate the eruption of MFR3. The slipping motion near LB3 was observed to begin from the eastern footpoint of LB3 (the first row of Figure 5), which depicts a hook-shaped part

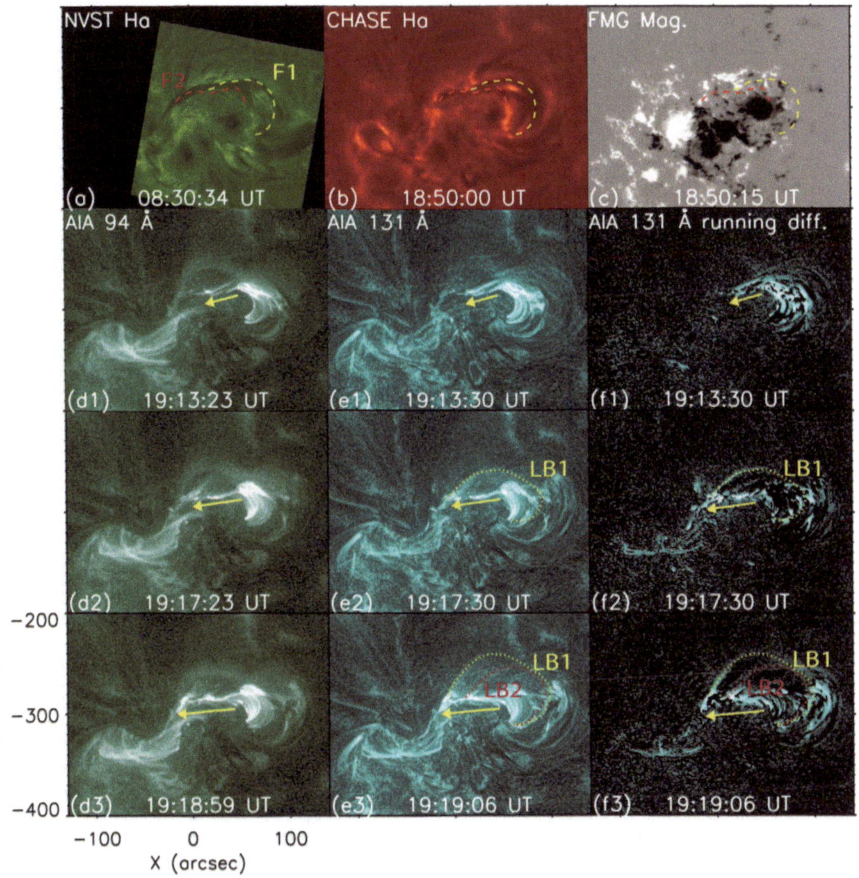

Figure 3 Pre-eruption observations and slipping process of MFR1. Panels (a) and (b) show the Hα observations from NVST and CHASE, respectively. Panel (c) displays the FMG magnetogram. The red and yellow dashed lines depict the location of Filament 1 (F1) and Filament 2 (F2), respectively. Panels (d)–(f) show the EUV observations of the slipping process of MFR1 in 94 Å (panel (d)), 131 Å (panel (e)), and 131 Å running difference with a 48-second cadence. The yellow arrow shows the slipping process of MFR1, while the yellow and red dotted lines represent the loop bundles of LB1 and LB2, respectively. The LB1 and LB2 might be associated with MFR1 and MFR2, respectively. (An animation of this figure is available online.)

of the positive-polarity flare ribbon (PR). At the same time, the western negative-polarity flare ribbon (NR) marked by blue arrows in panel (b2) appears. Note that the slipping of NR was bidirectional, as shown in panels (b1)–(b3). The slippage continued with the flare ribbon elongating southward approximately 70 Mm (panels (a1)–(a4) and (b1)–(b4)). Finally, the LB3 kept ascending and erupted, followed by a flare shown in panel (c4). From the online animation, the bidirectional slipping of NR corresponds to the slipping of PR and hook positive-polarity ribbon (HPR), respectively, which may indicate that this might be associated with both MFR2 and MFR3. Furthermore, the flare region covers the region of both MFR2 and MFR3, which is another evidence of the simultaneous slippage. Ultimately, the simultaneous eruptions of MFR2 and MFR3 contribute to an M9.8-class flare, as depicted in Figure 5(c4).

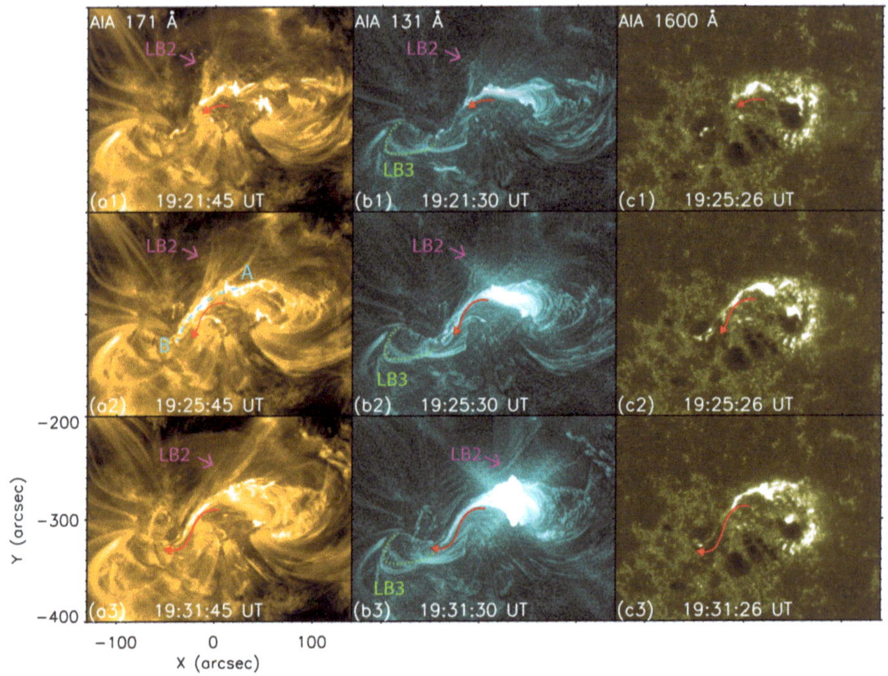

Figure 4 Slipping process of MFR2. Panels (a)-(c) display the observations in AIA 171 Å, 131 Å, and 1600 Å, respectively. The pink arrow denotes the location of the outflows resulting from the reconnection beneath the MFR, thereby aiding in the identification of MFR2. The red arrow shows the slipping process of MFR2. Additionally, the green dots indicate the loop bundle LB3, which might be associated with MFR3. Slice A–B in panel (a2) is specifically selected to illustrate the slipping process of MFR1 and MFR2 and the result is shown in Figure 6(a). The observed solar flare is the M3.4-class flare produced by the eruption of MFR1.

To characterize the kinetic dynamics of the slipping process and flare ribbon motions, we employed selected slices on the EUV images. Slice A–B in Figure 4(a2) illustrates the slipping process of MFR1 and MFR2, while the corresponding time–distance diagram is presented in Figure 6(a). Initially, MFR1 exhibits slipping motion at a velocity of 106 $\mathrm{km\,s^{-1}}$, shown in the distance from 60 to 30 Mm in panel (a). In this process, the slipping of MFR2 is covered by that of MFR1, making it difficult to distinguish. Following MFR1's slipping and subsequent eruption, MFR2 continues slipping southeastward at 40 $\mathrm{km\,s^{-1}}$ from 30 to 10 Mm in panel (a). This slipping extends toward the vicinity of MFR3, potentially initiating the slippage of MFR3. Next, slice C–D in Figure 5(a4) is placed to observe the southward slipping of MFR3, and the associated time–distance diagram is depicted in Figure 6(b). The starting time of MFR3's slipping process is corresponding to when the sequential brightening of MFR2 reaches near MFR3, suggesting that MFR3's slippage may be triggered by MFR2. Then, the eastern end of MFR3 moves southward at the speed of $\sim 137\ \mathrm{km\,s^{-1}}$ to a distance of ~ 70 Mm. Slice E–F in Figure 5(b4) elucidates the slipping direction of the negative-polarity ribbon of the double flare ribbon, and the corresponding time–distance diagram is shown in Figure 6(c). Bidirectional brightenings indicate simultaneous eruptions of MFR2 and MFR3. The brightening of MFR2 occurs earlier than that of MFR3, indicating that the buildup process of MFR2 is earlier. The slipping speeds of the PR are measured at 26 $\mathrm{km\,s^{-1}}$ towards the south, which probably corresponds to the western end of MFR2,

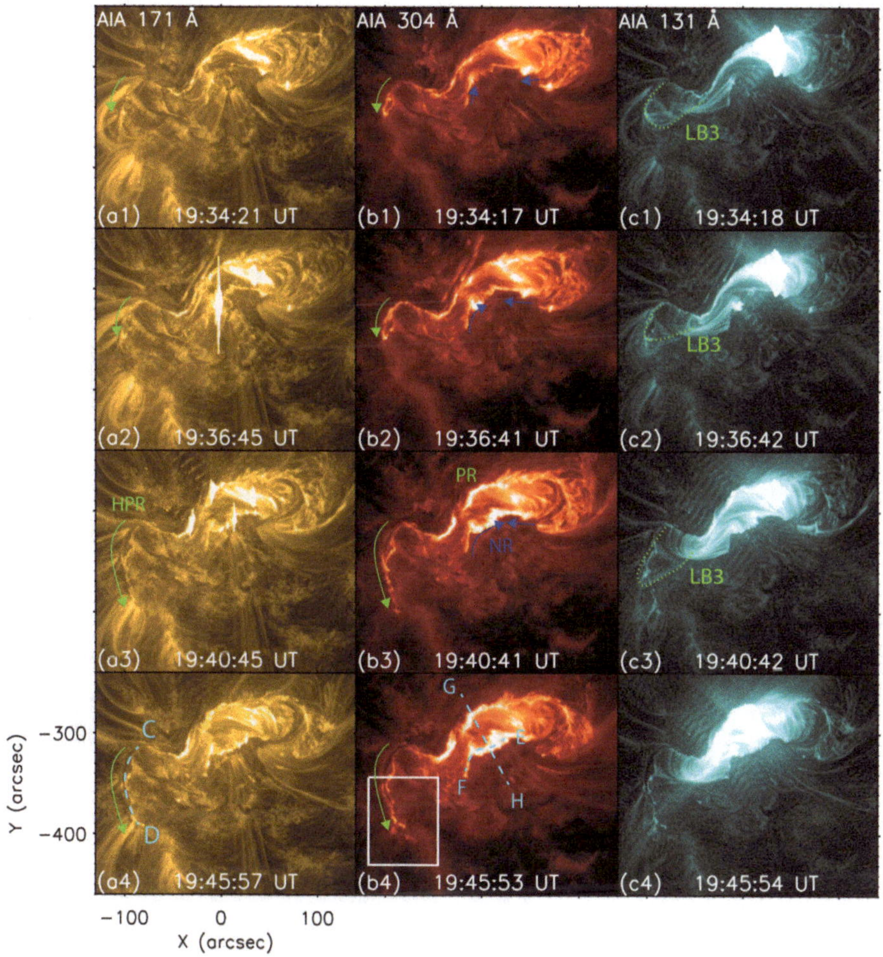

Figure 5 Slipping process of MFR3. Panels (a)–(c) depict the observations in AIA 171 Å, 304 Å, and 131 Å, respectively. The green arrow illustrates the slipping process of LB3, while the blue arrows indicate the bi-directional slipping of the second flare ribbon. Green dots in panels (c1)–(c3) mark the location of MFR3's loop bundle. The white rectangle in panel (b4) outlines the FOV depicted in Figure 7. Slices C–D, E–F, and G–H in the bottom row are specifically chosen to elucidate the slipping process of MFR3, bi-directional slipping of the second ribbon, and the separation of the double flare ribbons, respectively. The observed flare corresponds to the M9.8-class flare resulting from the eruption of MFR2 and MFR3.

and 51 $\mathrm{km\,s^{-1}}$ towards the north corresponding to the western end of the MFR3. Further examination of the time–distance diagram of Slice G–H in Figure 5(b4), represented in Figure 6(d), shows the separation of the double flare ribbons. Initially, around 19:15 UT, PR begins to separate due to the M3.4-class flare at the speed of 16 $\mathrm{km\,s^{-1}}$. Subsequently, around 19:35 UT, the negative-polarity ribbon starts to separate following the M9.8-class flare at a velocity of 18 $\mathrm{km\,s^{-1}}$. Notably, during this period, the northern ribbon remains relatively stable, exhibiting minimal movement perpendicular to the PIL.

Fine structures at the newly-formed eastern footpoint of MFR3 are highlighted within the white square in Figure 5(b4). The temporal evolution of this zoomed-in area is detailed in Figure 7. An analysis of AIA 304 Å and SUTRI 465 Å observations reveal that the elonga-

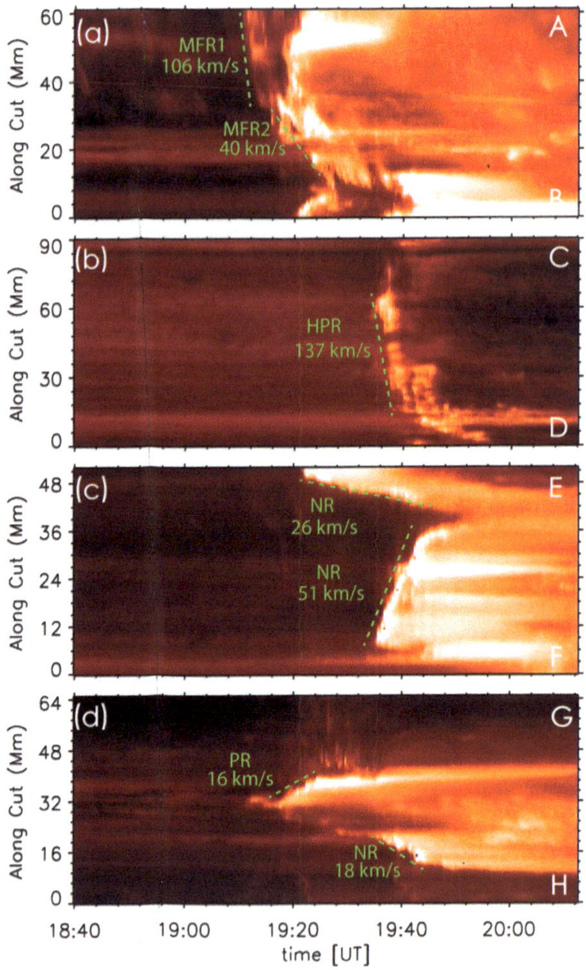

Figure 6 Time–distance diagrams. Slice A-B depicts the slipping process of MFR1 and MFR2, while slice C–D illustrates the slipping process of MFR3. Slice E–F shows the bi-directional slipping of the second ribbon, and slice G–H illustrates the separation of the double flare ribbons. Slice A-B and C–D are shown in Figure 4(a2) and Figure 5(a4), respectively, while Slice E–F and G–H are displayed in Figure 5(b4). The calculated velocity is denoted in green and all velocities are projected velocities.

tion of the HPR halts at $\sim$ 19:42 UT. Subsequently, the end of the flare ribbon undergoes a transformation to a curved shape, as observed in panels (a3) and (d3). Approximately 7 minutes later, it evolves into a pentagram shape (panels (a4), (a5), (b4), and (b5)). This observed evolution process may be attributed to the reconnection between the MFR and surrounding magnetic field lines, leading to the drifting of footpoints as predicted in Aulanier and Dudík (2019).

Large-scale EUV waves generated by the eruptions are shown in Figure 8. Panel (a) presents the EUV waves accompanying the M3.4-class flare, which is triggered by the eruption of MFR1. The wave propagates northward at a speed of 980 $\mathrm{km\,s^{-1}}$ with an angular width of approximately 120°. Panels (b) and (c) depict the waves generated by the M9.8-class flare, which results from the eruptions of MFR2 and MFR3. Notably, we observe the expansion of loops from MFR3, indicated by pink cross signs. The expansion of loops serves as an indicator for the ascent of MFR3, with observations revealing a slow rise at 45 $\mathrm{km\,s^{-1}}$ followed by a rapid increase to 390 $\mathrm{km\,s^{-1}}$. This slow-to-fast transition aligns with previous observations of MFR dynamics (Zhang, Cheng, and Ding, 2012; Kliem et al., 2014; Cheng et al., 2023). In front of the loop expansion, an EUV wave propagates southeastward

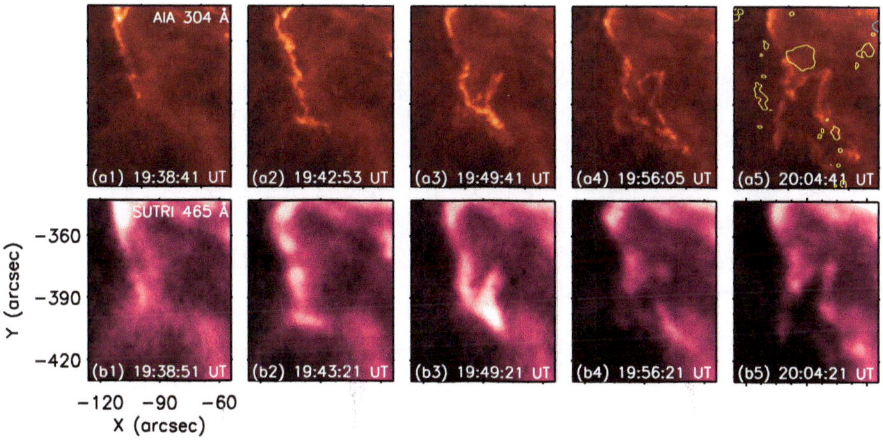

Figure 7 Fine structures of the end of the flare ribbon in the slipping process of MFR3, observed in AIA 304 Å (panels (a1)–(a5)) and SUTRI 465 Å (panels (b1)–(b5)). Panel (c) is superimposed by contours of the magnetic field at ± 400 Gauss levels, with yellow indicating positive values and blue representing negative values.

at a speed of 1080 $\mathrm{km\,s^{-1}}$. Additionally, another wavefront propagates northward at a speed of 1050 $\mathrm{km\,s^{-1}}$ (panel (c)), which is associated with the eruption of MFR2. Such a speed of these EUV waves should be explained as fast magnetoacoustic waves or shock waves (Patsourakos and Vourlidas, 2012; Warmuth, 2015; Sun et al., 2022).

Figure 9 displays white-light LASCO observations of the CMEs. LASCO captured two CMEs (CME1 and CME2) during this event, both documented by the SEEDS catalog.[1] At ∼ 21:30 UT, CME1 is observed propagating westward at a speed projected in the plane of sky of 535 $\mathrm{km\,s^{-1}}$. This CME is associated with the M3.4-class flare triggered by the eruption of MFR1. Approximately 24 minutes later, CME2 originates initially at the south limb of the Sun (panel (b)), evolving subsequently into a halo-CME (panels (c) and (d)), with a projected speed of 2227 $\mathrm{km\,s^{-1}}$ as recorded by SEEDS.

The Wind observations are depicted in Figure 10, where data is averaged over one-minute intervals in the Geocentric Solar Ecliptic (GSE) coordinate system. Analysis of the radial velocity data (panel (c)) reveals a background solar wind speed of approximately 350 $\mathrm{km\,s^{-1}}$. A notable sudden increase at ∼ 23:29 UT on November 30th (yellow dashed lines) indicates the arrival of a fast shock, followed by an ICME. After carefully checking of the EUV and coronagraph data, we found that there are two significant halo CMEs on November 27th and 28th, with the first being a quiescent filament eruption on November 27th and the second being the eruption analyzed in this study. The in-situ data on November 30th and December 1st only show two significant shocks, which are presented in Figure 10. So we could infer that these two significant shocks correspond to the two halo CMEs, respectively. As a result, the first shock should originate from the halo CME caused by a quiescent filament eruption on November 27th, the day prior to the analyzed eruption. After approximately 9 hours, another fast shock (orange dashed lines) arrived at ∼ 08:48 UT on December 1st. This ICME corresponds to the eruption under analysis. The main body of this ICME likely corresponds to the halo CME2 in Figure 9, however, we cannot exclude the possibility that CME2 caught up with CME1 and merged (Li et al., 2012; Lugaz et al., 2017). Following

[1] http://spaceweather.gmu.edu/seeds/.

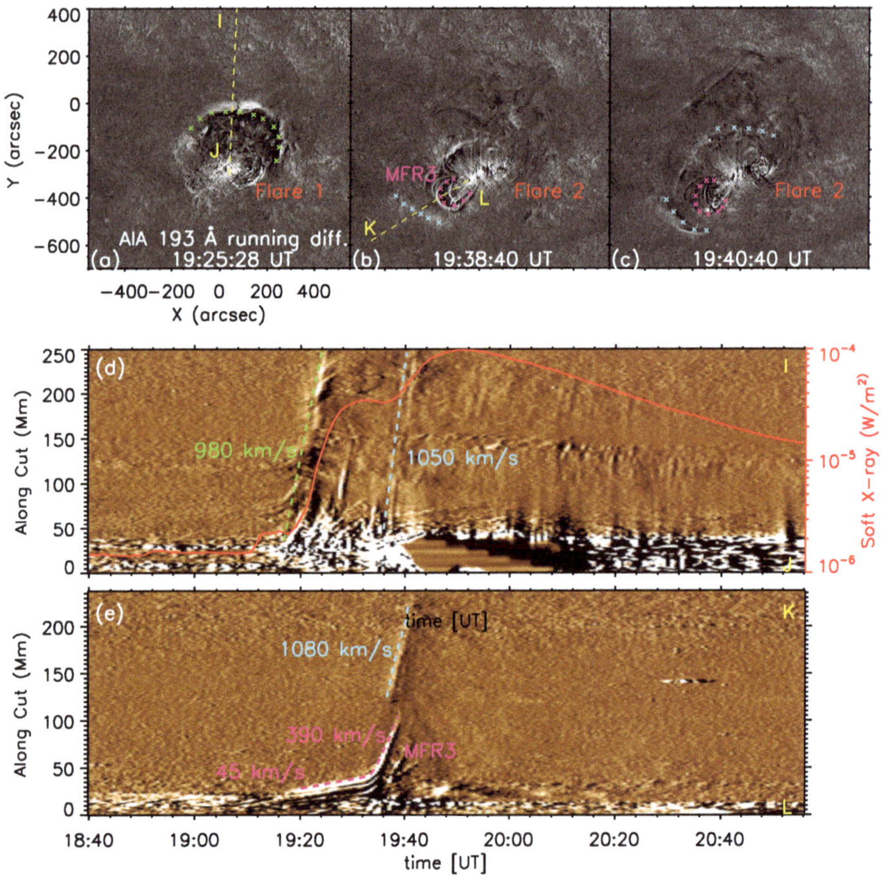

Figure 8 EUV waves produced by the eruption. Panels (a)–(c) show the EUV wavefronts propagating outward the AR 13500 in AIA 193 Å running difference with a 48-second cadence. The green and cyan cross signs indicate the observed wavefront, with the green one associated with Flare 1 (M3.4-class flare) and the others associated with Flare 2 (M9.8-class flare). The pink cross signs indicate the loops of MFR3. Yellow dashed lines are placed to generate of time–distance diagrams, as illustrated in panels (d)–(e). These slices are chosen to illustrate the kinetics of the EUV waves and MFR loops. Panels (d)–(e) show the time–distance diagrams of AIA 193 Å running difference. The green, cyan and pink dashed lines correspond to the fronts depicted in the same color in panels (a)–(c). Panel (d) is overlaid with the GOES SXR 1 – 8 Å curve in red.

the ICME shock-sheath region, the magnetic cloud (MC) passes, identified by systematic variations in the magnetic field components, indicative of a strong magnetic field associated with an MFR-like structure. The Bz component changes its sign from negative to positive. Based on these observations, we identified the time intervals for the MC as spanning from $\sim$ 19:48 UT on December 1st to $\sim$ 01:24 UT on December 2nd. Besides the clear MFR structure, there is a complex structure after December 3rd 00:00 UT, also with a magnetic component reversal. This might be a possible signal of another MFR structure.

Figure 11 presents hourly-averaged in-situ observations of geomagnetic conditions. Upon encountering the first ICME shock (yellow dashed lines), the Bz component increases to approximately -8 nT (where the minus indicates southward direction). Subsequently, upon the arrival of the second shock (orange dashed lines) produced by the eruption under

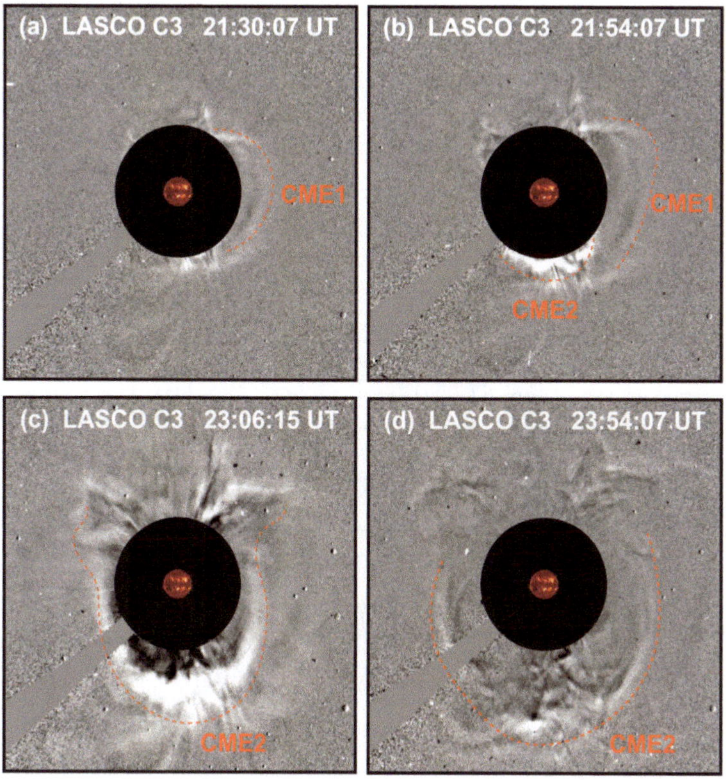

Figure 9 LASCO white-light observations within the near-Sun corona.

analysis, the Bz component further increases to around −23 nT. Such a significant southward magnetic field induces a rapid decrease in the Dst index, with the minimum reaching −105 nT, classifying the event as an intense geomagnetic storm (Dst ≤ -100 nT; Gonzalez et al., 1994; Echer et al., 2008; Lakhina and Tsurutani, 2016). Following the main phase of the geomagnetic storm, there is a long-duration recovery phase. Additionally, this event leads to an increase in the Kp index (panel (c)) to 6.7, categorized as G2 according to NOAA Space Weather Scales. Furthermore, this event triggers low-latitude auroras in the northern hemisphere, which falls beyond the scope of this paper.

4. Discussion

This paper investigates a solar eruption that resulted in an intense geomagnetic storm and a low-latitude aurora. The eruption involved three erupting MFRs, two flares and two CMEs. Throughout this event, we observed three slipping process during the buildup of the three MFRs. The possible scenario is as follows: MFR1 initiates its ascent, accompanied by eastward slipping of the its eastern end at the speed of 106 $\mathrm{km\,s^{-1}}$. Subsequently, MFR1 erupts, leading to the formation of the M3.4-class flare and CME1. Meanwhile, MFR2 undergoes slipping towards the southeast at 40 $\mathrm{km\,s^{-1}}$. As MFR2 slips approximately 30 Mm, it approaches the vicinity of MFR3, triggering the initiation of its destabilization. MFR3 then

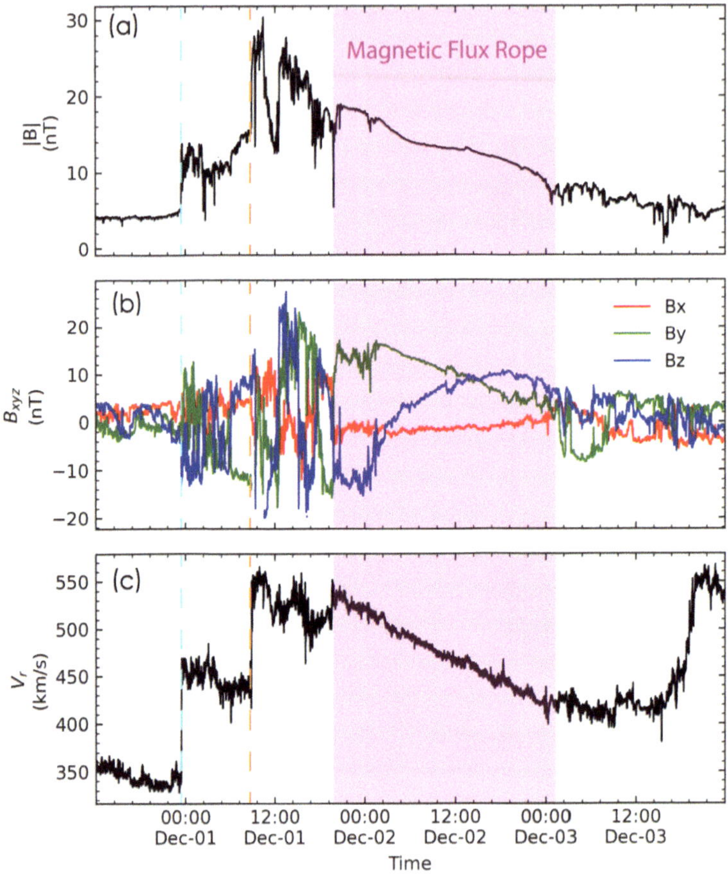

Figure 10 Wind in-situ observations of the ICME. The panels from top to bottom, display plots of magnetic field strength ($|B|$) and its components (Bx, By, Bz), as well as solar wind speed (Vr) as a function of time, respectively. The cyan and orange dashed lines indicate two instances of fast shock arrival, with the cyan line corresponding to the ICME produced one day before the eruption under analysis, and the orange line corresponding to the ICME being analyzed. The time period highlighted in pink denotes the duration of MFR passage.

begins slipping, with its flare ribbon elongating southward at the speed of 137 $\mathrm{km\,s^{-1}}$. Ultimately, the eruptions of MFR2 and MFR3 both contribute to the M9.8-class flare and CME2.

Slipping motions in the MFR eruptions have been simulated and observed in many works (Janvier et al., 2013; Dudík et al., 2014, 2016; Li and Zhang, 2014; Li et al., 2016). In the 3D extensions to the standard CSHKP model, MFRs are typically anchored at the hooks of double-J QSLs. There are numerous sheared arcades overlying the MFR, with one end anchored at the hook part of one J-shaped QSL and the other end at the straight part of the other J-shaped QSL (see in Figure 5 in Aulanier, Janvier, and Schmieder, 2012). During the pre-eruption or eruption phases of MFRs, the sheared arcades can form an HFT structure beneath the MFR and reconnect with each other. This reconnection process converts two sheared field lines into a magnetic flux rope line and a post-flare loop, thereby increasing the poloidal magnetic flux of the MFR and contributing to its buildup. The continuous slip-running reconnection of these field lines leads to observable slipping motions in the flare

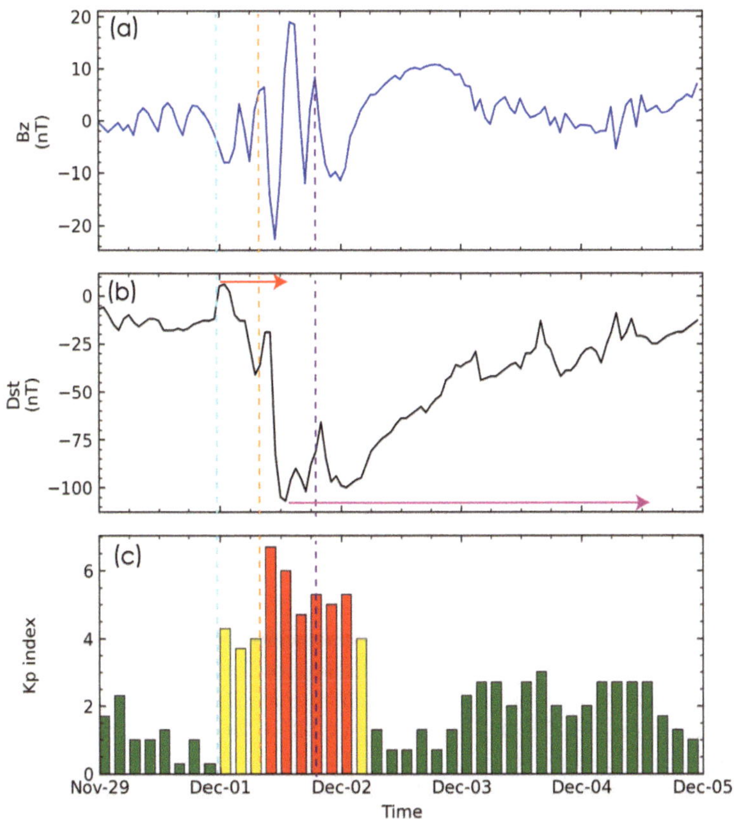

Figure 11 In-situ observations of the CME encountering Earth's atmosphere. Panel (a) presents the Bz in GSE coordinates averaged over one hour. Panel (b) displays the Dst index, reaching a minimum of −105 nT, indicating an intense geomagnetic storm. The cyan and orange dashed lines indicate the shock arrival times as in Figure 10. The purple dashed line indicates the MFR arrival time, also shown in Figure 10. The red and pink arrows denote the main phase and recovery phase of the geomagnetic storm, respectively. Panel (c) shows the histogram of the Kp index, with green, yellow, and red bars representing index values 0 – 3, 3 – 4.3, and 4.3 – 7, respectively. The Kp index values reach up to 6.7 during the storm's main phase, categorized as G2 according to NOAA Space Weather Scales.

ribbons during the buildup of the MFR (Janvier et al., 2013). The buildup process corresponds to the scenario where a preexisting seed MFR accumulates poloidal magnetic flux during the eruption phase. Our results reveal extrapolated MFR structures prior to the eruption and the subsequent buildup process, in agreement with the seed MFR eruption scenario.

Numerous observational case studies have confirmed the presence of slipping motions before and during MFR eruptions (Dudík et al., 2014, 2016; Li and Zhang, 2014; Li et al., 2016, 2018a). These studies have identified continuous brightening along the flare ribbons or QSLs during the buildup process of MFR eruptions. Despite many case studies demonstrating the slipping motions of individual MFR eruptions, few studies have investigated the slipping motions in multi-MFR systems. In our event, we observed the slipping motions of each MFR, providing observational supplements for the three-dimensional reconnection in the MFR eruption. Additionally, the multi-MFR eruption is commonly explained as the reduction of the restraining forces above (Schmieder, Démoulin, and Aulanier, 2013; Dhakal,

Chintzoglou, and Zhang, 2018; Green et al., 2018). However, in our event, the end of MFR2 was elongated to the vicinity of another flux rope MFR3 through the slipping process, which plausibly triggered the eruption of MFR3. Our observations imply that the slipping reconnection plays a key role in the multi-MFR eruptions.

During the slipping process, we observed that brightenings not only occurred in the immediate vicinity of the MFRs, but extended in length to a distance of up to 70 Mm. This phenomenon can also be clearly identified in Dudík et al. (2014), where an elongation of the flare ribbon is observed. The sequential brightening should occur along the hook part of the QSLs (i.e., the footpoints of the MFR), and thus, the direction and distance of the slipping should be associated with the specific configuration of MFRs (Aulanier et al., 2013; Li and Zhang, 2014). Priest and Longcope (2017) proposed a scenario for the buildup of MFR suggesting a two-step process: first, the 3D zipper reconnection of sheared arcades forms an MFR along the PIL, followed by quasi-2D main phase reconnection of unsheared flux. The former corresponds to the parallel extension of conjugate flare ribbons in the same direction, while the latter corresponds to the separation of double flare ribbons. However, in our event, we observed that the extension of conjugate flare ribbons HPR and NR occurs in opposite directions (see Figure 5), suggesting that slipping reconnection rather than zipper reconnection is involved in this event.

Numerous observations indicate that some intense geomagnetic storms are linked with multiple CMEs (Farrugia et al., 2006; Zhang et al., 2007; Liu et al., 2014a). CMEs in the near-Sun corona or interplanetary space can interact with each other, termed the CME–CME interaction. Studies have highlighted the significant geoeffectiveness of CME–CME interactions (Lugaz et al., 2017; Joshi et al., 2018; Scolini et al., 2020). The long-duration driving and compressed magnetic fields resulting from CME–CME interactions can contribute to intense geomagnetic storms (Lugaz et al., 2017). In our event, the halo CME2 includes eruptions possibly from both MFR2 and MFR3, indicating that CME2 itself is a product of CME–CME interaction. Moreover, CME1 preceding CME2 also suggests the possibility of another CME–CME interaction. Consequently, CME–CME interactions among these three MFRs' eruptions may significantly contribute to the observed geomagnetic effects. Additionally, when the ICME associated with this eruption reaches the Earth, it encounters the ICME produced on the day before (Figure 10), potentially compressing the prior ICME and further enhancing geoeffectiveness. Ongoing investigations with a focus on understanding how these CME–CME interactions influence the magnetosphere is crucial for the observed profound geoeffectiveness, which is beyond the scope of this paper.

5. Conclusion

This study investigated a multi-MFR eruption that occurred on 28 November 2023, involving MFR1, MFR2, and MFR3. The eruption of MFR1 led to the formation of an M3.4-class flare and CME1, while the eruptions of MFR2 and MFR3 collectively contributed to an M9.8-class flare and CME2. We observed distinct slipping processes during the buildup of the three MFRs. All three MFRs exhibited slipping motions, extending beyond their original locations. Notably, the slipping of MFR2 extends ~ 30 Mm and plausibly initiate the eruption of MFR3. To our knowledge, this is the first study to display successive slipping process in a multi-MFR system, showing how one MFR's slipping can trigger the slippage and eruption of another MFR.

The eruption resulted in an intense geomagnetic storm, with a Dst index reaching -105 nT on 1 December 2023. This event stands out as one of the six intense geomagnetic storms recorded since the start of Solar Cycle 25. Notably, the associated aurora reached the most southern latitudes in the northern hemisphere observed in the past two decades. Through analysis of the solar origin of the eruption, we consider that CME–CME interactions, triggered by multiple MFR eruptions, may be significant contributors to the observed geomagnetic effects. Further studies focusing on ICMEs and the Earth's magnetosphere are necessary to elucidate the detailed mechanisms behind the intense geomagnetic storm and the low-altitude aurora associated with this event.

Supplementary Information The online version contains supplementary material available at https://doi.org/10.1007/s11207-024-02329-4.

Acknowledgments We thank the referee and the editor for their careful work and thoughtful suggestions. We also thank Prof. Shuo Yao and Dr. Wensi Wang for fruitful discussions. The ASO-S mission is supported by the Strategic Priority Research Program on Space Science, Chinese Academy of Sciences. SUTRI is a collaborative project conducted by the National Astronomical Observatories of CAS, Peking University, Tongji University, Xi'an Institute of Optics and Precision Mechanics of CAS and the Innovation Academy for Microsatellites of CAS. The CHASE mission is supported by China National Space Administration (CNSA). AIA is a payload onboard SDO, a mission of NASA's Living With a Star Program. We thank the SDO, Wind, SOHO/LASCO, FY-3E/X-EUVI, SUTRI, CHASE, ASO-S, and NVST for providing data. This work is supported by the Strategic Priority Research Program of the Chinese Academy of Sciences (Grant No. XDB0560000 and XDB41000000), the National Key R&D Program of China (2019YFA0405000, 2021YFA1600500, 2022YFF0503800), the National Natural Science Foundations of China (11973056, 12222306, and 12273060, 12333009), and the Youth Innovation Promotion Association CAS (2023063).

Author contributions Z. Sun analyzed the data and wrote the first manuscript. T. Li proposed the initial idea and revised the manuscript. Y.J. Hou provided the NLFFF extrapolation results. H. Tian, Z.Q Wu, K. Li, Y.N. Zhang, Z.Y Hou and G.P. Zhou provided suggestions on the data analysis. Z.T Li and X.Y. Bai helped with the HXI and FMG data processing, respectively. L. Feng is Co-PI of LST. C. Li is of CHASE. Q. Song and J.S Wang contributed to the pipeline and release of X-EUVI data. All authors reviewed the manuscript.

Data availability No datasets were generated or analysed during the current study.

Declarations

Competing interests The authors declare no competing interests.

References

Archontis, V., Hood, A.W., Tsinganos, K.: 2014, Recurrent explosive eruptions and the "sigmoid-to-arcade" transformation in the Sun driven by dynamical magnetic flux emergence. *Astrophys. J. Lett.* **786**, L21. DOI. ADS.

Aulanier, G., Dudík, J.: 2019, Drifting of the line-tied footpoints of CME flux-ropes. *Astron. Astrophys.* **621**, A72. DOI. ADS.

Aulanier, G., Janvier, M., Schmieder, B.: 2012, The standard flare model in three dimensions. I. Strong-to-weak shear transition in post-flare loops. *Astron. Astrophys.* **543**, A110. DOI. ADS.

Aulanier, G., Pariat, E., Demoulin, P., DeVore, C.: 2006, Slip-running reconnection in quasi-separatrix layers. *Solar Phys.* **238**, 347.

Aulanier, G., Démoulin, P., Schrijver, C.J., Janvier, M., Pariat, E., Schmieder, B.: 2013, The standard flare model in three dimensions. II. Upper limit on solar flare energy. *Astron. Astrophys.* **549**, A66. DOI. ADS.

Awasthi, A.K., Liu, R., Wang, H., Wang, Y., Shen, C.: 2018, Pre-eruptive magnetic reconnection within a multi-flux-rope system in the solar corona. *Astrophys. J.* **857**, 124. DOI. ADS.

Bai, X., Tian, H., Deng, Y., Wang, Z., Yang, J., Zhang, X., Zhang, Y., Qi, R., Wang, N., Gao, Y., Yu, J., He, C., Shen, Z., Shen, L., Guo, S., Hou, Z., Ji, K., Bi, X., Duan, W., Yang, X., Lin, J., Hu, Z., Song, Q., Yang, Z., Chen, Y., Qiao, W., Ge, W., Li, F., Jin, L., He, J., Chen, X., Zhu, X., He, J., Shi, Q., Liu, L., Li, J., Xu, D., Liu, R., Li, T., Feng, Z., Wang, Y., Fan, C., Liu, S., Guo, S., Sun, Z., Wu, Y., Li, H., Yang,

Q., Ye, Y., Gu, W., Wu, J., Zhang, Z., Yu, Y., Ye, Z., Sheng, P., Wang, Y., Li, W., Huang, Q., Zhang, Z.: 2023, The Solar Upper Transition Region Imager (SUTRI) onboard the SATech-01 satellite. *Res. Astron. Astrophys.* **23**, 065014. DOI. ADS.

Berger, M.A., Prior, C.: 2006, The writhe of open and closed curves. *J. Phys. A, Math. Gen.* **39**, 8321. DOI.

Brueckner, G.E., Howard, R.A., Koomen, M.J., Korendyke, C.M., Michels, D.J., Moses, J.D., Socker, D.G., Dere, K.P., Lamy, P.L., Llebaria, A., Bout, M.V., Schwenn, R., Simnett, G.M., Bedford, D.K., Eyles, C.J.: 1995, The Large Angle Spectroscopic Coronagraph (LASCO). *Solar Phys.* **162**, 357. DOI. ADS.

Cai, Y.-F., Yang, X., Xiang, Y.-Y., Yan, X.-L., Jin, Z.-Y., Liu, H., Ji, K.-F.: 2022, The co-alignment of winged Hα data observed by the New Vacuum Solar Telescope. *Res. Astron. Astrophys.* **22**, 065010. DOI. ADS.

Canou, A., Amari, T.: 2010, A twisted flux rope as the magnetic structure of a filament in active region 10953 observed by Hinode. *Astrophys. J.* **715**, 1566. DOI. ADS.

Chen, P.F.: 2011, Coronal mass ejections: models and their observational basis. *Living Rev. Solar Phys.* **8**, 1. DOI. ADS.

Chen, H., Yang, J., Ji, K., Duan, Y.: 2019, Observational analysis on the early evolution of a CME flux rope: preflare reconnection and flux rope's footpoint drift. *Astrophys. J.* **887**, 118. DOI. ADS.

Chen, B., Zhang, X.-X., He, L.-P., Song, K.-F., Liu, S.-J., Ding, G.-X., Dun, J.-P., Li, J.-W., Li, Z.-H., Guo, Q.-F., et al.: 2022, Solar X-ray and EUV imager on board the FY-3E satellite. *Light: Sci. Appl.* **11**, 329. DOI.

Cheng, X., Guo, Y., Ding, M.: 2017, Origin and structures of solar eruptions I: magnetic flux rope. *Sci. China Earth Sci.* **60**, 1383. DOI. ADS.

Cheng, X., Zhang, J., Liu, Y., Ding, M.D.: 2011, Observing flux rope formation during the impulsive phase of a solar eruption. *Astrophys. J. Lett.* **732**, L25. DOI. ADS.

Cheng, X., Xing, C., Aulanier, G., Solanki, S.K., Peter, H., Ding, M.D.: 2023, Deciphering the slow-rise precursor of a major coronal mass ejection. *Astrophys. J. Lett.* **954**, L47. DOI. ADS.

Chintzoglou, G., Patsourakos, S., Vourlidas, A.: 2015, Formation of magnetic flux ropes during a confined flaring well before the onset of a pair of major coronal mass ejections. *Astrophys. J.* **809**, 34. DOI. ADS.

Demoulin, P., Henoux, J.C., Priest, E.R., Mandrini, C.H.: 1996, Quasi-separatrix layers in solar flares. I. Method. *Astron. Astrophys.* **308**, 643. ADS.

Demoulin, P., Bagala, L.G., Mandrini, C.H., Henoux, J.C., Rovira, M.G.: 1997, Quasi-separatrix layers in solar flares. II. Observed magnetic configurations. *Astron. Astrophys.* **325**, 305. ADS.

Deng, Y.-Y., Zhang, H.-Y., Yang, J.-F., Li, F., Lin, J.-B., Hou, J.-F., Wu, Z., Song, Q., Duan, W., Bai, X.-Y., Wang, D.-G., Lv, J., Ge, W., Wang, J.-N., Zheng, Z.-Y., Wang, C.-J., Wang, N.-G., Ni, H.-K., Zeng, Y.-Z., Zhang, Y., Yang, X., Sun, Y.-Z., Zhang, Z.-Y., Wang, X.-F.: 2019, Design of the full-disk MagnetoGraph (FMG) onboard the ASO-S. *Res. Astron. Astrophys.* **19**, 157. DOI. ADS.

Dhakal, S.K., Chintzoglou, G., Zhang, J.: 2018, A study of a compound solar eruption with two consecutive erupting magnetic structures. *Astrophys. J.* **860**, 35. DOI. ADS.

Ding, J.Y., Hu, Y.Q., Wang, J.X.: 2006, Catastrophic behavior of multiple coronal flux rope system. *Solar Phys.* **235**, 223. DOI. ADS.

Dudík, J., Janvier, M., Aulanier, G., Del Zanna, G., Karlický, M., Mason, H.E., Schmieder, B.: 2014, Slipping magnetic reconnection during an X-class solar flare observed by SDO/AIA. *Astrophys. J.* **784**, 144. DOI. ADS.

Dudík, J., Polito, V., Janvier, M., Mulay, S.M., Karlický, M., Aulanier, G., Del Zanna, G., Dzifčáková, E., Mason, H.E., Schmieder, B.: 2016, Slipping magnetic reconnection, chromospheric evaporation, implosion, and precursors in the 2014 September 10 X1.6-class solar flare. *Astrophys. J.* **823**, 41. DOI. ADS.

Echer, E., Gonzalez, W.D., Tsurutani, B.T., Gonzalez, A.L.C.: 2008, Interplanetary conditions causing intense geomagnetic storms (Dst ≤ -100 nT) during solar cycle 23 (1996 – 2006). *J. Geophys. Res. Space Phys.* **113**, A05221. DOI. ADS.

Fan, Y.: 2009, The emergence of a twisted flux tube into the solar atmosphere: sunspot rotations and the formation of a coronal flux rope. *Astrophys. J.* **697**, 1529. DOI. ADS.

Farrugia, C.J., Matsui, H., Kucharek, H., Jordanova, V.K., Torbert, R.B., Ogilvie, K.W., Berdichevsky, D.B., Smith, C.W., Skoug, R.: 2006, Survey of intense Sun–Earth connection events (1995 – 2003). *Adv. Space Res.* **38**, 498. DOI. ADS.

Gan, W.-Q., Zhu, C., Deng, Y.-Y., Li, H., Su, Y., Zhang, H.-Y., Chen, B., Zhang, Z., Wu, J., Deng, L., Huang, Y., Yang, J.-F., Cui, J.-J., Chang, J., Wang, C., Wu, J., Yin, Z.-S., Chen, W., Fang, C., Yan, Y.-H., Lin, J., Xiong, W.-M., Chen, B., Bao, H.-C., Cao, C.-X., Bai, Y.-P., Wang, T., Chen, B.-L., Li, X.-Y., Zhang, Y., Feng, L., Su, J.-T., Li, Y., Chen, W., Li, Y.-P., Su, Y.-N., Wu, H.-Y., Gu, M., Huang, L., Tang, X.-J.: 2019, Advanced Space-based Solar Observatory (ASO-S): an overview. *Res. Astron. Astrophys.* **19**, 156. DOI. ADS.

Gibson, S.E., Fan, Y., Török, T., Kliem, B.: 2006, The evolving sigmoid: evidence for magnetic flux ropes in the corona before, during, and after CMES. *Space Sci. Rev.* **124**, 131. DOI. ADS.

Gonzalez, W., Joselyn, J.-A., Kamide, Y., Kroehl, H.W., Rostoker, G., Tsurutani, B.T., Vasyliunas, V.: 1994, What is a geomagnetic storm? *J. Geophys. Res. Space Phys.* **99**, 5771. DOI.

Gou, T., Liu, R., Kliem, B., Wang, Y., Veronig, A.M.: 2019, The birth of a coronal mass ejection. *Sci. Adv.* **5**, 7004. DOI. ADS.

Green, L.M., Török, T., Vršnak, B., Manchester, W., Veronig, A.: 2018, The origin, early evolution and predictability of solar eruptions. *Space Sci. Rev.* **214**, 46. DOI. ADS.

Guo, Y., Schmieder, B., Démoulin, P., Wiegelmann, T., Aulanier, G., Török, T., Bommier, V.: 2010, Coexisting flux rope and dipped arcade sections along one solar filament. *Astrophys. J.* **714**, 343. DOI. ADS.

Hale, G.E., Nicholson, S.B.: 1938, Magnetic observations of sunspots, 1917 – 1924/ Washington.

Harten, R., Clark, K.: 1995, The design features of the GGS wind and polar spacecraft. *Space Sci. Rev.* **71**, 23. DOI. ADS.

Hou, Y.J., Zhang, J., Li, T., Yang, S.H., Li, X.H.: 2018, Eruption of a multi-flux-rope system in solar active region 12673 leading to the two largest flares in solar cycle 24. *Astron. Astrophys.* **619**, A100. DOI. ADS.

Hou, Y., Li, C., Li, T., Su, J., Qiu, Y., Yang, S., Yang, L., Li, L., Guo, Y., Hou, Z., Song, Q., Bai, X., Zhou, G., Ding, M., Gan, W., Deng, Y.: 2023, Partial eruption of solar filaments. I. Configuration and formation of double-decker filaments. *Astrophys. J.* **959**, 69. DOI. ADS.

Inoue, S., Kusano, K.: 2006, Three-dimensional simulation study of flux rope dynamics in the solar corona. *Astrophys. J.* **645**, 742. DOI. ADS.

Janvier, M., Aulanier, G., Pariat, E., Démoulin, P.: 2013, The standard flare model in three dimensions. III. Slip-running reconnection properties. *Astron. Astrophys.* **555**, A77. DOI. ADS.

Jiang, C., Feng, X., Wu, S.T., Hu, Q.: 2013, Magnetohydrodynamic simulation of a sigmoid eruption of active region 11283. *Astrophys. J. Lett.* **771**, L30. DOI. ADS.

Jiang, C., Zou, P., Feng, X., Hu, Q., Liu, R., Vemareddy, P., Duan, A., Zuo, P., Wang, Y., Wei, F.: 2018, Magnetohydrodynamic simulation of the X9.3 flare on 2017 September 6: evolving magnetic topology. *Astrophys. J.* **869**, 13. DOI. ADS.

Jiang, C., Duan, A., Feng, X., Zou, P., Zuo, P., Wang, Y.: 2019, Reconstruction of a highly twisted magnetic flux rope for an inter-active-region X-class solar flare. *Front. Astron. Space Sci.* **6**, 63. DOI. ADS.

Joshi, B., Ibrahim, M.S., Shanmugaraju, A., Chakrabarty, D.: 2018, A major geoeffective CME from NOAA 12371: initiation, CME-CME interactions, and interplanetary consequences. *Solar Phys.* **293**, 107. DOI. ADS.

Kliem, B., Török, T., Titov, V.S., Lionello, R., Linker, J.A., Liu, R., Liu, C., Wang, H.: 2014, Slow rise and partial eruption of a double-decker filament. II. A double flux rope model. *Astrophys. J.* **792**, 107. DOI. ADS.

Krall, J., Chen, J., Duffin, R.T., Howard, R.A., Thompson, B.J.: 2001, Erupting solar magnetic flux ropes: theory and observation. *Astrophys. J.* **562**, 1045. DOI. ADS.

Lakhina, G.S., Tsurutani, B.T.: 2016, Geomagnetic storms: historical perspective to modern view. *Geosci. Lett.* **3**, 5. DOI. ADS.

Leka, K.D., Canfield, R.C., McClymont, A.N., van Driel-Gesztelyi, L.: 1996, Evidence for current-carrying emerging flux. *Astrophys. J.* **462**, 547. DOI. ADS.

Lemen, J.R., Title, A.M., Akin, D.J., Boerner, P.F., Chou, C., Drake, J.F., Duncan, D.W., Edwards, C.G., Friedlaender, F.M., Heyman, G.F., Hurlburt, N.E., Katz, N.L., Kushner, G.D., Levay, M., Lindgren, R.W., Mathur, D.P., McFeaters, E.L., Mitchell, S., Rehse, R.A., Schrijver, C.J., Springer, L.A., Stern, R.A., Tarbell, T.D., Wuelser, J.-P., Wolfson, C.J., Yanari, C., Bookbinder, J.A., Cheimets, P.N., Caldwell, D., Deluca, E.E., Gates, R., Golub, L., Park, S., Podgorski, W.A., Bush, R.I., Scherrer, P.H., Gummin, M.A., Smith, P., Auker, G., Jerram, P., Pool, P., Soufli, R., Windt, D.L., Beardsley, S., Clapp, M., Lang, J., Waltham, N.: 2012, The Atmospheric Imaging Assembly (AIA) on the Solar Dynamics Observatory (SDO). *Solar Phys.* **275**, 17. DOI. ADS.

Li, T., Priest, E., Guo, R.: 2021, Three-dimensional magnetic reconnection in astrophysical plasmas. *Proc. R. Soc. Lond. A* **477**, 20200949. DOI. ADS.

Li, T., Zhang, J.: 2014, Slipping magnetic reconnection triggering a solar eruption of a triangle-shaped flag flux rope. *Astrophys. J. Lett.* **791**, L13. DOI. ADS.

Li, G., Moore, R., Mewaldt, R.A., Zhao, L., Labrador, A.W.: 2012, A twin-CME scenario for ground level enhancement events. *Space Sci. Rev.* **171**, 141. DOI. ADS.

Li, T., Yang, K., Hou, Y., Zhang, J.: 2016, Slipping magnetic reconnection of flux-rope structures as a precursor to an eruptive X-class solar flare. *Astrophys. J.* **830**, 152. DOI. ADS.

Li, T., Hou, Y., Yang, S., Zhang, J.: 2018a, Three-dimensional magnetic reconnection triggering an X-class confined flare in active region 12192. *Astrophys. J.* **869**, 172. DOI. ADS.

Li, T., Yang, S., Zhang, Q., Hou, Y., Zhang, J.: 2018b, Two episodes of magnetic reconnections during a confined circular-ribbon flare. *Astrophys. J.* **859**, 122. DOI. ADS.

Li, H., Chen, B., Feng, L., Li, Y., Huang, Y., Li, J.-W., Lu, L., Xue, J.-C., Ying, B.-L., Zhao, J., Yang, Y.-T., Gan, W.-Q., Fang, C., Song, K.-F., Wang, H., Guo, Q.-F., He, L.-P., Zhu, B., Zhu, C., Deng, L., Bao, H.-C., Cao, C.-X., Yang, Z.-G.: 2019, The Lyman-alpha Solar Telescope (LST) for the ASO-S mission – I. Scientific objectives and overview. *Res. Astron. Astrophys.* **19**, 158. DOI. ADS.

Li, C., Fang, C., Li, Z., Ding, M., Chen, P., Qiu, Y., You, W., Yuan, Y., An, M., Tao, H., Li, X., Chen, Z., Liu, Q., Mei, G., Yang, L., Zhang, W., Cheng, W., Chen, J., Chen, C., Gu, Q., Huang, Q., Liu, M., Han, C., Xin, H., Chen, C., Ni, Y., Wang, W., Rao, S., Li, H., Lu, X., Wang, W., Lin, J., Jiang, Y., Meng, L., Zhao, J.: 2022, The Chinese Hα Solar Explorer (CHASE) mission: An overview. *Sci. China, Phys. Mech. Astron.* **65**, 289602. DOI. ADS.

Liu, R.: 2020, Magnetic flux ropes in the solar corona: structure and evolution toward eruption. *Res. Astron. Astrophys.* **20**, 165. DOI. ADS.

Liu, R., Liu, C., Wang, S., Deng, N., Wang, H.: 2010, Sigmoid-to-flux-rope transition leading to a loop-like coronal mass ejection. *Astrophys. J. Lett.* **725**, L84. DOI. ADS.

Liu, Y.D., Luhmann, J.G., Kajdič, P., Kilpua, E.K.J., Lugaz, N., Nitta, N.V., Möstl, C., Lavraud, B., Bale, S.D., Farrugia, C.J., Galvin, A.B.: 2014a, Observations of an extreme storm in interplanetary space caused by successive coronal mass ejections. *Nat. Commun.* **5**, 3481. DOI. ADS.

Liu, Z., Xu, J., Gu, B.-Z., Wang, S., You, J.-Q., Shen, L.-X., Lu, R.-W., Jin, Z.-Y., Chen, L.-F., Lou, K., Li, Z., Liu, G.-Q., Xu, Z., Rao, C.-H., Hu, Q.-Q., Li, R.-F., Fu, H.-W., Wang, F., Bao, M.-X., Wu, M.-C., Zhang, B.-R.: 2014b, New vacuum solar telescope and observations with high resolution. *Res. Astron. Astrophys.* **14**, 705. DOI. ADS.

Liu, R., Kliem, B., Titov, V.S., Chen, J., Wang, Y., Wang, H., Liu, C., Xu, Y., Wiegelmann, T.: 2016, Structure, stability, and evolution of magnetic flux ropes from the perspective of magnetic twist. *Astrophys. J.* **818**, 148. DOI. ADS.

Liu, L., Cheng, X., Wang, Y., Zhou, Z., Guo, Y., Cui, J.: 2018, Rapid buildup of a magnetic flux rope during a confined X2.2 class flare in NOAA AR 12673. *Astrophys. J. Lett.* **867**, L5. DOI. ADS.

Liu, L., Cheng, X., Wang, Y., Zhou, Z.: 2019, Formation of a magnetic flux rope in the early emergence phase of NOAA active region 12673. *Astrophys. J.* **884**, 45. DOI. ADS.

Lugaz, N., Temmer, M., Wang, Y., Farrugia, C.J.: 2017, The interaction of successive coronal mass ejections: a review. *Solar Phys.* **292**, 64. DOI. ADS.

Manchester, I.W., Gombosi, T., DeZeeuw, D., Fan, Y.: 2004, Eruption of a buoyantly emerging magnetic flux rope. *Astrophys. J.* **610**, 588. DOI. ADS.

Nindos, A., Patsourakos, S., Vourlidas, A., Tagikas, C.: 2015, How common are hot magnetic flux ropes in the low solar corona? A statistical study of EUV observations. *Astrophys. J.* **808**, 117. DOI. ADS.

Patsourakos, S., Vourlidas, A.: 2012, On the nature and genesis of EUV waves: a synthesis of observations from SOHO, STEREO, SDO, and hinode (invited review). *Solar Phys.* **281**, 187. DOI. ADS.

Patsourakos, S., Vourlidas, A., Stenborg, G.: 2013, Direct evidence for a fast coronal mass ejection driven by the prior formation and subsequent destabilization of a magnetic flux rope. *Astrophys. J.* **764**, 125. DOI. ADS.

Priest, E.R., Démoulin, P.: 1995, Three-dimensional magnetic reconnection without null points. 1. Basic theory of magnetic flipping. *J. Geophys. Res.* **100**, 23443. DOI. ADS.

Priest, E.R., Forbes, T.G.: 1992, Magnetic flipping: reconnection in three dimensions without null points. *J. Geophys. Res.* **97**, 1521. DOI. ADS.

Priest, E.R., Longcope, D.W.: 2017, Flux-rope twist in eruptive flares and CMEs: due to zipper and main-phase reconnection. *Solar Phys.* **292**, 25. DOI. ADS.

Savcheva, A.S., Green, L.M., van Ballegooijen, A.A., DeLuca, E.E.: 2012, Photospheric flux cancellation and the build-up of sigmoidal flux ropes on the Sun. *Astrophys. J.* **759**, 105. DOI. ADS.

Savcheva, A., Pariat, E., McKillop, S., McCauley, P., Hanson, E., Su, Y., Werner, E., DeLuca, E.E.: 2015, The relation between solar eruption topologies and observed flare features. I. Flare ribbons. *Astrophys. J.* **810**, 96. DOI. ADS.

Schmieder, B., Démoulin, P., Aulanier, G.: 2013, Solar filament eruptions and their physical role in triggering coronal mass ejections. *Adv. Space Res.* **51**, 1967. DOI. ADS.

Scolini, C., Chané, E., Temmer, M., Kilpua, E.K.J., Dissauer, K., Veronig, A.M., Palmerio, E., Pomoell, J., Dumbović, M., Guo, J., Rodriguez, L., Poedts, S.: 2020, CME-CME interactions as sources of CME geoeffectiveness: the formation of the complex ejecta and intense geomagnetic storm in 2017 early September. *Astrophys. J. Suppl.* **247**, 21. DOI. ADS.

Song, H.Q., Chen, Y., Zhang, J., Cheng, X., Wang, B., Hu, Q., Li, G., Wang, Y.M.: 2015, Evidence of the solar EUV hot channel as a magnetic flux rope from remote-sensing and in situ observations. *Astrophys. J. Lett.* **808**, L15. DOI. ADS.

Sun, Z., Tian, H., Chen, P.F., Yao, S., Hou, Z., Chen, H., Chen, L.: 2022, Cross-loop propagation of a quasiperiodic extreme-ultraviolet wave train triggered by successive stretching of magnetic field structures during a solar eruption. *Astrophys. J. Lett.* **939**, L18. DOI. ADS.

Sun, Z., Li, T., Tian, H., Hou, Y., Hou, Z., Chen, H., Bai, X., Deng, Y.: 2023, Observation of two splitting processes in a partial filament eruption on the Sun: the role of breakout reconnection. *Astrophys. J.* **953**, 148. DOI. ADS.
Tian, H.: 2017, Probing the solar transition region: current status and future perspectives. *Res. Astron. Astrophys.* **17**, 110. DOI. ADS.
Titov, V.S., Démoulin, P.: 1999, Basic topology of twisted magnetic configurations in solar flares. *Astron. Astrophys.* **351**, 707. ADS.
Titov, V.S., Hornig, G., Démoulin, P.: 2002, Theory of magnetic connectivity in the solar corona. *J. Geophys. Res. Space Phys.* **107**, 1164. DOI. ADS.
Toriumi, S., Wang, H.: 2019, Flare-productive active regions. *Living Rev. Solar Phys.* **16**, 3. DOI. ADS.
Török, T., Leake, J.E., Titov, V.S., Archontis, V., Mikić, Z., Linton, M.G., Dalmasse, K., Aulanier, G., Kliem, B.: 2014, Distribution of electric currents in solar active regions. *Astrophys. J. Lett.* **782**, L10. DOI. ADS.
Veronig, A.M., Podladchikova, T., Dissauer, K., Temmer, M., Seaton, D.B., Long, D., Guo, J., Vršnak, B., Harra, L., Kliem, B.: 2018, Genesis and impulsive evolution of the 2017 September 10 coronal mass ejection. *Astrophys. J.* **868**, 107. DOI. ADS.
Wang, W., Liu, R., Wang, Y., Hu, Q., Shen, C., Jiang, C., Zhu, C.: 2017, Buildup of a highly twisted magnetic flux rope during a solar eruption. *Nat. Commun.* **8**, 1330. DOI. ADS.
Warmuth, A.: 2015, Large-scale globally propagating coronal waves. *Living Rev. Solar Phys.* **12**, 3. DOI. ADS.
Wheatland, M.S., Sturrock, P.A., Roumeliotis, G.: 2000, An optimization approach to reconstructing force-free fields. *Astrophys. J.* **540**, 1150. DOI. ADS.
Wiegelmann, T.: 2004, Optimization code with weighting function for the reconstruction of coronal magnetic fields. *Solar Phys.* **219**, 87. DOI. ADS.
Wiegelmann, T., Inhester, B., Sakurai, T.: 2006, Preprocessing of vector magnetograph data for a nonlinear force-free magnetic field reconstruction. *Solar Phys.* **233**, 215. DOI. ADS.
Wiegelmann, T., Thalmann, J.K., Inhester, B., Tadesse, T., Sun, X., Hoeksema, J.T.: 2012, How should one optimize nonlinear force-free coronal magnetic field extrapolations from SDO/HMI vector magnetograms? *Solar Phys.* **281**, 37. DOI. ADS.
Xie, H., Gopalswamy, N., Akiyama, S., Yashiro, S., Makela, P.: 2023, Magnetic flux rope structures associated with filament channels: two case studies. *J. Atmos. Solar-Terr. Phys.* **252**, 106154. DOI. ADS.
Zhang, J., Cheng, X., Ding, M.-D.: 2012, Observation of an evolving magnetic flux rope before and during a solar eruption. *Nat. Commun.* **3**, 747. DOI. ADS.
Zhang, J., Richardson, I.G., Webb, D.F., Gopalswamy, N., Huttunen, E., Kasper, J.C., Nitta, N.V., Poomvises, W., Thompson, B.J., Wu, C.-C., Yashiro, S., Zhukov, A.N.: 2007, Solar and interplanetary sources of major geomagnetic storms ($Dst \leq -100$ nT) during 1996–2005. *J. Geophys. Res. Space Phys.* **112**, A10102. DOI. ADS.
Zhang, Z., Chen, D.-Y., Wu, J., Chang, J., Hu, Y.-M., Su, Y., Zhang, Y., Wang, J.-P., Liang, Y.-M., Ma, T., Guo, J.-H., Cai, M.-S., Zhang, Y.-Q., Huang, Y.-Y., Peng, X.-Y., Tang, Z.-B., Zhao, X., Zhou, H.-H., Wang, L.-G., Song, J.-X., Ma, M., Xu, G.-Z., Yang, J.-F., Lu, D., He, Y.-H., Tao, J.-Y., Ma, X.-L., Lv, B.-G., Bai, Y.-P., Cao, C.-X., Huang, Y., Gan, W.-Q.: 2019, Hard X-ray imager (HXI) onboard the ASO-S mission. *Res. Astron. Astrophys.* **19**, 160. DOI. ADS.
Zhao, J., Gilchrist, S.A., Aulanier, G., Schmieder, B., Pariat, E., Li, H.: 2016, Hooked flare ribbons and flux-rope-related QSL footprints. *Astrophys. J.* **823**, 62. DOI. ADS.
Zheng, R., Zhang, L., Wang, B., Kong, X., Song, H., Wu, Z., Feng, S., Chen, H., Chen, Y.: 2021, Compound eruptions of twin flux ropes in a solar active region. *Astrophys. J. Lett.* **921**, L39. DOI. ADS.
Zhongxian, S., Jingxiu, W.: 1994, Delta-sunspots and X-class flares. *Solar Phys.* **149**, 105. DOI. ADS.

Authors and Affiliations

Zheng Sun[1] · Ting Li[2,3,4] · Yijun Hou[2,3,4] · Hui Tian[1,4] · Ziqi Wu[1] · Ke Li[1] · Yining Zhang[2] · Zhentong Li[5] · Xianyong Bai[2] · Li Feng[5,6] · Chuan Li[7,8] · Zhenyong Hou[1] · Qiao Song[9,10,11] · Jingsong Wang[9,10,11] · Guiping Zhou[2,3,4]

✉ T. Li
liting@nao.cas.cn

[1] School of Earth and Space Sciences, Peking University, Beijing 100871, People's Republic of China

[2] National Astronomical Observatories, Chinese Academy of Sciences, Beijing 100101, People's Republic of China

[3] School of Astronomy and Space Science, University of Chinese Academy of Sciences, Beijing 100049, People's Republic of China

[4] Key Laboratory of solar activity and space weather, National Space Science Center, Chinese Academy of Sciences, Beijing 100190, People's Republic of China

[5] Key Laboratory of Dark Matter and Space Astronomy, Purple Mountain Observatory, Chinese Academy of Sciences, Nanjing 210023, People's Republic of China

[6] School of Astronomy and Space Science, University of Science and Technology of China, Hefei 230026, People's Republic of China

[7] School of Astronomy and Space Science, Nanjing University, Nanjing 210023, People's Republic of China

[8] Key Laboratory for Modern Astronomy and Astrophysics (Nanjing University), Ministry of Education, Nanjing 210023, People's Republic of China

[9] National Satellite Meteorological Center (National Centre for Space Weather), China Meteorological Administration, Beijing 100081, People's Republic of China

[10] Innovation Center for FengYun Meteorological Satellite (FYSIC), Beijing 100081, People's Republic of China

[11] Key Laboratory of Space Weather, National Center for Space Weather, China Meteorological Administration, Beijing 100081, People's Republic of China

Solar Physics (2025) 300:26
https://doi.org/10.1007/s11207-024-02421-9

RESEARCH

Comparison of Relative Magnetic Helicity Flux Calculation Results Based on the Line-of-Sight Magnetograms of ASO-S/FMG and SDO/HMI

Shangbin Yang[1,2,3] · Suo Liu[1,2] · Jiangtao Su[1,2,3] · Yuanyong Deng[1,2,3]

Received: 11 March 2024 / Accepted: 22 December 2024 / Published online: 4 March 2025

Abstract
Magnetic helicity is a key geometrical parameter to describe the structure and evolution of solar coronal magnetic fields. The accumulation of magnetic helicity is correlated with the nonpotential magnetic field energy, which is released in the solar eruptions. Moreover, the relative magnetic helicity fluxes can be estimated only relying on the line-of-sight magnetic field (e.g. Démoulin and Berger 2003). The payload Full-disk MagnetoGraph (FMG) on the Advanced Space-based Solar Observatory (ASO-S) currently has been supplying the continuous evolution of line-of-sight magnetograms for the solar active regions, which can be used to estimate the magnetic helicity flux. In this study, we use eight-hour line-of-sight magnetograms of NOAA 13273, at which the Sun–Earth direction speed of the satellite is zero to avoid the oscillation of the magnetic field caused by the Doppler effect on polarization measurements. We obtain the helicity flux by applying fast Fourier transform (FFT) and local correlation tracking (LCT) methods to obtain the horizontal vector potential field and the motions of the line-of-sight polarities. We also compare the helicity flux derived using data from the Heliosesmic and Magnetic Imager (HMI) on board the Solar Dynamics Observatory (SDO) and the same method. It is found that the flux has the same sign and the correlation between measurements is 0.98. The difference of the absolute magnetic helicity normalized to the magnetic flux is less than 4%. This comparison demonstrates the reliability of ASO-S/FMG data and that it can be reliably used in future studies.

Keywords Helicity · Solar magnetic field · Corona

✉ S. Yang
yangshb@nao.cas.cn

S. Liu
lius@nao.cas.cn

J. Su
sjt@nao.cas.cn

Y. Deng
dyy@nao.cas.cn

1 Huairou Solar Observing Station, National Astronomical Observatories, Chinese Academy of Sciences, 100101 Beijing, China

2 State Key Laboratory of Solar Activity and Space Weather, 100190 Beijing, China

3 University of Chinese Academy of Sciences, 100049 Beijing, China

1. Introduction

It is well known that solar flares and coronal mass ejections (CMEs) are closely related to the sudden explosive release of stored free magnetic energy in the corona (see, e.g., Priest and Forbes 2002)). Unfortunately, magnetic fields can be observed only in the photosphere, and only few indirect measurements exist to determine the magnetic field in the corona (e.g. Lin, Kuhn, and Coulter 2004). Although linear and nonlinear mathematical approaches are used to extrapolate force-free magnetic fields from the photosphere into the corona, it is still difficult to estimate the free magnetic energy contained in the corona and the conditions of its release (Schrijver 2006).

Magnetic helicity is a quantity describing the twist, writhe, and torsion of magnetic field lines and magnetic configurations (Berger and Field 1984; Berger 1999). The concept of magnetic helicity has successfully been applied to characterize solar coronal processes; for a recent review about observations and computation of the photospheric magnetic helicity, see Démoulin and Pariat (2009). Zhang, Flyer, and Low (2006), Zhang and Flyer (2008) conjectured that there is an upper limit of the total magnetic helicity that a force-free magnetic field can contain before it erupts. It also has been found that the temporal evolution of the relative magnetic helicity is closely related to the evolution of the magnetic energy content of the corona (Santos, Büchner, and Otto 2011; Skála et al. 2015; Yang et al. 2013, 2018). For a better understanding of the energy storage and release in the corona, it is, therefore, essential to investigate the magnetic helicity accumulation. Park et al. (2008) investigated the variation of the magnetic helicity for the 11 X-class flares that occurred in seven active regions (ARs). They found that each of these major flares was preceded by a significant helicity accumulation at a nearly constant rate before the flares. Zhang, Tan, and Yan (2008) also found that there is a sharp jump of magnetic helicity change rate closely related to the solar eruption. Ninodos and Andrews (2004) investigated 133 events and showed that the amount of the stored preflare coronal helicity could determine if a flare is eruptive or confined. Tziotziou, Georgoulis, and Raouafi (2012) analyzed the free magnetic energy and relative magnetic helicity budgets in 162 vector magnetograms and found a statistically robust, monotonic correlation between the free magnetic energy and the relative magnetic helicity, which indicates that the relative magnetic helicity is an essential ingredient for major solar eruptions. Observations and simulations also show a threshold for current-carrying parts of relative magnetic helicity (Price et al. 2019; Gupta, Thalmann, and Veronig 2021). Liokati, Nindos, and Liu (2022), Liokati, Nindos, and Georgoulis (2023) also discussed the role of left- and right-handed accumulated helicity and found that the excess budget is cotemporal with the flare peaks. Both results discussed above also show that there is correlation between the evolution of magnetic helicity and solar eruptions. Moreover, the observation of magnetic helicity patterns over long periods could be used to analyze the so-called helicity hemispheric rule (Bao and Zhang 1998; Zhang et al. 2010), as well as to constrain the solar dynamo from the point of view of magnetic helicity conservation over large scales and modulation of the solar cycle strength (Yang et al. 2020). A comprehensive review about the investigation of magnetic helicity in geophysics and astrophysics can be found in Kuzanyan et al. (2023) and references therein.

As the first launched Chinese solar satellite, the Advanced Space-based Solar Observatory (ASO-S) is focused on observing solar magnetic fields, solar flares, and coronal mass ejections (CMEs) to study their relationships, which are key scientific questions in modern solar physics (Gan et al. 2023). Three payloads are therefore deployed on the ASO-S: the Full-disk vector MagnetoGraph (Deng et al. 2019, FMG), the Lyman-α Solar Telescope (Feng et al. 2019, LST), and the Hard X-ray Imager (Su et al. 2019, HXI). To study the correlation between the accumulated magnetic helicity and coronal dynamics, we analyze the helicity evolution using ASO-S/FMG data. In this study, we use line-of-sight ASO-S/FMG magnetograms to estimate the magnetic helicity flux and compare the results with those derived using the Heliosiesmic and Magnetic Imager (HMI) data on board the Solar Dynamics Observatory (SDO); this allows us to verify the feasibility of FMG data for helicity studies. In Section 2, we introduce the data processing and magnetic helicity flux estimation method. In Section 3, we present a comparison of magnetic field, velocity flows, magnetic helicity density maps and magnetic helicity accumulation computations between ASO-S/FMG and SDO/HMI data. A summary is presented in Section 4.

2. Data and Method

In this paper, we use the data from ASO-S/FMG and SDO/HMI for the comparison of magnetic helicity. The aperture of FMG is 14 cm. A CMOS camera with 4096 × 4096 pixels is used. FMG observes polarized Stokes images using a liquid crystal variable retarder (Hou et al., 2020). The routine observations of the FMG are taken at one wavelength position of the Fe I 5324.19 Å, and a linear calibration method is adopted. The longitudinal magnetic field sensitivity is 15 G (normal mode). The temporal resolution is 30 seconds (single component) and 2 minutes (vector magnetograms) under normal mode. The pixel size is about 0.5″, and the spatial resolution of FMG for the flight model (FM) is 1.04″ (Deng et al., 2019; Su et al., 2019; Gan et al., 2023). HMI is an instrument designed to study oscillations and the magnetic field at the solar photosphere. HMI is one of three instruments on the SDO. It observes the Sun nearly continuously and takes one terabyte of data per day. HMI observes the full solar disk at 6173 Å with a resolution of 1″ (Schou et al. 2012).

To calculate magnetic helicity fluxes across the photosphere S, we use the following formula:

$$\frac{dH_R}{dt} = -2\int (\vec{A}_p \cdot \vec{U}) B_n d\vec{S}, \tag{1}$$

where $\vec{U}$ denotes the horizontal velocity flows computed using local correlation tracking (LCT). The vector potential $\vec{A}_p$ is obtained using the fast Fourier transform (FFT) applied to the normal components of the photospheric magnetic field B_n (Chae 2001), which is estimated by multiplying the magnetic field line-of-sight component by $1/\cos\psi$, where ψ is the heliocentric angle of the region (Liu and Zhang 2006; Yang, Büchner, and Zhang 2009; Yang, Zhang, and Büchner 2009). According to the derivation of Démoulin and Berger (2003), Equation 1 includes the helicity flux from the emerging of magnetic flux and the motions of the magnetic field elements at the solar photosphere. Such approach has been widely used to calculate relative magnetic helicity flux across the solar photosphere to the corona (Green et al. 2002; Ninodos and Zhang 2002; Moon et al. 2002; Nindos, Zhang, and Zhang 2003). Currently, we cannot obtain the continuous vector magnetograms of solar active regions from the FMG data. The above equation indicates that we can evaluate the magnetic helicity accumulation in the solar corona evolution only based on a series of line-of-sight magnetic fields.

In this tudy, we choose the active region (AR) NOAA 13273 when it crossed the solar disk center on 10 April from 00:00 to 08:00 UTC. The longitude of this active region in the chosen time interval is less than E25, which reduced the projection effect. When processing the data, we have considered:

1) The resolution and spatial alignments. Since the optical diffraction limit of FMG and HMI are both close to 1″, we use the cubic convolution interpolation method to rescale the data to 1″ pixel resolution. Then, we crop the same region to 256 × 256.

2) The cadence and time alignments. The normal mode of ASO-S/FMG is to observe the polarization signal at the fixed off-band of Fe I 5324.19 Å. The Doppler shift caused by the satellite orbit changes the working wavelength position of Lyot narrow-band filter of FMG and takes the oscillation of measured polarization signals. This produces an uncertainty in solar magnetic field measurements. To avoid this problem, we use the orbit speed record in the header of FMG fits file and apply a cubic convolution interpolation method to obtain the processed FMG data in which the orbit Sun-satellite direction speed is zero. Then, we choose the HMI magnetic field data at the closet time and also apply the cubic convolution interpolation method to obtain the processed HMI data at the same time. Note that the new observing times of FMG and HMI data have been changed after the interpolation. The cadence after time alignment is 52 minutes.

3) The tracking parameter setting. Physically, significant transverse velocities of the photospheric elements are usually smaller than 1.5 $\mathrm{km\,s^{-1}}$ (Chae 2001). Hence the possible shift between two consequent magnetograms is less than 8 pixels in the estimated time cadence. For the Local Correlation Tracking (LCT) method, we choose 8″ for the FWHM (full width at half-maximum) of the apodizing function. To reduce the noise, we set the horizontal velocity to zero in regions where the magnetic field is small (< 10 G). To better track the emerging regions and to exclude the effect of relative quiet regions outside the emergence sites, we set the horizontal velocity to zero in regions of a weak cross-correlation (< 0.9) between two magnetograms.

After the above data processing, the final sizes of FMG and HMI data cubes is 256 × 256 × 10. The first two dimensions are the solar X and Y coordinate directions. The third dimension is for the new interpolation observing time. In the next section, we calculate the helicity flux and compare the results.

3. Results

Figure 1 depicts the evolution of the line-of-sight magnetic field of NOAA 13273 from ASO-S/FMG on 10 April 2023 from 01:00 UTC to 07:40 UTC. Figure 2 shows a similar evolution of NOAA 13273 observed by SDO/HMI. The distribution and evolution of the magnetic field is well consistent, and the average for the linear Pearson correlation coefficient is 0.93. Figures 3 and 4 show, respectively, the horizontal velocity using the LCT method applied to ASO-S/FMG and SDO/HMI. The two figures show consistent motions in general. We compute the separation between the leading positive polarity and the following negative polarity sunspots. A small emerging positive flux moved toward the leading sunspots and gradually canceled the negative polarity flux. Figures 5 and 6, respectively, depict the evolution of the magnetic helicity flux density map, which is the integration core of Equation 1 defined as G_A. The two density maps are also consistent in general. The majority of the sign of the magnetic helicity density maps is negative. This means that the accumulated magnetic helicity during the calculated time period is also negative. Note the mixture of positive and negative helicity density in the leading sunspots. This is due to the same trend of the horizontal velocity field and the reverse direction of the vector potential of the magnetic field around the same leading sunspot polarity, which creates some fake parity signals into the G_A. Figure 7 shows the evolution profile of the accumulated magnetic helicity of NOAA 12373 by using ASO-S/FMG (diamond) and SDO/HMI data (asterisk).

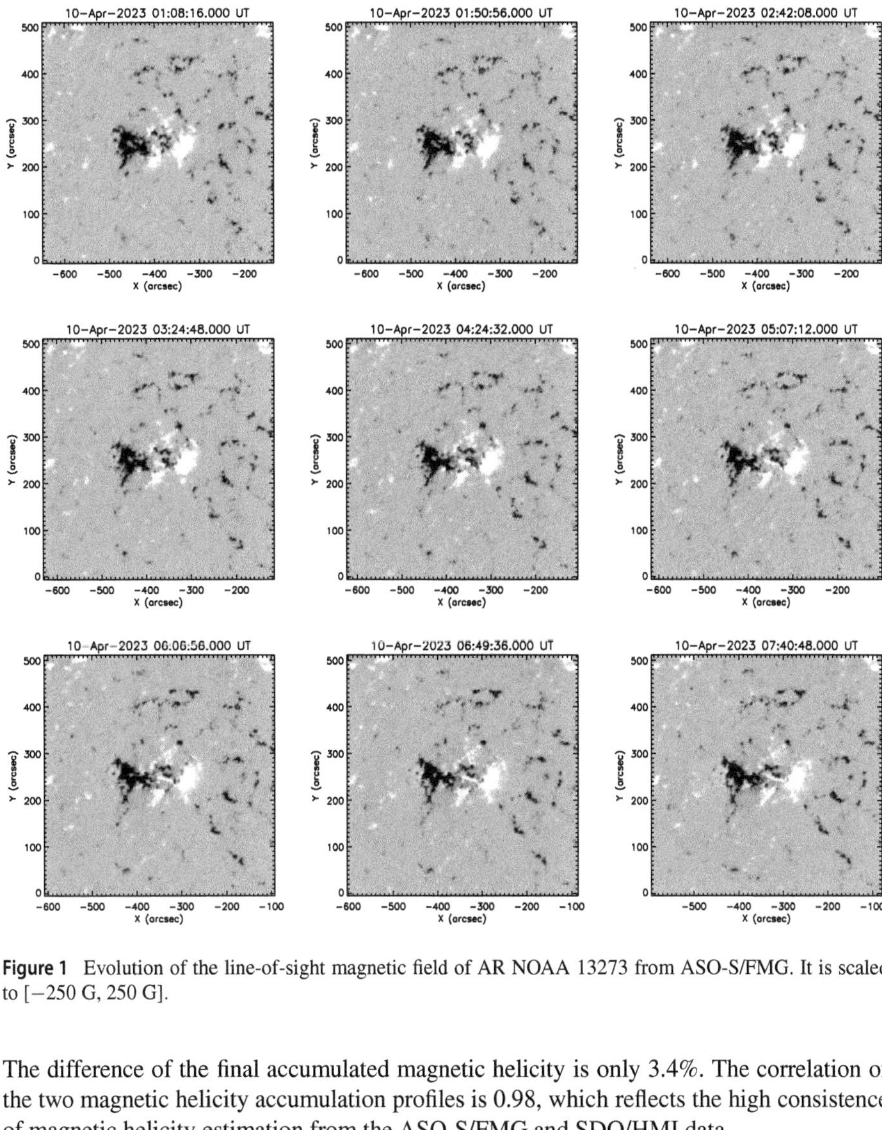

Figure 1 Evolution of the line-of-sight magnetic field of AR NOAA 13273 from ASO-S/FMG. It is scaled to [−250 G, 250 G].

The difference of the final accumulated magnetic helicity is only 3.4%. The correlation of the two magnetic helicity accumulation profiles is 0.98, which reflects the high consistence of magnetic helicity estimation from the ASO-S/FMG and SDO/HMI data.

4. Summary

Magnetic helicity plays an important role to understand the origin of solar eruptions, and even their long-term evolution could be used to constrain the solar dynamo process. Magnetic helicity can quantitatively be calculated using the continuous evolution of the magnetic field, which definitely helps to push forward the achievement of ASO-S scientific goals. After carefully choosing magnetic field measurements at the optimum moment, local-correlation-tracking parameters under appropriate resolution, and cadence with appropriate projection correction, we can obtain a highly trusted calculation of the magnetic helicity flux estimation by comparing with the results based on SDO/HMI data, which is widely used in

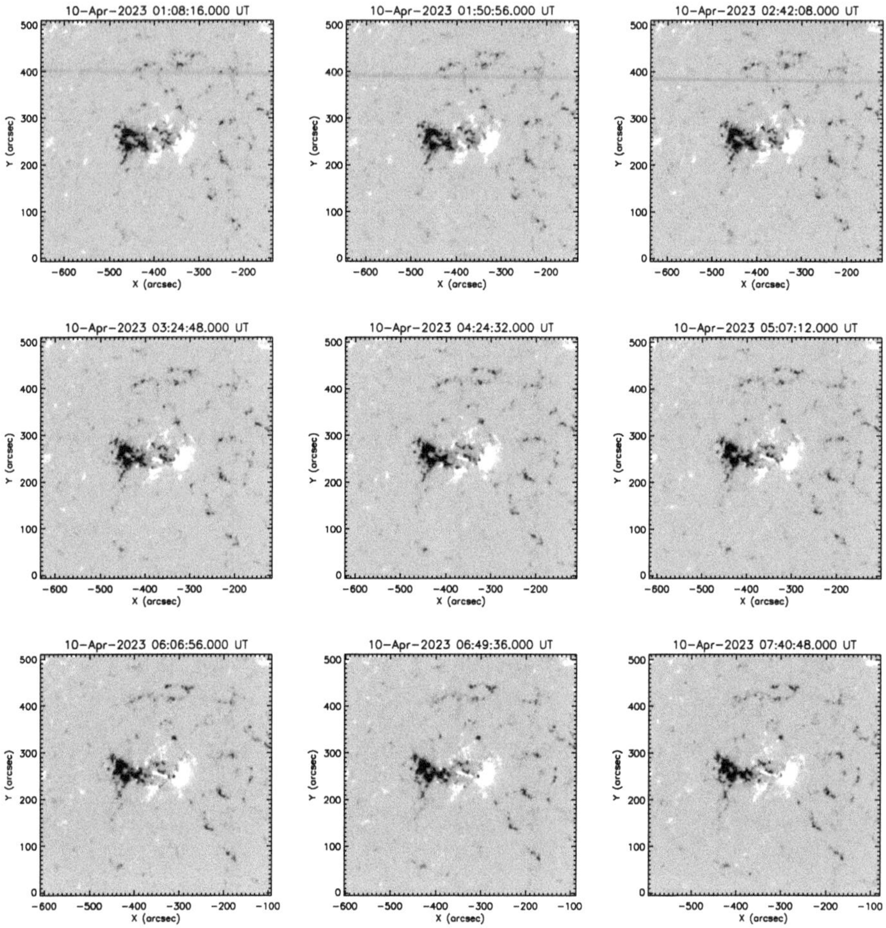

Figure 2 Evolution of the line-of-sight magnetic field of AR NOAA 13273 from SDO/HMI. It is scaled to [−250 G, 250 G].

previous magnetic helicity flux estimation research. The linear correlation between the profiles is 0.98, and the difference of magnetic helicity accumulation is less than 4% for the solar AR in this study. In the future, we will carry on a large statistically significant study of magnetic helicity evolution and analyze its relations with solar flares and CMEs during the solar maximum of Solar Cycle 25.

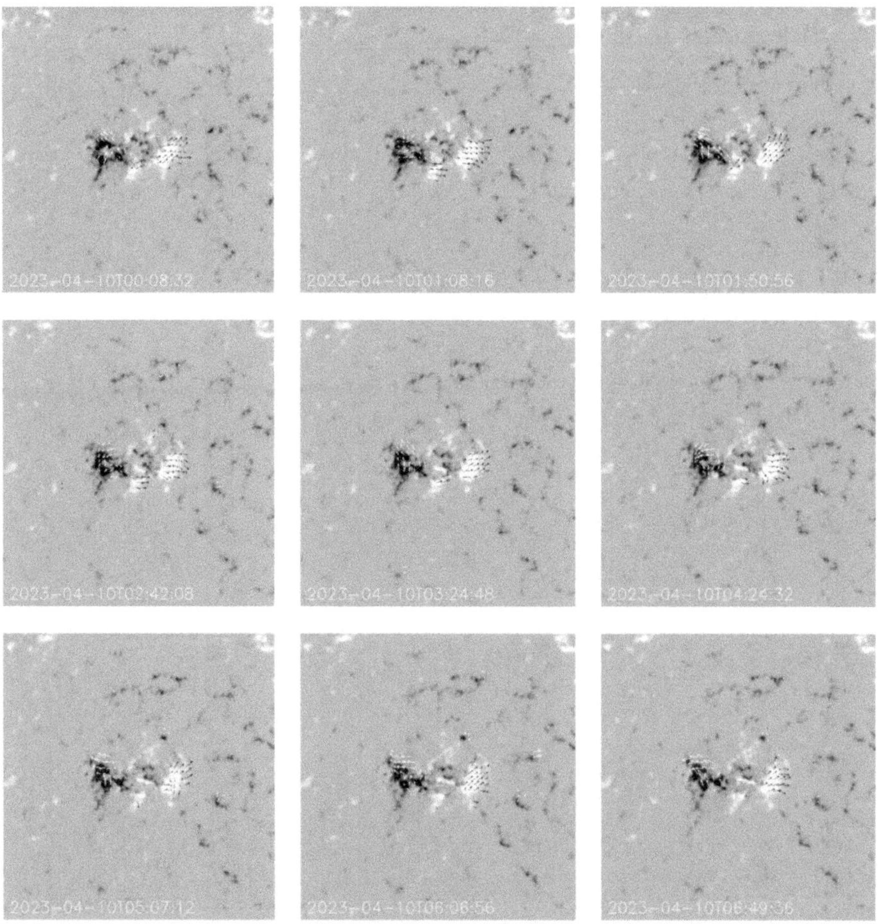

Figure 3 Evolution of the horizontal velocity of AR NOAA 13273 from ASO-S/FMG. The maximum arrow length measures a velocity of 0.8 km/s.

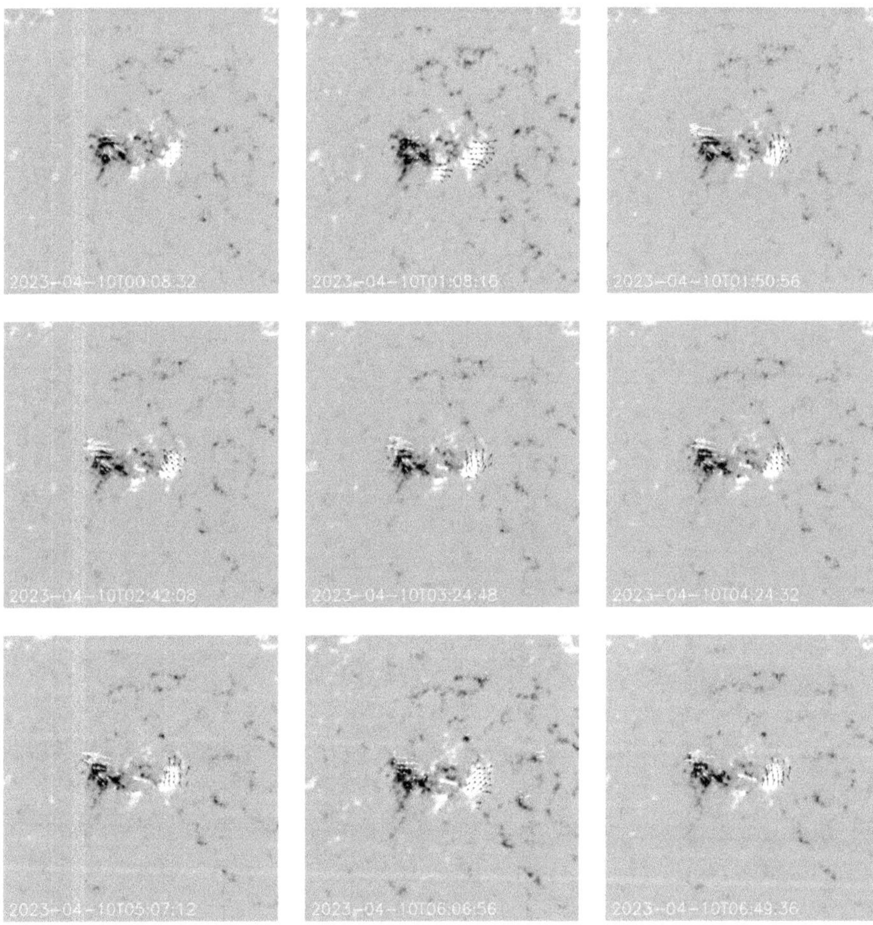

Figure 4 Evolution of the horizontal velocity of AR NOAA 13273 from SDO/HMI. The maximum arrow length measures a velocity of 0.8 km/s.

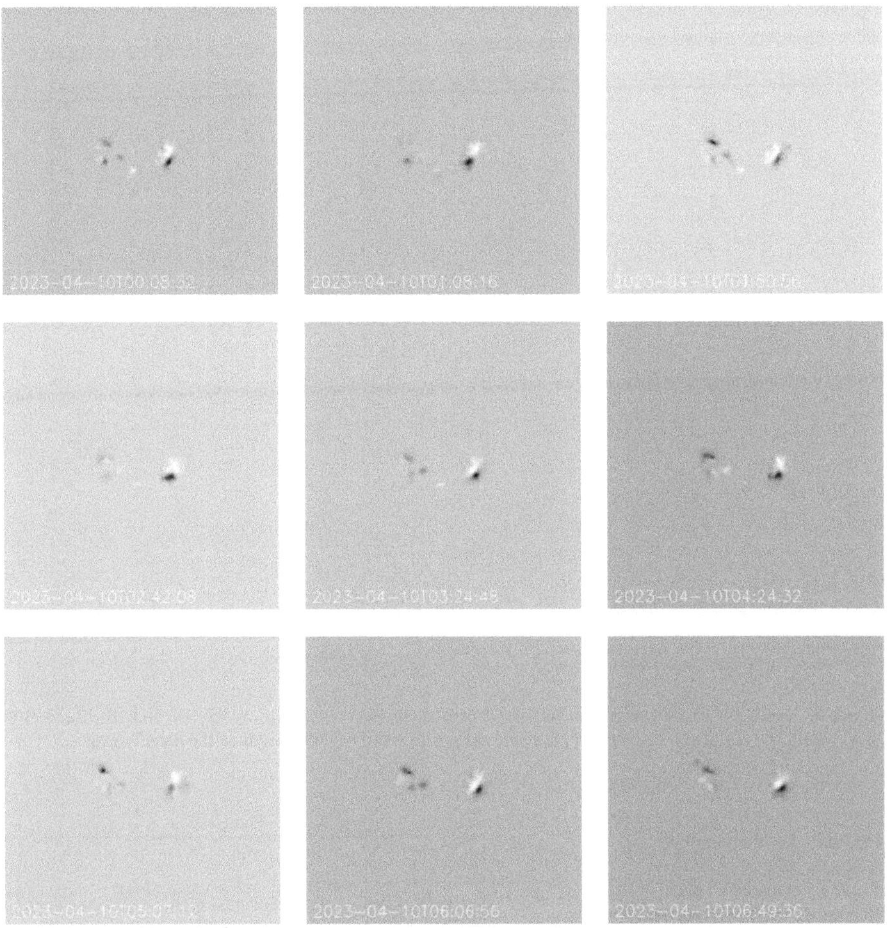

Figure 5 Evolution of the magnetic helicity density map $G_A = -2(A_p \cdot U)B_n$ of NOAA 13273 from ASO-S/FMG. The white and black colors indicate the positive and negative signs of the density map.

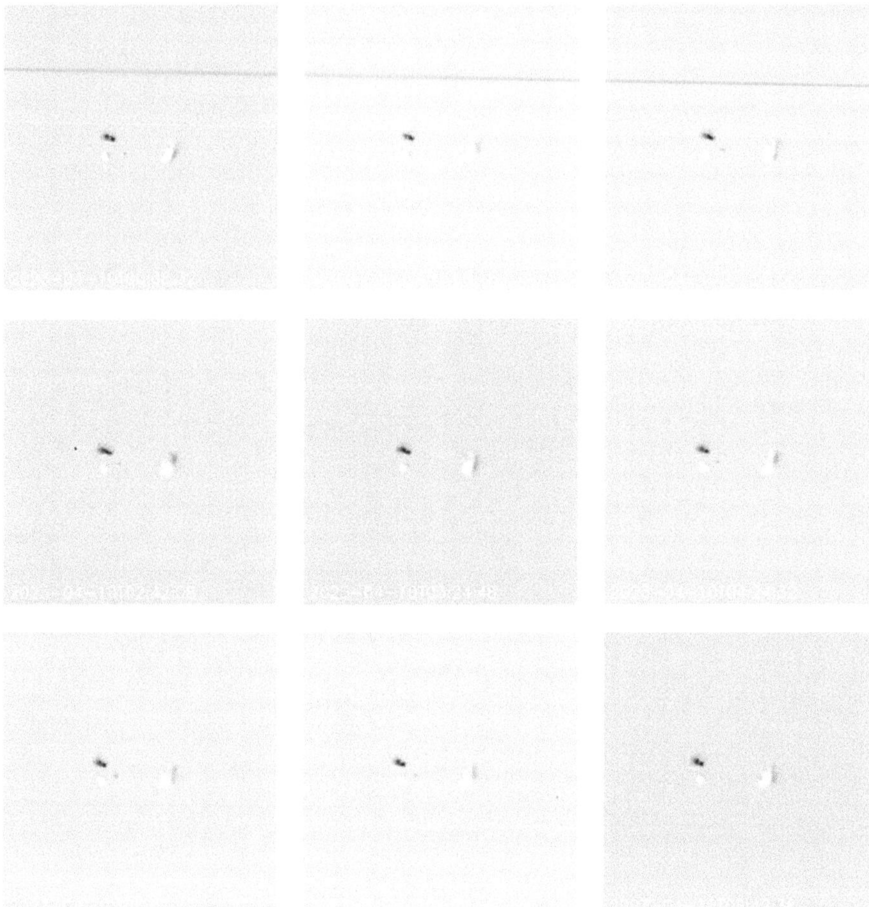

Figure 6 Evolution of the magnetic helicity density map $G_A = -2(A_p \cdot U)B_n$ of NOAA 13273 from SDO/HMI. The white and black colors indicate the positive and negative signs of the density map.

Figure 7 Accumulation of magnetic helicity of NOAA 13273 by using FMG data (diamond) and HMI (asterisk).

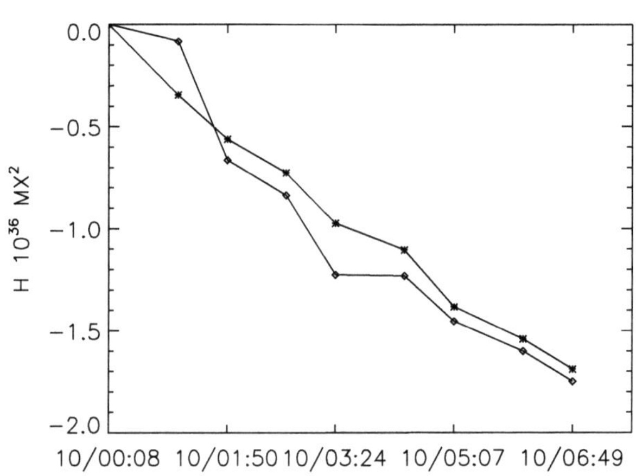

Acknowledgements We would like to thank the referee for carefully reading our manuscript and for giving constructive comments that substantially helped improving the paper. We would also like to thank the editor to significantly improve the language of the manuscript.

Author contributions Shangbin Yang prepared the main manuscript. Liu S. and Su J. T. supplied the data. Prof. Deng Yuanyuan presented the instruments issues.

Funding Information This research is supported by the National Key *R&D* Program of China No. 2022YFF0503800, 2021YFA1600500, 2022YFF0503001; National Natural Science Foundation of China (grants No. 12250005, 12073040, 12273059, 11973056, 12003051, 11573037, 12073041, 11427901, 11572005 and 11611530679); the Strategic Priority Research Program of the Chinese Academy of Sciences (grants No. XDB0560000, XDA15052200, XDB09040200, XDA15010700, XDB0560301, and XDA15320102); by the Chinese Meridian Project (CMP); and by Beijing Natural Science Foundation (grants No. IS23030, Z180007).

Data Availability No datasets were generated or analysed during the current study.

Declarations

Competing interests The authors declare no competing interests.

References

Bao, S.D., Zhang, H.Q.: 1998, *Astrophys. J.* **496**, L43.
Berger, M.A.: 1999, *Plasma Phys. Control. Fusion* **41**, 167.
Berger, M.A., Field, G.B.: 1984, *J. Fluid Mech.* **147**, 133.
Chae, J.: 2001, *Astrophys. J.* **560**, L95.
Démoulin, P., Berger, M.A.: 2003, *Solar Phys.* **215**, 203. DOI.
Démoulin, P., Pariat, E.: 2009, *Adv. Space Res.* **43**, 1013.
Deng, Y.Y., Zhang, H.Y., Yang, J.F., et al.: 2019, *Res. Astron. Astrophys.* **19**, 157.
Feng, L., Li, H., Chen, B., et al.: 2019, *Res. Astron. Astrophys.* **19** 16.
Gan, W.Q., Zhu, C., Deng, Y.Y., et al.: 2023, *Solar Phys.* **298**, 68.
Green, L.M., Lòpez fuentes, M.C., Mandrini, C.H., Démoulin, P., Van Driel-Gesztelyi, L., Culhane, J.L.: 2002, *Solar Phys.* **208**, 43.
Gupta, M., Thalmann, J.K., Veronig, A.M.: 2021, *Astron. Astrophys.* **653**, A69.
Kuzanyan, K., Yokoi, N., Georgoulis, M.K., Stepanov, R.: 2023, *Helicities in Geophysics, Astrophysics, and Beyond, Geophysical Monograph Series*. Wiley, New York, 1.
Lin, H., Kuhn, J.R., Coulter, R.: 2004, *Astrophys. J.* **613**, L177.
Liokati, E., Nindos, A., Georgoulis, M.K.: 2023, *Astron. Astrophys.* **672**, A38.
Liokati, E., Nindos, A., Liu, Y.: 2022, *Astron. Astrophys.* **662**, A6.
Liu, J., Zhang, H.: 2006, *Solar Phys.* **234**, 21.
Moon, Y.J., Chae, J., Wang, H., Choe, G.S., Park, Y.D.: 2002, **580**, 528.
Nindos, A., Zhang, J., Zhang, H.: 2003, *Astrophys. J.* **594**, 1033.
Ninodos, A., Andrews, M.D.: 2004, *Astrophys. J.* **616**, L175.
Ninodos, A., Zhang, H.: 2002, *Astrophys. J.* **573**, L133.
Park, S., Lee, J., Choe, G.S., Chae, J., Jeong, H., Guo, Y., Jing, J., Wang, H.: 2008, *Astrophys. J.* **686**, 1397.
Price, D.J., Pomoell, J., Lumme, E., Kilpua, E.K.J.: 2019, *Astron. Astrophys.* **628**, A114.
Priest, E.R., Forbes, T.G.: 2002, The magnetic nature of solar flares. *Astron. Astrophys. Rev.* **10**, 313.
Santos, J.C., Büchner, J., Otto, A.: 2011, *Astron. Astrophys.* **535**, A111.
Schou, J., Scherrer, P.H., Bush, R.I., et al.: 2012, *Solar Phys.* **275**, 229.
Schrijver, C.J.: 2006, *Solar Phys.* **235**, 161.
Skála, J., Baruffa, F., Büchner, J., RamppA, M.: 2015, *Astron. Astrophys.* **580**, A48.
Su, Y., Liu, W., Li, Y.P., et al.: 2019, *Res. Astron. Astrophys.* **19** 163.
Tziotziou, K., Georgoulis, M.K., Raouafi, N.E.: 2012, *Astrophys. J.* **759**, L4.
Yang, S., Büchner, J., Zhang, H.: 2009, *Astrophys. J.* **695**, L25.
Yang, S., Zhang, H., Büchner, J.: 2009, *Astron. Astrophys.* **502**, 333.
Yang, S., Büchner, J., Santos, J.C., Zhang, H.: 2013, *Solar Phys.* **283**, 369.
Yang, S., Büchner, J., Skala, J., Zhang, H.: 2018, *Astron. Astrophys.* **613**, A27.

Yang, S., Pipin, V.V., Sokoloff, D.D., Kuzanyan, K.M., Zhang, H.: 2020, *J. Plasma Phys.* **86**(3).
Zhang, M., Flyer, M.: 2008, *Astrophys. J.* **683**, 1160.
Zhang, M., Flyer, M., Low, B.: 2006, *Astrophys. J.* **644**, 575.
Zhang, Y., Tan, B.L., Yan, Y.H.: 2008, *Astrophys. J.* **682**, L133.
Zhang, H., Sakurai, T., Pevtsov, A., et al.: 2010, *Mon. Not. Roy. Astron. Soc.* **402**, L30.

Part 3: HXI Related Research

Solar Physics (2024) 299:158
https://doi.org/10.1007/s11207-024-02399-4

RESEARCH

A New Solar Hard X-ray Image Reconstruction Algorithm for ASO-S/HXI Based on Deep Learning

Yuehan Xia[1,2] · Yang Su[1] · Hui Liu[3] · Wenhui Yu[1,2] · Zhentong Li[1] · Wei Chen[1] · Yu Huang[1] · Weiqun Gan[1]

Received: 2 July 2023 / Accepted: 23 October 2024 / Published online: 15 November 2024

Abstract
Most solar hard X-ray (HXR) imagers in the past and current solar missions obtain X-ray images via Fourier transform imaging technology, which requires proper imaging algorithms to reconstruct images from spatially-modulated or temporally-modulated signals. A variety of algorithms have been developed during the last 50 years for the characteristics of respective instruments. In this work, we present a new imaging algorithm developed based on deep learning for the Hard X-ray Imager (HXI) onboard the Advanced Space-based Solar Observatory (ASO-S) and the preliminary test results of the algorithm with both simulated data and observations. We first created a training dataset by obtaining modulation data from simulated HXR images of single, double and loop-shaped sources, respectively, and the patterns of HXI sub-collimators. Then, we introduced machine-learning algorithm to develop a pattern-based deep learning network model: HXI_DLA, which can directly produce an image from modulation counts. After training the model with simple sources, we tested DLA for simple sources, extended sources, and double sources for imaging dynamic range. Finally, we compared CLEAN and DLA images reconstructed from HXI observations of three flares. Overall, these imaging tests revealed that the current HXI_DLA method produces comparable image result to those from the widely used imaging method CLEAN. In some cases, DLA images are even slightly better. Besides, HXI_DLA is super fast for imaging and parameter-free. Although this is only the first step towards a fully developed and practical DLA method, the tests have shown the potential of deep learning in the field of solar hard X-ray imaging.

Keywords X-ray bursts · Solar flares · Machine learning · Hard X-ray imaging

✉ Y. Su
yang.su@pmo.ac.cn

H. Liu
liuhui@ynao.ac.cn

1 Key Laboratory of Dark Matter and Space Astronomy, Purple Mountain Observatory, Chinese Academy of Sciences, 210023, Nanjing, China

2 School of Astronomy and Space Science, University of Science and Technology of China, Hefei 230026, People's Republic of China

3 Yunnan Observatories, Chinese Academy of Sciences, Kunming 650216, People's Republic of China

1. Introduction

Solar flares are a class of intense solar burst activity in which magnetic energy is released explosively through processes such as magnetic reconnection and energy conversion occurs in the forms of particle acceleration, plasma heating, etc. (see the review in Priest and Forbes 2002; Fletcher et al. 2011; Moore et al. 2011; Shibata and Magara 2011; Benz 2017). During this process, significant increase in hard X-ray radiation can be observed, which is a direct product of energetic electrons. Therefore, full-disk X-ray spectra can be used to derive the physical parameters of thermal plasma and accelerated particles, the amount of thermal and non-thermal energy. In particular, X-ray imaging observations have significantly advanced the studies and our knowledge of plasma heating, particle acceleration, reconnection, initial energy release site and spectral line diagnostics (see the review in Holman et al. 2011; Fletcher et al. 2011; Raymond et al. 2012). Furthermore, imaging spectra of an individual source contains important information on the acceleration, transportation, thermalization of energetic electrons (e.g., Krucker et al. 2008; Krucker and Battaglia 2014). Therefore, observation of solar hard X-ray sources is of great significance for studying flares and has been one of the focuses of solar observation since the very early stage of space age.

Since 1970s, a series of solar hard X-ray imagers have been launched. Most of these imagers are indirect imaging systems (e.g., Oda 1982), such as coded-aperture imaging (or coded-mask imaging) and Fourier-transform imaging, which is in nature a special type of coded-aperture system. According to the different modulation strategies, Fourier-transform imaging can be divided into two categories. One is space-modulation imaging (Oda 1965), such as HXT (Kosugi et al. 1991) onboard Yohkoh (Ogawara et al. 1991), the Spectrometer Telescope for Imaging X-rays (STIX, Krucker et al. 2020) on the ESA Solar Orbiter, and the Hard X-ray Imager (HXI, Zhang et al. 2019; Su et al. 2019, 2024) onboard the Advanced Space-based Solar Observatory (ASO-S, Gan, Feng, and Su 2022; Gan et al. 2023). The other is time-modulation imaging, such as RHESSI (the Ramaty High Energy Solar Spectroscopic Imager, Lin et al. 2002).

Along with these missions, a variety of imaging algorithms have been developed. For example, the most basic one is the back-projection (BP) algorithm (Mertz, Nakano, and Kilner 1986; Kilner and Nakano 1989; Hurford et al. 2002). There are also many widely used algorithms, such as: CLEAN (Högbom 1974), MEM_GE (Massa et al. 2020), Forward Fit (Aschwanden, Schmahl, and Team 2002), Pixon (Pina and Puetter 1993; Metcalf et al. 1996). These imaging algorithms have played an important role in reconstructing solar HXR images from modulated data.

However, they may have certain disadvantages and have to be carefully used for different scientific purposes (see the "Image Algorithm Summary" on the RHESSI website hesperia.gsfc.nasa.gov/rhessi3/software/imaging-software/image-algorithm-summary). Some require users to have good knowledge of both the corresponding instrument and parameters of the algorithms themselves (see also the tests in Dennis and Tolbert 2019).

For modulation-based indirect imaging, the HXR image is usually reconstructed from a limited number of modulated count data or visibilities. This is a typical underdetermined inverse problem, i.e., the inversion of high-dimensional manifold from projection of low-dimensional space, where machine learning can naturally contribute. In recent years, with the rapid development of computer technology, machine-learning methods have been applied in more and more aspects in the field of solar physics, such as, flare prediction based on deep neural networks and deep learning (Huang et al. 2018; Abed, Qahwaji, and Abed 2021), identification of preflare spectroscopic signatures (Woods, Sainz Dalda, and De Pontieu 2021), rapid recognition of solar images (Armstrong and Fletcher 2019), automatic de-

Table 1 Final grid configuration in the flight model of HXI. The position angles are the angles between grid slits and the X-axis of HXI FOV. The rows 3, 4, 5, 6 list the number of bi-grids for the corresponding phase of subcollimator.

Pitch/μm	36	52	76	108	156	224	344	524	800	1224
Thickness/mm	1.0	1.4	1.7	2.0	2.0	2.0	2.0	2.0	2.0	2.0
For phase/$\varphi = 0°$	4	5	5	5	5	5	5	5	3	3
For phase/$\varphi = 90°$	4	5	5	5	5	5	5	5	3	2
For phase/$\varphi = 120°$	0	0	0	0	0	0	0	0	0	1
For phase/$\varphi = 240°$	0	0	0	0	0	0	0	0	0	1
Spatial res/arcsec	3.1	4.5	6.5	9.3	13.4	19.3	29.6	45.1	68.8	105.2
Position angles/°		5	32	23	14	5	23	5		
	25	41	68	59	50	41	59	41	23	53
	70	77	104	95	86	77	95	77	83	113
	115	113	139	131	122	113	131	113	143	173
	160	149	176	167	158	149	167	149		

tection of occulted hard X-ray flares (Ishikawa et al. 2021), fine-grained solar flare forecasting based on the hybrid convolutional neural networks (Deng et al. 2021); solar radio-burst forecast based on a convolutional neural network (Ma et al. 2022); sparse representation of Hinode/SOT/SP spectra using convolutional neural networks (Flint and Milic 2021), etc.

In this article, we develop a neural network model based on CNN and MLP (Multi-Layer Perception) to reconstruct HXI images from input of modulation counts (or a set of visibilities). Here, the first version of the new Deep Learning Algorithm (HXI_DLA) is tested and compared with the widely used CLEAN method. The production of training dataset are described in Section 2. The development of the deep learning model, the model training and testing results are described in Section 3. The application of the method to the observational data of ASO-S/HXI is shown in Section 5. The summary and discussion are presented in Section 6.

2. Training Data Set

2.1. HXI Imaging Principle

ASO-S is China's first comprehensive solar space mission, which was successfully launched on 9 October 2022 (Gan et al. 2019, 2023). More details of the mission can be found at ASO-S website aso-s.pmo.ac.cn/en_index.jsp. The main scientific objective of ASO-S is to study solar magnetic fields, solar flares, coronal mass ejections (CMEs), and their relationships (Gan, Feng, and Su 2022). HXI is one of the three payloads onboard ASO-S designed to investigate non-thermal high-energy electrons accelerated in solar flares by providing HXR spectra and images in the energy range of $\sim 10-300$ keV (Su et al. 2024).

HXI is a spatial-modulation Fourier-transform imaging system that uses 91 sub-collimators and $LaBr_3$ detectors to obtain 91 modulated counts or 45 visibilities for imaging reconstruction with a spatial resolution as high as 3.1 arcsec. After performing a series of test and simulations (Su et al. 2019; Chen et al. 2020, 2021), HXI team determined the final Tungsten grid configuration for the flight model, summarized in Table 1.

2.2. Sample Simulation

First, we simulate a hypothetical hard X-ray source on the Sun to obtain an input image of size 204 × 204, a size that is large enough to cover a regular flare site, and small enough to boost training speed. The pixel size M is set to 1 arcsec. The intensity of the pixel $M(x, y)$ is physically meaningful: during the exposure time, $M(x, y)$ photons of corresponding energy hit the front grid (here we ignore the conversion between counts and photons, which is a coefficient depending on energy).

We consider three types of simulated sources: single source, double sources, and loop-shaped sources, all based on Gaussian distribution. These types of hard X-ray sources are relatively common and widely used in HXR analysis, (e.g. Dennis and Pernak 2009; Krucker and Battaglia 2014; Yu, Li, and Gan 2021; Krucker, Torre, and Schwartz 2021).

On the other hand, HXI patterns (Su et al. 2019; Jiang et al. 2023) are calculated (using hxi_get_patt.pro in HXI analysis software) for the same FOV and same pixel size, which are the 2D extension of the modulation function of each set of bi-grid along slit. Since HXI grids have not been fully calibrated, we used the design parameters in Table 1 but took into account the internal shadowing effect (due to the thickness of non-ideal grid, Su et al. 2019; Jiang et al. 2023). In a pattern P_d, d represents the detector ID (from 1 to 91). The pixel value (i.e., the bi-grid response) at (x, y) is $P_d(x, y)$ and the counts of detector d for simulated source M is C_d, then:

$$C_d = \sum_{x=1}^{204} \sum_{y=1}^{204} M(x, y) P_d(x, y). \tag{1}$$

Assuming an input image (M) with a certain type of source, we can get an array (C) of 91 counts from the above calculations. Here we also added statistical noise to the count data. The input image and the count array form a single, complete training sample.

In order to effectively train the model, a large number of training samples are required to cover the parameter space. This is done by randomly changing the parameters that describe the sources. For the case of a single source, five parameters are introduced, including the location of the source center (x, y), the width of the source sig along x and y and the peak value, which is realized through normalization of total flux. The distribution of intensities of the sources follows Gaussian distribution. The location (x, y) of the source randomly changes in the range: $2 < x, y \leq 199$. sig changes in the range $2 < sig \leq 10$ (unit is arcsec). Here $sig_x = sig_y$, and it turns into a circular source, where sig represents the radius of the source.

Each source in the case of double sources has the same number of parameters as those of single source. A selection formula is also introduced to ensure that the source is not too close to the boundary, which would result in an incomplete source.

We simulate the loop-shaped source (or elongated source) based on the simple quadratic function. After getting the expression of quadratic function, we use single sources to generate loop with a certain loop width. Finally, we use part of it as our loop. There are six parameters that need to be randomized. The apex coordinates of the loop (h, k), with an random range of: $15 < h, k \leq 185$. The opening size parameter a, with a range of $-0.08 < a \leq 0.08$. With the first three parameters, we can get a quadratic function.

$$y = a(x - h)^2 + k. \tag{2}$$

The fourth parameter is the size of each single source, which is randomly selected in the range of $2 < sig \leq 8$. The last two parameters are the cutting ranges, which have two parts

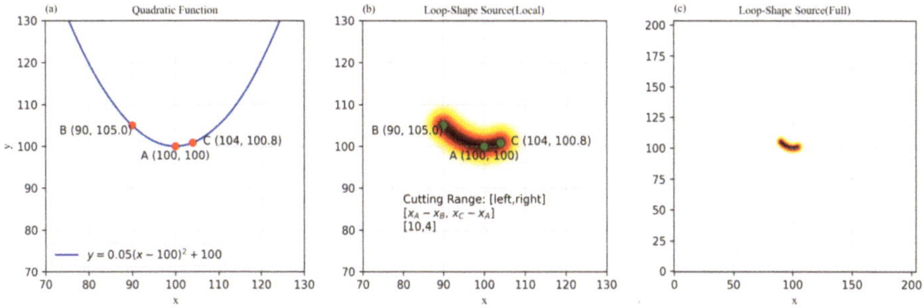

Figure 1 This is an example of how to get a loop-shaped source. (a) The quadratic function used for generating loop source. (b) The generated loop source and its parameters. (c) The loop source in the full image used as model input.

corresponding to the two legs of the loop source, left part and right part [left, right]. Their ranges were $5 < left \leq 15$ and $1 < right \leq 7$, respectively, which are selected to allow generation of asymmetric loops. Figure 1 shows how we get a loop-shaped source by parameters above.

Besides, we performed logarithmic processing on all the data before training. Because our simulation source is small relative to the field of view and most of pixel intensities are 0, so we added a fixed background of 0.001 to all the pixels before logarithmic processing on the whole image. The purpose of this process is to reduce the dynamic range of data, making the model converge faster. Finally, we introduce a parameter to limit the total flux within a range: $2.5 < lg(o) \leq 4.5$.

3. Method

3.1. Formulation of the Problem

The imaging principle of HXI is given by Equation 1. For the actual observation, we get the counted vector C of 91 detectors, and what the imaging algorithm needs to do is to reconstruct 2D image M from C. This is a typical underdetermined inverse problem, inversion of the high-dimensional manifold from the projection of the low-dimensional space. There are many studies using DNN (deep neural network) to solve this kind of inverse problem (e.g. Jin et al. 2017; Kelly, Matthews, and Anastasio 2017; Antholzer, Haltmeier, and Schwab 2019). This also inspired our idea of using DNN models to solve the hard X-ray image reconstruction problem.

3.2. Net Structure

The problem we address is how to get a 204×204 image M from 91 count values C. The imaging principle of HXI suggests that the intensity of each pixel of the target image contributes to the count array. Therefore, we constructed a hybrid deep neural network as an inversion model. The model first uses a fully connected MLP (Multilayer Perceptron) structure to build the inversion of the count value to the small-size image, and then interpolates the small-size image to the target image by CNN.

Specifically, the basic structure we take is MLP+Transition1 (transition block 1, denoted as Tran 1.0) + ResNet (He et al. 2016) + Transition2 (Transition Block 2, denoted Tran 2.0)

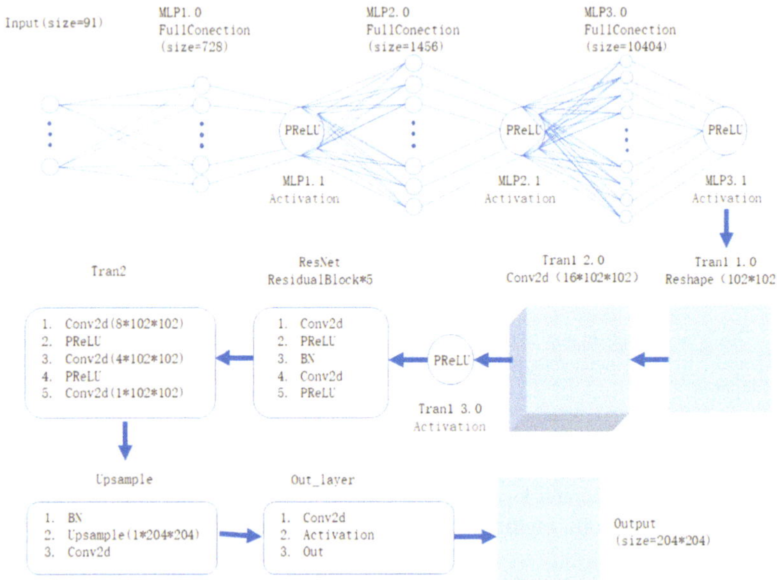

Figure 2 This is a schematic diagram of our network structure, from top left to bottom right. The input is 91 detector count array, and the output is a 204×204 image. The first row is the MPL part; the second row includes parts of Tran1.0, ResNet, and Tran2.0; the third row is the upsampling and output sections. The important parameters of each layer are also shown.

+ Up-sample. Tran1 is a transitional structure tasked with reshaping the output of MLP into a two-dimensional image. ResNet can increase the depth of the model to improve the fitting ability, while also speeding up convergence. The role of Tran2 is to merge hidden channels opened by Tran1. Up-sample layers are only enabled if the MLP output is not 204×204. Then the next problem is how to choose hyperparameters. The main hyperparameters to be selected are: the size of the convolution kernel, the output of MLP, the number of residual blocks, and the number of hidden channels.

For example, we train the model with different convolution kernel size using the same sample set and keep other parameters unchanged. There are two aspects to determine which convolution kernel size is better. One is obviously accuracy, which means we will select hyperparameters with the lower loss of the two types of losses, the training loss, which shows the fitting ability of model and the valid loss, which shows the generalization ability of model. The other aspect is at what epochs the loss converges (almost stable). Therefore, we select the model with a faster convergence speed by looking at the loss curve.

After training on 500 epochs with the same 92 samples, we determined the following parameters: convolution kernel size: $ks = 5$, linear layer output: 102×102, the number of residual blocks: $block = 5$, hidden channels: $channel = 16$. Figure 2 shows the model structure we finally adopted.

We acknowledge pytorch (Paszke et al. 2019) to achieve the net and CUDA (Chetlur et al. 2014) to accelerate the computational speed. For data analysis and processing, we used Astropy (Price-Whelan et al. 2018) and Numpy (van der Walt, Colbert, and Varoquaux 2011). All plots (except Figure 2) were done using Matplotlib (Hunter 2007).

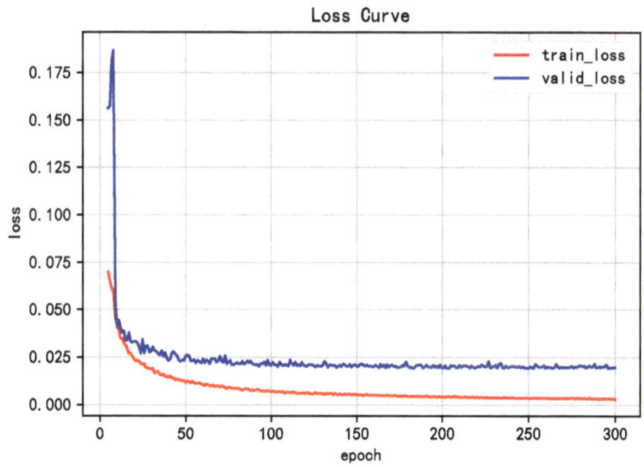

Figure 3 L2 loss curve with 300 epochs. The red curve is the loss of training set, and blue is the loss of valid set. The calculation of loss is introduced in Equation 3.

3.3. Training Process

The number of samples can greatly affect the model's ability. We conducted a simple evaluation of the sample size required for the model. The results show that as the sample size increases, the ability of the model is also constantly improving. Until the sample size reaches 1 million, this trend remains unchanged. Relevant literature also points out that: "Such supervised training needs a large number of data samples" (Kamyab et al. 2021). Finally, considering the time cost and our imaging quality requirements, we use 1,028,459 samples to train the model.

Besides, the sample size of the single, double and loop-shaped cases are kept same in the selected training samples to ensure the generalization performance of the model. We used the Adam optimizer (Kingma and Ba 2014) in our training with an initial learning rate of 0.0001, and the well-known L_2 loss function defined as the mean square error (MSE) between the output and the label, which in our situation is:

$$Loss = \sum_{x=1}^{n}\sum_{y=1}^{n}(M(x,y) - M'(x,y))^2/n^2, \tag{3}$$

where $n = 204$, M' is output of the model. We stop the training at the epoch of 300, when the loss of the training set stabilized at 0.003 (Figure 3 shows the loss curve). The valid set loss is stable at 0.02. It took almost two and a half months due to large amount of samples.

4. Test of DLA with Simulated Sources

We use the test set with a sample size of 10,310. The test results are mainly presented by four indicators (Figure 4): the loss L2 (Equation 3), the linear fitting slope K, the correlation coefficient (cc) and the quality index for X-ray image reconstruction (QuIX). Here smaller L2 means better result. It has an average of 0.028, which indicates that our training has achieved relatively good generalization ability. K is fitted by scatters, whose abscissa is taken from the intensity of the label M, and ordinate is the intensity of the corresponding position of the output image M'. The closer K is to 1, the better the result is. 'cc' can evaluate the linear correlation between the output and the label. QuIX is a new parameter dedicated

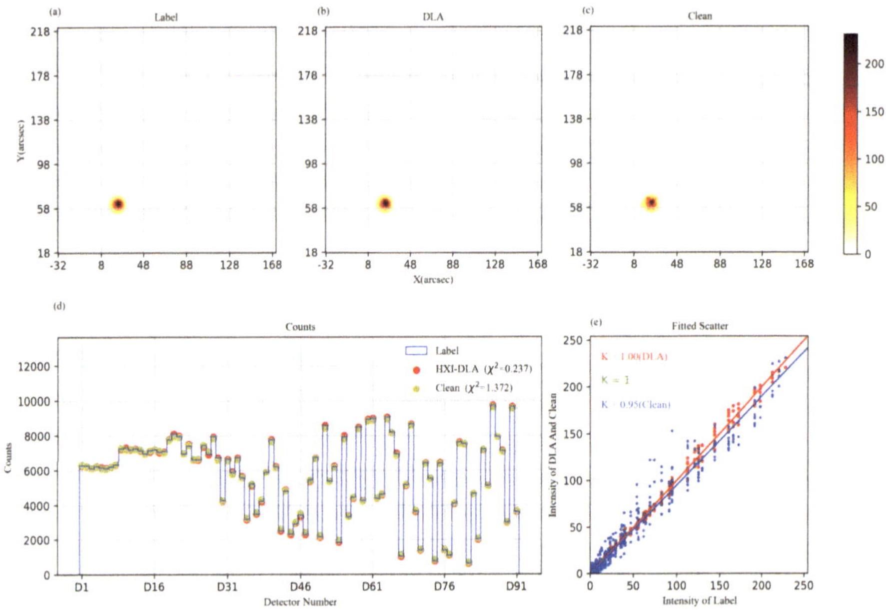

Figure 5 This figure shows the test result in a single-source case with K = 1. (a) The label. (b) The output (prediction) of the HXI_DLA. (c) The output of CLEAN. (d) The counts corresponding to the output (red for DLA, yellow for CLEAN) and the label (blue), where the count of the output is calculated by Equation 1. (e) A scatter plot of intensity at corresponding position of the above three images. The green line is y = x, which is covered by the red line for DLA with the same slope in the current situation. The blue line is for CLEAN result. The standard deviation of K is 0.0003 (DLA) and 0.0004 (CLEAN).

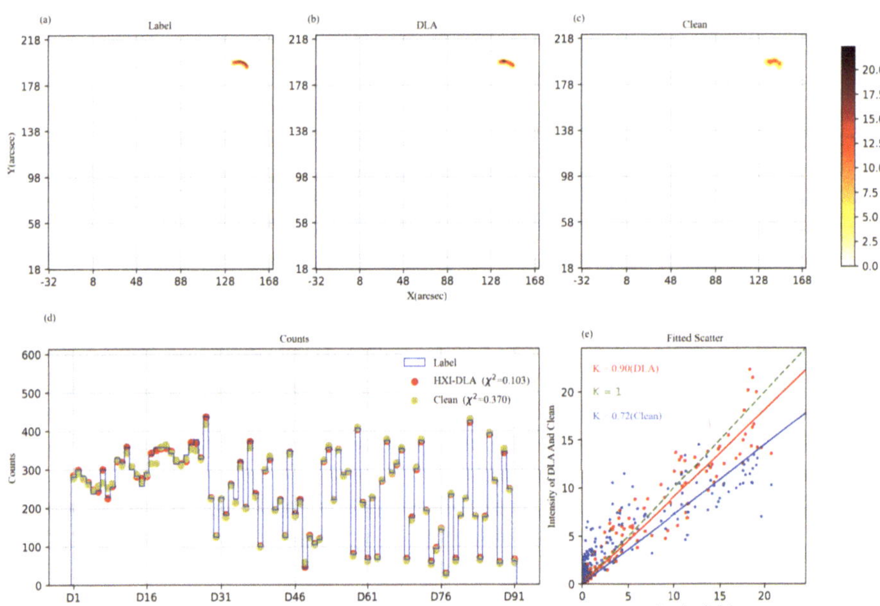

Figure 6 Same as Figure 5, but for a loop-shaped case with K = 0.9. The standard deviation of K in (e) is 0.0010 (DLA) and 0.0014 (CLEAN).

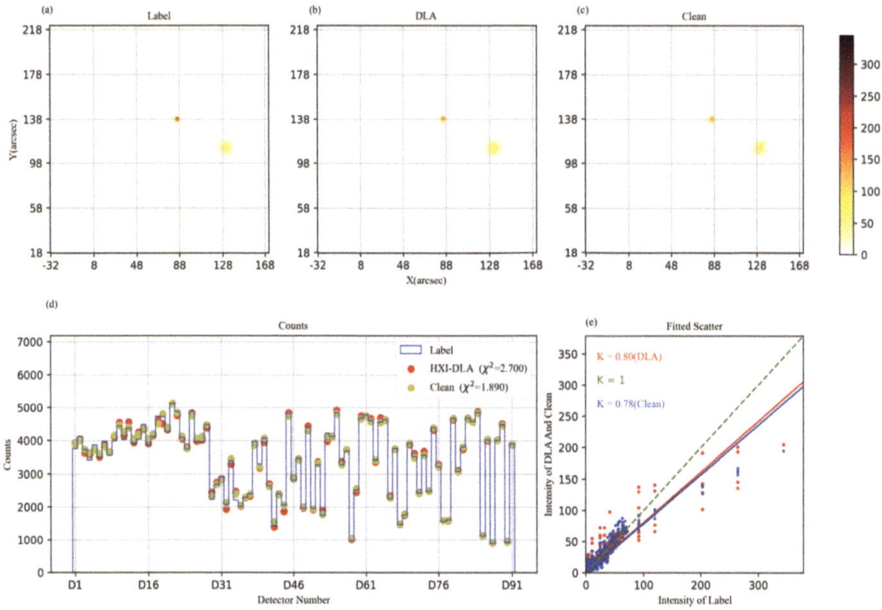

Figure 7 Same as Figure 5, but for a double-source case with K = 0.8. The standard deviation of K(DLA) and K(CLEAN) in (e) is 0.0014 for both.

are recorded as "unqualified", with a number of 753, accounting for only 7.3% of the total samples in the test set.

Overall, the testing result shows that the model produces good results for cases with simple source structures.

4.2. Dynamic Range Test

To evaluate the response of the model to different dynamic ranges of double-source, we designed the following simulations. We first generated ten double-source samples, whose only difference is the relative strength of the two single sources. The peak intensities (and fluxes) of the bright and weak sources ranges from 1:1 to 10:1. The ratio is also called as contrast factor in the following text. The rest parameters, such as positions, sizes, remain consistent.

The test result can be seen in Figure 8. From (a) to (d), we show the reconstruction image when the contrast factor is 1, especially for the weak source part. Next, we present the difference image of the weak source region when the contrast factor is 1, 4, 7, and 10 (second row in Figure 8). The difference image is calculated by:

$$diff = (prediction - label)/label \tag{5}$$

It can be seen that as the contrast factor increases, the difference becomes more significant, which also means that the reconstruction is getting worse. In Figure 8(i), we can clearly see from the graph that the fitting slope remains around 1 for contrast factors less than 5. After that, this indicator rapidly drops to a very poor level, and this change can also be clearly seen from the increasing chi-square. Note that these two indicators are both calcu-

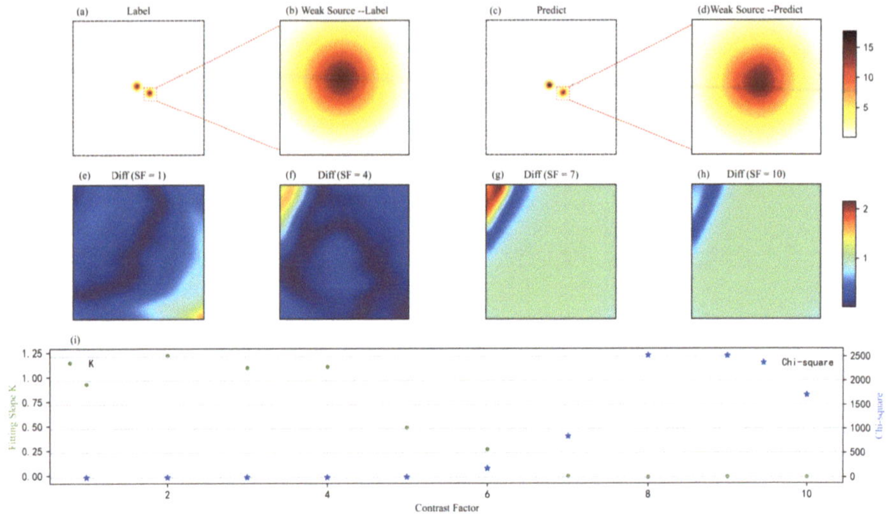

Figure 8 This figure highlights the imaging effects of the DLA model on dual sources with different dynamic ranges. (a), (c) The label (input) and prediction (output) when the double sources have the same intensity. (b), (d) The label (input) and prediction (output) of the weak source. (e) – (h) The difference images of the weak source region for the cases where the contrast factor between bright and weak sources is 1, 4, 7, 10, respectively. (i) Fitting slope K and Chi-square varies with the contrast factor from 1 to 10. These two indicators are both calculated based on the weak source area (as shown in (b) and (d)).

lated based on the weak source area (as shown in Figure 8(b) and (d)), to avoid the influence of the bright source.

4.3. Test on Extended Sources

To test the reconstruction of the DLA model on extended sources, we simulated two complex sources, which are combinations of compact and loop-shaped sources. The test results are displayed in Figure 9. DLA can reconstruct the overall shape of the source, but the details are not accurately reconstructed compared to CLEAN results, especially at both ends of the loop. CLEAN results are better here, but not as good as those in the cases of simple sources.

It is known that modulation-based indirect imaging of complex sources with very limited modulation data is difficult. But two more facts should be pointed out. First, the DLA model was not trained with samples of complex sources. Therefore, the DLA result can be improved with larger training set in the future. Second, there are a few free parameters (such as, different combination of grids, CLEAN beam width, etc.) in the CLEAN method. CLEAN images may also be improved through adjusting these parameters.

5. Test of DLA with ASO-S/HXI Observation

We further tested the HXI_DLA method by conducting imaging analysis for three flare events observed by ASO-S/HXI in 2023. Two of the three cases are compact flares with a single source, and the last one has double footpoint sources. Due to incomplete imaging calibration in the early stage of the ASO-S mission and the impact of jitter (or pointing drift) effects of the satellite platform, we decided to temporarily remove the finest sub-collimators

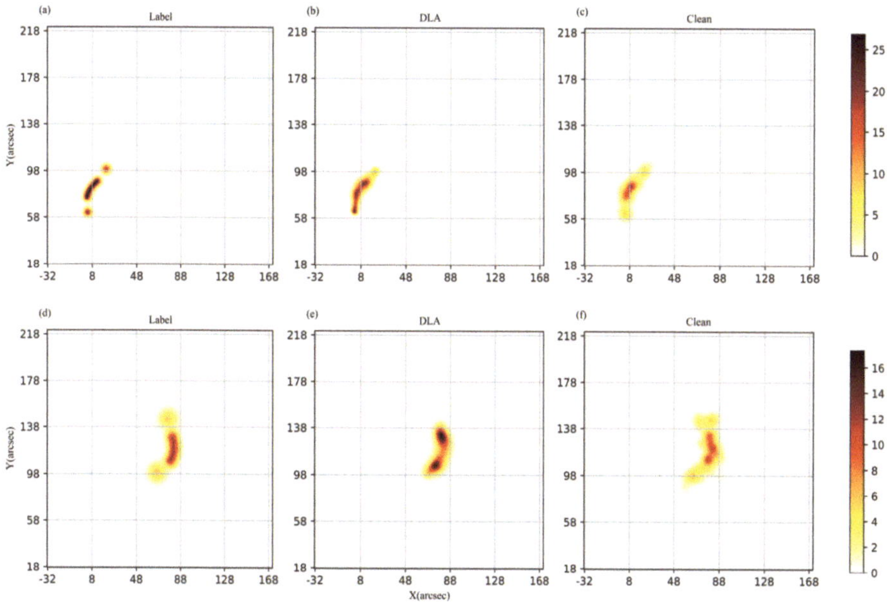

Figure 9 Test results of extended sources. The three images in each row are: label (a), DLA prediction (b) and CLEAN result (c). Here, CLEAN results are reconstructed without the first two groups of the fine grids.

with grid pitches of 36 μm or both 36 μm and 52 μm for the two compact flares. Therefore, the counts of up to 18 finest bi-grids are not used for the training of HXI_DLA model and the image reconstruction.

The image size is adjusted to 160×160 to reduce the training time. For comparison, we reconstructed CLEAN images in the field of view of HXI using HXI_CLEAN, provided in the HXI analysis software (HXI GUI v1.21beta). Imaging calibration and pointing corrections are not applied for both methods. The time ranges for imaging in each flare are selected to keep the maximum pointing drifts smaller than the finest resolution.

The first flare occurred on 7 Feb. 2023, with a GOES class of M6.3 and a peak time at 23:07 UT. The energy range for imaging is 20 – 25 keV, and the integration time is about 20 seconds from 23:06:40 – 23:07:00 UT. We compare the imaging results given by HXI_CLEAN and HXI_DLA in Figure 10. Both methods reconstructed a single, compact source at the same location. However, the HXI_DLA source is significantly smaller than that in the CLEAN image. The reconstructed modulation curves (yellow and green) suggest that HXI_DLA provides better result.

The second flare occurred on 9 May 2023, with GOES class of M6.5 and a peak time of 03:54:00 UT. The images (Figure 11) were made for the energy range of 30 – 40 keV with integration time of 9 seconds (03:51:40 – 03:51:49). Both the HXI_DLA image and the CLEAN image show clearly consistent double sources. The modulation curve of HXI_DLA is very similar to HXI_CLEAN in this case.

The third flare occurred on 23 May 2023, with GOES class of M3.1 and a peak time of 12:13:00 UT (GOES peak). The images (Figure 12) were made for the energy range of 15 – 25 keV with integration time of 14 seconds (12:12:20 – 12:12:34 UT). This is also a compact flare and the results are similar to the first one. The source location determined by the two methods are the same, but the source sizes are different. Like in the first case,

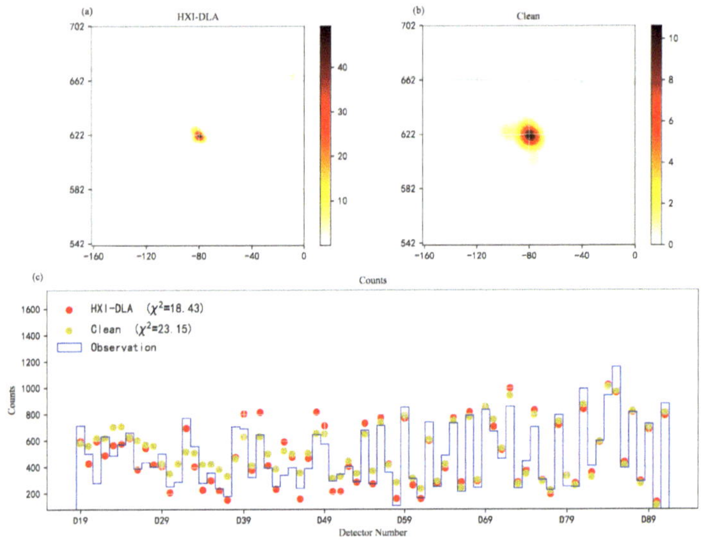

Figure 10 The comparison of the images made by HXI_DLA and HXI_CLEAN for the flare on 7 Feb. 2023. (a) the image obtained from the HXI_DLA algorithm. The coordinates in the figure are in arcsec relative to the center of HXI field of view. (b) the image obtained from the HXI_CLEAN algorithm. (c) The blue line is the modulated count data observed by HXI, the red scatters are the counts of detectors derived from the output HXI_DLA image, and the yellow is from the HXI_CLEAN image.

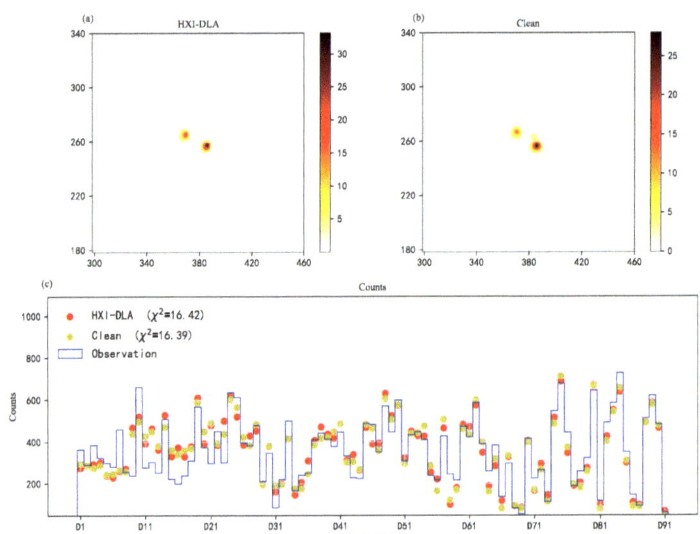

Figure 11 Same as Figure 10, but for the flare occurred on 9 May 2023.

the source size in HXI_DLA image is significantly smaller. The modulation curve of the reconstructed HXI_DLA image is also better than that of the HXI_CLEAN image.

Table 3 is the summary of the results for the three flares.

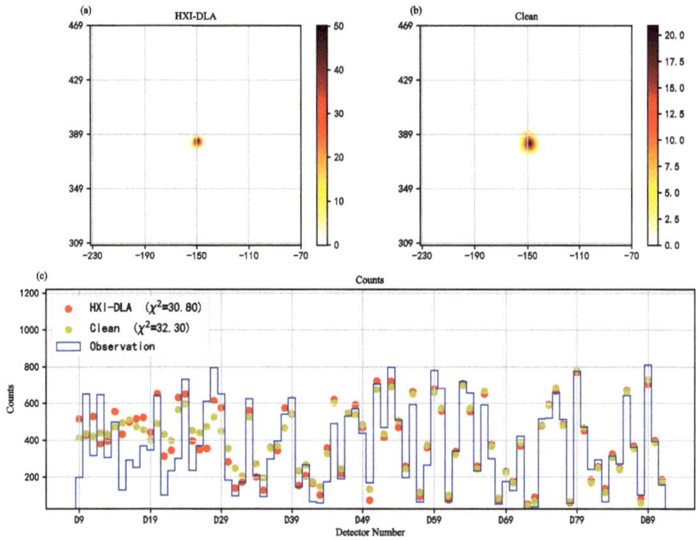

Figure 12 Same as Figure 10, but for the flare occurred on 23 May 2023. In this case, both images show compact HXR sources.

Table 3 The details of the three flares.

Date	7 Feb. 2023	9 May 2023	23 May 2023
Peak Time (HXI)	23:06	03:51	12:12
Peak Time (GOES)	23:07	03:54	12:13
GOES Class	M6.3	M6.5	M3.1
Integration Time (s)	20	9	14
Energy (keV)	20 – 25	30 – 40	15 – 25
$\mathcal{X}^2$ of DLA	18.43	16.42	30.8
$\mathcal{X}^2$ of CLEAN	23.15	16.39	32.3
SSIM of DLA	0.87	0.82	0.84
SSIM of CLEAN	0.79	0.82	0.76

6. Summary and Discussion

In this paper, we present a new imaging algorithm HXI_DLA, developed using deep learning for the HXI on board the ASO-S. Since the training for full-disk imaging requires extremely large volume of training samples, we only tested the method for a region that covers an usual flare site as the first step. The training and testing of the model is based on pattern and simulation data. We used three types of sources to generate simulated images, i.e., single sources, double sources, and loop-shaped sources. The model achieves the expected result under 1,028,459 training samples with 300 epochs. The test results revealed acceptable imaging quality that is comparable to CLEAN images or slightly better than CLEAN in some cases.

We also tested HXI_DLA for dynamic range and extended sources. The test of imaging dynamic range with a simple double-source case shows a value of 5:1 which is more or less expected for such kind of indirect imaging with very limited visibilities or subcollimators.

For example, the typical dynamic range of RHESSI is also ~ 5 (Krucker et al. 2014). The exact dynamic range of an imager is hard to determine as it also depends on counting statistics and source complexity (Krucker et al. 2014). We also tested DLA imaging for extended sources, but found DLA returns worse image quality than CLEAN. In fact, the image quality of both methods is not as good as those in the cases with simple sources. One possible reason for DLA is that the current model is not trained for complex sources.

We also notice that one of the important factors restricting the accuracy of the model is the sample size of training set. For sample size of nearly one million, it takes 4.5 h per epoch trained, which means 300 epochs would take more than two months (GPU: Tesla A100 40G; CPU: Intel(R) Xeon(R) Gold 6246R CPU @ 3.40 GHz *2). Therefore, the training time is also an important factor that we have to consider. It is foreseeable that as the sample size continues to increase, the performance of the model will most likely be further improved.

We also applied this method to HXI flare observations (M6.3 flare on 7 February 2023, M1.5 flare on 23 February 2023 and M6.5 flare on 9 May 2023), and obtained comparable images with those from the 'traditional' imaging method (CLEAN). According to the reconstructed modulation curves, the HXI_DLA images are more consistent with observation than the HXI_CLEAN images.

An interesting fact is that the size of the HXI_DLA source, which produces better modulation curve, is significantly smaller than that of the CLEAN source. In the current work, the beam size of HXI_CLEAN (an important parameter) by default is the spatial resolution of the finest sub-collimators selected for imaging, which is 6.5 arcsec of the sub-collimators (D19-D91) used for the two compact flares. This fact suggests that, (1) HXI_DLA may produce "over-resolved" but still reasonable sources, probably due to the fact that the training was done by a wide range of source sizes, including sub-resolution cases. This is an advantage of HXI_DLA method, which does not require pre-defined, fixed beam width as CLEAN does. (2) the default setting of beam width for HXI_CLEAN, i.e., the finest resolution of the sub-collimators selected for imaging, is larger than it should be. How to determine the beam width for CLEAN has been an open question. Now HXI_DLA could help us to answer this question.

In this work, the sub-collimators are not fully calibrated (meaning that design parameters are used for the phases and slit width). Only the effect of grid thickness is considered in the calculation of patterns. These un-calibrated grids does not affect the comparison of the two methods, since they are based on the same pattern.

In order to make HXI_DLA a practical method, some other aspects of this work need to be optimized:

i) Currently, the model is developed and trained for small regions. Full-disk imaging with HXI_DLA is only possible for low-resolution images. While it is also true for some transitional algorithms, HXI_DLA imaging requires training of the models for a large number of regions and energy ranges, and saving the results in advance. There are two reasons. First, the solution space for a full-disk image is too large to fit (the GPU is also more likely to be out of memory). The other one is that it will consume more computing resources and time.

   This leads to the fact that in application, we need to determine the approximate location of the source according to other information, and then train the model in this specific area, so that the obtained model has scientific imaging capabilities. This is doable as the imaging with other algorithms follows the same steps. Another solution is that we can train the model for different regions (and energies) separately and store the results in the database of the HXI software for fast applications.

ii) Satellite jitter problem or pointing shifts. The pointing of ASO-S platform is within a distance of 36 arcsec around the solar disk center, which means the center of the view can not be stably pointed at the disk center. Although the drifting speed is rather slow (usually <0.25 arcsec/s), it may still affect HXI imaging for a relatively long duration time and result in an elongated or curved source from a single compact source. Including a pointing correction in HXI_DLA would make the training process a more complicated task and a case-by-case work. However, it may be easier to apply the pointing correction for the visibilities-based DLA method.

iii) Since the data we use in the training and testing is pattern-based, and photons of different energy have different penetration capabilities, which result in different patterns. Our current models can only be used for imaging at one energy or a narrow energy range at a time, which is also a problem that limits the application of models. A solution to that is to train the models for different energy ranges in advance and save the results in the HXI database.

iv) At what level can we use the results for scientific research? This is a general, open question for a deep learning method used in science. For HXR imaging, the modulation curve consistent with observation is a good indicator, but may not be enough to evaluate the reconstruction quality of the result, as multiple solutions to the inversion problem may co-exist. Other self-testing indicator of image quality may be required to make DLA a stand-alone algorithm.

Despite these difficulties, the deep learning method have shown its potential in solar hard X-ray imaging. Based on the simulation tests of simulated sources and imaging test with observational data, HXI_DLA is super fast for imaging once properly trained and parameter-free, which requires no additional parameter setting from users. Most importantly, the current test version of HXI_DLA already provides comparable image quality to that of CLEAN. With calibrated parameters of the sub-collimators and improved deep learning model in the next steps, we expect that HXI_DLA will be more robust and produce better HXR images.

Author contributions Y.X. and H.L. developed the deep learning model and trained the model. Y.X. performed all the testing and made the figures. Y.S. proposed the study, calculated patterns, provided HXI data and CLEAN images. W.Y. and Z.L. helped with imaging and testing. Y.X., Y.S., and H.L. wrote the main manuscript text. All authors reviewed and helped improve the manuscript.

Funding This work was supported by the National Key R&D Program of China 2022YFF0503002, the National Natural Science Foundation of China (NSFC 12333010, 11873027, U2031140), and the Strategic Priority Research Program of the Chinese Academy of Sciences (Grant No. XDB0560000). ASO-S mission is supported by the Strategic Priority Research Program on Space Science, the Chinese Academy of Sciences, Grant No. XDA15320000.

Declarations

Competing interests The authors declare no competing interests.

References

Abed, A.K., Qahwaji, R., Abed, A.: 2021, The automated prediction of solar flares from SDO images using deep learning. *Adv. Space Res.* **67**, 2544. DOI. ADS.

Antholzer, S., Haltmeier, M., Schwab, J.: 2019, Deep learning for photoacoustic tomography from sparse data. *Inverse Probl. Sci. Eng.* **27**, 987. DOI.

Armstrong, J.A., Fletcher, L.: 2019, Fast solar image classification using deep learning and its importance for automation in solar physics. *Solar Phys.* **294**, 80. DOI. ADS.

Paszke, A., Gross, S., Massa, F., Lerer, A., Bradbury, J., Chanan, G., Killeen, T., Lin, Z., Gimelshein, N., Antiga, L., Desmaison, A., Kopf, A., Yang, E., DeVito, Z., Raison, M., Tejani, A., Chilamkurthy, S., Steiner, B., Fang, L., Bai, J., Chintala, S.: 2019, PyTorch: an imperative style, high-performance deep learning library. In: Wallach, H., Larochelle, H., Beygelzimer, A., d'Alche-Buc, F., Fox, E., Garnett, R. (eds.) *Advances in Neural Information Processing Systems* **32**. *33rd Conference on Neural Information Processing Systems (NeurIPS), Vancouver, Canada, Dec. 08 – 14, 2019.*

Pina, R.K., Puetter, R.C.: 1993, Bayesian image reconstruction: the pixon and optimal image modeling. *Publ. Astron. Soc. Pac.* **105**, 630. DOI. ADS.

Price-Whelan, A.M., Hogg, D.W., Rix, H.-W., De Lee, N., Majewski, S.R., Nidever, D.L., Troup, N., Fernández-Trincado, J.G., García-Hernández, D.A., Longa-Peña, P., Nitschelm, C., Sobeck, J., Zamora, O.: 2018, Binary companions of evolved stars in APOGEE DR14: search method and catalog of ~ 5000 companions. *Astron. J.* **156**, 18. DOI. ADS.

Priest, E.R., Forbes, T.G.: 2002, The magnetic nature of solar flares. *Astron. Astrophys. Rev.* **10**, 313. DOI. ADS.

Raymond, J.C., Krucker, S., Lin, R.P., Petrosian, V.: 2012, Observational aspects of particle acceleration in large solar flares. *Space Sci. Rev.* **173**, 197. DOI. ADS.

Shibata, K., Magara, T.: 2011, Solar flares: magnetohydrodynamic processes. *Living Rev. Solar Phys.* **8**, 6. DOI. ADS.

Su, Y., Liu, W., Li, Y.-P., Zhang, Z., Hurford, G.J., Chen, W., Huang, Y., Li, Z.-T., Jiang, X.-K., Wang, H.-X., Xia, F.-X.-Y., Chen, C.-X., Yu, W.-H., Yu, F., Wu, J., Gan, W.-Q.: 2019, Simulations and software development for the Hard X-ray Imager onboard ASO-S. *Res. Astron. Astrophys.* **19**, 163. DOI. ADS.

Su, Y., Zhang, Z., Chen, W., Chen, D., Yu, F., Hu, Y., et al.: 2024, The tests and calibrations of the Hard X-ray Imager aboard ASO-S. *Solar Phys.* **299**, 153. DOI.

van der Walt, S., Colbert, S.C., Varoquaux, G.: 2011, The NumPy array: a structure for efficient numerical computation. *Comput. Sci. Eng.* **13**, 22. DOI. ADS.

Woods, M.M., Sainz Dalda, A., De Pontieu, B.: 2021, Unsupervised machine learning for the identification of preflare spectroscopic signatures. *Astrophys. J.* **922**, 137. DOI. ADS.

Yu, W.-h., Li, Y.-p., Gan, W.-q.: 2021, Statistical studies on modified neupert effect. *Chin. Astron. Astrophys.* **45**, 82. DOI. ADS.

Zhang, Z., Chen, D.-Y., Wu, J., Chang, J., Hu, Y.-M., Su, Y., Zhang, Y., Wang, J.-P., Liang, Y.-M., Ma, T., Guo, J.-H., Cai, M.-S., Zhang, Y.-Q., Huang, Y.-Y., Peng, X.-Y., Tang, Z.-B., Zhao, X., Zhou, H.-H., Wang, L.-G., Song, J.-X., Ma, M., Xu, G.-Z., Yang, J.-F., Lu, D., He, Y.-H., Tao, J.-Y., Ma, X.-L., Lv, B.-G., Bai, Y.-P., Cao, C.-X., Huang, Y., Gan, W.-Q.: 2019, Hard X-ray Imager (HXI) onboard the ASO-S mission. *Res. Astron. Astrophys.* **19**, 160. DOI. ADS.

Solar Physics (2025) 300:56
https://doi.org/10.1007/s11207-025-02465-5

RESEARCH

Spectral Cross-Calibrations of ASO-S/HXI with Other X-ray Missions

Zhentong Li[1] · Yang Su[1,2] · Wei Liu[1] · Wei Chen[1] · Fu Yu[1] · Weiqun Gan[1,3]

Received: 15 March 2024 / Accepted: 2 April 2025 / Published online: 25 April 2025

Abstract
Cross-calibration of X-ray instruments is crucial for stereoscopic studies of solar hard X-ray directivity and three-dimensional morphology, where spectral and imaging differences observed from multiple perspectives provide critical constraints. We present the preliminary results of the spectral calibrations of ASO-S/HXI through internal consistency checks of its three total flux detectors and cross-calibrations with Solar Orbiter/STIX, Fermi/GBM, and Konus-Wind data. Both calibration tests suggest that the response matrix of one of the HXI total flux monitors, D94 (with a thin aluminum window), needs to be modified. The new response matrix is determined by testing the series of response matrices with changing effective aluminum thickness through simulations. Post-correction analysis reveals $\sim 88\%$ of co-observed events demonstrate electron spectral index consistency within ± 0.5, while comparisons of the total counts above 30 keV show $< 10\%$ differences for STIX-HXI flares with separation angles $< 20^\circ$ and Fermi-observed moderate flares. Though challenges persist regarding detector response uncertainties, the pile-up effect, and other nonideal factors, the achieved inter-instrument consistency enables quantitative electron anisotropy investigations through joint spectral analysis.

Keywords Spectrum, X-ray · Flares, spectrum · X-ray bursts, spectrum · Instrumental effects · Cross calibration

1. Introduction

Solar X-ray emissions provide critical insights into electron acceleration and plasma heating during flares (e.g., Brown 1971; Lin and Hudson 1976; Lin et al. 1981; Phillips 2004; Caspi and Lin 2010; Benz 2017). Through X-ray spectral fitting techniques, the properties

✉ Y. Su
yang.su@pmo.ac.cn

Z. Li
ztli@pmo.ac.cn

1 Key Laboratory of Dark Matter and Space Astronomy, Purple Mountain Observatory, CAS, Nanjing 210023, China

2 School of Astronomy and Space Science, University of Science and Technology of China, Hefei, 230026, China

3 University of Chinese Academy of Sciences (UCASNJ), Nanjing 211135, China

of thermal plasma and energetic electrons as well as their temporal evolution can be derived (e.g., Holman et al. 2003; Krucker et al. 2008; Holman et al. 2011; Grigis and Benz 2006).

In the last few decades, solar dedicated X-ray instruments have been persistently launched, e.g., Yohkoh (Ogawara et al. 1991), the Reuven Ramaty High Energy Solar Spectroscopic Imager (RHESSI; Lin et al. 2002), and the Spectrometer Telescope for Imaging X-rays (STIX; Krucker et al. 2020) aboard Solar Orbiter and the Hard X-ray Imager (HXI; Zhang et al. 2019; Su et al. 2019) aboard the Advanced Space-based Solar Observatory (ASO-S; Gan et al. 2023). Moreover, some X-ray missions for X-ray astrophysics, including Fermi (Meegan et al. 2009), Konus-Wind (Aptekar et al. 1995; Lysenko et al. 2022), and NuSTAR (Harrison et al. 2013), also observe the Sun with considerable observation time and have reported their solar flare list[1] or database[2].

The directivity of hard X-ray (HXR) emission, which is important for understanding the particle acceleration process and the anisotropy (pitch-angle distribution) of energetic electrons, is difficult to measure and poorly investigated (e.g., Kašparová, Kontar, and Brown 2007; Kontar et al. 2011; Dickson and Kontar 2013) mainly due to the lack of measurements, albedo effect, and large uncertainties in observed spectral data. Some early studies have even concluded that HXR performs near-isotropic distribution (e.g., Kontar and Brown 2006), based on limited measurements from a single viewing angle. A better and more direct approach to study HXR directivity is simultaneous observation of the same flare from multiple vantage points (Krucker et al. 2019; Jeffrey et al. 2024).

Before Solar Orbiter/STIX, the ULYSSES mission (Hurley et al. 1992) offered the opportunity to provide HXR spectral observations from a different view angle (Kane et al. 1998) other than near-earth orbit. Now, STIX and HXI can provide, for the first time, both X-ray images and spectra from various vantage points (see Krucker et al. 2019; Ryan et al. 2024, for more details) with a similar energy range (overlapped in $\sim$ 10 to 150 keV). The opportunity arises for HXR directivity studies with simultaneous X-ray imaging and spectroscopy from two different viewing angles. However, careful spectral cross-calibration is required before the direct comparison of the spectra/images obtained by the two instruments.

With the advent of the era of simultaneous, multi-instrument, multi-angle observation of solar X-rays, joint observations will not be limited only to complementary observations but will also be essential for verifying the reliability and comparability of data from different instruments. This makes the spectral cross-calibration among X-ray instruments increasingly important.

So far, only a few studies show detailed spectral calibration or comparison among different X-ray instruments in solar physics e.g. (Kane et al. 1998; Jeffrey et al. 2024; Bajnoková et al. 2024) and astrophysics e.g. (Nava et al. 2011; Kaneko et al. 2015). It is challenging to perform spectral cross-calibration due to various of factors, e.g., albedo effects in different levels (Kontar et al. 2006), different energy resolutions (Jeffrey et al. 2024), different detector designs, different background levels, different deadtimes, different instrumental effects, etc.

In this study, we present the preliminary result of the spectral cross-calibration of HXI by comparing its spectra with the spectra observed by various X-ray instruments including STIX, Fermi/GBM, and Konus-Wind, using selected sample events. In Section 2, we present the event selection and different approaches to perform spectral analysis for different instruments. In Section 3, we show the results of spectral self-calibrations of the HXI detectors

[1]Fermi/GBM flare list: https://hesperia.gsfc.nasa.gov/fermi/gbm/qlook/fermi_gbm_flare_list.txt. NuSTAR solar observations: http://ianan.github.io/nsigh_all/.

[2]Konus-Wind solar flare database: https://www.ioffe.ru/LEA/kwsun/index.html.

Table 1 Events Overview (✓means the corresponding data are used).

Time (near spec. fitting)	GOES class	HXI ($\theta_{HXI-flare}$)	$\theta_{HXI-STIX}$	STIX ($\theta_{STIX-flare}$)	Fermi	Konus-Wind
2023-01-06 00:56	X1.2	✓(81.9°)	23.6°	✓(63.7°)	✓	✓
2023-02-11 15:46	X1.1	✓(38.6°)	32.6°	✓(32.6°)	✓	✓
2023-02-28 17:42	M8.6	✓(44.3°)	32.3°	✓(66.1°)	✓	–
2023-07-16 15:07	M1.8	✓(31.3°)	–	–	✓	✓
2023-10-01 03:23	C9.6	✓(79.9°)	–	–	✓	✓
2024-03-20 07:32	M7.4	✓(48.64°)	0.6°	✓(48.66°)	✓	–
2024-03-25 06:43	M4.5	✓(23.8°)	15.9°	✓(8.6°)	✓	✓
2024-11-10 20:18	M4.9	✓(37.4°)	6.3°	✓(34.8°)	–	–
2023-12-14 17:06	X2.8	✓(53.4°)	13.6°	✓(65.6°, Att. set)	–	–
2024-02-22 22:28	X6.3	✓(37.0°)	26.6°	✓(19.6°, Att. set)	–	–
2024-10-31 21:16	X2.0	✓(30.3°)	9.6°	✓(20.7°, Att. set)	–	–

and cross-calibration based on joint comparisons with other instruments. The conclusions and discussion are given in Section 4.

2. Observations and Data Processing

2.1. Event Selection

The criteria of event selection is critical for the purpose of cross-calibration. First, the events should be simultaneously observed by two or three of the X-ray instruments including STIX, Fermi/GBM, and Konus-Wind. Second, the separation angle between HXI and STIX must be small enough ($\lesssim\sim 30°$ in this study) to minimize the differences in spectra caused by albedo or HXR directivity. Third, the count rate must be high enough to produce spectra with a good signal-to-noise ratio (SNR). However, it must also avoid severe pulse pile-up effect, which can distort the spectral shape. This effect occurs when multiple signals within a short period, which varies for different instruments, are detected as a single event. Finally, the flares shall be fully observed by STIX and HXI, meaning that occulted flares are not used here.

Table 1 shows the collected events. $\theta_{HXI-flare}$ and $\theta_{STIX-flare}$ are the angles between HXI/STIX viewing line and the flare normal direction, respectively. $\theta_{HXI-STIX}$ means the separation angle between HXI and STIX. The 2023-01-06 event is near the east limb of the Sun from Earth's perspective, but on disk for the STIX viewing point. The 2023-02-28 event is on disk from Earth's perspective, but near the west limb for the STIX viewing point. For the 2023-07-16 event and 2023-10-01 event, although they are M1.8 and C9.6 class flares, HXI recorded enough counts even at energies up to 300 keV. STIX reached a close viewing point (0.6° separation angle) with other near-Earth instruments during the event 2024-03-20, that diminished the X-ray directivity effect. It is preferable for the spectra to have broader energy coverage to facilitate spectral comparisons. The time ranges selected for spectral analysis are primarily around the impulsive peak or early impulsive peak for the top eight flares in Table 1, when the HXR photon fluxes reach the maximum and the spectra cover a wide energy range. Another advantage for selecting these time ranges is that there are less impacts from pile-up effects compared to the later phase.

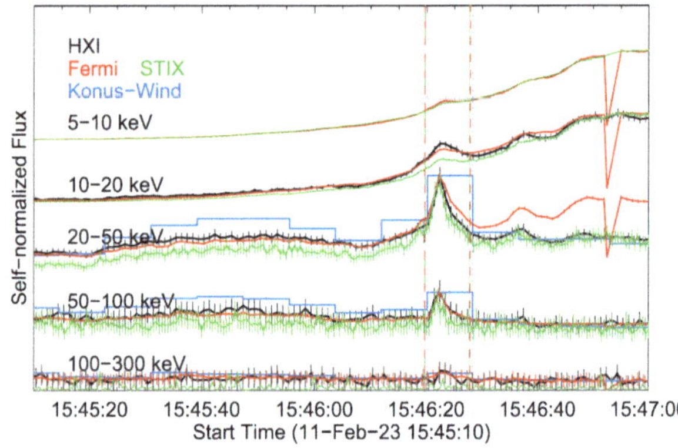

Figure 1 Normalized X-ray count fluxes of multiple energy ranges of the 2023-02-11 event observed by HXI (black curves), Fermi/GBM (red curves), STIX (green curves), and Konus-Wind (blue curves). The region in light gray and the dashed or dotted vertical lines in different colors indicate the time intervals used for the spectral fitting of corresponding instruments.

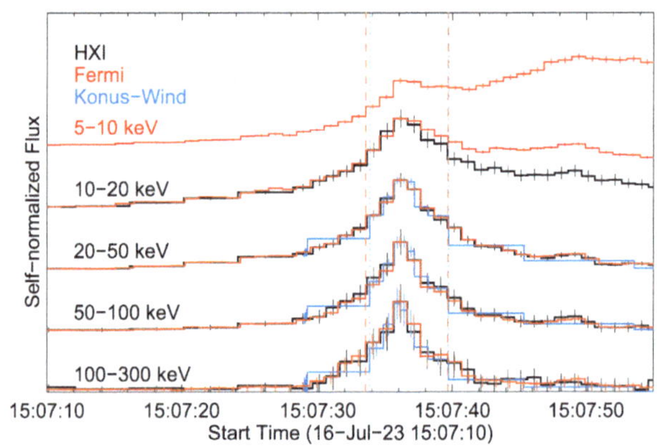

Figure 2 Normalized X-ray count fluxes of different energy ranges of the 2023-07-16 event observed by HXI (black curves), Fermi/GBM (red curves), and Konus-Wind (blue curves). The region in light gray and the dashed or dotted vertical lines in different colors indicate the time intervals of corresponding instruments.

In addition, we selected three extra events (the bottom three rows in Table 1) that are observed by STIX with the attenuator and discussed the comparison in Section 4. We also add another three events for the self-calibration among the HXI detectors in Section 3.1.

Here, we show the light curves of two events. Figure 1 displays the normalized light curves of the flare observed by HXI, STIX, Fermi/GBM, and Konus-Wind on 2023-02-11. The light curves of 20 – 300 keV from three instruments show a high consistency, despite that Konus-Wind has a lower time resolution apart from the peak. The Fermi 20 – 50 keV light curve shows relatively higher flux after the impulsive peak, which may be caused by a pile-up effect. Figure 2 shows the 2023-07-16 event observed by HXI, Fermi/GBM, and Konus-Wind. They also present good agreement near the peak time. The Fermi 10 – 20 keV light curve shows a relatively slower decrease, also probably due to a pile-up.

2.2. Spectral Analysis

In terms of spectra, HXI provides X-ray spectral data with energy resolution (FWHM/E_{peak}) of $\sim 16\%$ at 32 keV, in the range of ~ 10 to 300 keV (Su et al. 2024).[3] The spectra and response matrices can be generated by HXI Graphical User Interface (GUI)[4] for the total flux

[3]The spectral energy range has been extended after inflight testing phase. The low-energy limit of its spectra can reach 10 – 12 keV.

[4]Available online: http://aso-s.pmo.ac.cn/sodc/analysisSoftware.jsp.

monitors which is labeled with detector ID numbers, D92, D93, D94, separately, or for the combination of any two or three of them. Due to the severe level of charged particles in a part of the orbit of ASO-S satellite, background subtraction is crucial for analysis. Generally, there are three ways to select background: a) selecting the data observed 48 h before or after the flare time as background; b) using fitting profiles from non-flaring time intervals; and c) using data from background monitors. We recommend to firstly consider the first method because the satellite almost returns to the same location every 48 h, which means the background 48 h before or after the flare time is similar (in most cases) if there are no additional flare signals or particle events (usually caused by radiation belt, the South Atlantic Anomaly, magnetospheric disturbances, or solar particle events). The systematic error of background subtraction of this method for the flares selected in this study is roughly $\pm 2\%$ in most cases (see Appendix A). Otherwise, one can try the second method, which can be widely used when the background time profile is simple and can be properly fitted. The last option to subtract the background is the third method, which is ill-suited for scientific purpose for two reasons. First, the background monitors (D95 to D99) have different fluxes in the emission lines from the internal calibration sources. Second, they may still receive flare fluxes from high-energy photons in large flares penetrating the tungsten window, making their spectra higher than the true background. Here, we take the background of the data observed 48 h before for the events on 2023-01-06 and 2023-02-11, fitting profiles of non-flaring time intervals for the 2023-07-16 event and the background of 48 h after for the other events.

STIX is a solar X-ray spectrometer and imager containing 32 detectors that measure incident photon counts in 32 energy channels from 4 to 150 keV, with an energy resolution (FWHM) of 1 keV at 6 keV (Krucker et al. 2020). To avoid the detector fuzzy effect (Jeffrey et al. 2024) and for simplicity of joint calibration, we use the total flux of only large pixels of the coarsest imaging detectors from G6 to G10 to make the spectra. We use the STIX background data provided in the STIX data center website.[5] The fitting energy ranges start from 11 keV (the same starting energy for HXI), except for the last three events in Table 1 (from 12 keV). The version of STIX ground software (GSW) that we use in this study is v0.5.2.

The Gamma-Ray Burst Monitor (GBM) is one of the science payloads on the Fermi Gamma-Ray Space Telescope, which is designed mainly for research of gamma-ray bursts (Meegan et al. 2009). The 12 NaI(Tl) detectors on the GBM measure lower-energy (~ 8 keV to ~ 1 MeV) photons coming from 12 directions, with an energy resolution of $\sim 15\%$ at 100 keV, and 6 of them have relatively smaller pointing angles toward the Sun. Thus, GBM also triggers on solar flares and records photons in an overlapping energy range with typical solar photons. The NaI detectors can be easily affected by a pile-up effect for a solar event with high photon flux due to its large effective detector area. Therefore, the detector with the larger pointing angle is a better candidate for spectral analysis in this case. Background profiles are fitted from data selected during a non-flaring time within one orbit period before/after the flare.

Konus-Wind, designed to observe gamma-ray bursts and solar flares, consists of two NaI(Tl) detectors that can record solar X-ray and soft gamma-ray photons from ~ 20 to 1200 keV into three energy bands in waiting mode, and into 64 energy bins with higher temporal resolution in the triggered mode (e.g., 16 ms during ~ 30 s near trigger (see details in Aptekar et al. 1995; Lysenko et al. 2022)). The energy resolution is $\sim 20\%$ at 20 keV. The background profile is almost stable around the Sun–Earth Lagrange L1 point, and the background data are already provided in the spectra file.

[5] https://datacenter.stix.i4ds.net/.

The X-ray spectra are originally recorded in units of count, which is dependent on the detector response. Given this dependency, it is logical to compare the photon spectra between different instruments by normalizing them to a common physical quantity, such as photon flux. Our primary goal is to achieve a good fit of the spectra for accurate comparison, rather than to interpret the physical content directly from the fitting results. The fitted photon spectra from each instrument will then be used for inter-instrument comparison. Regarding the fitting model, we use a thermal bremsstrahlung emission model (`f_vth`) plus a thick target model (`f_thick2_vnorm`) combined with albedo correction provided in OSPEX packages (Schwartz et al. 2002) to fit all the events observed from each instrument individually.

All the best-fit parameters of each instrument are listed in Table 2, which also contains the fitting results of the HXI combined detectors with the newly adjusted response matrix (given in Section 3.2). Note that these instruments have different energy coverages and different energy resolutions, which could introduce different biases. We do not expect the thermal parameters and the nonthermal electron low-energy cutoffs to agree across different instruments due to varying energy ranges, the multi-thermal nature of flares, spectral resolutions, and pile-up levels. Such discrepancies are obviously present in Table 2. The pile-up levels can be indicated by the ratio of livetime to exposure time r_{live} (see the last column of Table 2). Overall, most of the fitting results achieve $\chi^2 < 2$, and all of the STIX fittings achieve $\chi^2 < 1.4$, while some spectra (mainly observed by Fermi) are difficult to fit due to the just-mentioned nonideal factors, resulting in larger χ^2 and structural residuals (see all the figures of spectral fitting in Appendix B). In the following sections, we focus mainly on the non-thermal electron distribution index and the flux at certain energies.

3. Results

3.1. Cross-Calibrations of HXI Detectors

The first step before spectral cross-calibration should be examining the consistency of the three HXI total flux detectors. These detectors are all covered by an attenuation layer that mainly consists of aluminum and Teflon to decrease the incident low-energy photon flux to avoid severe pile-up signals from low-energy channels. In addition, there are four multilayers in front of all detectors, whose effect is already taken into account in the response matrix. The aluminum thicknesses of D92 and D93 are thicker (0.85 mm) than that of D94 (0.6 mm). Henceforth, we call the total flux detectors D92, D93 detectors T1, T2, respectively, and D94 detector T3.[6]

By design, T3 observes higher counts that T1, T2, due to the lower attenuation from the thinner Al window, to allow a better response to low energies. Ideally, the photon models obtained from the best-fit parameters for each detector should be similar as the incident X-ray spectrum. However, if the detector response matrix (DRM) is inaccurate, the best-fit photon model may not be consistent with the incident spectrum. Therefore, we first evaluate the detector responses of the three detectors by comparing the best-fit photon spectra obtained from each detector individually (Figures 3(a – d)). It is clearly seen that there are obvious discrepancies between the fitted photon models of T3 and those for the other two detectors.

[6]The HXI detectors D92, D93, and D94 are total flux monitors which are distinct from the imaging detectors (D1 – D91). In the HXI GUI, they are labeled as T1, T2, and T3, respectively. Additionally, the background monitors (D95 – D99) are labeled as B1 – B5.

Table 2 Spectral fitting parameters of different events using different instruments.

Event date (GOES class)	Instrument (detector)	EM (10^{49} cm^{-3})	Temperature (MK)	Electron δ	E_C (keV)	r_{live}
2023-01-06 (X1.2)	HXI/T3 (0.6 mm)	$1.44_{\pm 0.37}$	$23.4_{\pm 1.0}$	$4.40_{\pm 0.04}$	$33.2_{\pm 2.3}$	> 0.990
	HXI/T123 (new)	$0.53_{\pm 0.15}$	$25.9_{\pm 1.4}$	$4.32_{\pm 0.03}$	$33.3_{\pm 2.3}$	> 0.992
	STIX	$0.51_{\pm 0.28}$	$24.3_{\pm 3.4}$	$4.29_{\pm 0.07}$	$21.3_{\pm 10.9}$	0.81
	Fermi/GBM	$0.08_{\pm 0.01}$	$32.3_{\pm 0.9}(*)$	$4.47_{\pm 0.02}$	$45.7_{\pm 1.3}$	0.64
	Konus-Wind	$0.06_{\pm 0.07}$	$41.5_{\pm 8.6}(*)$	$4.32_{\pm 0.03}$	$62.0_{\pm 3.7}$	0.44
2023-02-11 (X1.1)	HXI/T3 (0.6 mm)	$1.31_{\pm 0.62}$	$21.9_{\pm 1.8}$	$5.32_{\pm 0.15}$	$25.7_{\pm 7.6}$	> 0.995
	HXI/T123 (new)	$0.42_{\pm 0.18}$	$24.9_{\pm 2.2}$	$5.05_{\pm 0.12}$	$26.7_{\pm 8.6}$	> 0.996
	STIX	$0.21_{\pm 0.07}$	$27.4_{\pm 2.4}$	$4.96_{\pm 0.13}$	$23.8_{\pm 10.3}$	0.83
	Fermi/GBM	$0.057_{\pm 0.003}$	$34.0_{\pm 0.6}(*)$	$5.33_{\pm 0.06}$	$40.3_{\pm 1.8}$	0.75
	Konus-Wind	$0.004_{\pm 0.016}$	$49.5_{\pm 27.3}(*)$	$4.95_{\pm 0.14}$	$30.8_{\pm 22.2}$	0.74
2023-02-28 (M8.6)	HXI/T3 (0.6 mm)	$1.13_{\pm 0.41}$	$20.8_{\pm 1.2}$	$5.97_{\pm 0.21}$	$28.1_{\pm 3.0}$	> 0.996
	HXI/T123 (new)	$0.32_{\pm 0.09}$	$23.9_{\pm 1.3}$	$5.99_{\pm 0.18}$	$30.5_{\pm 2.4}$	> 0.997
	STIX	$0.55_{\pm 0.19}$	$20.7_{\pm 1.5}$	$5.79_{\pm 0.07}$	$19.5_{\pm 3.4}$	0.83
	Fermi/GBM	$0.027_{\pm 0.001}$	$37.3_{\pm 0.6}(*)$	$6.44_{\pm 0.14}$	$44.0_{\pm 1.6}$	0.80
2023-07-16 (M1.8)	HXI/T3 (0.6 mm)	$0.05_{\pm 0.03}$	$26.3_{\pm 3.3}$	$3.86_{\pm 0.05}$	$44.2_{\pm 4.0}$	> 0.996
	HXI/T123 (new)	$0.03_{\pm 0.02}$	$25.8_{\pm 3.7}$	$3.90_{\pm 0.04}$	$47.3_{\pm 2.8}$	> 0.996
	Fermi/GBM	$0.0018_{\pm 0.0002}$	$45.3_{\pm 2.4}(*)$	$4.26_{\pm 0.03}$	$72.4_{\pm 2.3}$	0.88
	Konus-Wind	$0.002_{\pm 0.002}$	$65.9_{\pm 17.3}(*)$	$4.34_{\pm 0.06}$	$91.3_{\pm 5.8}$	0.87
2023-10-01 (C9.6)	HXI/T3 (0.6 mm)	$0.12_{\pm 0.24}$	$21.3_{\pm 7.1}$	$4.05_{\pm 0.10}$	$28.0_{\pm 9.7}$	> 0.996
	HXI/T123 (new)	$0.04_{\pm 0.05}$	$23.9_{\pm 5.6}$	$4.08_{\pm 0.08}$	$33.9_{\pm 5.0}$	> 0.996
	Fermi/GBM	$0.0010_{\pm 0.0001}$	$59.2_{\pm 2.9}(*)$	$4.46_{\pm 0.06}$	$61.2_{\pm 3.4}$	0.94
	Konus-Wind	$0.012_{\pm 0.37}$	$39.3_{\pm 20.2}(*)$	$4.45_{\pm 0.32}$	$70.3_{\pm 14.2}$	0.79
2024-03-20 (M7.4)	HXI/T3 (0.6 mm)	$0.61_{\pm 0.22}$	$20.7_{\pm 1.2}$	$5.96_{\pm 0.4}$	$29.8_{\pm 4.0}$	> 0.998
	HXI/T123 (new)	$0.13_{\pm 0.04}$	$24.8_{\pm 1.3}$	$5.86_{\pm 0.36}$	$32.7_{\pm 3.7}$	> 0.998
	STIX	$0.17_{\pm 0.07}$	$22.5_{\pm 2.0}$	$6.01_{\pm 0.08}$	$23.8_{\pm 2.9}$	0.77
	Fermi/GBM	$0.013_{\pm 0.001}$	$37.9_{\pm 0.6}$	$6.83_{\pm 0.34}$	$46.5_{\pm 2.2}$	0.90
2024-03-25 (M4.5)	HXI/T3 (0.6 mm)	$0.27_{\pm 0.12}$	$22.0_{\pm 1.6}$	$4.11_{\pm 0.16}$	$47.2_{\pm 5.8}$	> 0.997
	HXI/T123 (new)	$0.10_{\pm 0.04}$	$24.4_{\pm 1.6}$	$4.22_{\pm 0.14}$	$53.5_{\pm 4.4}$	> 0.997
	STIX	$0.03_{\pm 0.01}$	$29.5_{\pm 1.4}$	$3.98_{\pm 0.10}$	$48.3_{\pm 4.7}$	0.85
	Fermi/GBM	$0.0056_{\pm 0.0003}$	$41.8_{\pm 0.9}(*)$	$4.91_{\pm 0.12}$	$85.3_{\pm 3.4}$	0.90
	Konus-Wind	$0.01_{\pm 0.01}$	$46.5_{\pm 11.9}(*)$	$4.44_{\pm 0.25}$	$85.7_{\pm 10.3}$	0.63
2024-11-10 (M4.9)	HXI/T3 (0.6 mm)	$1.33_{\pm 0.44}$	$20.3_{\pm 1.0}$	$5.44_{\pm 0.22}$	$32.0_{\pm 3.1}$	> 0.996
	HXI/T123 (new)	$0.33_{\pm 0.12}$	$23.4_{\pm 1.5}$	$5.27_{\pm 0.12}$	$30.4_{\pm 3.0}$	> 0.997
	STIX	$0.22_{\pm 0.12}$	$23.7_{\pm 3.0}$	$5.23_{\pm 0.10}$	$22.7_{\pm 6.2}$	0.86
2023-12-14 (X2.8)	HXI/T3 (0.6 mm)	$33.5_{\pm 12.2}$	$16.2_{\pm 0.8}$	$3.99_{\pm 0.08}$	$20.6_{\pm 131}$	> 0.992
	HXI/T123 (new)	$5.83_{\pm 0.95}$	$19.1_{\pm 0.5}$	$3.91_{\pm 0.06}$	$21.1_{\pm 79.6}$	> 0.994
	STIX (Att. set)	$2.62_{\pm 0.29}$	$21.8_{\pm 0.4}$	$3.62_{\pm 0.10}$	$20.8_{\pm 116}$	> 0.994

Table 2 (*Continued*)

Event date (GOES class)	Instrument (detector)	EM (10^{49} cm^{-3})	Temperature (MK)	Electron δ	E_C (keV)	r_{live}
2024-02-22 (X6.3)	HXI/T3 (0.6 mm)	$49.9_{\pm 14.6}$	$20.7_{\pm 1.0}$	$8.16_{\pm 0.07}$	$26.6_{\pm 1.1}$	0.91
	HXI/T123 (new)	$9.12_{\pm 2.62}$	$25.6_{\pm 1.5}$	$8.13_{\pm 0.06}$	$27.8_{\pm 1.4}$	0.93
	STIX (Att. set)	$4.82_{\pm 1.60}$	$27.5_{\pm 2.2}$	$8.17_{\pm 0.11}$	$25.9_{\pm 1.5}$	0.91
2024-10-31 (X2.0)	HXI/T3 (0.6 mm)	$2.04_{\pm 0.38}$	$24.5_{\pm 0.8}$	$4.78_{\pm 0.07}$	$47.0_{\pm 1.9}$	0.978
	HXI/T123 (new)	$0.51_{\pm 0.09}$	$29.3_{\pm 1.0}$	$4.83_{\pm 0.05}$	$50.4_{\pm 1.6}$	0.983
	STIX (Att. set)	$0.22_{\pm 0.05}$	$34.2_{\pm 1.7}$	$4.76_{\pm 0.10}$	$50.5_{\pm 3.1}$	0.97

Note: HXI/T3 (0.6 mm) indicates using HXI T3 DRM with 0.6-mm effective aluminum thickness. HXI/T123 (new) indicates using combined detector T1+T2+T3 and the new DRM after adjustment. r_{live} represents the averaged ratio of livetime to exposure time in each fitting time interval. The asterisk symbols (∗) mean the derived temperatures may be questionable (due to the largely reduced flux or missing measurement at low energies below 20 keV).

In order to compare the three spectra in count space, we plotted the observed count spectra of T1 and T2 (they are designed to be identical) in Figure 3(e), as well as the expected count spectra for T1 which is converted from the best-fit photon spectrum of T3 and T1's detector response matrix, following the general equation,

$$C = F \cdot \text{DRM}, \tag{1}$$

where C is the count flux recorded by an instrument with n energy bins, F is the X-ray photon emission flux divided into m energy bins, and DRM is the detector response matrix of the instrument with dimension $n \times m$ describing the correlation between the input photon spectrum and the output count spectrum.

After convolving the fitted photon spectra of T3 with T1 DRM, the three count spectra of three detectors (Figure 3(e)) are comparable and they are less model-dependent because only the best-fit model of T3 is used. The count fluxes recorded by T1 and T2 are indeed similar as expected, but they are significantly lower than the expected count spectrum from the T3 photon spectrum at low energies. To quantitatively describe the difference, similar to the spectral fitting residual, we introduce the comparison residual as $\frac{\text{Diff}}{\sigma_{\text{T1}}}$, where Diff means difference between the spectrum of T2 or T3 and T1 and σ_{T1} means the error of the observed T1 data. The differences between T1 and T2 are mostly within $\pm 2\sigma$ of the T1 data. Therefore, we can conclude for this case that T1 and T2 are consistent with each other, while the converted count fluxes of T3 are obviously larger at low energies than T1 (up to $5\sigma_{\text{T1}}$ at energies < 30 keV). The comparison results remain similar for other events (see Section 3.2). It is clear that the response matrix for T1(T2) or T3 needs adjustment to eliminate the discrepancies. In order to find out which detector(s) are the cause, we used the spectra observed from other instruments as valuable references.

Here we take the 2023-07-16 event as an example. Since the time resolutions are different for different instruments and different stages of the flare (see Figure 2), we take the most adjacent time intervals near the common peak for all the instruments to perform spectral analysis. For HXI, detector T1 and T2 are combined as detector T12 to improve SNR due to their consistency. We compare in Figure 4(a)(b) the observed T3 and T12 count spectra with the corresponding count spectra expected from the best-fit photon models of each instrument, respectively. It is evident that the expected T3 count flux from Fermi and KW

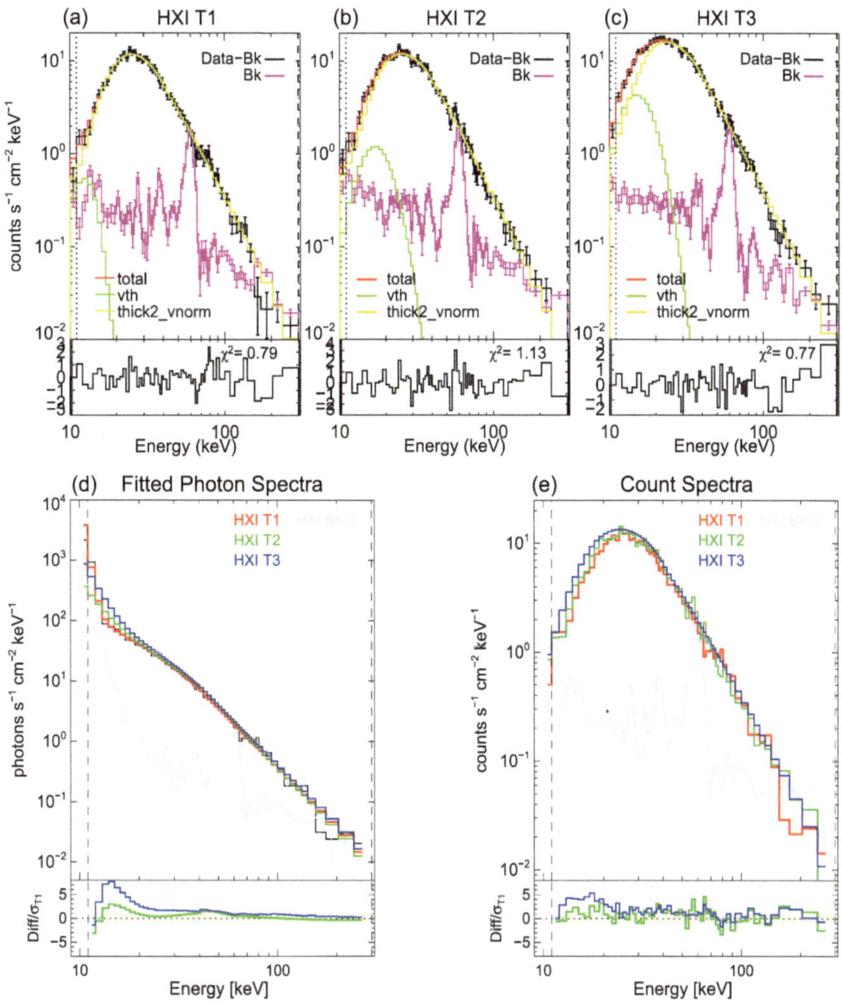

Figure 3 Comparison of HXI spectra of different detectors obtained from 2023-07-16 event data. Panel (a)–(c): Spectral fitting results of HXI detectors T1, T2, T3, respectively. The black curve in each panel is the observed background-subtracted count spectrum. Panel (d): Best-fit photon spectra derived from spectral fitting of T1, T2, and T3 data, respectively. Panel (e): Observed count spectra of T1 (red) and T2 (green) compared with the expected T1 count spectrum (blue curve) from T3 best-fit photon spectrum. Gray lines indicate the backgrounds selected by T1. Vertical dashed lines show the fitting energy range.

best-fit photon spectra are lower than the recorded T3 count flux in the low-energy range (11–30 keV). The differences in the same energy range are less significant for the case of T12. Another event (2024-03-25 in Figure 4(c)(d)) shows similar result that the expected T3 count flux from Fermi and STIX best-fit photon spectra are systematically lower than the recorded T3 count flux in the low-energy range (11–20 keV), which cannot be ignored.

After investigating the comparisons using the other events, we can conclude that T12 are more consistent with other instruments than T3 (see Figure 5). Although the results from other instruments are not claimed to be physically true due to possible pulse pile-up effect, albedo effect (STIX), and model-dependent fitting process, it is more reasonable to infer that

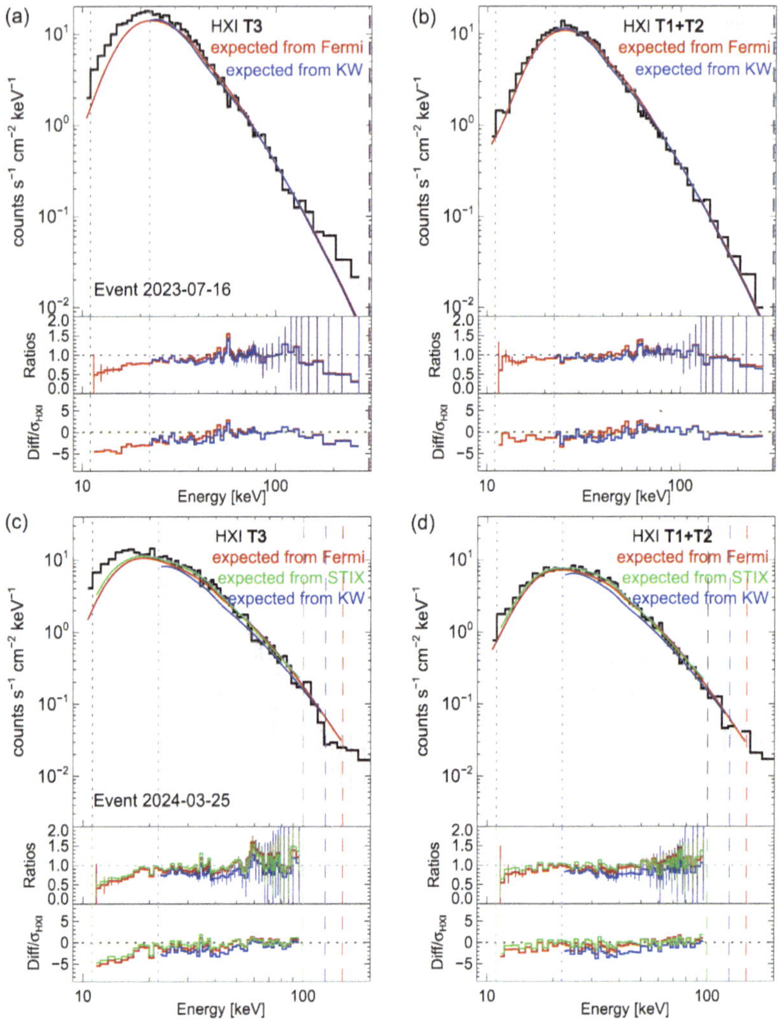

Figure 4 Spectral comparison between HXI detectors (T3 and combined detector T12) and other X-ray instruments for the 2023-07-16 event and 2024-03-25 event. Panel (a): Observed T3 count spectrum (black curve) compared with the expected T3 count spectra derived from the best-fit photon spectra of Fermi (red) and KW (blue) for the 2023-07-16 event. Gray curves indicate the HXI backgrounds. Vertical dotted and dashed lines show the fitting energy range. The differences between the observed and expected spectra are plotted below the spectra as ratios in percentages and normalized residuals. Panel (b): same as panel (a), but for combined detector T12 and its response matrix. Panel (c) and (d): same as panel (a) and (b) but for the 2024-03-25 event.

the HXI T3 detector needs to be further calibrated because the discrepancies between HXI T3 and T12 and those between HXI T3 and other instruments are clearly seen.

3.2. Adjustment of HXI Detector Response

The photon fluxes derived from HXI T12 and other instruments are temporarily assumed to be the true fluxes (as reference) due to their acceptable consistency with each other, ignoring

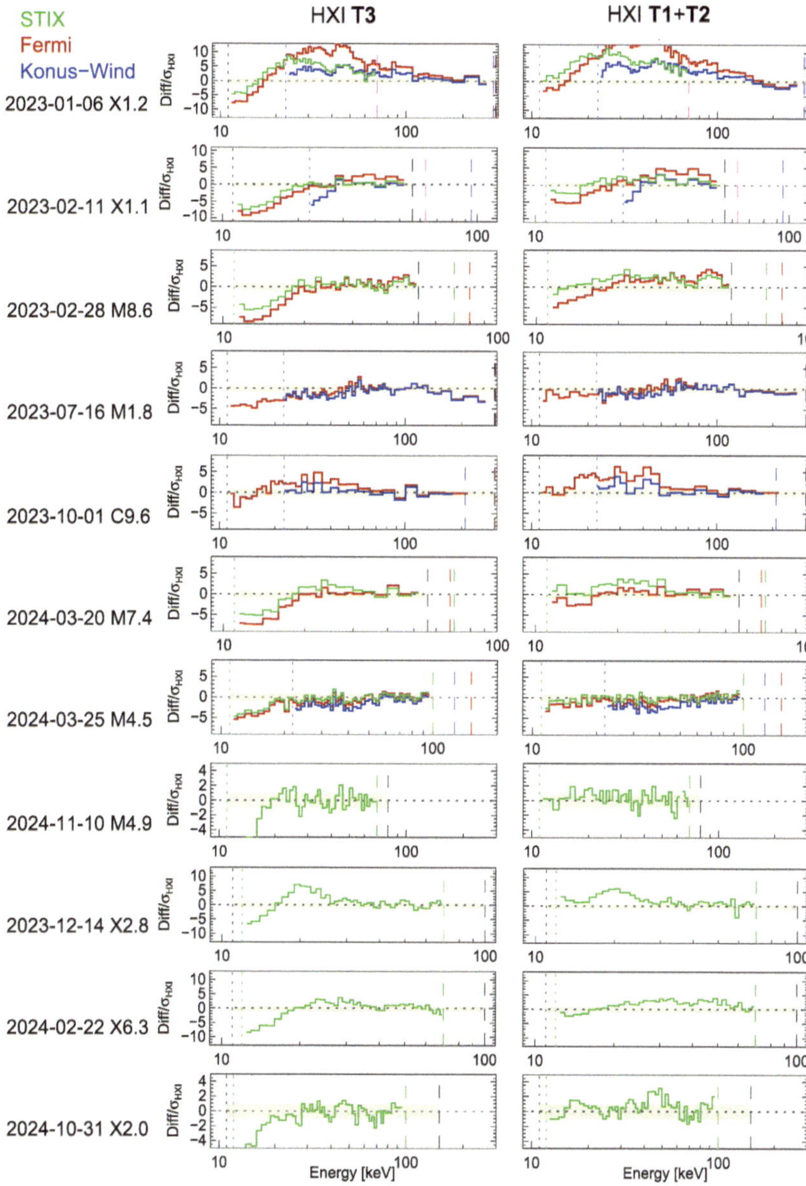

Figure 5 Spectral comparison residuals of 11 events observed by multiple instruments. Left panels: comparison residuals between the expected HXI T3 count spectra from other instruments and the observed T3 count spectrum. Right panels: comparison residuals for combined HXI detectors T12. The vertical dotted and dashed lines show the fitting energy range of each instrument in corresponding color labeled in the figure.

the fitting errors (shown in Appendix B) and instrumental effects, which will be discussed in Section 4. According to Figures 4 and 5, the measured T3 count fluxes at energies below ~ 20 keV are higher than the expected fluxes, which means that the effective thickness of

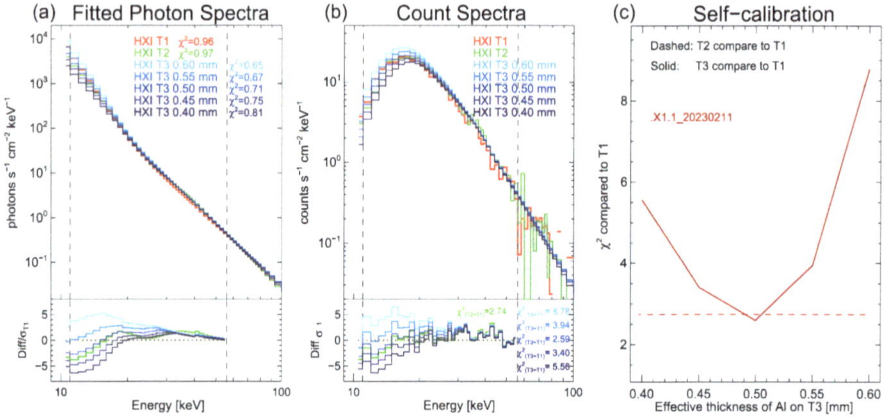

Figure 6 Comparisons of the best-fit photon spectra and count spectra for HXI detectors T1, T2, and T3 with multiple T3 response matrices based on the data of the 2023-02-11 event. Panel (a): Best-fit photon spectrum individually derived from T1 (red), T2 (green), and T3 (blue) with DRMs under various simulated window thicknesses (darker blue means thinner). Panel (b): T1 (solid red), T2 (solid green) observed counts, and the expected T1 count spectra derived from the best-fit photon spectrum for detector T3 under various simulated window thicknesses (blue lines with changing darkness). Dashed blue curve indicates the recorded counts spectrum of T3 detector. Vertical dotted and dashed lines show the fitting energy range. Panel (c): χ^2_{Diff} values in panel (b). The dashed and solid lines represent the χ^2_{Diff} value of T2 and T3 compared to T1, respectively. The X-axis shows the changing effective thickness of aluminum covered in front of T3.

the materials in front of T3 is larger than it should be according to Equation 1. Thus, the DRM of T3 should be adjusted to decrease/eliminate the discrepancies.

The main difference between the HXI detector T3 (D94, thin) and T1 (D92, thick) or T2 (D93, thick) is the thickness of the aluminum window, 0.60 mm and 0.85 mm, respectively. Other materials in front of the detectors include Teflon and multilayers. Their thickness were roughly estimated for DRM simulation. Here, we adjust the aluminum thickness to effectively modify the DRM of T3 by utilizing the GEANT4 simulation tool (Agostinelli et al. 2003). Initially, a mass model is established in GEANT4 based on the engineering parameters and the material composition of the HXI detection units. Subsequently, monoenergetic photons of varying energies are introduced at fixed intervals to interact physically with the mass model of the HXI detection unit, with the response energies within the HXI detection unit being recorded individually. This process enables the determination of the response matrix for the HXI detection unit.

According to the previous results, the effective thickness of the aluminum of the T3 (D94) should be smaller to gain a higher response at low energies. Therefore, we change the effective thickness from its original measurement (0.6 mm, which is used in Section 3.1) to smaller values with a step of 0.05 mm and obtain a sequence of DRMs of various thicknesses. Figure 6 shows the spectral comparisons among T1, T2, and T3, with varying effective window thickness for T3. The main variations are at the low energies in both the photon and count space. We apply the approach of calculating goodness-of-fit to quantify the difference between the converted count flux of one detector and recorded count flux of another detector,

$$\chi^2_{\mathrm{Diff}} = \sum_i^N \frac{(C_{i,\mathrm{conv}} - C_{i,\mathrm{obs}})^2}{(N-1)\sigma_{i,\mathrm{obs}}}, \tag{2}$$

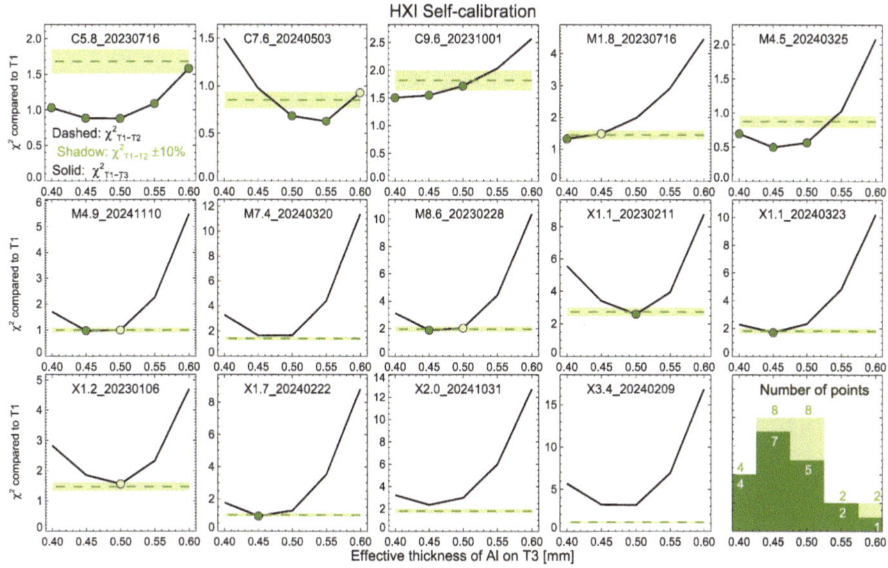

Figure 7 Self-calibration results of HXI detectors. The dashed and solid lines in each panel (except the last one) represent the $\chi^2_{\rm Diff}$ value of T2 and T3 compared to T1, respectively. The green shadow area indicates the $\pm 10\%$ range of $\chi^2_{\rm T1-T2}$. The marked dark green and light green points indicate $\chi^2_{\rm T1-T3} \leqslant \chi^2_{\rm T1-T2}$ and $\chi^2_{\rm T1-T3} \leqslant 110\% \cdot \chi^2_{\rm T1-T2}$, respectively. The X-axis in each panel shows the changing effective thickness of the aluminum cover in front of T3. The event information is labeled in the form of 'GOES class' plus 'event date'. The last panel shows the number of the dark and light green points in the previous panels.

where $C_{i,\rm conv}$ is converted count flux in ith energy bin and $C_{i,\rm obs}$ is the observed one, $\sigma_{i,\rm obs}$ is the random error of observed counts, and N is the number of energy bins within the fitting range. For detector T2, we directly use its original recorded count flux instead of $C_{i,\rm conv}$ due to its same response with T1. The $\chi^2_{\rm Diff}$ values of T1 and T3 with 5 different thicknesses are presented in the bottom right of Figure 6(b). The difference between T1 and T3 becomes smallest when T3 thickness changes to 0.5 mm according to the curves comparison and $\chi^2_{\rm Diff}$ values. In this case, the discrepancy between T3 and T1 ($\chi^2_{\rm T3-T1} = 2.59$) is even smaller than that between T2 and T1 ($\chi^2_{\rm T2-T1} = 2.74$).

We use 14 events to do the self-calibration by testing the discrepancy levels of T3 with changing effective window thicknesses. The results are shown in Figure 7. The dashed lines indicate the difference level between T2 and T1 ($\chi^2_{\rm T2-T1}$). The differences between T3 and T1 (solid lines) are considered to be acceptable if they are similar to or lower than $\chi^2_{\rm T2-T1}$ (i.e., the light and dark green points in Figure 7). For most of the events, $\chi^2_{\rm T3-T1}$ reaches the acceptable level when the aluminum thickness decreases to $\sim$0.5 mm. Although some of the events prefer smaller thickness to reach lower $\chi^2_{\rm T3-T1}$, we tend to minimize the change to the original thickness of T3, especially when the two effective thicknesses (0.5 mm and 0.45 mm) have a similar effect. Additionally, Figure 7 shows no strong correlation between flare classes and effective thicknesses, except that there is a weak trend for T3 that it is harder to reach the acceptable level when the flare is larger. According to the sample events in this work, the flare class should not affect the potential change of effective aluminum thickness of T3.

On the other hand, more tests on cross-calibrations among different satellites are needed to confirm that the selection of effective thickness is reasonable. Equation 2 can also be

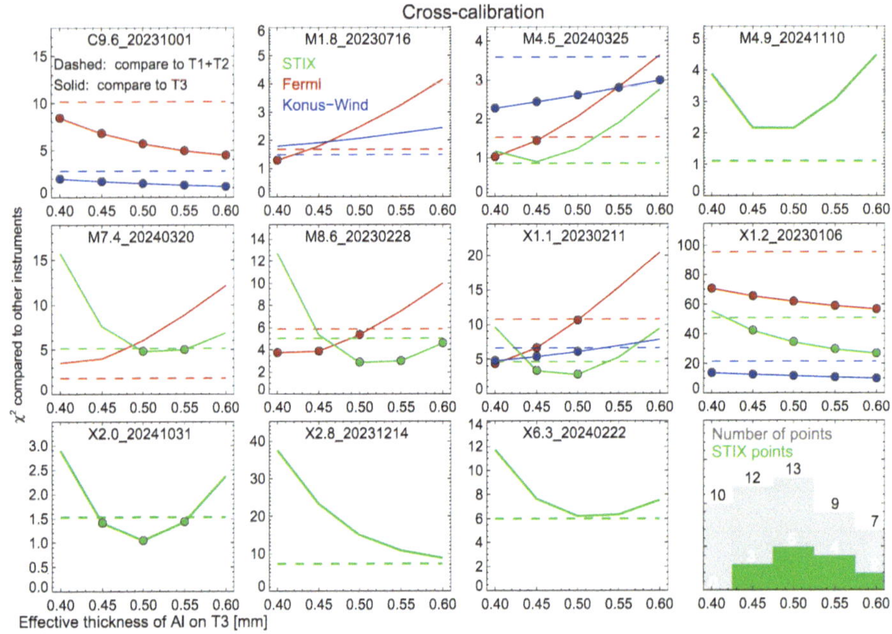

Figure 8 Cross-calibration results of HXI detectors with other X-ray instruments. The dashed and solid lines in each panel (except the last one) represent the χ^2_{Diff} value of each instrument compared to T1+T2 and T3, respectively. The marked points indicate where the solid lines are lower than the dashed lines. The X-axis in each panel shows the changing effective thickness of the aluminum cover in front of T3. The event information is labeled in the form of 'GOES class' plus 'event date'. The last panel shows the number of all the points and STIX points in the previous panels.

applied to calculate the χ^2 for cross-calibrations. Figure 8 shows the corresponding χ^2 for all the comparisons in Figure 5. The dashed lines represent the difference level between T1+T2 and other instruments, and they are unrelated to the effective thickness of T3. The differences between T3 and other instruments (solid lines) are considered to be acceptable if they are similar to, or lower than, those between T1+T2 and other instruments (dashed lines). These cases are shown as marked points in the last panel. Most of the points are recorded near 0.5 mm, indicating that the effective thickness of T3 should decrease to around 0.5 mm.

According to the cited self- and cross-calibration results, the best effective thickness in front of T3 can be determined to be 0.5 mm, and the DRM is consequently adjusted to the corresponding one. Be aware that, in HXI GUI v1.42, the DRM of T3 uses an effective thickness of 0.6 mm. The new DRM for detector T3 will be released in the latest version (v1.50beta) of HXI software packages.

With the latest energy calibration of HXI detectors and the new response matrix for T3, the three detectors T1, T2, T3 are well calibrated and consistent. Therefore, we suggest that readers can use HXI GUI to produce the combined spectra data from the three detectors for spectral analysis. An exception is that T3 may be affected by a (slight) pile-up effect around the peak time of very large flares with a GOES class larger than X5. In that case, T1 and T2 can be combined for spectral analysis. Due to the thick front window and the small deadtime (~ 1.85 μs) of HXI, they should not be affected by a pile-up effect. In order to effectively constrain the thermal component, we suggest a start energy of 10 or 11 keV for the fitting range (same as the lower observation limit of HXI).

3.3. Comparison Between HXI, STIX, Fermi, and Konus-Wind

The comparison between HXI and other instruments are performed with the new DRM in this section. We use the event 2023-02-11 and 2024-03-25 as examples. Figure 9 shows the spectral comparisons in count space after converting the best-fit photon spectra of other instruments. By comparing Figures 9(a) and (b), it is evident that the expected count spectra from STIX and Fermi in the low-energy range with the new T3 DRM are more consistent with the observed T3 spectrum. Similar results are found for event 2024-03-25 in Figures 9(c) and (d), where combined HXI detectors T1+T2+T3 are used. The change of DRM mainly affects the energies lower than 20 keV; thus, few impacts are shown in the expected spectra from Konus-Wind.

We add the best-fit parameters of the spectra of HXI combined detectors T1+T2+T3 with new DRM in Table 2. It is observed that the non-thermal electron indices δ fitted by HXI are more or less consistent with other instruments, but relatively large differences are found in the low-energy cutoff (which can be caused by a number of reasons). While E_c do affect study of thermal and non-thermal processes, they do not affect much in this study because the photon spectra are more important than the physical models here. On the other hand, since the distribution of fluxes as a function of energy is very important in the study of directivity, here we focus on δ.

For instance, the δ of the 2023-01-06 event derived from HXI and STIX are close to the values reported in Shi et al. (2024). For STIX, the slight differences in electron δ compared to other instruments with Earth's perspectives may be attributed to potential directivity of X-ray emissions or a pulse pile-up effect. Most of the fitted thermal components are not well-constrained due to several reasons: a) a single thermal function may not represent the multi-thermal nature of plasma emission in some flares; b) the responses below ~ 20 keV are relatively low for HXI than STIX; c) ~ 23 keV is the low-energy boundary for Konus-Wind (Lysenko et al. 2022) and $\sim$ 10 keV for HXI; d) the large difference of the low-energy cutoffs (E_c, which is also affected by the thermal component and pile-up effect); e) different levels of pulse pile-up effects; and f) the calibration at low energies for Fermi/GBM is not well understood. Besides, the thermal parameters derived from HXI T3 (with 0.6-mm DRM) and HXI T1+T2+T3 (with 0.5 mm-T3 DRM) are obviously different, which is reasonable because the change of T3 DRM mainly affect the low-energy part. Almost all the events show that the thermal parameters derived from combined HXI detectors with the new DRM are closer to that from STIX.

Figure 10 shows a summary of the spectral comparison of all the select events. The light blue regions indicate a relatively small difference of electron index δ (< 0.5) and photon flux ($< 10\%$) at certain energies (30 and 50 keV). The electron δ shows a better consistency (19 of 21 points are located within ± 0.5) than the photon flux at 50 keV (11 of 21 points are located within $\pm 10\%$) among different instruments. After excluding the worst cases that have a severe pile-up effect (with $r_{\rm live} < 0.7$) and the cases with maximum fitting energies near 50 keV, there are 15 of 17 points located within ± 0.5 δ difference and 11 of 17 points located within $\pm 10\%$ flux difference at 50 keV. For STIX, almost all the points are within ± 0.3 difference of electron δ.

Considering that the flux discrepancies among different instruments may be energy dependent, we examined the ratios of total counts of other instruments to those of HXI in

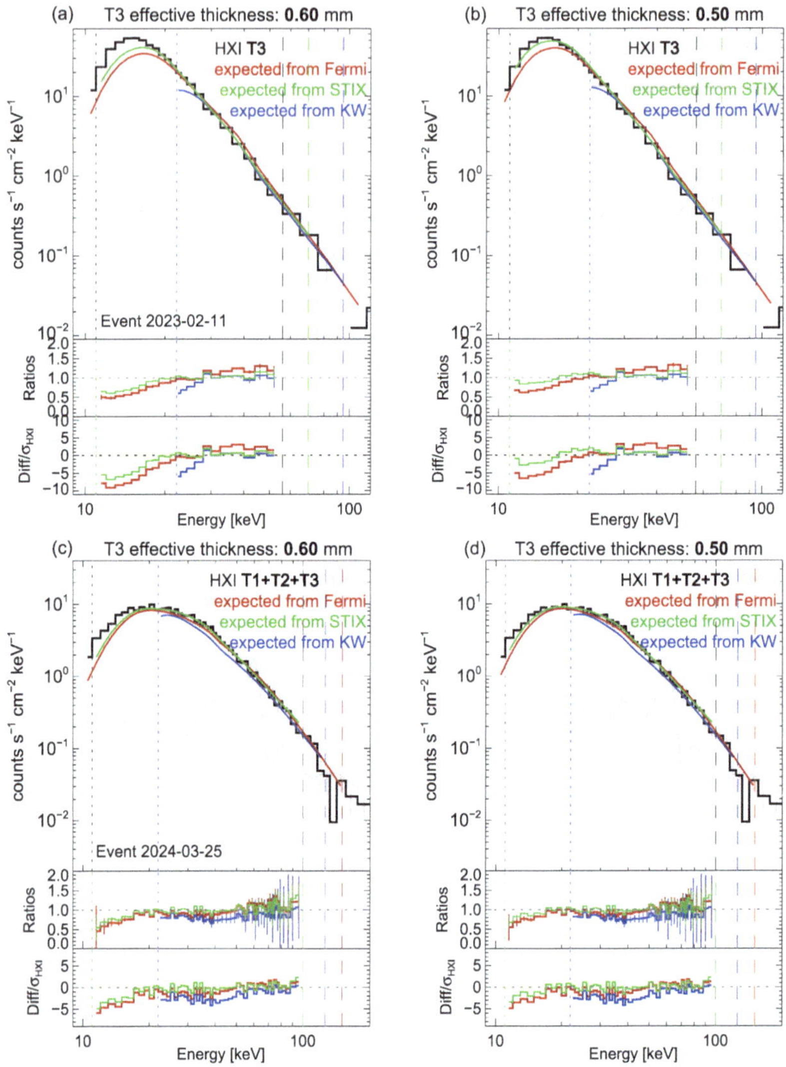

Figure 9 Spectral comparison with the original and new T3 response matrices for the 2023-02-11 event (top panels) and 2024-03-25 event (bottom panels). Panel (a): The green, red, and blue curves represent the expected count spectra derived from the best-fit photon spectra for STIX, Fermi, and Konus-Wind, respectively, using the DRM of HXI T3 with 0.6-mm effective aluminum thickness. Panel (b): same as panel (a) but using the DRM of T3 with 0.5-mm aluminum thickness. Black curves represent the observed background-subtracted count spectrum from the combined detectors T1+T2+T3. Gray curves indicate the HXI backgrounds. Panel (c) and (d): same as panel (a) and (b) but for combined T123 detectors and the 2024-03-25 event.

10 – 20 keV, 20 – 30 keV, and above 30 keV in Figure 11. Most of STIX events have a consistency of ±10% of the total counts in the 10 – 20 keV range. However, Fermi exhibits significant discrepancies within this range due to the influence of pile-up effects (a certain portion of low-energy photons has been recorded as higher-energy photons). In the 20 – 30 keV range, there are generally larger differences, which may be easily attributed to the effects of

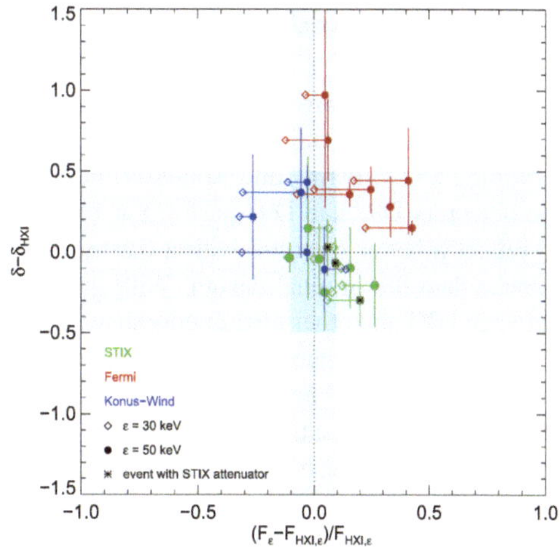

Figure 10 Spectral comparison summary of all the selected events. The X-axis values are the relative differences of the best-fit photon flux at certain energy ϵ between HXI and other instruments (shown in corresponding colors). The Y-axis values are the differences of electron indexes, and the uncertainties from spectral fitting are also plotted. For STIX, the three events with attenuator are indicated by an asterisk.

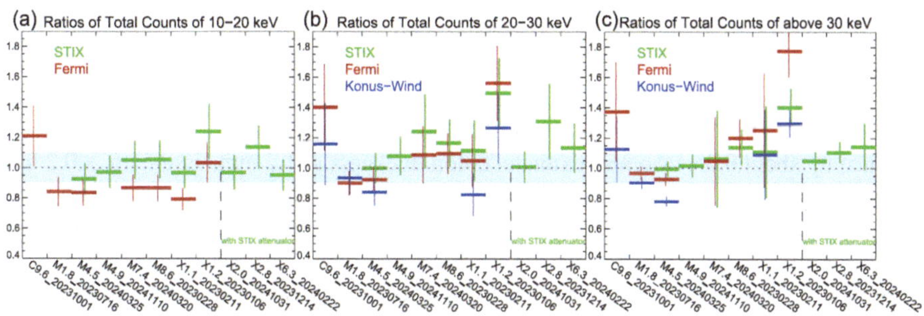

Figure 11 Ratios of total counts in various energy ranges of other instruments to those of HXI T123 (with new T3 DRM). The total counts in certain ranges of each instrument are derived from the expected HXI count spectrum using each best-fit photon spectrum. The flares labeled in the X-axis are in increasing order of GOES class. The last three flares in each panel are observed by STIX with attenuators.

albedo (related to X-ray directivity) and pile-up in this energy range. From 30 keV up to the upper limit of each fitting energy range, the three instruments are in relatively good agreement, except for events with substantial pile-up effects. In particular, STIX shows agreement mostly within $\pm 10\%$ difference from HXI for moderate flares and small separation angle events. For instance, there are 5 events with separation angles smaller than $20°$, and their > 30 keV total count ratios to HXI are 1.00 ± 0.05 (event 20240325), 1.02 ± 0.08 (event 20241110), 1.06 ± 0.32 (event 20240320), 1.05 ± 0.06 (event 20241031), and 1.11 ± 0.07 (event 20231214), respectively. Therefore, the difference above 30 keV of STIX and HXI can reach an averaged level of $\pm 4\%$, with an averaged random error of the ratios of $\sim 6\%$ after excluding event 20240320 which has a small number of counts above 50 keV. For Fermi, the worst-fitting events are those with severe pile-up effect (event 20230106, 20230211 and 20230228) according to Appendix B and r_{live} in Table 2. The difference in the Fermi to HXI ratio for the total counts above 30 keV can also reach a level of $\pm 10\%$ for the moderate flare events. Although the spectral fitting χ^2 values of Konus-Wind are usually small, the r_{live} values are relatively larger than other instruments, possibly resulting in a $> 10\%$ difference level of the total counts of above 30 keV.

4. Conclusion and Discussion

In this article, we carried out self-calibration of the HXI detector response matrix of three detectors and cross-calibration based on observations of HXI and other X-ray instruments, including STIX, Fermi, and Konus-Wind. Both tests suggested that the response matrix of the detectors D92 and D93 perform as expected and that of the D94 needs to be adjusted to reflect a better response with a 0.5-mm effective aluminum window, which is 0.1 mm thinner than the original design. With the new response matrix, the spectral comparisons between HXI and other instruments showed significantly improved consistency.

Although differences between the HXI spectra and the expected HXI spectra from other instruments (mainly Fermi/GBM) still exist at low energies below ~ 30 keV, a directivity study should not be affected since it focuses on non-thermal emissions at energies higher than ~ 30 keV in most cases, below which thermal emissions usually dominate. It is therefore possible to perform joint spectral study especially in the energy range of ~ 30 keV to ~ 80 keV which is the overlapped range within the responses of all the instruments mentioned.

It should be pointed out that the measured thicknesses of the aluminum window for D92 and D93 are 0.85 ± 0.03 mm and that for D94 is 0.59 ± 0.03 mm. These measurement errors are significantly smaller than the adjustment of the effective thickness of aluminum required to generate the new DRM. However, there are other materials (Teflon) that may affect the detection efficiency. At 20 keV, a 0.01-mm difference in aluminum thickness affects transmittance by about 0.05.[7] An approximate Teflon thickness at 1.2 mm, assuming ± 0.2 mm deviation, equivalently impacts the aluminum thickness by about 0.046 mm due to X-ray attenuation properties, leading to a possible total deviation of 0.07 to 0.08 mm (Hubbell and Seltzer 1995).

In total, $\sim 88\%$ of all the observations exhibit spectral indexes consistent within a ± 0.5 range. For STIX flares with $< 20^\circ$ separation angles and all the moderate flares observed by Fermi, the total counts above 30 keV show differences within $\pm 10\%$ when compared to HXI measurements, demonstrating strong potential for joint studies. Notably, STIX-HXI co-observed events within 20° separation angle display an averaged difference of $\pm 4\%$, with corresponding random error in ratios averaging $\sim 6\%$. These sub-10% discrepancies enable preliminary investigations of electron anisotropy (Jeffrey et al. 2024). Furthermore, the three events observed with STIX attenuator showed that the high-energy (> 30 keV) total counts exceeded those of HXI by $5-14\%$, constituting a significant improvement over large flares observed without STIX attenuator, where discrepancies exceeded $10-40\%$. Future joint spectral analyses will become statistically robust as HXI and STIX accumulate more observations of large flares with the same perspective.

It is known that achieving a high level of consistency among different instruments is challenging and that cross-calibrations are also influenced by various factors. In addition to the physical difference of X-ray emissions (mainly caused by X-ray directivity and albedo), here we list the factors that may affect cross-calibrations:

1. Detector response and grid transmission. The knowledge of detectors and imaging grids (how well we know the detector/pixel size, material composition of detector and attenuation window, grid transmissions, etc.) is the most important factor here. Since HXI has independent total flux detectors, all effects of these factors are related to the calibration

[7] According to the X-ray mass attenuation coefficient of aluminum: https://physics.nist.gov/PhysRefData/XrayMassCoef/ElemTab/z13.html.

accuracy of detector responses. Although STIX also has total flux monitors, their effective area is too small (~ 2.22 mm^2) to accumulate enough counts for moderate flares (Krucker et al. 2020). Total fluxes are mainly recorded by imaging detectors, which may induce fuzzy grid effect (Jeffrey et al. 2024). The uncertainty is $\sim 10\%$ if the 24 coarsest detectors are used and only a few percent if the 12 coarsest detectors are used. STIX detector response is also affected by the calibration status of the movable attenuator.

2. Different energy resolutions and coverage. As discussed in Jeffrey et al. (2024), the spectral energy accuracy is limited by different energy resolutions of different instruments and also different energy ranges. Consequently, the error ranges are different, and we can hardly conclude their consistency or discrepancy if some of the errors are large.
3. Energy calibration. The spectral energy calibration of HXI is primarily based on three emission lines: the 26.4 keV and 59.5 keV lines from ^{241}Am and the 356 keV line from ^{133}Ba. There is a potential issue with the 26.4 keV line since its profile could be influenced by the group of lines below 26 keV (with lower intensities) and the 31 keV line emitted by ^{133}Ba. Additionally, the activity ratios of ^{241}Am to ^{133}Ba vary across different detectors, which introduces uncertainty in pinpointing the exact position of the 26.4 keV line. Although all these factors are considered in energy calibration, the calibration result might be subject to bias at the lower end of the energy spectrum, particularly for energies below 15 keV, where no lines can be used for confirmation.
4. Albedo effect. The total X-ray flux and the spectral shape can be affected when observing from different viewing angles because of the albedo effect, which is model-dependent and sensitive in many cases (e.g., Kontar et al. 2006; Kontar and Jeffrey 2010). STIX mostly observes flares from a different perspective than the others. Although we added the simple albedo correction in spectral fitting functions, the electron distributions still need to be assumed (as isotropic or beamed). Thus, it would be a challenge to correctly construct/apply the albedo component to a flare observed from different perspectives for joint studies.
5. Pulse pile-up effect. Pile-up effect can easily alter the spectral shape, in some cases by introducing a second component as a bump in the spectra which affects the derived photon spectra. The determination of the pile-up effect depends on the instrument design (e.g., detector effective area, attenuator, deadtime) and the flux levels of the detection. For Fermi/GBM, the pile-up effect level is different for most sunward and less sunward detectors. For some events in our study (e.g., 2023-01-06 event), the spectral behavior is similar to some analytical models (Chaplin et al. 2013). For STIX, the most different scenarios are observing with or without attenuators, which significantly affects the fluxes at low energies and the calculation of livetimes (see the last column of Table 2). When STIX is observing with attenuators, the livetime should be closer to the exposure time. This is because more photons are blocked by the attenuators, which reduces the pile-up effect.
6. Different time resolutions. Although we can choose the time intervals to minimize the time difference, it may still introduce differences between average count rates (which is used in OSPEX) and averaged counts for changing livetimes during the time interval of interest. The discrepancy would be larger for fast-changing events with a low count rate observed by different instruments with various livetimes.
7. Background subtraction. Proper methods of background subtraction are crucial since they may directly impact the spectral shape and flux, especially at high energies. For different near-Earth orbiting instruments, they may gain severe signals from orbital background particles, which makes the background subtraction processes more complicated and uncertain. Although the background levels for STIX are relatively stable, the recommended

background files from the STIX data-center website can still be either overestimated or underestimated. This discrepancy may impact the accuracy of spectra for smaller flares.

8. Uncertainties of spectral fitting. The ultimate purpose of spectral fitting is to progressively approach the true X-ray emissions from the source. However, the spectral fitting result technically depends on the goodness of the fitted spectrum that is mostly evaluated by χ^2 which may still not describe the sufficiency of fitting all the information of the true spectrum, and physically depends on the selected fitting models, which may perform the equivalent total effect (Krucker and Lin 2008). Some spectral fitting residuals may contain systematic features, indicating incomplete fitting, even if the calculated χ^2 values are the same as that of a better fit. Different starting parameters could also affect the best-fit parameters.
9. Instrument temperature change. Temperature variations and different temperatures between the detectors at edge and center may have impacts on the detector response and detection efficiency. For HXI, the thermal control system plays a crucial role in monitoring and regulating the instrument temperature (Su et al. 2024), thereby mitigating the adverse effects on detector response and efficiency. However, for some other instruments, especially those affected by temperature changes due to variations in the distance from the Sun, the effects of such temperature variations may need to be considered.

Here, the most important factors include the knowledge of detector response, energy resolution, energy calibration, pulse pile-up, and albedo effect.

In summary, the preliminary results of our cross-calibration work suggest that directivity studies of X-rays, studies of occulted flares, and joint imaging can be carried out in the energy range of 30 – 80 keV with the adjusted response matrix of HXI detector T3. The differences between the two spectra of HXI and STIX will be acceptable (within $3\sigma_{\text{error}}$, $\pm4\%$ of δ, and $\pm4\%$ of counts) for drawing preliminary conclusions of X-ray directivity. In order to get more precise calibration results, more events that are suitable for cross-calibration are needed from new observations in the future. Better understanding of the STIX detector responses with attenuator and Fermi detector responses at low energies will also help to improve the cross-calibration results.

Appendix A: Systematic Error of HXI Background

For each investigated event, we compared the spectrum observed 2 days before (or after) with the one observed 4 days before (or after) during the same time interval for spectral fitting, both having no emission from flares. The results are shown in Figure 12. The averaged difference between each spectrum and its $\pm$ 48 h profile (which is regarded as background) is mainly in the $\pm$ 2% range. Some of the differences are even larger, primarily due to the short durations of the selected time intervals, which may introduce larger uncertainties.

Appendix B: Fitting Results of HXI, STIX, Fermi/GBM, and Konus-Wind

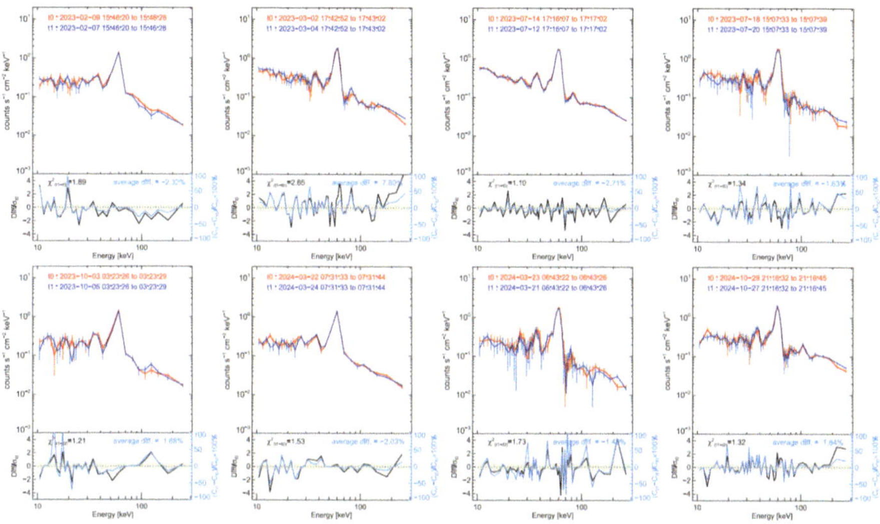

Figure 12 Systematic error of HXI background.

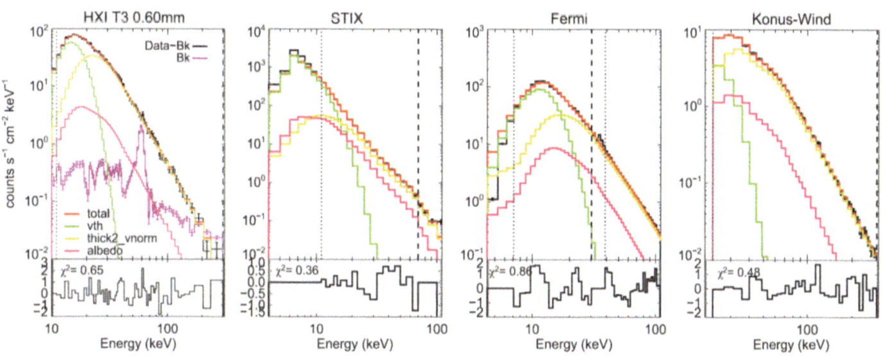

Figure 13 Fitting results of event 2023-01-06.

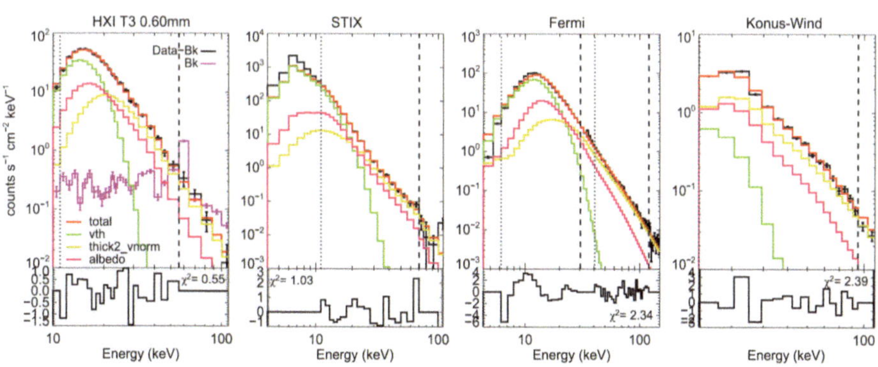

Figure 14 Fitting results of event 2023-02-11.

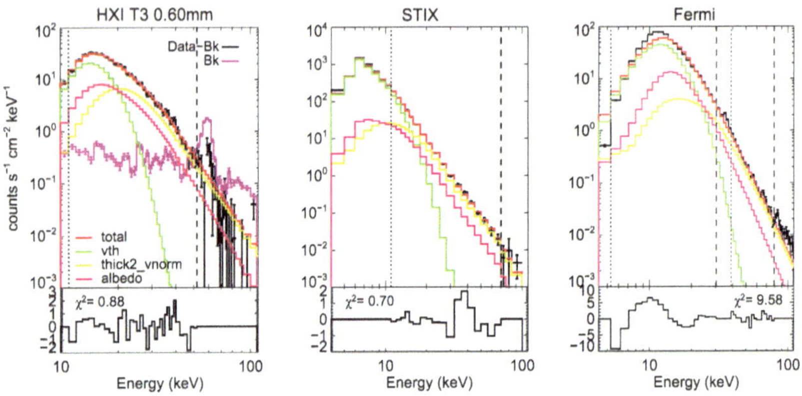

Figure 15 Fitting results of event 2023-02-28.

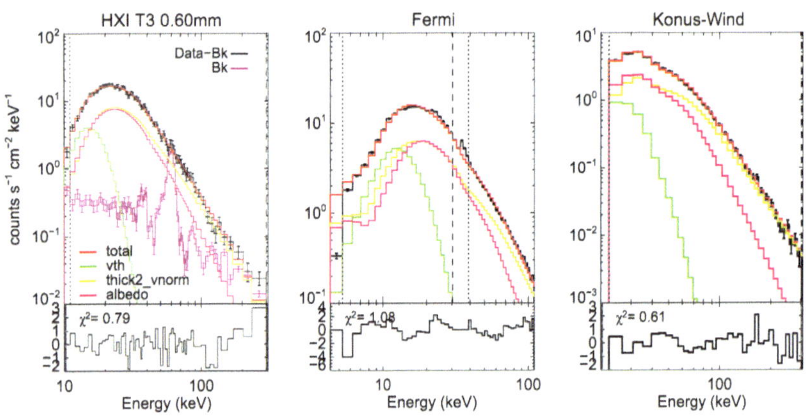

Figure 16 Fitting results of event 2023-07-16.

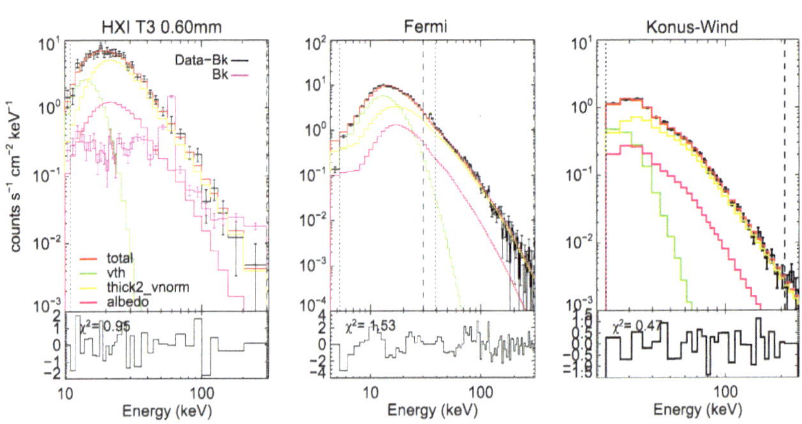

Figure 17 Fitting results of event 2023-10-01.

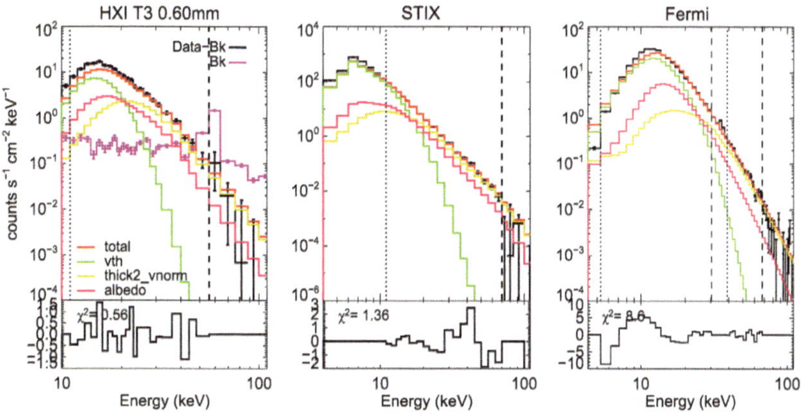

Figure 18 Fitting results of event 2024-03-20.

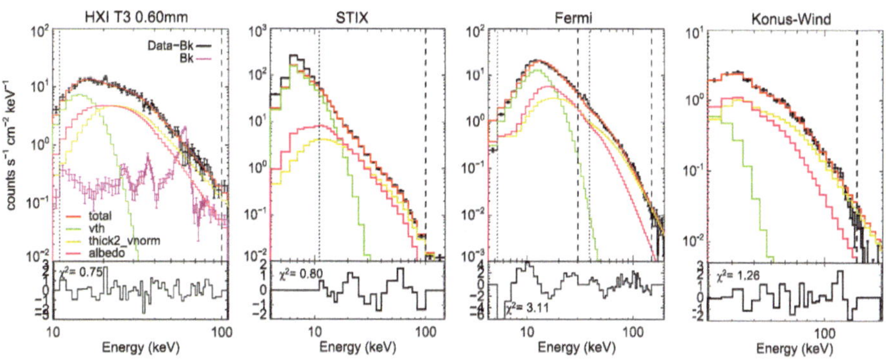

Figure 19 Fitting results of event 2024-03-25.

Figure 20 Fitting results of event 2024-11-10.

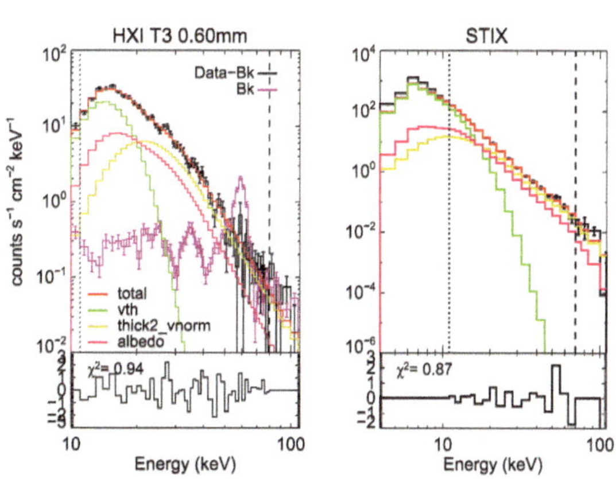

Figure 21 Fitting results of event 2023-12-14.

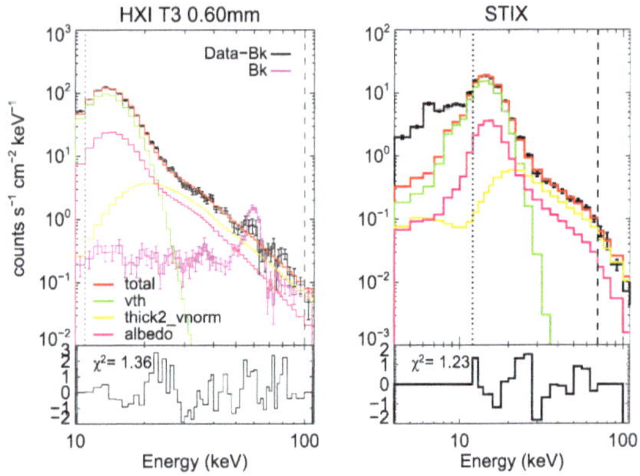

Figure 22 Fitting results of event 2024-02-22.

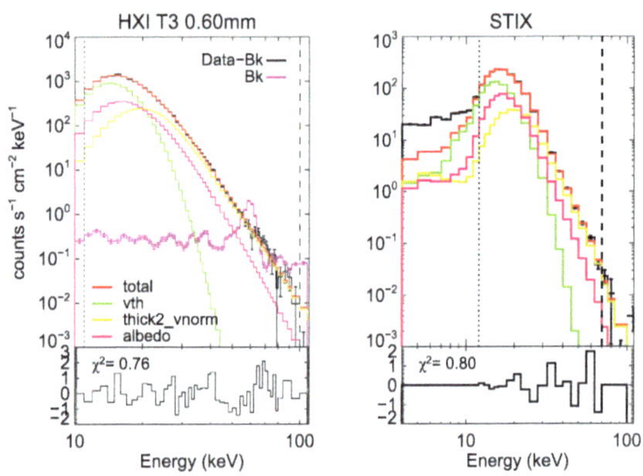

Figure 23 Fitting results of event 2024-10-31.

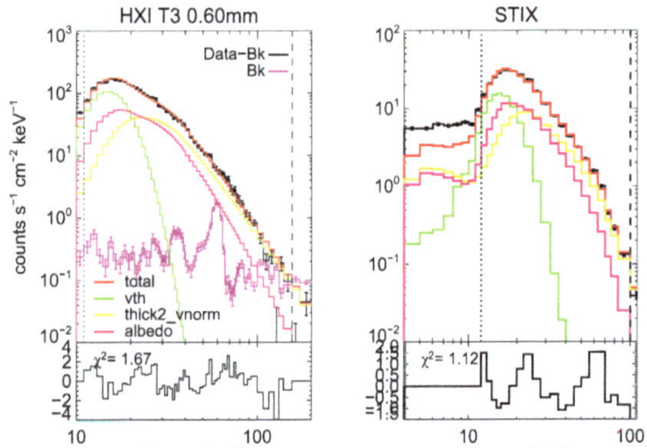

Acknowledgments The authors thank Säm Krucker, Tomasz Mrozek, and Karol Kulaga for providing suggestions on STIX spectral analysis. Many thanks to Alexandra Lysenko and Alexey Kuznetsov for interpretations of Konus-Wind data.

Author contributions Z.L. analyzed the data, performed almost all the tests, prepared all the figures, and wrote the main manuscript text. Y.S. proposed and guided this work, helped to improve and revise the manuscript, gave advices on data processing, calibration strategy, and spectral fitting, etc. Y.S. and W.C. completed the detector energy calibration and reproduced HXI science data . W.L. simulated the detector response matrices and helped to give explanations and interpretations on relative parts. W.C. helped to analyze HXI spectra and discuss the uncertainty caused by energy calibration. F.Y. helped to collect event samples and run the programs. W.G. led the ASO-S team and gave valuable suggestions. All authors reviewed the manuscript.

Funding Information The work is supported by the National Key R&D Program of China 2022YFF0503002, and the NSFC 12333010, 11921003, 11820101002. ZL is also supported by the Prominent Postdoctoral Project of Jiangsu Province (2023ZB304). ASO-S mission is supported by the Strategic Priority Research Program on Space Science, the Chinese Academy of Sciences, Grant No. XDA15320000.

Data Availability The datasets generated for this paper are available from the authors upon reasonable request.

Declarations

Competing interests The authors declare no competing interests.

References

Agostinelli, S., Allison, J., Amako, K.A., Apostolakis, J., Araujo, H., Arce, P., Asai, M., Axen, D., Banerjee, S., Barrand, G., et al.: 2003, GEANT4—a simulation toolkit. *Nucl. Instrum. Methods Phys. Res., Sect. A, Accel. Spectrom. Detect. Assoc. Equip.* **506**, 250.

Aptekar, R.L., Frederiks, D.D., Golenetskii, S.V., Ilynskii, V.N., Mazets, E.P., Panov, V.N., Sokolova, Z.J., Terekhov, M.M., Sheshin, L.O., Cline, T.L., Stilwell, D.E.: 1995, Konus-W gamma-ray burst experiment for the GGS Wind spacecraft. *Space Sci. Rev.* **71**, 265. DOI. ADS.

Bajnoková, N., Hannah, I.G., Cooper, K., Krucker, S., Grefenstette, B.W., Smith, D.M., Jeffrey, N.L.S., Duncan, J.: 2024, First joint X-ray solar microflare observations with NuSTAR and Solar Orbiter/STIX. *Mon. Not. Roy. Astron. Soc.* **533**, 3742. DOI. ADS.

Benz, A.O.: 2017, Flare observations. *Living Rev. Solar Phys.* **14**, 2. DOI. ADS.

Brown, J.C.: 1971, The deduction of energy spectra of non-thermal electrons in flares from the observed dynamic spectra of hard X-ray bursts. *Solar Phys.* **18**, 489. DOI. ADS.

Caspi, A., Lin, R.P.: 2010, RHESSI line and continuum observations of super-hot flare plasma. *Astrophys. J. Lett.* **725**, L161. DOI. ADS.

Chaplin, V., Bhat, N., Briggs, M.S., Connaughton, V.: 2013, Analytical modeling of pulse-pileup distortion using the true pulse shape; applications to Fermi-GBM. *Nucl. Instrum. Methods Phys. Res., Sect. A, Accel. Spectrom. Detect. Assoc. Equip.* **717**, 21. DOI. ADS.

Dickson, E.C.M., Kontar, E.P.: 2013, Measurements of electron anisotropy in solar flares using albedo with RHESSI X-ray data. *Solar Phys.* **284**, 405. DOI. ADS.

Gan, W., Zhu, C., Deng, Y., Zhang, Z., Chen, B., Huang, Y., Deng, L., Wu, H., Zhang, H., Li, H., Su, Y., Su, J., Feng, L., Wu, J., Cui, J., Wang, C., Chang, J., Yin, Z., Xiong, W., Chen, B., Yang, J., Li, F., Lin, J., Hou, J., Bai, X., Chen, D., Zhang, Y., Hu, Y., Liang, Y., Wang, J., Song, K., Guo, Q., He, L., Zhang, G., Wang, P., Bao, H., Cao, C., Bai, Y., Chen, B., He, T., Li, X., Zhang, Y., Liao, X., Jiang, H., Li, Y., Su, Y., Lei, S., Chen, W., Li, Y., Zhao, J., Li, J., Ge, Y., Zou, Z., Hu, T., Su, M., Ji, H., Gu, M., Zheng, Y., Xu, D., Wang, X.: 2023, The Advanced Space-based Solar Observatory (ASO-S). *Solar Phys.* **298**, 68. DOI. ADS.

Grigis, P.C., Benz, A.O.: 2006, Electron acceleration in solar flares: theory of spectral evolution. *Astron. Astrophys.* **458**, 641. DOI. ADS.

Harrison, F.A., Craig, W.W., Christensen, F.E., Hailey, C.J., Zhang, W.W., Boggs, S.E., Stern, D., Cook, W.R., Forster, K., Giommi, P., Grefenstette, B.W., Kim, Y., Kitaguchi, T., Koglin, J.E., Madsen, K.K., Mao, P.H., Miyasaka, H., Mori, K., Perri, M., Pivovaroff, M.J., Puccetti, S., Rana, V.R., Westergaard, N.J., Willis, J., Zoglauer, A., An, H., Bachetti, M., Barrière, N.M., Bellm, E.C., Bhalerao, V., Brejnholt, N.F., Fuerst, F., Liebe, C.C., Markwardt, C.B., Nynka, M., Vogel, J.K., Walton, D.J., Wik, D.R., Alexander, D.M., Cominsky, L.R., Hornschemeier, A.E., Hornstrup, A., Kaspi, V.M., Madejski, G.M., Matt, G., Molendi, S., Smith, D.M., Tomsick, J.A., Ajello, M., Ballantyne, D.R., Baloković, M., Barret, D., Bauer, F.E., Blandford, R.D., Brandt, W.N., Brenneman, L.W., Chiang, J., Chakrabarty, D., Chenevez, J., Comastri, A., Dufour, F., Elvis, M., Fabian, A.C., Farrah, D., Fryer, C.L., Gotthelf, E.V., Grindlay, J.E., Helfand, D.J., Krivonos, R., Meier, D.L., Miller, J.M., Natalucci, L., Ogle, P., Ofek, E.O., Ptak, A., Reynolds, S.P., Rigby, J.R., Tagliaferri, G., Thorsett, S.E., Treister, E., Urry, C.M.: 2013, The Nuclear Spectroscopic Telescope Array (NuSTAR) high-energy X-ray mission. *Astrophys. J.* **770**, 103. DOI. ADS.

Holman, G.D., Sui, L., Schwartz, R.A., Emslie, A.G.: 2003, Electron Bremsstrahlung hard X-ray spectra, electron distributions, and energetics in the 2002 July 23 solar flare. *Astrophys. J. Lett.* **595**, L97. DOI. ADS.

Holman, G.D., Aschwanden, M.J., Aurass, H., Battaglia, M., Grigis, P.C., Kontar, E.P., Liu, W., Saint-Hilaire, P., Zharkova, V.V.: 2011, Implications of X-ray observations for electron acceleration and propagation in solar flares. *Space Sci. Rev.* **159**, 107. DOI. ADS.

Hubbell, J.H., Seltzer, S.M.: 1995, Tables of X-ray mass attenuation coefficients and mass energy-absorption coefficients 1 keV to 20 MeV for elements Z = 1 to 92 and 48 additional substances of dosimetric interest. Technical report, National Inst. of Standards and Technology-PL, Gaithersburg. MD.

Hurley, K., Sommer, M., Atteia, J.-L., Boer, M., Cline, T., Cotin, F., Henoux, J.-C., Kane, S., Lowes, P., Niel, M.: 1992, The solar X-ray/cosmic gamma-ray burst experiment aboard ULYSSES. *Astron. Astrophys. Suppl. Ser.* **92**, 401. ADS.

Jeffrey, N.L.S., Krucker, S., Stores, M., Kontar, E.P., Saint-Hilaire, P., Battaglia, A.F., Hayes, L., Collier, H., Veronig, A., Su, Y., Tadepalli, S.P., Xia, F.: 2024, A modeling investigation for solar flare X-ray stereoscopy with Solar Orbiter/STIX and Earth-orbiting missions. *Astrophys. J.* **964**, 145. DOI. ADS.

Kane, S.R., Hurley, K., McTiernan, J.M., Boer, M., Niel, M., Kosugi, T., Yoshimori, M.: 1998, Stereoscopic observations of solar hard X-ray flares made by Ulysses and Yohkoh. *Astrophys. J.* **500**, 1003. DOI. ADS.

Kaneko, Y., Bostancı, Z.F., Göğüş, E., Lin, L.: 2015, Short gamma-ray bursts with extended emission observed with Swift/BAT and Fermi/GBM. *Mon. Not. Roy. Astron. Soc.* **452**, 824. DOI. ADS.

Kašparová, J., Kontar, E.P., Brown, J.C.: 2007, Hard X-ray spectra and positions of solar flares observed by RHESSI: photospheric albedo, directivity and electron spectra. *Astron. Astrophys.* **466**, 705. DOI. ADS.

Kontar, E.P., Brown, J.C.: 2006, Stereoscopic electron spectroscopy of solar hard X-ray flares with a single spacecraft. *Astrophys. J. Lett.* **653**, L149. DOI. ADS.

Kontar, E.P., Jeffrey, N.L.S.: 2010, Positions and sizes of X-ray solar flare sources. *Astron. Astrophys.* **513**, L2. DOI. ADS.

Kontar, E.P., MacKinnon, A.L., Schwartz, R.A., Brown, J.C.: 2006, Compton backscattered and primary X-rays from solar flares: angle dependent Green's function correction for photospheric albedo. *Astron. Astrophys.* **446**, 1157. DOI. ADS.

Kontar, E.P., Brown, J.C., Emslie, A.G., Hajdas, W., Holman, G.D., Hurford, G.J., Kašparová, J., Mallik, P.C.V., Massone, A.M., McConnell, M.L., Piana, M., Prato, M., Schmahl, E.J., Suarez-Garcia, E.: 2011, Deducing electron properties from hard X-ray observations. *Space Sci. Rev.* **159**, 301. DOI. ADS.

Krucker, S., Lin, R.P.: 2008, Hard X-ray emissions from partially occulted solar flares. *Astrophys. J.* **673**, 1181. DOI. ADS.

Krucker, S., Battaglia, M., Cargill, P.J., Fletcher, L., Hudson, H.S., MacKinnon, A.L., Masuda, S., Sui, L., Tomczak, M., Veronig, A.L., Vlahos, L., White, S.M.: 2008, Hard X-ray emission from the solar corona. *Astron. Astrophys. Rev.* **16**, 155. DOI. ADS.

Krucker, S., Hurford, G.J., Su, Y., Gan, W.-Q.: 2019, Joint hard X-ray observations with ASO-S/HXI and SO/STIX. *Res. Astron. Astrophys.* **19**, 167. DOI. ADS.

Krucker, S., Hurford, G.J., Grimm, O., Kögl, S., Gröbelbauer, H.-P., Etesi, L., Casadei, D., Csillaghy, A., Benz, A.O., Arnold, N.G., Molendini, F., Orleanski, P., Schori, D., Xiao, H., Kuhar, M., Hochmuth, N., Felix, S., Schramka, F., Marcin, S., Kobler, S., Iseli, L., Dreier, M., Wiehl, H.J., Kleint, L., Battaglia, M., Lastufka, E., Sathiapal, H., Lapadula, K., Bednarzik, M., Birrer, G., Stutz, S., Wild, C., Marone, F., Skup, K.R., Cichocki, A., Ber, K., Rutkowski, K., Bujwan, W., Juchnikowski, G., Winkler, M., Darmetko, M., Michalska, M., Seweryn, K., Białek, A., Osica, P., Sylwester, J., Kowalinski, M., Ścisłowski, D., Siarkowski, M., Stęślicki, M., Mrozek, T., Podgórski, P., Meuris, A., Limousin, O., Gevin, O., Le Mer, I., Brun, S., Strugarek, A., Vilmer, N., Musset, S., Maksimović, M., Fárník, F., Kozáček, Z., Kašparová, J., Mann, G., Önel, H., Warmuth, A., Rendtel, J., Anderson, J., Bauer, S., Dionies, F., Paschke, J., Plüschke, D., Woche, M., Schuller, F., Veronig, A.M., Dickson, E.C.M., Gallagher, P.T., Maloney, S.A., Bloomfield, D.S., Piana, M., Massone, A.M., Benvenuto, F., Massa, P., Schwartz, R.A., Dennis, B.R., van Beek, H.F., Rodríguez-Pacheco, J., Lin, R.P.: 2020, The Spectrometer/Telescope for Imaging X-rays (STIX). *Astron. Astrophys.* **642**, A15. DOI ADS.

Lin, R.P., Hudson, H.S.: 1976, Non-thermal processes in large solar flares. *Solar Phys.* **50**, 153. DOI. ADS.

Lin, R.P., Schwartz, R.A., Pelling, R.M., Hurley, K.C.: 1981, A new component of hard X-rays in solar flares. *Astrophys. J. Lett.* **251**, L109. DOI. ADS.

Lin, R.P., Dennis, B.R., Hurford, G.J., Smith, D.M., Zehnder, A., Harvey, P.R., Curtis, D.W., Pankow, D., Turin, P., Bester, M., Csillaghy, A., Lewis, M., Madden, N., van Beek, H.F., Appleby, M., Raudorf, T., McTiernan, J., Ramaty, R., Schmahl, E., Schwartz, R., Krucker, S., Abiad, R., Quinn, T., Berg, P., Hashii, M., Sterling, R., Jackson, R., Pratt, R., Campbell, R.D., Malone, D., Landis, D., Barrington-Leigh, C.P., Slassi-Sennou, S., Cork, C., Clark, D., Amato, D., Orwig, L., Boyle, R., Banks, I.S., Shirey, K., Tolbert, A.K., Zarro, D., Snow, F., Thomsen, K., Henneck, R., McHedlishvili, A., Ming, P., Fivian, M., Jordan, J., Wanner, R., Crubb, J., Preble, J., Matranga, M., Benz, A., Hudson, H., Canfield, R.C., Holman, G.D., Crannell, C., Kosugi, T., Emslie, A.G., Vilmer, N., Brown, J.C., Johns-Krull, C., Aschwanden, M., Metcalf, T., Conway, A.: 2002, The Reuven Ramaty High-Energy Solar Spectroscopic Imager (RHESSI). *Solar Phys.* **210**, 3. DOI. ADS.

Lysenko, A.L., Ulanov, M.V., Kuznetsov, A.A., Fleishman, G.D., Frederiks, D.D., Kashapova, L.K., Sokolova, Z.Y., Svinkin, D.S., Tsvetkova, A.E.: 2022, KW-Sun: the Konus-Wind solar flare database in hard X-ray and soft gamma-ray ranges. *Astrophys. J. Suppl.* **262**, 32. DOI. ADS.

Meegan, C., Lichti, G., Bhat, P.N., Bissaldi, E., Briggs, M.S., Connaughton, V., Diehl, R., Fishman, G., Greiner, J., Hoover, A.S., van der Horst, A.J., von Kienlin, A., Kippen, R.M., Kouveliotou, C., McBreen, S., Paciesas, W.S., Preece, R., Steinle, H., Wallace, M.S., Wilson, R.B., Wilson-Hodge, C.: 2009, The Fermi Gamma-Ray Burst Monitor. *Astrophys. J.* **702**, 791. DOI. ADS.

Nava, L., Ghirlanda, G., Ghisellini, G., Celotti, A.: 2011, Fermi/GBM and BATSE gamma-ray bursts: comparison of the spectral properties. *Mon. Not. Roy. Astron. Soc.* **415**, 3153. DOI. ADS.

Ogawara, Y., Takano, T., Kato, T., Kosugi, T., Tsuneta, S., Watanabe, T., Kondo, I., Uchida, Y.: 1991, The SOLAR-a mission – an overview. *Solar Phys.* **136**, 1. DOI. ADS.

Phillips, K.J.H.: 2004, The solar flare 3.8-10 keV X-ray spectrum. *Astrophys. J.* **605**, 921. DOI. ADS.

Ryan, D.F., Massa, P., Battaglia, A.F., Dickson, E.C.M., Su, Y., Chen, W., Krucker, S.: 2024, Triangulation of hard X-ray sources in an X-class solar flare with ASO-S/HXI and Solar Orbiter/STIX. *Solar Phys.* **299**, 114. DOI. ADS.

Schwartz, R.A., Csillaghy, A., Tolbert, A.K., Hurford, G.J., Mc Tiernan, J., Zarro, D.: 2002, RHESSI data analysis software: rationale and methods. *Solar Phys.* **210**, 165. DOI. ADS.

Shi, F., Li, D., Ning, Z., Warmuth, A., Chen, W., Su, Y., Li, Y., Xu, J., Song, Y., Yang, Y.: 2024, Multiwavelength observations of quasiperiodic pulsations in the impulsive phase of an eruptive flare with the hard X-ray imager on board ASO-S and other instruments. *Solar Phys.* **299**, 30. DOI.

Su, Y., Liu, W., Li, Y.-P., Zhang, Z., Hurford, G.J., Chen, W., Huang, Y., Li, Z.-T., Jiang, X.-K., Wang, H.-X., Xia, F.-X.-Y., Chen, C.-X., Yu, W.-H., Yu, F., Wu, J., Gan, W.-Q.: 2019, Simulations and software development for the Hard X-ray Imager onboard ASO-S. *Res. Astron. Astrophys.* **19**, 163. DOI. ADS.

Su, Y., Zhang, Z., Chen, W., Chen, D., Yu, F., Hu, Y., Zhang, Y., Xia, F., Chen, C., Li, Z., Jiang, X., Huang, Y., Zhang, Y., Liu, W., Ma, T., Li, D., Yu, W., Li, Y., Cai, M., Guo, J., Huang, Y., Wang, H., Liang, Y., Ma, M., Wang, J., Zhu, S., Tao, J., Yu, J., Yang, J., Wu, J., Gan, W.: 2024, The tests and calibrations of the hard X-ray imager aboard ASO-S. *Solar Phys.* **299**, 153. DOI. ADS.

Zhang, Z., Chen, D.-Y., Wu, J., Chang, J., Hu, Y.-M., Su, Y., Zhang, Y., Wang, J.-P., Liang, Y.-M., Ma, T., Guo, J.-H., Cai, M.-S., Zhang, Y.-Q., Huang, Y.-Y., Peng, X.-Y., Tang, Z.-B., Zhao, X., Zhou, H.-H., Wang, L.-G., Song, J.-X., Ma, M., Xu, G.-Z., Yang, J.-F., Lu, D., He, Y.-H., Tao, J.-Y., Ma, X.-L., Lv, B.-G., Bai, Y.-P., Cao, C.-X., Huang, Y., Gan, W.-Q.: 2019, Hard X-ray Imager (HXI) onboard the ASO-S mission. *Res. Astron. Astrophys.* **19**, 160. DOI. ADS.

Publisher's Note Springer Nature remains neutral with regard to jurisdictional claims in published maps and institutional affiliations.

Solar Physics (2024) 299:122
https://doi.org/10.1007/s11207-024-02366-z

RESEARCH

The Solar Aspect System of the Hard X-ray Imager Onboard ASO-S

Jirui Yu[1,2] · Yinghong He[1] · Jinyou Tao[1] · Yang Su[3,4] · Zhe Zhang[3] · Jianfeng Yang[1] · Nange Wang[1] · Song Guo[1] · Baogang Lv[1] · Xiaobo Chen[1] · Bin Xue[1] · Yiming Hu[3] · Dengyi Chen[3] · Fu Yu[3] · Mingde Ding[5] · Ping Ruan[1]

Received: 16 July 2023 / Accepted: 19 August 2024 / Published online: 2 September 2024

Abstract
The ASO-S/HXI is a bi-grids modulating instrument for solar hard X-ray imaging, whose collimator contains 91 pairs of tungsten grids. Since the solar disk is invisible in hard X-rays, a Solar Aspect System (SAS) is required to provide the pointing of hard X-ray imager (HXI) for locating X-ray sources on the solar disk. In addition, the knowledge of the alignment and relative twist of the corresponding front–rear grid pairs is important for image reconstruction as well as locating flares. Therefore, thc SAS system was designed to monitor the alignment status of HXI grids and to provide the pointing direction of the HXI collimator with two subsystems DM and SA during the whole life cycle of HXI. DM measures the centroids of the front frosted glasses and the solar disk. SA images the Sun and provides precise relative locations of the solar disk center. Both work in the visible light of 565–585 nm. With all the data together, we can solve with an inversion algorithm the alignment status of the front and rear grids, the relative twist, and the pointing direction. We tested and validated the SAS design with both the simulation model and ground coordinate measuring machine. Here we present the detailed system design, the testing results, the inversion algorithm, and the in-orbit status of the SAS. Currently, the SAS has realized the rotational measurement accuracy of about 4 arcsec, and a translational measurement accuracy of about 15 μm, and the SAS pointing data has been used in both imaging calibration for flare locations and imaging corrections for the platform drifting effect. The high-cadence precise measurement (better than 0.3 arcsec) of the pointing will allow the study of source locations at different energies and therefore help us to understand electron acceleration and transportation in flares.

Keywords Solar flare imaging · Hard X-ray · Visual monitoring system · Inverse problem · Accuracy test

1. Introduction

Over the past few decades, solar exploration continued to develop and scientists have made many promising achievements in this field. After the exploration in recent years, China hopes to make a greater contribution in the field of solar physics (Gan, Yan, and Huang 2019). Therefore, China has actively launched two dedicated exploration satellites into orbit to provide the latest solar observation data for scholars around the world. As China's

Extended author information available on the last page of the article

first comprehensive solar exploration satellite, the Advanced Space-based Solar Observatory (ASO-S) was developed to observe the solar eruptive activities (Gan et al. 2019a,b; Huang et al. 2019), including solar magnetic field (Deng et al. 2019), solar flares (Zhang et al. 2019), and coronal mass ejections (Chen et al. 2019; Li et al. 2019). One of the three payloads, the Hard X-ray Imager (HXI), is a hard X-ray instrument that uses a group of bi-grids to modulate photon transmission and recover flare images by Fourier reconstruction (Chen et al. 2021; Jiang et al. 2022; Krucker et al. 2019; Liu et al. 2022; Su et al. 2019; Zhang et al. 2019). It is well known that HXR imagers cannot determine X-ray source positions relative to the solar disk solely by X-ray imaging, since the disk is invisible in HXRs. An optical system is therefore required to provide solar aspect solution. Besides, according to the principle of modulation-based imaging, the factors affecting imaging result mainly from two aspects: firstly, it experiences the change of force and thermal conditions in the space environment, which leads to the deformation of the collimator and causes the grid slit to misalign and twist. This will affect the accuracy of the reconstructed images. Secondly, the pointing towards the Sun may change slowly due to the drifting effect of the satellite platform (within ± 36 arcsec by design), which causes some blurring of the X-ray images. Therefore, a monitoring device is essential in monitoring the alignment of the grid in real-time, and also providing accurate pointing information during imaging. Thus, for photon-modulated imaging instruments, how to achieve high-performance monitoring has always been a hot research topic of such payloads.

In this context, SAS was born, and the design details will be introduced in this paper. The SAS dataset includes information on the deformation of the HXI collimator, specifically the relative shift and rotation between the front and rear substrates, as well as the HXI's pointing (measurements of the location of the solar disk center in SAS FOV). The variations of the pointing with time can be used in correcting the pointing drift/jitter effect of the platform during imaging, which may produce blurry images. Furthermore, the pointing data are valuable for locating flare X-ray sources in the reconstructed images on the solar disk. This is accomplished by establishing the relationship between the imaging axis of HXI and the optical axis of the SAS camera, using flare X-ray observations and known locations of the sources from multiwavelength observations.

Due to the limitations of HXR imaging discussed above, optical aspect systems were part of previous HXR imagers. For instance, SMM/HXIS achieved a solar limb measurement accuracy of 2 arcsecs by the holes in the SLS system (Van Beek et al. 1980). The pointing system in the Yohkoh/HXT payload consisted of two identical systems, each contained a lens, a set of fiducial marks, and a CCD, and determined the pointing with an accuracy of approximately 1 arcsec by measuring the brightness distribution of the white light from the Sun (Kosugi et al. 1991). RHESSI used three optical systems and three line array CCDs to fit the center of the Sun and achieved an accuracy better than an RMS error of 0.4 arcsecs (Fivian et al. 2002; Fivian, Henneck, and Zehnder 2003; Schwartz et al. 2002; Zehnder et al. 2003). STIX achieves a pointing accuracy of ± 4 arcsecs by using a plate with four orthogonal radial arms cross-shape apertures to project the solar limb image on the photodiode (Krucker et al. 2020; Warmuth et al. 2020). The GRIPS balloon payload's pointing system used a telephoto lens to image the Sun and achieved solar pointing measurements with arcsec accuracy (Duncan et al. 2016; Shih et al. 2012). Each payload has its unique characteristics and technologies, which are worthy of in-depth study to provide a basis for higher performance monitoring in the future.

For the HXI system, the function of the SAS is realized with the method of optical measurement. First, we regard the 91 grids and the supported substrate as a whole to establish the correspondence between the grid and the substrate fiducials, and then the position of the

Table 1 Main specifications of the HXI.

Parameter	Specification									
Grid pitch / μm	36	52	76	108	156	224	344	524	800	1224
Spatial resolution / ″	3.1	4.5	6.5	9.3	13.4	19.3	29.6	45.1	68.8	105.2
Required pointing accuracy / ″	2									

front substrate is determined by the SAS, which subsequently determines the grid slit position. Second, the SAS can simultaneously take into account the solar pointing information during observation. Finally, although the optical measurement method is easy in data acquisition, the optical system aberration and the inevitable image processing errors will result in data that requires complex processing to ensure accuracy and stability.

The overall structure of the study takes the form of three sections, including the hardware design, the software design, and operational performance. Our study provides a new methodological idea for in-orbit monitoring of solar hard X-ray modulation instruments in the future, and likewise, the method could also be implemented in several other potential regions that require aspect monitoring.

2. Hardware Design of the SAS

In this section, we elaborate on the design of the SAS in three aspects, including optical, electronic, and mechanical considerations. As mentioned above, the SAS is a Simultaneous Localization and Mapping (SLAM) system based on optical vision (Boal, Sánchez-Miralles, and Arranz 2014; Fuentes-Pacheco, Ruiz-Ascencio, and Rendón-Mancha 2015). It is assembled with the collimator, and one of its subsystems, called the Deformation Monitor (DM), images some specific objects, involving the Sun and in total three frosted glasses mounted on the front substrate. The frosted glass is a flat panel glass with a scattering surface and a smooth surface, which can be illuminated by solar light and create a uniform circle surface light source. They are used as the feature targets for optical measurement. As the incident direction of the solar light shifts, or the position relation between the front and rear substrates changes, the images on the detector will move accordingly, then the observation orientation and the collimator deformation can be deduced computationally with the image movement. Since the ill-posed characteristics of the DM calculation function decreases the inversion accuracy, another subsystem of SAS named Solar Aspect (SA) is introduced to correct the DM data. Its high accuracy promotes the performance of the deformation computation, simultaneously improving the accuracy of solar center pointing. As the finest of the 91 groups of grids provide the highest spatial resolution of about 3.1 arcsecs, the expected pointing accuracy of SA is set to 2 arcsecs, as shown in Table 1. Since the SAS monitors the displacement of objects between two states, so actually, it achieves the relative deviation from one state to another, but not the absolute value in a specific state. The layout of the SAS and its assembly position is shown in Figure 1.

2.1. Optical Design

The optical components, as the basis for the realization of the SAS functions, require lenses with good imaging quality, reasonable filter and attenuation film to adjust the light intensity,

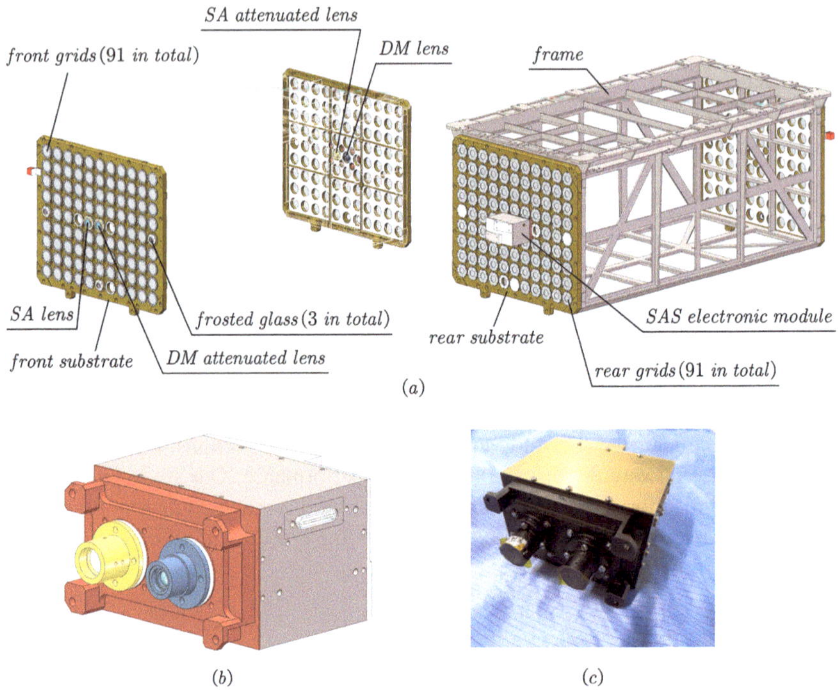

Figure 1 Layout of the HXI collimator and the SAS, containing Solar Aspect (SA) and Deformation Monitor (DM). (a) Left: front view and the frame structure has been set transparent for clarity; right: back view. (b) Overview of the SAS module. (c) The SAS photograph.

Table 2 Optical design specifications for the SAS.

Parameter	SA	DM
Focal length	1214.16 mm	32.8 mm
Sensor pixel number	2048×2048	2048×2048
Sensor pixel size	5.5 μm	5.5 μm
Field of view	45′ (diagonal)	27.8° (diagonal)
Spatial resolution	0.934″ / pixel	34.6″ / pixel
Pupil	25 mm	3 mm
Optical distortion	<0.05%	<0.3%
Wavelength coverage	565∼585 nm	565∼585 nm
Lens attenuation	−30 dB	−50 dB

and uniform and stable source targets for imaging. The main design specifications of the SAS optical assembly are summarized in Table 2, and its performance has been verified in the subsequent tests.

2.1.1. SA Subsystem

The SA is comprised of a single planoconvex lens and a CMOS detector. It directly images the Sun and provides the pointing information from the solar image. For the requirement of radiation resistance, the planoconvex lens uses fused silica material with a diameter of

Figure 2 Light spots' morphological characteristics and positions on the detectors.

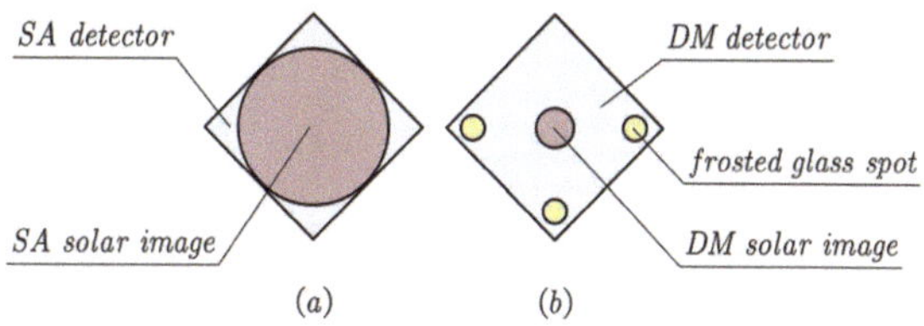

Figure 3 The optical path diagram of the SA.

25 mm. The focal length of the lens is nearly 1215 mm and the flat surface of the lens is placed outside so that it can be used as a reference for optical axis calibration. The detector is a CMOS sensor located at a 45° rotation through the optical axis and soldered to the printed circuit board (PCB). To satisfy the expected pointing accuracy, SA was designed with the longest possible focal length to ensure high spatial resolution, but it also requires a relatively big field of view (FOV) to cover the full solar disk to implement the center extraction algorithm, thus we set the diagonal FOV of the SA optical system to about 45′, which is slightly bigger than the collimator FOV of about 40′. At this time, the solar image occupies almost the whole detector, leaving blanks at the four corners of the CMOS, as shown in Figure 2 (a). The optical path diagram of the SA is shown in Figure 3. A stable and sharp solar image guarantees the accuracy of the solar center extraction, thus the wavelength window for the SA should be narrow enough without strong chromospheric lines that possibly affect the sharpness of the solar image. After a comprehensive assessment, we select a window of 565-585 nm that contains only a number of weak metallic lines. The photospheric image at this wavelength is stable and does not change much even during solar flares, and the 20 nm bandwidth passes through sufficient light to allow a fast exposure. The flat surface of the SA lens is covered with this narrow-band filter film, and the transmittance curve of the filter film is demonstrated in Figure 4.

2.1.2. DM Subsystem

The DM contains a triplet optical lens and a CMOS detector. The focal length of the DM lens is approximately 32.8 mm. The DM utilizes a CMOS sensor identical to the SA sensor, which is also located at a 45° rotation through the optical axis. The optical system of the DM has a diagonal field of view of about 27.8°, allowing it to simultaneously image the Sun and the three frosted glasses. The optical path diagram of the DM is shown in Figure 5. Similar to the SA, an incident window, which is a plane glass, is placed on the front grid substrate. This incident window is covered with the same narrow-band filter film as the SA, located on its first surface. When solar light passes through this incident window, it creates a centered solar image on the detector, covering an area of about 55 pixels in diameter. Additionally, the three frosted glasses scatter the incident light, resulting in three surface light sources. These sources are imaged by the DM and also cover an area of about 55 pixels on the CMOS, as shown in Figure 2 (b). It is important to note that the size of the incident window limits the

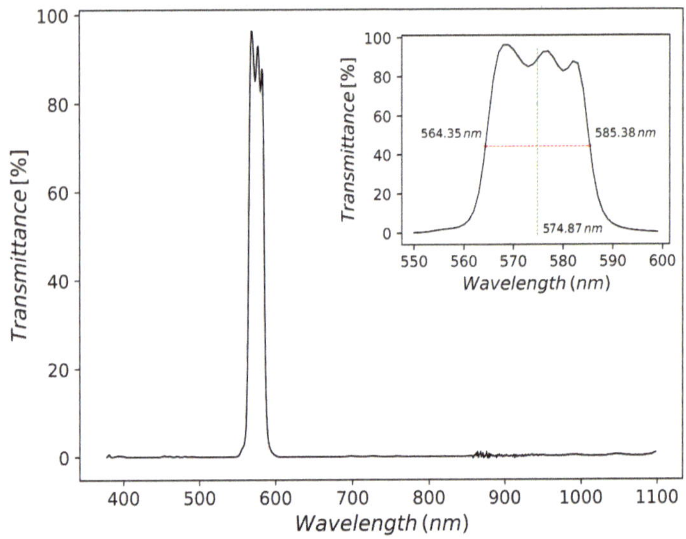

Figure 4 The actual transmittance curve of the narrow-band filter film. The upper right shows the full width at half maximum (FWHM) of the film.

Figure 5 The optical path diagram of the DM.

FOV of solar imaging in a range of not less than 50′, i.e., if the solar pointging deviates by more than 25′, the solar image may be blocked.

2.1.3. Frosted Glasses

The frosted glasses are all made of 11 mm diameter and 4 mm thickness panel glass with the scattering surface towards the DM lens, while the incident plane towards the Sun is smooth and covered with a filter film identical to the SA lens, which results in a constant uniform diffused light and minimizes the influence of chromatic aberration of the DM. The three frosted glasses are all first assembled in the titanium alloy structures and then placed separately on the edge of the front substrate so that they are more sensitive to rotation around the center of the substrate. Consequently, the frosted glasses are imaged in the corner of the detector.

2.1.4. Solar Attenuated Lens

In Figure 2, there are five light spots in total on the two detectors. However, the light that reaches the detector directly through the lens has a higher intensity than the light scattered by the frosted glass, so the intensity of the five light spots is different. To simplify the procedure

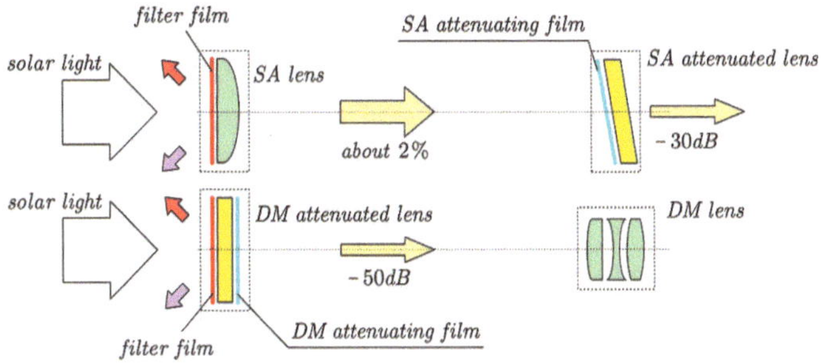

Figure 6 The location of the attenuating films.

of calculating the center position of the five light spots, it is designed to approximate the digital number (DN) of the five light spots on the detector at an identical exposure time. Therefore, the solar attenuated lens is crucial to modulate the intensity balance of the five light spots. The attenuation rate is determined by a solar observation test that reflects the actual observation situation in orbit.

The attenuated lens is different in the two subsystems. For the SA, the attenuated lens is a tilted plate covered with an attenuating film on the front surface, which is set in a titanium frame and mounted on the shell of the SAS electronic unit. The film attenuates the incident solar light by about 30 dB, which enables the system to obtain a more appropriate brightness of the solar spot at a moderate exposure time, without being too dark or overexposed. This consideration guarantees sufficient redundancy to accommodate variations in sunlight intensity during orbital implementation. For the DM, the attenuated lens is a plate incident window with the identical 565–585 nm filter film on the first surface and an attenuating film on the second surface. It is mounted on the front substrate and reduces the incident solar light intensity by about 50 dB.

The attenuating film would absorb the vast majority of the incident energy, which would be stored to heat the particular portion of the front grid substrate and cause unwanted deformation, it is advantageous to reflect the out-of-band incident light before it is absorbed by the attenuating film. Therefore, in the SA and DM subsystems, we both set the layout that the filter film is located in front of the attenuating film in the light path, it would reflect most of the incident energy so that the attenuation film does not absorb too much heat, as shown in Figure 6.

2.2. Electronic Module

The electronic module is the key part of the SAS in which all electrical processes are determined. It is connected to the HXI Electronic Control Box (HXI-E). All operations such as data transfer, signal transmission, and power supply are carried out entirely through the cable. In this section, the electronic design and some specific considerations are described.

2.2.1. Circuit Design

The SAS electronic module consists of three parts according to its function: the main control circuit, the detector circuit, and the power and interface circuit. The main control circuit is the hardware core of the SAS electronic module, which essentially implements all driving, controlling and signaling functions. The detector circuit mainly implements the image

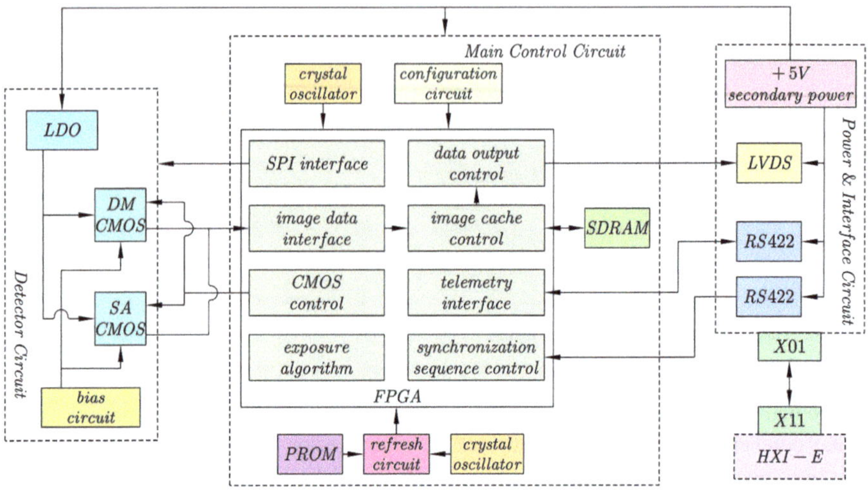

Figure 7 Block diagram of the electronic module.

Table 3 Overview of the electronic module specifications.

Items	Specification
Input voltage	+5 V
Voltage of the Sensor power supply	+3.3 V/+3.0 V/+2.1 V
Power consumption	6.5 W
Sensor type	CMOS
Sensor pixel number	2048×2048
Sensor pixel size	5.5 μm
Exposure time	106 μs to12856 μs (50 μs / step)
SA image frame frequency	4 fps
Data transmission interface	LVDS
Command & Telemetry interface	RS-422
Synchronization sequence interface	RS-422
Data transmission rate	25 Mbps
Command transmission rate	115.2 Kbps

sensor configuration and image data acquisition functions. The power and interface circuit generates the power supply for SAS and provides the input/output transmission path for various types of signals. The block diagram of the electronic module is depicted in Figure 7, while a summary of its specifications is listed in Table 3. For thermal control demands, the components with power consumption greater than 0.3 W, such as FPGA and LDOs, are covered with a thermally conductive insulating layer and then attached to an aluminum made heat sink, which conduct the heat to the rear face of the electronic box, and finally, the heat is transmitted to the radiating surface by heat pipes fixed outside, as shown in Figure 8.

The main control circuit consists of a Field Programmable Gate Array (FPGA) and its various peripheral and configuration circuits, which realize the following functions:

- CMOS sensor parameters configuration, including gain and readout channel setting, and exposure time setting, etc.;

Figure 8 Thermal design for the high-power components. Left: the FPGA and the LDOs are covered with a thermally conductive insulating layer; Right: the aluminum heat sink before it is glued to the back of the electronic box.

- Driving the sensor for image sampling;
- Image data caching;
- Image data output through LVDS interface;
- Telemetry signal acquisition and transmission through RS422 with HXI-E;
- Synchronization sequence generation and timing control;

The SA and DM each possess a CMOS sensor, which is soldered side-by-side on the detector circuit board to achieve compactness. The CMOS specifications are demonstrated in Table 3. As mentioned above, both CMOS sensors are rotated 45° through their normal vector to guarantee the longest focal length of DM while imaging.

The power and interface circuit comprises mainly of modules that generate the required voltage, connectors that establish the external transmission interface, and some peripheral circuitry. The SAS is connected to the HXI-E using a cable, and all power and data transfer occurs through the cable. The HXI-E provides +5 V secondary power, and the SAS has a total power consumption of 6.5 W. The image data is transmitted via the Low Voltage Differential Signal (LVDS) interface, while the telemetry information, parameters, commands, and synchronization pulses are sent via the RS-422 interface.

2.2.2. Software Design

The software for SAS implements five functions: 1) CMOS sensor drivers; 2) exposure algorithm; 3) image processing algorithm; 4) image data transmission; 5) communication control.

The FPGA software is responsible for controlling the two CMOS sensors of the SAS, enabling the DM or SA subsystem to operate independently. The exposure time can be configured in two modes: Manual exposure mode and automatic exposure mode, the exposure time of the system is adjustable in steps, with a minimum of approximately 106 μs, each step increasing linearly by 50 μs, and totally 255 steps, i.e., the time range is from 106 μs to 12.856 ms. In this time range, adding the data transmission time, the high cadence of the SA measurement can be achieved, which is 0.25 s (4 fps).

After the image data is transferred to the Synchronous Dynamic Random Access Memory (SDRAM), the FPGA performs two types of online image processing algorithms: Calculating the centroid position of the solar image and the three frosted glass spots in the DM subsystem (processed with the method in Section 4.1.1); and calculating the total number of pixels covered by the solar image in the four corner areas of the detector in the SA subsystem (used in four quadrant method (FQM), mentioned in Section 4.1.4). To reduce the computational burden in orbit, the FPGA performs only the two online processes above, while the complex inversion algorithm is computed on the ground.

SAS science data can be classified into two categories: calibration mode data and observation mode data. In either mode, the scientific data package includes header information,

Table 4 Data specification in different working modes.

Mode		SA	DM
Observation	Image data type	14 row data	Null
	Image Transmission	LVDS	Null
	Online processing	Four quadrant pixel number statistic	Light spot center calculation
	Result Data Transmission	LVDS	LVDS/RS-422
	Engineering parameter	RS-422	RS-422
Calibration	Image data type	Raw image	Raw image
	Image Transmission	LVDS	LVDS
	Online processing	Four quadrant pixel number statistic	Light spot center calculation
	Result data Transmission	LVDS	LVDS/RS-422
	Engineering parameter	RS-422	RS-422

identification information, status information, image data, and its online processing result. During calibration mode, the image data acquired by both SA and DM are outputted in full, allowing researchers on the ground to process and apply the data. As a comparison, during observation mode, only SA works, and only a specific 14 rows of the image data are binarized and outputted instead of the entire image (used for the least square method, mentioned in Section 4.1.3). This is done to reduce the pressure of data transmission.

The scientific data package is transmitted through the LVDS interface at a rate of 25 Mbps, and the various commands, telemetry data, and engineering parameters of SAS are exchanged with the HXI-E through a bidirectional asynchronous RS-422 interface with a transfer rate of 115.2 Kbps. For clarity, we have listed the data types and transmission channels for the different operating modes in Table 4.

Then, a summary of the SAS operational process is demonstrated as follows and in Figure 9:

1) The SAS is powered up by the HXI-E;
2) Telemetry commands, including reset signals, operation mode settings, and exposure parameters settings, are received by the SAS;
3) Two working modes are available: observation mode, and calibration mode. In both modes, the SAS receives a synchronization pulse signal, and both DM and SA start imaging and then cache the images;
4) During the processes 2) and 3), the SAS defaults to automatic exposure mode. If the resulting images are not satisfactory, an artificial exposure time command can be sent through the HXI-E, the SAS adjusts the exposure time immediately after recciving the command and simultaneously sends a response signal;
5) Executing online image processing based on the cached images. For the SA image, it calculates the number of pixels occupied by the Sun in the four corner areas of the CMOS, and for the DM image, it calculates the centroid positions of the four light spots;
6) In observation mode, the scientific data package consists of 14 rows of data which come from the SA imge, the online processing results in 5), and some other status information;
7) In calibration mode, the scientific data package consists of raw image data from DM or SA (could be chosen), the online processing results in 5), and some other status information;
8) The scientific data package is transferred through the LVDS interface, and it is worth noting that to facilitate monitoring the status of the collimator, the four light spot center

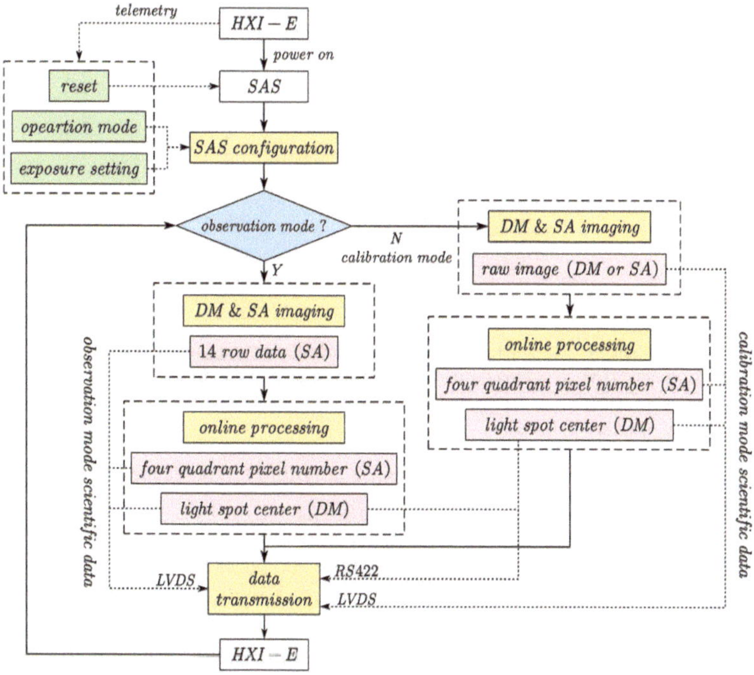

Figure 9 Operational process chart of the SAS. White boxes refer to the instruments, yellow boxes to operations, green boxes to commands, the blue box to judgment, red boxes to data, the dashed line to data flow, the solid line to process orientation.

in DM calculated in 5) is packaged into engineering parameters, which are constantly transmitted and accessed through the RS-422 periodically.

2.3. Assembly and Calibration

The components of the SAS are situated in various locations on the HXI. The layout of the SAS necessitates self-calibration of the two subsystems as well as co-calibration with the HXI during the assembly process. This calibration procedure ensures accurate and coordinated functioning of the entire system.

2.3.1. Mechanical Assembly

As a monitoring system, the SAS electronic module with two CMOSs are mounted on the rear central beam of the HXI collimator support frame, without contact to the rear substrate, which benefits the stability of the rear grids. The CMOSs are considered to have the same motion mode as the rear substrate. Moreover, for stability demands, it was determined that the SAS needed to operate within the temperature range of 17 °C to 27 °C. To achieve this, a polyimide film electric heater was used to heat the side of the SAS electronic box, and the two bevels on the rear face of the SAS are used to attach the heat pipe. This design is due to the need for the heat pipes to be placed at an angle to avoid the HXI Spectrometer (HXI-S), which is placed at the rear end of the collimator. Both heat pipes have a radiant heat sink attached to the end of the heat pipe, as shown in Figure 10. Finally, the entire

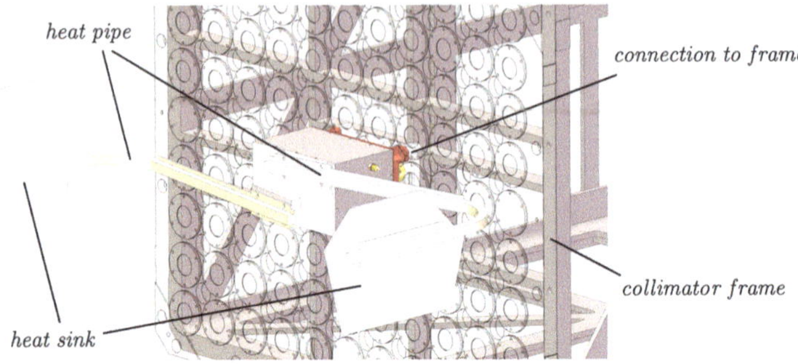

Figure 10 The electronic box is mounted on the collimator frame and has no contact with the rear substrate. The HXI Spectrometer (HXI-S) is directly behind the collimator, and the model of the HXI-S is hidden here for clarity of presentation.

exterior of the electronic box is covered with a polyimide multi-layer material to improve thermal insulation.

In the SA subsystem, the planoconvex lens is separately mounted on the front substrate and remotely aligned with the detector through the optical axis. To prevent stray light, the SA attenuated lens is mounted in front of the detector with an inclination of 12°, which causes a 0.268 mm displacement of the image.

The DM lens is mounted on the electronic box directly and adjacent to the detector. The DM attenuated lens is placed on the front grid substrate without tilt, aligned with the DM lens. Three frosted glasses occupy the array vacancy at the edge of the front substrate. To meet thermal stability requirements, the SA lens, DM attenuated lens, and three frosted glasses are all thermally insulated from the front grid substrate by placing a 3 mm thick glass fiber reinforced polymer (GFRP) ring pad.

We can adjust the focus by slightly moving the lens with some aluminum gaskets of different thickness. For DM, to balance the infinite distance Sun imaging and the finite distance frosted glass imaging, the CMOS sensor is set to the 1.8 m focal plane, which is the intermediate position between the two objects. This causes the objects to be off-focus and the light spot area to expand slightly, but it does not affect the graphical accuracy. But for the SA, its long focal length guarantees that it is insensitive to off-focus.

As the SAS is required to realize the observation function with graphic and visual processing, the stability of the detector is the crucial factor to guarantee the observation accuracy. Thermal dissipation by the CMOS sensor would heat the detector's electronic board during operation, and the heating flux makes the electronic board expand and produces a micro displacement of the detector, which leads to a deviation of the light spot position on the detector. It triggers false information and affects the accuracy of the inversion algorithm. Therefore, we set up a low coefficient of thermal expansion (CTE) invar alloy shielding structure, which is mounted on the inside of the electronic box to cover the CMOS. As shown in Figure 11, the shield wraps around the edge of the detector, leaving a slit filled with silicone rubber to constrain the position of the CMOS and accelerate the heat flux conduction. The structure also prevents redundancy, such as the dust to adhere to the CMOS surface, or stray light interference.

2.3.2. Axis Calibration

The pointing information of the collimator axes is critical, therefore, it is necessary to precisely calibrate the direction and relationship of each axis within the system during the

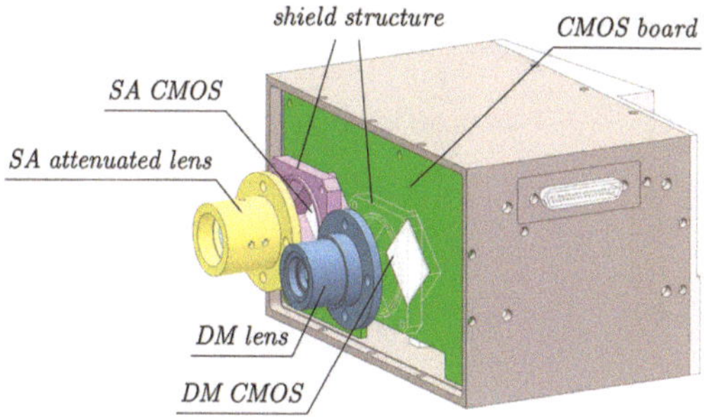

Figure 11 The invar alloy shield structrure. The front panel has been hidden to exhibit the interior layout, and one of the shield structure is hyalinized.

assembly process. The HXI contains a total of three axes, one being the collimator axis, and the other two being the optical axes of the SAS subsystems.

In the assembly process, the outer surface of the DM lens is not a plane, so we need to extract the perpendicular vector of the CMOS surface as the optical axis. We first adjust a parallel light pipe to the perpendicular direction to the CMOS sensor by theodolite, then mount the DM lens and adjust it until the light spot from the pipe is located near the central pixel of the detector. Finally, the light spot deviation from the center of the detector of ± 20 pixels is deemed acceptable. For the SA subsystem, the optical axis is defined as the perpendicular vector of the lens' first flat surface, which is distinct from the DM axis definition. The optical axis is briefly to be extracted in this setting, it is the reason why to place the lens' flat surface outside.

The collimator is an array-type modulator with 91 groups of bi-grids, each of which has an axis that is the connection line between the center points of the two grids. The 91 axes are not overlapped, and it is not necessary to define all 91 axes for the collimator. Moreover, arbitrarily selecting one axis from the 91 axes to represent the main axis of the collimator is also considered unrepresentative. We tend to define a collective axis that can take into account all the bi-grids. Before the definition, we need to consider the grids' assembly procedure. While the front and rear substrates are assembled and all the grids are installed on the corresponding vacancies, the two corresponding grid slits were precisely aligned at this moment with an image coordinate measuring machine (CMM). Then we create some joint fiducials for all grids, that is, a total of eight small standard platforms are polished on the two substrates with an accuracy of 2 μm, as shown in Figure 12. It is obvious that once the front and rear grid substrates have been aligned and mounted on the support frame, the collimator structure relationship is determined, and then the collimator reference coordinate system could be created with the Leica AT901 laser tracker. We create a horizontal plane that is multi-point fitted by the four horizontal fiducials firstly with its normal vector representing the Z-axis. Meanwhile, using the four vertical platforms, we obtain another plane that intersects the first plane and creates a line representing the Y-axis. The Y-axis is parallel to the two fitting planes and thus be regarded as mostly parallel to all the central point connecting lines of bi-grids in an ideal situation. It is considered the most appropriate definition that could represent all the bi-grids' axes and reveals the collimator's zero field of view at which the X-ray penetrates the bi-grids at the zero-phase transmittance of the modulated curve. The X-axis is determined by the Cartesian coordinate definition, so far as demonstrated in Figure 13, both the reference coordinate system and the collimator axis orientation have been defined. Subsequently, we take advantage of the laser tracker to extract the normal vector of the flat surface of the SA lens according to the reference coordinate system, and then adjust its mounting azimuth by grinding the GFRP pad under the lens frame flange until the normal vector is parallel to the Y (the collimator axis) orientation. In addition, with a parallel light

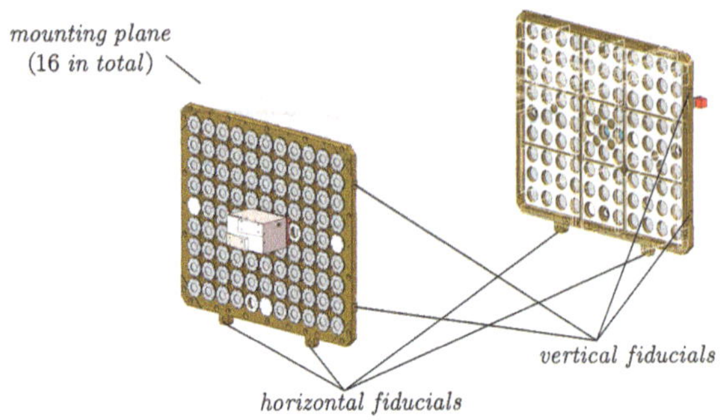

Figure 12 The joint fiducials on the two substrates.

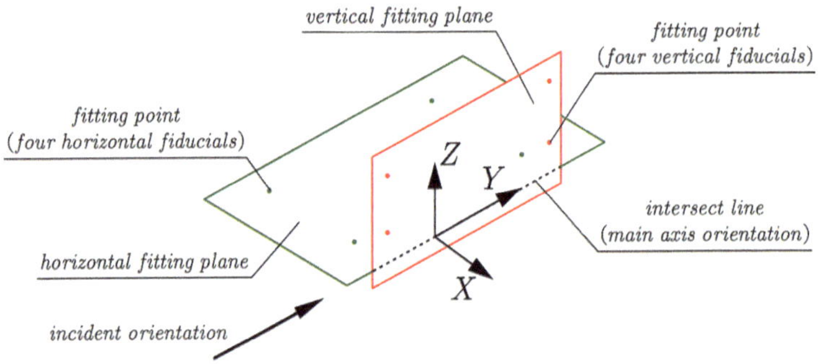

Figure 13 Definition of the reference coordinate system and the main axis of the collimator.

pipe illuminating the SA through the Y orientation, the position of the SA lens must also be translated to ensure that the light spot from the zero field of view is located in the center of the CMOS, of course with a tolerance of ±20 pixels.

The above operation actually aligns the collimator axis with the SA optical axis, and the FOV of the SA and HXI have a definite positional relationship. In inflight applications, it is necessary to re-calibrate the position of the SA CMOS center in relation to the HXI FOV by means of other images such as the SDO/AIA. Unfortunately, we have to neglect the DM optical axis direction due to the non-coincidence of the SA and the DM optical axis. Although both the SA and DM can visually represent the orientation of the collimator axis by imaging the Sun, the SA has a higher accuracy in axis positioning due to the focal length discrepancy, so it is more appropriate to express the collimator axis in terms of the SA optical axis. In this case, the position of the light spot created by the illumination through the Y-orientation needs to be recorded, which will be used as the original input for the inversion algorithm in the future.

3. Function Realization

The previous section described the hardware design of the SAS. In this section, we will interpret how SAS achieves its monitoring functions. The SAS is a sampling device that detects the light spot positions corresponding to the collimator status. By comparing the light spot positions under two different states, we can monitor the changes of the collimator observation states in real-time with the inversion algorithms whose equations are constructed

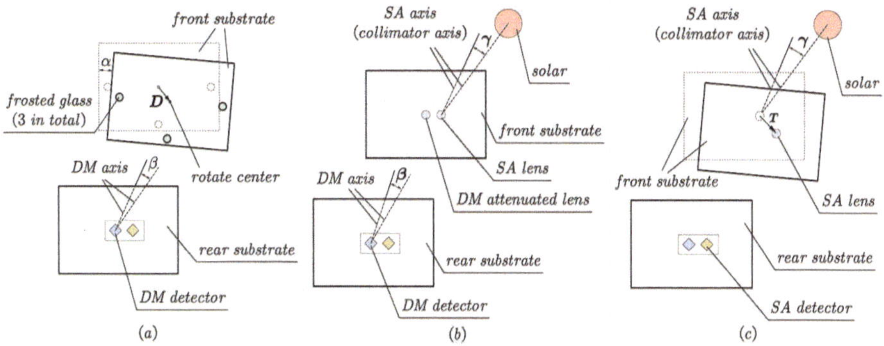

Figure 14 Physical model of the SAS.

by the physical model of the SAS. In the following, we will interpret the functions of the SAS including several aspects: 1) the physical model supporting the algorithm; 2) the optical parameter calibration affecting the inverse equation; 3) the inversion algorithm calculating all the state information of the HXI collimator.

3.1. Physical Model of the SAS

To interpret the SAS inverse procedure clearly, we need to investigate the physical model of the SAS. As described in Section 2, there are a total of three types of light spots imaged on the detectors, including three frosted glass spots and one solar image on the DM, one solar image on the SA. If the collimator frame deforms, or the satellite pointing changes, or the DM axis itself shifts, these five light spots would take a movement on CMOS. After analysis, four variables would influence the results.

- The relative rotation α between the two substrates on the collimator;
- The relative translation $\boldsymbol{D}$ between the two substrates on the collimator;
- The deviation $\boldsymbol{\beta}$ of the DM optical axis orientation toward the initial state;
- The pointing deviation $\boldsymbol{\gamma}$ of the collimator axis (also the SA optical axis with an offset) from the solar center.

SAS uses the very traditional SLAM operating mode of monitoring the positions of the feature points. Because the measurement is insensitive to collimator deformation along the axis, meanwhile, the change in distance to the Sun resulting from the orbit operation is relatively insignificant compared to the distance itself. We, therefore, simplify the calculation process by neglecting the main axis dimension and reducing the equations from 3 dimensions to 2 dimensions. Obviously, the different variables contribute separately to the different modes of motion of the light spots. We show the relationships in Figure 14 and illustrate how the three types of motion of the light spots are triggered separately.

As the positions of the two substrates change due to the deformation of the collimator frame, the three frosted glasses would move away from their original position, causing the three light spot positions on the CMOS of the DM to change. Assuming that the rear substrate is fixed, the motion mode consists of two parts, the relative rotation at the centroid of the front substrate, and its monolithic displacement. In addition, when the optical axis of the DM changes, the light spots also move. The physical principle is illustrated in Figure 14 (a).

Table 5 The relationship between the HXI states and the move mode of the five light spots. The variables are all demonstrated in Figure 14.

Variables	Refer to light spots	Effect on X-ray imaging
α	Frosted glass SA solar image	Affects the shape and peak-to-peak value of the slit modulation curve, causing out-of-focus effects in reconstructed flares
$\boldsymbol{D}$	Frosted glass SA solar image	Similar effects to $\boldsymbol{\gamma}$, a 12 µm shift is equivalent to a pointing change of about 2 arcsec
$\boldsymbol{\beta}$	Frosted glass DM solar image	No effect, but builds the inversion Equations 9, 10, and 13 to solve the unknown variables α, $\boldsymbol{D}$, $\boldsymbol{\beta}$, and $\boldsymbol{\gamma}$
$\boldsymbol{\gamma}$	SA solar image DM solar image	Affects the pointing to the solar center

The Sun can also be imaged on the DM detector center, the position of the solar image is affected by the offset of the DM optical axis and the orientation of the collimator axis simultaneously, as shown in Figure 14 (b).

As the SA lens and the detector are mounted separately on the two substrates of the collimator, the factors affecting the movement of the solar image are complicated. Firstly, since the SA subsystem is an integral part, the change in the orientation of the collimator axis, which is equivalent to the change of the solar light's incident angle, directly prompts the spot's movement. Moreover, the SA lens would move passively with the movement of the front substrate due to the collimator deformation, causing a misalignment between the SA lens and its detector, which contributes to a displacement on the solar image, as described in Figure 14 (c).

In summary, the movement of the light spots is a complex coupled process, as listed in Table 5.

3.2. Parameter Calibration

Before the SAS starts working, some prior calibrations are essential to obtain the precise optical parameters that are crucial to guarantee the accuracy of the inversion algorithm. Generally, for an optical measurement system, the orientation elements need to be calibrated, but especially for the SAS, not all elements need to be concerned. When in orbit, the satellite maintains a relatively stable temperature state due to effective thermal control, and the collimator would not have severe deformation, the five light spots of the SAS always occupy a relatively fixed position around the initial regions, so the elements such as focal length, longitudinal magnification, and optical distortion have little effect on the algorithm. Considering the exclusive function of the SAS, we only need to calibrate some requisite parameters for the inversion algorithm, which are listed below.

- The transverse magnification B, which reflects the relationship between the objects and the images in DM.
- The relationship L between the solar light's incident angle offset and the light spots' displacement on the detectors. The relationships for DM and SA are distinct due to the different focal length, symbolized as L_D and L_S respectively.
- The relationship L_G between the frosted glasses light spots' movement and the offset of the DM optical axis orientation.

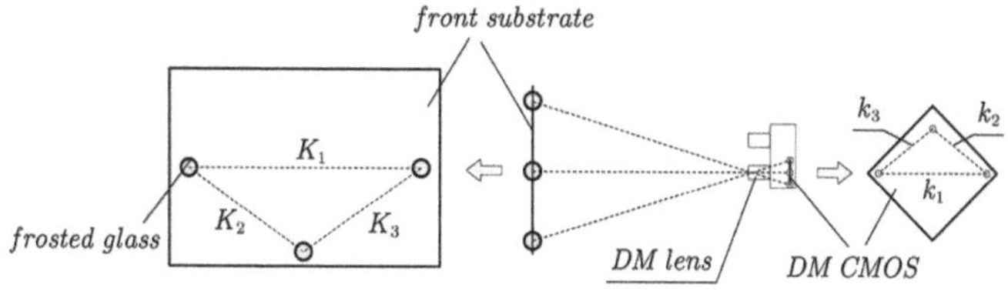

Figure 15 The principle of transverse magnification calibration. The three light spot values are averaged.

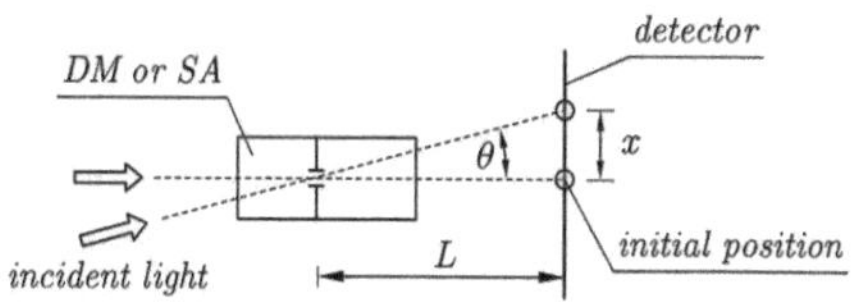

Figure 16 Schematic of the calibration of L.

The parameters above are calibrated with the conventional method. The transverse magnification B is obtained by image inspection. As shown in Figure 15, we illuminate the three frosted glasses and record the corresponding coordinates on the DM CMOS, then calculate their pixel distance k. Meanwhile, the three frosted glasses' real-world distances K are measured with precision coordinate metrology equipment, then B can be obtained with the formula:

$$B = \frac{1}{3}\sum_{i=1}^{3} k_i / K_i. \tag{1}$$

By changing the angle of the incident light and recording the position of the light spots, the value of the parameter L could be calculated, as shown in Figure 16. We use a parallel light pipe to generate the incident light and rotate it at θ angle, which is monitored by a high-precision self-collimated theodolite. L can be calculated from the formula:

$$L_D = x_D / \tan\theta_D, \tag{2}$$

$$L_S = x_S / \tan\theta_S, \tag{3}$$

where x refers to the distance that the light spots on the detector move over due to the angle θ, and the subscripts D and S indicate DM and SA, respectively. Taking into account the different focal lengths of the SA and DM lenses, the relationship between x and θ is distinct in the two subsystems, with a difference of about 36-fold.

When we calibrate the relationship L_G, for the frosted glasses in DM, because the frosted glass is of finite object distance, which is different from the infinite solar incident light, its value is distinguished from the L_D. As the frosted glasses are imaged, their light spots' positions satisfy the relationship:

$$l = L_D \cdot \tan\theta_G, \tag{4}$$

where θ_G represents the angle of view of the frosted glass in the DM, and l is the distance from the light spot to the zero field of view pixel on the detector. If the DM optical axis has a skewness of $d\theta_G$, the movement dl of the light spot follows the relationship:

$$L_G = dl / d\theta_G = L_D \cdot \sec^2\theta_G. \tag{5}$$

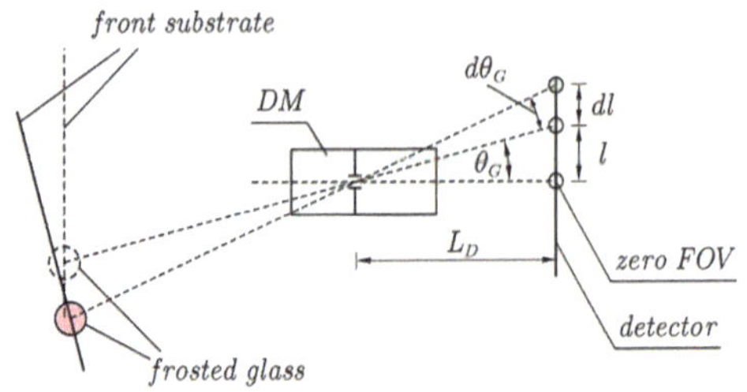

Figure 17 Definition of the parameter L_G.

L_G is similar to L_D in sense, and the coefficient $\sec^2\theta_G$ can be considered as a correction to L_D to accommodate measurements on finitely distant objects, as shown in Figure 17.

For higher accuracy, the parameter calibration process is repeatedly executed several times along different coordinate directions.

3.3. Inversion Algorithm

Combining with the physical model of the SAS, an inversion algorithm is designed to realize the function of monitoring the four variables: α, $\boldsymbol{D}$, $\boldsymbol{\beta}$, and $\boldsymbol{\gamma}$, which are considered to influence the accuracy of the collimator imaging result. The algorithm would not be implemented in orbit as the complexity is beyond the capacity of the FPGA. We simply take the light spot positions in orbit as input data and correspond them to the time of the flare, the inverse equation based on the physical model is then subsequently solved in the ground application system, thus obtaining the instantaneous states of the HXI collimator in the flare duration and making it realizable to carry out a data modification to increase the image reliability.

In practice, the algorithm needs two types of input data. One is the original position fiducials of the light spots, while the other is the new positions of the light spots when some changes happen in the HXI system at a certain time. We have made a theoretical and experimental study of the physical models and found that the movements of all the light spots are derived from coupled quantities, which result in the complexity of the inverse equations. Although it is possible to solve the equations with a certain numerical method, the computational burden and time complexities have no advantages. For simplicity, we divide the inverse problem into two sub-processes. The first is to deduce the rotation and translation of the two substrates through the three frosted glass positions. The latter is to calculate the other variables by adding the information of the two solar spot positions on top of the first step.

In the first sub-procedure, which is a common visual SLAM model that projects the object locations onto the camera coordinate, as shown in Figure 14 (a), the motion mode of the frosted glasses is expressed as:

$$\boldsymbol{p}' = [\mathcal{K}] \cdot \boldsymbol{p}. \tag{6}$$

In the approximate case of neglecting the Z direction, $\boldsymbol{p} = \begin{bmatrix} x & y & 1 \end{bmatrix}^T$ represents the original positions of the three frosted glasses, while $\boldsymbol{p}' = \begin{bmatrix} x' & y' & 1 \end{bmatrix}^T$ represents the new positions after the deformation event, both recorded by the DM. The $[\mathcal{K}]$ shows the projection matrix of the form:

$$[\mathcal{K}] = \begin{bmatrix} R & \boldsymbol{C} \\ O & 1 \end{bmatrix} = \begin{bmatrix} \cos\alpha & -\sin\alpha & c_x \\ \sin\alpha & \cos\alpha & c_y \\ 0 & 0 & 1 \end{bmatrix}, \tag{7}$$

where α represents the rotation of the two grid substrates, and

$$R = \begin{bmatrix} \cos\alpha & -\sin\alpha \\ \sin\alpha & \cos\alpha \end{bmatrix} \tag{8}$$

is the rotation matrix. According to the physical model, $\boldsymbol{C} = \begin{bmatrix} c_x & c_y \end{bmatrix}^T$ is a synthetic value that takes into account the same effect of $\boldsymbol{D}$ and $\boldsymbol{\beta}$ on the light spot shift, it can be written directly as the combination of the two parts:

$$\boldsymbol{D} + \boldsymbol{\beta} \cdot L_G = \boldsymbol{C}. \tag{9}$$

Combining the three light spots, Equation 6 can be decomposed into six formulae along two coordinate directions. It is an overdetermined problem with three unknown variables α, c_x, and c_y, and cannot achieve absolute accuracy solutions, but we can obtain a moderate solution that adapts to the measurement errors using methods such as least squares (LS) (Arun, Huang, and Blostein 1987). By solving Equation 6, the rotation α and the coupled displacement $\boldsymbol{C}$ are calculated, but the variables $\boldsymbol{D}$, $\boldsymbol{\beta}$, and $\boldsymbol{\gamma}$ are still unknown. Another method is then established in the second sub-procedure.

For the solar image on the DM, its motion is also composed of two factors, $\boldsymbol{\beta}$ and $\boldsymbol{\gamma}$. The movement can be expressed as:

$$(\boldsymbol{\beta} + \boldsymbol{\gamma}) \cdot L_D = \boldsymbol{m}' - \boldsymbol{m}, \tag{10}$$

where $\boldsymbol{m} = \begin{bmatrix} m_x & m_y \end{bmatrix}^T$ is the initial position of the solar image and $\boldsymbol{m}' = \begin{bmatrix} m'_x & m'_y \end{bmatrix}^T$ is its position in a new state. It is worth mentioning that $\boldsymbol{m}$ is calibrated with the incident light in the +Y direction, as we described in Section 2.3.2.

In the case of the last one, that is, the solar image on the SA, its motion mode also needs to concatenate two impact factors, i.e. the offset of the collimator axis and the misalignment between the SA lens and its detector. Thus, the movement of the image can be synthesized as:

$$\boldsymbol{T} + \boldsymbol{\gamma} \cdot L_S = \boldsymbol{s}' - \boldsymbol{s}, \tag{11}$$

where $\boldsymbol{s} = \begin{bmatrix} s_x & s_y \end{bmatrix}^T$ is the initial coordinate of the solar light spot, and $\boldsymbol{s}' = \begin{bmatrix} s'_x & s'_x \end{bmatrix}^T$ is the new coordinate in another state. Similarly, $\boldsymbol{s}$ is calibrated using the incident light along the +Y direction. The variable $\boldsymbol{T}$ represents the misalignment between the SA lens and the detector in this situation. Since the SA lens is mounted on the front substrate, it has the same move mode as the front substrate, thus $\boldsymbol{T}$ can be expressed with the rotation and translation of the front substrate, as illustrated in Equation 12:

$$\boldsymbol{T} = -\frac{\boldsymbol{t}' - \boldsymbol{t}}{B} = -\frac{(R \cdot \boldsymbol{t} + \boldsymbol{D}) - \boldsymbol{t}}{B}, \tag{12}$$

where $\boldsymbol{t} = \begin{bmatrix} t_x & t_y \end{bmatrix}^T$ is the coordinate of the SA lens in the DM subsystem, as it can be considered as a target, therefore the coordinate can be obtained indirectly according to the layout of the front substrate, and $\boldsymbol{t}' = (R \cdot \boldsymbol{t} + \boldsymbol{D})$ is the calculated new coordinate of the SA lens after it undergoing the rotation R and the translation $\boldsymbol{D}$ of the substrate. The negative sign means that the misalignment effect in this case influences the movement of the light spots oppositely compared to the imaging process.

It is worth mentioning that since the deformation values are small, to linearize the equations to simplify the solution process, the tangent values in Equation 9 to Equation 11 are simplified to the arc angle values. Combining Equation 9 to Equation 12, we obtain the linear equation set of the second sub-procedure in the form:

$$H\boldsymbol{y} = \boldsymbol{g}, \tag{13}$$

where the coefficient

$$H = \begin{bmatrix} 1 & L_G & 0 \\ 0 & L_D & L_D \\ -1/B & 0 & L_S \end{bmatrix} \tag{14}$$

is a 3×3 matrix constructed with the previously defined parameters, $\boldsymbol{y} = \begin{bmatrix} \boldsymbol{D} & \boldsymbol{\beta} & \boldsymbol{\gamma} \end{bmatrix}^T$ constitutes the variables set to be calculated, and the right side

$$\boldsymbol{g} = \begin{bmatrix} \boldsymbol{C} \\ \boldsymbol{m}' - \boldsymbol{m} \\ \boldsymbol{s}' - \boldsymbol{s} + \frac{R \cdot t - t}{B} \end{bmatrix} \tag{15}$$

is formed with known data derived from the light spots image. By solving the linear equation 13, the variables $\boldsymbol{D}$, $\boldsymbol{\beta}$, and $\boldsymbol{\gamma}$ can be calculated.

So far, the functional principle of SAS, the calibration of the parameters and the implementation of the inversion algorithm has been fully described. Here, due to space constraints, we present only the fundamental principles of the processing algorithm in this section. However, there exist numerous methods for ensuring data stability and regularization processing, which we will describe in detail in a separate article (Yu et al. 2024).

4. Performance and Result

The SAS operates throughout the whole life cycle, from HXI assembly to in-orbit application. Before application, we carry out data pre-processing as the light spot images need to translate into position data. Several methods are applied to adapt to multiple types of light spots, including frosted glass spots, solar spots, and simulated laser light spots. In the following, we present the actual performance of the SAS during the ground and in-flight tests. The performance of the SAS will be evaluated by another metrology instrument on the ground or by the restored flare image quality in orbit. Based on the requirements, we expect that the SAS should achieve a rotation measurement accuracy of 5 arcsec, a displacement measurement accuracy of 15 μm, a pointing accuracy of less than 2 arcsec and a pointing stability of less than 0.5 arcsec / min.

4.1. Accuracy of Center Extraction Method

All the input data used in the SAS algorithm are derived from the light spot images, so a high-accuracy position-extracting method is the basis of the algorithm. The principle of defining the light spot position is to calculate the centroid of the spots, there have been many mature theories that realize the center extraction efficiently with excellent adaptivity and robustness (Chen et al. 2016; Fitzgibbon, Pilu, and Fisher 1999; Tang et al. 2009; Yuan et al. 2018; Zhang et al. 2018). The solar illumination conditions in orbit are stable, which

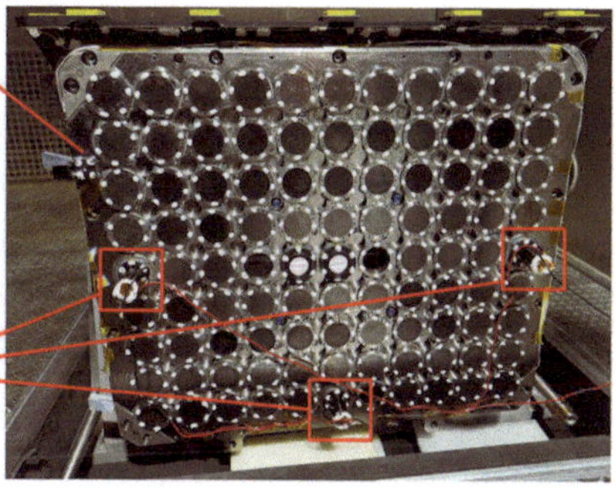

Figure 18 The LED lights mounted on the frosted glasses in ground test. Before departure, the LED lights would be uninstalled.

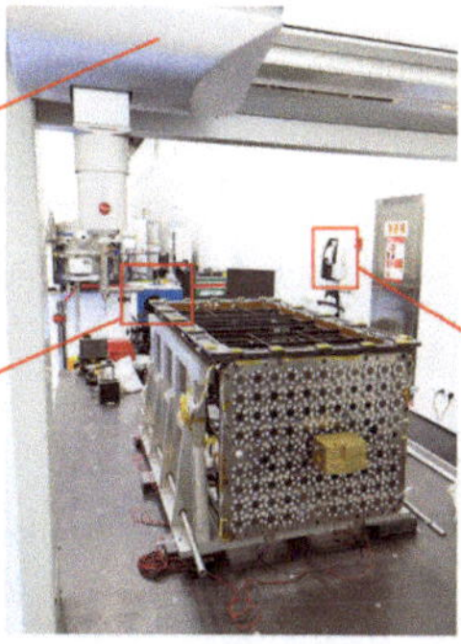

Figure 19 1.5-meter focal parallel pipe used in ground test.

makes the SAS five light spots uniform and immutable. Meanwhile, the optical design with low aberration ensured that the light spots are imaged with tiny deformation. Both make the extraction job relatively easy to achieve. However, in the ground test, after the total HXI components have been integrated, it caused great difficulties in driving such an enormous instrument to track the Sun and carrying out actual solar observation. Therefore, we designed exclusive schemes to simplify the ground test procedure. First, we use LED lights with stable intensity instead of solar light to illuminate the frosted glasses, as shown in Figure 18. Secondly, a single-mode fiber (SMF) assembled on a 1.5-meter focal parallel pipe emits a laser beam at a wavelength of 577 ± 2 nm with power stability of 1.5% to produce a steady laser light spot simulating the incident solar light, as shown in Figure 19. In summary, the various types of light spots in application, whose morphological structures are shown in Figure 20 and Figure 21, necessitate the utilization of different center extraction methods. It is important to note that a weird light spot lies upper right to the centroid of the DM image. This light spot has an arrow-like shape that points to the DM optical axis. This is because the solar incident light may illuminate the fringe of the SA lens structure, creating a stray light and passing through the filter film. The stray light would be just imaged along the arrow direction by the DM, however, it does not affect the SA and DM imaging and is discarded during data processing.

4.1.1. Shape Centroid Method

In a satellite system, the computing resources of FPGA do not represent much redundancy in orbit, so the most traditional and widely used method of shape centroid algorithm has more advantages. It is easy to realize by binarizing operation during the test, as the frosted glass spots, the solar image on DM, and the laser light spots on SA are small circular spots of

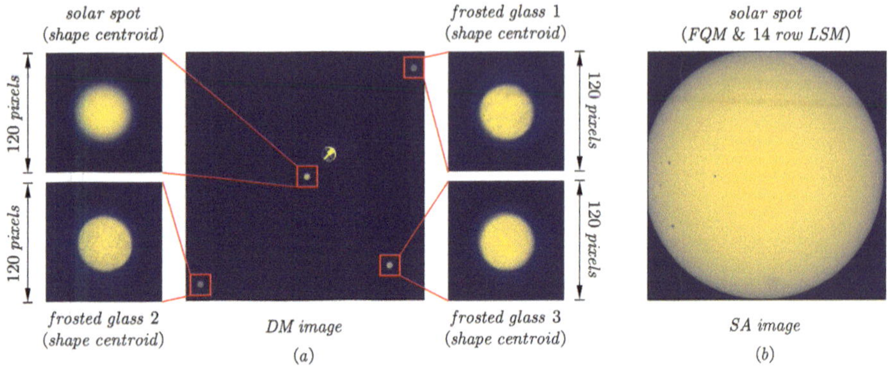

Figure 20 Morphological structures of light spots in orbit.

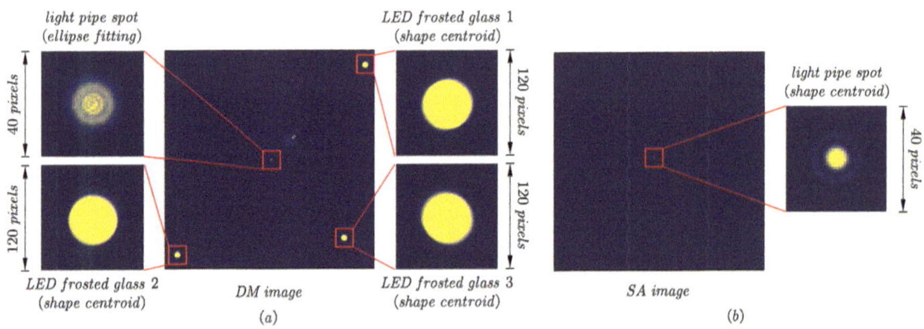

Figure 21 Morphological structures of light spots in ground test.

uniform or Gaussian intensity, it is wise to use this method. The shape centroid is calculated as:

$$X_{shape} = \sum_i p_i x_i / \sum_i p_i, \tag{16}$$

$$Y_{shape} = \sum_i p_i y_i / \sum_i p_i, \tag{17}$$

where the p_i demonstrates the pixel intensity after binarization process, the x and y is the pixel coordinate. In the simulation test, we shift the frosted glass spot artificially by 1-5 pixels and add random noise points to the image in a certain proportion, whose intensity is set to the maximum value in the image area, using the method to calculate the amount of its movement and compared with the theoretical value, it can be seen that the standard deviation of spot center positioning for 10,000 times of the test with the additional 0.1% random noise is about 0.04 pixels, as shown in Figure 22.

4.1.2. Ellipse Fitting

Due to the off-focus effect, a multi-circular spot is formed on the DM when the laser parallel light tube simulates solar light in ground tests, as depicted in Figure 21. The morphological

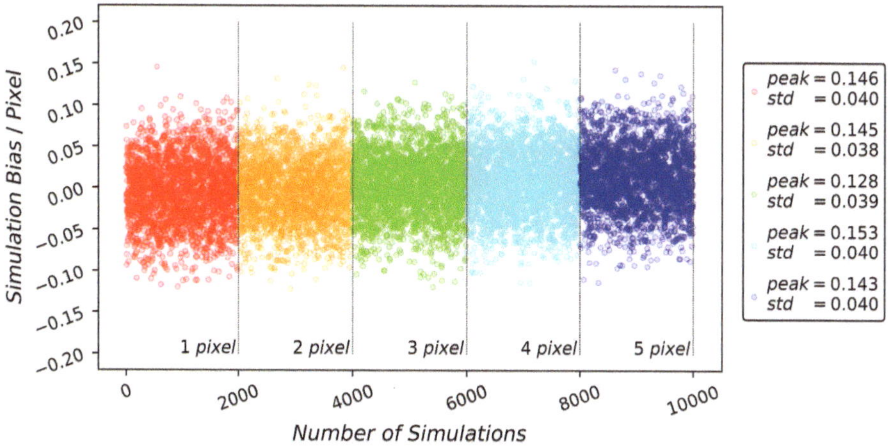

Figure 22 Accuracy test of the shape centroid method.

characteristics of this spot are influenced by several factors, including the stability of the light source, vibrations, airflow disturbances, and others, leading to subtle variations in the shape of the multi-circular ring and reducing the accuracy of position extraction. To address this issue, we use the ellipse fitting method to determine the center of the relatively stable inner dark ring. The ellipse-fitting equation is:

$$F(A, B, C, D, E) = \sum_i \left(x_i^2 + Ax_i y_i + By_i^2 + Cx_i + Dy_i E\right)^2, \tag{18}$$

where x and y are the sampling points with the extreme value of the inner dark ring. The objective function is taken to be optimal at $\partial F = 0$, and the solution can obtain the ellipse center position.

$$X_{ellipse} = \frac{2BC - AD}{A^2 - 4B}, \tag{19}$$

$$Y_{ellipse} = \frac{2D - AC}{A^2 - 4B}. \tag{20}$$

In the simulation test, 10,000 tests with a 0.1% random error achieved a positioning standard deviation of approximately 0.03 pixels, as shown in Figure 23. The high concentration of the simulation results obtained from this method is due to the fact that the algorithm only calculates the characteristics of the spot and does not involve all image pixels. Therefore, it is not particularly sensitive to random errors within the whole surface. However, if noise points appear on the spot, they may affect the position of the sampling points, resulting in a few values that deviate significantly from the expected results. In practical applications, these values should be eliminated in advance during the long-range averaging process.

4.1.3. 14-row Least Square Method (LSM)

As outlined in Section 2.2.2, in observation mode, the scientific data transmitted by SAS is not the complete original image, but a sequence of data formed by the data from a specific 14 lines of pixels in the image binarized. This method yields data that is in principle equivalent

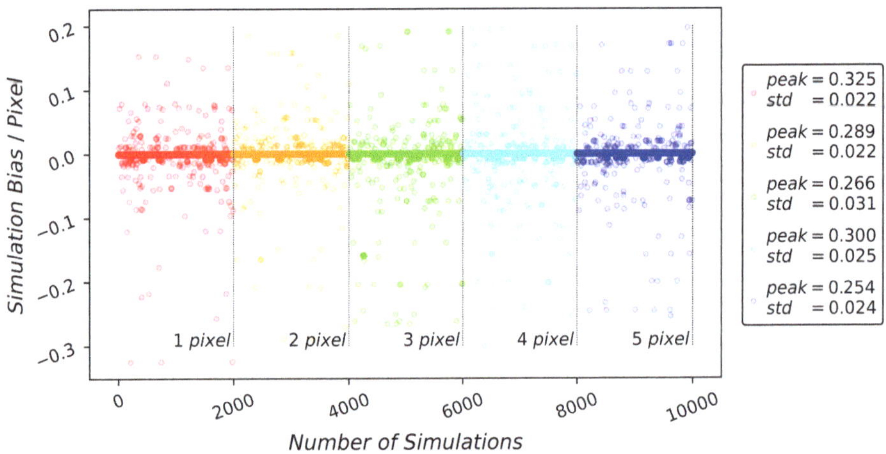

Figure 23 Accuracy test of ellipse fitting.

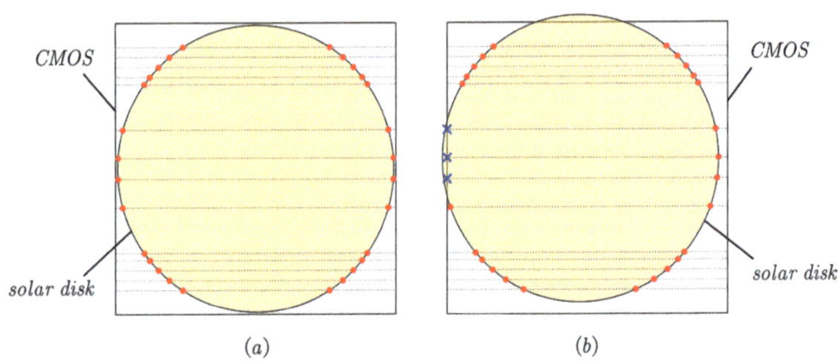

Figure 24 The 14-row horizontal lines on the solar disk and their intersections with the edge of the image. (a) Indicates that the solar image is just near the center of the detector, at this time, 28 edge-fitting points are formed after the binarization of 14 image elements, as shown by the red dots. (b) When the solar image is offset to a certain extent, the edge of the solar image exceeds the detector range, at this time there are some failed edge-fitting points, as shown by the blue crosses.

to the intersection of the solar image with 14 horizontal lines, and the intersection point of each row of data represents the theoretical edge of the solar image. By utilizing this method, up to 28 solar edge points can be obtained, which can then be fitted to the center of the solar spot using the least squares method, as demonstrated in Figure 24. The 14 rows are predetermined, which are the 171, 241, 311, 381, 431, 751, 951, 1301, 1501, 1611, 1681, 1751, 1821, and 1871 rows in the total 2048 rows of data. The solar limb points satisfy the circular fitting equation:

$$f(x, y, x_0, y_0) = (x - x_0)^2 + (y - y_0)^2 = r_0^2, \tag{21}$$

and considering the 28 fitting points in total, define the residual equation:

$$M(\hat{x}, \hat{y}, \hat{r}) = \sum_{x_i \in E, y_i \in E} \left[\hat{r}^2 - f(x_i, y_i, \hat{x}, \hat{y})\right]^2, \tag{22}$$

where E is the set of the edge points, according to LSM, the center of the solar image $(\hat{x}, \hat{y})$ to be fitted should satisfy the objective function:

$$arg \min M\left(\hat{x}, \hat{y}, \hat{r}\right), \tag{23}$$

that is,

$$\frac{\partial M}{\partial \hat{x}} = \frac{\partial M}{\partial \hat{y}} = \frac{\partial M}{\partial \hat{r}} = 0. \tag{24}$$

Solving for Equation 24 yields $\hat{x}$ and $\hat{y}$, and of course the $\hat{r}$, which can be used as the solar disk radius assessment. More details of this algorithm are described in detail by Yu et al. (2020).

However, due to the continuously changing observation pointing and the fact that the solar image is not always located near the center of the detector, there may be scenarios where the solar image is far away from the detector center. When the distance is significant enough, some of the 28 edge points may be beyond the range of the detector, causing some of the edge points to fail, resulting in fewer fit points available for the entire algorithm. This, in turn, leads to an increase in the error of the centering.

Yu et al. (2020) reported that, in algorithm simulation, the test images are generated by artificially shifting one solar disk image, the center localization accuracy of the 14-row least squares method is around 0.11 pixel at a 0.05% image random noise level, and the localization error is uniformly distributed and not easily impacted by noise. Furthermore, due to its small data volume, this method is highly valuable for in-orbit applications.

4.1.4. Four Quadrant Method (FQM)

The solar spot on the SA occupies almost the entire detector and it is not appropriate to use the shape centroid algorithm due to incomplete edges, as shown in Figure 20 (b). Therefore, we use a special fitting method to find the morphological center of the solar image. Firstly the image is binarized, then the principle is that the area (or the number of pixels) where the solar disk intersects the four corners of the CMOS is directly related to the center and radius of the disk, so that the center $\left(\hat{x}, \hat{y}\right)$ and the disk radius $\hat{r}$ can be fitted from the four area values S_1, S_2, S_3, S_4, and some other related parameters h, b, m, n, θ, ϕ, φ. To ensure that the four corner regions can still cover the solar image after it has shifted, we chose four regions of 445×445 pixels from the edge, as shown in Figure 25. Taking the calculation for the first quadrant as an example, we assume that the upper left corner is the detector coordinate origin, at which point the area S has the relationship:

$$S = \frac{\theta}{2}\hat{r}^2 - \frac{mb}{2} - \frac{nh}{2}, \tag{25}$$

according to trigonometric functions and CMOS coordinates, the expressions of each intermediate variable can be obtained:

$$\begin{cases} \theta = \pi/2 - \phi - \varphi \\ \sin\phi = h/\hat{r} \\ \sin\varphi = b/\hat{r} \\ m = L - (\hat{y} - \sqrt{\hat{r}^2 - b^2}) \\ n = L - (\hat{x} - \sqrt{\hat{r}^2 - h^2}) \\ b = \hat{x} - L \\ h = \hat{y} - L \end{cases}. \tag{26}$$

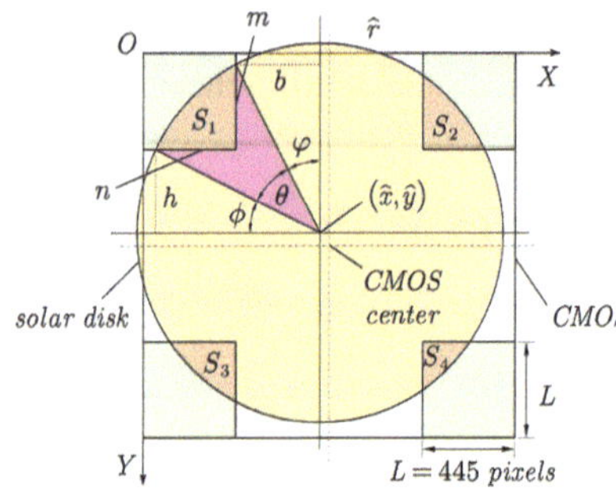

Figure 25 Principle of the four quadrant method. The different areas are represented by different colors.

Through this approach, it is possible to establish the correlation between the area S and the variables $\hat{x}$, $\hat{y}$ and $\hat{r}$, which are to be determined. Likewise, by integrating the remaining three area measurements, a set of four equations can be formulated. Thus, the residual equation can be defined:

$$F\left(\hat{x}, \hat{y}, \hat{r}\right) = \sum_{i=1}^{4} \left[S_i - \left(\frac{\theta_i}{2} \hat{r}^2 - \frac{m_i b_i}{2} - \frac{n_i h_i}{2} \right) \right]^2, \tag{27}$$

the values $\hat{x}$, $\hat{y}$, and $\hat{r}$ can be determined by solving the optimization function:

$$arg \min F\left(\hat{x}, \hat{y}, \hat{r}\right). \tag{28}$$

Equation 28 is solved by the Powell method, with the initial value of the detector center coordinates, a convergence accuracy of 10 times the machine accuracy, and 10,000 iterations of aborts. Additional information regarding the accuracy simulation of the method can be found in Yu et al. (2020), which provides a more comprehensive overview.

The accuracy of the spot center calculation is influenced by both the choice of the binarization threshold and random noise. After verifying the algorithm, the preferred threshold is determined by selecting the point with the largest gradient at the solar image's edge. As the solar center is fitted using the pixel number occupied by the solar image in the four corners of the detector, random noise may result in a lower weighting of the effective area in that region, thereby increasing the error.

According to the conclusion of Yu et al. (2020), overall, the measurement error of the method tends to increase as the spot offset increases. With a random noise level of 0.05%, the center extraction accuracy is superior to 0.3 pixels in standard deviation, corresponding to an instrumental optical axis measurement accuracy of approximately 0.3 arcsec. Moreover, the comparison verification also found that, while FQM is somewhat more sensitive to errors compared to LSM, under a suitable threshold, FQM offers higher localization accuracy. This will be demonstrated later in the in-orbit performance introduction. In the initial design considerations, LSM and FQM were deemed as mutually verifiable algorithms.

4.2. In Ground Test Application

To verify the usability of the HXI collimator, it should be monitored over a long period before the satellite's departure. From the moment the HXI payload begins to be assembled to the satellite platform, there is no room to monitor the collimator state with the external metrology instrument, such as the CMM, for this, the SAS takes over the monitoring mission from then on. On the basis of this, we must also establish correspondence between the initial

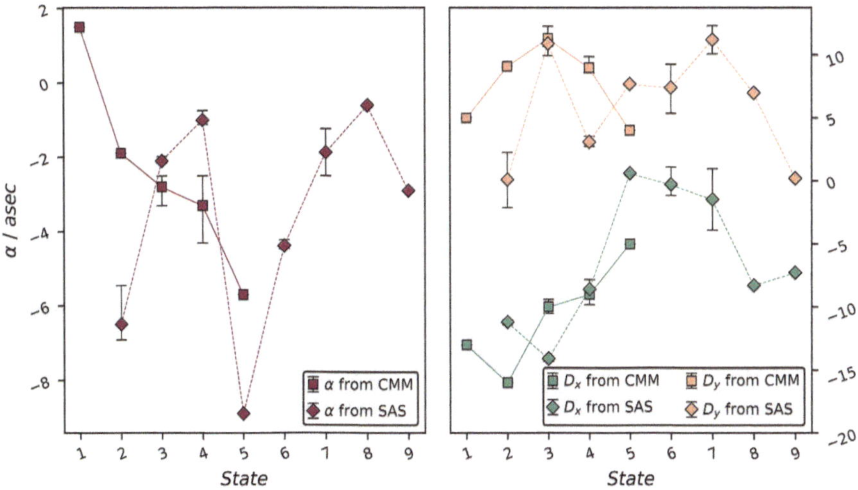

Figure 26 Measurement result in key times. State 1: HXI assembly completed, initial state of SAS; State 2: after HXI vibration test; State 3: after HXI thermal cycle test; State 4: after HXI transport vibration test; State 5: completion of all HXI tests, from then on the CMM was unavailable; State 6: HXI assembled onto the satellite; State 7: the whole satellite assembly completed; State 8: after satellite environmental test; State 9: before satellite launch.

state of the collimator, as measured by the CMM, and the initial position of the light spots recorded by SAS. The inverse value of the state change is then added to the initial state to obtain the current state of the collimator. It is important to note that when recording the initial spot position, the laser incidence angle must be parallel to the calibrated axis of the collimator to accurately capture the direction of the Sun center.

In practical applications, a large number of comparison tests were conducted with CMM, and it was found that SAS achieved a rotational measurement accuracy better than 4 arcsec and a translational measurement accuracy better than 15 microns. The states between the two substrates measured at key times are listed in Figure 26. The results are averaged over multiple measurements, and the error bar indicates the fluctuation range of the measurement results. As shown, the SAS and CMM data are well articulated, indicating the high reliability of the SAS, and the small range of data fluctuations throughout the test process indicates that the collimator was also always maintained in a good state of high accuracy and stability as well.

4.3. Performance in Orbit Application

Unlike ground tests, where the collimator state calibration is the main task of SAS, the solar pointing becomes the main task of SAS during in-orbit operation. But the circumstances of the satellite launch is a severe test for the HXI collimator, we recommended for the SAS to execute a state calibration firstly to determine whether the payload meets the working state in orbit. The ASO-S satellite was launched on October 9th, 2022, and after several days of adjustment, it reached a favorable state and was switched on. The three main payloads, including the HXI, were prepared to begin working. At an opportune moment, the SAS took its first image exhibited in Figure 27. Through the inverse computation, we determined that the collimator was in good state: the grid substrates with a relative rotation α of 0.5 arcsec, a relative displacement D_x of -11 microns and D_y of 14 microns. The

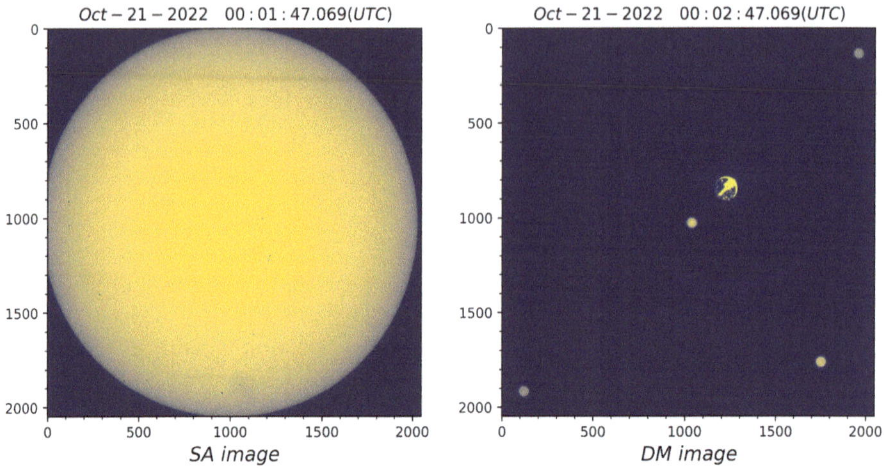

Figure 27 Image of the SAS on October 21, 2022. Due to the actual light spots in DM being too small, we refer to its enlargement in Figure 20, as it is also the actual image collected in orbit.

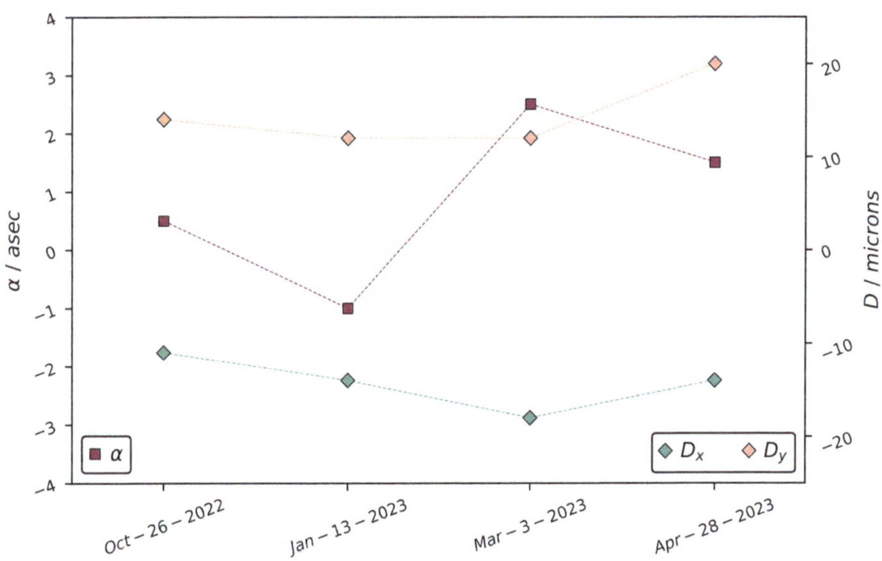

Figure 28 Collimator state calibrations in orbit. The values of the α curve in the figure refer to the left coordinate, the D curve refers to the right coordinate, and all values are the absolute offsets from the initial state calibrated on the ground.

stability of the orbital environment ensures that the collimator does not undergo deformation. As a result, frequent instrument calibration is not required, and periodic calibration is sufficient. Figure 28 presents the results of previous calibrations in orbit, which demonstrate the consistent stability of the relative rotations and translations between the grid substrates. This indicates that the collimator remains in good condition during its time in orbit. Flare imaging always needs pointing information. To achieve this, solar pointing measurements are performed simultaneously with each flare flux sampling.

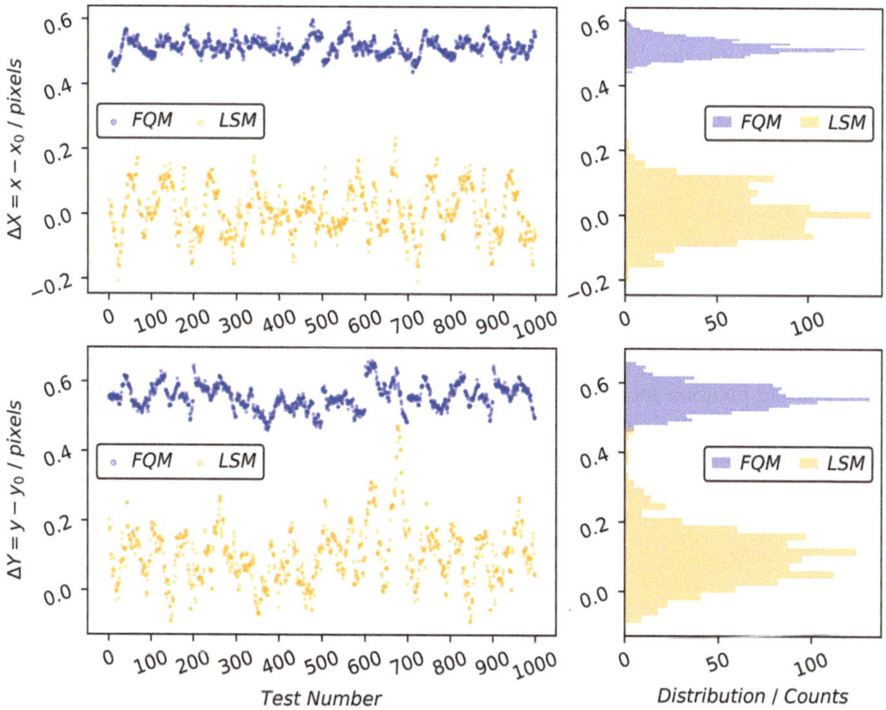

Figure 29 Pointing accuracy test of LSM and FQM in orbit use. x_0 and y_0 come from multi-point fitting and are deemed as the ideal position of the solar disk center.

The solar pointing measurement can be considered a simplified version of the inversion algorithm. Since we assume that the collimator frame would not deform in orbit, the variables α, $\boldsymbol{D}$, and $\boldsymbol{\beta}$ are zero, so the inverse equations will take a simple form:

$$\boldsymbol{\gamma} \cdot L_S = \boldsymbol{s}' - \boldsymbol{s}. \tag{29}$$

The in-orbit pointing of the SAS can be calculated using Equation 29. The key factor that affects the pointing accuracy is the positioning accuracy of the spot centers $\boldsymbol{s}'$ and $\boldsymbol{s}$. The primary techniques for extracting the centers of solar spots are the LSM and FQM. In contrast to simulations, the ideal center position of the solar disk is obtained by fitting multiple edge points in a real observational solar image. We compare the offsets of the LSM and FQM results to this ideal center position to verify their accuracy in practical applications. The rationale for treating the fitting center as the true position of the solar disk is based on the large number of edge points available from the complete solar image, which guarantees a high and stable theoretical accuracy of center fitting. Specifically, we extract 100 points uniformly from the solar disk edge to fit the center position.

We selected 10 actual SA images with the same exposure time and determined the binarization threshold as the grayscale value at the maximum gradient of the solar disk edge. To verify the influence of the threshold on the calculation accuracy, we obtained a total of 1000 sets of data by taking 100 values within the threshold range of ± 50 for each of the 10 images. The results are depicted in Figure 29. Both the LSM and FQM methods achieve high localization accuracy, with FQM achieving a localization accuracy of about 0.6 pixels in both directions and LSM better than 0.3 pixels. Within reasonable limits, the threshold value has less influence on FQM, as the fluctuation of the curve on the same image (every 100 points) is basically within 0.1 pixel, resulting in a higher concentration of distribution in FQM than in LSM. Overall, both methods satisfy the expected accuracy of better than 2 arcsec (1 pixel corresponds to about 0.94 arcsec according to the system parameters), demonstrating their feasibility in practical applications.

Since the satellite platform points to the solar disk center with an accuracy of better than 2 arcsec and a drift rate of 1 arcsec/min, additionally the SAS with an accuracy of 0.3 arcsec and a stability of 0.1 arcsec, which means the satisfaction of the expected pointing accuracy within 2 arcsec in application considerations.

Due to the high accuracy of solar disk center measurement of SAS, the imaging results are optimized to a large extent. Of course, the correction of flare images is a complex process, and the pointing measurement of SAS is only an important part of the flare image correction. Therefore, the SAS correction alone cannot guarantee the accuracy of flare images, but needs to be combined with the actual calibration data of many other indicators in the HXI system. As mentioned before, the imaging axis of the collimator and the one for the SA are different. We obtained the offset during the preliminary calibration by comparing HXI images with AIA EUV images. Here, we used the observations of the flare that occurred on May 9, 2023. The HXI image was made in HXI FOV for the energy range of 30-40 keV to reveal the clear double footpoint sources. The source locations in EUV are determined in an AIA 94 Å image, which also clearly presented two bright sources. Combining the two results we found an offset [15.20, −3.10] arcsec of the SA center in HXI FOV. The details can be found in Su et al. (2024). This offset has been used in HXI GUI since February 2024 (in hxi_get_image.pro). It is working well so far. We estimate an uncertainty of <2 arcsec for most of the flares (in fact we do not need manually shift the sources for most of the cases). Of course, more accurate offset requires more flare observations of clear sources in both HXR and EUV. We have listed it as one of the most important tasks in the near future.

The aspect is determined from both the instant location of the solar disk center on SA CMOS and the relative location of the CMOS center in HXI FOV (relative to HXI imaging axis). To elaborate, the source locations are determined in HXI FOV through imaging and then converted to coordinates relative to the SA CMOS center. This conversion is done with the knowledge of the relative location of the CMOS center in HXI FOV, which was determined through axis calibration. Subsequently, the source coordinates are further transformed to the solar disk center, which is determined relative to the SA CMOS center using the pointing data. It is worth mentioning that the platform is three-axis stabilized, with one plane in the ecliptic and an accuracy within 36 arcsec, which does not cause significant changes of X-ray source locations. Therefore, we only need consider the solar rotational obliquity angle to the ecliptic from Earth view, which can be calculated from our procedure 'sun_north_angle.pro' in HXI software or by some SSW routines.

Figure 30 shows an example of HXI Clean image reconstructed for the flare observed on February 23, 2023. The two panels compare the HXR source in HXI field of view without any correction of HXI pointing and the HXI source overplotted on the solar disk in solar coordinates after locating it with the pointing data recorded during this flare and preliminary calibration based on SAS data and flare data recorded during other events. The details can be consulted in Su et al. (2024). After complete imaging calibration with the high-precision measurements of solar disk center, more flare data, and other measurements of HXI pointing, HXR source locations will be more accurately determined.

These confirm the high performance of the HXI collimator and the effectiveness of the SAS monitoring. In future work, the science team will continue to optimize the imaging

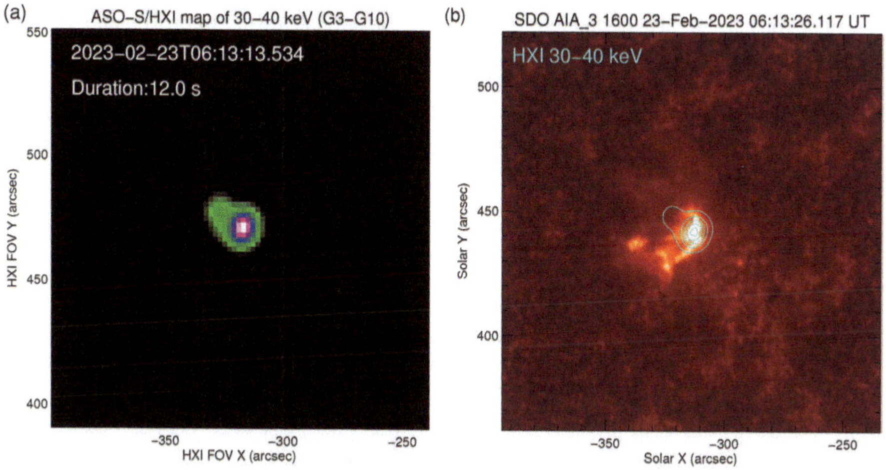

Figure 30 Reconstructed flare images without (a) and with (b) SAS aspect data. This figure displays the reconstructed Clean image of the February 23, 2023 flare observed by ASO-S/HXI. The time interval is from 06:13:13 to 06:13:25 UT and the energy range for imaging is 30-40 keV. (a) the HXI image reconstructed in the HXI field of view with sub-collimator groups from G3 to G10. This image is obtained without any knowledge of the pointing direction of the imager system. (b) The HXI contours overplotted on the SDO/AIA 1600 Å image. It is noticed that the automatically determined location of the HXI source agrees with the bright source in AIA 1600 Å image. This is done with the preliminary imaging calibration (for flare locating) using SAS data (recorded during other flares) and the high-cadence pointing data of SAS-SA camera recorded during this flare.

capability of the HXI instrument by combining other calibration tools, and the complete calibration and correction method for flare imaging will be described in detail in a separate article (Su et al. 2024).

5. Conclusion

In this paper, we have elaborated on the SAS of the HXI payload in the ASO-S mission, including its optical design, mechanical design, electronical and software design, calibration method, algorithm principle, and its performance results. The SAS realizes the two main functions of collimator state monitoring and solar pointing.

In ground test application, compared to the external metrology instrument measurement data, we estimate that the accuracy of the state calibration is better than 4 arcsec in rotation and 15 microns in translation.

In orbit application, the data from several state calibration tests indicate that the HXI instrument can maintain a good condition. Calculation and verification through the actual solar image taken in orbit demonstrate that the SAS realizes an accuracy of better than 0.3 arcsec for the measurements of the solar disk center. The SAS data was used for aspect calibration and determination of the source location within 2 arcsec of the correct flare site in most of the cases.

In conclusion, although we have used the optical measurements of SAS to provide a method for monitoring and calibrating collimators, we still face significant challenges in improving the accuracy and robustness of the algorithms. To address this, we have integrated some advanced data processing methods. However, due to space constraints, we cannot describe them in detail in this paper. In the future, monitoring systems like this one will remain essential for hard X-ray-modulated imaging. We encourage further improvements by optimizing the forms and parameters of the optical system to make the inverse physical model more reasonable, reducing solving difficulties and errors, and increasing robustness against perturbations. Additionally, more advanced algorithms need to be thoroughly researched and extended to accommodate more complex demands. Ultimately, we also look forward to advancing solar exploration technology in the future.

Acknowledgments The work is supported by the National Key R&D Program of China 2022YFF0503002, the NSFC 12333010, and the Strategic Priority Research Program of the Chinese Academy of Sciences, Grant No. XDB0560000.

Author contributions All authors listed in the manuscript provided significant contributions to the article. The contributions made by all authors are: Jirui Yu was responsible for writing the manuscript, simulating and analyzing the data, producing Figure 1-Figure 30; Yinghong He assisted in writing Chapter 2.1 and Chapter 3; Jinyou Tao assisted in writing Chapter 2.2, Chapter 3, and Chapter 4.1; Yang Su assisted in writing Chapter 4.3, producing Figure 31, and assisting in data analysis and usage; Zhe Zhang led the design of the HXI and was responsible for proofreading the article; Jianfeng Yang was responsible for proofreading the article; Nange Wang assisted in writing Chapter 2.2; Song Guo was responsible for the organization of the experimental data; Baogang Lv assisted in writing Chapter 2.2.2; Xiaobo Chen assisted in writing Chapter 2.2.1; Bin Xue assisted in data simulation and design; Yiming Hu & Dengyi Chen participated in the HXI collimator design and experiments; Fu Yu made the SA accuracy simulation and its data calibration; Mingde Ding designed the observation wavelength; Ping Ruan was responsible for article architecture design and manuscript review.

Declarations

Competing interests The authors declare no competing interests.

References

Arun, K.S., Huang, T.S., Blostein, S.D.: 1987, Least-squares fitting of two 3-D point sets. *IEEE Transactions on pattern analysis machine intelligence*, 698. DOI.

Boal, J., Sánchez-Miralles, A., Arranz, A.: 2014, Topological simultaneous localization and mapping: a survey. *Robotica* **32**, 803. DOI.

Chen, H., Yang, Z.-H., Guo, P., Zhang, Y., Chen, S.-Y.: 2016, Research of the high precision laser spot center location algorithm. *Trans. Beijing Inst. Techol.* **36**, 181. DOI.

Chen, B., Li, H., Song, K.-F., Guo, Q.-F., Zhang, P.-J., He, L.-P., Dai, S., Wang, X.-D., Wang, H.-F., Liu, C.-L.: 2019, The Lyman-alpha Solar Telescope (LST) for the ASO-S mission–II. Design of LST. *Res. Astron. Astrophys.* **19**, 159. DOI.

Chen, D.-Y., Hu, Y.-M., Ma, T., Su, Y., Yang, J.-F., Wang, J.-P., Xu, G.-Z., Jiang, X.-K., Guo, J.-H., Zhang, Y.-Q.: 2021, Design and verification of the HXI collimator on the ASO-S mission. *Res. Astron. Astrophys.* **21**, 136. DOI.

Deng, Y.-Y., Zhang, H.-Y., Yang, J.-F., Li, F., Lin, J.-B., Hou, J.-F., Wu, Z., Song, Q., Duan, W., Bai, X.-Y.: 2019, Design of the full-disk MagnetoGraph (FMG) onboard the ASO-S. *Res. Astron. Astrophys.* **19**, 157. DOI.

Duncan, N., Saint-Hilaire, P., Shih, A., Hurford, G., Bain, H., Amman, M., Mochizuki, B., Hoberman, J., Olson, J., Maruca, B.: 2016, First flight of the Gamma-Ray Imager/Polarimeter for Solar flares (GRIPS) instrument. *Space Telescopes and Instrumentation 2016: Ultraviolet to Gamma Ray, SPIE*, 876. DOI.

Fitzgibbon, A., Pilu, M., Fisher, R.B.: 1999, Direct least square fitting of ellipses. *IEEE Trans. Pattern Anal. Mach. Intell.* **21**, 476. DOI.

Fivian, M., Henneck, R., Zehnder, A.: 2003, RHESSI aspect system and in-flight calibration. Innovative Telescopes and Instrumentation for Solar Astrophysics. *SPIE*, 60. DOI.

Fivian, M., Hemmeck, R., Mchedlishvili, A., Zehnder, A.: 2002, RHESSI aspect reconstruction. *Solar Phys.* **210**, 87. DOI.

Fuentes-Pacheco, J., Ruiz-Ascencio, J., Rendón-Mancha, J.M.: 2015, Visual simultaneous localization and mapping: a survey. *Artif. Intell. Rev.* **43**, 55. DOI.

Gan, W., Yan, Y., Huang, Y.: 2019, Prospect for space solar physics in 2016-2030. *Scientia Sinica Physica, Mechanica Astronomica* **49**, 059602. DOI.

Gan, W.-Q., Ding, M.-D., Huang, Y., Su, Y.-N.: 2019a, Preface: advanced space-based solar observatory (ASO-S). *Res. Astron. Astrophys.* **19**, 155. DOI.

Gan, W.-Q., Zhu, C., Deng, Y.-Y., Li, H., Su, Y., Zhang, H.-Y., Chen, B., Zhang, Z., Wu, J., Deng, L.: 2019b, Advanced space-based solar observatory (ASO-S): an overview. *Res. Astron. Astrophys.* **19**, 156. DOI.

Huang, Y., Li, H., Gan, W.-Q., Li, Y.-P., Su, J.-T., Deng, Y.-Y., Feng, L., Su, Y., Chen, W., Lei, S.-J.: 2019, The science operations and data center (SODC) of the ASO-S mission. *Res. Astron. Astrophys.* **19**, 164. DOI.

Jiang, X.-K., Jian, W., Chen, D., Hu, Y., Wang, H.X., Liu, W., Zhang, Z.: 2022, Construction and validation of a geometry-based mathematical model for the hard X-ray imager. *Res. Astron. Astrophys.* **23**, 025004. DOI.

Kosugi, T., Makishima, K., Murakami, T., Sakao, T., Dotani, T., Inda, M., Kai, K., Masuda, S., Nakajima, H., Ogawara, Y.: 1991, The hard X-ray telescope (HXT) for the Solar-A mission. *The Yohkoh Mission*, 17. DOI.

Krucker, S., Hurford, G.J., Su, Y., Gan, W.-Q.: 2019, Joint hard X-ray observations with ASO-S/HXI and SO/STIX. *Res. Astron. Astrophys.* **19**, 167. DOI.

Krucker, S., Hurford, G.J., Grimm, O., Kögl, S., Gröbelbauer, H.-P., Etesi, L., Casadei, D., Csillaghy, A., Benz, A.O., Arnold, N.G.: 2020, The spectrometer/telescope for imaging X-rays (STIX). *Astron. Astrophys.* **642**, A15. DOI.

Li, H., Chen, B., Feng, L., Li, Y., Huang, Y., Li, J.-W., Lu, L., Xue, J.-C., Ying, B.-L., Zhao, J.: 2019, The Lyman-alpha Solar Telescope (LST) for the ASO-S mission—I. Scientific objectives and overview. *Res. Astron. Astrophys.* **19**, 158. DOI.

Liu, W., Chen, D.-Y., Jiang, X.-K., Wu, J., Zhang, Z., Hu, Y.-M., Su, Y., Chen, W., Ma, T.: 2022, Research on the on-orbit background of the hard X-ray imager onboard ASO-S. *Res. Astron. Astrophys.* **22**, 095011. DOI.

Schwartz, R., Csillaghy, A., Tolbert, A., Hurford, G., Mc Tiernan, J., Zarro, D.: 2002, RHESSI data analysis software: rationale and methods. *Solar Phys.* **210**, 165. DOI.

Shih, A.Y., Lin, R.P., Hurford, G.J., Duncan, N.A., Saint-Hilaire, P., Bain, H.M., Boggs, S.E., Zoglauer, A.C., Smith, D.M., Tajima, H.: 2012, The gamma-ray imager/polarimeter for solar flares (GRIPS). *Space Telescopes and Instrumentation 2012: Ultraviolet to Gamma Ray, SPIE*, 1297. DOI.

Su, Y., Liu, W., Li, Y.-P., Zhang, Z., Hurford, G.J., Chen, W., Huang, Y., Li, Z.-T., Jiang, X.-K., Wang, H.-X.: 2019, Simulations and software development for the hard X-ray imager onboard ASO-S. *Res. Astron. Astrophys.* **19**, 163. DOI.

Su, Y., Zhang, Z., Chen, W., Chen, D., Yu, F., Hu, Y., Zhang, Y., Xia, F., Chen, C., Li, Z., Jiang, X., Huang, Y., Zhang, Y., Liu, W., Ma, T., Li, D., Yu, W., Li, Y., Cai, M., Guo, J., Huang, Y., Wang, H., Liang, Y., Ma, M., Wang, J., Zhu, S., Tao, J., Yu, J., Yang, J., Wu, J., Gan, W.: 2024, The tests and calibrations of the Hard X-ray Imager onboard ASO-S. *Solar Phys.* submitted.

Tang, S., Zhou, Z., Guo, X., Xiao, Y.: 2009, Improved iteration centroid algorithm based on linear CCD light-spot location. In: *2009 9th International Conference on Electronic Measurement & Instruments*, IEEE 4. DOI.

Van Beek, H.F., Hoyng, P., Lafleur, B., Simnett, G.M.: 1980, The hard X-ray imaging spectrometer (HXIS). *Solar Phys.* **65**, 39. DOI.

Warmuth, A., Önel, H., Mann, G., Rendtel, J., Strassmeier, K.G., Denker, C., Hurford, G., Krucker, S., Anderson, J., Bauer, S.-M.: 2020, The STIX aspect system (SAS): the optical aspect system of the spectrometer/telescope for imaging X-rays (STIX) on solar orbiter. *Solar Phys.* **295**, 1. DOI.

Yu, F., Su, Y., Zhang, Z., Huang, Y.: 2020, Research on the solar aspect system algorithm of ASO-S/HXI. *Acta Astron. Sin.* **61**, 79. DOI.

Yu, J.-R., Ruan, P., Su, Y., He, Y.-H., Tao, J.-Y., Zhang, Z., Guo, S., Xue, B., Yang, J.-F.: 2024, Inverse calculation and regularization process for the solar aspect system (SAS) of HXI payload on ASO-S spacecraft. *Res. Astron. Astrophys.* **24**, 045003. DOI.

Yuan, X., Li, G., Tang, X., Gao, X., Huang, G., Li, Y.: 2018, Centroid automatic extraction of spaceborne laser spot image. *Acta Geodaetica et Cartographica Sinica* **47**, 135. DOI.
Zehnder, A., Bialkowski, J., Burri, F., Fivian, M., Henneck, R., Mchedlishvili, A., Ming, P., Welte, J., Thomsen, K., Clark, D.: 2003, RHESSI imager and aspect systems. Keil, S.L., Avakyan, S.V. (eds.): Innovative Telescopes and Instrumentation for Solar Astrophysics, Proc. SPIE 41. DOI.
Zhang, H., Li, G., Zhang, Y., Hu, S.: 2018, Extracting the centroid of the light spot by two-dimensional sigmoid fitting. *Opt. Eng.* **57**, 104113. DOI.
Zhang, Z., Chen, D.-Y., Wu, J., Chang, J., Hu, Y.-M., Su, Y., Zhang, Y., Wang, J.-P., Liang, Y.-M., Ma, T., Guo, J.-H., Cai, M.-S., Zhang, Y.-Q., Huang, Y.-Y., Peng, X.-Y., Tang, Z.-B., Zhao, X., Zhou, H.-H., Wang, L.-G., Song, J.-X., Ma, M., Xu, G.-Z., Yang, J.-F., Lu, D., He, Y.-H., Tao, J.-Y., Ma, X.-L., Lv, B.-G., Bai, Y.-P., Cao, C.-X., Huang, Y., Gan, W.-Q.: 2019, Hard X-ray Imager (HXI) onboard the ASO-S mission. *Res. Astron. Astrophys.* **19**, 160. DOI.

Authors and Affiliations

Jirui Yu[1,2] · Yinghong He[1] · Jinyou Tao[1] · Yang Su[3,4] · Zhe Zhang[3] · Jianfeng Yang[1] · Nange Wang[1] · Song Guo[1] · Baogang Lv[1] · Xiaobo Chen[1] · Bin Xue[1] · Yiming Hu[3] · Dengyi Chen[3] · Fu Yu[3] · Mingde Ding[5] · Ping Ruan[1]

✉ P. Ruan
ruanp@opt.ac.cn

J. Yu
yujirui@opt.ac.cn

1 Xi'an Institute of Optics and Precision Mechanics of Chinese Academy of Sciences, Xi'an, 710119, China

2 University of Chinese Academy of Sciences, Beijing, 100049, China

3 Key Laboratory of Dark Matter and Space Astronomy, Purple Mountain Observatory, Chinese Academy of Sciences, Nanjing, 210023, China

4 School of Astronomy and Space Science, University of Science and Technology of China, Hefei, 230026, China

5 School of Astronomy and Space Science, Nanjing University, Nanjing 210023, China

Solar Physics (2024) 299:30
https://doi.org/10.1007/s11207-024-02272-4

RESEARCH

Multiwavelength Observations of Quasiperiodic Pulsations in the Impulsive Phase of an Eruptive Flare with the *Hard X-Ray Imager* On Board ASO-S and Other Instruments

Fanpeng Shi[1,2] · Dong Li[1,2] · Zongjun Ning[1,2] · Alexander Warmuth[3] · Wei Chen[1,2] · Yang Su[1,2] · Ying Li[1,2] · Jun Xu[1,2] · Yuxiang Song[1,2] · Yuzhi Yang[1,2]

Received: 10 December 2023 / Accepted: 8 February 2024 / Published online: 12 March 2024

Abstract
We investigated the quasiperiodic pulsations (QPPs) of the X1.2 solar flare (SOL2023-01-06T00:57) based on multi-instrument observations, especially the *Hard X-ray Imager* (HXI) on board the *Advanced Space-based Solar Observatory* (ASO-S). The quasiperiod of ≈ 27 s was identified in hard X-rays (HXR) and microwaves using the Markov Chain Monte Carlo method, while no corresponding oscillation was found in soft X-rays. The HXI imaging demonstrates that HXR pulsations arise from the double-footpoint HXR sources. Moreover, the fluctuation of ≈ 27 s period was also present in the nonthermal electron power-law index derived from the X-ray spectra, which is the signature of periodic electron acceleration/precipitation during the impulsive phase. The electron spectral indices not only exhibited the well-known "soft–hard–soft" evolution, but also showed the negative correlation with the HXR pulsations. These results suggest that QPPs directly originate from quasiperiodic injections of accelerated electrons into flare-loop footpoints. We also discuss the possible generation mechanisms for QPPs.

Keywords Oscillations · Solar · Flares · Impulsive phase · X-ray bursts · Association with flares · Radio emission

1. Introduction

Quasiperiodic pulsations (QPPs), the fluctuation of electromagnetic emissions (light curves), are common phenomena in solar and stellar flares. The statistical study of solar flares QPPs

✉ F. Shi
shifp@pmo.ac.cn

Y. Su
yang.su@pmo.ac.cn

1 Key Laboratory of Dark Matter and Space Science, Purple Mountain Observatory, Chinese Academy of Sciences, Nanjing 210023, China

2 School of Astronomy and Space Science, University of Science and Technology of China, Hefei 230026, China

3 Leibniz-Institut für Astrophysik Potsdam (AIP), An der Sternwarte 16, 14482 Potsdam, Germany

clearly demonstrates that QPPs are present in many flares (Simões, Hudson, and Fletcher, 2015; Inglis et al., 2016; Pugh, Broomhall, and Nakariakov, 2017; Dominique et al., 2018; Hayes et al., 2020). QPPs are observed across the entire electromagnetic spectrum, with characteristic periods ranging from subseconds to several minutes (Nakariakov and Melnikov, 2009; Van Doorsselaere, Kupriyanova, and Yuan, 2016; McLaughlin et al., 2018; Kupriyanova et al., 2020; Zimovets et al., 2021).

Many mechanisms have been proposed to explain the observations for QPPs. In general, two main categories can be found, i.e. the modulation by magnetohydrodynamic (MHD) waves in coronal magnetic loops and quasiperiodic magnetic reconnection. In the interpretations by MHD waves, the emissions are modulated by eigenmodes of flare-loop oscillations, in which the sausage (Melnikov et al., 2005; Tian et al., 2016; Li et al., 2020), kink (Kolotkov et al., 2015; Yuan and Van Doorsselaere, 2016a,b; Li et al., 2018; Nakariakov et al., 2021), and slow modes (Wang et al., 2003; Reznikova and Shibasaki, 2011; Wang et al., 2021) may all be candidates for generating QPPs. In the scenario of quasiperiodic magnetic reconnection, charged particles are accelerated periodically to produce enhanced emissions. The reconnection processes could be spontaneous (McLaughlin, Thurgood, and MacTaggart, 2012; Thurgood, Pontin, and McLaughlin, 2017, 2019; Karampelas et al., 2022) or periodically triggered by external waves (Nakariakov et al., 2006; Chen and Priest, 2006; Inglis and Nakariakov, 2009; Li et al., 2022).

The imaging observation of QPPs allows us to analyze the spatial locations of pulsations, which is crucial for understanding the nature of QPPs. With the HXR imaging analysis, it was found that pulsations may originate from flare-loop tops or footpoints, or both of them (Asai et al., 2001; Li and Gan, 2008; Huang et al., 2016; Yuan et al., 2019; Clarke et al., 2021; Li and Chen, 2022). Additionally, the microwave imaging also reveals important information about QPP sources, such as the flare loop and footpoints (Melnikov et al., 2005; Kupriyanova et al., 2010; Reznikova and Shibasaki, 2011; Zimovets, Kuznetsov, and Struminsky, 2013; Kolotkov et al., 2015; Huang et al., 2016) and even the reconnection current sheet (Kou et al., 2022).

In this article, we analyzed the QPPs in an X1.2 flare (SOL2023-01-06T00:57) using multiwavelength observations, including extreme ultraviolet (EUV), white-light, soft X-rays (SXR), hard X-rays (HXR), and microwaves. Based on the HXR imaging and X-ray spectral analysis, the evidence of periodic electron acceleration/precipitation was obtained. Section 2 presents the observations and instruments, followed by the data analysis in Section 3. The discussion and conclusion are shown in Section 4.

2. Observations and Instruments

An X1.2 solar flare (SOL2023-01-06T00:57) occurred in active region NOAA 13182 (S20 E81) located near the eastern limb of the solar disk. According to SXR fluxes recorded by the *X-ray Sensors* (XRS) on board the *Geostationary Operational Environmental Satellite* (GOES), it started at 00:43, peaked at around 00:57, and ended at 01:07 UT. This event was well detected by several instruments, as shown in Table 1, including the *Atmospheric Imaging Assembly* (AIA; Lemen et al., 2012) on board the *Solar Dynamics Observatory* (SDO; Pesnell, Thompson, and Chamberlin, 2012), the *Hard X-ray Imager* (HXI; Su et al., 2019; Zhang et al., 2019), the *White-light Solar Telescope* (WST) of the *Lyman-alpha (Lyα) Solar Telescope* (LST; Li et al., 2019) on board the *Advanced Space-based Solar Observatory* (ASO-S; Gan et al., 2019, 2023), the *Spectrometer/Telescope for Imaging X-rays* (STIX; Krucker et al., 2020) on board the *Solar Orbiter* (SolO; Müller et al., 2020), the

Table 1 Observational instruments and data used in this study.

Instrument	Wavelength	Cadence (s)	Pixel size (arcsec)	Description	Dimension
GOES/XRS	1 – 8 Å	≈ 1	–	SXR	1D
SDO/AIA	94, 131 Å	12	0.6	EUV imaging	2D
	1600 Å	24	0.6	UV imaging	2D
ASO-S/HXI	15 – 25, 25 – 50, 50 – 84 keV	0.5	–	HXR/SXR	1D
	15 – 20, 25 – 50, 50 – 84 keV	–	2	Imaging	2D
	10 – 300 keV	4	–	Spectra	1D
ASO-S/LST/WST	360 nm	120	0.5	White-light imaging	2D
SolO/STIX	15 – 25, 25 – 50, 50 – 84 keV	0.5	–	HXR/SXR	1D
	4 – 150 keV	4	–	Spectra	1D
Fermi/GBM	50 – 101 keV	1	–	HXR	1D
KW	20 – 80 keV	0.5	–	HXR	1D
NoRP	17, 35 GHz	1	–	Microwave	1D
CBS	35.25 – 39.25 GHz	≈ 0.53	–	Microwave	1D

Fermi Gamma-ray Burst Monitor (GBM; Meegan et al., 2009), the *Konus-Wind* (KW; Lysenko et al., 2022), the *Nobeyama Radio Polarimeters* (NoRP; Nakajima et al., 1985), and the *Chashan Broadband Solar millimeter spectrometer* (CBS; Shang et al., 2022, 2023; Yan et al., 2023; note that it is the first solar-dedicated radio spectrometer in the millimeter regime).

ASO-S/HXI has a higher temporal cadence in the burst mode and hence the light curves were interpolated to the uniform 0.5 s in this study. Likewise, the time series of Fermi/GBM and KW were interpolated to lower cadences of 1 s and 0.5 s, respectively.

In contrast to these near-Earth assets, SolO/STIX was at a distance of 0.95 AU from the Sun, and its position with respect to solar longitude was $\approx 23.6^\circ$ east from the Sun–Earth line at the time of the event.

Figures 1a–c show the AIA 94 Å, 131 Å, and 1600 Å images taken at the impulsive phase of the flare. The AIA 94 Å and 131 Å passbands reflect the flaring plasma with the characteristic temperature of ≈ 6 MK and ≈ 10 MK, respectively. The AIA 1600 Å flare excess emission is of chromospheric origin (Simões et al., 2019). From the AIA movie (aia94_movie.mp4), we can see the underlying bright compact "postflare" loops accompanied by the overlying large-scale eruption and the fall of matter. This event was a double-ribbon flare, the two ribbons were connected by flare loops with the prominent cusp structure above the loops. Furthermore, the contours of HXI images of three energy bands are overlaid, the contour levels represent 25%, 50%, and 75% of the peak intensity. At the beginning of the eruption (panel a), the 25 – 50 keV band (cyan) shows double HXR sources, while the 50 – 84 keV (magenta) shows a single source. Subsequently, both 25 – 50 keV and 50 – 84 keV bands show double sources, which coincide with the footpoints of EUV compact loops. The lower energy band 15 – 20 keV (red) exhibits some ejected structures, as well as the footpoints and loop structures. The HXR images are reconstructed by the HXI Clean method (HXI pattern-based Clean algorithm) using the detectors from D19 to D91, since the fine grids (i.e. D1 – D18) still need to be carefully calibrated. Figure 1d illustrates that the white-light brightening shows the same location as the HXR sources. This flare has also been confirmed as a white-light flare by Jing et al. (2024).

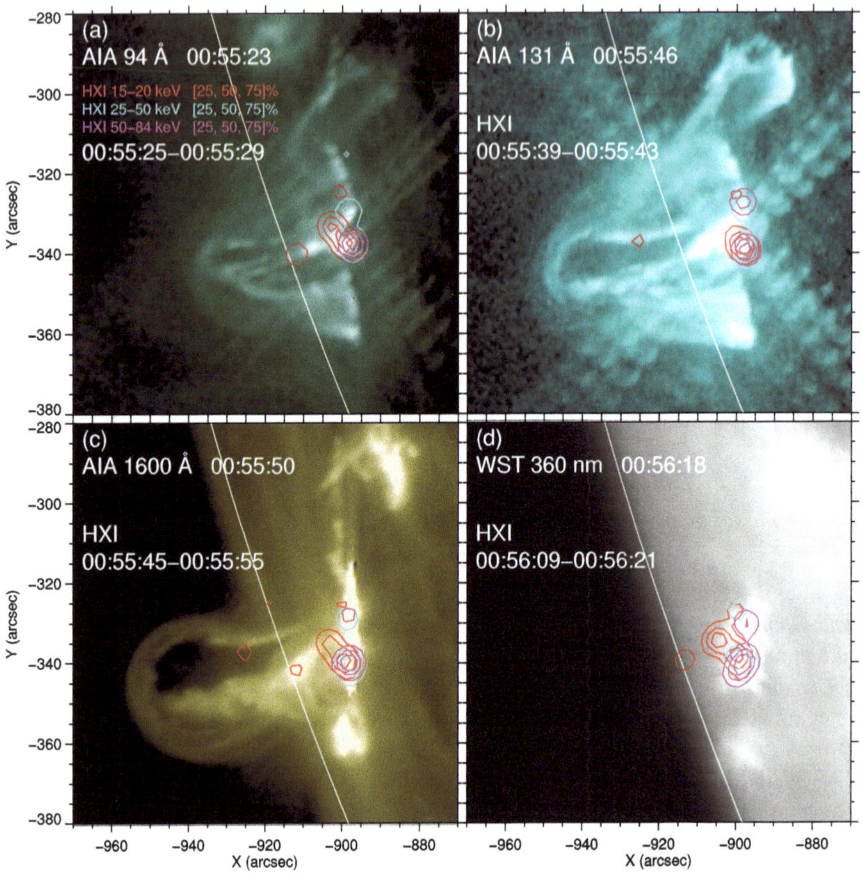

Figure 1 (**a**) AIA 94 Å, (**b**) 131 Å, and (**c**) 1600 Å images taken during the impulsive phase of the flare. The HXR contours of three energy bands are overlaid. The contour levels are 25%, 50%, and 75% of the peak intensity. (**d**) LST/WST 360 nm white-light image with overplotted HXR contours.

3. Data Analysis

3.1. The Identification of QPPs

QPPs were identified as intensity fluctuations above the noise level. In this study, the Markov Chain Monte Carlo (MCMC) method was used to recognize the oscillatory signals. The details of the MCMC method for detecting periodic signals can be seen in previous literature (Vaughan, 2005, 2010; Inglis, Ireland, and Dominique, 2015; Inglis et al., 2016; Hayes et al., 2019, 2020; Liang et al., 2020; Guo et al., 2023; Shi et al., 2023). This method fits the Fourier power spectral density (PSD) of a time series with a model and determines the best-fit parameters. We selected the three-parameter model, which is a power-law plus a constant, i.e. $M = Af^{-\alpha} + C$, where f is the Fourier frequency, $Af^{-\alpha}$ and C reflect the properties of red and white noise in Fourier PSD, respectively.

Figure 2 plots the time series and their MCMC simulation results for multiwavelengths, including the HXI 50–84 keV (panel a), 25–50 keV (panel c), 15–25 keV (panel e), STIX 50–84 keV (panel b), 25–50 keV (panel d), 15–25 keV (panel f),

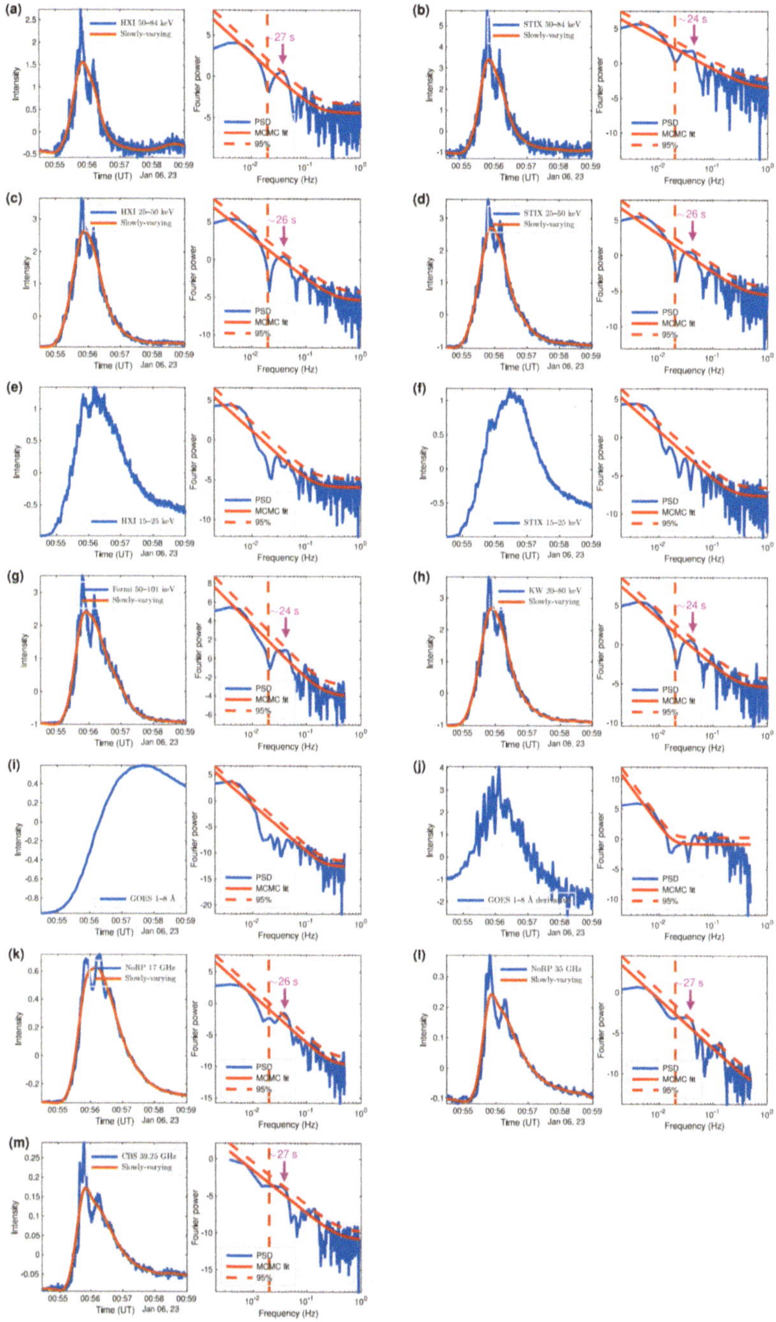

Figure 2 (**a**) The HXI 50 – 84 keV intensity (blue) and its slowly varying components (orange) in the left panel. The Fourier PSD (blue), best-fit (red solid), and the 95% confidence level (red dashed) in the right panel. The vertical orange line represents the cutoff threshold between rapidly (high-) and slowly varying (low-frequency) components. The prominent peak of ≈ 27 s is also marked. (**b**)–(**m**) Similar to (**a**), results for other instruments and wavelengths. For the (**e**)–(**f**) and (**i**)–(**j**), slowly varying components were not plotted because the ≈ 27 s QPPs are not detected in these wavelengths.

Fermi 50 – 101 keV (panel g), KW 20 – 80 keV (panel h), GOES 1 – 8 Å (panel i), 1 – 8 Å derivative (panel j), NoRP 17 GHz (panel k), NoRP 35 GHz (panel l), and the CBS 39.25 GHz (panel m). As shown in Figure 2a, the intensity (blue) and its slowly varying components (orange) are plotted in the left panel, while the Fourier PSD (blue), best-fit (red solid), and the 95% confidence level (red dashed) are plotted in the right panel (periodogram). The PSD at the position of ≈ 27 s shows a prominent peak above the 95% confidence level, indicating that the period of QPPs in HXI 50 – 84 keV is about 27 s. Figure 2c shows a similar peak of ≈ 26 s in HXI 25 – 50 keV, but it seems to have just reached the 95% confidence level. Likewise, STIX 50 – 84 keV (≈ 24 s), STIX 50 – 84 keV (≈ 26 s), Fermi 50 – 101 keV (≈ 24 s), Fermi 50 – 101 keV (≈ 24 s), KW 20 – 80 keV (≈ 24 s), NoRP 35 GHz (≈ 27 s), and CBS 39.25 GHz (≈ 27 s) all appear to have just reached the 95% confidence level, while NoRP 17 GHz (≈ 26 s) shows a peak above that. However, GOES 1 – 8 Å and its derivative (an approximation for the HXR), HXI 15 – 25 keV, and STIX 15 – 25 keV, do not show any definite periodic signals above the red-noise level, which suggests that the absence of ≈ 27 s QPPs in SXR and HXR below 25 keV. Therefore, our observation confirms the existence of ≈ 27 s QPPs in HXR above 25 keV and microwaves.

The presence of a period of about 27 s is convincingly shown. It is also notable that the PSD in Figure 2 show peaks with a frequency of about ≈ 10 s, near the 95% confidence level above the red-noise level, which may be another quasiperiodic signal. Additionally, as shown in the GOES 1 – 8 Å derivative curve, there are some clear pulsations with a period of ≈ 10 s. Although these oscillations are higher than the random white noise in Fourier space, rather than the red noise, they are indeed likely to be the ≈ 10 s QPPs signals.

3.2. QPPs in Nonthermal Emissions

After identifying the ≈ 27 s QPPs, it is natural to decompose the light curves into rapidly and slowly varying components via the fast Fourier transform (FFT) method. A value of 50 s (0.02 Hz) was adopted as the cutoff threshold between rapidly (high-) and slowly varying (low-frequency) components, as marked by the vertical orange line in the periodogram in Figure 2. Figure 3a plots the self-normalized light curves between 00:54 and 00:59 UT, including the GOES 1 – 8 Å (purple), HXI 15 – 25 keV (dark blue), 25 – 50 keV (black), 50 – 84 keV (blue), STIX 15 – 25 keV (orange), 25 – 50 keV (red), 50 – 84 keV (green), Fermi 50 – 101 keV (gold), and the KW 20 – 80 keV (pink). Figure 3b plots the CBS spectrum with the microwave light curves (NoRP 17 GHz, 35 GHz, and CBS 39.25 GHz) overplotted. The HXRs exhibit similar pulsations as the microwaves during the impulsive phase, i.e. 00:55 – 00:57 UT, as marked by the gray background. Figure 3c shows the corresponding rapidly varying components that are obtained from the FFT method mentioned above. Morlet-wavelet analysis was performed on the rapidly varying components to determine the periodicity. This routine was provided by Torrence and Compo (1998). Figure 3d plots the wavelet power spectra of the detrended HXI 50 – 84 keV emission, and the 95% significance level is marked. The significant period of ≈ 27 s is found in the global power spectrum, which is consistent with the periodicity found in the MCMC simulation. As for the ≈ 10 s signal mentioned before, the corresponding weak signatures are also visible on the wavelet spectrum, almost simultaneously with the ≈ 27 s oscillation, which could be the sign of second or third harmonics.

Both the HXR and microwave emissions here are caused by nonthermal electrons – by nonthermal bremsstrahlung for HXR, and gyrosynchrotron emission for microwaves. Figure 4 gives the crosscorrelation diagrams between the rapidly varying components of HXI

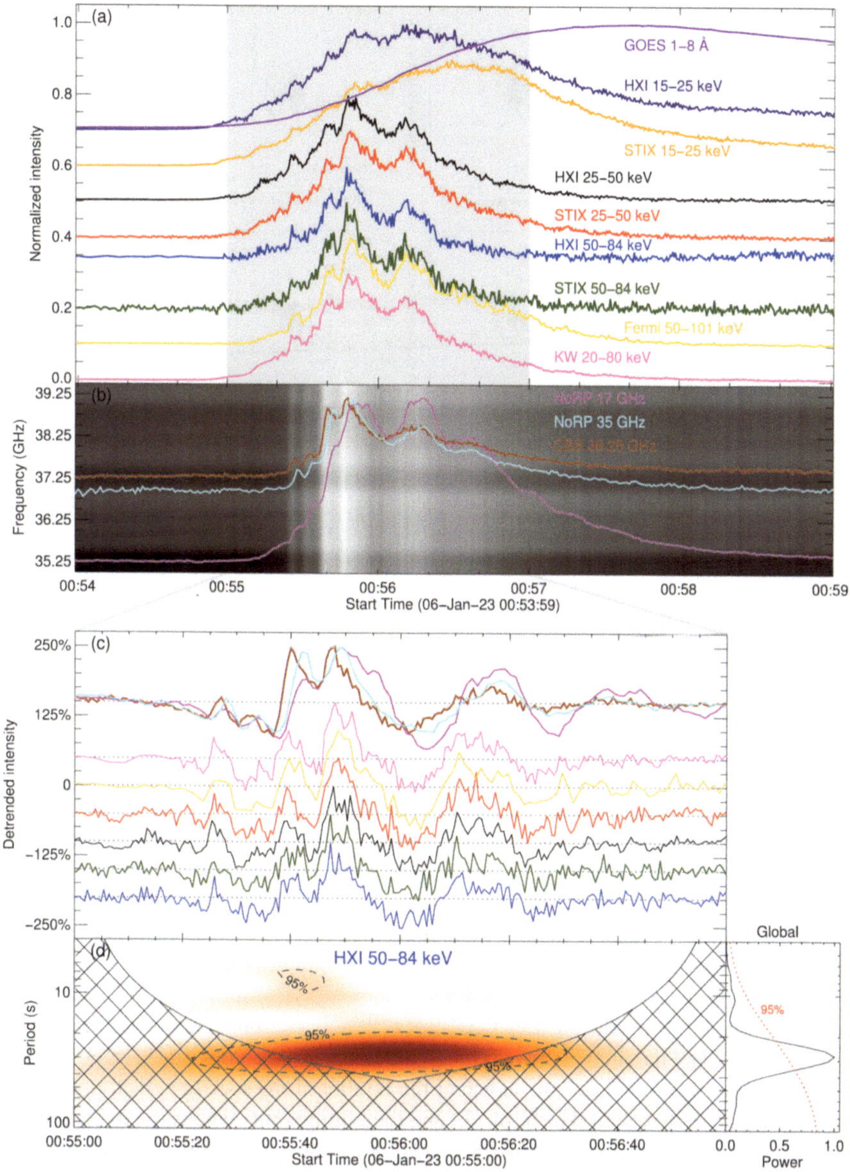

Figure 3 (**a**) The self-normalized light curves between 00:54 and 00:59 UT, including the GOES 1 – 8 Å (purple), HXI 15 – 25 keV (dark blue), 25 – 50 keV (black), 50 – 84 keV (blue), STIX 15 – 25 keV (orange), 25 – 50 keV (red), 50 – 84 keV (green), Fermi 50 – 101 keV (gold), and the KW 20 – 80 keV (pink). (**b**) The CBS spectrum with the microwave light curves (NoRP 17 GHz, 35 GHz, and CBS 39.25 GHz) are overplotted. (**c**) The rapidly varying components of light curves at 00:55 – 00:57 UT. (**d**) The Morlet-wavelet spectra of the detrended HXI 50 – 84 keV emission. The 95% significance level is also marked.

50 – 84 keV and that of other wavelengths at 00:55 – 00:57 UT. Their correlation coefficients (CCs) are also labeled. The HXI 50 – 84 keV data are interpolated to the same temporal resolution as Fermi, NoRP, and CBS. These diagrams show good positive correlation, with the

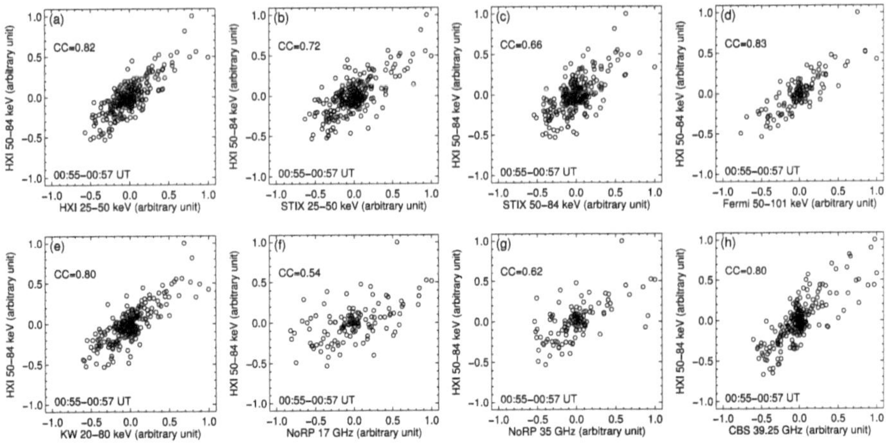

Figure 4 The intensity fluctuations at HXI 50 – 84 keV vs. at (**a**) HXI 25 – 50 keV, (**b**) STIX 25 – 50 keV, (**c**) STIX 50 – 84 keV, (**d**) Fermi 50 – 101 keV, (**e**) KW 20 – 80 keV, (**f**) NoRP 17 GHz, (**g**) NoRP 35 GHz, and (**h**) CBS 39.25 GHz at 00:55 – 00:57 UT. The correlation coefficient (CC) is also added.

CCs all larger than 0.54 and the maximum as high as 0.83, which is sufficient to illustrate the almost inphase oscillations between HXRs and microwaves.

3.3. QPPs of Nonthermal Electron Spectral Index

We also performed X-ray spectral analysis on the HXI and STIX data during the impulsive phase (i.e. 00:55 – 00:57 UT) using the Object Spectral Analysis Executive (OSPEX) software package within the Solar SoftWare (SSW). These spectra were obtained over time intervals of 4 s and fitted by single-temperature thermal bremsstrahlung (vth) plus a nonthermal thick-target single power-law (thick2_vnorm) model. These fits also included the isotropic albedo correction (Kontar et al., 2006), i.e. the actual fit model was "vth+thick2_vnorm+albedo". Figure 5 shows four examples of fitting results around the time of the HXR peak, in which panels a and b show the HXI spectral fitting, and panels c and d show the STIX result. The black profiles are the background-subtracted spectral data with error bars, and the fitting thermal (blue) and nonthermal components (green) are overplotted. The reduced residuals are shown at the bottom of the spectra, which represent the deviation between the spectral data (black) and total fitting (red). The vertical red lines indicate the upper and lower limits of energy bands for fitting. The derived physical parameters are also given.

In terms of thermal parameters, the differences in plasma temperature (T) between HXI and STIX were not significant, with values of ≈ 19 MK and ≈ 21 MK, respectively, but there were minor differences in the emission measure (EM). The EM was approximately $(3.96 - 4.42) \times 10^{49}$ cm^{-3} for HXI, while it was $(0.73 - 0.90) \times 10^{49}$ cm^{-3} for STIX during the same time interval. This discrepancy in the thermal component is most likely due to the fact that the STIX spectral fits go down to 4 keV and also have a higher energy resolution at low energies as compared to HXI, for which 12 keV was used as the minimum energy. In contrast, the fitted nonthermal parameters showed a very good agreement. The electron fluxes for HXI and STIX were estimated to be $(5.11 - 5.95)$ and $(4.90 - 5.72) \times 10^{33}$ s^{-1} keV^{-1}. Additionally, the nonthermal electron spectral indices (δ)

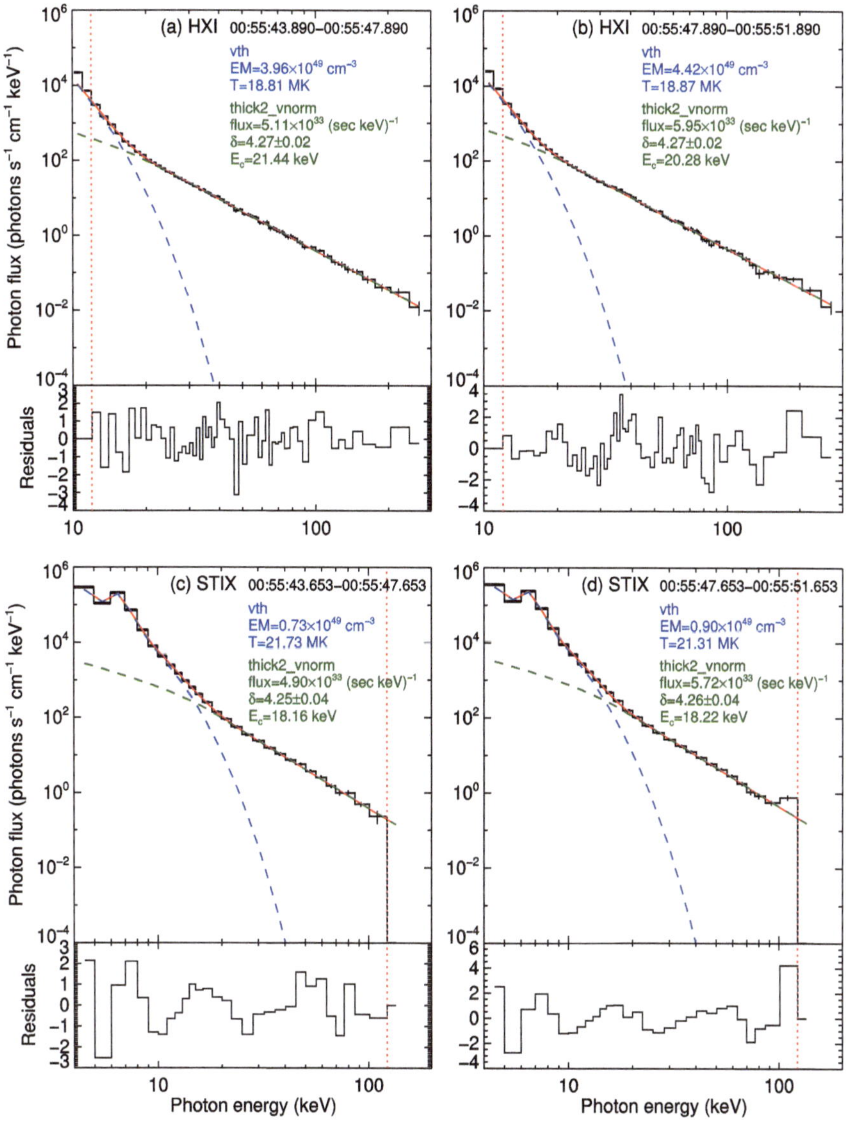

Figure 5 (**a**) and (**b**) HXI spectra and fitting results with the "vth+thick2_vnorm+albedo" model. Shown is the background-subtracted spectral data (black) with error bars, the thermal (blue) and nonthermal fitting components (green), and the total fit (red). The reduced residuals are shown at the bottom of the spectra. The vertical red lines indicate the lower limit of the energy range used for fitting. The fitting parameters are also given. (**c**) and (**d**) Similar to (**a**) and (**b**), for the STIX spectra and fitting results.

were almost the same considering the sigma (σ) error, and the low-energy cutoffs (E_c) of HXI ($\approx$20 keV) were only marginally larger than that of STIX ($\approx$18 keV).

Figure 6a shows the temporal evolution of nonthermal electron spectral indices (red for HXI, blue for STIX) from 00:55 to 00:57 UT, with the light curve of HXI 50–84 keV overplotted. The spectral indices of HXI and STIX not only show the well-known "soft–hard–soft" trend (Benz, 1977; Brown and Loran, 1985; Lin and Schwartz, 1987; Fletcher

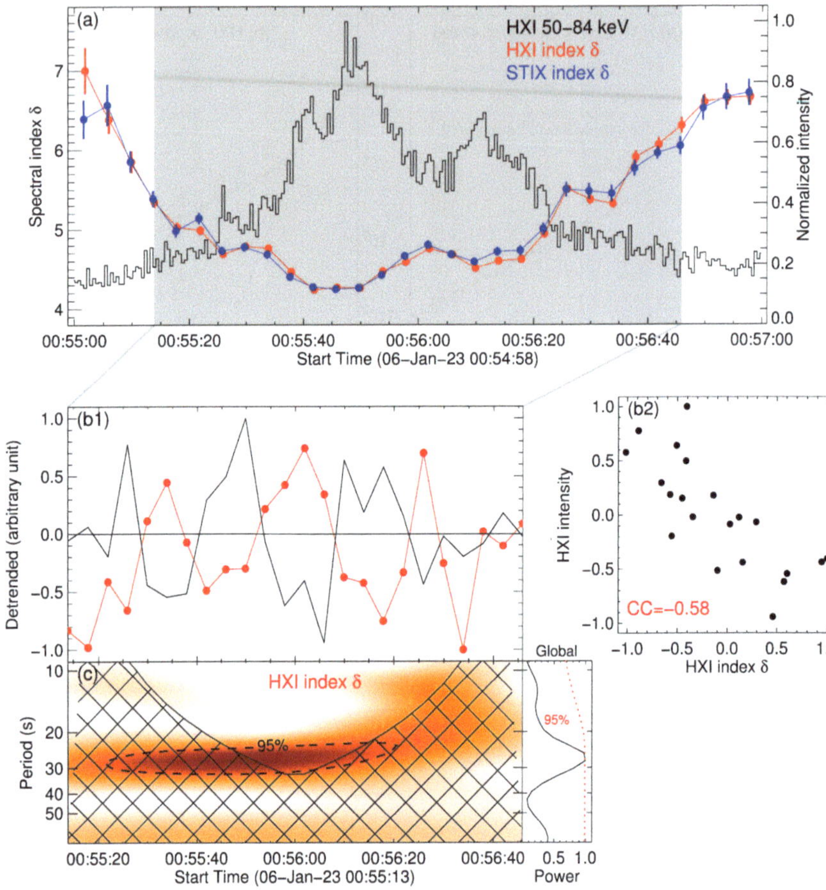

Figure 6 (**a**) The temporal evolution of nonthermal electron spectral indices (red for HXI, blue for STIX) from 00:55 to 00:57 UT, and the normalized intensity of HXI 50 – 84 keV. (**b1**) and (**b2**) The correlation between the detrended spectral index and the detrended HXR intensity. (**c**) The Morlet-wavelet spectra of the detrended spectral index. The 95% significance level is also marked.

and Hudson, 2002; Grigis and Benz, 2004), but also show the negative correlation with the HXR emission. In order to investigate this correlation, Figure 6b1 plots the detrended HXI spectral index and the detrended HXR intensity, which were derived with the same FFT method as above. To compare with the spectral index, the detrended HXR intensity had been interpolated to the temporal resolution of 4 s. Figure 6b2 plots the correlation between the detrended spectral index and HXR intensity. Although panel b2 exhibits a good negative correlation, the CC is only -0.58, which could be caused by the calculation using only 24 points. We also performed the wavelet analysis on the detrended spectral index, as shown in Figure 6c. The Morlet-wavelet spectra show a peak at ≈ 27 s, reaching the 95% significance level, which is consistent with the period of QPPs analyzed before. The same dominant period implies the origin of QPPs, i.e. the periodic electron acceleration/precipitation during the impulsive phase.

4. Discussion and Conclusion

We studied the QPPs of the X1.2 solar flare (SOL2023-01-06T00:57) based on multi-instrument observations. The ≈ 27 s QPPs were identified in HXRs and microwaves during the impulsive phase, but not found in the SXR thermal emission. The weak signatures of another quasiperiod of ≈ 10 s were also present, which may be the sign of second or third harmonics. Spatial analysis of HXI imaging demonstrates that HXR pulsations originate from the double-footpoint HXR sources. Furthermore, the nonthermal electron spectral indices also show the ≈ 27 s oscillation via the X-ray spectral analysis using HXI and STIX data. Our results support the scenario of quasiperiodic injections of accelerated electrons, in which periodic electrons spiral along magnetic field lines in flare loops to generate microwaves, and then precipitate into the chromosphere at the double footpoints to produce HXR emissions.

Figure 1 indicates that there are remote hook-like flare ribbons and in general the development of this flare resembles a picture of the three-dimensional (3D) standard flare model (e.g. Janvier et al., 2014). Although after around 01:05 UT, well after the end of QPPs, the eruption observed in EUV stopped, i.e. the flare was accompanied by a confined eruption and a coronal mass ejection was not observed in white light in this event. In this scenario, a slipping reconnection may occur in the current layer, which may be accompanied by a displacement of HXR sources along the magnetic polarity-inversion line. There are observations indicating that such magnetic reconnection can occur in a quasiperiodic mode in some flares (Li and Zhang, 2015; Li et al., 2016). A simulation work performed by Thurgood, Pontin, and McLaughlin (2017) demonstrated that reconnection at 3D magnetic null points can proceed naturally in a time-dependent and periodic fashion, illustrating that oscillatory reconnection likely plays a role in the observed QPPs.

QPPs accompanied by motions of HXR sources along a polarity-inversion line can be also interpreted in terms of quasiperiodic triggering or modulation of magnetic reconnection in a current sheet by slow magnetoacoustic waves (Nakariakov and Zimovets, 2011) or by flapping oscillations (Artemyev and Zimovets, 2012). However, due to the observation perspective of this flare arcade, or due to limited angular resolution of the used HXI collimating grids, motion of HXR sources was hard to observe in this event. This does not allow us to conclude which of these mechanisms is the more probable as the origin of QPPs.

There are other theoretical works (models) showing the possibility of quasiperiodic modulation of physical parameters in the cusp region, the signs of which are observed in this flare. Takasao and Shibata (2016) revealed the above-the-loop-top oscillation controlled by the backflow of the reconnection outflow, which then excited quasiperiodic propagating fast-mode magnetoacoustic waves (QPFs), a model known as a "magnetic tuning fork". They also found that the above-the-loop-top region presented shocks and waves, as well as the periodic variation of the termination-shock strength. The termination shock is a promising driver for particle acceleration in flares (Chen et al., 2015). This means that electrons could be accelerated periodically through this physical process located at the above-the-loop-top region.

QPPs with short periods were generally interpreted as sausage modes in terms of MHD waves (Nakariakov, Melnikov, and Reznikova, 2003; Melnikov et al., 2005; Inglis, Nakariakov, and Melnikov, 2008; Van Doorsselaere et al., 2011; Kolotkov et al., 2015). Tian et al. (2016) found strong evidence for trapped sausage modes, which explained the ≈ 25 s QPPs. However, in this study, due to the explosive and highly dynamic nature of the impulsive phase, the standing MHD waves may not be effectively established in the flare loops.

We report here some of the first observations obtained with new instruments, which show the capabilities to investigate QPPs using the joint observations of HXI and STIX. In the future, we will apply this to more events to obtain statistically sound results. For example, Collier et al. (2023) found periods across the range of 4 – 128 s in a study of several STIX flares, and the joint observation may help to investigate the causes of the variation of the QPP periods.

Supplementary Information The online version contains supplementary material available at https://doi.org/10.1007/s11207-024-02272-4.

Acknowledgment We thank the referee for valuable comments. ASO-S mission is supported by the Strategic Priority Research Program on Space Science, the Chinese Academy of Sciences, Grant No. XDA15320000. Solar Orbiter is a space mission of international collaboration between ESA and NASA, operated by ESA. We appreciate the teams of GOES, SDO, Fermi, KW, NoRP, and CBS for their open data-use policy.

Author contributions F.S., D.L., and Z.N. wrote the main manuscript text, A.W. and W.C. prepared Figures 5 – 6, Y.S. and Y.L. prepared Figure 1. J.X., Y.S., and Y.Y. participated in the discussion. All authors reviewed the manuscript.

Funding This work is supported by the Strategic Priority Research Program of the Chinese Academy of Sciences, Grant No. XDB0560000. This work is also funded by the National Key R&D Program of China2022YFF0503002 (2022YFF0503000), NSFC under grants 12073081, 12333010, and 11820101002.

Data Availability ASO-S data: http://aso-s.pmo.ac.cn/sodc/dataArchive.jsp. SolO data: https://soar.esac.esa.int/soar/. SDO data: http://jsoc.stanford.edu. KW data: http://www.ioffe.ru/LEA/kwsun/. NoRP data: http://solar.nro.nao.ac.jp/norp/index.html. GOES and Fermi data can be downloaded via SSW. CBS data can be obtained by contacting the instrument team.

Materials Availability The movie (aia94_movie.mp4) of Figure 1 is available.

Declarations

Competing interests The authors declare no competing interests.

References

Artemyev, A., Zimovets, I.: 2012, Stability of current sheets in the solar corona. *Solar Phys.* **277**, 283. DOI. ADS.

Asai, A., Shimojo, M., Isobe, H., Morimoto, T., Yokoyama, T., Shibasaki, K., Nakajima, H.: 2001, Periodic acceleration of electrons in the 1998 November 10 solar flare. *Astrophys. J. Lett.* **562**, L103. DOI. ADS.

Benz, A.O.: 1977, Spectral features in solar hard X-ray and radio events and particle acceleration. *Astrophys. J.* **211**, 270. DOI. ADS.

Brown, J.C., Loran, J.M.: 1985, Possible evidence for stochastic acceleration of electrons in solar hard X-ray bursts observed by SMM. *Mon. Not. Roy. Astron. Soc.* **212**, 245. DOI. ADS.

Chen, P.F., Priest, E.R.: 2006, Transition-region explosive events: reconnection modulated by p-mode waves. *Solar Phys.* **238**, 313. DOI. ADS.

Chen, B., Bastian, T.S., Shen, C., Gary, D.E., Krucker, S., Glesener, L.: 2015, Particle acceleration by a solar flare termination shock. *Science* **350**, 1238. DOI. ADS.

Clarke, B.P., Hayes, L.A., Gallagher, P.T., Maloney, S.A., Carley, E.P.: 2021, Quasi-periodic particle acceleration in a solar flare. *Astrophys. J.* **910**, 123. DOI. ADS.

Collier, H., Hayes, L.A., Battaglia, A.F., Harra, L.K., Krucker, S.: 2023, Characterising fast-time variations in the hard X-ray time profiles of solar flares using Solar Orbiter's STIX. *Astron. Astrophys.* **671**, A79. DOI. ADS.

Dominique, M., Zhukov, A.N., Dolla, L., Inglis, A., Lapenta, G.: 2018, Detection of quasi-periodic pulsations in solar EUV time series. *Solar Phys.* **293**, 61. DOI. ADS.

Fletcher, L., Hudson, H.S.: 2002, Spectral and spatial variations of flare hard X-ray footpoints. *Solar Phys.* **210**, 307. DOI. ADS.

Gan, W.-Q., Zhu, C., Deng, Y.-Y., Li, H., Su, Y., Zhang, H.-Y., Chen, B., Zhang, Z., Wu, J., Deng, L., Huang, Y., Yang, J.-F., Cui, J.-J., Chang, J., Wang, C., Wu, J., Yin, Z.-S., Chen, W., Fang, C., Yan, Y.-H., Lin, J., Xiong, W.-M., Chen, B., Bao, H.-C., Cao, C.-X., Bai, Y.-P., Wang, T., Chen, B.-L., Li, X.-Y., Zhang, Y., Feng, L., Su, J.-T., Li, Y., Chen, W., Li, Y.-P., Su, Y.-N., Wu, H.-Y., Gu, M., Huang, L., Tang, X.-J.: 2019, Advanced Space-based Solar Observatory (ASO-S): an overview. *Res. Astron. Astrophys.* **19**, 156. DOI. ADS.

Gan, W., Zhu, C., Deng, Y., Zhang, Z., Chen, B., Huang, Y., Deng, L., Wu, H., Zhang, H., Li, H., Su, Y., Su, J., Feng, L., Wu, J., Cui, J., Wang, C., Chang, J., Yin, Z., Xiong, W., Chen, B., Yang, J., Li, F., Lin, J., Hou, J., Bai, X., Chen, D., Zhang, Y., Hu, Y., Liang, Y., Wang, J., Song, K., Guo, Q., He, L., Zhang, G., Wang, P., Bao, H., Cao, C., Bai, Y., Chen, B., He, T., Li, X., Zhang, Y., Liao, X., Jiang, H., Li, Y., Su, Y., Lei, S., Chen, W., Li, Y., Zhao, J., Li, J., Ge, Y., Zou, Z., Hu, T., Su, M., Ji, H., Gu, M., Zheng, Y., Xu, D., Wang, X.: 2023, The Advanced Space-Based Solar Observatory (ASO-S). *Solar Phys.* **298**, 68. DOI. ADS.

Grigis, P.C., Benz, A.O.: 2004, The spectral evolution of impulsive solar X-ray flares. *Astron. Astrophys.* **426**, 1093. DOI. ADS.

Guo, Y., Liang, B., Feng, S., Yuan, D., Nakariakov, V.M., Dai, W., Yang, Y.: 2023, Feature identification and statistical characteristics of quasi-periodic pulsation in solar flares using the Markov-chain-Monte-Carlo approach. *Astrophys. J.* **944**, 16. DOI. ADS.

Hayes, L.A., Gallagher, P.T., Dennis, B.R., Ireland, J., Inglis, A., Morosan, D.E.: 2019, Persistent quasi-periodic pulsations during a large X-class solar flare. *Astrophys. J.* **875**, 33. DOI. ADS.

Hayes, L.A., Inglis, A.R., Christe, S., Dennis, B., Gallagher, P.T.: 2020, Statistical study of GOES X-ray quasi-periodic pulsations in solar flares. *Astrophys. J.* **895**, 50. DOI. ADS.

Huang, J., Kontar, E.P., Nakariakov, V.M., Gao, G.: 2016, Quasi-periodic acceleration of electrons in the flare on 2012 July 19. *Astrophys. J.* **831**, 119. DOI. ADS.

Inglis, A.R., Ireland, J., Dominique, M.: 2015, Quasi-periodic pulsations in solar and stellar flares: re-evaluating their nature in the context of power-law flare Fourier spectra. *Astrophys. J.* **798**, 108. DOI. ADS.

Inglis, A.R., Nakariakov, V.M.: 2009, A multi-periodic oscillatory event in a solar flare. *Astron. Astrophys.* **493**, 259. DOI. ADS.

Inglis, A.R., Nakariakov, V.M., Melnikov, V.F.: 2008, Multi-wavelength spatially resolved analysis of quasi-periodic pulsations in a solar flare. *Astron. Astrophys.* **487**, 1147. DOI. ADS.

Inglis, A.R., Ireland, J., Dennis, B.R., Hayes, L., Gallagher, P.: 2016, A large-scale search for evidence of quasi-periodic pulsations in solar flares. *Astrophys. J.* **833**, 284. DOI. ADS.

Janvier, M., Aulanier, G., Bommier, V., Schmieder, B., Démoulin, P., Pariat, E.: 2014, Electric currents in flare ribbons: observations and three-dimensional standard model. *Astrophys. J.* **788**, 60. DOI. ADS.

Jing, Z., Li, Y., Feng, L., Li, H., Huang, Y., Li, Y., Su, Y., Chen, W., Tian, J., Song, D., Li, J., Xue, J., Zhao, J., Lu, L., Ying, B., Zhang, P., Su, Y., Zhang, Q., Li, D., Ge, Y., Li, S., Li, Q., Li, G., Liu, X., Shi, G., Shan, J., Tian, Z., Zhou, Y., Gan, W.: 2024, A statistical study of solar white-light flares observed by the White-Light Solar Telescope of the Lyman-Alpha Solar Telescope on the Advanced Space-Based Solar Observatory (ASO-S/LST/WST) at 360 nm. *Solar Phys.* **299**, 11. DOI. ADS.

Karampelas, K., McLaughlin, J.A., Botha, G.J.J., Régnier, S.: 2022, Oscillatory reconnection of a 2D X-point in a hot coronal plasma. *Astrophys. J.* **925**, 195. DOI. ADS.

Kolotkov, D.Y., Nakariakov, V.M., Kupriyanova, E.G., Ratcliffe, H., Shibasaki, K.: 2015, Multi-mode quasi-periodic pulsations in a solar flare. *Astron. Astrophys.* **574**, A53. DOI. ADS.

Kontar, E.P., MacKinnon, A.L., Schwartz, R.A., Brown, J.C.: 2006, Compton backscattered and primary X-rays from solar flares: angle dependent Green's function correction for photospheric albedo. *Astron. Astrophys.* **446**, 1157. DOI. ADS.

Kou, Y., Cheng, X., Wang, Y., Yu, S., Chen, B., Kontar, E.P., Ding, M.: 2022, Microwave imaging of quasi-periodic pulsations at flare current sheet. *Nat. Commun.* **13**, 7680. DOI. ADS.

Krucker, S., Hurford, G.J., Grimm, O., Kögl, S., Gröbelbauer, H.-P., Etesi, L., Casadei, D., Csillaghy, A., Benz, A.O., Arnold, N.G., Molendini, F., Orleanski, P., Schori, D., Xiao, H., Kuhar, M., Hochmuth, N., Felix, S., Schramka, F., Marcin, S., Kobler, S., Iseli, L., Dreier, M., Wiehl, H.J., Kleint, L., Battaglia, M., Lastufka, E., Sathiapal, H., Lapadula, K., Bednarzik, M., Birrer, G., Stutz, S., Wild, C., Marone, F., Skup, K.R., Cichocki, A., Ber, K., Rutkowski, K., Bujwan, W., Juchnikowski, G., Winkler, M., Darmetko, M., Michalska, M., Seweryn, K., Białek, A., Osica, P., Sylwester, J., Kowalinski, M., Ścisłowski, D., Siarkowski, M., Stęślicki, M., Mrozek, T., Podgórski, P., Meuris, A., Limousin, O., Gevin, O., Le Mer, I., Brun, S., Strugarek, A., Vilmer, N., Musset, S., Maksimović, M., Fárník, F., Kozáček, Z., Kašparová, J., Mann, G., Önel, H., Warmuth, A., Rendtel, J., Anderson, J., Bauer, S., Dionies, F., Paschke, J., Plüschke, D., Woche, M., Schuller, F., Veronig, A.M., Dickson, E.C.M., Gallagher, P.T., Maloney, S.A., Bloomfield, D.S., Piana, M., Massone, A.M., Benvenuto, F., Massa, P., Schwartz, R.A., Dennis, B.R., van Beek, H.F., Rodríguez-Pacheco, J., Lin, R.P.: 2020, The spectrometer/telescope for imaging X-rays (STIX). *Astron. Astrophys.* **642**, A15. DOI. ADS.

Kupriyanova, E.G., Melnikov, V.F., Nakariakov, V.M., Shibasaki, K.: 2010, Types of microwave quasi-periodic pulsations in single flaring loops. *Solar Phys.* **267**, 329. DOI. ADS.
Kupriyanova, E., Kolotkov, D., Nakariakov, V., Kaufman, A.: 2020, Quasi-periodic pulsations in solar and stellar flares. Review. *Solar-Terr. Phys.* **6**, 3. DOI. ADS.
Lemen, J.R., Title, A.M., Akin, D.J., Boerner, P.F., Chou, C., Drake, J.F., Duncan, D.W., Edwards, C.G., Friedlaender, F.M., Heyman, G.F., Hurlburt, N.E., Katz, N.L., Kushner, G.D., Levay, M., Lindgren, R.W., Mathur, D.P., McFeaters, E.L., Mitchell, S., Rehse, R.A., Schrijver, C.J., Springer, L.A., Stern, R.A., Tarbell, T.D., Wuelser, J.-P., Wolfson, C.J., Yanari, C., Bookbinder, J.A., Cheimets, P.N., Caldwell, D., Deluca, E.E., Gates, R., Golub, L., Park, S., Podgorski, W.A., Bush, R.I., Scherrer, P.H., Gummin, M.A., Smith, P., Auker, G., Jerram, P., Pool, P., Soufli, R., Windt, D.L., Beardsley, S., Clapp, M., Lang, J., Waltham, N.: 2012, The Atmospheric Imaging Assembly (AIA) on the Solar Dynamics Observatory (SDO). *Solar Phys.* **275**, 17. DOI. ADS.
Li, D., Chen, W.: 2022, Quasi-periodic accelerations of energetic particles during a solar flare. *Astrophys. J. Lett.* **931**, L28. DOI. ADS.
Li, Y.P., Gan, W.Q.: 2008, Observational studies of the X-ray quasi-periodic oscillations of a solar flare. *Solar Phys.* **247**, 77. DOI. ADS.
Li, T., Zhang, J.: 2015, Quasi-periodic slipping magnetic reconnection during an X-class solar flare observed by the Solar Dynamics Observatory and Interface Region Imaging Spectrograph. *Astrophys. J. Lett.* **804**, L8. DOI. ADS.
Li, T., Yang, K., Hou, Y., Zhang, J.: 2016, Slipping magnetic reconnection of flux-rope structures as a precursor to an eruptive X-class solar flare. *Astrophys. J.* **830**, 152. DOI. ADS.
Li, D., Yuan, D., Su, Y.N., Zhang, Q.M., Su, W., Ning, Z.J.: 2018, Non-damping oscillations at flaring loops. *Astron. Astrophys.* **617**, A86. DOI. ADS.
Li, H., Chen, B., Feng, L., Li, Y., Huang, Y., Li, J.-W., Lu, L., Xue, J.-C., Ying, B.-L., Zhao, J., Yang, Y.-T., Gan, W.-Q., Fang, C., Song, K.-F., Wang, H., Guo, Q.-F., He, L.-P., Zhu, B., Zhu, C., Deng, L., Bao, H.-C., Cao, C.-X., Yang, Z.-G.: 2019, The Lyman-alpha Solar Telescope (LST) for the ASO-S mission — I. Scientific objectives and overview. *Res. Astron. Astrophys.* **19**, 158. DOI. ADS.
Li, B., Antolin, P., Guo, M.-Z., Kuznetsov, A.A., Pascoe, D.J., Van Doorsselaere, T., Vasheghani Farahani, S.: 2020, Magnetohydrodynamic fast sausage waves in the solar corona. *Space Sci. Rev.* **216**, 136. DOI. ADS.
Li, D., Shi, F., Zhao, H., Xiong, S., Song, L., Peng, W., Li, X., Chen, W., Ning, Z.: 2022, Flare quasi-periodic pulsation associated with recurrent jets. *Front. Astron. Space Sci.* **9**, 1032099. DOI. ADS.
Liang, B., Meng, Y., Feng, S., Yang, Y.: 2020, Estimating red noise in quasi-periodic signals with MCMC-based Bayesian. *Astrophys. Space Sci.* **365**, 40. DOI. ADS.
Lin, R.P., Schwartz, R.A.: 1987, High spectral resolution measurements of a solar flare hard X-ray burst. *Astrophys. J.* **312**, 462. DOI. ADS.
Lysenko, A.L., Ulanov, M.V., Kuznetsov, A.A., Fleishman, G.D., Frederiks, D.D., Kashapova, L.K., Sokolova, Z.Y., Svinkin, D.S., Tsvetkova, A.E.: 2022, KW-Sun: the konus-wind solar flare database in hard X-ray and soft gamma-ray ranges. *Astrophys. J. Suppl. Ser.* **262**, 32. DOI. ADS.
McLaughlin, J.A., Thurgood, J.O., MacTaggart, D.: 2012, On the periodicity of oscillatory reconnection. *Astron. Astrophys.* **548**, A98. DOI. ADS.
McLaughlin, J.A., Nakariakov, V.M., Dominique, M., Jelínek, P., Takasao, S.: 2018, Modelling quasi-periodic pulsations in solar and stellar flares. *Space Sci. Rev.* **214**, 45. DOI. ADS.
Meegan, C., Lichti, G., Bhat, P.N., Bissaldi, E., Briggs, M.S., Connaughton, V., Diehl, R., Fishman, G., Greiner, J., Hoover, A.S., van der Horst, A.J., von Kienlin, A., Kippen, R.M., Kouveliotou, C., McBreen, S., Paciesas, W.S., Preece, R., Steinle, H., Wallace, M.S., Wilson, R.B., Wilson-Hodge, C.: 2009, The Fermi gamma-ray burst monitor. *Astrophys. J.* **702**, 791. DOI. ADS.
Melnikov, V.F., Reznikova, V.E., Shibasaki, K., Nakariakov, V.M.: 2005, Spatially resolved microwave pulsations of a flare loop. *Astron. Astrophys.* **439**, 727. DOI. ADS.
Müller, D., St. Cyr, O.C., Zouganelis, I., Gilbert, H.R., Marsden, R., Nieves-Chinchilla, T., Antonucci, E., Auchère, F., Berghmans, D., Horbury, T.S., Howard, R.A., Krucker, S., Maksimovic, M., Owen, C.J., Rochus, P., Rodriguez-Pacheco, J., Romoli, M., Solanki, S.K., Bruno, R., Carlsson, M., Fludra, A., Harra, L., Hassler, D.M., Livi, S., Louarn, P., Peter, H., Schühle, U., Teriaca, L., del Toro Iniesta, J.C., Wimmer-Schweingruber, R.F., Marsch, E., Velli, M., De Groof, A., Walsh, A., Williams, D.: 2020, The Solar Orbiter mission. Science overview. *Astron. Astrophys.* **642**, A1. DOI. ADS.
Nakajima, H., Sekiguchi, H., Sawa, M., Kai, K., Kawashima, S.: 1985, The radiometer and polarimeters at 80, 35, and 17 GHz for solar observations at Nobeyama. *Publ. Astron. Soc. Japan* **37**, 163. ADS.
Nakariakov, V.M., Melnikov, V.F.: 2009, Quasi-periodic pulsations in solar flares. *Space Sci. Rev.* **149**, 119. DOI. ADS.
Nakariakov, V.M., Melnikov, V.F., Reznikova, V.E.: 2003, Global sausage modes of coronal loops. *Astron. Astrophys.* **412**, L7. DOI. ADS.

Nakariakov, V.M., Zimovets, I.V.: 2011, Slow magnetoacoustic waves in two-ribbon flares. *Astrophys. J. Lett.* **730**, L27. DOI. ADS.
Nakariakov, V.M., Foullon, C., Verwichte, E., Young, N.P.: 2006, Quasi-periodic modulation of solar and stellar flaring emission by magnetohydrodynamic oscillations in a nearby loop. *Astron. Astrophys.* **452**, 343. DOI. ADS.
Nakariakov, V.M., Anfinogentov, S.A., Antolin, P., Jain, R., Kolotkov, D.Y., Kupriyanova, E.G., Li, D., Magyar, N., Nisticò, G., Pascoe, D.J., Srivastava, A.K., Terradas, J., Vasheghani Farahani, S., Verth, G., Yuan, D., Zimovets, I.V.: 2021, Kink oscillations of coronal loops. *Space Sci. Rev.* **217**, 73. DOI. ADS.
Pesnell, W.D., Thompson, B.J., Chamberlin, P.C.: 2012, The Solar Dynamics Observatory (SDO). *Solar Phys.* **275**, 3. DOI. ADS.
Pugh, C.E., Broomhall, A.-M., Nakariakov, V.M.: 2017, Significance testing for quasi-periodic pulsations in solar and stellar flares. *Astron. Astrophys.* **602**, A47. DOI. ADS.
Reznikova, V.E., Shibasaki, K.: 2011, Flare quasi-periodic pulsations with growing periodicity. *Astron. Astrophys.* **525**, A112. DOI. ADS.
Shang, Z., Xu, K., Liu, Y., Wu, Z., Lu, G., Zhang, Y., Zhang, L., Su, Y., Chen, Y., Yan, F.: 2022, A broadband solar radio dynamic spectrometer working in the millimeter-wave band. *Astrophys. J. Suppl. Ser.* **258**, 25. DOI. ADS.
Shang, Z., Wu, Z., Liu, Y., Bai, Y., Lu, G., Zhang, Y., Zhang, L., Su, Y., Chen, Y., Yan, F.: 2023, The calibration of the 35-40 GHz solar radio spectrometer with the new Moon and a noise source. *Astrophys. J. Suppl. Ser.* **268**, 45. DOI. ADS.
Shi, F., Li, D., Ning, Z., Guo, Y., Feng, S., Xu, J.: 2023, Are quasi-periodic pulsations independent of loop oscillations in solar flare? *Astrophys. J.* **958**, 39. DOI. ADS.
Simões, P.J.A., Hudson, H.S., Fletcher, L.: 2015, Soft X-ray pulsations in solar flares. *Solar Phys.* **290**, 3625. DOI. ADS.
Simões, P.J.A., Reid, H.A.S., Milligan, R.O., Fletcher, L.: 2019, The spectral content of SDO/AIA 1600 and 1700 Å filters from flare and plage observations. *Astrophys. J.* **870**, 114. DOI. ADS.
Su, Y., Liu, W., Li, Y.-P., Zhang, Z., Hurford, G.J., Chen, W., Huang, Y., Li, Z.-T., Jiang, X.-K., Wang, H.-X., Xia, F.-X.-Y., Chen, C.-X., Yu, W.-H., Yu, F., Wu, J., Gan, W.-Q.: 2019, Simulations and software development for the Hard X-ray Imager onboard ASO-S. *Res. Astron. Astrophys.* **19**, 163. DOI. ADS.
Takasao, S., Shibata, K.: 2016, Above-the-loop-top oscillation and quasi-periodic coronal wave generation in solar flares. *Astrophys. J.* **823**, 150. DOI. ADS.
Thurgood, J.O., Pontin, D.I., McLaughlin, J.A.: 2017, Three-dimensional oscillatory magnetic reconnection. *Astrophys. J.* **844**, 2. DOI. ADS.
Thurgood, J.O., Pontin, D.I., McLaughlin, J.A.: 2019, On the periodicity of linear and nonlinear oscillatory reconnection. *Astron. Astrophys.* **621**, A106. DOI. ADS.
Tian, H., Young, P.R., Reeves, K.K., Wang, T., Antolin, P., Chen, B., He, J.: 2016, Global sausage oscillation of solar flare loops detected by the Interface Region Imaging Spectrograph. *Astrophys. J. Lett.* **823**, L16. DOI. ADS.
Torrence, C., Compo, G.P.: 1998, A practical guide to wavelet analysis. *Bull. Am. Meteorol. Soc.* **79**, 61. DOI. ADS.
Van Doorsselaere, T., Kupriyanova, E.G., Yuan, D.: 2016, Quasi-periodic pulsations in solar and stellar flares: an overview of recent results (invited review). *Solar Phys.* **291**, 3143. DOI. ADS.
Van Doorsselaere, T., De Groof, A., Zender, J., Berghmans, D., Goossens, M.: 2011, LYRA observations of two oscillation modes in a single flare. *Astrophys. J.* **740**, 90. DOI. ADS.
Vaughan, S.: 2005, A simple test for periodic signals in red noise. *Astron. Astrophys.* **431**, 391. DOI. ADS.
Vaughan, S.: 2010, A Bayesian test for periodic signals in red noise. *Mon. Not. Roy. Astron. Soc.* **402**, 307. DOI. ADS.
Wang, T.J., Solanki, S.K., Innes, D.E., Curdt, W., Marsch, E.: 2003, Slow-mode standing waves observed by SUMER in hot coronal loops. *Astron. Astrophys.* **402**, L17. DOI. ADS.
Wang, T., Ofman, L., Yuan, D., Reale, F., Kolotkov, D.Y., Srivastava, A.K.: 2021, Slow-mode magnetoacoustic waves in coronal loops. *Space Sci. Rev.* **217**, 34. DOI. ADS.
Yan, F., Wu, Z., Shang, Z., Wang, B., Zhang, L., Chen, Y.: 2023, The first flare observation with a new solar microwave spectrometer working in 35 – 40 GHz. *Astrophys. J. Lett.* **942**, L11. DOI. ADS.
Yuan, D., Van Doorsselaere, T.: 2016a, Forward modeling of standing kink modes in coronal loops. I. Synthetic views. *Astrophys. J. Suppl. Ser.* **223**, 23. DOI. ADS.
Yuan, D., Van Doorsselaere, T.: 2016b, Forward modeling of standing kink modes in coronal loops. II. Applications. *Astrophys. J. Suppl. Ser.* **223**, 24. DOI. ADS.
Yuan, D., Feng, S., Li, D., Ning, Z., Tan, B.: 2019, A compact source for quasi-periodic pulsation in an M-class solar flare. *Astrophys. J. Lett.* **886**, L25. DOI. ADS.
Zhang, Z., Chen, D.-Y., Wu, J., Chang, J., Hu, Y.-M., Su, Y., Zhang, Y., Wang, J.-P., Liang, Y.-M., Ma, T., Guo, J.-H., Cai, M.-S., Zhang, Y.-Q., Huang, Y.-Y., Peng, X.-Y., Tang, Z.-B., Zhao, X., Zhou, H.-H.,

Wang, L.-G., Song, J.-X., Ma, M., Xu, G.-Z., Yang, J.-F., Lu, D., He, Y.-H., Tao, J.-Y., Ma, X.-L., Lv, B.-G., Bai, Y.-P., Cao, C.-X., Huang, Y., Gan, W.-Q.: 2019, Hard X-ray Imager (HXI) onboard the ASO-S mission. *Res. Astron. Astrophys.* **19**, 160. DOI. ADS.

Zimovets, I.V., Kuznetsov, S.A., Struminsky, A.B.: 2013, Fine structure of the sources of quasi-periodic pulsations in "single-loop" solar flares. *Astron. Lett.* **39**, 267. DOI. ADS.

Zimovets, I.V., McLaughlin, J.A., Srivastava, A.K., Kolotkov, D.Y., Kuznetsov, A.A., Kupriyanova, E.G., Cho, I.-H., Inglis, A.R., Reale, F., Pascoe, D.J., Tian, H., Yuan, D., Li, D., Zhang, Q.M.: 2021, Quasi-periodic pulsations in solar and stellar flares: a review of underpinning physical mechanisms and their predicted observational signatures. *Space Sci. Rev.* **217**, 66. DOI. ADS.

Solar Physics (2024) 299:57
https://doi.org/10.1007/s11207-024-02299-7

RESEARCH

A Statistical Investigation of the Neupert Effect in Solar Flares Observed with ASO-S/HXI

Dong Li[1,2,3] · Hanyang Dong[1,2] · Wei Chen[1,2] · Yang Su[1,2] · Yu Huang[1,2] · Zongjun Ning[1,2]

Received: 31 January 2024 / Accepted: 4 April 2024 / Published online: 9 May 2024

Abstract
The Neupert effect refers to the strong correlation between the soft X-ray (SXR) light curve and the time-integrated hard X-ray (HXR) or microwave flux, which is frequently observed in solar flares. In this article, we therefore utilized the newly launched *Hard X-ray Imager* (HXI) on board the *Advanced Space-based Solar Observatory* to investigate the Neupert effect during solar flares. By checking the HXR light curves at 20 – 50 keV, a sample of 149 events that cover the flare impulsive phase were selected. Then, we performed a crosscorrelation analysis between the HXR fluence (i.e., the time integral of the HXR flux) and the SXR 1 – 8 Å flux measured by the *Geostationary Operational Environmental Satellite*. All the selected flares show high correlation coefficients (>0.90), which seem to be independent of the flare location and class. The HXR fluences tend to increase linearly with the SXR peak fluxes. Our observations indicate that all the selected flares obey the Neupert effect.

Keywords Flares · Neupert effect · X-ray emission · Radio emission

1. Introduction

Neupert (1968) first noted that the shape of the soft X-ray (SXR) flux appeared to resemble the time-integrated microwave light curve during the impulsive phase of a solar flare. Such a close relationship was also observed between the SXR light curve and the cumulative time integral of the hard X-ray (HXR) flux, this was dubbed the 'Neupert effect' (e.g. Hudson, 1991; Yu, Li, and Gan, 2020; Li et al., 2023c). The empirical correlation suggests that a causal relationship between the nonthermal and thermal emissions could exist in a

✉ D. Li
lidong@pmo.ac.cn

Z. Ning
ningzopngjun@pmo.ac.cn

1 Key Laboratory of Dark Matter and Space Astronomy, Purple Mountain Observatory, CAS, Nanjing 210023, China

2 School of Astronomy and Space Science, University of Science and Technology of China, Hefei 230026, China

3 Yunnan Key Laboratory of the Solar physics and Space Science, Kunming 650216, P.R. China

solar flare. That is, the instantaneous radiation in the SXR waveband depends on the accumulated energy that is deposited by the electron beam during the flare impulsive phase (e.g. Neupert, 1968; Hudson, 1991). Based on the thick-target flare model, the HXR emission is produced by the nonthermal bremsstrahlung via Coulomb collisions, whereas the SXR emission is the cumulative energy produced by the thermal bremsstrahlung that is also related to the same electron beam. During this process, the heated plasma, via the thermal bremsstrahlung, could rapidly expand up into the upper atmosphere, called 'chromospheric evaporation' (e.g. Fisher, Canfield, and McClymont, 1985; Battaglia, Fletcher, and Benz, 2009; Tian et al., 2015; Tian and Chen, 2018; Li, Ning, and Zhang, 2015; Li, Hong, and Ning, 2022). Therefore, the Neupert effect could be regarded as evidence of the chromospheric evaporation driven by nonthermal electrons (e.g. Lee, Petrosian, and McTiernan, 1995; McTiernan, Fisher, and Li, 1999; Mann, Aurass, and Warmuth, 2006; Ning, 2009; Ning and Cao, 2010a; Li et al., 2023b).

The Neupert effect has been studied by many authors since it was reported by Neupert (1968). Some authors estimated the nonthermal and thermal energies using X-ray spectroscopic imager observations to study the Neupert effect in solar flares (e.g. Datlowe, Elcan, and Hudson, 1974; Starr et al., 1988; Feldman, 1990; Dennis et al., 2003; Ning, 2008b; Ning and Cao, 2010b). This is because the SXR radiation is related to the thermal energy accumulated in the solar corona, while the HXR emission is linked to the nonthermal energy during the solar flare. However, those observations gave conflicting results. Dennis (1988) proposed that the causal correlation between SXR and HXR emissions could be attributed to different flare types or phases (see also Dennis and Zarro, 1993). The Neupert-type flare, which followed the Neupert effect, was usually found in the model when the flare energy was released primarily by nonthermal electrons (e.g. Brown, 1971; Ning, 2009). Based on a statistical investigation of the time difference between the SXR maximum and the end of the HXR emission in solar flares, Veronig et al. (2002a) concluded that the chromospheric evaporation driven by electron beams plays a major role in Neupert-type flares (see also Veronig et al., 2002b, 2005). The Neupert effect was also observed in microflare data from the *Reuven Ramaty High Energy Solar Spectroscopic Imager* (RHESSI) (e.g. Ning, 2008a).

The recently launched *Hard X-ray Imager* (HXI: Su et al., 2019; Zhang et al., 2019) on board the *Advanced Space-based Solar Observatory* (ASO-S: Gan et al., 2019, 2023; Huang et al., 2019) provides a good chance to study the relationship between the SXR light curve and the time-integrated HXR flux during the impulsive phase of solar flares. In this study, we statistically investigated the Neupert effect in solar flares, which were simultaneously measured by ASO-S/HXI and the *Geostationary Operational Environmental Satellite* (GOES: Loto'aniu et al., 2019). The microwave emissions of some flares were measured by the *Chashan Broadband Solar millimeter spectrometer* (CBS: Shang et al., 2022; Yan et al., 2023) and the *Nobeyama Radio Polarimeters* (NoRP), which were used to confirm the Neupert effect. The article is organized as follows: Section 2 introduces the observations and data analysis, Section 3 shows our primary results, Section 4 presents our discussion, and a brief summary is given in Section 5.

2. Observations and Data Analysis

2.1. Observations

HXI is one of three payloads on board ASO-S. It provides flare observations in the HXR energy range of $\approx 15-300$ keV. The time cadence is 4 s in the regular mode, and it can be as high as 0.125 s in the flare mode. Two types of data products are provided by the HXI team, Level Q1 and Level 1. In our study, the full-disk light curves in the energy channel of 20 – 50 keV are used. The light curve is the sum of three flux monitors (D92, D93, and D94), which are derived from the data production of Level Q1. The time cadence of these data is 4 s in the regular mode and becomes 2 s in the flare mode.

GOES is designed to continuously monitor atmospheric conditions, space weather, and solar activity. The *X-Ray Sensor* (XRS) on board GOES provides SXR light curves at 1 – 8 Å and 0.5 – 4 Å, which can be used for detecting solar flares (Loto'aniu et al., 2019). They have a uniform time cadence of 1 s. We also used the microwave flux measured by CBS and NoRP to confirm the causal relationship between the thermal and nonthermal emissions. CBS is a newly built solar radio spectrograph at the Chashan Solar Observatory (CSO). It is the first solar-dedicated radio spectrometer in the millimeter regime of 35 – 40 GHz, which has a time cadence of ≈0.53 s (Shang et al., 2023; Yan et al., 2023). NoRP provides the solar radio flux at six frequencies, and it has a time cadence of 1 s.

2.2. Data Analysis

ASO-S/HXI has captured hundreds of solar flares since its first flight in the HXR energy range, which gives us an opportunity to statistically study the HXR emission in solar flares. However, the HXI data can be affected when ASO-S crosses the South Atlantic Anomaly (SAA) and the radiation belt (RB), resulting in incomplete flare data. Hence, we first selected solar flares that were observed completely by HXI, namely, the flares that showed significant enhancements during their impulsive phases at HXI 20 – 50 keV, as shown in Figures 1 and 2. The main reason for using HXI 20 – 50 keV flux is to maintain a significant number of flare events. Moreover, the selected flares had to be recorded by GOES at 1 – 8 Å, because in this way we could study the flare thermal emission in SXR. We finally chose 149 flares from the one-year HXI observations, i.e., from 11 November 2022 to 11 November 2023 during the increasing period of Cycle 25, as listed in Table 1. Here, there are mainly two data biases in our event selection: (1) the 149 flares were completely recorded by GOES, (2) HXI 20 – 50 keV fluxes showed obviously flare radiation during the impulsive phase.

In order to investigate the Neupert effect, we compared the causal relationship between the HXR fluence (Γ_{HXR}) and SXR flux (F_{SXR}) during the flare impulsive phase, as expressed in previous articles (e.g. Lee, Petrosian, and McTiernan, 1995; Veronig et al., 2002a):

$$F_{\mathrm{SXR}} \approx k \cdot \Gamma_{\mathrm{HXR}}, \tag{1}$$

$$\Gamma_{\mathrm{HXR}} = \int_{t_1}^{t_2} (F_{\mathrm{HXR}} - F_{\mathrm{BKG}}) \cdot dt, \tag{2}$$

where F_{HXR} and F_{BKG} denote the HXR light curve and its background flux, which were observed by ASO-S/HXI at 20 – 50 keV. t_1 and t_2 are the start and end time of the HXR light curve. The HXR fluence is defined as the time integral of the HXR light curve after removing its background emission during the impulsive phase of a solar flare. According to the theoretical expectation that the HXR emission ends and the SXR maximum occurs at almost the same time for the Neupert-type flare, the end time of the HXR light curve is regarded as the maximum time of the SXR flux.

For a solar flare recorded by GOES, the start, peak, and stop times are automatically provided by the GOES team. In this study, we obtained this information from the Solar Flare Finder Widget, which is available via the Solar Software (Freeland and Handy, 1998; Milligan and Ireland, 2018). This widget also provides the flare locations, which are calculated

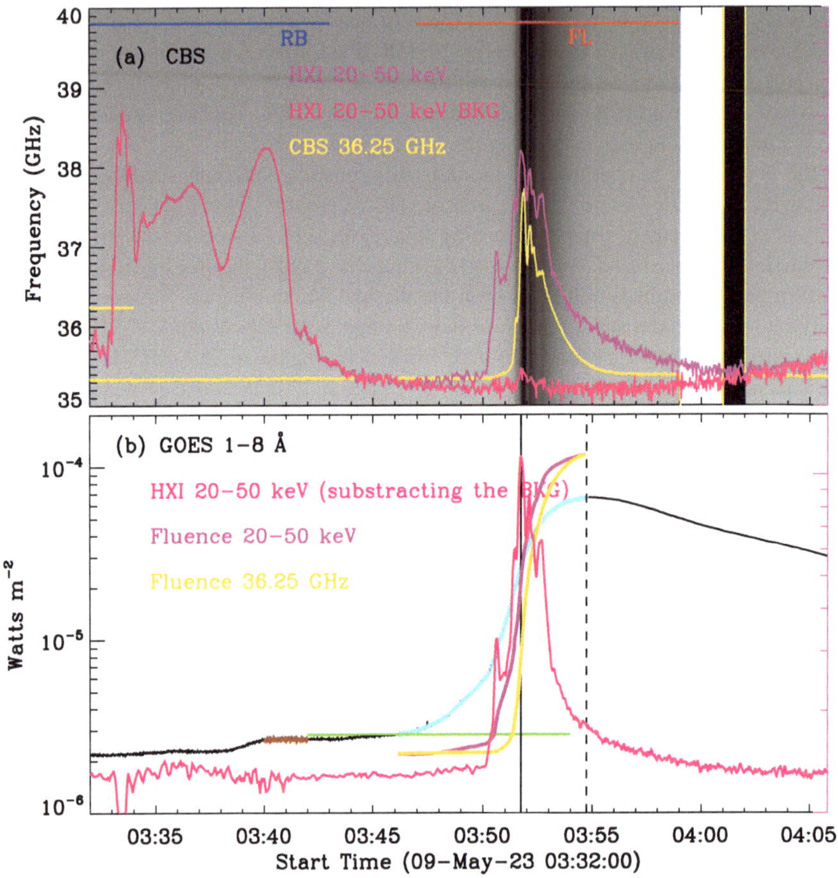

Figure 1 Overview of the solar flare on 09 May 2023. (**a**): HXR and microwave light curves measured by ASO-S/HXI (magenta and deep pink) and CBS (gold), respectively. The context image is the radio dynamic spectrum in the high-frequency range of ≈35 – 40 GHz. (**b**): The SXR light curve recorded by GOES (black) and the HXR flux (deep pink) after removing the background (BKG) emission. HXR and microwave fluences are integrated over the HXI and CBS fluxes after removing their background emissions. The vertical lines mark the HXR peak (solid) and the SXR maximum (dashed). The horizontal line outlines the duration from the start and peak time given by the GOES team.

from the *Atmospheric Imaging Assembly* images. Figure 1b presents the SXR flux recorded by GOES 1 – 8 Å, the horizontal green line outlines the duration of the impulsive phase; the start (t_s) and peak (t_p) times are extracted from the Solar Flare Finder Widget. The vertical lines mark the time when the HXR (solid) and SXR (dashed) emissions reach their maximum, respectively. Obviously, the SXR maximum time such as t_2 is different from t_p given by the GOES team; t_2 is regarded as the end of the HXR emission. Similarly, t_1 is not equal to t_s. In order to obtain t_1, the SXR emission with a duration of two minutes before the solar flare (t_s) is first chosen as the background, as marked by the brown pluses. Then, a threshold is determined by the average value (avg) plus a statistically significant ($9\,\sigma$) of the background emission, as shown by the green line. That is, t_1 occurs when the SXR flux exceeeds the threshold ($avg + 9\,\sigma$). We note that t_1 and t_2 are identified from the SXR flux rather than the HXR light curve, because the HXR light curve measured by HXI has a high

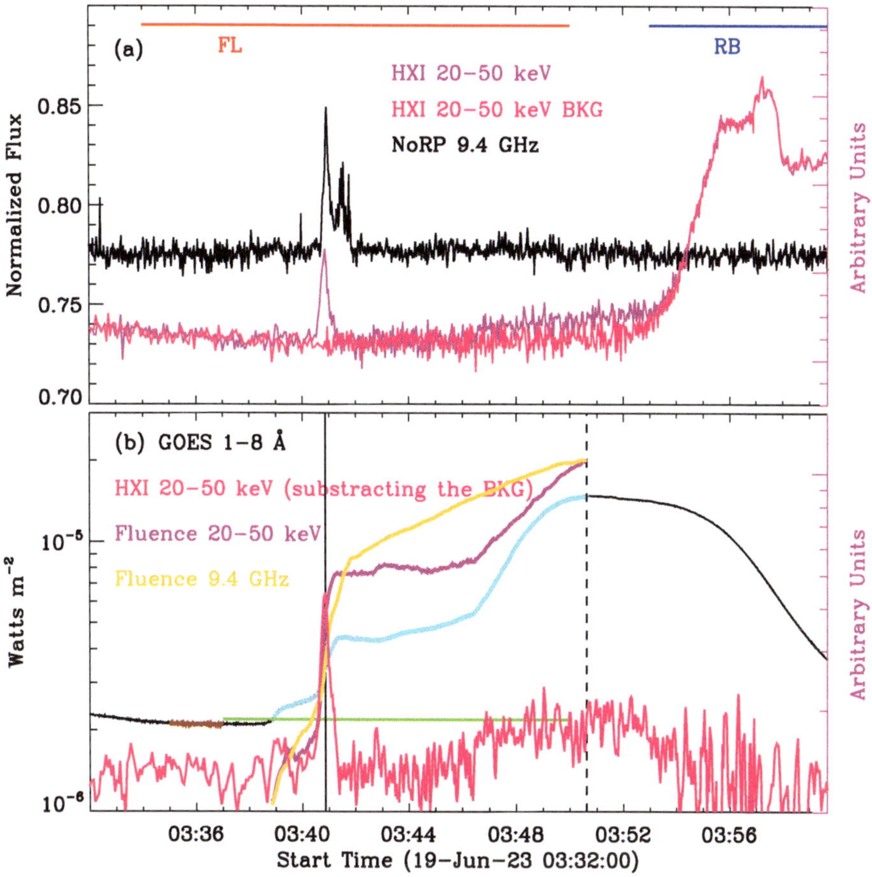

Figure 2 Overview of the solar flare on 19 June 2023. (**a**): HXR and microwave light curves measured by ASO-S/HXI (magenta and deep pink) and NoRP (black), respectively. (**b**): The SXR light curve recorded by GOES (black), and the HXR flux (deep pink) after removing the background emission (BKG). HXR and microwave fluence are integrated over the HXI and NoRP fluxes, respectively. The vertical lines mark the HXR peak (solid) and the SXR maximum (dashed). The horizontal line outlines the duration from the start and peak times given by the GOES team.

noise, especially for the nonflare flux. In such a case, the fluctuation of the background is too large, and it is impossible to use a unique standard value. Figure 2b shows another flare recorded by GOES 1 – 8 Å; similar temporal properties confirm our selection.

3. Results

To show our statistical results, we first describe two events, as examples.

3.1. Examples in the Sample List

Figure 1 shows an example of the solar flare that took place on 09 May 2023, which was an M6.5-class flare. Panel a draws the HXR light curves measured by HXI 20 – 50 keV,

Table 1 Key parameters of the solar flares in this study.

No.	Date	Time		Flux (GOES)	Fluence (HXI)	Position	• cc.
		t_1	t_2	(W m^{-2})	(Counts)	X, Y	
1	11 Nov 2022	01:31:14 UT	01:35:59 UT	4.85×10^{-6}	4.15×10^{3}	100″, 130″	0.987
2	11 Nov 2022	03:15:28 UT	03:17:16 UT	2.54×10^{-6}	3.74×10^{3}	116″, 130″	0.957
3	11 Nov 2022	11:33:13 UT	11:40:41 UT	1.28×10^{-5}	1.11×10^{4}	182″, 132″	0.991
4	12 Nov 2022	18:01:55 UT	18:03:53 UT	6.94×10^{-6}	5.31×10^{3}	419″, 138″	0.967
5	02 Dec 2022	09:15:20 UT	09:20:54 UT	8.65×10^{-6}	1.86×10^{3}	−924″, 266″	0.978
6	15 Dec 2022	01:29:03 UT	01:37:33 UT	1.67×10^{-5}	1.94×10^{3}	678″, −341″	0.979
7	15 Dec 2022	22:23:52 UT	22:40:16 UT	5.84×10^{-5}	2.24×10^{4}	805″, −342″	0.986
8	16 Dec 2022	15:35:11 UT	15:41:00 UT	1.30×10^{-5}	2.47×10^{3}	862″, −328″	0.914
9	16 Dec 2022	19:04:49 UT	19:11:14 UT	9.35×10^{-6}	3.53×10^{3}	877″, −329″	0.904
10	20 Dec 2022	04:09:19 UT	04:10:17 UT	4.77×10^{-6}	2.50×10^{3}	−709″, 365″	0.922
11	20 Dec 2022	13:52:46 UT	14:06:49 UT	1.18×10^{-5}	3.81×10^{3}	−674″, 383″	0.921
12	24 Dec 2022	04:09:41 UT	04:14:16 UT	4.21×10^{-6}	3.15×10^{3}	259″, 302″	0.932
13	25 Dec 2022	06:56:45 UT	07:00:59 UT	8.32×10^{-6}	3.70×10^{3}	485″, 300″	0.975
14	27 Dec 2022	20:35:16 UT	20:37:21 UT	4.28×10^{-6}	1.10×10^{3}	−922″, −317″	0.940
15	29 Dec 2022	18:15:04 UT	18:20:37 UT	2.23×10^{-5}	1.14×10^{4}	−922″, 319″	0.992
16	30 Dec 2022	19:08:04 UT	19:09:43 UT	4.53×10^{-6}	1.51×10^{3}	−144″, 379″	0.996
17	06 Jan 2023	00:47:58 UT	00:57:40 UT	1.25×10^{-4}	1.48×10^{5}	−906″, −324″	0.999
18	07 Jan 2023	00:44:25 UT	00:52:07 UT	1.66×10^{-5}	6.38×10^{3}	−840″, −327″	0.997
19	09 Jan 2023	04:49:50 UT	04:51:18 UT	5.02×10^{-6}	5.56×10^{2}	−921″, −256″	0.988
20	09 Jan 2023	08:46:30 UT	09:01:18 UT	2.16×10^{-5}	3.74×10^{3}	381″, −227″	0.994
21	09 Jan 2023	13:17:47 UT	13:22:14 UT	1.04×10^{-5}	3.70×10^{3}	394″, −244″	0.987
22	10 Jan 2023	02:13:06 UT	02:16:18 UT	1.09×10^{-5}	4.86×10^{3}	−886″, −178″	0.932
23	10 Jan 2023	14:45:51 UT	14:47:36 UT	9.23×10^{-6}	1.06×10^{3}	−825″, −186″	0.968
24	10 Jan 2023	17:47:35 UT	17:48:31 UT	1.34×10^{-5}	2.01×10^{3}	−816″, −185″	0.971
25	14 Jan 2023	20:08:31 UT	20:21:22 UT	3.62×10^{-5}	3.11×10^{4}	626″, −231″	0.958
26	15 Jan 2023	14:20:12 UT	14:31:34 UT	4.92×10^{-5}	3.12×10^{4}	−809″, −214″	0.990
27	18 Jan 2023	10:28:35 UT	10:34:58 UT	1.83×10^{-5}	8.42×10^{3}	−322″, −194″	0.998
28	18 Jan 2023	20:44:36 UT	20:47:23 UT	9.73×10^{-6}	1.94×10^{3}	−212″, 347″	0.993
29	07 Feb 2023	23:05:24 UT	23:07:36 UT	6.69×10^{-5}	2.21×10^{4}	−44″, 580″	0.997
30	08 Feb 2023	02:48:26 UT	02:53:40 UT	2.13×10^{-5}	3.93×10^{3}	−15″, 580″	0.968
31	09 Feb 2023	02:49:43 UT	03:10:37 UT	3.42×10^{-5}	4.31×10^{3}	−937″, −91″	0.908
32	09 Feb 2023	07:15:59 UT	07:17:59 UT	1.28×10^{-5}	9.43×10^{2}	−898″, −112″	0.986
33	09 Feb 2023	14:50:47 UT	14:56:46 UT	1.53×10^{-5}	2.65×10^{3}	250″, 563″	0.996
34	10 Feb 2023	02:49:51 UT	03:03:55 UT	3.74×10^{-5}	2.91×10^{4}	330″, 573″	0.998
35	10 Feb 2023	08:03:34 UT	08:05:44 UT	1.64×10^{-5}	1.43×10^{3}	379″, 584″	0.991
36	10 Feb 2023	10:47:02 UT	10:49:21 UT	9.72×10^{-6}	1.13×10^{3}	−788″, −88″	0.966
37	11 Feb 2023	11:31:47 UT	11:34:36 UT	1.52×10^{-5}	1.99×10^{3}	800″, 502″	0.985
38	11 Feb 2023	15:44:14 UT	15:48:45 UT	1.21×10^{-4}	7.27×10^{4}	−604″, −82″	0.982
39	12 Feb 2023	08:39:05 UT	08:48:04 UT	3.37×10^{-5}	1.55×10^{4}	−466″, −70″	0.996
40	12 Feb 2023	15:35:20 UT	15:37:57 UT	1.11×10^{-5}	2.21×10^{3}	−406″, −67″	0.963
41	17 Feb 2023	19:46:53 UT	20:16:54 UT	2.29×10^{-4}	4.46×10^{5}	−811″, 449″	0.995
42	18 Feb 2023	18:33:17 UT	18:35:52 UT	7.31×10^{-6}	2.15×10^{3}	423″, 239″	0.981

Table 1 (*Continued*)

No.	Date	Time		Flux (GOES)	Fluence (HXI)	Position	• cc.
		t_1	t_2	(W m^{-2})	(Counts)	X, Y	
43	20 Feb 2023	14:54:29 UT	14:58:15 UT	4.51×10^{-5}	1.93×10^{4}	−870″, 429″	0.997
44	21 Feb 2023	20:04:26 UT	20:16:59 UT	5.12×10^{-5}	3.49×10^{4}	−785″, 469″	0.993
45	23 Feb 2023	06:12:32 UT	06:13:55 UT	1.68×10^{-5}	4.06×10^{3}	−313″, 437″	0.996
46	23 Feb 2023	23:11:18 UT	23:13:38 UT	6.85×10^{-6}	1.13×10^{3}	−176″, 426″	0.995
47	24 Feb 2023	17:13:04 UT	17:15:41 UT	1.23×10^{-5}	5.63×10^{3}	−16″, 444″	0.988
48	28 Feb 2023	17:38:57 UT	17:50:28 UT	8.66×10^{-5}	7.54×10^{4}	420″, 532″	0.992
49	02 Mar 2023	04:42:49 UT	04:50:17 UT	9.38×10^{-6}	5.87×10^{3}	751″, 409″	0.987
50	04 Mar 2023	02:24:15 UT	02:25:56 UT	5.99×10^{-6}	8.69×10^{2}	−703″, 217″	0.994
51	05 Mar 2023	21:30:19 UT	21:35:47 UT	5.19×10^{-5}	3.84×10^{4}	798″, 355″	0.999
52	06 Mar 2023	02:11:46 UT	02:28:13 UT	5.86×10^{-5}	3.96×10^{4}	821″, 350″	0.988
53	07 Mar 2023	04:00:29 UT	04:05:01 UT	4.79×10^{-6}	9.94×10^{2}	906″, 317″	0.947
54	19 Mar 2023	02:14:36 UT	02:15:42 UT	3.12×10^{-6}	2.20×10^{3}	−754″, −316″	0.933
55	19 Mar 2023	03:55:16 UT	03:56:38 UT	2.51×10^{-6}	5.32×10^{2}	−740″, −331″	0.981
56	19 Mar 2023	06:31:30 UT	06:35:31 UT	4.62×10^{-6}	1.96×10^{3}	−740″, −331″	0.972
57	19 Mar 2023	18:05:44 UT	18:08:48 UT	5.12×10^{-6}	4.64×10^{3}	−656″, −317″	0.988
58	21 Mar 2023	13:51:33 UT	13:53:57 UT	5.56×10^{-6}	1.63×10^{3}	−619″, −362″	0.979
59	14 Apr 2023	23:22:59 UT	23:27:51 UT	1.64×10^{-5}	2.13×10^{3}	−721″, 242″	0.985
60	15 Apr 2023	03:02:56 UT	03:03:26 UT	4.01×10^{-6}	3.88×10^{2}	618″, −112″	0.991
61	16 Apr 2023	17:33:40 UT	17:43:46 UT	9.15×10^{-6}	4.84×10^{3}	−872″, 391″	0.990
62	21 Apr 2023	17:48:07 UT	18:11:44 UT	1.79×10^{-5}	7.63×10^{3}	153″, −296″	0.980
63	27 Apr 2023	11:11:22 UT	11:13:53 UT	2.18×10^{-5}	3.81×10^{3}	−77″, −287″	0.996
64	01 May 2023	13:05:45 UT	13:09:25 UT	7.72×10^{-5}	3.44×10^{4}	617″, −343″	0.995
65	03 May 2023	10:07:31 UT	10:14:16 UT	3.19×10^{-5}	7.05×10^{3}	−625″, 246″	0.985
66	03 May 2023	10:39:40 UT	10:45:20 UT	7.54×10^{-5}	1.71×10^{4}	−625″, 246″	0.991
67	08 May 2023	08:22:43 UT	08:25:48 UT	3.13×10^{-6}	6.09×10^{3}	193″, 269″	0.966
68	08 May 2023	14:16:48 UT	14:21:03 UT	9.85×10^{-6}	7.24×10^{3}	209″, 268″	0.949
69	08 May 2023	20:16:11 UT	20:25:19 UT	2.33×10^{-5}	1.35×10^{4}	255″, 283″	0.999
70	09 May 2023	03:46:10 UT	03:54:44 UT	6.65×10^{-5}	6.37×10^{4}	407″, 263″	0.996
71	09 May 2023	06:07:42 UT	06:13:32 UT	1.27×10^{-5}	3.33×10^{3}	378″, 264″	0.983
72	09 May 2023	20:34:22 UT	20:52:38 UT	5.03×10^{-5}	2.98×10^{4}	546″, 257″	0.995
73	11 May 2023	08:51:30 UT	09:01:38 UT	2.21×10^{-5}	8.98×10^{3}	622″, −60″	0.972
74	16 May 2023	16:34:19 UT	16:43:26 UT	9.68×10^{-5}	2.87×10^{4}	−873″, −370″	0.989
75	18 May 2023	06:15:26 UT	06:26:14 UT	1.03×10^{-5}	2.03×10^{3}	−906″, 278″	0.977
76	18 May 2023	20:17:07 UT	20:23:36 UT	4.04×10^{-5}	1.40×10^{4}	−884″, 285″	0.999
77	20 May 2023	04:41:04 UT	04:50:06 UT	7.03×10^{-6}	2.66×10^{3}	−790″, −419″	0.975
78	20 May 2023	14:56:09 UT	14:59:48 UT	6.16×10^{-5}	1.01×10^{4}	−744″, 297″	0.959
79	22 May 2023	13:26:08 UT	13:27:28 UT	7.66×10^{-6}	2.39×10^{3}	−370″, 305″	0.973
80	23 May 2023	12:08:52 UT	12:13:53 UT	3.99×10^{-5}	7.37×10^{3}	−174″, 305″	0.983
81	27 May 2023	04:33:28 UT	04:35:42 UT	8.80×10^{-6}	5.01×10^{3}	611″, 277″	0.947
82	30 May 2023	10:17:38 UT	10:21:58 UT	1.41×10^{-5}	9.11×10^{3}	877″, −354″	0.994
83	30 May 2023	13:36:55 UT	13:38:53 UT	1.66×10^{-5}	6.25×10^{3}	528″, −298″	0.983
84	09 Jun 2023	14:31:54 UT	14:40:00 UT	4.61×10^{-6}	1.27×10^{3}	−128″, −250″	0.987

Table 1 (*Continued*)

No.	Date	Time		Flux (GOES)	Fluence (HXI)	Position	cc.
		t_1	t_2	(W m^{-2})	(Counts)	X, Y	
85	09 Jun 2023	16:52:22 UT	17:11:26 UT	2.51×10^{-5}	1.37×10^{4}	−642″, −358″	0.993
86	18 Jun 2023	00:28:06 UT	00:31:37 UT	1.37×10^{-5}	9.98×10^{3}	−388″, −358″	0.977
87	19 Jun 2023	03:38:54 UT	03:50:37 UT	1.49×10^{-5}	1.82×10^{3}	−923″, −197″	0.947
88	19 Jun 2023	12:08:12 UT	12:14:18 UT	1.21×10^{-5}	2.73×10^{3}	−912″, −200″	0.948
89	20 Jun 2023	10:49:23 UT	11:06:05 UT	8.74×10^{-6}	2.65×10^{3}	806″, −349″	0.973
90	24 Jun 2023	12:13:23 UT	12:17:03 UT	1.17×10^{-5}	2.93×10^{3}	580″, 315″	0.910
91	26 Jun 2023	16:13:09 UT	16:22:22 UT	1.65×10^{-5}	3.04×10^{3}	536″, 342″	0.996
92	27 Jun 2023	15:01:32 UT	15:14:25 UT	1.29×10^{-5}	1.63×10^{3}	657″, 345″	0.954
93	28 Jun 2023	08:32:00 UT	08:44:47 UT	1.99×10^{-5}	6.26×10^{3}	759″, 335″	0.997
94	02 Jul 2023	08:47:59 UT	08:52:50 UT	8.96×10^{-6}	3.39×10^{3}	672″, 245″	0.935
95	03 Jul 2023	06:47:15 UT	06:53:14 UT	1.40×10^{-5}	4.65×10^{3}	−588″, −282″	0.997
96	04 Jul 2023	12:23:54 UT	12:35:53 UT	1.49×10^{-5}	1.23×10^{4}	907″, 259″	0.935
97	04 Jul 2023	19:11:32 UT	19:27:51 UT	8.12×10^{-6}	2.45×10^{3}	915″, 227″	0.911
98	05 Jul 2023	18:14:26 UT	18:17:05 UT	4.45×10^{-6}	1.88×10^{3}	−886″, −324″	0.990
99	11 Jul 2023	02:11:18 UT	02:14:13 UT	8.90×10^{-6}	2.11×10^{3}	−886″, 320″	0.972
100	15 Jul 2023	09:49:43 UT	09:53:15 UT	1.09×10^{-5}	3.84×10^{3}	−591″, 289″	0.993
101	15 Jul 2023	10:07:25 UT	10:09:53 UT	2.35×10^{-5}	8.88×10^{3}	−580″, 288″	0.984
102	16 Jul 2023	08:33:14 UT	08:40:48 UT	5.05×10^{-6}	2.46×10^{3}	732″, −346″	0.983
103	16 Jul 2023	15:06:32 UT	15:09:02 UT	1.92×10^{-5}	1.43×10^{4}	−388″, 277″	0.985
104	16 Jul 2023	17:12:35 UT	17:19:24 UT	5.90×10^{-6}	2.99×10^{3}	700″, −365″	0.977
105	16 Jul 2023	17:38:41 UT	17:48:13 UT	4.04×10^{-5}	1.82×10^{5}	740″, −418″	0.960
106	17 Jul 2023	22:44:44 UT	22:55:20 UT	2.74×10^{-5}	2.28×10^{4}	845″, −408″	0.947
107	17 Jul 2023	23:19:24 UT	23:34:30 UT	5.02×10^{-5}	6.45×10^{4}	870″, −348″	0.996
108	18 Jul 2023	20:19:40 UT	20:28:07 UT	2.16×10^{-5}	7.20×10^{3}	880″, −341″	0.999
109	18 Jul 2023	22:40:54 UT	22:44:12 UT	8.74×10^{-6}	2.48×10^{3}	892″, −309″	0.962
110	25 Jul 2023	21:11:37 UT	21:16:47 UT	1.67×10^{-5}	3.99×10^{3}	−600″, −229″	0.993
111	26 Jul 2023	15:18:53 UT	15:24:03 UT	8.17×10^{-6}	2.61×10^{3}	−467″, −239″	0.922
112	26 Jul 2023	15:53:19 UT	15:59:41 UT	2.08×10^{-5}	7.36×10^{3}	840″, 312″	0.993
113	29 Jul 2023	16:16:11 UT	16:24:14 UT	1.45×10^{-5}	8.42×10^{3}	146″, −269″	0.970
114	30 Jul 2023	13:12:59 UT	13:15:51 UT	4.40×10^{-6}	1.25×10^{3}	303″, −267″	0.979
115	01 Aug 2023	14:03:30 UT	14:09:16 UT	1.45×10^{-5}	3.98×10^{3}	664″, −198″	0.913
116	02 Aug 2023	10:47:22 UT	10:49:48 UT	1.38×10^{-5}	2.84×10^{3}	808″, −211″	0.997
117	02 Aug 2023	14:50:28 UT	14:52:27 UT	1.77×10^{-5}	9.21×10^{3}	823″, −208″	0.994
118	02 Aug 2023	16:20:12 UT	16:22:19 UT	1.40×10^{-5}	3.68×10^{3}	831″, −207″	0.976
119	03 Aug 2023	11:49:19 UT	11:55:10 UT	2.16×10^{-5}	1.02×10^{4}	893″, −206″	0.981
120	06 Aug 2023	18:25:18 UT	18:40:32 UT	5.53×10^{-5}	1.04×10^{5}	936″, 129″	0.991
121	08 Aug 2023	09:25:10 UT	09:31:16 UT	3.70×10^{-5}	1.12×10^{4}	895″, 280″	0.996
122	25 Aug 2023	01:00:29 UT	01:09:15 UT	1.49×10^{-5}	5.52×10^{3}	−713″, −269″	0.984
123	26 Aug 2023	21:47:48 UT	21:50:20 UT	2.69×10^{-6}	3.42×10^{3}	942″, 113″	0.992
124	03 Sep 2023	00:15:25 UT	00:23:00 UT	1.14×10^{-5}	8.50×10^{3}	933″, 176″	0.968
125	06 Sep 2023	00:48:50 UT	00:50:29 UT	5.23×10^{-6}	1.63×10^{3}	208″, 132″	0.992
126	06 Sep 2023	17:54:58 UT	17:56:29 UT	5.66×10^{-6}	8.65×10^{2}	377″, 122″	0.928

Table 1 (*Continued*)

No.	Date	Time		Flux (GOES)	Fluence (HXI)	Position	• cc.
		t_1	t_2	($W\ m^{-2}$)	(Counts)	X, Y	
127	11 Sep 2023	06:00:03 UT	06:01:16 UT	1.07×10^{-5}	6.62×10^{2}	$-796'', -164''$	0.986
128	11 Sep 2023	21:09:46 UT	21:14:02 UT	8.37×10^{-6}	1.88×10^{3}	$-93'', 244''$	0.990
129	12 Sep 2023	06:49:39 UT	07:02:15 UT	2.58×10^{-5}	1.18×10^{4}	$487'', 163''$	0.971
130	14 Sep 2023	19:28:21 UT	19:31:52 UT	2.05×10^{-5}	6.18×10^{3}	$-148'', 48''$	0.996
131	16 Sep 2023	00:45:53 UT	00:49:49 UT	2.99×10^{-5}	8.01×10^{3}	$131'', 48''$	0.990
132	16 Sep 2023	05:28:22 UT	05:38:45 UT	3.45×10^{-5}	2.89×10^{4}	$180'', 49''$	0.958
133	16 Sep 2023	15:50:13 UT	15:51:00 UT	2.72×10^{-6}	7.38×10^{2}	$-942'', 144''$	0.987
134	16 Sep 2023	15:55:40 UT	15:56:57 UT	4.13×10^{-6}	5.45×10^{2}	$260'', 52''$	0.999
135	19 Sep 2023	15:41:42 UT	15:52:32 UT	4.32×10^{-6}	3.92×10^{2}	$658'', 249''$	0.971
136	20 Sep 2023	12:10:33 UT	12:12:49 UT	3.57×10^{-6}	7.02×10^{2}	$-446'', 29''$	0.967
137	20 Sep 2023	14:14:12 UT	14:19:06 UT	8.31×10^{-5}	1.33×10^{5}	$-546'', 20''$	0.998
138	26 Sep 2023	16:16:37 UT	16:18:08 UT	4.54×10^{-6}	1.42×10^{3}	$253'', -385''$	0.971
139	28 Sep 2023	09:05:38 UT	09:07:30 UT	1.33×10^{-5}	5.21×10^{3}	$-828'', -354''$	0.969
140	01 Oct 2023	03:22:27 UT	03:24:07 UT	9.75×10^{-6}	4.25×10^{3}	$898'', -273''$	0.998
141	05 Oct 2023	19:09:32 UT	19:13:25 UT	4.84×10^{-6}	1.05×10^{3}	$0'', 159''$	0.957
142	07 Oct 2023	18:03:04 UT	18:06:19 UT	1.86×10^{-5}	1.14×10^{3}	$-607'', -263''$	0.972
143	07 Oct 2023	18:27:57 UT	18:29:59 UT	3.50×10^{-6}	4.97×10^{2}	$474'', 75''$	0.917
144	07 Oct 2023	23:17:22 UT	23:25:00 UT	5.33×10^{-6}	1.18×10^{3}	$532'', 62''$	0.939
145	19 Oct 2023	12:08:40 UT	12:10:47 UT	1.53×10^{-6}	4.54×10^{2}	$820'', -351''$	0.911
146	02 Nov 2023	12:21:09 UT	12:22:32 UT	2.09×10^{-5}	6.22×10^{3}	$461'', -359''$	0.993
147	05 Nov 2023	11:35:57 UT	11:43:06 UT	2.18×10^{-5}	2.32×10^{3}	$-755'', -192''$	0.989
148	05 Nov 2023	14:31:06 UT	14:32:49 UT	1.91×10^{-5}	5.25×10^{3}	$-745'', -193''$	0.945
149	11 Nov 2023	03:55:28 UT	03:59:20 UT	7.65×10^{-6}	1.56×10^{3}	$566'', 90''$	0.985

• cc. is the correlation coefficient between the HXR fluence and the SXR flux.

the magenta and deep pink curves are the HXI data observed by three flux monitors and the corresponding background monitors, respectively. It can be immediately noted that the radiation enhancement in the approximate period of ≈03:33 – 03:44 UT is from the background rather than from the solar flare, since the magenta and deep pink curves overlap. This is mainly because ASO-S goes through the radiation belt (RB) during that time interval, as indicated by the blue line. On the other hand, the radiation enhancement between about 03:45 UT and 03:55 UT is the flare emission in the HXR channel and the background curve does not reveal any significant enhancement. The microwave flux at the frequency of 36.25 GHz (gold) further confirms that the radiation is from a solar flare. The background image is the radio dynamic spectrum in the frequency range of 35 – 40 GHz observed by CBS, which clearly shows a group of microwave bursts accompanying the solar flare.

In Figure 1b, we show the SXR flux recorded by GOES at 1 – 8 Å, which starts at about 03:42 UT and peaks at around 03:54 UT,[1] as outlined by the green line. We note that both the SXR and HXR (deep pink curve) fluxes do not exhibit a significant enhancement at about 03:42 UT, which is not suitable for the start time (t_1) of the HXR fluence. Here, the

[1] https://www.solarmonitor.org/?date=20230509.

time when the SXR flux exceeds the threshold ($avg + 9\,\sigma$) is regarded as t_1, while the end time (t_2) of the HXR fluence is the time when the SXR flux reaches its maximum, based on the assumption that the SXR emission reaches its maximum when the HXR ends (Veronig et al., 2002a). Then, the HXR fluence can be calculated by integrating over the HXI flux after removing the background emission between the time interval t_1 and t_2, as shown by the magenta curve. It matches well the SXR flux, as indicated by the cyan curve. Using the same method, the microwave fluence is estimated by integrating over the CBS flux from t_1 to t_2, as shown by the gold curve. It seems to be a little different from the SXR flux, which could be because the microwave flux starts later than the HXR flux, as shown by the magenta and gold line in Figure 1a.

Figure 2 presents another solar flare that occurred on 19 June 2023, which was an M1.4-class flare. The HXR light curves were observed by HXI 20 – 50 keV, and the microwave flux was detected by NoRP 9.4 GHz. They both show a nonthermal pulse during the M1.4 flare, as shown in panel a. The SXR flux recorded by GOES 1 – 8 Å suggests that the flare starts at about 03:37 UT and peaks at around 03:50 UT.[2] Similarly, these are not the HXR start and end times. Moreover, the SXR flux reveals two primary peaks during the flare impulsive phase, as shown in panel b. Using the same method, the HXR/microwave fluence is estimated via integrating over the HXI and NoRP fluxes after subtracting their background from t_1 to t_2, as indicated by the magenta and gold curves. Those two curves appear to match the SXR flux, particularly for the HXR fluence.

To look closely at their causal relationship, we perform a crosscorrelation analysis between the SXR flux and the HXR and microwave fluences in the two flares, as shown in Figure 3. Here, the time cadences of the SXR and microwave light curves are both interpolated to the HXI time resolution, so they have the same points during the same time interval. A very high correlation coefficient (cc.) is obtained for the M6.5 flare on 09 May 2023, implying that it is completely consistent with the Neupert effect. Similarly to the observational result in Figure 1b, the correlation coefficient between the SXR flux and HXR fluence is higher than that between the SXR flux and microwave fluence. Conversely, the correlation coefficient for the M1.4 flare on 19 June 2023 is a little lower, which may suggest that it weakly agrees with the Neupert effect.

3.2. Statistical Results

Based on the same method, we investigated the simple correlation between the HXR fluence and the SXR flux for 149 flares measured by ASO-S/HXI and GOES. Figure 4 presents the distribution of correlation coefficients, namely, the correlation coefficients as a function of time (panel a) and the histogram of the correlation coefficients (panel b). We can see that all these solar flares show high correlation coefficients, >0.90. Moreover, there are 123 flares that show correlation coefficients above 0.95, and only 26 flares show correlation coefficients below 0.95, as shown in the histogram of the correlation coefficients. That is, about 82.5% (123/149) of the studied flares show a strong correlation between the SXR flux and the time integral of the HXR emission. Here, a threshold of 0.95 (magenta line) is chosen, since it is a middle value for the correlation coefficients. We also considered the confidence interval at the level of 95% in wavelet analysis (e.g. Scargle, 1982; Torrence and Compo, 1998).

Figure 5 shows the locations of all these studied flares. The colors represent the different ranges of the correlation coefficients. As shown in Figure 5, these studied flares appear to be

[2] https://www.solarmonitor.org/?date=20230619.

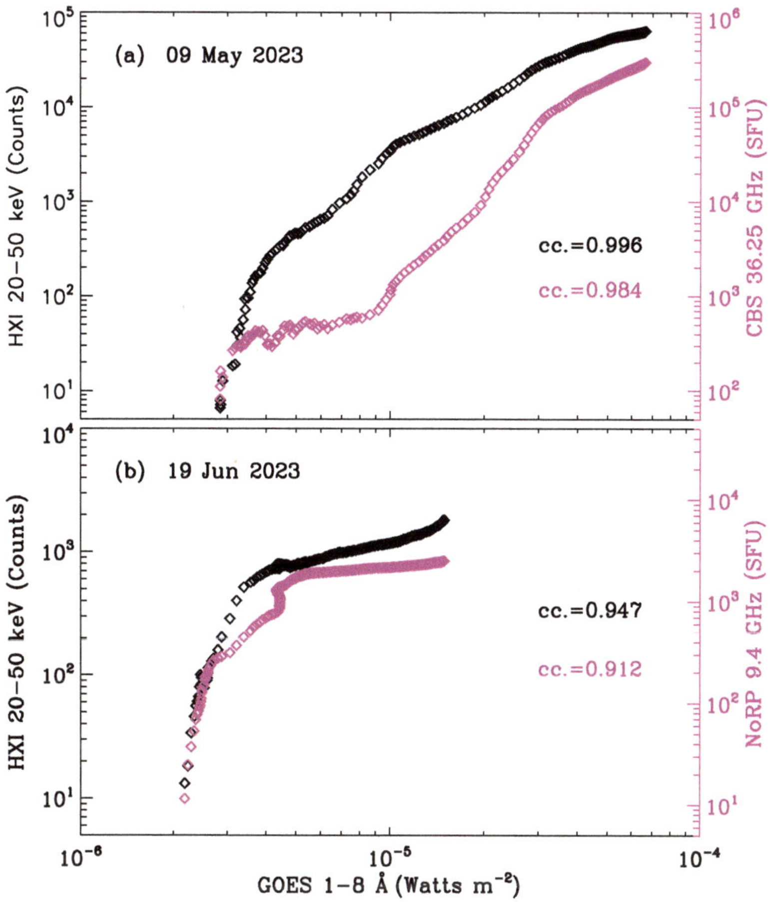

Figure 3 Correlation coefficients between the HXR and microwave fluences and the SXR flux for the sampled flares on 09 May 2023 and 19 June 2023, respectively. The correlation coefficient values appear in the inset.

located at the low and middle latitudes, including all the flares no matter if they have higher ($\geq$0.95) or lower ($<$0.95) correlation coefficients. On the other hand, those flares that show lower correlation coefficients seem to occur at the region that is close to the solar limb, east or west solar limbs, while those flares that have higher correlation coefficients appear in any low- and middle-latitude region, i.e., the solar limb and solar center.

Figure 6 presents four scatter plots between some key parameters. Panels a and b show the HXR peak flux and fluence depending on the SXR peak flux. Comparing with the HXR peak flux, the HXR fluence exhibits a better linear relationship with the SXR peak flux, as indicated by the black line in panel b. Note that the statistical result is based on all the selected flares, whether they have higher (cyan) or lower (magenta) correlation coefficients. The linear fitting is also performed for those flares that have higher and lower correlation coefficients, as indicated by the cyan and magenta lines, respectively. The two fitting results (black and magenta lines) are consistent with that for all the flares (green line), which may suggest that all these flares are obeying the Neupert effect. Figure 6c displays the SXR peak flux or GOES class depending on the correlation coefficient. It seems that these lower

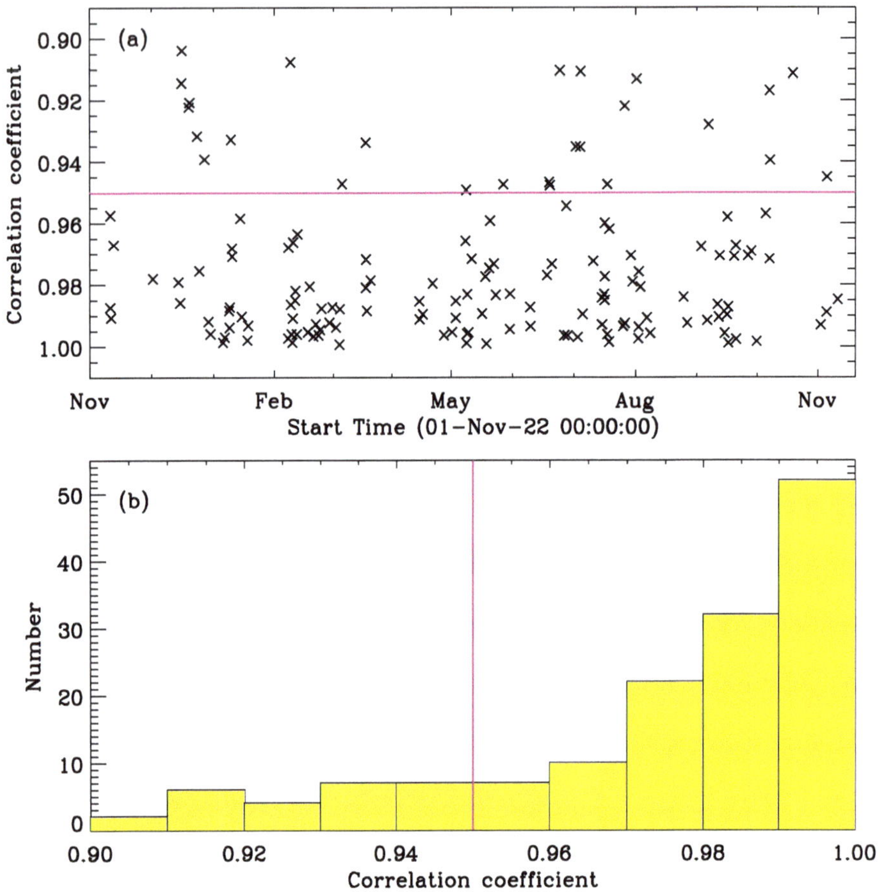

Figure 4 (**a**) Correlation coefficients as a function of time. (**b**) Histogram of correlation coefficients. The magenta line marks the correlation coefficient of 0.95.

correlation coefficients tend to be related to small flares, while major flares tend to exhibit a higher correlation. However, we should state that this is mainly because the major flares are very rare in our study, i.e., only three X-class flares. Panel d presents the ratio between the time difference and lifetime depending on the correlation coefficient. Here, the time difference refers to the time delay between the HXR peak and the SXR maximum, as marked by the vertical solid and dashed lines in Figures 1b and 2b. The lifetime refers to the duration of the HXR fluence, i.e., $t_2 - t_1$. The lower correlation coefficients seem to appear in those flares that have a large ratio between the time difference and the lifetime. For instance, the ratio of those correlation coefficients below 0.95 is close to 1.0.

4. Discussion

In this study, 149 events were selected to investigate the causal relationship between the HXR fluence and the SXR flux, since the impulsive phases of these flares were completely observed by ASO-S/HXI in the energy range of 20 – 50 keV. The SXR fluxes were recorded

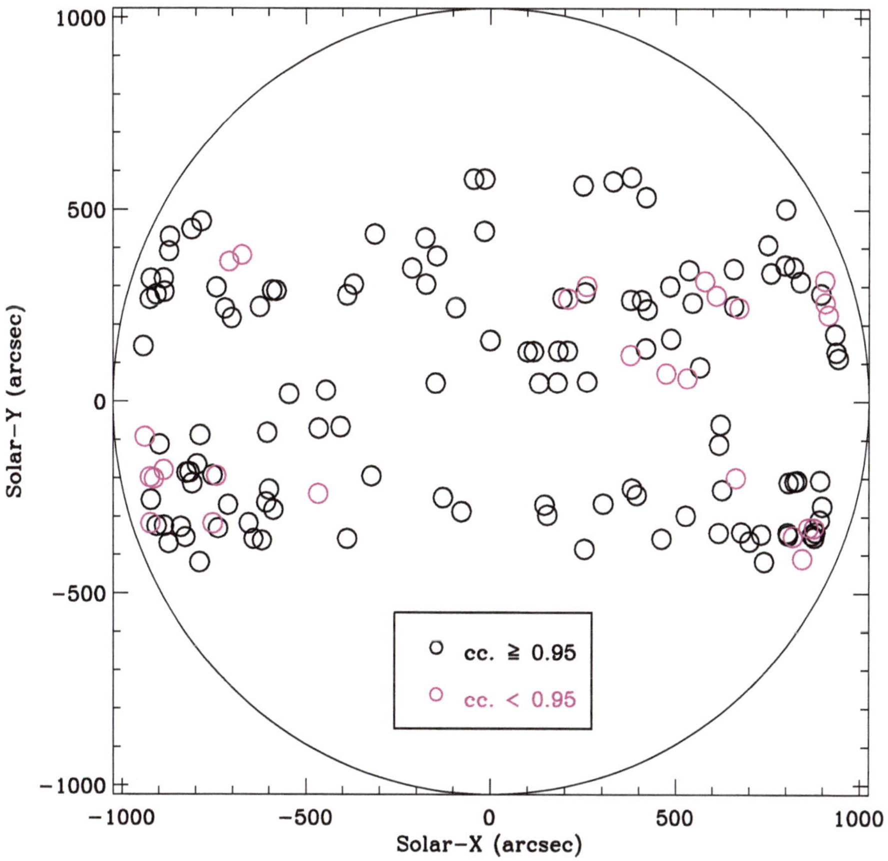

Figure 5 Positions of the studied flares on the Sun. The colors represent the different ranges of correlation coefficients.

by GOES at 1 – 8 Å, and the HXR fluence was defined as the time integral of HXR light curves measured by HXI at 20 – 50 keV. The microwave fluxes, which were measured by CBS and NoRP, were also used to examine the correlation between the microwave fluence and the SXR flux. For those selected solar flares, their HXR fluences were found to closely match the SXR fluxes, implying that the studied flares could follow the Neupert effect.

In this work, the causal correlation between the time integral of the HXR flux and the SXR emission is primarily identified as the correlation coefficient between the HXR fluence and the SXR flux during the flare impulsive phase. This is mainly because the HXR emission produced by the nonthermal electron is strongly related to the magnetic reconnection during the flare impulsive phase (e.g. Krucker et al., 2008; Shibata and Magara, 2011; Jiang et al., 2021; Li et al., 2023a). The duration of the time integral is determined from the SXR flux observed by GOES, since the HXR flux measured by ASO-S/HXI reveals a high noise level, which could increase the uncertainty in the start and end times. The start time (t_1) is regarded as the statistically significant excess of 9 σ above the background emission, similarly to the triggered mode of Konus-Wind.[3] The end time (t_2) is defined as the maximum of the SXR

[3]http://www.ioffe.ru/LEA/kwsun/.

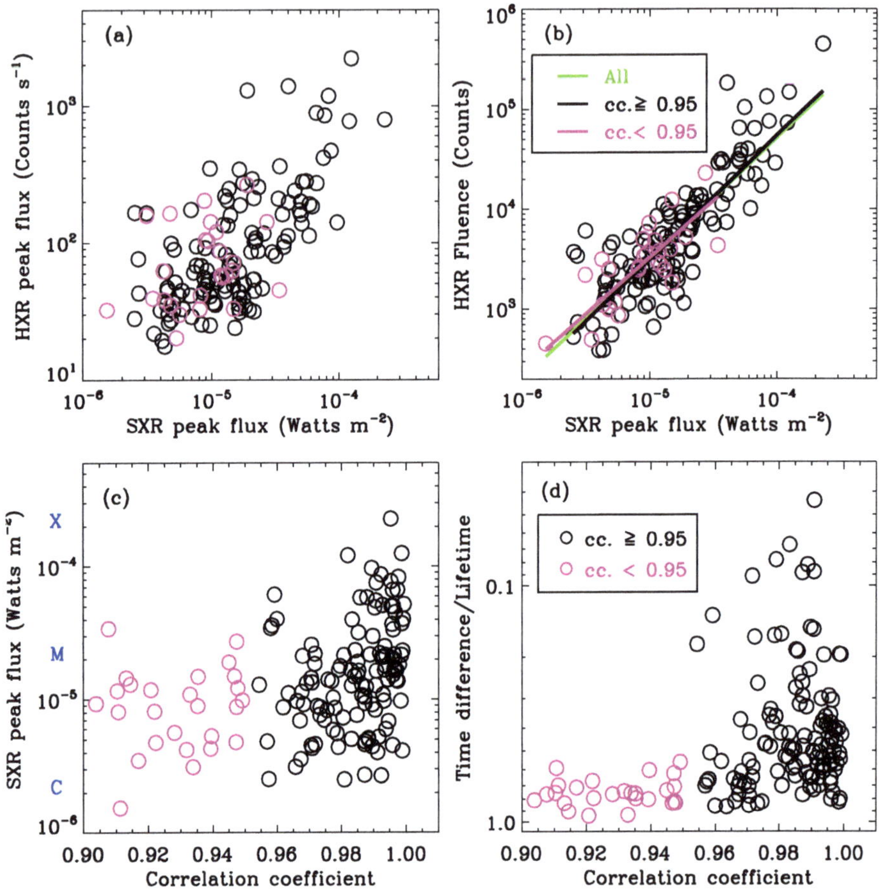

Figure 6 Scatter plots. (**a**) HXR peak flux vs. SXR peak flux. (**b**) HXR fluence vs. SXR peak flux. The color lines indicate linear fittings. (**c**) SXR peak flux (or GOES class) vs. correlation coefficient. (**d**) The ratio between the time difference of the HXR peak and the SXR maximum and the lifetime of the HXR fluence vs. the correlation coefficients.

flux, according to the fact that the maximum of SXR light curves and the end of HXR fluxes occur at nearly the same time (Veronig et al., 2002a). Then, the lifetime of the HXR fluence can be obtained as $t_2 - t_1$. We also estimated the time difference between the peak time of HXR pulse and the maximum time of the SXR flux. We note that the time difference studied here is not the same as that reported by Veronig et al. (2002a), who investigated the time delay between the SXR maximum and the HXR end.

The correlation coefficients of all these 149 flares are greater than 0.90, which may suggest a close relationship between the HXR fluence and the SXR flux. The similar fitting results for these flares confirm the strong correlation. Moreover, about 82.5% of the studied flares show much higher correlation coefficients, ≥ 0.95. The statistical results agree with the standard flare model (e.g. Masuda et al., 1994; Su et al., 2013; Yan et al., 2018, 2022) or the CSHKP model (e.g. Carmichael, 1964; Sturrock, 1966; Hirayama, 1974; Kopp and Pneuman, 1976), indicating that the characterization we used in this study is reasonable. Twenty six flares show lower correlation coefficients, <0.95, but they are still greater than

0.90. All the lower correlation coefficients tend to show a large ratio between the time difference of the HXR peak and the SXR maximum and its lifetime ($t_2 - t_1$). For instance, their ratios could be as high as 1.0, indicating that the acceleration process of nonthermal electrons is much shorter than the heating process of thermal plasmas. Our observations suggest that those solar flares might have an additional heating process to the nonthermal electrons, which is consistent with the previous results (e.g. Veronig et al., 2002a; Li et al., 2021; Yu, Li, and Gan, 2020).

For all these 149 flares, the correlation coefficients seem to be independent of the flare location. They appear to take place in the middle- and low-latitude regions and not in the high-latitude region, while for these 26 flares that have lower correlation coefficients, they appear to be distributed close to the solar limbs, i.e., the east and west limbs. The correlation coefficients do not exhibit a significant relation with the flare class or the SXR peak flux. The lower correlation coefficients seem to appear in small flares (C and M), which may be because the intense flares (X) are very rare in our study, only three X-class flares. Similarly to previous observations (cf. Veronig et al., 2002a), the HXR fluence displays a linear relationship with the SXR peak flux, suggesting that the flare energy released during the impulsive phase is primarily from the nonthermal electron beam (Priest and Forbes, 2002; Chen et al., 2017; Warmuth and Mann, 2016, 2020). Finally, we note that the HXI and GOES data are selected during the increasing period of Solar Cycle 25, which might motivate the curiosity to perform a similar study for the later time of the solar cycle.

5. Summary

Using the HXR and SXR light curves observed by ASO-S/HXI and GOES, we statistically investigated the Neupert effect during solar flares. Our primary results are summarized as follows:

(1) 149 flares are selected from the HXI database. These flares exhibit apparently radiation enhancements in the energy range of 20 – 50 keV, and their impulsive phases are completely recorded by HXI.

(2) The selected flares show high correlation coefficients (>0.90) between the HXR fluence and the SXR flux. 123 (82.5%) of them exhibit much higher correlation coefficients, ≥0.95.

(3) The HXR fluence displays a linear relationship with the SXR peak flux, and the correlation coefficient between the time integral of the HXR flux and the SXR light curve appears to be independent of the flare location and class. Our observational facts suggest that all the selected flares could follow the Neupert effect.

Author contributions Z. J. Ning provided the idea, D. Li led this work and wrote the manuscript. H. Y. Dong participated in data analysis, W. Chen, Y. Su and Y. Huang participated the HXI data correction and some discussions. All authors reviewed the manuscript.

Funding This work is funded by the National Key R&D Program of China 2022YFF0503002 (2022YFF0503000), NSFC under grants 12073081, 12333010. D. Li is supported by Yunnan Key Laboratory of Solar Physics and Space Science under the number YNSPCC202207. This work is also supported by the Strategic Priority Research Program of the Chinese Academy of Sciences, Grant No. XDB0560000. ASO-S mission is supported by the Strategic Priority Research Program on Space Science, the Chinese Academy of Sciences, Grant No. XDA15320000.

Data Availability No datasets were generated during the current study.

Declarations

Competing interests The authors declare no competing interests.

References

Battaglia, M., Fletcher, L., Benz, A.O.: 2009, Observations of conduction driven evaporation in the early rise phase of solar flares. *Astron. Astrophys.* **498**, 891. DOI. ADS.
Brown, J.C.: 1971, The deduction of energy spectra of non-thermal electrons in flares from the observed dynamic spectra of hard X-ray bursts. *Solar Phys.* **18**, 489. DOI. ADS.
Carmichael, H.: 1964, A Process for Flares. NASA Special Publication 50 451. ADS.
Chen, Y., Wu, Z., Liu, W., Schwartz, R.A., Zhao, D., Wang, B., Du, G.: 2017, Double-coronal X-ray and microwave sources associated with a magnetic breakout solar eruption. *Astrophys. J.* **843**, 8. DOI. ADS.
Datlowe, D.W., Elcan, M.J., Hudson, H.S.: 1974, OSO-7 observations of solar X-rays in the energy range 10 100 keV. *Solar Phys.* **39**, 155. DOI. ADS.
Dennis, B.R.: 1988, Solar flare hard X-ray observations. *Solar Phys.* **118**, 49. DOI. ADS.
Dennis, B.R., Zarro, D.M.: 1993, The Neupert effect: what can it tell up about the impulsive and gradual phases of solar flares. *Solar Phys.* **146**, 177. DOI. ADS.
Dennis, B.R., Veronig, A., Schwartz, R.A., Sui, L., Tolbert, A.K., Zarro, D.M., Rhessi Team: 2003, The Neupert effect and new RHESSI measures of the total energy in electrons accelerated in solar flares. *Adv. Space Res.* **32**, 2459. DOI. ADS.
Feldman, U.: 1990, The beam-driven chromospheric evaporation model of solar flares: a model not supported by observations from nonimpulsive large flares. *Astrophys. J.* **364**, 322. DOI. ADS.
Fisher, G.H., Canfield, R.C., McClymont, A.N.: 1985, Flare loop radiative hydrodynamics. V. Response to thick-target heating dynamics of the thick-target heated chromosphere. *Astrophys. J.* **289**, 414. DOI. ADS.
Freeland, S.L., Handy, B.N.: 1998, Data analysis with the SolarSoft system. *Solar Phys.* **182**, 497. DOI. ADS.
Gan, W.-Q., Zhu, C., Deng, Y.-Y., Li, H., Su, Y., Zhang, H.-Y., Chen, B., Zhang, Z., Wu, J., Deng, L., Huang, Y., Yang, J.-F., Cui, J.-J., Chang, J., Wang, C., Wu, J., Yin, Z.-S., Chen, W., Fang, C., Yan, Y.-H., Lin, J., Xiong, W.-M., Chen, B., Bao, H.-C., Cao, C.-X., Bai, Y.-P., Wang, T., Chen, B.-L., Li, X.-Y., Zhang, Y., Feng, L., Su, J.-T., Li, Y., Chen, W., Li, Y.-P., Su, Y.-N., Wu, H.-Y., Gu, M., Huang, L., Tang, X.-J.: 2019, Advanced Space-based Solar Observatory (ASO-S): an overview. *Res. Astron. Astrophys.* **19**, 156. DOI. ADS.
Gan, W., Zhu, C., Deng, Y., Zhang, Z., Chen, B., Huang, Y., Deng, L., Wu, H., Zhang, H., Li, H., Su, Y., Su, J., Feng, L., Wu, J., Cui, J., Wang, C., Chang, J., Yin, Z., Xiong, W., Chen, B., Yang, J., Li, F., Lin, J., Hou, J., Bai, X., Chen, D., Zhang, Y., Hu, Y., Liang, Y., Wang, J., Song, K., Guo, Q., He, L., Zhang, G., Wang, P., Bao, H., Cao, C., Bai, Y., Chen, B., He, T., Li, X., Zhang, Y., Liao, X., Jiang, H., Li, Y., Su, Y., Lei, S., Chen, W., Li, Y., Zhao, J., Li, J., Ge, Y., Zou, Z., Hu, T., Su, M., Ji, H., Gu, M., Zheng, Y., Xu, D., Wang, X.: 2023, The Advanced Space-Based Solar Observatory (ASO-S). *Solar Phys.* **298**, 68. DOI. ADS.
Hirayama, T.: 1974, Theoretical model of flares and prominences. I: evaporating flare model. *Solar Phys.* **34**, 323. DOI. ADS.
Huang, Y., Li, H., Gan, W.-Q., Li, Y.-P., Su, J.-T., Deng, Y.-Y., Feng, L., Su, Y., Chen, W., Lei, S.-J., Li, Y., Ge, Y.-Y., Su, Y.-N., Liu, S.-M., Zang, J.-J., Xu, Z.-L., Bai, X.-Y., Li, J.-W.: 2019, The Science Operations and Data Center (SODC) of the ASO-S mission. *Res. Astron. Astrophys.* **19**, 164. DOI. ADS.
Hudson, H.S.: 1991, Differential emission-measure variations and the "Neupert effect". In: *Bulletin of the American Astronomical Society* **23**, 1064. ADS.
Jiang, C., Feng, X., Liu, R., Yan, X., Hu, Q., Moore, R.L., Duan, A., Cui, J., Zuo, P., Wang, Y., Wei, F.: 2021, A fundamental mechanism of solar eruption initiation. *Nat. Astron.* **5**, 1126. DOI. ADS.
Kopp, R.A., Pneuman, G.W.: 1976, Magnetic reconnection in the corona and the loop prominence phenomenon. *Solar Phys.* **50**, 85. DOI. ADS.
Krucker, S., Battaglia, M., Cargill, P.J., Fletcher, L., Hudson, H.S., MacKinnon, A.L., Masuda, S., Sui, L., Tomczak, M., Veronig, A.L., Vlahos, L., White, S.M.: 2008, Hard X-ray emission from the solar corona. *Astron. Astrophys. Rev.* **16**, 155. DOI. ADS.
Lee, T.T., Petrosian, V., McTiernan, J.M.: 1995, The Neupert effect and the chromospheric evaporation model for solar flares. *Astrophys. J.* **448**, 915. DOI. ADS.
Li, D., Hong, Z., Ning, Z.: 2022, Simultaneous observations of chromospheric evaporation and condensation during a C-class flare. *Astrophys. J.* **926**, 23. DOI. ADS.

Li, D., Ning, Z.J., Zhang, Q.M.: 2015, Observational evidence of electron-driven evaporation in two solar flares. *Astrophys. J.* **813**, 59. DOI. ADS.
Li, D., Warmuth, A., Lu, L., Ning, Z.: 2021, An investigation of flare emissions at multiple wavelengths. *Res. Astron. Astrophys.* **21**, 066. DOI. ADS.
Li, D., Hou, Z., Bai, X., Li, C., Fang, M., Zhao, H., Wang, J., Ning, Z.: 2023c, Simultaneous detection of flare-associated kink oscillations and extreme-ultraviolet waves. *Sci. China, Technol. Sci.*. DOI. ADS.
Li, D., Li, C., Qiu, Y., Rao, S., Warmuth, A., Schuller, F., Zhao, H., Shi, F., Xu, J., Ning, Z.: 2023b, Observational signatures of electron-driven chromospheric evaporation in a white-light flare. *Astrophys. J.* **954**, 7. DOI. ADS.
Li, D., Warmuth, A., Wang, J., Zhao, H., Lu, L., Zhang, Q., Dresing, N., Vainio, R., Palmroos, C., Paassilta, M., Fedeli, A., Dominique, M.: 2023a, Global energetics of solar powerful events on 2017 September 6. *Res. Astron. Astrophys.* **23**, 095017. DOI. ADS.
Loto'aniu, T.M., Redmon, R.J., Califf, S., Singer, H.J., Rowland, W., Macintyre, S., Chastain, C., Dence, R., Bailey, R., Shoemaker, E., Rich, F.J., Chu, D., Early, D., Kronenwetter, J., Todirita, M.: 2019, The GOES-16 spacecraft science magnetometer. *Space Sci. Rev.* **215**, 32. DOI. ADS.
Mann, G., Aurass, H., Warmuth, A.: 2006, Electron acceleration by the reconnection outflow shock during solar flares. *Astron. Astrophys.* **454**, 969. DOI. ADS.
Masuda, S., Kosugi, T., Hara, H., Tsuneta, S., Ogawara, Y.: 1994, A loop-top hard X-ray source in a compact solar flare as evidence for magnetic reconnection. *Nature* **371**, 495. DOI. ADS.
McTiernan, J.M., Fisher, G.H., Li, P.: 1999, The solar flare soft X-ray differential emission measure and the Neupert effect at different temperatures. *Astrophys. J.* **514**, 472. DOI. ADS.
Milligan, R.O., Ireland, J.: 2018, On the performance of multi-instrument solar flare observations during solar cycle 24. *Solar Phys.* **293**, 18. DOI. ADS.
Neupert, W.M.: 1968, Comparison of solar X-ray line emission with microwave emission during flares. *Astrophys. J. Lett.* **153**, L59. DOI. ADS.
Ning, Z.: 2008a, RHESSI microflares with quiet microwave emission. *Astrophys. J.* **686**, 674. DOI. ADS.
Ning, Z.: 2008b, RHESSI observations of the Neupert effect in three solar flares. *Solar Phys.* **248**, 99. DOI. ADS.
Ning, Z.: 2009, The investigation of the Neupert effect in two solar flares. *Sci. China, Phys. Mech. Astron.* **52**, 1686. DOI. ADS.
Ning, Z., Cao, W.: 2010a, Investigation of chromospheric evaporation in a Neupert-type solar flare. *Astrophys. J.* **717**, 1232. DOI. ADS.
Ning, Z., Cao, W.: 2010b, Investigation of the Neupert effect in the various intervals of solar flares. *Solar Phys.* **264**, 329. DOI. ADS.
Priest, E.R., Forbes, T.G.: 2002, The magnetic nature of solar flares. *Astron. Astrophys. Rev.* **10**, 313. DOI. ADS.
Scargle, J.D.: 1982, Studies in astronomical time series analysis. II. Statistical aspects of spectral analysis of unevenly spaced data. *Astrophys. J.* **263**, 835. DOI. ADS.
Shang, Z., Xu, K., Liu, Y., Wu, Z., Lu, G., Zhang, Y., Zhang, L., Su, Y., Chen, Y., Yan, F.: 2022, A broadband solar radio dynamic spectrometer working in the millimeter-wave band. *Astrophys. J. Suppl.* **258**, 25. DOI. ADS.
Shang, Z., Wu, Z., Liu, Y., Bai, Y., Lu, G., Zhang, Y., Zhang, L., Su, Y., Chen, Y., Yan, F.: 2023, The calibration of the 35-40 GHz solar radio spectrometer with the new moon and a noise source. *Astrophys. J. Suppl.* **268**, 45. DOI. ADS.
Shibata, K., Magara, T.: 2011, Solar flares: magnetohydrodynamic processes. *Living Rev. Solar Phys.* **8**, 6. DOI. ADS.
Starr, R., Heindl, W.A., Crannell, C.J., Thomas, R.J., Batchelor, D.A., Magun, A.: 1988, Energetics and dynamics of simple impulsive solar flares. *Astrophys. J.* **329**, 967. DOI. ADS.
Sturrock, P.A.: 1966, Model of the high-energy phase of solar flares. *Nature* **211**, 695. DOI. ADS.
Su, Y., Veronig, A.M., Holman, G.D., Dennis, B.R., Wang, T., Temmer, M., Gan, W.: 2013, Imaging coronal magnetic-field reconnection in a solar flare. *Nat. Phys.* **9**, 489. DOI. ADS.
Su, Y., Liu, W., Li, Y.-P., Zhang, Z., Hurford, G.J., Chen, W., Huang, Y., Li, Z.-T., Jiang, X.-K., Wang, H.-X., Xia, F.-X.-Y., Chen, C.-X., Yu, W.-H., Yu, F., Wu, J., Gan, W.-Q.: 2019, Simulations and software development for the hard X-ray imager onboard ASO-S. *Res. Astron. Astrophys.* **19**, 163. DOI. ADS.
Tian, H., Chen, N.-H.: 2018, Multi-episode chromospheric evaporation observed in a solar flare. *Astrophys. J.* **856**, 34. DOI. ADS.
Tian, H., Young, P.R., Reeves, K.K., Chen, B., Liu, W., McKillop, S.: 2015, Temporal evolution of chromospheric evaporation: case studies of the M1.1 flare on 2014 September 6 and X1.6 flare on 2014 September 10. *Astrophys. J.* **811**, 139. DOI. ADS.
Torrence, C., Compo, G.P.: 1998, A practical guide to wavelet analysis. *Bull. Am. Meteorol. Soc.* **79**, 61. DOI. ADS.

Veronig, A., Vršnak, B., Dennis, B.R., Temmer, M., Hanslmeier, A., Magdalenić, J.: 2002a, Investigation of the Neupert effect in solar flares. I. Statistical properties and the evaporation model. *Astron. Astrophys.* **392**, 699. DOI. ADS.

Veronig, A., Vršnak, B., Temmer, M., Hanslmeier, A.: 2002b, Relative timing of solar flares observed at different wavelengths. *Solar Phys.* **208**, 297. DOI. ADS.

Veronig, A.M., Brown, J.C., Dennis, B.R., Schwartz, R.A., Sui, L., Tolbert, A.K.: 2005, Physics of the Neupert effect: estimates of the effects of source energy, mass transport, and geometry using RHESSI and GOES data. *Astrophys. J.* **621**, 482. DOI. ADS.

Warmuth, A., Mann, G.: 2016, Constraints on energy release in solar flares from RHESSI and GOES X-ray observations. II. Energetics and energy partition. *Astron. Astrophys.* **588**, A116. DOI. ADS.

Warmuth, A., Mann, G.: 2020, Thermal-nonthermal energy partition in solar flares derived from X-ray, EUV, and bolometric observations. Discussion of recent studies. *Astron. Astrophys.* **644**, A172. DOI. ADS.

Yan, X.L., Yang, L.H., Xue, Z.K., Mei, Z.X., Kong, D.F., Wang, J.C., Li, Q.L.: 2018, Simultaneous observation of a flux rope eruption and magnetic reconnection during an X-class solar flare. *Astrophys. J. Lett.* **853**, L18. DOI. ADS.

Yan, X., Xue, Z., Jiang, C., Priest, E.R., Kliem, B., Yang, L., Wang, J., Kong, D., Song, Y., Feng, X., Liu, Z.: 2022, Fast plasmoid-mediated reconnection in a solar flare. *Nat. Commun.* **13**, 640. DOI. ADS.

Yan, F., Wu, Z., Shang, Z., Wang, B., Zhang, L., Chen, Y.: 2023, The first flare observation with a new solar microwave spectrometer working in 35-40 GHz. *Astrophys. J. Lett.* **942**, L11. DOI. ADS.

Yu, W.H., Li, Y.P., Gan, W.Q.: 2020, Statistical studies on modified Neupert effect. *Acta Astron. Sin.* **61**, 53. ADS.

Zhang, Z., Chen, D.-Y., Wu, J., Chang, J., Hu, Y.-M., Su, Y., Zhang, Y., Wang, J.-P., Liang, Y.-M., Ma, T., Guo, J.-H., Cai, M.-S., Zhang, Y.-Q., Huang, Y.-Y., Peng, X.-Y., Tang, Z.-B., Zhao, X., Zhou, H.-H., Wang, L.-G., Song, J.-X., Ma, M., Xu, G.-Z., Yang, J.-F., Lu, D., He, Y.-H., Tao, J.-Y., Ma, X.-L., Lv, B.-G., Bai, Y.-P., Cao, C.-X., Huang, Y., Gan, W.-Q.: 2019, Hard X-ray Imager (HXI) onboard the ASO-S mission. *Res. Astron. Astrophys.* **19**, 160. DOI. ADS.

Solar Physics (2024) 299:80
https://doi.org/10.1007/s11207-024-02331-w

RESEARCH

Spatial and Spectral Evolution of Microwave and X-Ray Sources During the Solar Limb Flare on February 5, 2023

Julia N. Shamsutdinova[1] · Larisa K. Kashapova[1] · Zhentong Li[2] · Yang Su[2,3]

Received: 15 March 2024 / Accepted: 23 May 2024 / Published online: 10 June 2024

Abstract
We present an empirical scenario of the energy release process during a solar limb flare on February 5, 2023. This event was observed by the Siberian Radioheliograph (SRH) within the 3 – 12 GHz range and the Advanced Space-based Solar Observatory / Hard X-ray Imager (ASO-S/HXI) within 10 – 300 keV range. The combination of these data allowed us to use information not only about the spectral features but also about the spatial evolution of the flare. The main source of the energy released was a small compact loop which was revealed in both the X-ray and microwave ranges. During the early phases of the flare evolution, a spectral analysis of microwave emission showed that thermal gyrosynchrotron emission turned to gyrosynchrotron emission of nonthermal electrons. This indicated the transition from the heating process to the acceleration processes. Spectral indices derived from hard X-ray and microwave data closely agree with each other and show the classical soft–hard–soft scenario of acceleration. The hardening of the average microwave spectrum at the end of the impulsive phase was caused by the contribution of jet emission to microwaves rather than by peculiarities of the acceleration processes.

Keywords Flares · Flares, dynamics · Flares, energetic particles · Flares, models · X-ray bursts, association with flares · Heating, in flares radio bursts, microwave (mm, cm)

✉ J.N. Shamsutdinova
yulia@iszf.irk.ru

L.K. Kashapova
lkk@iszf.irk.ru

Z. Li
ztli@pmo.ac.cn

Y. Su
yang.su@pmo.ac.cn

1 Institute of Solar-Terrestrial Physics SB RAS, 664033 Irkutsk, Russia

2 Key Laboratory of Dark Matter and Space Astronomy, Purple Mountain Observatory, CAS, Nanjing 210023, China

3 School of Astronomy and Space Science, University of Science and Technology of China, Hefei, 230026, China

1. Introduction

Solar flares are powerful transient events in the solar atmosphere where we see transformation of energy supplied by magnetic field into heating, radiation, and particle acceleration, followed by energy propagation to different levels of the solar atmosphere (see Fletcher et al., 2011; Hudson, 2011; Benz, 2017, and references therewith). The quantitative description of solar flare plasma evolution could help both in improving forecast of geoeffective events and better understanding energy release and propagation processes in the solar atmosphere. There are two spectral regions which can provide direct tracing of flare plasma parameter evolution, namely, X-ray and microwave. These types of emission have different mechanisms of generation, advantages, and disadvantages, and they supplement each other (White et al., 2011).

The mechanism of X-ray emission generation is bremsstrahlung (see Kontar et al., 2011, and the references therein). The less energetic electrons with thermal velocity distribution lose energy at higher levels of the solar atmosphere and produce emission at the level of flare loop top. The electrons with energy of several dozen keV, instead, precipitate towards the lower levels of the solar atmosphere and are responsible for the generation of the emission coming from the footpoints of the flare loop.

The most widespread mechanism forming microwave emission of solar flares is gyrosynchrotron or magnetobremsstrahlung (Dulk, 1985). We note that bremsstrahlung is also present among mechanisms of flare emission. But its contribution is not significant relative to the contribution of the gyrosynchrotron emission (Bastian, Benz, and Gary, 1998), except for the cases when a large amount of heated plasma fills the flare loops (posteruptive arcade phase, for example). Flare emission in the microwave range could also be generated by the plasma mechanism, which is mostly associated with decimeter and meter range (see, for example, Benz, 2017, and the references therein). The appearance of this type of emission is usually caused by high plasma density and low level of magnetic field that is usually related to peculiarities in active region topology. The gyrosynchrotron emission could be generated by both electrons with thermal distribution as well as nonthermal electrons. The index obtained from the spectral range above the spectrum maximum indicates the distribution of electrons with energy (or velocities) and it linearly relates to the electron spectral index. Emission of this spectral range produced in the regions associated with flare loop legs. The lower frequency range is traditionally associated with emission of the flare loop top and nearby levels. If we assume that electrons generating both types of emission come from the same energy release site, then plasma parameters at the same flare source location taken from the different types of emission should coincide or closely agree (Silva, Wang, and Gary, 2000; Kawate et al., 2012). However, observations show that temporal, spectral, and spatial features taken from these ranges could differ or demonstrate complicate agreement. It is well known that the differences between the electron spectral indexes and the maximum of microwave (MW) emission have a little delay relative to the hard X-ray (HXR) emission (see, for example, Cornell et al., 1984). There are two main possible reasons for the delay between HXR and MW time profile maxima (see, for example, Lu and Petrosian, 1990). The acceleration during a flare can develop according to the "soft-hard-soft" scenario. In this case the nonthermal electrons with higher energies that produce MW emission will appear later than the nonthermal electrons generating HXR emission. The other reason is a magnetic trapping where electrons are trapped by magnetic mirrors formed by a converging magnetic field. A model that includes the scenario of energy release and particle acceleration is required to interpret the observations. This model should account for the magnetic topology and density within the active region. The spatial and height structure of flare sources also could differ as a result of magnetic topology features. However, the plasma parameters obtained from different spectral ranges have to be consistent with each other in order to provide a reliable interpretation and confirm the results obtained from a self-consistent model (see an example in Ruan, Xia, and Keppens, 2020). So, a numerically consistent flare scenario based on multiwavelength observations is still important for observation interpretation, theoretical model improvement, and space weather forecast development.

The ideal case for designing such a model is a compact flare located near the solar limb, which helps avoid projection effects. The event should have significant levels of HXR and MW emission which allow doing not only spectral but spatial analysis. We present a study of spectral and spatial evolution of a limb event to form a consistent empirical scenario of energy release during solar flares.

2. Observational Data

We used data obtained by the Hard X-ray Imager (HXI; Su et al., 2019) onboard Advanced Space-based Solar Observatory (ASO-S; Gan et al., 2023) for spectral and spatial study of X-ray emission of the event. HXI observes solar X-ray emission of 10 – 300 keV energy range with high temporal and spectral resolution. Spectral analysis of HXI data allowed us to determine the cutoff energy of accelerated electrons, as well as estimate parameters such as spectral index and electron fluxes. In the current study we used the detector response matrix of HXI software v1.30bcta. The best spatial resolution of the images obtained by HXI is about 3.1″, which allows us to study the spatial distribution of accelerated electrons and plasma heating process. As HXI detects X-ray flux above 10 keV, we obtained information about the soft X-ray emission of the flare using the data from the Extreme Ultraviolet and X-ray Irradiance Sensors (EXIS) on the GOES–18 Series satellites (Thiemann et al., 2019).

Observations from the Siberian Radioheliograph (SRH; Altyntsev et al., 2020) provided data for spectral and spatial analysis of microwave (MW) emission within the 3 – 12 GHz frequency range. SRH is a complex consisting of three separate T-shaped antenna arrays for 3 – 6, 6 – 12, and 12 - 24 GHz frequency bands. The radioheliograph takes the R and L circularly polarized components and transforms them into I and V Stokes parameters. The temporal resolution of the SRH is 0.1 seconds for taking an image at one frequency. As images are taken in series for the same antenna array, the temporal resolution depends on the number of frequencies, and it is 3 – 4 seconds on average. The number of antennas in each array defines the spatial resolution and sensitivity. There are 129, 192, and 207 antennas for the 3 – 6, 6 – 12, and 12 – 24 GHz antenna arrays, respectively. We used observations of 3 – 6 and 6 – 12 GHz frequency bands, having sensitivity of 40 – 70 $\times 10^{-4}$ solar flux units (sfu). The spatial resolutions are 30 to 15 arcseconds for the 3 – 6 GHz band and 24 – 12 arcseconds for the 6 – 12 GHz band. We used both the imaging spectroscopy techniques and fluxes from SRH as spectropolarimeters. The fluxes at 3 and 9 GHz obtained by the Nobeyama Radio Polarimeters (NoRP; Torii et al., 1979) with 1 second temporal resolution were used to verify the flux values from SRH.

Information about the magnetic field structure was obtained from the line-of-sight (LOS) magnetograms by the Helioseismic and Magnetic Imager (HMI; Scherrer et al., 2012) onboard of the Solar Dynamics Observatory (SDO; Pesnell, Thompson, and Chamberlin, 2012). We used HMI magnetograms to coalign SRH images to define MW sunspot source positions, which were then applied for solar disks center correction. We used ultraviolet (UV) and extreme ultraviolet (EUV) images obtained by the Atmospheric Imaging Assembly (AIA; Lemen et al., 2012) onboard SDO to study flare topology evolution both in the corona and at the chromospheric level.

3. Data Analysis

3.1. General Analysis of Flare

The studied flare took place on February 5, 2023, and was of C6.6 class according to GOES classification (the background level of 7×10^{-7} W m^{-2} was not subtracted). The maximum of the flare at $1-8$ Å was at 03:26 UT. The event located on the eastern limb of the Sun at [−950, −225] arcseconds. The flare occurred in a decaying active region which did not get a number after appearing on the solar disk. The EUV observation by SDO/AIA reveal the onset of jet eruption at about 03:25 UT.

Figure 1 shows a comparison of the flare's temporal profiles in microwaves and X-ray emission. The temporal profiles in hard X-rays (HXR) and MW show a close correlation for the first flare peak according to data obtained by various instruments. This relationship indicated the generation of both HXR and MW emission by the same population of electrons. However, during the decay phase we see a second peak only in the MW temporal profiles but not in the HXR temporal profiles at the energies higher than 50 keV. An additional marker of accelerated electrons is the derivative of the soft X-ray (SXR) emission temporal profiles (the Neupert effect: Neupert, 1968). The derivative taken from GOES X-ray flux in the $1-8$ Å energy band revealed evidence of the second peak coinciding with the one seen

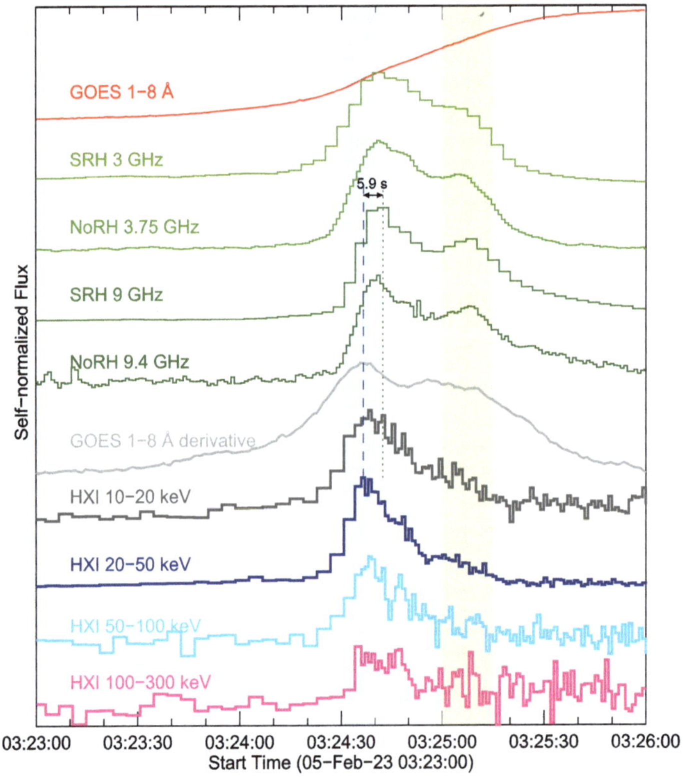

Figure 1 Temporal profiles of the event in soft X-rays, microwaves and hard X-ray emission detected by different instruments. The blue shaded area indicates the first peak in HXRs. The green shaded area indicates the second peak of MWs.

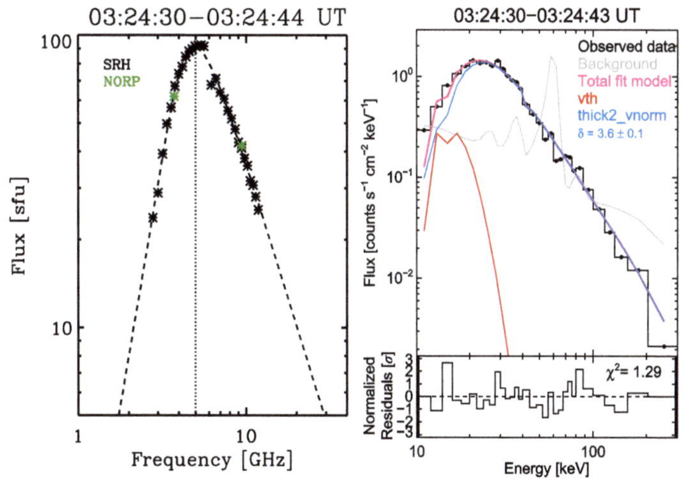

Figure 2 The X-ray and MW spectra at the first flare peak. (Left) The MW spectrum obtained from observations with SRH as a spectropolarimeter and the NoRP fluxes at 3.7 and 9.4 GHz. The dashed line shows the peak frequency at about 5 GHz. (Right) The HXI averaged count spectrum fitted with thermal model (red color, `f_vth`) plus thick target model (blue color, `thick2_vnorm`). The observed count spectrum data are marked by black dots, and the background is shown in a solid gray curve. The bottom plot shows the normalized residuals.

in MW time profiles. In our case, there is a good agreement of MW time profiles with the derivative of SXR time profile, that indicates the existence of electrons and their impact in the generation of radio emission. The weak increase seen in the HXI 20 – 50 keV temporal profile during the second peak provides further evidence of the presence of low-energy electrons only. One can see that the maximum in HXR and MW emission took place at about 03:24:40 UT. The HXR and MW fluxes decreased to preflare level before the maximum in soft X-rays.

Figure 2 shows the obtained spectra for the main peak in two spectral ranges. The left panel of Figure 2 presents the MW spectrum obtained combining data by SRH and NoRP within the 3 – 12 GHz range. We used the SRH fluxes that were integrated over the whole Sun at each frequency. The data from NoRP and SRH shows an excellent agreement between the two. The form of the spectrum is typical for the gyrosynchrotron emission of non-thermal electrons. We obtained a peak frequency of 5.2 GHz using the fitting method from Nita, Gary, and Lee (2004). The MW emission above peak frequency corresponds to emission of an optically thin source and the presence of "pure" emission by accelerated electrons forming in the legs of the flare loop. The height of the source seen at these frequencies depends on various factors, but in any case, the electron spectral index derived from this part should be correlated with the HXR electron spectral index. The MW sources seen at frequencies below 5.2 GHz correspond to the optically thick emission and are mainly forming near to the loop top of a flare.

The amount of plasma concentrated in the upper part of a flare loop can cause a distortion in the flux distribution at that part of the microwave spectrum. For this reason, the spectral and temporal analyses of emissions from this part is difficult to interpret. But the images at these frequencies provide useful information about the position and shape of the upper part of the loop. We chose 3.4 and 9 GHz for further analysis of the flare source structure.

The right panel of Figure 2 shows the ASO-S/HXI spectral fitting results utilizing the OSPEX package (Schwartz et al., 2002). We use the isothermal model (`f_vth`) combined

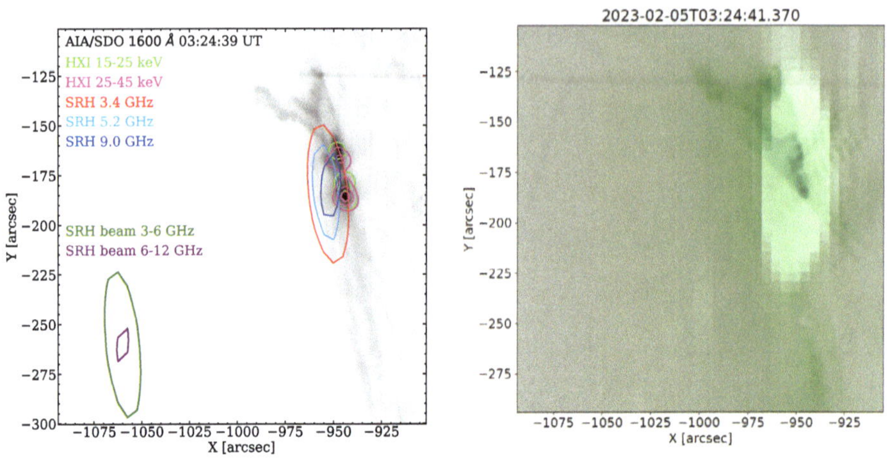

Figure 3 (Left) The SDO/AIA 1600 Å image overlaid with contours of the MW and HXR flare sources obtained by SRH and ASO-S/HXI, respectively. SRH contours present the difference between images taken at 3:24:38 UT and the image taken at the moment without flare activity. The contours present emission at 80% of maximum for MW and 20%, 50%, and 80% for X-ray emission. The SRH beam is shown for correct interpretation of the MW sources at 80% of the maximum. Right panel: Combination of the SDO/AIA 94 Å image at 03:24:36 UT and the 11.8 GHz SRH image at 03:24:41 UT shown by means of different transparency. UV images at 1600 and 94 Å are negative.

with the thick target model (`thick2_vnorm`) to fit HXI spectrum averaged during the first peak time (03:24:30 – 03:24:43 UT). Background is subtracted by fitting the nonflaring times instead of using the orbit data 48 h before or ahead due to the relatively large fluctuations during the time range. We used the 'This E Profile' ratio method from the OSPEX software package for background fitting. The low-energy edge of HXI is about 10 keV, which allows us to use only continuum emission and exclude lines related to hot plasma. Therefore, the thermal component would only be loosely constrained, although we can still roughly estimate the temperature and emission measure. The thermal emission contribution is also substantial to determine the parameters of the accelerated electrons including the low-energy cutoff. One can see from the plot of normalized residuals and χ^2 value that the chosen model can sufficiently explain the observed X-ray spectrum. The results of the fitting show that non-thermal emissions dominate above ~ 20 keV in comparison to thermal part contribution. This indicates that emissions with energies above ~ 20 keV are reasonable for analyzing non-thermal temporal profiles and X-ray source topologies. The most reasonable upper energy level for temporal profile analysis is ~ 40–45 keV, because the subtracted flux is above the background level. We note that the 10 – 20 keV X-ray emission, according to the fitting results, is a mixture of thermal and nonthermal components.

The left panel of Figure 3 demonstrates the MW and HXR sources obtained at the maximum of the flare and overlaid as contours on the 1600 Å image by SDO/AIA. Sometimes, the center of the solar disk in MW images does not precisely coincide with the center of the image. This could result in errors in the position of the source. The method used for coaligning SRH images and correcting for this artifact is described in the Appendix. To remove the contribution of quasistationary emission from sunspot sources, we subtracted an image taken before the flare at 3:15 UT from the images taken during the microwave burst.

The contours of this sources at several frequencies are shown in the left panel of Figure 3. The SRH beams for two spectral ranges are also shown in the figure for correct inter-

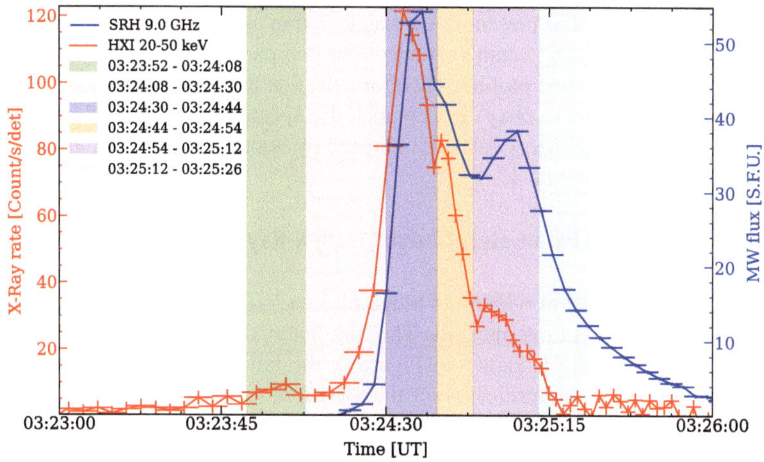

Figure 4 Comparison of microwave and X-ray normalized time profiles. Color blocks demonstrate selected time intervals used for obtaining MW spectra. Error bars are obtained for time, X-Ray rate and MW flux.

pretation of the MW source shape. The 15 – 25 keV and 25 – 45 keV images by HXI were reconstructed for the temporal interval from 03:24:26.757 UT to 03:24:43.257 UT by the `HXI_Clean` algorithm using sub-collimator groups G3 through G10, which have a spatial resolution of 6.5 arcsec. The finer grids are not used due to the incomplete calibration.

One can see that the HXR sources in both energy ranges show two footpoints of a flare loop. The centers of the MW sources demonstrate a decrease in height over the solar limb as the frequency increases. This shows the transformation from optically thick emission to optically thin emission, or from the flare loop top to the flare footpoints. However, the MW sources do not demonstrate a loop shape, but they have the same shape as the SRH beams. This fact means that the flare loop was small and compact. The right panel of Figure 3 showing the 94 Å EUV image from SDO/AIA overlapped with the 11.8 GHz image from SRH without subtracting a nonflaring MW image also confirms this. Both panels in Figure 3 demonstrate the presence of a bright cusp-like structure related to MW and HXR flare sources. These structures are usually associated with current sheets and a place of particle acceleration. The MW and HXR sources are located only in the southern part of the structure. The 94 Å image shows a small flare loop at this place. This small loop indicates that the main energy release and particle precipitation took place in this location.

Figure 4 shows a comparison of the MW and HXR time profiles. Vertical error bars show the uncertainty in flux, and horizontal error bars represent the time uncertainty. A weak increase at 03:24:40 UT during preflare phase, as seen in the HXR data, is valid, according to error estimates. We see that the delay between HXR and MW temporal profiles during the rise phase is within the errors. Thus, the fluxes of both spectral ranges rose without a delay to each other. The time profile at a frequency of 9.0 GHz indicates the presence of accelerated electrons observed in the SRH. The time profile of ASO-S/HXI in the energy range of 20 – 50 keV is also associated with bremsstrahlung emission from nonthermal electrons. The MW time profile shows that the maximum of the event was around 03:24:38 UT. The HXR time profile has a maximum about 4 seconds earlier, and this value is longer than the temporal error bars. As mentioned in the Introduction, there are two possible explanations of the delay (see Silva, Wang, and Gary, 2000, and the references therein). It can be the manifestation of a gradual hardening of the electron acceleration or the magnetic trapping of accelerated electrons. The presence of a magnetic trap usually leads to a slower decrease in the MW flux generated by nonthermal electrons compared to the HXR flux. We did not observe this effect in our time profiles. Therefore, the first hypothesis should be tested in the event under study. In order to carry out a comparative spectral analysis of the event, we split the period of nonthermal energy release, indicated by both MW and HXR emissions, into time intervals shown in Figure 4.

3.2. Evolution of Flare Parameters Derived from X-Rays and Microwaves

Figure 5 shows MW spectra within the range of 3 – 12 GHz, using fluxes from the SRH spectrograph for selected time intervals. The time interval of 03:24:30 – 03:24:54 UT for the main peak of the flare is shown in Figure 2, the left panel. The NoRP data at 3.75 and 9.4 GHz were used for verification of the SRH flux values. One can see that the SRH and NoRP fluxes agree within errors, and the agreement becomes better as the flux value

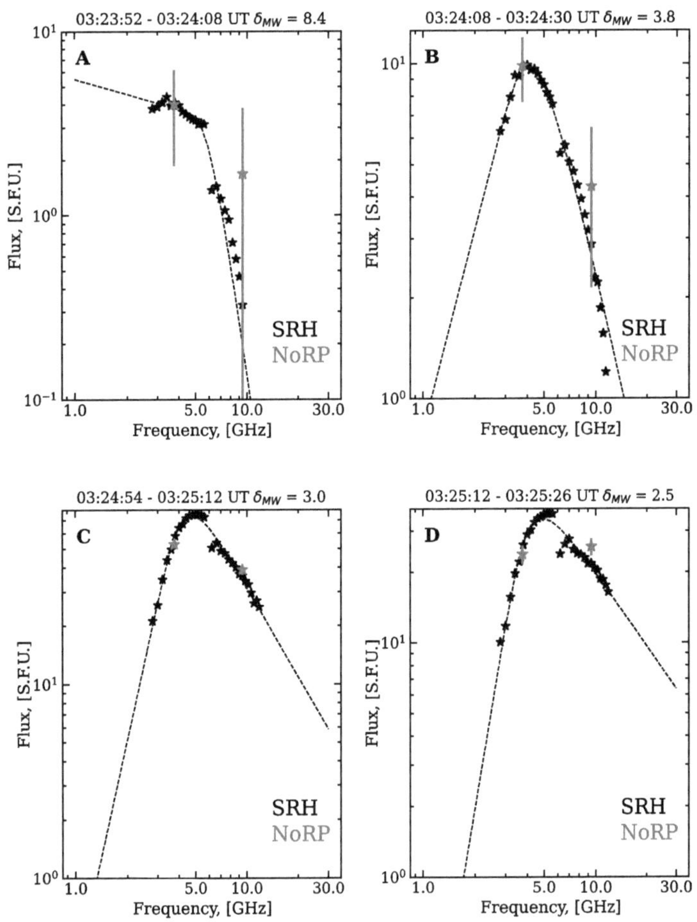

Figure 5 The MW spectra obtained from the averaged fluxes of SRH at 3 – 12 GHz during the selected time intervals, as shown in Figure 4. The green color marks NoRP data at 3.75 and 9.4 GHz as a verification of the SRH flux values at the same frequencies.

Table 1 Comparison of microwave and hard X-ray spectral parameters.

Time intervals UT	δ_{MW}	δ_{HXI}	T_{HXI}, MK	EM_{HXI}, 10^{47} cm^{-3}	T_{GOES}, MK	EM_{GOES}, 10^{47} cm^{-3}
03:23:52 – 03:24:08	8.4 ± 1.3[1]	–	–	–	11.9 ± 0.7	3.1 ± 2.0
03:24:08 – 03:24:30	3.8 ± 0.3[1]	4.4 ± 0.3	15.0 ± 5.3	1.9 ± 55.0	12.8 ± 0.7	5.0 ± 2.0
03:24:30 – 03:24:44	3.3 ± 0.2[1]	3.6 ± 0.1	31.5 ± 0.8	0.1 ± 0.2	13.3 ± 0.7	10.0 ± 2.0
03:24:44 – 03:24:54	3.3 ± 0.3[1]	3.5 ± 0.1	25.2 ± 0.6	0.4 ± 0.5	13.7 ± 0.7	13.2 ± 2.0
03:24:54 – 03:25:12	3.0 ± 0.4[1]	3.4 ± 0.1	24.3 ± 0.9	0.3 ± 0.6	14.2 ± 0.7	16.7 ± 2.0
03:25:12 – 03:25:26	$[2.5 \pm 0.1]$[1] 4.2 ± 0.1[2]	3.7 ± 0.5	13.3 ± 0.9	10.8 ± 79.0	14.4 ± 0.7	20.2 ± 2.0

[1] The electron index obtained from the average MW spectrum.

[2] The electron index obtained from imaging spectroscopy.

increases. This is a result of the better sensitivity of the SRH relative to the spectropolarimeter. The spectra were fitted using the method from Nita, Gary, and Lee (2004) to obtain the peak frequency and photon index of the high-frequency spectral region of a spectrum (α). The obtained MW photon spectral indices were transformed to the electron spectral index (δ_{MW}) using the formula $\delta_{mw} = 1.1(\alpha - 1.2)$ from (Dulk, 1985). The results are shown in the second column of Table 1. The table presents the results of the analysis of X-ray emission from ASO-S/HXI and GOES-18/EXIS. The fitting model of HXI spectra is described in Section 3.1. The third column is the electron spectral index as deduced from HXI (δ_{HXI}). The fourth and fifth columns show the electron temperature and emission measure. As the lowest energy in the HXI spectrum is about 10 keV, we lose a very important region (6 – 10 keV) for thermal emission fitting. This can result in ambiguous thermal function fitting and not sufficiently reliable values of electron temperature and emission measure. Therefore, we added the thermal plasma parameters from GOES to Table 1. The temperature and emission measure from the GOES data were derived with a standard code of the SSW IDL package. This code is based on the technique described in White, Thomas, and Schwartz (2005).

The time interval of 03:23:52 – 03:24:08 UT was analyzed because one can see a small burst in the 20 – 50 keV energy band temporal profile. Due to the low number of counts in X-rays during this time interval, we were unable to fit the HXI spectrum. However, we had the opportunity to check for the presence of accelerated electrons during the preflare phase using MW data. The MW spectrum had the shape of gyrosynchrotron emission spectrum with electron index equal to $\delta = 8.4$ (see panel A, Figure 5). This value corresponds to the emission of electrons with a thermal distribution of energies (Dulk, 1985). Therefore, we can conclude that during this phase, thermal electrons dominated. The temperature derived from GOES data for this time interval was about 12 MK, which is close to the average electron temperature at a flare maximum. All these facts indicate heating during the preflare phase before the acceleration process begins. The same electron temperature value is characteristic of hot-onset flares (Hudson et al., 2021; da Silva et al., 2023; Battaglia et al., 2023). The emission measure value obtained from the GOES data agrees, within errors, with the values obtained by da Silva et al. (2023) and indicates a small amount of hot plasma. We also note that the height of the MW flare source and the flare loop seen in the EUV range show low flare loops. This fact agrees with the results of the previous study by Hudson et al. (2021).

The next time interval of 03:24:08 – 03:24:30 UT shows emission of the rise phase. The MW spectrum has a classical gyrosynchrotron shape, with an electron spectral index equal to 3.8 (see panel B, Figure 5). According to fitting of the X-ray spectrum, the electron spectral

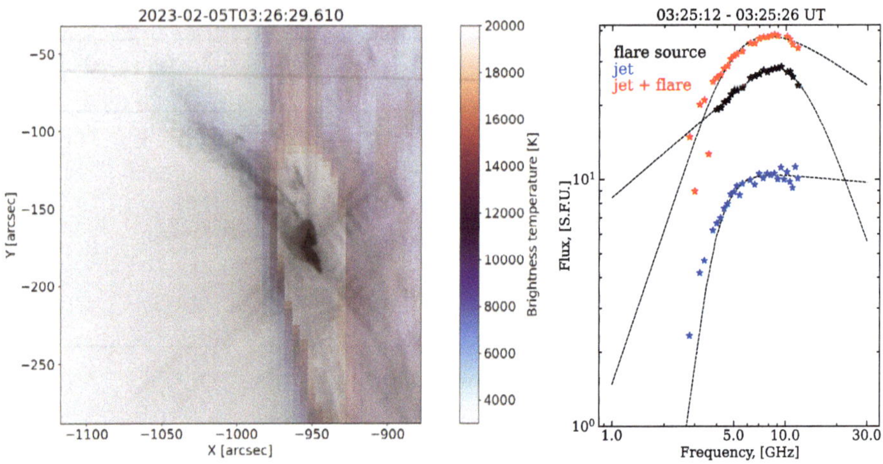

Figure 6 (Left) Overlay of the SRH image at 11.8 GHz (color palette) and an SDO/AIA image in the 131 Å band (black) shows a flare loop in the EUV and an MW flare source, combined with a jet seen in both images. (Right) Corrected spectrum obtained from a selected area of the flare source using the SRH image.

index is 4.4. The HXR electron spectral index is slightly softer than the microwave electron spectral index, but both electron spectral indices are agreed within two sigma errors of each other. These index values indicate the presence of accelerated electrons. Furthermore, in Table 1, the electron spectral indices both in the MW and HXR bands become harder, and they are consistent within the error bars for the next three intervals. This qualitatively and quantitatively agrees with earlier results (see Silva, Wang, and Gary, 2000, and the references therein).

The MW spectral index obtained for the latest time interval is 2.5, and it has become harder (panel D, Figure 5). This behavior is unusual because the HXR photon spectral index has become softer at this time. There are two possible reasons for this disagreement. One possible reason is any process that could prevent the precipitation of high-energy electrons into the atmospheric levels, where the HXR emission is generated. But this possibility is unlikely. We used the averaged fluxes from the whole Sun for spectrum reconstruction. However, one can see the jet in the left panel of Figure 6 during this time interval, and it is visible not only in the EUV but also in microwave range. So, the most likely explanation can be the contribution to the spectrum from emission of a new transient source. We note that the 15 – 25 keV image from HXI does not show a source that is spatially associated with the jet. However, this could be a result of the application of the HXI_CLEAN algorithm to image reconstruction in the presence of very bright and very weak sources in the field of view.

Figure 6, left panel, shows the overlaid SRH image at 11.8 GHz on the 131 Å image of SDO/AIA at 03:26:29 UT. We see that the emission at 11.8 GHz consists of two components. The first is the flare source, which coincides with the flare loop. The second component is the jet. To get a spectrum of the flare source only, we summarized SRH images at each frequency during the time interval 03:25:12 – 03:25:26 UT. For this, we selected and summarized the brightness temperature values of pixels related to the microwave flare source and jet separately. The brightness temperatures were recalculated into a flux using the Rayleigh–Jeans formula. Figure 6, right panel, shows the MW spectra of the flare source, the jet and the sum of the fluxes. The new value of the MW spectral index for the flare source is equal to 4.2, and it agrees within the uncertainty to the photon spectral index from X-rays. In the MW spectrum obtained for the jet, the flux distribution shows a flat behavior above the peak frequency. Such an MW spectrum shape is a feature of the thermal bremsstrahlung or the free-free radio emission (see Dulk, 1985; Nindos, 2020). This emission is associated with Coulomb collisions or interactions between electrons and ions. The free–free emission becomes optically thin at high frequencies, and its brightness temperature depends on frequency as $T_b \propto \nu^{-2}$. The MW flux relates to brightness temperature according to the Rayleigh–Jeans formula, as $I \propto T_b \times \sigma \times \nu^2$, where sigma is the area of the source. If the area of the source does not change significantly with frequency, then the flux behavior will be flat at frequencies above the peak frequency. Several flare-related phenomena characterized by large amounts of heated plasma, such as posteruptive arcades, coronal mass ejections (see the previous observations by SRH in Uralov et al., 2023) and jets, can show dominance of optically thin free–free radio emission in the MW spectrum. Thus, the thermal bremsstrahlung emission from the jet caused the hardening of the MW photon spectral index derived from microwaves. The flare-related MW electron spectral index agrees with the softening of the nonthermal electron spectral index revealed by HXR observations. Moreover, the second peak seen in microwave range, but poorly seen in the HXR time profiles, can be explained by the contribution of jet emission at high frequencies.

4. Discussion and Conclusions

We carried out a comparative analysis of the flare plasma parameters obtained from microwave and X-ray emission in order to form an empirical consistent scenario of acceleration processes. The subject of the study was the C6.6 solar flare that occurred on February 5, 2023 on the eastern limb. Spatially resolved observations from ASOS/HXI and SRH have allowed us to not only analyze the spectral evolution of plasma parameters, but also to add information about the flare topology. The flare consisted of a small flare loop where both HXR and MW flare sources were found and a higher structure seen only in UV and EUV which transformed into a cusp-like structure during the flare maximum and later into a jet. A cusp-like structure assumes a classical CSHKP scenario (Carmichael, 1964; Sturrock, 1966; Hirayama, 1974; Kopp and Pneuman, 1976) with flare ribbons and HXR sources related to the footpoints of the flare loop arcade (see a description in Benz, 2017). We are unable to analyze the magnetic field configuration because the flare being located near the limb and any measurements of the magnetic field have large errors. Nevertheless, both HXR and MW data indicate that the main energy release occurred in a small compact loop.

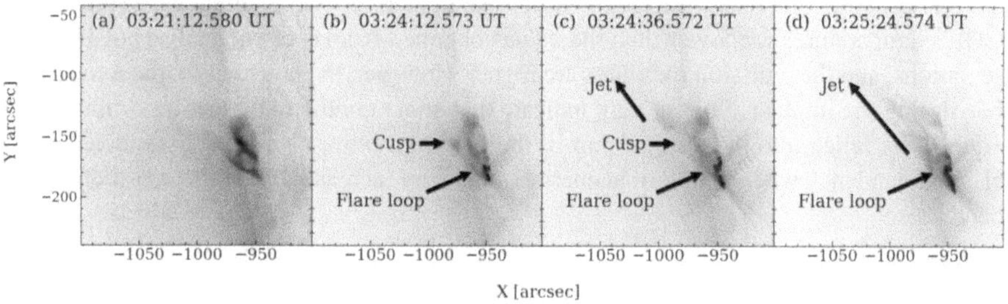

Figure 7 The evolution of flare topology in the SDO/AIA 94 Å band (negative images).

The 94 Å SDO/AIA images of the flare region evolution from before the flare to the decaying part of the impulsive phase are presented in Figure 7. One can see two bright low loops at 03:21:11 UT, before the flare onset (panel (a) in Figure 7). In the next moment, associated with the preflare stage, the two loops transformed into a system consisting of a small bright flare loop and a cusp-like structure. The cusp-like structure rose and developed over time, and after 03:25 UT it transformed into a powerful jet (panels (c) and (d) in Figure 7). As this evolution of the cusp-like structure coincides with a rise of the electron temperature and emission measure, it looks like the process of jet onset due to violation of balance between gas pressure and magnetic pressure.

So the seen cusp-like structure is not a feature of the CSHKP model but the event topology is similar to that of a two-loop flare model (Hanaoka, 1997). A two-loop flare model assumes interaction between a larger overlying loop and a smaller emerging loop. A feature of this model is absence of coronal mass ejections, but presence of such flare eruptions like jets and surges. Based on observations, the author noted that the smaller loop or the preceding region should be related to HXR, MW, and the chromospheric sources of a flare, as being close to the location of loop interaction and initial energy release, although the following kernel associated with the footpoint of the large loop did not show any HXR source. The flare studied here took place in a decaying active region that did not get a number after its appearance on the solar disk. However, a system of loops could exist, and new emerging magnetic flux could appear here. The small flare loop observed during our event had both microwave and HXR sources. We did not find any response in microwaves at the source associated with the large loop footpoint. This fact can be the evidence of a low density of accelerated electrons combined with a low value of the magnetic field strength. We can estimate the magnetic field strength at the observed MW flare source by using the MW photon spectral index, the visible size of the source, and the flux density from the HXR spectrum fitting, applying the formula from Dulk (1985). The height of the small flare loop estimated using the distance between the centers of HXR sources is about 9 – 14 Mm. This is in good agreement with the source size about 30 arcseconds obtained from UV and EUV images using for magnetic field strength estimations. If the photon spectral index was 3.3 and the electron flux was $3 \times 10^{27}\ \mathrm{erg\,s^{-1}}$ recorded during the time interval of 4 seconds, and the peak frequency of the MW spectrum was 5.2 GHz, then the magnetic field value is approximately 130 G. The magnetic field strength in the long loop's leg should be weaker, and thus it could not produce a detectable MW flux. In general, this flare topology looks like the event studied by Fleishman et al. (2016) but with a "hot-onset" of the event.

Figure 8 shows the evolution of plasma parameters during the flare compared to its temporal profiles. The estimation of thermal energy is very rough because we did not have the possibility to obtain the exact values of electron temperature and emission measure. The comparison between the temperature and the emission measure values derived from HXI and GOES show the usual difference in the parameters of two techniques. Typically, the GOES temperatures are lower than the values obtained from spectral analysis of HXR observations, and the emission measures are higher. However, the unusually rapid temperature rise during the impulsive phase may indicate that some nonthermal electrons with low energies have fallen into the thermal part of the spectrum fitting. Nevertheless, one can see a classical tendency when the thermal energy of the flare increases while the non-thermal flux decreases.

In general, the behavior of the electron spectral indices corresponds to the classic "soft–hard–soft" scenario for particle acceleration during a flare. We note that the electron spectral indices derived from the X-ray spectra by HXI and the MW spectra by SRH show a very close agreement within the error bars. The apparent hardening of the average MW electron

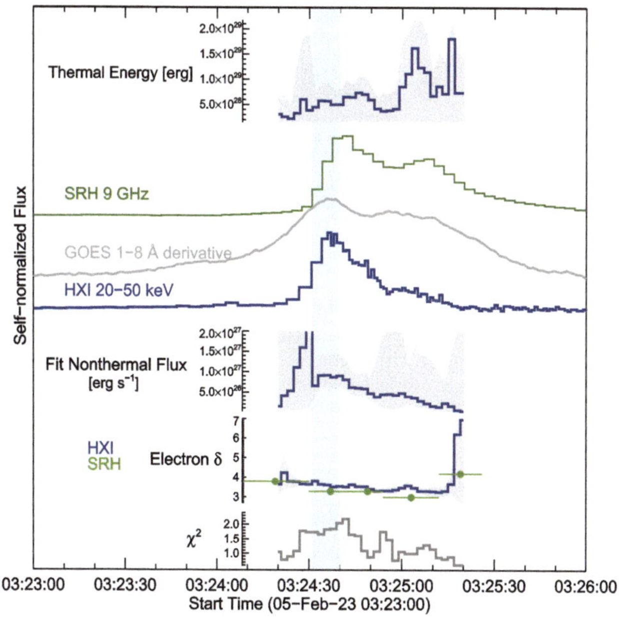

Figure 8 Evolution of plasma parameters during the flare and its time profiles. The dark blue curves are HXI related data, green curve and dots are SRH related data. Thermal energy and nonthermal flux evolutions are derived by HXI spectral fitting parameters. Electron δ panel shows electron indexes fitted by HXI and SRH spectra, respectively. The gray shadow region in each panel indicates the 1σ error range. The normalized SRH 9 GHz, HXI 20 – 50 keV, and GOES 1 – 8 Å derivative light curves are also plotted in between.

spectral index at the end of the impulsive phase turned out to be the jet emission contribution to flare microwave emission. This is an additional argument for the absence of magnetic trapping and an indicator of synchronization between the production of MW and HXR emissions, caused by the small size of the flare loop and the close location of the initial energy release site to the layers where the source emission is generated.

The studied event was a flare with a "hot-onset" (da Silva et al., 2023), according to the GOES data analysis. The MW flare source position indicates a low flare loop as it was found in a study by Hudson et al. (2021). Although because of a low signal the X-ray source was not reconstructed, unlike in previous studies (Battaglia, Fletcher, and Benz, 2009; Battaglia et al., 2023). However, the mechanism of emission generation was revealed through MW spectral analysis. We found a domination of gyrosynchrotron emission of thermal plasma. Thus, evidence shows that the onset of the acceleration process was preceded by plasma heating.

Appendix

The method described below is used for coaligning SRH images and correcting the center of solar disk. The correction technique used the fact that the projection effect is negligible at the center of the disk. Thus, the MW sunspot sources located in the center of the solar disk should not be shifted relative to each other at two nearby frequencies. It was assumed that the center of mass of the MW source is the center of the source. We chose the centers of the MW sunspot sources that are associated with several active regions located on the

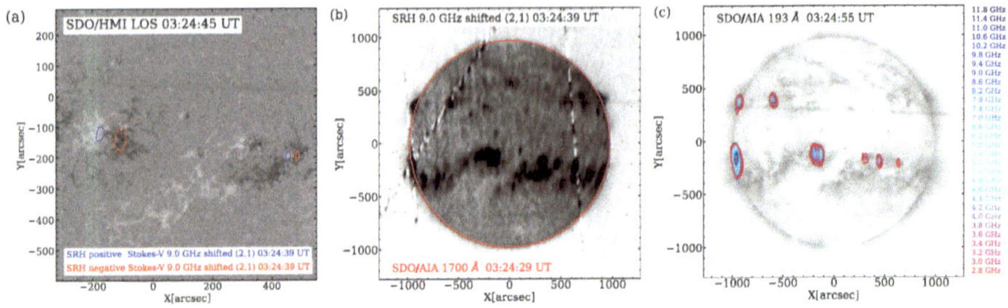

Figure 9 (a) SDO/HMI LOS magnetogram overlaid with contours of the MW sources obtained from SRH Stokes V images at 9.0 GHz. The contours show emission at 20% of maximum of negative and positive polarization. (b) SRH image at 9.0 GHz with overlaid contour obtained from SDO/AIA 1700 Å at 10% of maximum. (c) SDO/AIA 194 Å image overlaid with contours obtained from SRH images at 2.8 – 11.4 GHz. The contours present emission at 50% of maximum.

solar disk as reference points. We corrected the positions of images at one frequency using information about the sunspot center and the solar limb position from the SDO/HMI line-of-sight magnetogram and SRH Stokes V polarization images at 9.0 GHz. We corrected the position of the MW solar disk center, assuming that positive Stokes V polarization should be located over positive magnetic polarity, and negative Stokes V has to be located over negative magnetic polarity. Panel (a) of Figure 9 shows the corrected positions of the MW sunspot sources and their shift values. To test the correctness of the shift for MW solar disk center, we compared the positions of solar limb from SDO/AIA 1700 Å image and SRH 9.0 GHz image. In case of insufficient or wrong correction, the limb of the MW solar disk will be asymmetric relative to the solar disk as seen in the 1700 Å image. Results of this stage are presented in panel (b) of Figure 9.

The shifted image was accepted as the reference image (in this case, the image is at 9.0 GHz). Then, shifts for images were found at other frequencies. We shifted the image relative to the reference image until maximum agreement between the center of the sunspot was achieved. As a result of this correction, the MW emission contours at all frequencies must be positioned over each other for the sunspot sources located at the center of the solar disk, as shown in panel (c) of Figure 9.

Acknowledgments J.Sh. acknowledges the ASO-S Guest Investigator Program. L.K. and J.Sh. thank the Purple Mountie Observatory, Chinese Academy of Sciences for hospitality and facilities provided during the visits. The work of Y.S. and Z.L. is supported by the National Key R&D Program of China 2022YFF0503002, and the NSFC 12333010, 11820101002. L.K. and J.Sh. acknowledge the Ministry of Science and Higher Education of the Russian Federation. The authors thank the SRH team for help with data processing and the teams of the ASO-S, GOES, SDO, and NoRP for open access to the observational data. We would also like to thank the referee for useful comments and suggestions on improving our paper.

Author contributions J.S. and L.K. wrote the main manuscript text. Y.S. provided the software HXI_Clean for data processing. Z.L. and Y.S. prepared HXI data. J.S. and L.K. prepared SRH data. Z.L. prepared Figures 1, 2b, 8. J.S. prepared Figures 3a, 4, 5, 6b, 7, 9. L.K. prepared Figures 2a, 3b, 6a. All authors reviewed the manuscript.

Funding This work was funded by the Ministry of Science and Higher Education of the Russian Federation. J.Sh. was also supported by ASOS-S Guest Investigator Program. Y.S. and Z.L. were supported by the National Key R&D Program of China 2022YFF0503002, and the NSFC 12333010, 11820101002.

Data availability http://aso-s.pmo.ac.cn/en_index.jsp, https://badary.iszf.irk.ru/, https://www.ngdc.noaa.gov/stp/satellite/goes-r.html, https://solar.nro.nao.ac.jp/norp/, http://jsoc.stanford.edu/ajax/lookdata.html.

Declarations

Competing interests The authors declare no competing interests.

References

Altyntsev, A., Lesovoi, S., Globa, M., Gubin, A., Kochanov, A., Grechnev, V., Ivanov, E., Kobets, V., Meshalkina, N., Muratov, A., Prosovetsky, D., Myshyakov, I., Uralov, A., Fedotova, A.: 2020, Multiwave Siberian Radioheliograph. *Sol.-Terr. Phys.* **6**, 30. DOI. ADS.

Bastian, T.S., Benz, A.O., Gary, D.E.: 1998, Radio emission from solar flares. *Annu. Rev. Astron. Astrophys.* **36**, 131. DOI. ADS.

Battaglia, M., Fletcher, L., Benz, A.O.: 2009, Observations of conduction driven evaporation in the early rise phase of solar flares. *Astron. Astrophys.* **498**, 891. DOI. ADS.

Battaglia, A.F., Hudson, H., Warmuth, A., Collier, H., Jeffrey, N.L.S., Caspi, A., Dickson, E.C.M., Saqri, J., Purkhart, S., Veronig, A.M., Harra, L., Krucker, S.: 2023, The existence of hot X-ray onsets in solar flares. *Astron. Astrophys.* **679**, A139. DOI. ADS.

Benz, A.O.: 2017, Flare observations. *Living Rev. Solar Phys.* **14**, 2. DOI. ADS.

Carmichael, H.: 1964, A process for flares. *NASA Spec. Publ.* **50**, 451. ADS.

Cornell, M.E., Hurford, G.J., Kiplinger, A.L., Dennis, B.R.: 1984, The relative timing of microwaves and hard X-rays in solar flares. *Astrophys. J.* **279**, 875. DOI. ADS.

da Silva, D.F., Hui, L., Simões, P.J.A., Valio, A., Costa, J.E.R., Hudson, H.S., Fletcher, L., Hayes, L.A., Hannah, I.G.: 2023, Statistical analysis of the onset temperature of solar flares in 2010 – 2011. *Mon. Not. Roy. Astron. Soc.* **525**, 4143. DOI. ADS.

Dulk, G.A.: 1985, Radio emission from the sun and stars. *Annu. Rev. Astron. Astrophys.* **23**, 169. DOI. ADS.

Fleishman, G.D., Pal'shin, V.D., Meshalkina, N., Lysenko, A.L., Kashapova, L.K., Altyntsev, A.T.: 2016, A cold flare with delayed heating. *Astrophys. J.* **822**, 71. DOI. ADS.

Fletcher, L., Dennis, B.R., Hudson, H.S., Krucker, S., Phillips, K., Veronig, A., Battaglia, M., Bone, L., Caspi, A., Chen, Q., Gallagher, P., Grigis, P.T., Ji, H., Liu, W., Milligan, R.O., Temmer, M.: 2011, An observational overview of solar flares. *Space Sci. Rev.* **159**, 19. DOI. ADS.

Gan, W., Zhu, C., Deng, Y., Zhang, Z., Chen, B., Huang, Y., Deng, L., Wu, H., Zhang, H., Li, H., Su, Y., Su, J., Feng, L., Wu, J., Cui, J., Wang, C., Chang, J., Yin, Z., Xiong, W., Chen, B., Yang, J., Li, F., Lin, J., Hou, J., Bai, X., Chen, D., Zhang, Y., Hu, Y., Liang, Y., Wang, J., Song, K., Guo, Q., He, L., Zhang, G., Wang, P., Bao, H., Cao, C., Bai, Y., Chen, B., He, T., Li, X., Zhang, Y., Liao, X., Jiang, H., Li, Y., Su, Y., Lei, S., Chen, W., Li, Y., Zhao, J., Li, J., Ge, Y., Zou, Z., Hu, T., Su, M., Ji, H., Gu, M., Zheng, Y., Xu, D., Wang, X.: 2023, The Advanced Space-Based Solar Observatory (ASO-S). *Solar Phys.* **298**, 68. DOI. ADS.

Hanaoka, Y.: 1997, Double-loop configuration of solar flares. *Solar Phys.* **173**, 319. DOI. ADS.

Hirayama, T.: 1974, Theoretical model of flares and prominences. I: Evaporating flare model. *Solar Phys.* **34**, 323. DOI. ADS.

Hudson, H.S.: 2011, Global properties of solar flares. *Space Sci. Rev.* **158**, 5. DOI. ADS.

Hudson, H.S., Simões, P.J.A., Fletcher, L., Hayes, L.A., Hannah, I.G.: 2021, Hot X-ray onsets of solar flares. *Mon. Not. Roy. Astron. Soc.* **501**, 1273. DOI. ADS.

Kawate, T., Nishizuka, N., Oi, A., Ohyama, M., Nakajima, H.: 2012, Hard X-ray and microwave emissions from solar flares with hard spectral indices. *Astrophys. J.* **747**, 131. DOI. ADS.

Kontar, E.P., Brown, J.C., Emslie, A.G., Hajdas, W., Holman, G.D., Hurford, G.J., Kašparová, J., Mallik, P.C.V., Massone, A.M., McConnell, M.L., Piana, M., Prato, M., Schmahl, E.J., Suarez-Garcia, E.: 2011, Deducing electron properties from hard X-ray observations. *Space Sci. Rev.* **159**, 301. DOI. ADS.

Kopp, R.A., Pneuman, G.W.: 1976, Magnetic reconnection in the corona and the loop prominence phenomenon. *Solar Phys.* **50**, 85. DOI. ADS.

Lemen, J.R., Title, A.M., Akin, D.J., Boerner, P.F., Chou, C., Drake, J.F., Duncan, D.W., Edwards, C.G., Friedlaender, F.M., Heyman, G.F., Hurlburt, N.E., Katz, N.L., Kushner, G.D., Levay, M., Lindgren, R.W., Mathur, D.P., McFeaters, E.L., Mitchell, S., Rehse, R.A., Schrijver, C.J., Springer, L.A., Stern, R.A., Tarbell, T.D., Wuelser, J.-P., Wolfson, C.J., Yanari, C., Bookbinder, J.A., Cheimets, P.N., Caldwell, D., Deluca, E.E., Gates, R., Golub, L., Park, S., Podgorski, W.A., Bush, R.I., Scherrer, P.H., Gummin, M.A., Smith, P., Auker, G., Jerram, P., Pool, P., Soufli, R., Windt, D.L., Beardsley, S., Clapp, M., Lang, J., Waltham, N.: 2012, The Atmospheric Imaging Assembly (AIA) on the Solar Dynamics Observatory (SDO). *Solar Phys.* **275**, 17. DOI. ADS.

Lu, E.T., Petrosian, V.: 1990, The relative timing of microwaves and X-rays from solar flares. *Astrophys. J.* **354**, 735. DOI. ADS.

Neupert, W.M.: 1968, Comparison of solar X-ray line emission with microwave emission during flares. *Astrophys. J. Lett.* **153**, L59. DOI. ADS.
Nindos, A.: 2020, Incoherent solar radio emission. *Front. Astron. Space Sci.* **7**, 57. DOI. ADS.
Nita, G.M., Gary, D.E., Lee, J.: 2004, Statistical study of two years of solar flare radio spectra obtained with the Owens Valley Solar Array. *Astrophys. J.* **605**, 528. DOI. ADS.
Pesnell, W.D., Thompson, B.J., Chamberlin, P.C.: 2012, The Solar Dynamics Observatory (SDO). *Solar Phys.* **275**, 3. DOI. ADS.
Ruan, W., Xia, C., Keppens, R.: 2020, A fully self-consistent model for solar flares. *Astrophys. J.* **896**, 97. DOI. ADS.
Scherrer, P.H., Schou, J., Bush, R.I., Kosovichev, A.G., Bogart, R.S., Hoeksema, J.T., Liu, Y., Duvall, T.L., Zhao, J., Title, A.M., Schrijver, C.J., Tarbell, T.D., Tomczyk, S.: 2012, The Helioseismic and Magnetic Imager (HMI) Investigation for the Solar Dynamics Observatory (SDO). *Solar Phys.* **275**, 207. DOI. ADS.
Schwartz, R.A., Csillaghy, A., Tolbert, A.K., Hurford, G.J., Mc Tiernan, J., Zarro, D.: 2002, RHESSI data analysis software: Rationale and methods. *Solar Phys.* **210**, 165. DOI. ADS.
Silva, A.V.R., Wang, H., Gary, D.E.: 2000, Correlation of microwave and hard X-ray spectral parameters. *Astrophys. J.* **545**, 1116. DOI. ADS.
Sturrock, P.A.: 1966, Model of the high-energy phase of solar flares. *Nature* **211**, 695. DOI. ADS.
Su, Y., Liu, W., Li, Y.-P., Zhang, Z., Hurford, G.J., Chen, W., Huang, Y., Li, Z.-T., Jiang, X.-K., Wang, H.-X., Xia, F.-X.-Y., Chen, C.-X., Yu, W.-H., Yu, F., Wu, J., Gan, W.-Q.: 2019, Simulations and software development for the Hard X-ray Imager onboard ASO-S. *Res. Astron. Astrophys.* **19**, 163. DOI. ADS.
Thiemann, E.M.B., Eparvier, F.G., Woodraska, D., Chamberlin, P.C., Machol, J., Eden, T., Jones, A.R., Meisner, R., Mueller, S., Snow, M., Viereck, R., Woods, T.N.: 2019, The GOES-R EUVS model for EUV irradiance variability. *J. Space Weather Space Clim.* **9**, A43. DOI. ADS.
Torii, C., Tsukiji, Y., Kobayashi, S., Yoshimi, N., Tanaka, H., Enome, S.: 1979, Full-automatic radiopolarimeters for solar patrol at microwave frequencies. *Proc. Res. Inst. Atmos. Nagoya Univ.* **26**, 129. ADS.
Uralov, A.M., Grechnev, V.V., Lesovoi, S.V., Globa, M.V.: 2023, Plasma heating in an erupting prominence detected from microwave observations with the Siberian Radioheliograph. *Solar Phys.* **298**, 117. DOI. ADS.
White, S.M., Thomas, R.J., Schwartz, R.A.: 2005, Updated expressions for determining temperatures and emission measures from GOES soft X-ray measurements. *Solar Phys.* **227**, 231. DOI. ADS.
White, S.M., Benz, A.O., Christe, S., Fárník, F., Kundu, M.R., Mann, G., Ning, Z., Raulin, J.-P., Silva-Válio, A.V.R., Saint-Hilaire, P., Vilmer, N., Warmuth, A.: 2011, The relationship between solar radio and hard X-ray emission. *Space Sci. Rev.* **159**, 225. DOI. ADS.

Solar Physics (2024) 299:99
https://doi.org/10.1007/s11207-024-02333-8

RESEARCH

High-Resolution Observation of Blowout Jets Regulated by Sunspot Rotation

Tingyu Gou[1,2] · Rui Liu[1] · Yang Su[3] · Astrid M. Veronig[4] · Hanya Pan[1] · Runbin Luo[1] · Weiqun Gan[3,5]

Received: 13 March 2024 / Accepted: 30 May 2024 / Published online: 9 July 2024

Abstract
Coronal jets are believed to be the miniature version of large-scale solar eruptions. In particular, the eruption of a minifilament inside the base arch is suggested to be the trigger and even driver of blowout jets. Here, we propose an alternative triggering mechanism, based on high-resolution Hα observations of a blowout jet associated with a minifilament and an M1.2-class flare. The minifilament remains largely stationary during the blowout jet, except that it is straddled by flare loops connecting two flare ribbons, indicating that the magnetic arcade embedding the minifilament has been torn into two parts, with the upper part escaping with the blowout jet. In the wake of the flare, the southern end of the minifilament fans out like neighboring fibrils, indicative of mass and field exchanges between the minifilament and the fibrils. The blowout jet is preceded by a standard jet. With Hα fibrils moving toward the single-strand spire in a sweeping fashion, the standard jet transitions to the blowout jet. A similar pattern of standard-to-blowout jet transition occurs in an earlier C-class flare before the minifilament forms. The spiraling morphology and sweeping direction of these fibrils are suggestive of their footpoints being dragged by the leading sunspot that undergoes clockwise rotation for over two days. Soon after the sunspot rotation reaches a peak angular speed as fast as 10 deg h^{-1}, the dormant active region becomes flare productive, and the minifilament forms through the interaction of moving magnetic features from the rotating sunspot with satellite spots/pores. Hence, we suggest that the sunspot rotation plays a key role in building up free energy for flares and jets and in triggering blowout jets by inducing sweeping motions of fibrils.

Keywords Flares · Coronal jets · Filaments · Magnetic fields · X-ray emission

1. Introduction

Jets are ubiquitous in the solar atmosphere, characterized by a collimated ejection of plasma and an inverted-Y morphology with a base arch and a spire (Raouafi et al., 2016; Shen, 2021). Coronal jets are often dichotomized into standard jets and blowout jets (Moore et al., 2010; Sterling et al., 2015). The latter are different from the former in that the base arch does not remain static and closed but explodes to open up, hence is considered as a miniature version of the large-scale eruptions that are associated with coronal mass ejections (CMEs). A

Extended author information available on the last page of the article

minifilament is often found inside the base arch of coronal jets, and apparently the failed or successful eruption of the minifilament corresponds to the standard- or blowout-jet morphology, respectively (Sterling et al., 2015). Similar to their large-scale counterpart (Wang et al., 2000), minifilaments may serve as a marker of a sheared or twisted magnetic field (Mackay et al., 2010). Hence, it has been argued that the dichotomy of coronal jets might result from "the dichotomy of base arches that do not have and base arches that do have enough shear and twist to erupt open" (Moore et al., 2010). If this is true, there must exist a critical threshold of magnetic shear or magnetic twist, or magnetic helicity, to determine if a base arch is capable of eruption or not. The existence of such a threshold remains contentious for active regions (Toriumi and Wang, 2019), and its existence for coronal jets is also elusive.

Nevertheless, it is important to understand how the magnetic free energy of coronal jets is built up through the evolution of the photosphere. It has been noted that coronal jets often occur in regions in which satellite spots continually collide with the main sunspot, with the presence of coronal null points, or in regions where magnetic elements of opposite polarities cancel each other through converging and/or shearing motions (Shen, 2021).

On the other hand, nature seldom agrees with a simple dichotomy. For example, there is a continuum spectrum of filament eruptions (Gilbert, Alexander, and Liu, 2007), ranging from failed eruptions in which both the filament mass and the associated magnetic structure are confined in the lower corona, to partial eruptions in which either the filament mass (e.g., Tripathi et al., 2009) or the associated magnetic structure partly escape (e.g., Liu, Alexander, and Gilbert, 2007), and to complete eruptions, not to mention similarly diverse eruptive behaviors of double-decker filaments (see a review in Liu, 2020). This raises questions as to whether there also exists a continuum spectrum of minifilament eruptions, and how their eruptive behaviors affect the associated jets.

Below, we present high-resolution observations of a minifilament on 2022 November 11 using the 1-m New Vacuum Solar Telescope (NVST; Liu et al., 2014) in China, and investigate its evolution in association with the sunspot rotation in the NOAA active region (AR) 13141 and the role of sunspot rotation in a series of jets that occurred in the same AR.

2. Data and Instruments

In this study, we use both the line-of-sight and vector magnetograms observed by the Helioseismic and Magnetic Imager (HMI; Scherrer et al., 2012) onboard the Solar Dynamics Observatory (SDO; Pesnell, Thompson, and Chamberlin, 2012). The line-of-sight magnetograms are obtained every 45 s with a pixel scale of 0.6″. The vector magnetograms for space-weather HMI active-region patch (SHARP) are produced every 12 min and remapped with the cylindrical equal area (CEA) projection. With HMI vector magnetograms, we examine the injection rate of magnetic helicty [$\mathrm{Mx}^2\ \mathrm{s}^{-1}$] across the photosphere in the active region, and use the Differential Affine Velocity Estimator for Vector Magnetograms (DAVE4VM; Schuck, 2008) to obtain photospheric velocities. We further calculated the relative helicity flux across the photospheric boundary S (Berger, 1984):

$$\left.\frac{dH}{dt}\right|_S = 2\int_S (\mathbf{A}_p \cdot \mathbf{B}_t) V_{\perp n}\, dS - 2\int_S (\mathbf{A}_p \cdot \mathbf{V}_{\perp t}) B_n\, dS, \tag{1}$$

where $\mathbf{A}_p$ is the vector potential of the reference potential field $\mathbf{B}_p$; t and n refer to the tangential and normal directions, respectively. Obtained from DAVE4VM, $\mathbf{V}_\perp$ is the photospheric velocity that is perpendicular to the magnetic field (Liu and Schuck, 2012). The first term on the right of the equation describes the emergence of twisted flux tubes into the corona; the second term describes photospheric motions that shear and braid magnetic-field lines.

The flares in AR 13141 are observed by the Atmospheric Imaging Assembly (AIA; Lemen et al., 2012) onboard SDO and by NVST in Hα. AIA takes full-disk images of the Sun with a pixel resolution of 0.6$''$ and a temporal cadence of 12 s in seven EUV channels. We mainly use images in 131 Å and 304 Å passbands to study the flare activities and jets in AR 13141. The NVST Hα images are taken from 01:40 to 08:40 UT on 2022 November 11, and we use the linecenter (6562.81 Å) images with a temporal cadence of 72 s and a pixel scale of 0.165$''$.

The hard X-ray (HXR) emissions from the flares are observed by the Hard X-ray Imager (HXI; Zhang et al., 2019; Su et al., 2019) onboard the Advanced Space-based Solar Observatory (ASO-S; Gan et al., 2019) and by the Spectrometer Telescope for Imaging X-rays (STIX; Krucker et al., 2020) onboard Solar Orbiter (Müller et al., 2020), both of which provide HXR imaging of the Sun using a Fourier-transform indirect-imaging technique but observing from different locations with a separation angle of $\sim 22^\circ$ on 2022 November 11. HXI images the full solar disk in the energy range of $\sim$10 – 300 keV (Su et al., 2024) with a spatial resolution as high as 3.1$''$. The flares studied in this work were observed during the early stage of HXI and the HXR images are not available below 20 keV. HXR images by HXI are reconstructed using HXI_Clean method with the preliminary calibration parameters and the subcollimator groups G3-G10, which result in a spatial resolution of $\sim 6.5''$. When on its Sun-synchronous orbit ASO-S passes through the South Atlantic Anomaly region or the radiation belt, we use the HXR data from STIX, which provides HXR spectroscopy in the 4 – 150 keV energy range. On 2022 November 11, Solar Orbiter was separated from the Earth by about -22° in longitude, $+4.5^\circ$ in latitude, at a heliocentric distance of 0.62 AU from the Sun. The observation times of STIX light curves and images are corrected to account for the time difference of light traveling to Earth and Solar Orbiter. We reconstruct the STIX HXR images in the energy ranges of 6 – 10 keV, 15 – 25 keV, and 25 – 45 keV (if there is any enhanced emission) with a time integration of 1 min. The HXR images are primarily reconstructed using the CLEAN algorithm, while MEM_GE and EM algorithms (Massa et al., 2019, 2020) are also performed for a further consistency check especially for the HXR sources in the high-energy energy band. The STIX images are reprojected to the Earth's perspective and are overlaid on AIA and NVST images to study the flare X-ray emissions.

3. Analyses and Results

3.1. Overview of AR 13141

AR 13141 shows up at the solar northeast limb on 2022 November 4 and rotates to the backside of the Sun after 2022 November 17. The AR produces a series of C-class and a few M-class flares during its pass over the visible solar disk, which makes it the dominant flaring region on the Sun (Figure 1). Especially on 2022 November 11, it produced 17 flares of GOES C-class and above, including two moderate M-class flares (Table 1); on November 12, it produced 10 more flares, including an M1.1 one; while it was relatively dormant during earlier days (e.g., November 7 – 10), as shown by the GOES soft X-ray (SXR) fluxes

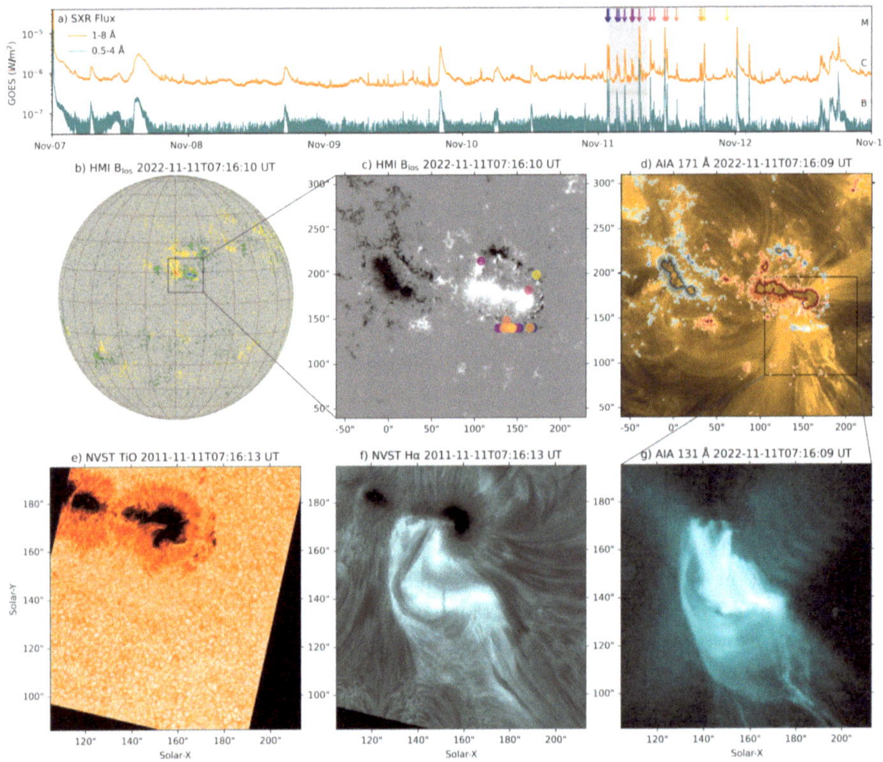

Figure 1 Overview of solar activities in AR 13141. a) GOES 1 – 8 Å (orange) and 0.5 – 4 Å (green) SXR fluxes. The colored arrows mark the peak time of flares on 2022 November 11 in Table 1. The light gray shade indicates the observation period of NVST. b) SDO/HMI full-disk line-of-sight magnetogram on 2022 November 11. The black rectangle indicates the FOV of panels (c and d). c) Zoom-in view of AR 13141. The colored dots mark the locations (after the correction of solar rotation) of flares on November 11 as listed in Table 1, whose peak times are marked by arrows with the same colors in panel (a). d) SDO/AIA 171 Å image overplotted with HMI B_{los} contours at ±300, ±600, and ±1000 G. The black rectangle indicates the FOV of panels (e–g), which show NVST TiO and Hα as well as AIA 131 Å images, respectively.

(Figure 1a). All of these flares occurred in the western part of the AR, mostly between the main spots with positive polarities and satellite spots with negative polarities (Figure 1c).

The AR consists of a leading sunspot with positive polarity and a trailing one with negative polarity as it rotates toward the solar disk center (Figure 2a). Both become elongated and transform into complex sunspot groups as time progresses. In particular, the leading spot eventually separates into two major spots, surrounded by satellite spots and pores with negative polarities (Figure 2b).

Overall, the AR's evolution is characterized by flux cancelation until mid-November 9 (Figure 3a) and injection of positive helicity [$\mathrm{Mx}^2\ \mathrm{s}^{-1}$] on November 8 and 9 (Figure 3b), which is dominated by the plasma flows, i.e., the shear term of helicity injection (2nd term in Eq. 1; Liu and Schuck, 2012).

3.2. Sunspot Rotation and Separation

The westernmost sunspot in the leading sunspot group rotates clockwise as it separates from the rest of the sunspot group. The rotation is visible not only in the maps of B_z but also in

Table 1 List of flares in AR 13141 on 2022 November 11.

No.	Start Time	Peak Time	GOES Class	Location
1	01:30	01:35	C6.8	N11W07
2	01:44	01:50	C6.5	N11W07
3	03:10	03:20	C2.5	N11W07
4	03:36	03:41	C1.3	N11W07
5	04:33	04:41	C3.3	N11W07
6	05:46	05:56	C1.9	N11W07
7	06:04	06:09	C1.3	N16W06
8	07:00	07:14	M1.2	N11W09
9	08:58	09:09	C5.9	N14W11
10	09:44	09:47	C1.7	N11W10
11	11:27	11:40	M1.2	N11W11
12	12:02	12:06	C3.6	N12W11
13	13:35	13:43	C1.0	N11W12
14	17:45	17:52	C1.5	N11W14
15	18:03	18:09	C1.4	N11W16
16	18:22	18:37	C4.7	N11W15
17	22:21	22:32	C1.1	N15W19

the maps of J_z, the vertical component of electric current density (see Figures 2b and c and the accompanying animation). The rotation is observed to start at around 9 UT on November 9, when the sunspot is notched in the northwestern quadrant by weak magnetic fields (gray colors) in the B_z map. This notch serves as a reference to mark the clockwise rotation of the sunspot (Figure 2b). A pair of current ribbons with opposite signs but large magnitude $\geqslant 60$ mA/m^2 is found to make a similar V shape as the notch, and together they rotate with the sunspot (Figure 2c). As shown by the DAVE method, in the sunspot region the typical rotating flow speed is about 0.3 km s^{-1} (Figure 4; averaged over 2 h), which, however, does not represent the overall speed of the sunspot rotation.

The sunspot is originally dominated by injection of positive magnetic-helicity flux [Mx^2 cm^{-2} s^{-1}] before November 10 (Figures 4b1 and 5d). Starting from around 12:00 UT on November 10, negative helicities are injected into the sunspot (Figure 4b2). Eventually, it is the negative-helicity injection that begins to dominate in the sunspot by about 13 UT on November 11 (Figure 4d1). We identify the centroid locations of the positive and negative helicity flux injections inside the sunspot (diamond and square symbols in Figure 4e), weighted by $H_{\rm pos} \geqslant 3 \times 10^9$ Mx^2 cm^2 s^{-1}, $H_{\rm neg} \leqslant -2.5 \times 10^9$ Mx^2 cm^2 s^{-1}, respectively. Associated with the sunspot motion, the centroid of positive-helicity fluxes moves from north to south (diamond symbols) and that of negative-helicity fluxes from east to west (square symbols) as demonstrated in Figure 4e. This is consistent with our impression that the patches of enhanced helicity fluxes of both signs seem to rotate about each other, along with the rotation of the sunspot notch, especially on late November 10 and early November 11 (Figures 4a and b).

To quantify the sunspot rotation, we measure the rotation angles (with respect to the solar north) of different features. First, we measure the rotation angle of the sunspot notch in B_z maps, by linearly fitting four manually selected points along the southern edge of the notch as it appears (only the central straight part is used as it curves outward). We take B_z maps with 1-h cadence for the manual measurement and repeat the measurement

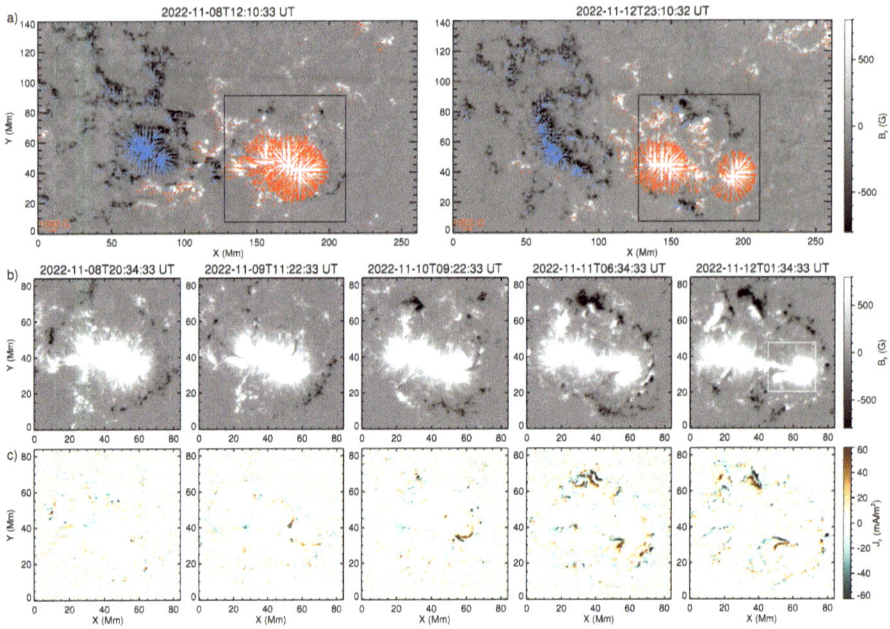

Figure 2 Evolution of magnetic field and current density. a) Vector magnetograms of AR 13141 on November 8 and 12. The FOV in panel (a) is selected for the DAVE4VM calculation. The black rectangles indicate the FOV of the bottom panels (b and c). Orange and blue arrows show the tangential magnetic-field components in the positive and negative polarities, respectively. b and c) Vertical component of magnetic fields and current densities at different times. The white rectangle in the middle right panel indicates the FOV plotted in Figure 4. An animation of this figure is available.

three times to obtain the average value and measurement errors. The results are plotted as purple dots with error bars in Figure 5a. Secondly, we measure the rotation of the current ribbons in J_z maps (Figure 2c). We use the contour of $J_z = 40\ \mathrm{mA/m^2}$ to obtain the pattern of the positive current ribbon. Although the current ribbon is overall curved, its inner 2/3 portion close to the center of the sunspot is almost straight, and this section is used for linear fitting to obtain the angle. The results are plotted with green dots in Figure 5a. Since the positive current ribbon is observed to lag behind the rotating notch in B_z (Figure 2 and its animation), the measured angles are generally smaller than those measured from B_z maps. In addition, we measure the angle subtended by the centriods of magnetic flux and positive-helicity flux (plus and diamond symbols, respectively, in Figure 4e), and the angle subtended by the centroids of positive- and negative-helicity fluxes (diamond and square symbols, respectively, in Figure 4e), and the results are shown as light blue and magenta dots in Figure 5a, respectively.

To derive the rotation speed of the sunspot, we first smooth the measured data points with the Savitzky–Golay method (solid lines in Figure 5a). The angular rotation speeds derived from the four different datasets are generally consistent with each other (Figure 5b), showing a long-duration rising phase lasting about 36 h, which reaches a peak value of about $10^\circ\ \mathrm{h}^{-1}$ at the end of November 10, and a following decay phase of about 12 h until the speed decreases to zero at about 12 UT on November 11. The sunspot rotates over 240° within two days. Both the total rotational angle and the transient angular speed of the sunspot are larger than many previously reported results (e.g., Brown et al., 2003; Zhang, Li, and Song,

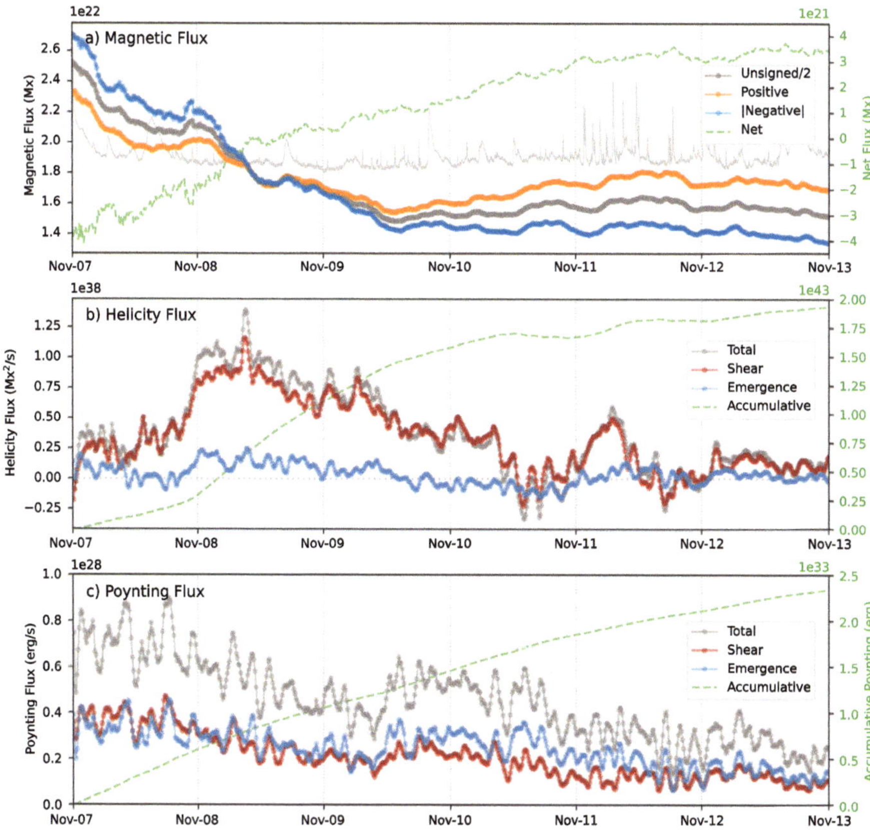

Figure 3 Long-term evolution of AR 13141. a) Magnetic flux within the FOV in Figure 2a, scaled by the left y-axis and shown in orange (positive flux), blue (negative flux in magnitude), and dark gray (unsigned flux divided by 2); net flux is shown as dashed green curve and scaled by the right y-axis. The GOES 1 – 8 Å SXR flux (light gray) are plotted in an arbitrary unit. b) Magnetic-helicity injection in AR 13141. The total (gray), shear term (red), and emergence term (blue) of helicity injection rates are scaled by the left y-axis. The accumulative helicity (dashed green) is scaled by the right y-axis. c) Same as those in panel (b) but for the Poynting fluxes.

2007; Yan and Qu, 2007; Török et al., 2013; Bi et al., 2016). By identifying the sunspot centroids weighted by the magnetic flux ($B_z \geqslant 500$ G; colored plus symbols in Figure 4e), one can see that the sunspot starts to move westward at about 12 UT on November 10, which roughly coincides with a steep rise in the angular rotation speed.

We obtain the total vertical electric currents and the rate of helicity injection (shear term) [$\mathrm{Mx}^2\ \mathrm{s}^{-1}$] inside the rotating sunspot (within the $B_z = 500$ G contour in Figure 4) and study the temporal evolution (Figures 5c and d). Both the positive and negative currents increase with time as the sunspot rotates rapidly (Figure 5c). On November 11 after the sunspot's angular speed peaks, the net current flowing through the sunspot reaches a higher level than on the preceding days (the green curve in Figure 5c). For the helicity injection by shearing motions, the sunspot is dominated by positive-helicity injection in the beginning (Figure 5d). However, from about 12 UT on November 10, negative helicities start to be injected into the sunspot (see also the helicity-density maps in Figure 4b), which coincides with the steep increase in the angular rotation speed (Figure 5b) and the westward motion (Figure 4e) of

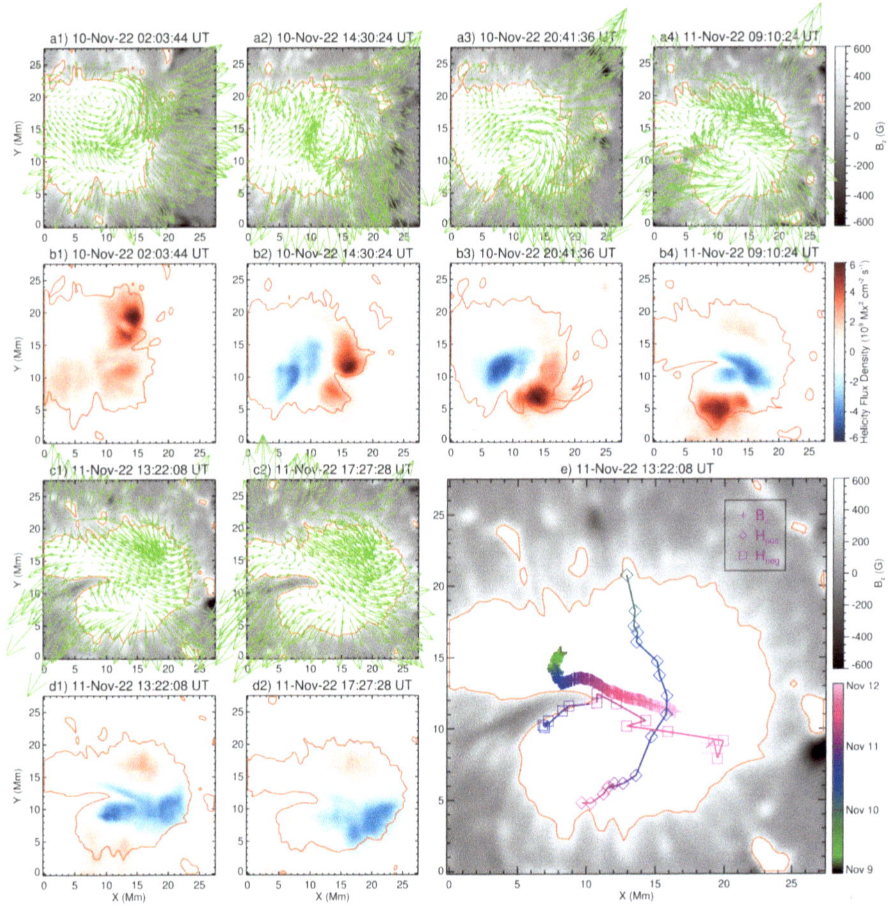

Figure 4 Evolution of DAVE flow fields and injection of magnetic-helicity flux densities associated with the sunspot. The results are averaged over 2 h around the time of each map. The orange contours outline locations where $B_z = 500$ G, which serves as an indication of the sunspot. The green arrows in panels (a and c) indicate tangential DAVE velocities, which are overlaid on the closest HMI B_z maps. The red and blue colors in the helicity flux density maps in panels (b and d) represent the positive and negative values of injected helicities (only shear term plotted in the figure). Panel (e) shows the temporal evolution of identified centroid locations during Nov 9 – 12, overlaid on a B_z map on Nov 11, coded by colors in two color bars in the right, respectively. The plus symbols indicate the centroid locations of the sunspot weighted by normal magnetic fluxes of $B_z \geqslant 500$ G. The diamond and square symbols indicate the centriod locations weighted by injections of positive and negative helicity fluxes, respectively.

the sunspot. The sunspot is dominated by negative-helicity injection from about 12 UT on November 11 onward (see the green dotted curve in Figure 5d), when the sunspot rotation is almost halted (Figure 5b).

3.3. Flare Activities

A series of flares occur in AR 13141 (e.g., Figure 6 and its animation), after the sunspot rotation reaches its peak angular speed and starts to move westward. Most flares originate from the peripheral regions of the rotating sunspot (Figure 1c; except for Flare #7 in Table 1

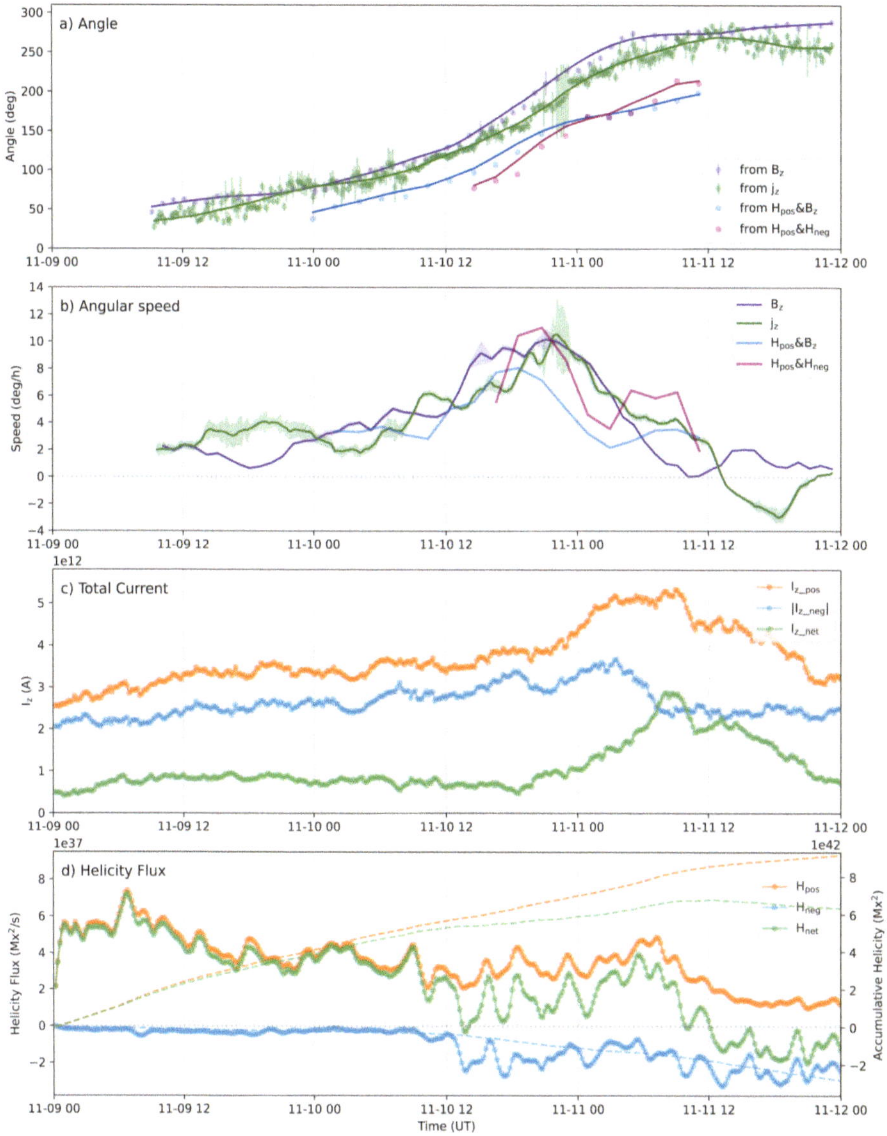

Figure 5 Evolution of the sunspot. a) Rotation angles with respect to the solar north, measured from the patterns of B_z, J_z, and from the centroids of B_z and positive helicity flux, centroids of positive- and negative-helicity fluxes, respectively (see the text for details). The colored solid curves are the corresponding smoothed results after applying the Savitzky–Golay smoothing filter. b) Angular speeds of the sunspot rotation derived from the smoothed angles in panel (a). c) Total vertical currents inside the sunspot region. d) Injection of helicity fluxes (shear term only) into the sunspot. The temporal injection rates (dotted lines) are scaled by the left y-axis, and the accumulative helicity flux injections (dashed curves) are scaled by the right y-axis.

in the northeast) and show similar morphologies, often associated with a jet (Figure 6). Bright flare loops form at the base of coronal jets and connect to two flare ribbons, where HXR footpoint sources with energies >25 keV are observed (Figures 6 and 7). The northern HXR footpoint is located near the sunspot penumbra associated with strong magnetic-field

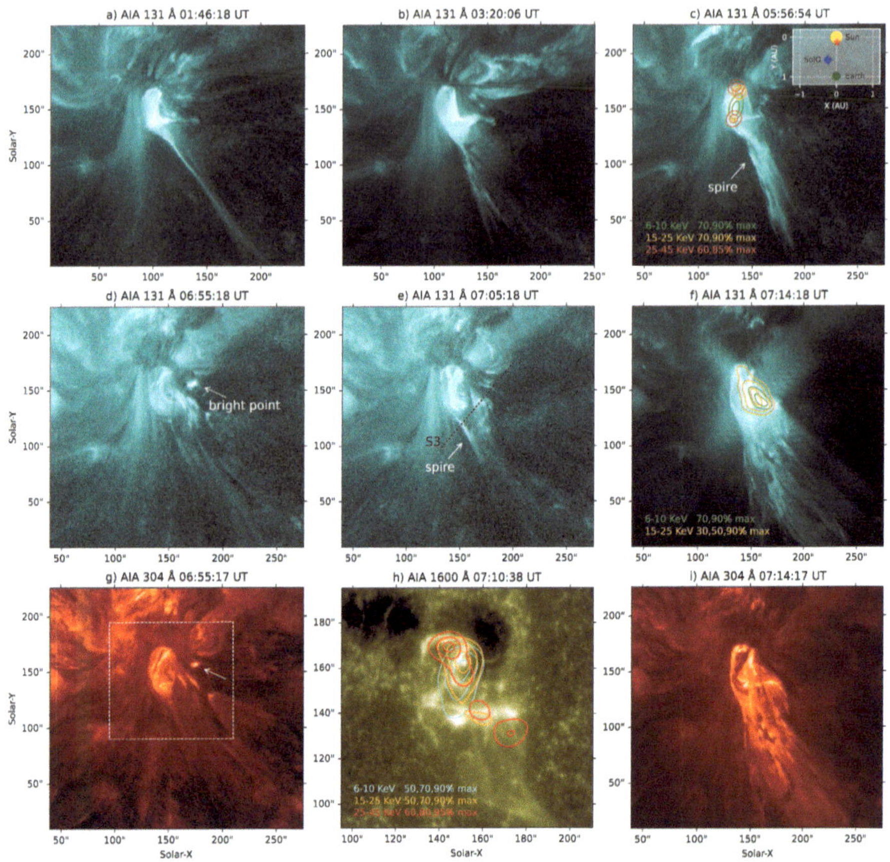

Figure 6 SDO/AIA observations of flares on 2022 November 11. Panels (a–c) show the C-class flares #2, #3, and #6 listed in Table 1, respectively. Panels (d–i) show the M1.2 flare (#8). The inserted plot in panel (c) shows the location of Solar Orbiter (SO) on November 11 relative to the Sun and Earth. The red star symbol indicates the location of flares on the Sun. The STIX HXR sources in different energy bands obtained using the CLEAN algorithm are overlaid on AIA images in panels (c, f, h). The black dotted line in panel (e) indicates the virtual slit S3 to generate the stack plot in Figure 8e. The white dotted rectangle in the panel (g) indicates the FOV of panel (h) and the NVST Hα images in Figure 7. An animation of this figure is available.

strengths, but does not show a significant asymmetry in HXR fluxes compared with the other footpoint (Figure 7), which is the usual case in many other flares (e.g., Yang et al., 2012). The HXR emissions from the footpoints are similar to the cases in Saqri et al. (2024), which includes a microflare in the same AR on November 10 showing unusually hard energy spectra and efficient particle acceleration. In between the two footpoints there is a third source at lower X-ray energies, which is interpreted as the coronal HXR source located at the top of the flare loops (e.g., Figures 6c, f, and h). In particular, flares #6 and #8 are each associated with a blowout jet. The M1.2-class flare (#8) well conforms with the classic, two-ribbon-flare morphology, with both ribbons visible in AIA 1600 Å (Figure 6h). The ribbon associated with positive polarities is partially located in the sunspot; the other ribbon associated with negative polarities is located to the south of the rotating sunspot. Flare loops are observed to connect the two ribbons (Figure 6f).

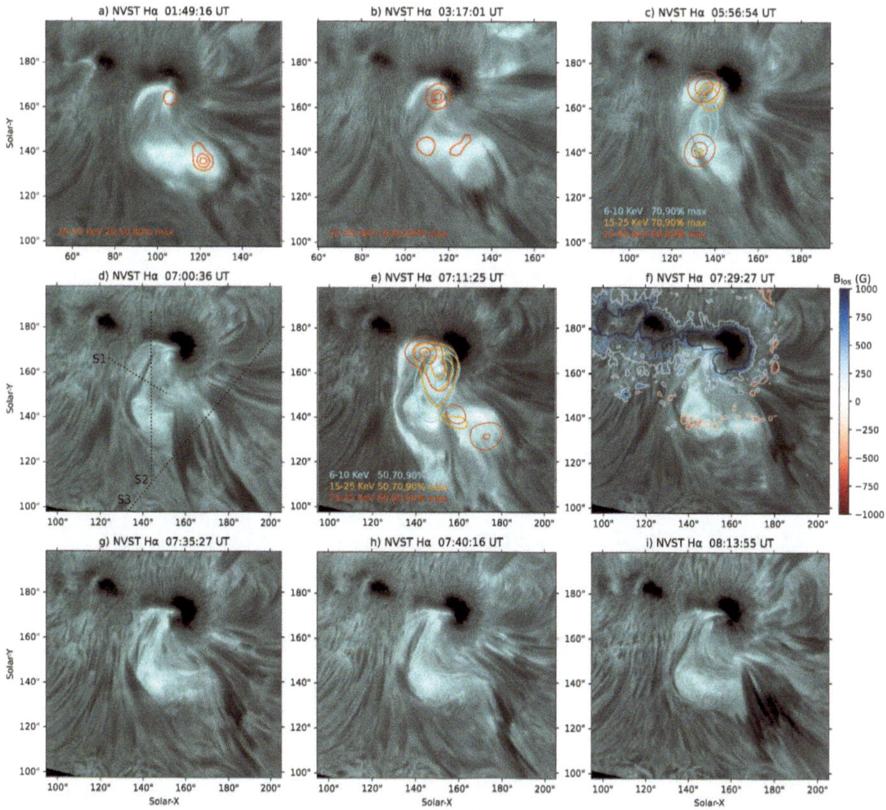

Figure 7 Snapshots of NVST Hα observation on 2022 November 11. The HXR sources reconstructed using the CLEAN algorithm from the HXI observation are overlaid in panels (a and b). The STIX HXR sources are overlaid in panels (c and e), which are the same as those in Figures 6(c and h). The dotted lines in panel (d) indicate the virtual slits used to generate the stack plots in Figure 8, where labels S1–S3 mark the origins of slits. The colored contours in panel (f) show the simultaneous HMI magnetogram at $B_{\text{los}} = \pm 300$, ± 600, and ± 1000 G. An animation of this figure is available.

High-resolution NVST Hα observations show a minifilament connecting the sunspot and the southern negative-polarity patches (Figures 1f and 7 and the accompanying animation). The northern footpoint of the minifilament seems to be rooted in around the tip of the notch of the rotating sunspot (Figures 1e, f and 7f), where the injection of negative magnetic helicities becomes dominant since about 12 UT on November 10 (Figures 4 and 5d), the day before the flares. We put a virtual slit (S1 in Figure 7d) to cut across the northern portion of the minifilament to investigate its dynamic evolution (see the stack plot in Figure 8b). In Hα, the minifilament becomes prominent after about 6 UT on November 11 (Figure 8b), with threads apparently twisting about each other in a left-handed sense, in the wake of a C-class flare (#6; see the accompanying animation of Figure 7). During the subsequent M1.2-class flare (#8), the filament becomes thicker and darker (Figure 8b), and the left-handed twist becomes more evident (e.g., Figure 7g and its animation).

Flare #6 at 05:56 UT is associated with a small blowout jet featured by a curtain-like spire (Figure 6c and its animation), but in the wake of the blowout jet a single-strand spire remains and persists until the early phase of the M1.2-class flare (Flare #8; Figure 6d and e).

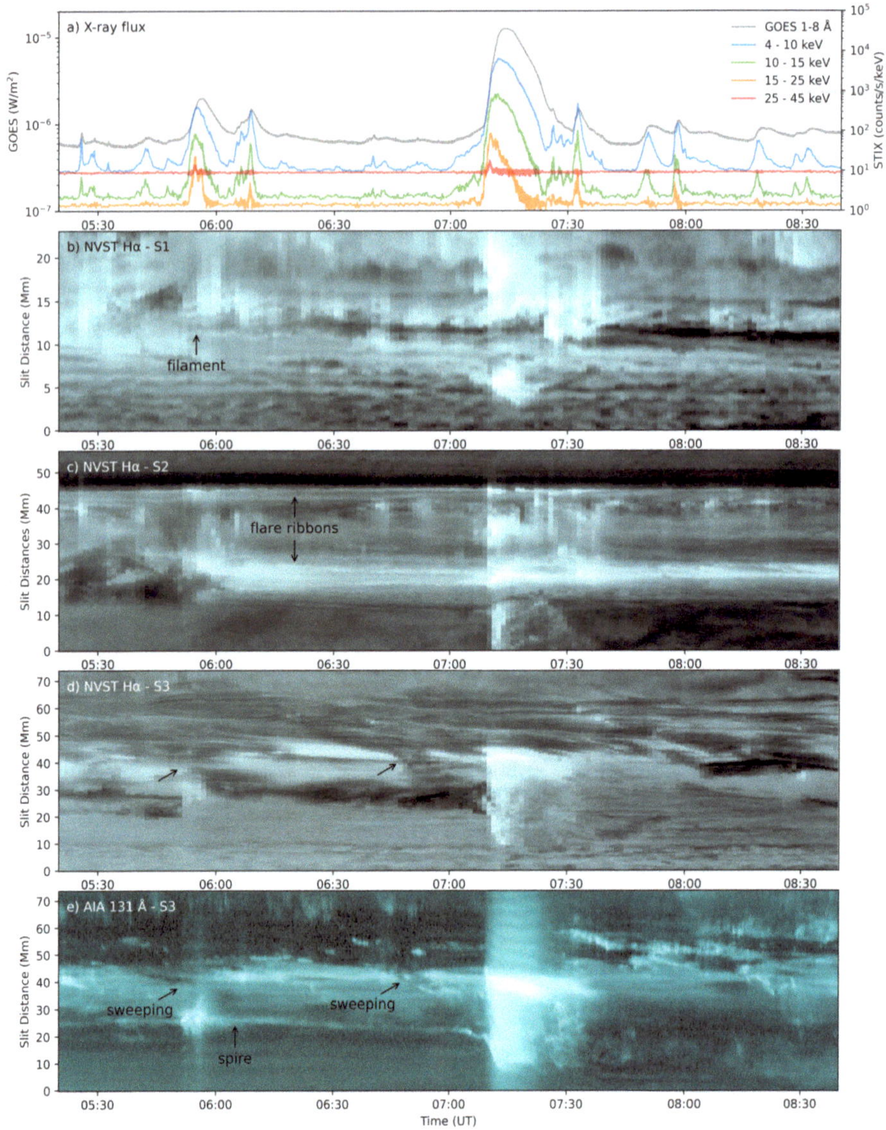

Figure 8 Temporal evolution of the minifilament and flares. a) GOES 1 – 8 Å SXR flux (gray; scaled by the left y-axis) and STIX HXR light curves (colored lines, scaled by the right y-axis). b–e) Stack plots of slits S1–S3 in Figures 7(d) and 6(e). In panels (d) and (e), arrows mark the sweeping motions of fibrils preceding two blowout jets at 05:50 and 07:10 UT, respectively.

Hence, the M-class flare initiates as a standard jet, which is featured by the single-strand spire located side by side with a thick bundle of fibrils (Figures 6g and 7d) and a miniature flare-arcade bright point located to the west of the base arch (e.g., at 06:55 UT, Figure 6(d and g)). Inside the base arch, a minifilament is clearly seen in 304 Å. By 07:05 UT, while the bright point has dimmed, multiple bright loops across the minifilament appear in 131 Å (Figure 6e and its animation), indicating that the magnetic fields overlying the minifilament have started to participate in magnetic reconnection. Meanwhile, the dark, spiraling Hα fibrils located to the west of the filament move clockwise (especially after 06:30 UT) to approach the filament in a sweeping manner (see the animation of Figure 7). We adopt two different slits (S2 and S3 in Figure 7d) to show this dynamic evolution in Hα. Slit S2 is made to cut across the two flare ribbons. Slit S3 is made to cut across both the dark fibrils and the jet spire. In Figure 8d, one can see that starting from about 06:45 UT, the thick bundle of Hα fibrils (slit position ~ 30 Mm) immediately neighboring the single-strand jet spire become darker and thicker, while farther away fibrils (slit position $\sim 35-50$ Mm) move rapidly toward the spire, as indicated by the steepening of dark strips in the stack plot, which is also associated with the further darkening of the filament (Figure 8b). At about 07:10 UT, around the peak of HXRs at $15-25$ and $25-45$ keV, the thick bundle of fibrils suddenly disappear, and in AIA 304 Å it is replaced by a bright curtain-like spire, along which materials spray outward (see the animation accompanying Figure 7). These features are consistent with the occurrence of a blowout jet. The earlier, smaller blowout jet associated with Flare #6 is similarly preceded by an episode of clockwise fibril sweeping at about 05:50 UT. Both episodes of fibril motions are also registered in the stack plot made from AIA 131 Å images (Figure 8e).

With the onset of the blowout jet at 07:10 UT, two flare ribbons appear in Hα on both sides of the minifilament (Figure 8c), with flare loops connecting the two ribbons in AIA 131 Å. These features indicate that the magnetic structure embedding the minifilament also participates in the blowout jet. However, the minifilament does not erupt as in typical blowout jets. Instead, it remains largely stationary, except that its southern end rooted in the negative-polarity region becomes less well defined than before (Figures 7e–i and the animation). At about 07:35 UT, the southern portion of the filament seems to fan out like neighboring fibrils (Figure 7g), indicating that interchange reconnections might be ongoing between the filament field and the fibril field.

3.4. Magnetic Configuration of the Flaring Region

To investigate the magnetic structure associated with the minifilament and its evolution during the M-class flare, we studied the magnetic-field configuration of the flaring region by extrapolating coronal magnetic fields, using HMI SHARP data as the boundary condition. These vector magnetograms are "preprocessed" to best match the force-free condition, before being fed into the "weighted optimization" nonlinear force-free field (NLFFF) code (Wiegelmann et al., 2006). Here, NLFFF is constructed over a uniform grid of $640 \times 360 \times 360$ pixels (pixel size 0.36 Mm), covering the major magnetic elements of AR13141 (cf. Figure 1c), and magnetic connectivities are investigated by calculating the map of squashing factor Q (Titov, Hornig, and Démoulin, 2002). We distinguished open from closed field lines. The former have one end at the upper or side boundaries of the box region that is selected for the NLFFF extrapolation, i.e., they are locally open; the latter have both ends at the bottom of the box region. We also obtained the maps of twist number $T_w = \int \nabla \times \mathbf{B} \cdot \mathbf{B}/4\pi B^2\, ds$, where the line integration is performed along each individual field lines (Liu et al., 2016).

Using the T_w maps, we identify a magnetic flux rope (MFR) colocated with the minifilament, by tracing magnetic-field lines from T_w-enhanced regions (dark green curves in Figure 9). This MFR is formed as early as ~02 UT on November 11, coinciding with the formation of a filament channel, where the chromospheric fibrils are roughly aligned along the PIL (Figure 7a and b). Soon afterwards, the minifilament is formed in the channel as evidenced in high-resolution NVST Hα images (Figure 7 and its animation). At about 07 UT a

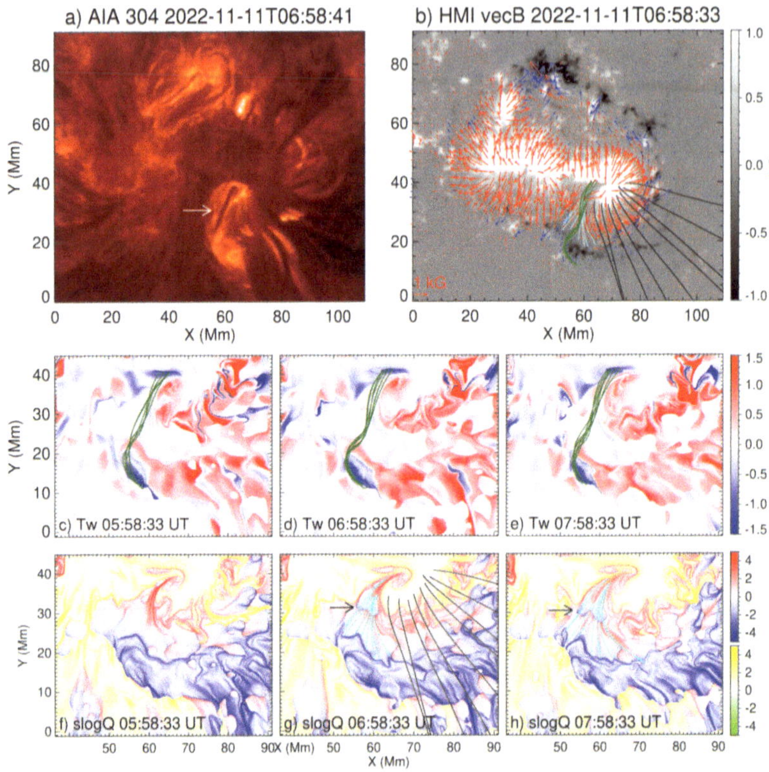

Figure 9 Magnetic configuration related to the M1.2-class flare on 2022 November 11. a) AIA 304 Å image taken just before the flare onset, which is remapped with the CEA projection to match the HMI SHARP Vector magnetogram (b) taken at the same time. The arrow in (a) marks the filament of interest. In (b) the background image shows B_z as scaled by ± 1000 G; overplotted vectors are the transverse field components originating from positive (red) and negative (blue) polarities. (c–e) maps of twist number T_w, where representative twisted-field lines traced from T_w-enhanced regions are shown in dark green. (f–h) maps of signed squashing factor, slog $Q = \text{sign}(B_Z) \times \log Q$, where representative field lines that graze the surface at the bald patch segment (marked by the arrows in (g and h)) are shown in cyan. The green-yellow and blue-red color tables indicate open- and closed-field regions, respectively. In open-field regions, the green (yellow) color indicates negative (positive) polarities, while in closed-field region the blue (red) color indicates negative (positive) polarities. In (g) a few locally open field lines that have similar orientations as the fibrils undergoing sweeping motions are shown in black. The field lines in (d) and (g) are reproduced in (b).

bald-patch (BP; Titov, Priest, and Demoulin, 1993) segment appears below the MFR in the Q-map (Figure 9g, marked by a black arrow). Representative field lines that are tangential to the surface at this BP segment are shown in cyan colors. However, both the MFR and the BP survives the M-class flare that peaks at 07:14 UT (Flare #8; Figures 9(e and h)), without exhibiting any obvious changes.

In addition, from the maps of signed squashing factor, slog $Q = \text{sign}(B_Z) \times \log Q$ (Figures 9(f–h)), one can see that the fibrils undergoing sweeping motions are likely rooted in the open-field regions surrounding the northern end of the minifilament, where magnetic-field lines have similar footpoint locations and projected orientations as the fibrils. However, one must keep in mind that the cold materials of fibrils may not show the full length but only segments of the field lines. On the other hand, these open-field lines are not associated with a null point or a bald-patch separatrix surface, as revealed in some jets originating from the edges of active regions (e.g., Schmieder et al., 2013; Chandra et al., 2017; Wyper, DeVore, and Antiochos, 2019). This could be partly attributed to the magnetic complexity of the active region under investigation as well as to the limit posed by the NLFFF modeling.

4. Conclusion and Discussion

The energy buildup for coronal jets in active regions is often associated with a satellite spot (e.g., Sakaue et al., 2017). In our case, the sunspot rotation appears to be the dominant driving force of the active-region evolution. Since a sunspot is a monolithic region of concentrated magnetic flux, any rotation would act to twist up a large amount of flux. With the twist being transported into the corona via Alfvén waves, the active-region field would hold more free energy to be released in the form of jets, flares, and CMEs (Brown et al., 2003; Tian, Alexander, and Nightingale, 2008). As expected, the active region becomes flare productive soon after the angular rotation speed of the sunspot peaks by the end of Nov 10 (Figure 5b), and most of the flares occur in the periphery of the rotating sunspot (Figure 1c). In particular, the minifilament channel associated with the blowout jet and M1.2-class flare (#8) is developed where moving magnetic features originating from the rotating sunspot continually advance into the satellite spots/pores with negative polarities and are subsequently canceled therein (Figure 10).

The evolution of the minifilament during the M1.2-class flare could be easily misidentified as a failed eruption. Only in high-resolution observations is the minifilament revealed to be interchanging mass and field with the neighboring fibrils, which is evidenced by the fanning-out of its southern end, exhibiting similar morphological features as the neighboring fibrils (Figure 7g). However, since it is largely stationary, the minifilament cannot be the driver of the blowout jet, as envisaged in some cartoon models (e.g., Sterling et al., 2015).

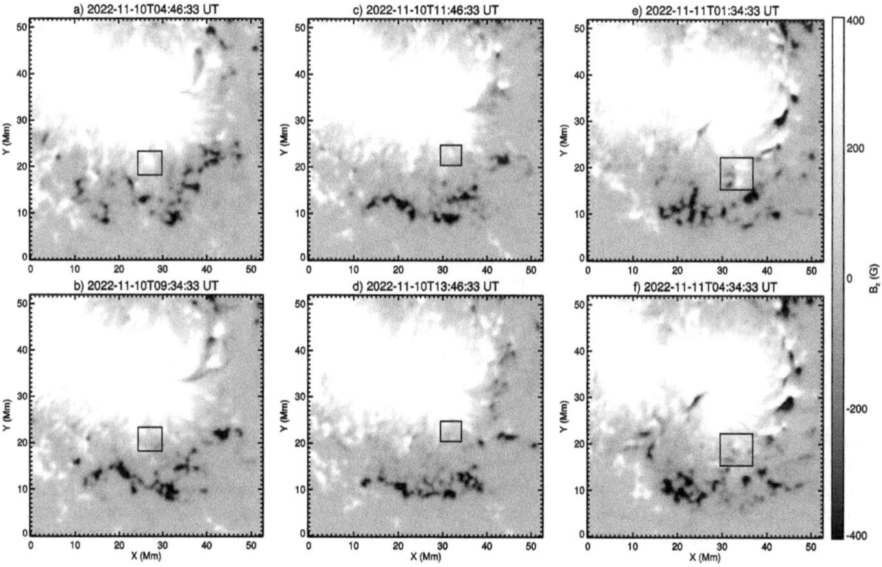

Figure 10 Evolution of HMI magnetograms around the flare region. Panels (a, b) (c, d), and (e, f) show three examples of the evolution of moving magnetic features originating from the rotating sunspot, respectively, as indicated by the black rectangles.

Nevertheless, the observation of the two flare ribbons sandwiching, and the flaring loops striding, the minifilament suggests that the magnetic arcade embedding the minifilament has been torn into two parts, most likely through internal reconnection (Gibson and Fan, 2006). The upper part erupts with the blowout jet, and the lower part still holds the minifilament in the low atmosphere, which is probably aided by the bald-patch field (§3.4). The internal reconnection might start early, as demonstrated by bright 131 Å loops straddling the minifilament during the standard-jet phase before the M1.2-class flare (Figure 6e). During the flare, the southern HXR footpoint source (25 – 45 keV) is apparently located below the fibrils involved in the blowout jet (Figure 7e), which could be attributed to the interchange reconnection between the sheared arcade overlying the minifilament and the fibrils. Thus, the minifilament partially erupts in two aspects: 1) the filament itself exchanges mass and field with neighboring fibrils, and 2) only part of the magnetic arcade overlying the filament escapes with the jet.

On the other hand, the transition from the standard to blowout jet clearly involves both the mass and fields of the fibrils located in the 'upstream' of the single-strand spire of the standard jet, with these fibrils moving toward the spire. The fibril motions may trigger the blowout jet as suggested by the close temporal and spatial relation between them (Figure 8), and by the absence of obvious changes in other relevant features prior to the blowout jet, including the base arch and the single-strand spire. In particular, the spiraling morphology and sweeping direction of these fibrils are consistent with the clockwise rotation of the sunspot. Despite the difficulty of following the evolution of each individual fibrils, we suggest that the fibril motions are an indication that the magnetic field associated with the fibrils changes orientation, which could be driven by the rotation of the sunspot. Such sweeping motions and changes of field orientations may make reconnections at the inverted Y-point of the single spire more favorable and therefore proceed faster, potentially triggering the eruption. In this regard, the sunspot rotation is not only crucial to the energy buildup but also to the trigger of the blowout jet.

Supplementary Information The online version contains supplementary material available at https://doi.org/10.1007/s11207-024-02333-8.

Author contributions T.G. and R.L. led the study, performed the analysis, and wrote the manuscript. Y.S. provided and analyzed the HXI data and contributed to the discussion. A.M.V. contributed to the interpretation and discussion. H.P. contributed to the STIX data analysis. R.B.L contributed to the NVST observation and data. W.G. is the PI of ASO-S mission and contributed to the discussion. All authors reviewed the manuscript.

Funding R.L. and Y.S. acknowledge the support from the National Key R&D Program of China 2022YFF0503002. R.L., H.P., and R.B.L. acknowledge the support by NSFC (11925302, 42188101, and 42274204) and by the Strategic Priority Program of the Chinese Academy of Sciences (XDB41000000). Y.S. and W.G. also acknowledge the support by NSFC (12333010, 11820101002, 11921003, and 12233012) and by the Strategic Priority Research Program of the Chinese Academy of Science (Grant No. XDB0560000). A.M.V acknowledges support by the Austrian Science Fund (FWF) 10.55776/I4555.

Data Availability The solar data used in the study are publicly available for download from the mission archives. The SDO data are available at http://jsoc.stanford.edu/. The NVST data are available at https://fso.ynao.ac.cn/DataService/query. The STIX data are available at https://datacenter.stix.i4ds.net/. The HXI data used in this study are during the mission commissioning phase and are available upon reasonable request, and the data after April 2023 are publicly available for download at http://aso-s.pmo.ac.cn/sodc/dataArchive.jsp.

Declarations

Competing interests The authors declare no competing interests.

References

Berger, M.A.: 1984, Rigorous new limits on magnetic helicity dissipation in the solar corona. *Geophys. Astrophys. Fluid Dyn.* **30**, 79. DOI. ADS.

Bi, Y., Jiang, Y., Yang, J., Hong, J., Li, H., Yang, B., Xu, Z.: 2016, Observation of a reversal of rotation in a sunspot during a solar flare. *Nat. Commun.* **7**, 13798. DOI. ADS.

Brown, D.S., Nightingale, R.W., Alexander, D., Schrijver, C.J., Metcalf, T.R., Shine, R.A., Title, A.M., Wolfson, C.J.: 2003, Observations of rotating sunspots from TRACE. *Solar Phys.* **216**, 79. DOI. ADS.

Chandra, R., Mandrini, C.H., Schmieder, B., Joshi, B., Cristiani, G.D., Cremades, H., Pariat, E., Nuevo, F.A., Srivastava, A.K., Uddin, W.: 2017, Blowout jets and impulsive eruptive flares in a bald-patch topology. *Astron. Astrophys.* **598**, A41. DOI. ADS.

Gan, W.-Q., Zhu, C., Deng, Y.-Y., Li, H., Su, Y., Zhang, H.-Y., Chen, B., Zhang, Z., Wu, J., Deng, L., Huang, Y., Yang, J.-F., Cui, J.-J., Chang, J., Wang, C., Wu, J., Yin, Z.-S., Chen, W., Fang, C., Yan, Y.-H., Lin, J., Xiong, W.-M., Chen, B., Bao, H.-C., Cao, C.-X., Bai, Y.-P., Wang, T., Chen, B.-L., Li, X.-Y., Zhang, Y., Feng, L., Su, J.-T., Li, Y., Chen, W., Li, Y.-P., Su, Y.-N., Wu, H.-Y., Gu, M., Huang, L., Tang, X.-J.: 2019, Advanced space-based solar observatory (ASO-S): an overview. *Res. Astron. Astrophys.* **19**, 156. DOI. ADS.

Gibson, S.E., Fan, Y.: 2006, The partial expulsion of a magnetic flux rope. *Astrophys. J. Lett.* **637**, L65. DOI. ADS.

Gilbert, H.R., Alexander, D., Liu, R.: 2007, Filament kinking and its implications for eruption and reformation. *Solar Phys.* **245**, 287. DOI. ADS.

Krucker, S., Hurford, G.J., Grimm, O., Kögl, S., Gröbelbauer, H.-P., Etesi, L., Casadei, D., Csillaghy, A., Benz, A.O., Arnold, N.G., Molendini, F., Orleanski, P., Schori, D., Xiao, H., Kuhar, M., Hochmuth, N., Felix, S., Schramka, F., Marcin, S., Kobler, S., Iseli, L., Dreier, M., Wiehl, H.J., Kleint, L., Battaglia, M., Lastufka, E., Sathiapal, H., Lapadula, K., Bednarzik, M., Birrer, G., Stutz, S., Wild, C., Marone, F., Skup, K.R., Cichocki, A., Ber, K., Rutkowski, K., Bujwan, W., Juchnikowski, G., Winkler, M., Darmetko, M., Michalska, M., Seweryn, K., Białek, A., Osica, P., Sylwester, J., Kowalinski, M., Ścisłowski, D., Siarkowski, M., Steślicki, M., Mrozek, T., Podgórski, P., Meuris, A., Limousin, O., Gevin, O., Le Mer, I., Brun, S., Strugarek, A., Vilmer, N., Musset, S., Maksimović, M., Fárník, F., Kozáček, Z., Kašparová, J., Mann, G., Önel, H., Warmuth, A., Rendtel, J., Anderson, J., Bauer, S., Dionies, F., Paschke, J., Plüschke, D., Woche, M., Schuller, F., Veronig, A.M., Dickson, E.C.M., Gallagher, P.T., Maloney, S.A., Bloomfield, D.S., Piana, M., Massone, A.M., Benvenuto, F., Massa, P., Schwartz, R.A., Dennis, B.R., van Beek, H.F., Rodríguez-Pacheco, J., Lin, R.P.: 2020, The spectrometer/telescope for imaging X-rays (STIX). *Astron. Astrophys.* **642**, A15. DOI. ADS.

Lemen, J.R., Title, A.M., Akin, D.J., Boerner, P.F., Chou, C., Drake, J.F., Duncan, D.W., Edwards, C.G., Friedlaender, F.M., Heyman, G.F., Hurlburt, N.E., Katz, N.L., Kushner, G.D., Levay, M., Lindgren, R.W., Mathur, D.P., McFeaters, E.L., Mitchell, S., Rehse, R.A., Schrijver, C.J., Springer, L.A., Stern, R.A., Tarbell, T.D., Wuelser, J.-P., Wolfson, C.J., Yanari, C., Bookbinder, J.A., Cheimets, P.N., Caldwell, D., Deluca, E.E., Gates, R., Golub, L., Park, S., Podgorski, W.A., Bush, R.I., Scherrer, P.H., Gummin, M.A., Smith, P., Auker, G., Jerram, P., Pool, P., Soufli, R., Windt, D.L., Beardsley, S., Clapp, M., Lang, J., Waltham, N.: 2012, The Atmospheric Imaging Assembly (AIA) on the Solar Dynamics Observatory (SDO). *Solar Phys.* **275**, 17. DOI. ADS.

Liu, R.: 2020, Magnetic flux ropes in the solar corona: structure and evolution toward eruption. *Res. Astron. Astrophys.* **20**, 165. DOI. ADS.

Liu, R., Alexander, D., Gilbert, H.R.: 2007, Kink-induced catastrophe in a coronal eruption. *Astrophys. J.* **661**, 1260. DOI. ADS.

Liu, Y., Schuck, P.W.: 2012, Magnetic energy and helicity in two emerging active regions in the sun. *Astrophys. J.* **761**, 105. DOI. ADS.

Liu, Z., Xu, J., Gu, B.-Z., Wang, S., You, J.-Q., Shen, L.-X., Lu, R.-W., Jin, Z.-Y., Chen, L.-F., Lou, K., Li, Z., Liu, G.-Q., Xu, Z., Rao, C.-H., Hu, Q.-Q., Li, R.-F., Fu, H.-W., Wang, F., Bao, M.-X., Wu, M.-C., Zhang, B.-R.: 2014, New vacuum solar telescope and observations with high resolution. *Res. Astron. Astrophys.* **14**, 705. DOI. ADS.

Liu, R., Kliem, B., Titov, V.S., Chen, J., Wang, Y., Wang, H., Liu, C., Xu, Y., Wiegelmann, T.: 2016, Structure, stability, and evolution of magnetic flux ropes from the perspective of magnetic twist. *Astrophys. J.* **818**, 148. DOI. ADS.

Mackay, D.H., Karpen, J.T., Ballester, J.L., Schmieder, B., Aulanier, G.: 2010, Physics of solar prominences: II—magnetic structure and dynamics. *Space Sci. Rev.* **151**, 333. DOI. ADS.

Massa, P., Piana, M., Massone, A.M., Benvenuto, F.: 2019, Count-based imaging model for the spectrometer/telescope for imaging X-rays (STIX) in solar orbiter. *Astron. Astrophys.* **624**, A130. DOI. ADS.

Massa, P., Schwartz, R., Tolbert, A.K., Massone, A.M., Dennis, B.R., Piana, M., Benvenuto, F.: 2020, MEM_GE: a new maximum entropy method for image reconstruction from solar X-ray visibilities. *Astrophys. J.* **894**, 46. DOI. ADS.

Moore, R.L., Cirtain, J.W., Sterling, A.C., Falconer, D.A.: 2010, Dichotomy of solar coronal jets: standard jets and blowout jets. *Astrophys. J.* **720**, 757. DOI. ADS.

Müller, D., St. Cyr, O.C., Zouganelis, I., Gilbert, H.R., Marsden, R., Nieves-Chinchilla, T., Antonucci, E., Auchère, F., Berghmans, D., Horbury, T.S., Howard, R.A., Krucker, S., Maksimovic, M., Owen, C.J., Rochus, P., Rodriguez-Pacheco, J., Romoli, M., Solanki, S.K., Bruno, R., Carlsson, M., Fludra, A., Harra, L., Hassler, D.M., Livi, S., Louarn, P., Peter, H., Schühle, U., Teriaca, L., del Toro Iniesta, J.C., Wimmer-Schweingruber, R.F., Marsch, E., Velli, M., De Groof, A., Walsh, A., Williams, D.: 2020, The solar orbiter mission. Science overview. *Astron. Astrophys.* **642**, A1. DOI. ADS.

Pesnell, W.D., Thompson, B.J., Chamberlin, P.C.: 2012, The solar dynamics observatory (SDO). *Solar Phys.* **275**, 3. DOI. ADS.

Raouafi, N.E., Patsourakos, S., Pariat, E., Young, P.R., Sterling, A.C., Savcheva, A., Shimojo, M., Moreno-Insertis, F., DeVore, C.R., Archontis, V., Török, T., Mason, H., Curdt, W., Meyer, K., Dalmasse, K., Matsui, Y.: 2016, Solar coronal jets: observations, theory, and modeling. *Space Sci. Rev.* **201**, 1. DOI. ADS.

Sakaue, T., Tei, A., Asai, A., Ueno, S., Ichimoto, K., Shibata, K.: 2017, Observational study on the fine structure and dynamics of a solar jet. I. Energy build-up process around a satellite spot. *Publ. Astron. Soc. Japan* **69**, 80. DOI.

Saqri, J., Veronig, A.M., Battaglia, A.F., Dickson, E.C.M., Gary, D.E., Krucker, S.: 2024, Efficiency of solar microflares in accelerating electrons when rooted in a sunspot. *Astron. Astrophys.* **683**, A41. DOI. ADS.

Scherrer, P.H., Schou, J., Bush, R.I., Kosovichev, A.G., Bogart, R.S., Hoeksema, J.T., Liu, Y., Duvall, T.L., Zhao, J., Title, A.M., Schrijver, C.J., Tarbell, T.D., Tomczyk, S.: 2012, The helioseismic and magnetic imager (HMI) investigation for the solar dynamics observatory (SDO). *Solar Phys.* **275**, 207. DOI. ADS.

Schmieder, B., Guo, Y., Moreno-Insertis, F., Aulanier, G., Yelles Chaouche, L., Nishizuka, N., Harra, L.K., Thalmann, J.K., Vargas Dominguez, S., Liu, Y.: 2013, Twisting solar coronal jet launched at the boundary of an active region. *Astron. Astrophys.* **559**, A1. DOI. ADS.

Schuck, P.W.: 2008, Tracking vector magnetograms with the magnetic induction equation. *Astrophys. J.* **683**, 1134. DOI. ADS.

Shen, Y.: 2021, Observation and modelling of solar jets. *Proc. Roy. Soc. London Ser. A, Math. Phys. Sci.* **477**, 217. DOI. ADS.

Sterling, A.C., Moore, R.L., Falconer, D.A., Adams, M.: 2015, Small-scale filament eruptions as the driver of X-ray jets in solar coronal holes. *Nature* **523**, 437. DOI. ADS.

Su, Y., Liu, W., Li, Y.-P., Zhang, Z., Hurford, G.J., Chen, W., Huang, Y., Li, Z.-T., Jiang, X.-K., Wang, H.-X., Xia, F.-X.-Y., Chen, C.-X., Yu, W.-H., Yu, F., Wu, J., Gan, W.-Q.: 2019, Simulations and software development for the hard X-ray imager onboard ASO-S. *Res. Astron. Astrophys.* **19**, 163. DOI. ADS.

Su, Y., Zhang, Z., Chen, D., Hu, Y., Zhang, Y., Jiang, X., et al.: 2024, The tests and calibrations of the Hard X-ray Imager onboard ASO-S. *Solar Phys.*, submitted.

Tian, L., Alexander, D., Nightingale, R.: 2008, Origins of coronal energy and helicity in NOAA 10030. *Astrophys. J.* **684**, 747. DOI. ADS.

Titov, V.S., Hornig, G., Démoulin, P.: 2002, Theory of magnetic connectivity in the solar corona. *J. Geophys. Res. Space Phys.* **107**, 1164. DOI. ADS.

Titov, V.S., Priest, E.R., Demoulin, P.: 1993, Conditions for the appearance of "bald patches" at the solar surface. *Astron. Astrophys.* **276**, 564. ADS.

Toriumi, S., Wang, H.: 2019, Flare-productive active regions. *Living Rev. Solar Phys.* **16**, 3. DOI. ADS.

Török, T., Temmer, M., Valori, G., Veronig, A.M., van Driel-Gesztelyi, L., Vršnak, B.: 2013, Initiation of coronal mass ejections by sunspot rotation. *Solar Phys.* **286**, 453. DOI. ADS.

Tripathi, D., Gibson, S.E., Qiu, J., Fletcher, L., Liu, R., Gilbert, H., Mason, H.E.: 2009, Partially-erupting prominences: a comparison between observations and model-predicted observables. *Astron. Astrophys.* **498**, 295. DOI. ADS.

Wang, J., Li, W., Denker, C., Lee, C., Wang, H., Goode, P.R., McAllister, A., Martin, S.F.: 2000, Minifilament eruption on the quiet sun. I. Observations at Hα central line. *Astrophys. J.* **530**, 1071. DOI. ADS.

Wiegelmann, T., Inhester, B., Kliem, B., Valori, G., Neukirch, T.: 2006, Testing non-linear force-free coronal magnetic field extrapolations with the Titov-Démoulin equilibrium. *Astron. Astrophys.* **453**, 737. DOI. ADS.

Wyper, P.F., DeVore, C.R., Antiochos, S.K.: 2019, Numerical simulation of helical jets at active region peripheries. *Mon. Not. Roy. Astron. Soc.* **490**, 3679. DOI. ADS.

Yan, X.L., Qu, Z.Q.: 2007, Rapid rotation of a sunspot associated with flares. *Astron. Astrophys.* **468**, 1083. DOI. ADS.

Yang, Y.-H., Cheng, C.Z., Krucker, S., Hsieh, M.-S., Chen, N.-H.: 2012, Asymmetry of hard X-ray emissions at conjugate footpoints in solar flares. *Astrophys. J.* **756**, 42. DOI. ADS.

Zhang, J., Li, L., Song, Q.: 2007, Interaction between a fast rotating sunspot and ephemeral regions as the origin of the major solar event on 2006 December 13. *Astrophys. J. Lett.* **662**, L35. DOI. ADS.
Zhang, Z., Chen, D.-Y., Wu, J., Chang, J., Hu, Y.-M., Su, Y., Zhang, Y., Wang, J.-P., Liang, Y.-M., Ma, T., Guo, J.-H., Cai, M.-S., Zhang, Y.-Q., Huang, Y.-Y., Peng, X.-Y., Tang, Z.-B., Zhao, X., Zhou, H.-H., Wang, L.-G., Song, J.-X., Ma, M., Xu, G.-Z., Yang, J.-F., Lu, D., He, Y.-H., Tao, J.-Y., Ma, X.-L., Lv, B.-G., Bai, Y.-P., Cao, C.-X., Huang, Y., Gan, W.-Q.: 2019, Hard X-ray imager (HXI) onboard the ASO-S mission. *Res. Astron. Astrophys.* **19**, 160. DOI. ADS.

Authors and Affiliations

Tingyu Gou[1,2] · Rui Liu[1] · Yang Su[3] · Astrid M. Veronig[4] · Hanya Pan[1] · Runbin Luo[1] · Weiqun Gan[3,5]

✉ T. Gou
tygou@ustc.edu.cn; tingyu.gou@cfa.harvard.edu

R. Liu
rliu@ustc.edu.cn

1 CAS Key Laboratory of Geospace Environment, University of Science and Technology of China, Hefei 230026, China

2 Present address: Center for Astrophysics | Harvard & Smithsonian, 60 Garden Street, Cambridge, MA 02138, USA

3 Key Laboratory of Dark Matter and Space Astronomy, Purple Mountain Observatory, Chinese Academy of Sciences, Nanjing 210023, China

4 Institute of Physics & Kanzelhöhe Observatory for Solar and Environmental Research, University of Graz, 8010 Graz, Austria

5 University of Chinese Academy of Sciences, Nanjing 211135, China

Solar Physics (2024) 299:81
https://doi.org/10.1007/s11207-024-02325-8

RESEARCH

Kink-and-Disconnection Failed Eruption in 3D

Tomasz Mrozek[1] · **Zhentong Li**[2] · **Marian Karlický**[3] · **Nicolina Chrysaphi**[4] · **Yang Su**[2,5] · **Wei Chen**[2,5] · **Weiqun Gan**[2,6]

Received: 29 February 2024 / Accepted: 15 May 2024 / Published online: 11 June 2024

Abstract

We present a case study of a failed eruption that accompanied an M1.5 GOES class solar flare. It was observed by STIX onboard Solar Orbiter, HXI onboard the Advanced Space-based Solar Observatory, AIA onboard Solar Dynamics Observatory, and WAVES onboard the STEREO-A. The important input is from stereoscopic hard X-ray (HXR) observations obtained by HXI and STIX, whose vantage points were separated by 31.5°, allowing us to unfold the 3D geometry of the event. The eruption was a two-phase event. First, it started with the rope helical kink and then was slowed down, but with the structure still unstable, it erupted two minutes later due to ongoing reconnection in the interacting legs of the kinked structure. A Type III burst was observed in association with the eruption, indicating the acceleration of semirelativistic electrons into the heliosphere. During the second phase, a hot cloud was disconnected and confined in the overlying magnetic field, where the overlying loops connected two adjacent active regions. The estimated and corrected for real geometry velocities are in the range of $385-400\ \mathrm{km\,s^{-1}}$, whereas acceleration reached $4.78-6.33\ \mathrm{km\,s^{-2}}$. These extreme values are much more demanding from a perspective of conditions that are needed to stop the eruption. Images obtained simultaneously by HXI and STIX located in different vantage points showed that flare-related sources are not lying along a normal to the solar surface. The understanding of the eruption analyzed here has been highly enriched thanks to the stereoscopic information about HXR source locations.

Keywords Solar flares · X-ray emission · X-ray stereoscopic observations · Failed eruption

✉ T. Mrozek
tmrozek@cbk.pan.wroc.pl

[1] Space Research Centre, Polish Academy of Sciences, ul. Bartycka 18a, 00-716 Warszawa, Poland

[2] Key Laboratory of Dark Matter and Space Astronomy, Purple Mountain Observatory, Chinese Academy of Sciences, Nanjing 210023, China

[3] Astronomical Institute ASCR, Fričova 298, 251 65, Ondřejov, Czech Republic

[4] Sorbonne Université, École Polytechnique, Institut Polytechnique de Paris, CNRS, Laboratoire de Physique des Plasmas (LPP), 4 Place Jussieu, 75005 Paris, France

[5] School of Astronomy and Space Science, University of Science and Technology of China, Hefei 230026, China

[6] University of Chinese Academy of Sciences, Nanjing 211135, China

1. Introduction

Coronal mass ejections (CME) are phenomena that can potentially affect Earth's magnetosphere when properly directed, gaining a lot of attention. However, predictions of arrival times, real direction of movement, and severeness of effects induced in the Earth's surroundings are still far from acceptance levels. One of the possible reasons is the evolution of the eruption at its initial phase before it reaches the middle corona. Before reaching that distance ($4-6\ R_{\odot}$), an eruption experiences a lot of interactions with, still strong, coronal magnetic fields, and its structure is prone to various deformations and disruptions. The eruptions are often observed to change direction, partially erupt, or even stop at some distance from the Sun. Therefore the analysis of the initial phase of their development, with special care given to strong events that stop, may give important constraints to our understanding (and better predicting) of geoeffective CMEs.

Failed eruptions (Gilbert, Alexander, and Liu, 2007) are particular cases of events accompanying the confined flares (Švestka, 1986). They start as typical eruptions that develop into a CME, but for some reason, after initial acceleration, they stop in the corona. One of the first observed (Ji et al., 2003) and widely analyzed (Török and Kliem, 2005; Alexander, Liu, and Gilbert, 2006; Hassanin and Kliem, 2016) failed eruptions was that accompanying the M2.0 flare observed on 27 May 2002. The eruption started to develop and shortly after it was observed to kink and stop. Analysis of that event further showed that one of the important parameters determining the eruption or confinement is the overlying horizontal magnetic field decay index (Cheng et al., 2011; Nindos, Patsourakos, and Wiegelmann, 2012; Baumgartner, Thalmann, and Veronig, 2018; Jing et al., 2018). Wang and Zhang (2007) showed that the location within the active region is also important. Namely, eruptions that started closer to a central part of the active region (stronger horizontal component) were stopped more often than the ones located on the side (stronger radial component). This result shows that when the horizontal component of a magnetic field is strong, the eruption is more prone to stop.

The role of the overlying field in stopping an eruption is investigated from two perspectives. One is the tension of the overlying field, which is like an elastic cage that prevents the eruption from moving higher (Myers et al., 2015; Chen, Ma, and Zhang, 2013; Mrozek, 2011). The second is when the overlying field reconnects with the moving filament causing the disintegration of its magnetic skeleton (Ding, Zhang, and Hong, 2022; Chen et al., 2023a,b; Jiang et al., 2023) and stop.

Amari and Luciani (1999) modeled the eruption of a magnetic flux rope and showed the possibility of destroying its magnetic skeleton due to interaction with an overlying magnetic field. Hassanin and Kliem (2016) modeled the 27 May 2002 eruption providing evidence for ongoing reconnection ahead of the eruption. In addition, as the eruption developed, a vertical current sheet formed behind. Both reconnection sites decreased a hoop force acting on the erupting structure. Such a scenario was further investigated in several papers revealing places for possible reconnection (Chen et al., 2023a; Jiang et al., 2023; Wang et al., 2022). Both reconnections, ahead of and behind an eruption, are possible areas for accelerating particles and direct heating of coronal plasma to high temperatures. Therefore, we expect hot and/or nonthermal sources present in the corona as features accompanying failed eruptions.

Here we present a case study of a failed eruption, which shows helical kinking, disconnection, and interaction with the overlying field as observed by instruments located at different points in space. It enabled us to reconstruct the three-dimensional structure of processes under study. The paper is organized as follows. In Section 2, we present observational data. In Section 3, these data are analyzed. Section 4 summarizes the results together with their discussion. Finally, in Section 5, we present our conclusions.

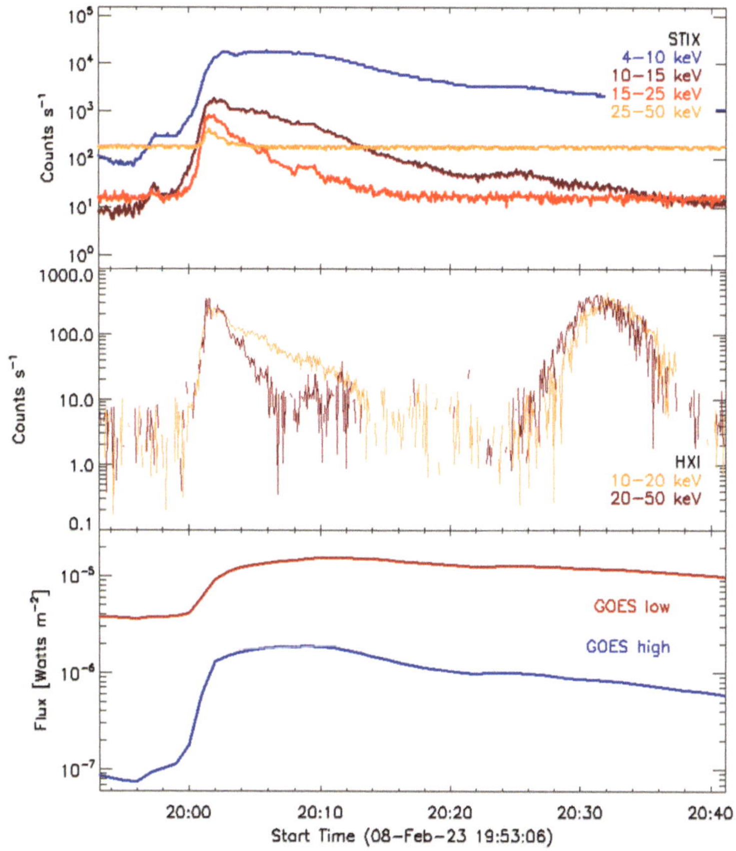

Figure 1 X-ray light curves measured by SO/STIX (top), ASO-S/HXI (middle), and GOES (bottom). The HXI count rates are the summed count rates of the total flux monitors (D92, D93, D94), subtracted by the summed count rates of the corresponding three background monitors (D95, D96, D99). The second bump seen in the HXI after 20:25 UT is related to a particle background, which was not fully subtracted.

2. Observational Data

The event analyzed (SOL20230208T20:05) was a moderate M1.5 GOES class flare (Figure 1). It was preceded by a precursor with a maximum around 19:57:30 UT. From Earth's perspective it was located close to the solar limb in the NOAA13217 active region. The flare has been registered by several instruments/observatories. In this paper, we used data from the following sources:

- The Spectrometer Telescope for Imaging X-rays (STIX; Krucker et al., 2020) onboard Solar Orbiter (Müller et al., 2020) is a coded aperture telescope equipped with the set of 32 pixelated detectors measuring energies of counts in 32 energy channels from 4 to 150 keV. A present instrument calibration allows us to reconstruct images with an angular resolution of 14.6 arcsec.
- The Hard X-ray Imager (HXI; Zhang et al., 2019) onboard ASO-S (Gan et al., 2023) can generate solar X-ray spectra from ~ 10 to 300 keV by the total flux monitors (D92, D93, D94) and reconstruct images starting from ~ 14 keV using detectors from D1 to D91 (but

we usually use detectors from D19 or D29 to D91 in practical analysis because the fine grids corresponding to detectors from D1 to D18 need to be further calibrated).

- Atmospheric Imaging Assembly (AIA, Lemen et al., 2012) onboard the Solar Dynamics Observatory (SDO, Pesnell, Thompson, and Chamberlin, 2012). The instrument provides EUV and UV images with 12-s cadence and 1.2-arcsec angular resolution.
- The S/WAVES instrument on board the Solar Terrestrial Relations Observatory (STEREO-A, Kaiser, 2005) includes radio receivers that measure radio wave intensity in the frequency range from 16 MHz to 40 kHz, corresponding to source distances of about 1 $R_\odot$ to 1 AU.

X-ray data were collected by coded-aperture telescopes, so images have to be reconstructed. STIX images were obtained with the MARLIN algorithm (Siarkowski et al., 2020). For several times and energies, we also used other algorithms, the Expectation Maximisation (Massa et al., 2019), MEM_GE (Massa et al., 2020) and VIS_FWDFIT_PSO (Volpara et al., 2022), which we cross-checked with MARLIN for consistency. At the time of the flare occurrence, STIX was in a test mode during which several detector parameters and onboard configurations were changing. Therefore the energy borders for images are slightly more uncertain but still useful for the analysis we performed. Similarly, spectra were also affected, and thus, for spectral analysis, we used only the power law index of the nonthermal emission as it is not affected by the type of tests performed. HXI images were reconstructed by dedicated HXI_CLEAN algorithm (Su et al., 2019) using detectors from D19 to D91 after subtracting the background of orbit data 48 h before.

3. Data Analysis

3.1. Height-Time Evolution

For the height-time analysis, we used the photometric cut marked in Figure 2. The dynamics of the structures along this cut are presented for two AIA filters (171 and 131 Å) in Figure 3. In the 171 Å filter, we see cold structures, overlying magnetic loops, whereas the 131 Å shows hot structures of temperatures around 10 MK.

The eruption was visible only in the hot channels of AIA (see online movie). Two phases have been observed. The structure starts to rise slowly at around 19:46 UT and becomes fast at 19:47 UT, moving with increasing speeds. After reaching the maximum of 260 $\mathrm{km\,s^{-1}}$, the kinked structure was visible, and the velocity dropped to the level of 60 – 90 $\mathrm{km\,s^{-1}}$. Simultaneously, a small burst of HXRs has been recorded by STIX and GOES. After three minutes, around 20:01 UT, the second phase started. The velocity jumped abruptly to about 250 $\mathrm{km\,s^{-1}}$ and was almost stable for one minute, after which it again dropped down to values less than 100 $\mathrm{km\,s^{-1}}$. After slowing down, the structure continued to rise up until $\sim$ 20:37 UT when we observed stabilization and it stopped completely.

Simultaneously observed cool structures along the cut in 171 Å images show a different evolution. Namely, these structures were observed initially at the height of 70 Mm and were almost stable. Some fluctuation, by 5 Mm, is visible around 19:59 UT, which is two minutes after the maximum velocity was reached during the first stage of the eruption. This could reflect a disturbance/wave traveling through the coronal structures with a projected velocity of 450 $\mathrm{km\,s^{-1}}$. The estimated velocity is based on the assumption that the disturbance is related to a compression on the front of the eruption, which continued to travel up after the eruption stopped during the first stage. The second stage is visible in the heights of overlying

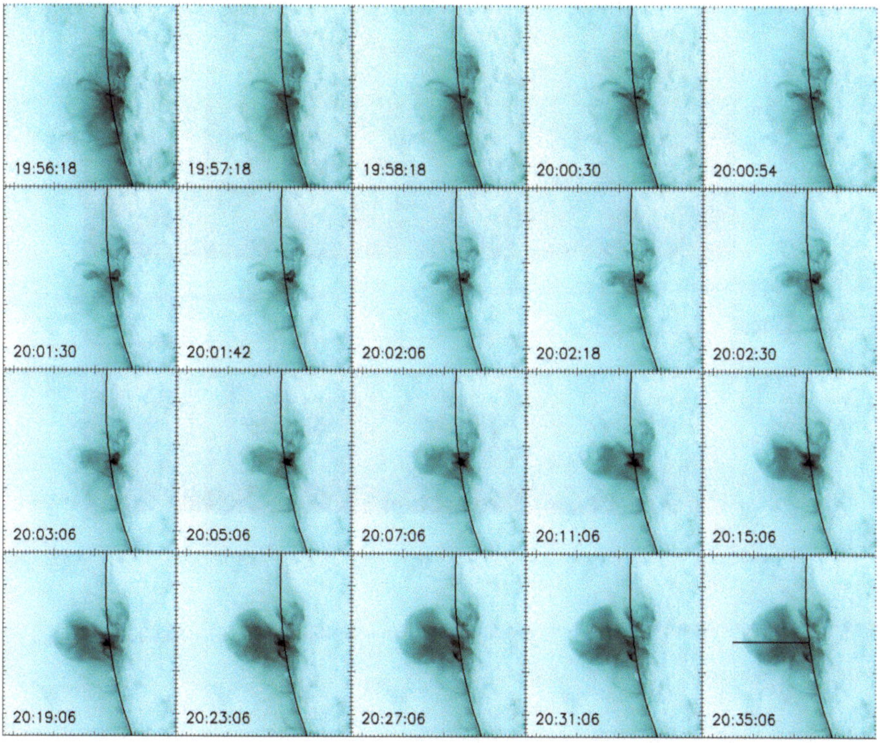

Figure 2 Snapshots of SDO/AIA 131 Å images presenting the event during the rising phase and shortly after the maximum of the flare. On the bottom right panel a line is plotted showing the photometric cut line used for the kinematics analysis.

loops. They started to grow up simultaneously with the velocity peak, without any time delay being observed. From that moment, the height evolution of hot and cold structures is similar. A slowing down hot structure observed simultaneously with cold structures rising is a clear evidence of interaction between an erupting structure and an overlying magnetic field.

Simultaneously with the second phase of velocity increase, we observed a burst of HXRs with energies up to 50 keV. The X-rays measured by HXI in 10 – 20 keV dropped to the background level around 20:14 UT, whereas STIX recorded energies of 10 – 15 keV until 20:36 UT.

STIX and HXI spectra were analyzed with the standard OSPEX[1] package. We used a thermal model (`f_vth`) and nonthermal model (`f_thick2`) to fit the spectra. We note that the HXI spectra is contaminated when passing through the particle region (from ~20:00 UT to ~20:02 UT). It is a technical hazard to analyze HXI spectra by subtracting the background from the background monitor data, which is not always at the same level as – or has a consistent trend with – the real background, also because of the different lifetimes between open flux monitors and background monitors. Moreover, due to instrumental effects (e.g., cross-calibration) and other physical effects (e.g., X-ray directivity and albedo effect), we do not expect perfect agreement between the parameters obtained from spectral fitting.

[1] https://hesperia.gsfc.nasa.gov/rhessi3/software/spectroscopy/spectral-analysis-software/index.html.

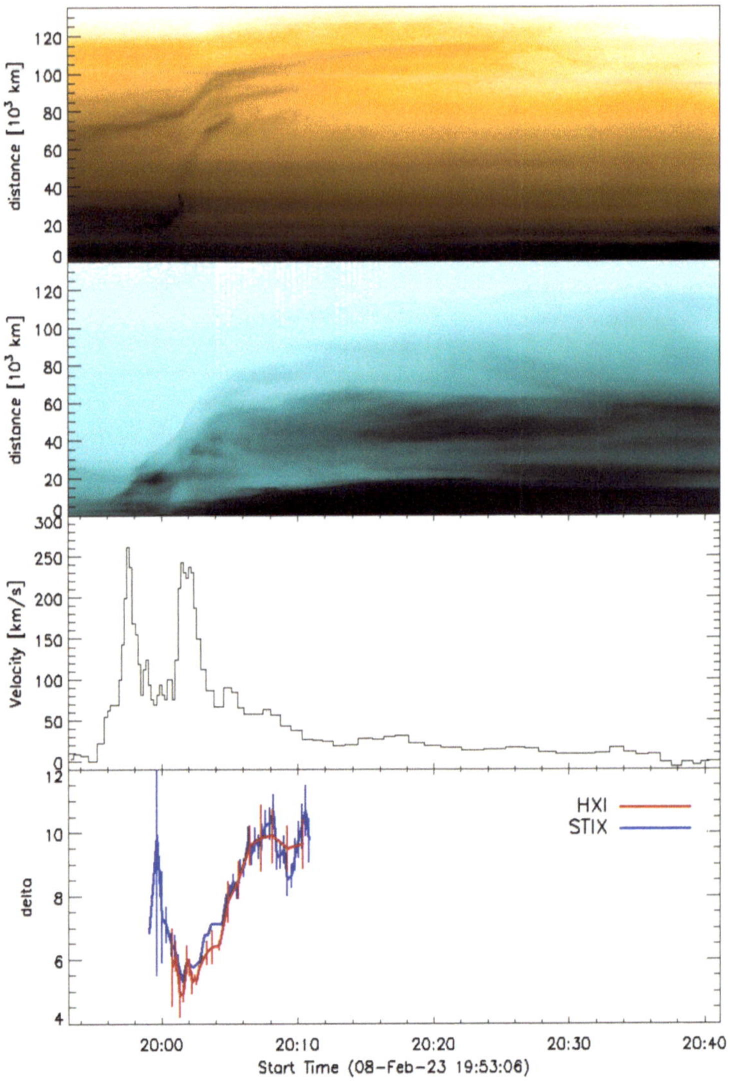

Figure 3 Temporal evolution of the entire event. The top two panels (SDO/AIA 171 and 131 Å, respectively) show brightness evolution (negative) along a cut marked in Figure 2. The third panel presents the velocity of the eruption as measured by SDO/AIA in 131 Å (shown in the panel above). The bottom panel contains the non-thermal electron power-law spectrum index estimated for STIX (blue) and HXI (red).

However, the power law index on nonthermal electron spectra should be consistent at the very basic level of data generation. The comparison is shown in Figure 3. Given that STIX data has been gathered during onboard tests, the overall agreement is satisfying. The power law is correlated to the velocity of the second phase, which supports that this phase is driven by reconnection.

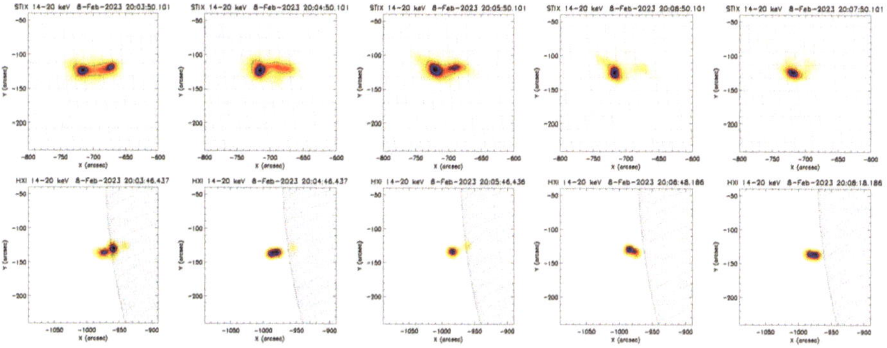

Figure 4 Top panel: STIX images (14 – 20 keV) reconstructed for consecutive time intervals when observations overlapped with HXI. Bottom panel: HXI images reconstructed in the same time and energy intervals. Maps have been distance-corrected to a common 1 AU perspective.

3.2. Location of X-ray Sources in the 3D Space

The analyzed event was observed in common by two HXR imagers located in different vantage points. The ecliptic latitude difference was 3.7°, and the angular separation, projected on the ecliptic plane was 31.5°, allowing for stereoscopic observations. Therefore we may try to unfold the real positions of X-ray sources in the solar corona. For this purpose, we modified SolarSoftWare (SSW) function *SCC_TRIANGULATE*, which was originally used for triangulation based on STEREO observations. Providing helioprojective Cartesian coordinates of two sources (centroids), we can obtain 3D coordinates in the heliocentric inertial system.

The main problem we met was the identification of pairs of the same sources when more than one was visible on the STIX and HXI reconstructed images. For this purpose, we analyzed a double-source morphology, which was observed around the maximum of the flare measured in the STIX 4 – 10 keV light curve. We reconstructed STIX and HXI images in the same time intervals (60-s long) and the same energy bands (14 – 20 keV). Despite small differences, the observed morphology is similar. For the first time interval, around 20:03:50 UT, we observe two sources (Figure 4). In the HXI images, the bottom source (H1) appears slightly brighter than the top one (H2). The STIX image also reveals a double source morphology, with the western source (s1) appearing slightly weaker and the eastern one (S2) brighter.

We must remember that the STIX and HXI images are reconstructed in the count space. This means that the source brightness is affected by the response of the instrument, and comparison between two instruments is not straightforward. In addition, we expect that sources may have elliptical shapes, which, when projected on the sky from two perspectives, can be reconstructed slightly differently by the used algorithms because of differences in surface brightness. More complication is introduced in the identification procedure from HXI images. They start with a double source, then only one is visible around 20:05 UT, and a double structure is seen again from 20:06 UT, but with different orientation. This is unclear, but we have to remember that these are reconstructed images. If two sources strongly overlap, then it might be a result of the reconstruction parameters. For example, a small change in the CLEAN beam would smear the source, visible after 20:06 UT, into one elliptical source.

Taking the above into account, we decided to only investigate the double source seen around 20:03:50 UT. The different instrumental responses of STIX and HXI affect the

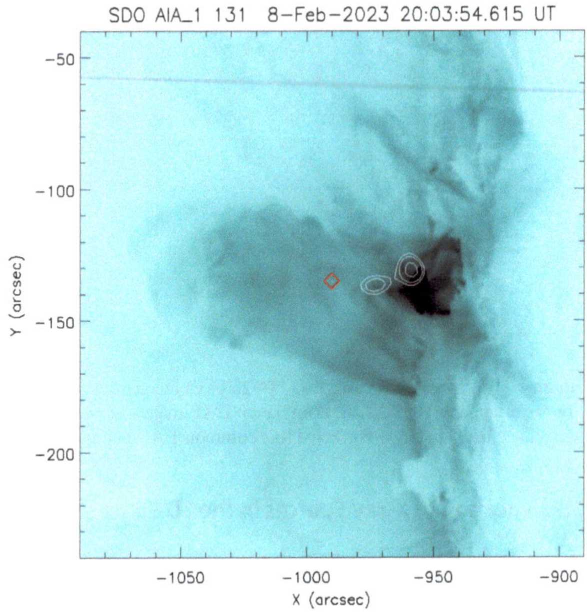

Figure 5 ASO-S/HXI contours (14–20 keV) overplotted on the AIA 131 Å image taken at the same moment. The red diamond marks the position of the X-point visible in the kinked structure several minutes before.

brightness. However, the relative brightness of two sources should be preserved when reconstructed for the same time interval, but different energies. We reconstructed such images and found that the S1 and H2 sources and the S2 and H1 sources behaved similarly as pairs, that is, with increasing energy, the brightness of the S1 and H2 sources increased, whereas the S2 and H1 pair of sources became dimmer. From this observation we concluded that source S1 visible in the STIX image corresponds to source H2 from HXI.

Images in Figure 4 have been rescaled to the same distance from the Sun. In the first column, where a double structure is clear, we can see that the distances between the sources are significantly different. Namely, from the HXI perspective they are separated by 17 arcsec. If they were located one above another, on the same line perpendicular to the solar surface, then the distance visible from the STIX position would be smaller, given that from STIX perspective the flare is located on the disk. However, we see that in STIX images the separation of sources is significantly larger, close to 32 arcsec. This means that in reality the sources are horizontally separated, and the separation is large. Interpreting such a configuration from only one instrument may be highly misleading. For example, assuming that we have only HXI observations, we would interpret the data as a double source located close to the reconnection site, which is illustrated in Figure 5. However, following triangulation, we can conclude that sources are located on a line that is about 26.6° inclined to the normal. The line is inclined toward the observer on Earth's orbit, meaning that the high-lying source H2 is closer to the observer.

An independent estimation of the 3D location of the X-ray source can be performed from another interesting phase seen in the STIX images. Namely, it is the occurrence of a large X-ray source located far from the two sources, S1 and S2, registered in the vicinity of the maximum phase (Figure 6). The source is visible in the STIX 6–10 keV images reconstructed after 20:22 UT. It is also visible on the solar disk. The source is located about 100 arcsec to the east comparing to the location of S1 and S2. HXI is unable to reconstruct images below 13 keV when using detectors D19 to D91 due to the lack of photon records.

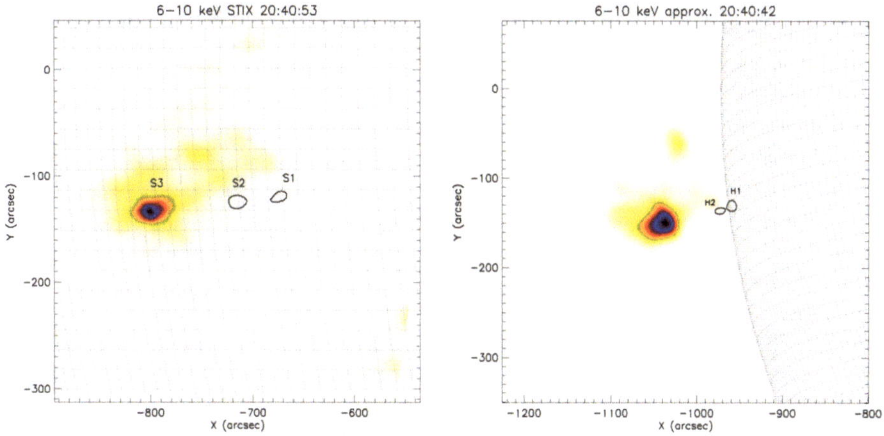

Figure 6 Left panel: STIX image (6 – 10 keV) reconstructed during the decay phase of the flare. Contours of 14 – 20 keV STIX sources, visible during the flare maximum phase, have been added. Right panel: the X-ray 6 – 10 keV emission approximated from DEM maps obtained at the same time as the left panel reconstruction. Contours of 14 – 20 keV HXI sources, visible during the flare maximum phase, have been added.

The question arises whether this source is visible from Earth's perspective or whether it is behind the solar limb. To answer this question, we may compare GOES and STIX total fluxes. The GOES class estimated from the STIX 4 – 10 keV flux (Xiao et al., 2023) is M1 (C7 – M2 range), whereas measurements by GOES gives M1.5. Such closeness of values is typical when both instruments observe the entire flare emission. If from one perspective (of either STIX or GOES), a flare is behind the limb, then the estimated and observed classes will differ significantly. Thus we assume that the STIX source visible during the decay phase is located in the corona and observed from Earth's orbiting instruments. We perform DEM analysis in the next section to investigate the hot thermal source in the post-eruption structure.

3.3. Differential Emission Measure Evolution

To identify the possible location of the hot plasma source, we use the AIA EUV images rebinned by 3×3 pixels and apply the differential emission measure (DEM) method, utilizing the sparse algorithm developed by Cheung et al. (2015) and Su et al. (2018) (with improved settings in Li et al., 2022), which better constrains the unreasonably high plasma temperature over ~ 10 MK. The DEM maps show hot plasma sites more clearly than single AIA images. This is due to the typical two-temperature response of AIA filters. The difference is noticeable in the top two panels of Figure 7 presenting the AIA 131 Å map (panel (a)) and the plasma emission measure (EM) map with temperatures above 5.6 MK in panel (b). Although both maps are similar, there are differences due to the response of AIA 131 Å, which has a maximum around 10^7 K and a second one close to $10^{5.7}$ K. Panel (c) of Figure 7 presents the evolution of the EM-weighted temperature ($T_{\mathrm{EM}} = \frac{\sum \mathrm{EM}_i \cdot T_i}{\sum \mathrm{EM}_i}$, where the temperature T is in linear scale, and i denotes the ith temperature bin) distribution of the hot plasma along the slit shown in the top panels. The first signature of hot loops disconnected from the reconnection site is detected before 20:01 UT and is correlated with the burst of HXRs and the maximum velocity of the eruption's second phase. For several minutes (20:01 – 20:25 UT), we observe a rising hot region as expected in a standard flare

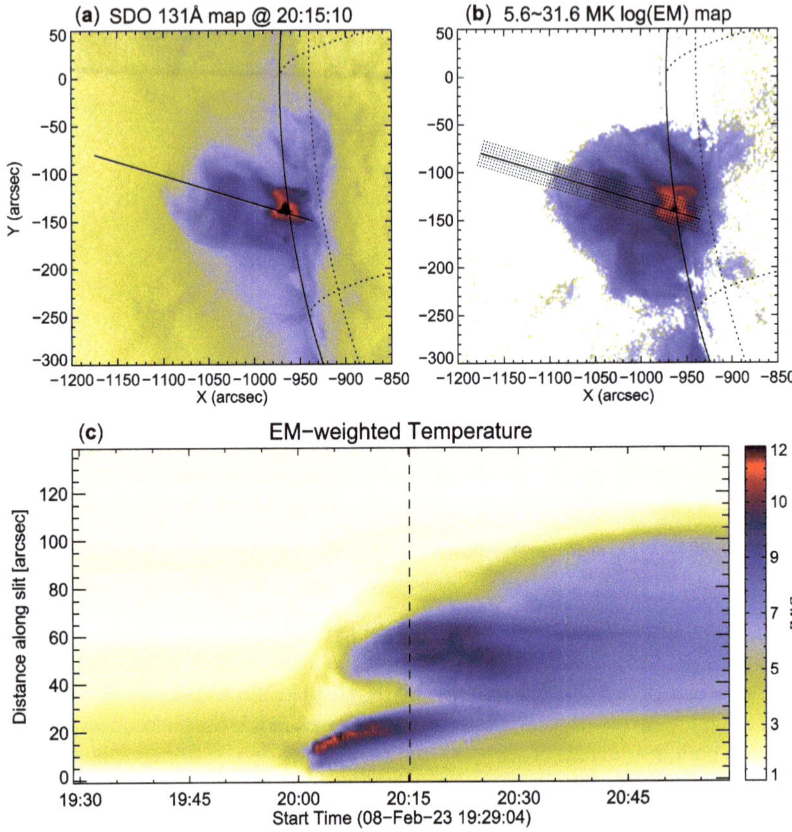

Figure 7 Panel (a): SDO/AIA 131 Å enhanced map observed at 20:15:10 UT. Panel (b): spatial distribution of the plasma emission measure (in log scale) of temperature ranges from 5.6 to 31.6 MK for the same time. The solid black line indicates the slit used to show the time–distance diagram of temperature evolution, and the black dots around the slit represent the width of slit used for averaging. Panel (c): time evolution of the EM-weighted average plasma temperature along the slit marked in panels (a) and (b).

scenario. However, as was noticed in the 3D analysis of the hot source, there is a significant horizontal separation of the X-ray sources, meaning that the evolution of the hot plasma sites seen along the photometric cut is more complicated. Assuming the 3D geometry estimated from the HXRs, we can conclude that we do not see a change of altitudes, but rather an ongoing reconnection along an inclined line, which, projected on the sky, produces the observed changes.

In the later phase, after the eruption stops, a hot cloud forms around 20:15 UT. Figures 8(c) and (d) show the time evolution of the temperature distribution of the EM averaged in the hot cloud and the underlying flare loop region, respectively. Around 20:02 UT, the plasma inside the flare loop region is rapidly heated, and a temperature distribution up to ~ 22 MK is obtained, whereas the temperature of the separate source in the corona is relatively lower. Later, the high-temperature plasma component of the flaring region is decreasing, corresponding to cooling or less heating, whereas for the coronal source, the high-temperature component (near 12 MK) is slowly increasing. After $\sim 20{:}20$ UT, the dominant temperature in the coronal source becomes higher than in the flaring region (from the amount of EMs) and drops by only 3 MK during the next ~ 40 min.

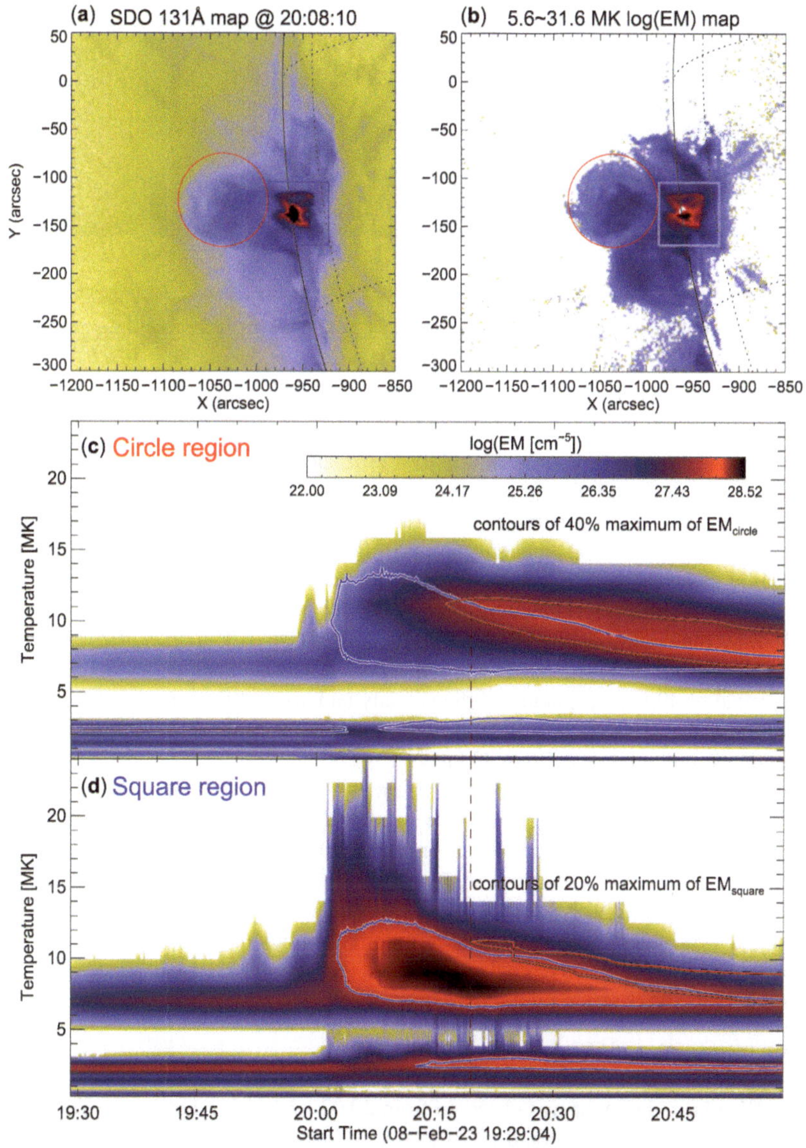

Figure 8 Panel (a): SDO/AIA 131 Å enhanced map observed in 20:08:10 UT. Panel (b): spatial distribution of plasma emission measure (in log scale) of temperature ranges from 5.6 to 31.6 MK for the same time. The red circle and blue square mark the regions used to calculate the averaged emission measure (EM) in the following panels. Panels (c) and (d): time evolution of temperature distribution of averaged EM of pixels inside the circle region (red) and square region (blue), respectively. The minimum value of the diagrams is set to 10^{22} cm^{-5}. The blue and red contours are the contours of EM diagrams of square and circle regions, in 40% maximum of EM of circle region in panel (c) and in 20% maximum of EM of square region in panel (d), respectively.

Having DEM maps, we can estimate the locations of the thermal X-ray emission in a given energy range. The result for 6 – 10 keV is presented in the right panel of Figure 6. The high-lying source is clearly visible, which we interpret as the STIX source reconstructed

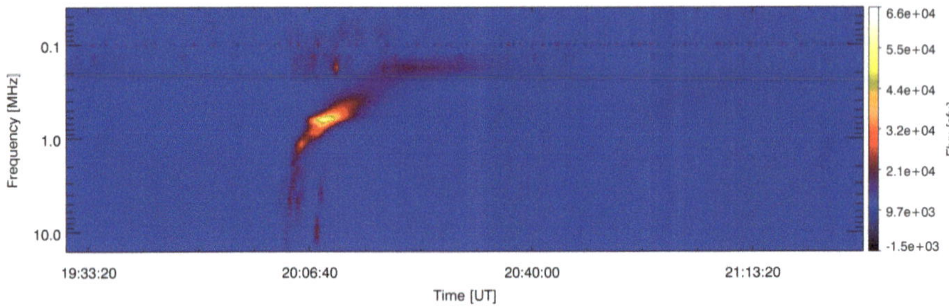

Figure 9 Radio spectrum of a Type III burst recorded by STEREO-A/WAVES during the event.

for the same time and energy (Figure 6, left panel). For completeness, the contours of 14 – 20 keV sources observed around the flare maximum have been overplotted.

The observed 6 – 10 keV source and DEM estimated can also be triangulated to obtain the location of source S3. Comparing the real 3D location of source S3 to source S1 (H2) and S2 (H1), we are able to estimate the approximate direction of the eruption, which started close to S1 (H2) and S2 (H1) and then stopped, forming the hot cloud S3. We have limited information about the real trajectory of the eruption, so we assumed that the range of possible eruption directions is defined by straight lines connecting source S3 to sources S1 and S2. The line connecting sources S2 and S3 is inclined about 70.3° to the normal, whereas for the S1 – S3 line, we obtained 28.9°. The actual direction is between these values. Assuming a mean value of 49.6°, we calculated the velocity and acceleration corrected for projection effects. The real maximum velocity of the eruption is in the range 385 – 400 $\mathrm{km\,s^{-1}}$, and the acceleration is in the range 4.78 – 6.33 $\mathrm{km\,s^{-1}}$.

Another interesting feature seen in the AIA images and the AIA-based DEMs is the shape of the regions containing hot plasma. Namely, after the eruption stops, we can observe that the high-lying hot cloud is separated from the bottom hot structures, which are chalice-shaped. This is interesting because such a structure is typical for a kink failed eruption modeled by Hassanin and Kliem (2016), Figure 6c. The chalice-shaped structure is loops formed when reconnection cuts the kinked eruption. In our event, we see such a structure, and the hot cloud visible above could be a disconnected part of the kinked structure, which was confined by the overlying field.

3.4. Radio Emission

A Type III solar radio burst was observed from 20:02 – 20:40 UT between ∼ 0.1 – 15 MHz by the radio instruments onboard STEREO-A, Wind, and Solar Orbiter. Type III bursts are understood to result from semirelativistic electron beams, which have velocities in the range ∼ 0.1 – 0.6 c (where c is the speed of light; e.g., Suzuki and Dulk, 1985). Figure 9 shows the Type III burst recorded by STEREO-A/WAVES, where the frequency-dependent background was subtracted. At the highest frequencies (above ∼ 2 MHz), there are clear signatures of multiple electron-beam injections, some of which lead to the Type III burst. Such multiple injections are frequently observed in low-frequency Type III bursts recorded by space-based instruments, as well as by higher-frequency Type III bursts recorded by ground-based detectors (e.g., Reid and Kontar, 2015). We note that below ∼ 0.2 MHz and in temporal relation to the Type III burst, there exists auroral kilometric radiation (AKR; e.g. Panchenko et al., 2014), characterized by its high degree of circular polarization (Stokes V; measured by STEREO-A), suggesting a reaction of the Earth's atmosphere to the energetic electron beams.

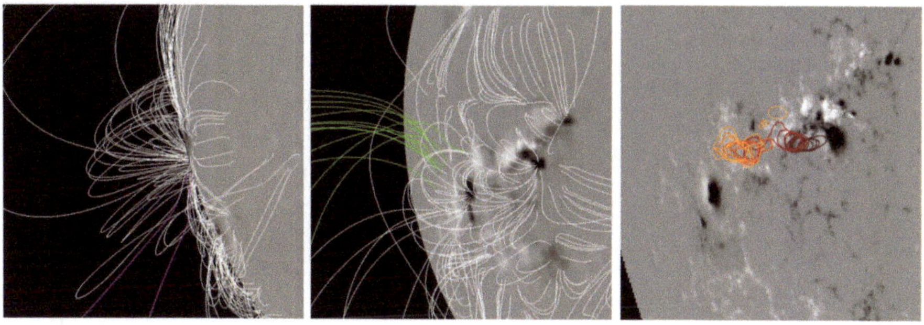

Figure 10 PFSS extrapolation of magnetic fields during the event (left panel). The middle panel presents the extrapolation based on observations obtained 60 hours after the flare occurred, which is related to a 31.1-degree difference in the ecliptic longitudes of SO and ASO-S. Right panel: STIX contours (6 – 10 keV) have been overplotted with color coded contours (brighter means later phase) on the SDO/HMI image collected 60 hours after the flare.

3.5. Magnetic Field Context

There are no available magnetograms from SO/PHI observations for the analyzed event. Therefore, due to the flare location being close to the limb, on the basis of SDO/HMI observations, we are not able to obtain much information about coronal magnetic fields via extrapolation of photospheric fields. Nevertheless, we used PFSS extrapolation to get an idea of the magnetic configuration. The plot in Figure 10 reveals the existence of a large structure of closed loops that might have confined the eruption. A similar extrapolation was also prepared (Figure 10) for a time 2.5 days later. The idea behind this is the following: at the moment the event occurred, the separation between Earth and Solar Orbiter was 31.5°, which can be translated to roughly 2.5 days of solar rotation. The magnetogram from 06:00 UT on 11 February 2023 is related to what is seen by Solar Orbiter at 18:00 UT on 8 February 2024. Here we assumed that the potential magnetic field at high altitudes may not change significantly during such a period; however, this assumption is strong. On the HMI image recorded at 06:00 UT on 11 February 2023, we overplot the STIX 6 – 10 keV sources reconstructed for 2-min intervals covering the entire flare evolution. The result is shown in the right panel of Figure 10. The brightness of contours is related to time. The darkest contours show the rise phase, whereas the brightest ones are related to the decay phase. We can see that the high-lying decay phase source is located approximately between two active regions. This means that the highly inclined eruption was stopped by overlying magnetic loops, which were rooted in two adjacent active regions. This conclusion is not definitive as we have to remember that 2.5 days is a long time during which a new active region may occur. Therefore comparing STIX with HMI image obtained 2.5 later has to be treated with care.

Nevertheless, there is one STEREO-A 171 Å image taken at 20:09 UT. The separation between Earth and STEREO-A was −13.4°, meaning that we are able to see some structures behind the limb defined by the Earth's perspective. In that image, we can recognize structures from the second active region and loops that are connecting them, supporting the scenario of the eruption being confined by trans-region loops.

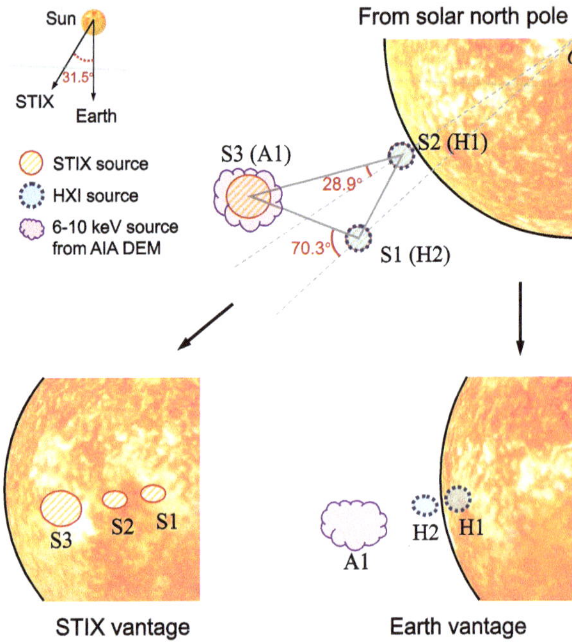

Figure 11 Sketch of the 3D locations of X-ray sources as observed from the points of view of STIX and HXI (which reflects what is seen from Earth).

4. Results and Discussion

Figure 11 summarizes the main observational findings for the analyzed event and the proposed scenario on the basis of stereoscopic HXI-STIX observations. The leading axis of the eruption is on average inclined to the normal by about 49.6°. Regardless of geometric details, the eruption front is moving along the inclined axis which changes the estimated velocities of the event. The velocities estimated from height–time analysis of EUV images in the plane of the sky give values of 250 – 260 $\mathrm{km\,s^{-1}}$, which are smaller than the real velocity estimated around 385 – 400 $\mathrm{km\,s^{-1}}$. This is an important finding as eruptions are frequently modeled on the basis of observations that cannot provide the real 3D geometry. Therefore, the assumed radial movement may, for example, be less demanding (smaller velocities) for model parameters that lead to eruption confinement. This is especially important for cases where gravity seems to play an important role in stopping the eruption, as the measured velocities are low. The estimated real velocities were reached in 36 – 48 s, corresponding to an acceleration of 4.78 – 6.33 $\mathrm{km\,s^{-2}}$. This value is large, exceeding the largest acceleration values recorded for CMEs (Bein et al., 2011). However, even higher accelerations of 7 $\mathrm{km\,s^{-2}}$ have been reported by Song et al. (2014), who estimated this acceleration as projected in the plane of the sky, without any corrections.

Wang and Zhang (2007) compared the locations of strong flares and related them to the eruptive or confined scenario. They found that events located closer to the center of the active region are more prone to be confined due to the strong horizontal component of magnetic fields. Here the stereoscopic analysis of HXI, STIX, and AIA DEMs shows a different scenario. Namely, the flare occurred close to flanks of one active region, but the eruption was confined by loops connecting that region to another one. It shows that even if the eruption occurs in strong radially oriented fields, it still can be stopped and confined by larger scale structures. To some extent, this is observed in statistical analyses of failed eruptions (Mrozek et al., 2020), which are more common close to the solar equator, where we expect large systems of trans-equatorial loops (Chase et al., 1976) connecting active regions.

The analyzed event revealed two distinct phases. Namely, we first observed the eruption, which was stopped after experiencing a kink instability. There are several similar events reported in the literature (Ji et al., 2003; Gilbert, Alexander, and Liu, 2007; Xu et al., 2020). Such eruptions may be stopped after reaching some new stable configuration (Török and Kliem, 2005), or due to reconnections lowering the hoop force acting on the unstable filament (Hassanin and Kliem, 2016). In the case analyzed here, the structure did not reach a stable configuration after the kink, and instead continued to move with a decreased velocity. Its velocity, projected on the sky plane, decreased to less than $100\ \mathrm{km\,s^{-1}}$. In that phase, we see (Figure 2) that the magnetic rope is not only kinked, but its legs touch each other, as shown by Karlický and Kliem (2010), where at the moment, these legs touched a 17-GHz radio source, which appeared at the apparent crossing of these legs. The microwave source indicated that during this touch, magnetic reconnection between the rope legs started, and electrons were accelerated. The same process has likely occurred in the present eruption, where a Type III burst is produced. A similar situation was also observed in HXRs, where sources were reported (Liu and Alexander, 2009; Netzel et al., 2012) to exist in the interacting legs of an eruption.

The above scenario is supported by the temporal correlation of the eruption velocity and the power-law index of nonthermal electrons derived from HXI and STIX data (Figure 2). Once the eruption slowed down following its first phase, we observe the occurrence of a nonthermal component in X-rays with an electron index lowering from 10 to 5. This is a clear evidence of reconnection, which is temporally correlated to the second phase of eruption's acceleration. Moreover, the observation of the Type III burst (Figure 9 indicates the existence of open magnetic fields near the kinked magnetic field structure of the failed eruption, along which the beam of accelerated electrons propagates into the heliosphere). Indeed, such open field structures can be seen in the left panel of Figure 10.

Disconnection observed during the second phase of the event divided the magnetic structure into two parts that are typically predicted in theoretical models of such events (e.g., Hassanin and Kliem, 2016). At lower heights, there is a leftover magnetic field structure, initially of chalice shape. Above it, we see a hot cloud that is spatially correlated with the STIX 6 – 10 keV source. Both structures are separated by a low-temperature region. The X-ray source is seen for about 20 minutes, but in AIA images taken in hot channels (94 and 131 Å), it persisted for more than 2.5 hours. Such structures in failed eruptions were reported earlier (Song et al., 2014; Liu et al., 2015). They are high lying, hot clouds observed in events with a strong EUV late phase. The process leading to such a hot structure surviving high in the corona for 2 hours or more is unclear. Song et al. (2014) proposed that the hot cloud is heated directly, without chromospheric evaporation. In addition, some energy release was expected as the decay phase was long, demanding additional energy release. Finally, in the failed eruption the kinetic energy was significant, and its dissipation is another source of heat for such coronal clouds. Another possibility is that after the abrupt stop of the eruption, a significant turbulence is present in the entire hot cloud. As was shown by Bian et al. (2016), turbulence may completely cut out the conduction flux allowing hot loops persist for even 10,000 s or longer. In the case analyzed here, the hot cloud cooled from 10 MK to 0.7 MK in about 2.5 hours, which demands strong turbulence or/and some additional heating during the decay phase.

5. Conclusions

In this paper, we present a case study of a failed eruption observed by SO/STIX, ASO-S/HXI, SDO/AIA, and STEREO-A/WAVES. We are able to reveal, for the first time, the real 3D geometry of the eruption on the basis of stereoscopic HXR observations. Images obtained simultaneously by HXI and STIX located in different vantage points showed that flare-related sources are not located along a normal to the solar surface. Apart from the vertical separation seen in HXI images, they are also separated horizontally (STIX perspective). Our main findings regarding this event are as follows:

1. Two phases of the failed eruption have been observed. The first failing was related to the helical kink after which the eruption was slowed down. It continued to grow with a much lower velocity, which was possibly the cause for the start of reconnection between interacting legs and the start of the second episode of eruption acceleration. During that phase, we observed disconnection of a large part of the eruption due to reconnection ongoing below. A disconnected cloud was confined in the overlying magnetic field.
2. The velocities of both phases projected on the sky plane were in the range $250-260\ \mathrm{km\,s^{-1}}$ with acceleration $3.1-4.1\ \mathrm{km\,s^{-2}}$. After correcting for the real geometry, we estimated that the eruption was moving at speeds up to $385-400\ \mathrm{km\,s^{-1}}$. The related acceleration reached a level of $4.78-6.33\ \mathrm{km\,s^{-2}}$.
3. The disconnected part of the eruption formed a hot cloud confined by the overlying large-scale magnetic loops. From comparison with STIX thermal images and HMI magnetograms, we conclude that the overlying field is loops connecting two adjacent active regions.

ASO-S has been launched on 9 October 2022. Since HXI started to operate on 21 October 2022, the flare data base covers a few hundreds of flares with more than 200 observed simultaneously with STIX. It gives a chance for a completely new perspective on HXR observations in general and stereoscopy in particular. The advantage is that usually HXR emission is in the form of a few sources whose identification on images obtained from two vantage points should not be complicated. Therefore the 3D geometry may be resolved with high confidence. A statistical analysis of events similar to the one presented here is underway.

Supplementary Information The online version contains supplementary material available at https://doi.org/10.1007/s11207-024-02325-8.

Acknowledgments First, we are grateful to the anonymous referee for a fruitful discussion and many remarks that significantly helped to improve the paper. Solar Orbiter is a mission of international cooperation between ESA and NASA, operated by ESA. The STIX instrument is an international collaboration between Switzerland, Poland, France, Czech Republic, Germany, Austria, Ireland, and Italy. We would like to give special thanks to the STIX Team for their hard work in constructing and understanding the instrument. Ewan Dickson and Olivier Limousin helped us to make correct calibration of the STIX test data. The ASO-S is the first Chinese solar space observatory. We would like to thank the HXI Team for their excellent work in building and understanding the instrument. We would like to give special thanks to the HXI Team for preparing great data analysis software which made this paper possible.

Author contributions T.M. wrote the main manuscript text and prepared Figures 1–6, 10. Z.L. wrote part of the mauscript related to HXI and DEM analysis and prepared Figures 7,8,11. MK wrote parts of the manuscript related to Introduction and radio data. N.C. wrote part of the paper related to radio data nad prepared Figure 9. YS and WC wrote the software used for HXI image reconstruction. WG supervised work on every stage. All authors reviewed the manuscript.

Funding ASO-S mission is supported by the Strategic Priority Research Program on Space Science, the Chinese Academy of Sciences, Grant No. XDA15320000. This research was funded in part by the National Science Centre, Poland grant No. 2020/39/B/ST9/01591. For the purpose of Open Access, the author has applied a CC-BY public copyright licence to any Author Accepted Manuscript (AAM) version arising from this submission. MK acknowledges support from GAČR grant 22-34841S. ZL is supported by the Prominent Postdoctoral Project of Jiangsu Province (2023ZB304). NC acknowledges funding support from the Initiative Physique des Infinis (IPI), a research training program of the Idex SUPER at Sorbonne Universitė. The Chinese authors are also supported by the Strategic Priority Research Program of the Chinese Academy of Sciences (XDB0560000), the National Key R&D Program of China (2022YFF0503002), and National Natural Science Foundation of China (12233012, 1233301, 11921003).

Data Availability No datasets were generated or analysed during the current study.

Declarations

Competing interests The authors declare no competing interests.

References

Alexander, D., Liu, R., Gilbert, H.R.: 2006, Hard X-ray production in a failed filament eruption. *Astrophys. J.* **653**, 719. DOI. ADS.

Amari, T., Luciani, J.F.: 1999, Confined disruption of a three-dimensional twisted magnetic flux tube. *Astrophys. J. Lett.* **515**, L81. DOI. ADS.

Baumgartner, C., Thalmann, J.K., Veronig, A.M.: 2018, On the factors determining the eruptive character of solar flares. *Astrophys. J.* **853**, 105. DOI. ADS.

Bein, B.M., Berkebile-Stoiser, S., Veronig, A.M., Temmer, M., Muhr, N., Kienreich, I., Utz, D., Vršnak, B.: 2011, Impulsive acceleration of coronal mass ejections. I. Statistics and coronal mass ejection source region characteristics. *Astrophys. J.* **738**, 191. DOI. ADS.

Bian, N.H., Watters, J.M., Kontar, E.P., Emslie, A.G.: 2016, Anomalous cooling of coronal loops with turbulent suppression of thermal conduction. *Astrophys. J.* **833**, 76. DOI. ADS.

Chase, R.C., Krieger, A.S., Svestka, Z., Vaiana, G.S.: 1976, Skylab observations of X-ray loops connecting separate active regions. In: *Space Research* **XVI**, 917. ADS.

Chen, H., Ma, S., Zhang, J.: 2013, Overlying extreme-ultraviolet arcades preventing eruption of a filament observed by AIA/SDO. *Astrophys. J. Lett.* **778**, 70. DOI. ADS.

Chen, Y., Cheng, X., Chen, J., Dai, Y., Ding, M.: 2023b, Observations of a failed solar filament eruption involving external reconnection. *Astrophys. J.* **959**, 67. DOI. ADS.

Chen, J., Cheng, X., Kliem, B., Ding, M.: 2023a, A model for confined solar eruptions including external reconnection. *Astrophys. J. Lett.* **951**, L35. DOI. ADS.

Cheng, X., Zhang, J., Ding, M.D., Guo, Y., Su, J.T.: 2011, A comparative study of confined and eruptive flares in NOAA AR 10720. *Astrophys. J.* **732**, 87. DOI. ADS.

Cheung, M.C.M., Boerner, P., Schrijver, C.J., Testa, P., Chen, F., Peter, H., Malanushenko, A.: 2015, Thermal diagnostics with the atmospheric imaging assembly on board the solar dynamics observatory: a validated method for differential emission measure inversions. *Astrophys. J.* **807**, 143. DOI. ADS.

Ding, T., Zhang, J., Hong, J.: 2022, Are the magnetic field directions of surrounding loops a key parameter for confining a solar filament eruption? *Astrophys. J. Lett.* **933**, L38. DOI. ADS.

Gan, W., Zhu, C., Deng, Y., Zhang, Z., Chen, B., Huang, Y., Deng, L., Wu, H., Zhang, H., Li, H., Su, Y., Su, J., Feng, L., Wu, J., Cui, J., Wang, C., Chang, J., Yin, Z., Xiong, W., Chen, B., Yang, J., Li, F., Lin, J., Hou, J., Bai, X., Chen, D., Zhang, Y., Hu, Y., Liang, Y., Wang, J., Song, K., Guo, Q., He, L., Zhang, G., Wang, P., Bao, H., Cao, C., Bai, Y., Chen, B., He, T., Li, X., Zhang, Y., Liao, X., Jiang, H., Li, Y., Su,

Y., Lei, S., Chen, W., Li, Y., Zhao, J., Li, J., Ge, Y., Zou, Z., Hu, T., Su, M., Ji, H., Gu, M., Zheng, Y., Xu, D., Wang, X.: 2023, The Advanced Space-Based Solar Observatory (ASO-S). *Solar Phys.* **298**, 68. DOI. ADS.

Gilbert, H.R., Alexander, D., Liu, R.: 2007, Filament kinking and its implications for eruption and re-formation. *Solar Phys.* **245**, 287. DOI. ADS.

Hassanin, A., Kliem, B.: 2016, Helical kink instability in a confined solar eruption. *Astrophys. J.* **832**, 106. DOI. ADS.

Ji, H., Wang, H., Schmahl, E.J., Moon, Y.-J., Jiang, Y.: 2003, Observations of the failed eruption of a filament. *Astrophys. J. Lett.* **595**, L135. DOI. ADS.

Jiang, C., Duan, A., Zou, P., Zhou, Z., Bian, X., Feng, X., Zuo, P., Wang, Y.: 2023, A model of failed solar eruption initiated and destructed by magnetic reconnection. *Mon. Not. Roy. Astron. Soc.* DOI. ADS.

Jing, J., Liu, C., Lee, J., Ji, H., Liu, N., Xu, Y., Wang, H.: 2018, Statistical analysis of torus and kink instabilities in solar eruptions. *Astrophys. J.* **864**, 138. DOI. ADS.

Kaiser, M.L.: 2005, The STEREO mission: an overview. *Adv. Space Res.* **36**, 1483. DOI. ADS.

Karlický, M., Kliem, B.: 2010, Reconnection of a kinking flux rope triggering the ejection of a microwave and hard X-ray source I. Observations and interpretation. *Solar Phys.* **266**, 71. DOI. ADS.

Krucker, S., Hurford, G.J., Grimm, O., Kögl, S., Gröbelbauer, H.-P., Etesi, L., Casadei, D., Csillaghy, A., Benz, A.O., Arnold, N.G., Molendini, F., Orleanski, P., Schori, D., Xiao, H., Kuhar, M., Hochmuth, N., Felix, S., Schramka, F., Marcin, S., Kobler, S., Iseli, L., Dreier, M., Wiehl, H.J., Kleint, L., Battaglia, M., Lastufka, E., Sathiapal, H., Lapadula, K., Bednarzik, M., Birrer, G., Stutz, S., Wild, C., Marone, F., Skup, K.R., Cichocki, A., Ber, K., Rutkowski, K., Bujwan, W., Juchnikowski, G., Winkler, M., Darmetko, M., Michalska, M., Seweryn, K., Białek, A., Osica, P., Sylwester, J., Kowalinski, M., Ścisłowski, D., Siarkowski, M., Stęślicki, M., Mrozek, T., Podgórski, P., Meuris, A., Limousin, O., Gevin, O., Le Mer, I., Brun, S., Strugarek, A., Vilmer, N., Musset, S., Maksimović, M., Fárník, F., Kozáček, Z., Kašparová, J., Mann, G., Önel, H., Warmuth, A., Rendtel, J., Anderson, J., Bauer, S., Dionies, F., Paschke, J., Plüschke, D., Woche, M., Schuller, F., Veronig, A.M., Dickson, E.C.M., Gallagher, P.T., Maloney, S.A., Bloomfield, D.S., Piana, M., Massone, A.M., Benvenuto, F., Massa, P., Schwartz, R.A., Dennis, B.R., van Beek, H.F., Rodríguez-Pacheco, J., Lin, R.P.: 2020, The Spectrometer/Telescope for Imaging X-rays (STIX). *Astron. Astrophys.* **642**, A15. DOI. ADS.

Lemen, J.R., Title, A.M., Akin, D.J., Boerner, P.F., Chou, C., Drake, J.F., Duncan, D.W., Edwards, C.G., Friedlaender, F.M., Heyman, G.F., Hurlburt, N.E., Katz, N.L., Kushner, G.D., Levay, M., Lindgren, R.W., Mathur, D.P., McFeaters, E.L., Mitchell, S., Rehse, R.A., Schrijver, C.J., Springer, L.A., Stern, R.A., Tarbell, T.D., Wuelser, J.-P., Wolfson, C.J., Yanari, C., Bookbinder, J.A., Cheimets, P.N., Caldwell, D., Deluca, E.E., Gates, R., Golub, L., Park, S., Podgorski, W.A., Bush, R.I., Scherrer, P.H., Gummin, M.A., Smith, P., Auker, G., Jerram, P., Pool, P., Soufli, R., Windt, D.L., Beardsley, S., Clapp, M., Lang, J., Waltham, N.: 2012, The Atmospheric Imaging Assembly (AIA) on the Solar Dynamics Observatory (SDO). *Solar Phys.* **275**, 17. DOI. ADS.

Li, Z., Su, Y., Veronig, A.M., Kong, S., Gan, W., Chen, W.: 2022, Detailed thermal and nonthermal processes in an A-class microflare. *Astrophys. J.* **930**, 147. DOI. ADS.

Liu, R., Alexander, D.: 2009, Hard X-ray emission in kinking filaments. *Astrophys. J.* **697**, 999. DOI. ADS.

Liu, K., Wang, Y., Zhang, J., Cheng, X., Liu, R., Shen, C.: 2015, Extremely large EUV late phase of solar flares. *Astrophys. J.* **802**, 35. DOI.

Massa, P., Piana, M., Massone, A.M., Benvenuto, F.: 2019, Count-based imaging model for the Spectrometer/Telescope for Imaging X-rays (STIX) in solar orbiter. *Astron. Astrophys.* **624**, A130. DOI. ADS.

Massa, P., Schwartz, R., Tolbert, A.K., Massone, A.M., Dennis, B.R., Piana, M., Benvenuto, F.: 2020, MEM_GE: a new maximum entropy method for image reconstruction from solar X-ray visibilities. *Astrophys. J.* **894**, 46. DOI. ADS.

Mrozek, T.: 2011, Failed eruption of a filament as a driver for vertical oscillations of coronal loops. *Solar Phys.* **270**, 191. DOI. ADS.

Mrozek, T., Kołomański, S., Stęślicki, M., Gronkiewicz, D.: 2020, Catalog of solar failed eruptions and other dynamic features registered by SDO/AIA. *Astrophys. J. Suppl.* **249**, 21. DOI. ADS.

Müller, D., St. Cyr, O.C., Zouganelis, I., Gilbert, H.R., Marsden, R., Nieves-Chinchilla, T., Antonucci, E., Auchère, F., Berghmans, D., Horbury, T.S., Howard, R.A., Krucker, S., Maksimovic, M., Owen, C.J., Rochus, P., Rodriguez-Pacheco, J., Romoli, M., Solanki, S.K., Bruno, R., Carlsson, M., Fludra, A., Harra, L., Hassler, D.M., Livi, S., Louarn, P., Peter, H., Schühle, U., Teriaca, L., del Toro Iniesta, J.C., Wimmer-Schweingruber, R.F., Marsch, E., Velli, M., De Groof, A., Walsh, A., Williams, D.: 2020, The solar orbiter mission. Science overview. *Astron. Astrophys.* **642**, A1. DOI. ADS.

Myers, C.E., Yamada, M., Ji, H., Yoo, J., Fox, W., Jara-Almonte, J., Savcheva, A., Deluca, E.E.: 2015, A dynamic magnetic tension force as the cause of failed solar eruptions. *Nature* **528**, 526. DOI. ADS.

Netzel, A., Mrozek, T., Kołomański, S., Gburek, S.: 2012, Extreme-ultraviolet and hard X-ray signatures of electron acceleration during the failed eruption of a filament. *Astron. Astrophys.* **548**, A89. DOI. ADS.

Nindos, A., Patsourakos, S., Wiegelmann, T.: 2012, On the role of the background overlying magnetic field in solar eruptions. *Astrophys. J. Lett.* **748**, L6. DOI. ADS.

Panchenko, M., Macher, W., Rucker, H.O., Fischer, G., Oswald, T.H., Cecconi, B., Maksimovic, M.: 2014, In-flight calibration of STEREO-B/WAVES antenna system. *Radio Sci.* **49**, 146. DOI. ADS.

Pesnell, W.D., Thompson, B.J., Chamberlin, P.C.: 2012, The Solar Dynamics Observatory (SDO). *Solar Phys.* **275**, 3. DOI. ADS.

Reid, H.A.S., Kontar, E.P.: 2015, Stopping frequency of type III solar radio bursts in expanding magnetic flux tubes. *Astron. Astrophys.* **577**, A124. DOI. ADS.

Siarkowski, M., Mrozek, T., Sylwester, J., Litwicka, M., Dąbek, M.: 2020, The non-Fourier image reconstruction method for the STIX instrument. *Open Astron. J.* **29**, 220. DOI. ADS.

Song, H.Q., Zhang, J., Cheng, X., Chen, Y., Liu, R., Wang, Y.M., Li, B.: 2014, Temperature evolution of a magnetic flux rope in a failed solar eruption. *Astrophys. J.* **784**, 48. DOI. ADS.

Su, Y., Veronig, A.M., Hannah, I.G., Cheung, M.C.M., Dennis, B.R., Holman, G.D., Gan, W., Li, Y.: 2018, Determination of differential emission measure from solar extreme ultraviolet images. *Astrophys. J. Lett.* **856**, L17. DOI. ADS.

Su, Y., Liu, W., Li, Y.-P., Zhang, Z., Hurford, G.J., Chen, W., Huang, Y., Li, Z.-T., Jiang, X.-K., Wang, H.-X., Xia, F.-X.-Y., Chen, C.-X., Yu, W.-H., Yu, F., Wu, J., Gan, W.-Q.: 2019, Simulations and software development for the hard X-ray imager onboard ASO-S. *Res. Astron. Astrophys.* **19**, 163. DOI. ADS.

Suzuki, S., Dulk, G.A.: 1985, Bursts of type III and type V. In: McLean, D.J., Labrum, N.R. (eds.) *Solar Radiophysics: Studies of Emission from the Sun at Metre Wavelengths*, 289. ADS.

Švestka, Z.: 1986, On the varieties of solar flares. In: Neidig, D.F., Machado, M.E. (eds.) *The Lower Atmosphere of Solar Flares*, 332. ADS.

Török, T., Kliem, B.: 2005, Confined and ejective eruptions of kink-unstable flux ropes. *Astrophys. J. Lett.* **630**, L97. DOI. ADS.

Volpara, A., Massa, P., Perracchione, E., Francesco Battaglia, A., Garbarino, S., Benvenuto, F., Krucker, S., Piana, M., Massone, A.M.: 2022, Forward fitting STIX visibilities. *Astron. Astrophys.* **668**, A145. DOI. ADS.

Wang, Y., Zhang, J.: 2007, A comparative study between eruptive X-class flares associated with coronal mass ejections and confined X-class flares. *Astrophys. J.* **665**, 1428. DOI. ADS.

Wang, C., Chen, F., Ding, M., Lu, Z.: 2022, Radiative magnetohydrodynamic simulation of the confined eruption of a magnetic flux rope: magnetic structure and plasma thermodynamics. *Astrophys. J. Lett.* **933**, L29. DOI. ADS.

Xiao, H., Maloney, S., Krucker, S., Dickson, E., Massa, P., Lastufka, E., Francesco Battaglia, A., Etesi, L., Hochmuth, N., Schuller, F., Ryan, D.F., Limousin, O., Collier, H., Warmuth, A., Piana, M.: 2023, The data center for the Spectrometer and Telescope for Imaging X-rays (STIX) on board solar orbiter. *Astron. Astrophys.* **673**, A142. DOI. ADS.

Xu, H., Su, J., Chen, J., Ruan, G., Awasthi, A.K., Zhang, H., Zhang, M., Ji, K., Zhang, Y., Liu, J.: 2020, Multiwavelength observation of a failed eruption from a helical kink-unstable prominence. *Astrophys. J.* **901**, 121. DOI.

Zhang, Z., Chen, D.-Y., Wu, J., Chang, J., Hu, Y.-M., Su, Y., Zhang, Y., Wang, J.-P., Liang, Y.-M., Ma, T., Guo, J.-H., Cai, M.-S., Zhang, Y.-Q., Huang, Y.-Y., Peng, X.-Y., Tang, Z.-B., Zhao, X., Zhou, H.-H., Wang, L.-G., Song, J.-X., Ma, M., Xu, G.-Z., Yang, J.-F., Lu, D., He, Y.-H., Tao, J.-Y., Ma, X.-L., Lv, B.-G., Bai, Y.-P., Cao, C.-X., Huang, Y., Gan, W.-Q.: 2019, Hard X-ray imager (HXI) onboard the ASO-S mission. *Res. Astron. Astrophys.* **19**, 160. DOI. ADS.

Solar Physics (2024) 299:114
https://doi.org/10.1007/s11207-024-02341-8

RESEARCH

Triangulation of Hard X-Ray Sources in an X-Class Solar Flare with ASO-S/HXI and Solar Orbiter/STIX

Daniel F. Ryan[1,2] · Paolo Massa[1] · Andrea F. Battaglia[1,3,4] · Ewan C.M. Dickson[5] · Yang Su[6,7] · Wei Chen[6] · Säm Krucker[1,8]

Received: 1 March 2024 / Accepted: 20 June 2024 / Published online: 21 August 2024

Abstract
HXI on ASO-S and STIX onboard Solar Orbiter are the first simultaneously operating solar hard X-ray imaging spectrometers. ASO-S's low Earth orbit and Solar Orbiter's periodic displacement from the Sun–Earth line enables multi-viewpoint solar hard X-ray spectroscopic imaging analysis for the first time. Here, we demonstrate the potential of this new capability by reporting the first results of 3D triangulation of hard X-ray sources in the SOL2023-12-31T21:55 X5 flare. HXI and STIX observed the flare near the east limb with an observer separation angle of 18°. We triangulated the brightest regions within each source, which enabled us to characterise the large-scale hard X-ray geometry of the flare. The footpoints were found to be in the chromosphere within uncertainty, as expected, while the thermal looptop source was centred at an altitude of 15.1 ± 1 Mm. Given the footpoint separation, this implies a more elongated magnetic-loop structure than predicted by a semi-circular model. These results show the strong diagnostic power of joint HXI and STIX observations for understanding the 3D geometry of solar flares. We conclude by discussing the next steps required to fully exploit their potential.

Keywords Spectrum, X-ray · Flares, impulsive phase · Flares, spectrum · Instrumentation and data management · Integrated Sun observations

1. Introduction

Hard X-ray emission (HXR; photon energies $\gtrsim 10$ keV) is a crucial diagnostic of some of the most energetic processes in solar flares. These include non-thermal electron acceleration and plasma heating to tens of megakelvin, both of which are thought to be among the earliest observable consequences of magnetic reconnection in the solar atmosphere. Our understanding of these processes has been revolutionised (Fletcher et al. 2011; Holman et al. 2011; Kontar et al. 2011) since the launch in 2002 of the Reuven Ramaty High Energy Solar Spectroscopic Imager (RHESSI; Lin et al. 2002). However, further understanding was eventually restricted by limitations including imaging dynamic range and projection effects from the single observing angle.

RHESSI has since been succeeded by two more solar HXR spectroscopic imagers: the Spectrometer/Telescope for Imaging X-rays (STIX; Krucker et al. 2020) onboard Solar Orbiter (Müller et al. 2020), and the Hard X-ray Imager (HXI; Zhang et al. 2019; Su et al.

Extended author information available on the last page of the article

2019, 2024) onboard the Advanced Space-based Solar Observatory (ASO-S; Gan et al. 2019). Critically, Solar Orbiter operates beyond Earth orbit, while ASO-S maintains a low-Earth orbit. Their launches in 2020 and 2022, respectively, have ushered in the first age of multi-viewpoint solar HXR imaging spectroscopy. The 3D location, morphology and dynamics of HXR sources, revealed by such observations, are integral components of the evolution and energetics of solar flares. For example, rising thermal X-ray sources are often attributed to plasma heating on new flare loops, formed at increasingly higher altitudes by ongoing magnetic reconnection. Meanwhile outward migration of HXR footpoints is thought to be the chromospheric signatures of non-thermally accelerated electrons, conduction fronts, and/or magnetohydrodynamic (MHD) waves propagating downwards along those same newly formed loops. The thermal and non-thermal energies contained within HXR sources are linked to their density and volume, which cannot be accurately estimated from single-viewpoint studies due to projection effects. Multi-viewpoint solar HXR imaging spectroscopy therefore has the potential to shed new light on the initiation, evolution and energetics of solar eruptive events and the high-energy processes that mediate them.

The first multi-viewpoint solar X-ray imaging was performed by Ryan et al. (2024). They combined STIX with the Earth-orbiting Hinode/X-ray Telescope (XRT; Golub et al. 2007) to perform the first 3D reconstruction of an X-ray thermal looptop source. There are several benefits of using XRT in this way, including its high angular resolution and direct-imaging design. These make XRT's images more straighforward to interpret than the indirect imaging of STIX and HXI. However, while STIX is a spectroscopic imager, XRT uses imaging filters to image plasma at different temperatures. This restricts multi-viewpoint analysis to the thermal X-ray range, and does not allow specific spectral ranges to be extracted from both instruments' observations. Consequently, combining XRT and STIX for 3D reconstruction is only valid when the sources seen by both instruments are dominated by the same plasma population, which is sometimes difficult to determine. By contrast, HXI is a spectroscopic X-ray imager whose spectral range overlaps significantly with that of STIX. This should make it possible to more reliably, and more often, isolate the same plasma population for 3D reconstruction and other multi-viewpoint X-ray analyses. Moreover, such analyses can be performed as a function of energy, including at higher energies where non-thermal processes dominate. Combining STIX with HXI therefore promises to provide a richer 3D understanding of various X-ray sources in solar flares and the physical mechanisms underpinning them.

In this paper, we take the first steps in multi-viewpoint solar X-ray spectroscopic imaging analysis. We present some of the first joint HXI–STIX observations of an X-class flare, and triangulate the brightest locations in the observed thermal and non-thermal sources. In Section 2, we discuss the HXI and STIX observations and the triangulation methodology. In Section 3, we combine these to derive the 3D locations of the X-ray sources and the properties of the flare loop. In Section 4, we highlight issues that should be addressed by future studies to maximise the potential of solar HXR 3D analyses. Finally, we provide our conclusions in Section 5.

2. Observations and Analysis

2.1. Observational Summary

The X5 flare examined in this study occurred in NOAA active region 13536 on 31 December 2023, and peaked in the GOES $1-8$ Å channel at 21:55 UT. Figure 1 shows an overview of

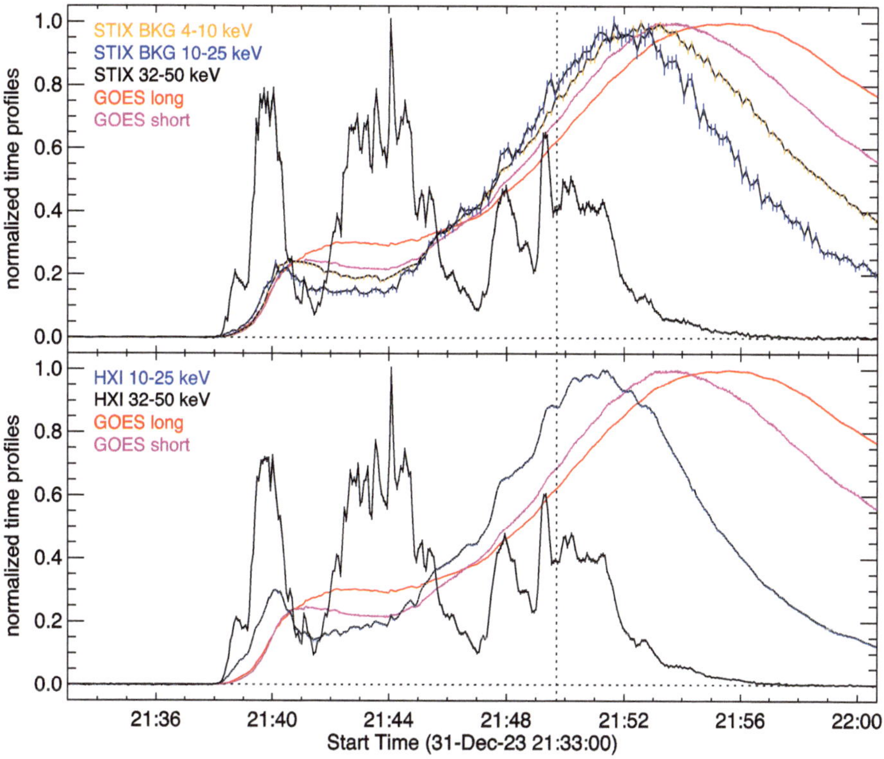

Figure 1 Time evolution of SOL2023-12-31T21:55 in X-rays as seen by GOES, STIX and HXI. The lightcurves are normalised count fluxes with the pre-flare background subtracted. The STIX profiles have been shifted by 16.0 s to account for the different light travel time to Solar Orbiter compared to Earth. Contributions to the HXI lightcurves from particle events during the impulsive phase are negligible. The dashed vertical line gives the time of the AIA images shown in Figure 2.

its soft- and hard X-ray time evolution. As the STIX attenuator moved in around 21:40:23 UT, we used the STIX background detector (Krucker et al. 2020) to derive the time profiles below 25 keV. As its effective area is low, the lightcurves become noisy at the highest energies. The STIX time profile in the non-thermal range (32 – 76 keV) is derived by summing all pixels of the 24 detectors used for imaging. As the non-thermal counts are integrated over many more detectors, it has much smaller error bars. In order to account for the difference in light travel time from the flare to HXI and STIX (16.0 s), we have adjusted the STIX times to coincide with light arrival time at Earth. The HXI time profiles are taken from two detectors with the thick entrance window (D92 and D93) and not covered by imaging grids. To reduce the error bars, all HXR time profiles are summed to a 4 s resolution. The GOES soft X-ray profiles are also shown for context. In order to show all lightcurves on the same scale, they are normalised to their peak flux value.

The impulsive phase of the flare lasts about 16 min with enhanced emission in the non-thermal range ($\gtrsim 30$ keV). The thermal time profiles below 30 keV show the often-observed behaviour of lower energies peaking later, clearly demonstrating the flare's multi-thermal nature. The non-thermal emission drastically reduces after the thermal peak, and there are no significant non-thermal bursts during the decay phase.

In order to demonstrate the feasibility of reconstructing the flare geometry by triangulating HXR sources, we investigated a single time interval in the later part of the impulsive phase. The interval is centred around the unsaturated 193 Å image recorded at 21:49:42.6 UT by the Atmospheric Imaging Assembly onboard the Solar Dynamics Observatory (SDO/AIA; Lemen et al. 2012). It was chosen because the HXR geometry at that time exhibited two footpoints, which simplifies co-alignment of the HXR images with the ultraviolet (UV) ribbons observed by AIA (Section 2.2).

2.2. Imaging

HXI and STIX provide spectroscopic imaging in the hard X-ray ranges 10 – 300 keV, and 4 – 150 keV, respectively, with angular resolutions down to $3.1''$ and $7''$. Due to the difficulty of focusing X-rays above $\gtrsim 10$ keV, both HXI and STIX employ indirect imaging techniques.[1] They use pairs of absorbing grids (bi-grids) to modulate the incoming X-ray flux before it is recorded by spectroscopic X-ray detectors (one detector per bi-grid; see Meuris et al. 2012; Zhang et al. 2019). The modulation is sensitive to the position of the X-ray source on the plane of the sky, as well as its angular extent perpendicular to the long axis of the bi-grids. The subcollimators (bi-grid plus detector) have different grid orientations and periods (pitches). They therefore sample different angular Fourier components of the source, more conveniently referred to as (complex) visibilities (Prince et al. 1988; Hurford 2013). The phase and amplitude of each STIX visibility are encoded into a Moiré pattern produced by the X-rays' transmission through the bi-grid, and recorded by the subcollimator's pixellated detector (Massa et al. 2023). By contrast, HXI uses non-pixellated detectors, and so employs two subcollimators per visibility to measure the sine and cosine components separately (Zhang et al. 2019). Once measured, the visibilities can be combined to reconstruct an image of the X-ray source using various well-established Fourier-based imaging algorithms (e.g. Piana et al. 2022).

Figure 2 shows the 2D image reconstructions of the thermal (15 – 18 keV, red contours, 20 – 95%) and non-thermal (32 – 76 keV, blue contours, 50 – 90%) emission seen by HXI and STIX at 21:49:42 UT.[2] The thermal and non-thermal images were integrated over intervals of 10 s and 1 min, respectively, centred around this time. This resulted in total imaged thermal counts above background of 116,174 and 19,910 and non-thermal counts above background of 626,282 and 65,441 for HXI and STIX, respectively. All image reconstructions only utilise subcollimators corresponding to angular scales $\geq 14''$, because the calibration of both instruments' finest subcollimators is not yet complete (Massa et al. 2022). In both cases, the Clean image reconstruction algorithm was used, which included a convolution with a Gaussian beam with a full width at half-maximum (FWHM) of $14''$, i.e. the resolution as the finest subcollimators used. HXI background subtraction was performed using a 6-min pre-flare integration centred on 21:33:20 UT. Note that care must be taken in selecting the HXI background due to its high time variability (Liu et al. 2022). By contrast,

[1] Solar-optimised directly focused hard X-ray observations have been achieved up to ~ 20 keV by the Focusing Optics X-ray Solar Imager (FOXSI) sounding rocket program (Krucker et al. 2009; Christe et al. 2016). A satellite version of FOXSI would provide directly focused observations up to ~ 55 keV (Christe et al. 2023; Reid et al. 2023), without the severe count-rate issues intrinsic to NuSTAR solar observing campaigns (Nuclear Spectroscopic Telescope Array; Harrison et al. 2013). Direct focusing has improved capabilities in sensitivity and imaging dynamic range, but remains challenging at higher energies.

[2] The HXI and STIX images shown in this paper were reconstructed with the HXI GUI Beta Version 1.40 and the STIX SSWIDL software v0.5.2 (Dickson et al. 2023, Note that a DOI for v0.5.2 is not yet available, so this citation represents v0.5.1).

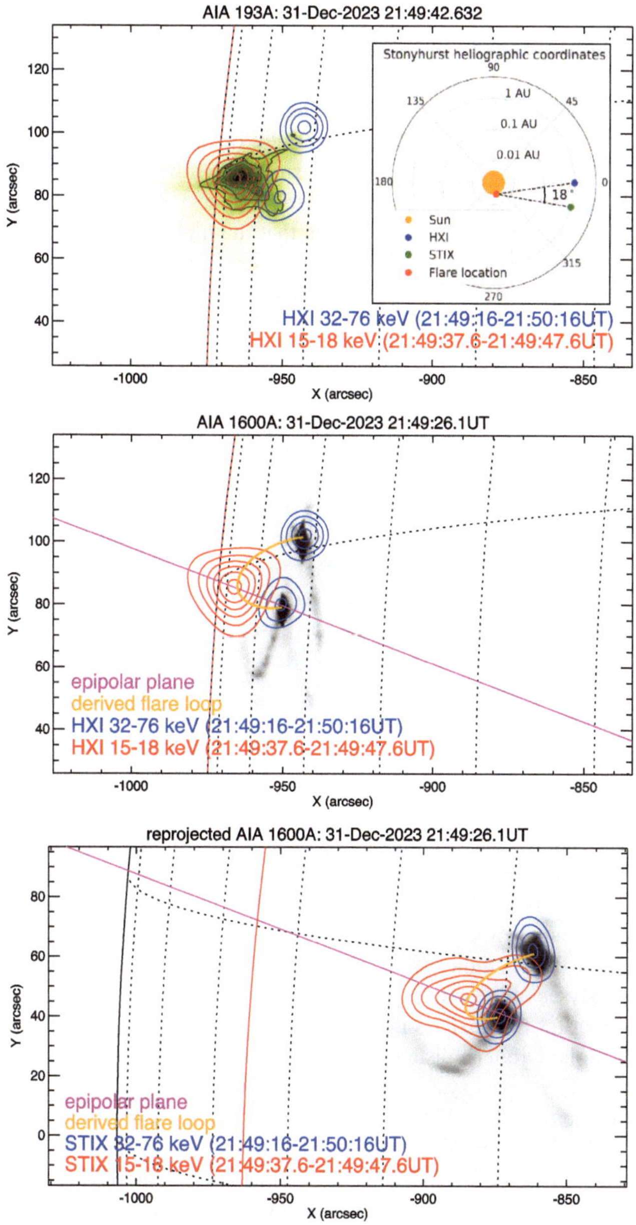

Figure 2 Imaging results in X-rays, EUV and UV around 21:49:42UT. The locations of HXI, STIX, the Sun and the flare are shown in the right inset of the top panel. The distances are plotted on a logarithmic scale to more clearly show the flare location on the Sun. The top and middle panels show thermal (red contours) and non-thermal (blue contours) observed by HXI overlaid on the AIA 193 Å and 1600 Å images, respectively. The bottom panel (see on the next page) shows the STIX HXR sources (thermal, red and non-thermal, blue) overlaid on the AIA 1600 Å image reprojected to the STIX vantage point. The magenta line is the projection of the epipolar plane intersecting Earth, Solar Orbiter and the brightest location of the thermal source (see Section 2.3). It appears as a line because each observer is located on the plane. The epipolar planes through the footpoint sources are not shown. The orange curve is the projection of the 3D loop reconstructed from the source locations. The red curve shows the solar limb as seen from Earth.

the STIX background is much more stable in time, and therefore a background estimate was obtained from observations over a non-flaring period on 28 December 2023 between 16:02 to 16:50 UT.

The top and middle panels show the resulting HXI contours overlaid on the AIA 193 Å and 1600 Å images, respectively. The morphology of the thermal HXI reconstruction is consistent with that of the hot ($\sim$ 10 MK) plasma shown by AIA 193 Å. The maximum intensity of the X-ray emission is slightly offset with respect to the core of the extreme ultraviolet (EUV) emission. This may indicate that the higher temperatures to which HXI is more sensitive reside at slightly higher altitudes than those seen by AIA 193 Å. However, as the displacement is less than $2''$, it could instead represent a combination of the alignment and image reconstruction inaccuracies. The next upgrade of the HXI imaging calibration may enable a closer examination of this question. The morphology of the HXI non-thermal footpoints is consistent with the dimension, relative orientation and separation of the UV ribbons (Figure 2, middle panel). The bottom panel shows the AIA 1600 Å map reprojected to Solar Orbiter's vantage point[3] and overlaid with the STIX contours. As part of the reprojection, the plate scale of the 1600 Å image has been corrected for the slightly different observer distances. The STIX non-thermal footpoints are also consistent with the reprojected UV flare ribbons, and all STIX sources appear consistent with those seen by HXI, considering the different viewing angles.

Figure 3 shows the count spectra observed during the time interval used for STIX and HXI thermal images. The STIX attenuator was inserted at this time, which drastically reduces the counts at lower energies and hence shifts the peak of the count spectrum to 15 keV. The peak of the HXI count spectrum is at roughly 16 keV, similar to STIX, but is broadened by HXI's coarser spectral resolution. The spectra were fitted with a model including isothermal, non-thermal thick target and albedo components (red, blue and grey histograms, respectively).[4] While a detailed discussion of the fitting results is beyond the scope of this paper, it is important to note two things relevant to 3D triangulation. First, both count spectra are dominated by the same model component in the imaged energy ranges (thermal for 15 – 18 keV and non-thermal for 32 – 76 keV). Secondly, the parameters of the components are sufficiently similar for both instruments. Together, this suggests that the brightest subregions of the sources imaged by both instruments' most likely represent the same 3D location. That said, future studies attempting more detailed 3D reconstructions may require more forensic analyses of the count spectra in the imaged spectral ranges (see Section 4).

A crucial aspect for the analysis in this study is the co-alignment of the STIX and HXI reconstructions. Solar Orbiter's distance from the Sun (0.95 AU) exceeded the 0.75 AU limit within which the STIX aspect system provides reliable pointing information (Warmuth et al. 2020). To overcome this, the STIX footpoints were co-aligned with the UV ribbons in the reprojected AIA 1600 Å map. This resulted in a shift of $(\Delta x, \Delta y) = (-17'', 29'')$ to the coordinates of the centre of the thermal and non-thermal images, producing a new STIX pointing accuracy of $2''$. The new uncertainty is based on the assumption that the UV ribbons and HXR footpoints were co-spatial, and the fact that shifts beyond $2''$ clearly showed, by eye, a misalignment between the cores of the HXR and UV sources. A similar process was applied

[3]Reprojection was performed assuming the flare ribbons are on the solar surface, and using the sunpy Map class, which leverages astropy reproject (SunPy Community et al. 2020; Mumford et al. 2020; Astropy Collaboration et al. 2022).

[4]The spectral fit of both instruments has been performed using the Object Spectral Executive (OSPEX; Schwartz et al. 2002) software available in SSWIDL. For the STIX spectral fit, a spectral response matrix with finer energy binning was used. At present, this feature is not available in the STIX software and it is not yet user-friendly. However, it will be included in a future release.

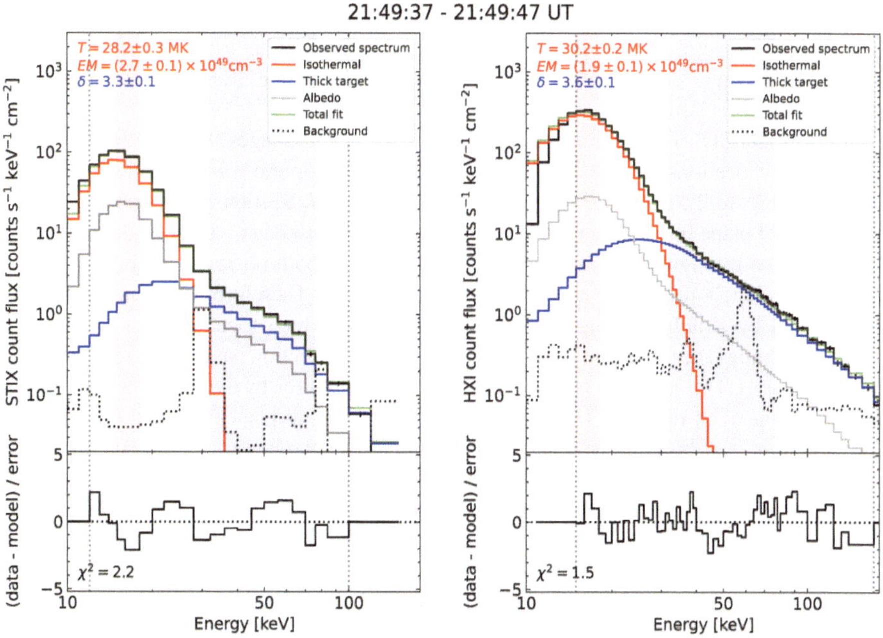

Figure 3 Count spectra from STIX (left) and HXI (right) of the 10 s time interval corresponding to the thermal images in Figure 2. The black histograms show the observed count spectra, while the red (thermal), blue (non-thermal), grey (albedo) and dotted (background) histograms show the components of the best-fit model (green). The lower panels show the residuals between the fit and observations. The vertical dotted lines show the range over which the fit was performed, while the red and blue shaded areas show the spectral ranges corresponding to the thermal and non-thermal images in Figure 2, respectively. The non-thermal spectral indices, δ, are the same within uncertainty, and the peaks of both the count spectra are within the thermal imaging range, where potential differences in the instruments temperature responses are minimised. This suggests that the corresponding HXI and STIX image pairs show the same plasma populations.

to co-align the HXI images with the UV flare ribbons. However, since the HXI pointing information was more reliable at this time, a much smaller shift of $(\Delta x, \Delta y) = (3'', 2'')$ was required. While the nominal pointing accuracy for the HXI observations used at the time of study was $< 1''$ (Zhang et al. 2019), a more detailed study suggests that calibration of HXI's aspect system to $< 0.3''$ should be feasible in the future (Su et al. 2024).

2.3. 3D Triangulation

The SOL2023-12-31T21:55 X-class flare was observed from Heliographic Stonyhurst positions (longitude, latitude and radial distance from Sun centre) $0°$, $-2.9°$ and 0.983 AU (HXI, low Earth orbit) and $-16.9°$, $3.3°$and 0.953 AU (STIX, beyond Earth orbit). These positions are shown in the inset in the top panel of Figure 2, along with the corresponding $18°$ observer separation relative to the flare site. Ryan et al. (2024) showed that the uncertainties associated with full 3D volumetric reconstructions grow rapidly for observer-separation angles less than $70°$. (See Appendix B of that paper, specifically Figure B.5.) Therefore, we derived the 3D scalar locations of the brightest subregions of the HXR sources via triangulation. Triangulation does not suffer the same complications as volumetric reconstructions because they depend more on the accuracy with which specific features can be identified from both viewing angles, rather than the extent of their 3D boundaries. The optically thin nature of the X-ray emission and similar viewing angles, and similarities between the instruments' count spectra (Section 2.2 and Figure 3) suggest that the brightest location in each image should represent the same subregion of the X-ray source. (See Section 4 for further discussion of this issue.)

Performing 3D triangulation requires knowledge of the images' coordinate frames and the observer locations. The first step is to produce a co-temporal pair of images from STIX and HXI (Figure 2), corrected for pointing misalignments (Section 2.2). Next, an epipolar plane – a 2D plane in 3D space that passes through both observers – must be defined so that it also passes through the feature being triangulated. As both observers sit in the epipolar plane, the plane's projection onto their images is a line. Therefore, the epipolar plane is determined by simply altering the arbitrary 3rd point, which together with the observer locations defines the plane, until the line of the plane's projection passes through the feature in both images. This is shown for the case of the thermal source by the magenta lines marked "epipolar plane" in Figure 2. (For a diagram of how epipolar planes relate to the observers and a feature, see Figure 2 of Ryan et al. 2024.) Next, the pixel locations of the feature along the plane are identified in both images. If we then recall that any pixel represents an infinite line of sight stretching away from the observer, it becomes clear that the 3D location of the feature must correspond to the intersection of those lines of sight.[5] The mathematical details of how to calculate the intersection within the epipolar plane and transform it to 3D real-world coordinates are outlined in Section 4 of Inhester (2006).

3. Results

The above methodology (Section 2.3) was applied to the HXI and STIX images (Figure 2) to calculate the 3D coordinates of the brightest locations of the thermal looptop and non-thermal footpoint sources. The coordinate components are shown in Table 1, as well as the sources' altitudes above the solar surface (1 $R_\odot$). The altitude of the thermal source is similar to estimates at similar stages of previous events (e.g. Gallagher et al. 2002; Milligan, McAteer, and Dennis 2010; Ryan et al. 2024), which gives confidence in the reliability of our triangulation calculations. The 1σ uncertainties of the 3D positions[6] and altitudes were estimated to be $\pm$ 2 Mm and $\pm$ 1 Mm, respectively. This was determined by assuming that the distribution of pointing uncertainty for each instrument was a Gaussian with a 3σ extent equal to the estimated error of the coalignment technique, i.e. $2''$. Random shifts from this pointing distribution were applied to the feature in the STIX and HXI images separately, and the 3D locations were recalculated along with their associated altitudes. The above-quoted uncertainties reflect the standard deviation of the resulting positional distributions.

This study represents the first determination of the 3D positions of flare footpoints via triangulation of multi-viewpoint HXR spectroscopic imaging observations. Within the above-stated uncertainty of $\pm$ 1 Mm, their altitudes above 1 $R_\odot$ place them approximately at the solar surface or slightly above. This is consistent with the footpoints being located in the chromosphere, as expected. This is consistent, within uncertainties, with previous studies that analysed single-viewpoint HXR observations of limb events. They found HXR

[5] As the lines of sight have been chosen to lie in the same 2D epipolar plane, they must intersect unless they are parallel. They cannot be parallel if the same feature has been identified in both images and the observers' separation angle is not 0° or 180°.

[6] In this analysis, we provide an uncertainty on the 3D location in terms of the radial distance from the derived 3D location. Therefore, we do not provide uncertainties on the coordinates of the location separately.

Table 1 3D locations and altitudes of the brightest regions in the X-ray sources. 3D coordinates are in Heliographic Stonyhurst, while altitudes are defined as above (positive) or below (negative) 1 $R_\odot$ (1 $R_\odot = 6.957 \times 10^8$ m, in accorrdance with IAU Resolution B3; Mamajek et al. 2015). The uncertainty in the 3D position is ± 2 Mm.

Source	Longitude [deg]	Latitude [deg]	Radius [$R_\odot$]	Altitude [Mm above 1 $R_\odot$]
Thermal looptop	−76.57	4.27	1.022	15.1 ± 1
Northern footpoint	−76.33	5.32	0.998	-1.2 ± 1
Southern footpoint	−77.09	4.00	1.003	1.8 ± 1

Table 2 Parameters of the elliptical flare loop defined by the thermal looptop source and footpoints.

Parameter	Value
Loop length	42 Mm
Loop inclination (from surface normal)	22.8°
Footpoint separation (minor axis)	18.8 Mm
Looptop to footpoint midpoint (semi-major axis)	16.1 Mm
Semi-axis ratio (minor/major)	0.58

footpoint altitudes in the ranges 0 – 2.5 Mm (at 30 keV; Brown, Aschwanden, and Kontar 2002; Aschwanden, Brown, and Kontar 2002), 0.8 – 1.7 Mm (Battaglia and Kontar 2011), 0.125 – 0.475 Mm (Martínez Oliveros et al. 2012), and 0.58 – 1.05 Mm (Krucker et al. 2015), including uncertainties.

It may be suggested that the footpoint altitudes found in this study are dictated by the co-alignment process, which requires that the HXR footpoints be aligned with the 1600 Å emission from the flare ribbon in the lower atmosphere. However, we have reason to believe that the derived altitudes are physically representative. In order for the footpoints in each image to correspond to the same point in 3D space, their projections must lie on the same epipolar plane. The co-alignment process was performed without prior knowledge of the 3D position of the sources. Despite this, for each source, an epipolar plane was found that passed through the brightest pixel in both images to within half a pixel. If this were not the case, it would imply an error in the image alignment. However, no adjustments were made post-facto, and the same instrument-specific pointing shifts were used for both the non-thermal and thermal source, which clearly resides at a higher altitude. This strongly suggests that the co-alignments, and hence the derived 3D positions, are representative of the true 3D source locations.

Using the 3D positions in Table 1, we parameterised the flare loop as a semi-ellipse rooted at the footpoints, with its apex in the thermal source (orange curve in Figure 2). The loop properties are shown in Table 2. The loop is significantly elongated into the corona (semi-axis ratio of 0.58, compared to 1 for a semi-circle) and tilted relative to vertical (22.8°). This is generally consistent with higher-resolution EUV images of loops in flaring active regions, which sometimes appear to be substantially inclined. The loop inclination leads to a longer derived loop length (42 Mm) and potentially different apex altitude than might otherwise be inferred from single-viewpoint studies. This may have consequences for the importance of different thermodynamic processes, such as the balance between radiative and conductive cooling (e.g. Cargill, Mariska, and Antiochos 1995; Klimchuk, Patsourakos, and Cargill 2008; Ryan et al. 2013).

4. Considerations for Future Studies

The results presented in Section 3 clearly demonstrate the potential of stereoscopic HXR imaging spectroscopy for characterising the geometry of high-energy processes in solar flares. In the future, we expect HXI and STIX to jointly observe many more flares, including at greater separation angles better suited to full 3D reconstruction. However, before more detailed studies are performed, a number of issues should be considered.

At present, the uncertainties in the HXR footpoint altitudes do not allow a more detailed consideration of the consequences for models of chromospheric energy deposition in flares than those provided by previous studies. However, triangulation uncertainties would be improved by better knowledge of the absolute source positions in the HXI and STIX images. This could be achieved in the future with improved calibration and a greater separation angle between HXI and STIX. Moreover, triangulation of footpoint sources further from the limb would not be subject to the ambiguities of foreshortening and absorption by optically thick plasma in the lower atmosphere, which had to be considered by those previous studies. Nevertheless, the preferred accuracy for addressing these science questions is of the order 0.1 Mm, which will be challenging to achieve with HXR triangulation.

While indirect imagers, such as HXI and STIX, are well suited to reconstructing large-scale features of X-ray sources, they are much less suited to reconstructing fine details. Care should therefore be given to not over-interpret such fine-scale structure, especially as part of a 3D reconstruction. When comparing HXI and STIX images of the same events, it may be advantageous to only use imaging subcollimators that result in both instruments' images having a similar spatial resolution, as was done in this study. HXI samples roughly 1.5 times as many visibilites as STIX, including some at finer angular scales ($3.1''$ for HXI compared to $7''$ for STIX). This should enable HXI to reconstruct slightly more morphological detail than STIX. However, when STIX is within ~ 0.45 AU of the Sun, its angular resolution corresponds to a finer spatial resolution than that of HXI. Therefore, which subcollimators result in the most comparable images will change from event to event.

Multi-viewpoint HXR analysis may also prove useful in disentangling the contribution of albedo in HXR imaging. This topic has proven challenging in the past with previous studies providing inconclusive results (e.g. Battaglia, Kontar, and Hannah 2011). While this topic is beyond the scope of this paper, it should be explored in future studies.

HXI's spectral 7 keV resolution contrasts starkly with the 1 keV resolution of STIX. When producing HXI images over a spectral range of a few keV, some of the counts will be due to photons outside the selected spectral range. When imaging close to the boundary of a spectral regime – e.g. the transition from thermal to non-thermal – counts associated with sources from the neighbouring spectral range can contaminate the image. This may alter the morphology of the imaged source, or the location of its centroid. The energy range of the thermal images in this paper (15 – 18 keV) are safely distant from this transition, as the non-thermal component only dominates over the thermal above 30 keV.

A similar effect can be induced by the instruments' different spectral responses, even within the selected spectral range. The spectral response defines the shape of the count spectrum as a function of energy for a given incident photon spectrum. The slope of the count spectrum, combined with the spectral resolution, dictates the weighted average energy of the imaged spectral range. If these are sufficiently different between HXI and STIX, different behaviour may be revealed by their images. For example, if the source is a multi-thermal plasma, the instrument whose image has the lower weighted-average energy will be sensitive to slightly lower temperatures. If these are spatially offset from the higher temperatures, the instruments may show a different source location or morphology. They may also show a different time evolution, because lower temperatures tend to decay slower, as the flare cools first from higher temperatures and then through lower ones. One way of minimising this phenomenon is to image at the peak of the count spectrum, as was done in this study. This was possible because the presence of the attenuator shifted the peak of STIX's count spectrum to a similar energy to that of HXI (~ 16 keV). In this scenario, the different spectral responses and resolutions have a smaller impact on the weighted average energies, as their effects are more evenly distributed between lower and higher energies within the spectral range. A more flexible approach, however, may be to image slightly different spectral ranges for the two instruments that nonetheless result in the same weighted-average energy. This would enable a greater exploration of the X-ray source's 3D structure throughout broader energy and time ranges.

Future 3D analyses that examine the temporal and spectral evolution of HXR sources should be performed in light of a greater understanding of the cross-calibration between HXI and STIX. Such efforts are currently underway by the relevant instrument teams, and should greatly benefit studies like this in the future.

5. Summary and Conclusions

We have presented calculations of the 3D locations of the thermal X-ray looptop and non-thermal X-ray footpoint sources during the impulsive phase of the SOL2023-12-31T21:55 X5-class flare. This was achieved by combining observations from ASO-S/HXI and Solar Orbiter/STIX, and represents the first triangulation of solar flare X-ray sources using HXR spectroscopic imaging observations. The footpoints were found to be separated by 18.8 Mm and have altitudes of -1.2 ± 1 Mm and 1.8 ± 1 Mm above one solar radius. The brightest point in the thermal source was found to have an altitude of 15.1 ± 1 Mm. These results are consistent with previous studies that used alternative observations and techniques for estimating the source altitudes. The loop joining the three sources was parameterised as a semi-ellipse and found to be elongated compared to a semi-circle. It had a length of 42 Mm and an inclination of 22.8°.

These results confirm the viability of 3D triangulation of HXR sources using HXI and STIX. More comprehensive 3D volumetric reconstructions were not attempted because the narrow observer separation angle (18°) leads to large uncertainties in the reconstructed volumes of the individual sources. Uncertainties in the triangulation of scalar 3D locations are not so severe as they depend more on the accuracy with which the same features can be identified between images, rather than the 3D source boundaries. We expect HXI and STIX to jointly observe more flares in the future, including many at substantially wider viewing angles, which will enable more comprehensive 3D reconstructions.

In Section 4, we outlined a number of complications that should be considered as part of more detailed studies in the future. These will, in part, be resolved by a better understanding of the calibration and cross-calibration of HXI and STIX. Nonetheless, this paper has demonstrated the current potential of multi-viewpoint HXR observations for scientific exploitation. Ongoing efforts by the HXI and STIX instrument teams to provide the community with better calibrations will only increase that potential. This promises to enable future studies to reconstruct the 3D HXR geometry of flares and eruptions in greater detail than previously possible, and hence elucidate the high-energy processes that underpin them.

Acknowledgments Solar Orbiter is a space mission operated by ESA and enabled by an international collaboration between ESA and NASA. The STIX instrument is an international collaboration between Switzerland, Poland, France, Czech Republic, Germany, Austria, Ireland and Italy.

Author contributions PM and SK produced the STIX images and PM and YS produced the HXI images. AFB, ECMD and WC performed the spectral analysis. 3D triangulation was performed DFR, who also led the writing of the manuscript. All authors provided expertise as part of scientific discussions that informed the content of the paper.

Funding Open access funding provided by FHNW University of Applied Sciences and Arts Northwestern Switzerland. AFB, DFR and SK are supported by the Swiss National Science Foundation Grant 200021L_189180 and the PRODEX experiment agreement No. 4000140488. PM is supported by the Swiss National Science Foundation in the framework of the project Robust Density Models for High Energy Physics and Solar Physics (rodem.ch), CRSII5_193716. ECMD is supported by the Austrian Science Fund (FWF) 10.55776/I4555. YS and WC are supported by the National Key R&D Program of China (2022YFF0503002), the National Natural Science Foundation of China (12333010,11820101002) and the Strategic Priority Research Program of the Chinese Academy of Sciences (XDB0560000). The ASO-S mission is supported by the Strategic Priority Research Program on Space Science, the Chinese Academy of Sciences, Grant No. XDA15320000.

Data availability The HXI and STIX data analysed in the current study are available at the ASO-S Data Archive (http://aso-s.pmo.ac.cn/sodc/dataArchive.jsp) and at the Solar Orbiter Archive (https://soar.esac.esa.int/soar/), respectively.

Declarations

Competing interests The authors declare no competing interests.

References

Aschwanden, M.J., Brown, J.C., Kontar, E.P.: 2002, Chromospheric height and density measurements in a solar flare observed with RHESSI II. Data analysis. *Solar Phys.* **210**, 383. DOI. ADS.

Astropy Collaboration, Price-Whelan, A.M., Lim, P.L., Earl, N., Starkman, N., Bradley, L., Shupe, D.L., Patil, A.A., Corrales, L., Brasseur, C.E., Nöthe, M., Donath, A., Tollerud, E., Morris, B.M., Ginsburg, A., Vaher, E., Weaver, B.A., Tocknell, J., Jamieson, W., van Kerkwijk, M.H., Robitaille, T.P., Merry, B., Bachetti, M., Günther, H.M., Aldcroft, T.L., Alvarado-Montes, J.A., Archibald, A.M., Bódi, A., Bapat, S., Barentsen, G., Bazán, J., Biswas, M., Boquien, M., Burke, D.J., Cara, D., Cara, M., Conroy, K.E., Conseil, S., Craig, M.W., Cross, R.M., Cruz, K.L., D'Eugenio, F., Dencheva, N., Devillepoix, H.A.R., Dietrich, J.P., Eigenbrot, A.D., Erben, T., Ferreira, L., Foreman-Mackey, D., Fox, R., Freij, N., Garg, S., Geda, R., Glattly, L., Gondhalekar, Y., Gordon, K.D., Grant, D., Greenfield, P., Groener, A.M., Guest, S., Gurovich, S., Handberg, R., Hart, A., Hatfield-Dodds, Z., Homeier, D., Hosseinzadeh, G., Jenness, T., Jones, C.K., Joseph, P., Kalmbach, J.B., Karamehmetoglu, E., Kaluszyński, M., Kelley, M.S.P., Kern, N., Kerzendorf, W.E., Koch, E.W., Kulumani, S., Lee, A., Ly, C., Ma, Z., MacBride, C., Maljaars, J.M., Muna, D., Murphy, N.A., Norman, H., O'Steen, R., Oman, K.A., Pacifici, C., Pascual, S., Pascual-Granado, J., Patil, R.R., Perren, G.I., Pickering, T.E., Rastogi, T., Roulston, B.R., Ryan, D.F., Rykoff, E.S., Sabater, J., Sakurikar, P., Salgado, J., Sanghi, A., Saunders, N., Savchenko, V., Schwardt, L., Seifert-Eckert, M., Shih, A.Y., Jain, A.S., Shukla, G., Sick, J., Simpson, C., Singanamalla, S., Singer, L.P., Singhal, J., Sinha, M., SipHocz, B.M., Spitler, L.R., Stansby, D., Streicher, O., Sumak, J., Swinbank, J.D., Taranu, D.S., Tewary, N., Tremblay, G.R., Val-Borro, M.D., Van Kooten, S.J., Vasović, Z., Verma, S., de Miranda Cardoso, J.V., Williams, P.K.G., Wilson, T.J., Winkel, B., Wood-Vasey, W.M., Xue, R., Yoachim, P., Zhang, C., Zonca, A., Astropy Project Contributors: 2022, The Astropy Project: sustaining and growing a community-oriented open-source project and the latest major release (v5.0) of the core package. *Astrophys. J.* **935**, 167. DOI. ADS.

Battaglia, M., Kontar, E.P.: 2011, Height structure of X-ray, EUV, and white-light emission in a solar flare. *Astron. Astrophys.* **533**, L2. DOI. ADS.

Battaglia, M., Kontar, E.P., Hannah, I.G.: 2011, The influence of albedo on the size of hard X-ray flare sources. *Astron. Astrophys.* **526**, A3. DOI. ADS.

Brown, J.C., Aschwanden, M.J., Kontar, E.P.: 2002, Chromospheric height and density measurements in a solar flare observed with RHESSI I. Theory. *Solar Phys.* **210**, 373. DOI. ADS.

Cargill, P.J., Mariska, J.T., Antiochos, S.K.: 1995, Cooling of solar flare plasmas. I. Theoretical considerations. *Astrophys. J.* **439**, 1034. DOI. ADS.

Christe, S., Glesener, L., Buitrago-Casas, C., Ishikawa, S.-N., Ramsey, B., Gubarev, M., Kilaru, K., Kolodziejczak, J.J., Watanabe, S., Takahashi, T., Tajima, H., Turin, P., Shourt, V., Foster, N., Krucker, S.: 2016, FOXSI-2: upgrades of the focusing optics X-ray solar imager for its second flight. *J. Astron. Instrum.* **5**, 1640005. DOI. ADS.

Christe, S., Alaoui, M., Allred, J., Battaglia, M., Baumgartner, W., Buitrago-Casas, J.C., Chen, B., Chen, T., Dennis, B., Drake, J., Glesener, L., Hannah, I., Hudson, H., Inglis, A., Ireland, J., Klimchuk, J., Kowalski, A., Krucker, S., Massone, A.M., Musset, S., Piana, M., Ryan, D., Shih, A., Veronig, A., Vilmer, N., Warmuth, A., White, S., Caspi, A., Hayes, L.A.: 2023, The Focusing Optics X-ray Solar Imager (FOXSI). *Bull. Am. Astron. Soc.* **55**, 065. DOI. ADS.

Dickson, E., Schwartz, R.A., Etesi, L.I., Massa, P., Schuller, F., Maloney, S., Battaglia, A.F., Hochmuth, N., Volpara, A., Benvenuto, F., Mrozek, T.: 2023, i4Ds/STIX-GSW: release of v0.5.1 – bugfix release. Zenodo. DOI.

Fletcher, L., Dennis, B.R., Hudson, H.S., Krucker, S., Phillips, K., Veronig, A., Battaglia, M., Bone, L., Caspi, A., Chen, Q., Gallagher, P., Grigis, P.T., Ji, H., Liu, W., Milligan, R.O., Temmer, M.: 2011, An observational overview of solar flares. *Space Sci. Rev.* **159**, 19. DOI. ADS.

Gallagher, P.T., Dennis, B.R., Krucker, S., Schwartz, R.A., Tolbert, A.K.: 2002, Rhessi and trace observations of the 21 April 2002 x1.5 flare. *Solar Phys.* **210**, 341. DOI. ADS.

Gan, W.-Q., Zhu, C., Deng, Y.-Y., Li, H., Su, Y., Zhang, H.-Y., Chen, B., Zhang, Z., Wu, J., Deng, L., Huang, Y., Yang, J.-F., Cui, J.-J., Chang, J., Wang, C., Wu, J., Yin, Z.-S., Chen, W., Fang, C., Yan, Y.-H., Lin, J., Xiong, W.-M., Chen, B., Bao, H.-C., Cao, C.-X., Bai, Y.-P., Wang, T., Chen, B.-L., Li, X.-Y., Zhang, Y., Feng, L., Su, J.-T., Li, Y., Chen, W., Li, Y.-P., Su, Y.-N., Wu, H.-Y., Gu, M., Huang, L., Tang, X.-J.: 2019, Advanced Space-based Solar Observatory (ASO-S): An overview. *Res. Astron. Astrophys.* **19**, 156. DOI. ADS.

Golub, L., DeLuca, E., Austin, G., Bookbinder, J., Caldwell, D., Cheimets, P., Cirtain, J., Cosmo, M., Reid, P., Sette, A., Weber, M., Sakao, T., Kano, R., Shibasaki, K., Hara, H., Tsuneta, S., Kumagai, K., Tamura, T., Shimojo, M., McCracken, J., Carpenter, J., Haight, H., Siler, R., Wright, E., Tucker, J., Rutledge, H., Barbera, M., Peres, G., Varisco, S.: 2007, The X-Ray Telescope (XRT) for the Hinode Mission. *Solar Phys.* **243**, 63. DOI. ADS.

Harrison, F.A., Craig, W.W., Christensen, F.E., Hailey, C.J., Zhang, W.W., Boggs, S.E., Stern, D., Cook, W.R., Forster, K., Giommi, P., Grefenstette, B.W., Kim, Y., Kitaguchi, T., Koglin, J.E., Madsen, K.K., Mao, P.H., Miyasaka, H., Mori, K., Perri, M., Pivovaroff, M.J., Puccetti, S., Rana, V.R., Westergaard, N.J., Willis, J., Zoglauer, A., An, H., Bachetti, M., Barrière, N.M., Bellm, E.C., Bhalerao, V., Brejnholt, N.F., Fuerst, F., Liebe, C.C., Markwardt, C.B., Nynka, M., Vogel, J.K., Walton, D.J., Wik, D.R., Alexander, D.M., Cominsky, L.R., Hornschemeier, A.E., Hornstrup, A., Kaspi, V.M., Madejski, G.M., Matt, G., Molendi, S., Smith, D.M., Tomsick, J.A., Ajello, M., Ballantyne, D.R., Baloković, M., Barret, D., Bauer, F.E., Blandford, R.D., Brandt, W.N., Brenneman, L.W., Chiang, J., Chakrabarty, D., Chenevez, J., Comastri, A., Dufour, F., Elvis, M., Fabian, A.C., Farrah, D., Fryer, C.L., Gotthelf, E.V., Grindlay, J.E., Helfand, D.J., Krivonos, R., Meier, D.L., Miller, J.M., Natalucci, L., Ogle, P., Ofek, E.O., Ptak, A., Reynolds, S.P., Rigby, J.R., Tagliaferri, G., Thorsett, S.E., Treister, E., Urry, C.M.: 2013, The Nuclear Spectroscopic Telescope Array (NuSTAR) high-energy X-Ray Mission. *Astrophys. J.* **770**, 103. DOI. ADS.

Holman, G.D., Aschwanden, M.J., Aurass, H., Battaglia, M., Grigis, P.C., Kontar, E.P., Liu, W., Saint-Hilaire, P., Zharkova, V.V.: 2011, Implications of X-ray observations for electron acceleration and propagation in solar flares. *Space Sci. Rev.* **159**, 107. DOI. ADS.

Hurford, G.J.: 2013, X-ray imaging with collimators, masks and grids. In: *Observing Photons in Space*, Springer, Berlin, 243.

Inhester, B.: 2006, Stereoscopy basics for the STEREO mission. DOI. arXiv. ADS.

Klimchuk, J.A., Patsourakos, S., Cargill, P.J.: 2008, Highly efficient modeling of dynamic coronal loops. *Astrophys. J.* **682**, 1351. DOI. ADS.

Kontar, E.P., Brown, J.C., Emslie, A.G., Hajdas, W., Holman, G.D., Hurford, G.J., Kašparová, J., Mallik, P.C.V., Massone, A.M., McConnell, M.L., Piana, M., Prato, M., Schmahl, E.J., Suarez-Garcia, E.: 2011, Deducing electron properties from hard X-ray observations. *Space Sci. Rev.* **159**, 301. DOI. ADS.

Krucker, S., Christe, S., Glesener, L., McBride, S., Turin, P., Glaser, D., Saint-Hilaire, P., Delory, G., Lin, R.P., Gubarev, M., Ramsey, B., Terada, Y., Ishikawa, S.-N., Kokubun, M., Saito, S., Takahashi, T., Watanabe, S., Nakazawa, K., Tajima, H., Masuda, S., Minoshima, T., Shomojo, M.: 2009, The Focusing Optics X-ray Solar Imager (FOXSI). In: O'Dell, S.L., Pareschi, G. (eds.) *Optics for EUV, X-Ray, and Gamma-Ray Astronomy IV, Society of Photo-Optical Instrumentation Engineers (SPIE) Conference Series* **7437**, 743705. DOI. ADS.

Krucker, S., Saint-Hilaire, P., Hudson, H.S., Haberreiter, M., Martinez-Oliveros, J.C., Fivian, M.D., Hurford, G., Kleint, L., Battaglia, M., Kuhar, M., Arnold, N.G.: 2015, Co-spatial white light and hard X-ray flare footpoints seen above the solar limb. *Astrophys. J.* **802**, 19. DOI. ADS.

Krucker, S., Hurford, G.J., Grimm, O., Kögl, S., Gröbelbauer, H.-P., Etesi, L., Casadei, D., Csillaghy, A., Benz, A.O., Arnold, N.G., et al.: 2020, The Spectrometer/Telescope for Imaging X-rays (STIX). *Astron. Astrophys.* **642**, A15.

Lemen, J.R., Title, A.M., Akin, D.J., Boerner, P.F., Chou, C., Drake, J.F., Duncan, D.W., Edwards, C.G., Friedlaender, F.M., Heyman, G.F., Hurlburt, N.E., Katz, N.L., Kushner, G.D., Levay, M., Lindgren, R.W., Mathur, D.P., McFeaters, E.L., Mitchell, S., Rehse, R.A., Schrijver, C.J., Springer, L.A., Stern, R.A., Tarbell, T.D., Wuelser, J.-P., Wolfson, C.J., Yanari, C., Bookbinder, J.A., Cheimets, P.N., Caldwell, D., Deluca, E.E., Gates, R., Golub, L., Park, S., Podgorski, W.A., Bush, R.I., Scherrer, P.H., Gummin, M.A., Smith, P., Auker, G., Jerram, P., Pool, P., Soufli, R., Windt, D.L., Beardsley, S., Clapp, M., Lang, J., Waltham, N.: 2012, The Atmospheric Imaging Assembly (AIA) on the Solar Dynamics Observatory (SDO). *Solar Phys.* **275**, 17. DOI.

Lin, R.P., Dennis, B.R., Hurford, G.J., Smith, D.M., Zehnder, A., Harvey, P.R., Curtis, D.W., Pankow, D., Turin, P., Bester, M., Csillaghy, A., Lewis, M., Madden, N., van Beek, H.F., Appleby, M., Raudorf, T., McTiernan, J., Ramaty, R., Schmahl, E., Schwartz, R., Krucker, S., Abiad, R., Quinn, T., Berg, P., Hashii, M., Sterling, R., Jackson, R., Pratt, R., Campbell, R.D., Malone, D., Landis, D., Barrington-Leigh, C.P., Slassi-Sennou, S., Cork, C., Clark, D., Amato, D., Orwig, L., Boyle, R., Banks, I.S., Shirey, K., Tolbert, A.K., Zarro, D., Snow, F., Thomsen, K., Henneck, R., Mchedlishvili, A., Ming, P., Fivian, M., Jordan, J., Wanner, R., Crubb, J., Preble, J., Matranga, M., Benz, A., Hudson, H., Canfield, R.C., Holman, G.D., Crannell, C., Kosugi, T., Emslie, A.G., Vilmer, N., Brown, J.C., Johns-Krull, C., Aschwanden, M., Metcalf, T., Conway, A.: 2002, The Reuven Ramaty High-Energy Solar Spectroscopic Imager (RHESSI). *Solar Phys.* **210**, 3. DOI. ADS.

Liu, W., Chen, D.-Y., Jiang, X.-K., Wu, J., Zhang, Z., Hu, Y.-M., Su, Y., Chen, W., Ma, T.: 2022, Research on the on-orbit background of the Hard X-Ray Imager Onboard ASO-S. *Res. Astron. Astrophys.* **22**, 095011. DOI. ADS.

Mamajek, E.E., Prsa, A., Torres, G., Harmanec, P., Asplund, M., Bennett, P.D., Capitaine, N., Christensen-Dalsgaard, J., Depagne, E., Folkner, W.M., Haberreiter, M., Hekker, S., Hilton, J.L., Kostov, V., Kurtz, D.W., Laskar, J., Mason, B.D., Milone, E.F., Montgomery, M.M., Richards, M.T., Schou, J., Stewart, S.G.: 2015, IAU 2015 Resolution B3 on recommended nominal conversion constants for selected solar and planetary properties. arXiv e-prints. DOI. arXiv. ADS.

Martínez Oliveros, J.-C., Hudson, H.S., Hurford, G.J., Krucker, S., Lin, R.P., Lindsey, C., Couvidat, S., Schou, J., Thompson, W.T.: 2012, The height of a white-light flare and its hard X-ray sources. *Astrophys. J. Lett.* **753**, L26. DOI. ADS.

Massa, P., Battaglia, A.F., Volpara, A., Collier, H., Hurford, G.J., Kuhar, M., Perracchione, E., Garbarino, S., Massone, A.M., Benvenuto, F., Schuller, F., Warmuth, A., Dickson, E.C.M., Xiao, H., Maloney, S.A., Ryan, D.F., Piana, M., Krucker, S.: 2022, First hard X-ray imaging results by solar orbiter STIX. *Solar Phys.* **297**, 93. DOI. ADS.

Massa, P., Hurford, G.J., Volpara, A., Kuhar, M., Battaglia, A.F., Xiao, H., Casadei, D., Perracchione, E., Garbarino, S., Guastavino, S., Collier, H., Dickson, E.C.M., Emslie, A.G., Ryan, D.F., Maloney, S.A., Schuller, F., Warmuth, A., Massone, A.M., Benvenuto, F., Piana, M., Krucker, S.: 2023, The STIX imaging concept. *Solar Phys.* **298**, 114. DOI. ADS.

Meuris, A., Hurford, G., Bednarzik, M., Limousin, O., Gevin, O., Le Mer, I., Martignac, J., Horeau, B., Grimm, O., Resanovic, R., Krucker, S., Orleański, P.: 2012, Caliste-SO X-ray micro-camera for the STIX instrument on-board Solar Orbiter space mission. *Nucl. Instrum. Methods Phys. Res., Sect. A, Accel. Spectrom. Detect. Assoc. Equip.* **695**, 288. DOI. ADS.

Milligan, R.O., McAteer, R.T.J., Dennis, B.R., Young, C.A.: 2010, Evidence of a plasmoid-looptop interaction and magnetic inflows during a solar flare/coronal mass ejection eruptive event. *Astrophys. J.* **713**, 1292. DOI. ADS.

Müller, D., St. Cyr, O.C., Zouganelis, I., Gilbert, H.R., Marsden, R., Nieves-Chinchilla, T., Antonucci, E., Auchère, F., Berghmans, D., Horbury, T.S., Howard, R.A., Krucker, S., Maksimovic, M., Owen, C.J., Rochus, P., Rodriguez-Pacheco, J., Romoli, M., Solanki, S.K., Bruno, R., Carlsson, M., Fludra, A., Harra, L., Hassler, D.M., Livi, S., Louarn, P., Peter, H., Schühle, U., Teriaca, L., del Toro Iniesta, J.C.,

Wimmer-Schweingruber, R.F., Marsch, E., Velli, M., De Groof, A., Walsh, A., Williams, D.: 2020, The Solar Orbiter mission. Science overview. *Astron. Astrophys.* **642**, A1.

Mumford, S., Freij, N., Christe, S., Ireland, J., Mayer, F., Hughitt, V., Shih, A., Ryan, D., Liedtke, S., Pérez-Suárez, D., Chakraborty, P., KI, V., Inglis, A., Pattnaik, P., Sipőcz, B., Sharma, R., Leonard, A., Stansby, D., Hewett, R., Hamilton, A., Hayes, L., Panda, A., Earnshaw, M., Choudhary, N., Kumar, A., Chanda, P., Haque, M., Kirk, M., Mueller, M., Konge, S., Srivastava, R., Jain, Y., Bennett, S., Baruah, A., Barnes, W., Charlton, M., Maloney, S., Chorley, N., Himanshu, M.S., Mason, J., Naman Campos Rozo, J.I., Manley, L., Chatterjee, A., Evans, J., Malocha, M., Bobra, M., Ghosh, S., Airmansmith, Stańczak, D., De Visscher, R., Verma, S., Agrawal, A., Buddhika, D., Sharma, S., Park, J., Bates, M., Goel, D., Taylor, G., Cetusic, G., Jacob, I.M., Dacie, S., Dubey, S., Sharma, D., Bray, E., Rideout, J., Zahniy, S., Meszaros, T., Bose, A., Chicrala, A., Ankit, G.C., D'Avella, D., Williams, D., Ballew, J., Murphy, N., Lodha, P., Robitaille, T., Krishan, Y., Hill, A., Eigenbrot, A., Mampaey, B., Wiedemann, B., Molina, C., Keşkek, D., Habib, I., Letts, J., Bazán, J., Arbolante, Q., Gomillion, R., Kothari, Y., Sharma, Y., Stevens, A., Price-Whelan, A., Mehrotra, A., Kustov, A., Stone, B., Dang, T., Arias, E., Dover, F., Verstringe, F., Kumar, G., Mathur, H., Babuschkin, I., Wimbish, J., Buitrago-Casas, J., Krishna, K., Hiware, K., Mangaonkar, M., Mendero, M., Schoentgen, M., Gyenge, N., Streicher, O., Mekala, R., Mishra, R., Srikanth, S., Jain, S., Yadav, T., Wilkinson, T., Pereira, T., Agrawal, Y., Yasintoda, Murray, S.: 2020, SunPy: a Python package for solar physics. *J. Open Sour. Softw.* **5**, 1832. DOI. ADS.

Piana, M., Emslie, A.G., Massone, A.M., Dennis, B.R.: 2022, *Hard X-ray Imaging of Solar Flares*, Springer, Berlin.

Prince, T.A., Hurford, G.J., Hudson, H.S., Crannell, C.J.: 1988, Gamma-ray and hard X-ray imaging of solar flares. *Solar Phys.* **118**, 269. DOI. ADS.

Reid, H.A.S., Musset, S., Ryan, D.F., Andretta, V., Auchère, F., Baker, D., Benvenuto, F., Browning, P., Buchlin, E., Calcines Rosario, A., Christe, S.D., Corso, A.J., Dahlin, J., Dalla, S., Del Zanna, G., Denker, C., Dudík, J., Erdélyi, R., Ermolli, I., Fletcher, L., Fludra, A., Green, L.M., Gordovskyy, M., Guglielmino, S.L., Hannah, I., Harrison, R., Hayes, L.A., Inglis, A.R., Jeffrey, N.L.S., Kašparová, J., Kerr, G.S., Kintziger, C., Kontar, E.P., Krucker, S., Laitinen, T., Laurent, P., Limousin, O., Long, D.M., Maloney, S.A., Massa, P., Massone, A.M., Matthews, S., Mrozek, T., Nakariakov, V.M., Parenti, S., Piana, M., Polito, V., Pesce-Rollins, M., Romano, P., Rouillard, A.P., Sasso, C., Shih, A.Y., Steslicki, M., Orozco Suárez, D., Teriaca, L., Verma, M., Veronig, A.M., Vilmer, N., Vocks, C., Warmuth, A.: 2023, The Solar Particle Acceleration Radiation and Kinetics (SPARK) mission concept. *Aerospace* **10**, 1034. DOI. https://www.mdpi.com/2226-4310/10/12/1034.

Ryan, D.F., Chamberlin, P.C., Milligan, R.O., Gallagher, P.T.: 2013, Decay-phase cooling and inferred heating of M- and X-class solar flares. *Astrophys. J.* **778**, 68. DOI. ADS.

Ryan, D.F., Laube, S., Nicula, B., Krucker, S., Maloney, S.A., Battaglia, A.F., Warmuth, A., Csillaghy, A., Müller, D.: 2024, 3D evolution of a solar flare thermal X-ray loop-top source. *Astron. Astrophys.* **681**, A61. DOI. ADS.

Schwartz, R.A., Csillaghy, A., Tolbert, A.K., Hurford, G.J., Mc Tiernan, J., Zarro, D.: 2002, RHESSI data analysis software: rationale and methods. *Solar Phys.* **210**, 165. DOI. ADS.

Su, Y., Liu, W., Li, Y.-P., Zhang, Z., Hurford, G.J., Chen, W., Huang, Y., Li, Z.-T., Jiang, X.-K., Wang, H.-X., Xia, F.-X.-Y., Chen, C.-X., Yu, W.-H., Yu, F., Wu, J., Gan, W.-Q.: 2019, Simulations and software development for the Hard X-ray Imager onboard ASO-S. *Res. Astron. Astrophys.* **19**, 163. DOI. ADS.

Su, Y., et al.: 2024, Solar Phys. In prep.

SunPy Community, Barnes, W.T., Bobra, M.G., Christe, S.D., Freij, N., Hayes, L.A., Ireland, J., Mumford, S., Perez-Suarez, D., Ryan, D.F., Shih, A.Y., Chanda, P., Glogowski, K., Hewett, R., Hughitt, V.K., Hill, A., Hiware, K., Inglis, A., Kirk, M.S.F., Konge, S., Mason, J.P., Maloney, S.A., Murray, S.A., Panda, A., Park, J., Pereira, T.M.D., Reardon, K., Savage, S., Sipőcz, B.M., Stansby, D., Jain, Y., Taylor, G., Yadav, T., Rajul, D.T.K.: 2020, The SunPy project: open source development and status of the version 1.0 core package. *Astrophys. J.* **890**, 68. DOI. ADS.

Warmuth, A., Önel, H., Mann, G., Rendtel, J., Strassmeier, K.G., Denker, C., Hurford, G.J., Krucker, S., Anderson, J., Bauer, S.-M., Bittner, W., Dionies, F., Paschke, J., Plüschke, D., Sablowski, D.P., Schuller, F., Senthamizh Pavai, V., Woche, M., Casadei, D., Kögl, S., Arnold, N.G., Gröbelbauer, H.-P., Schori, D., Wiehl, H.J., Csillaghy, A., Grimm, O., Orleanski, P., Skup, K.R., Bujwan, W., Rutkowski, K., Ber, K.: 2020, The STIX Aspect sYstem (SAS): the optical aspect system of the Spectrometer/Telescope for Imaging X-Rays (STIX) on Solar Orbiter. *Solar Phys.* **295**, 90. DOI. ADS.

Zhang, Z., Chen, D.-Y., Wu, J., Chang, J., Hu, Y.-M., Su, Y., Zhang, Y., Wang, J.-P., Liang, Y.-M., Ma, T., et al.: 2019, Hard X-ray Imager (HXI) onboard the ASO-S mission. *Res. Astron. Astrophys.* **19**, 160.

Authors and Affiliations

Daniel F. Ryan[1,2] · **Paolo Massa**[1] · **Andrea F. Battaglia**[1,3,4] · **Ewan C.M. Dickson**[5] · **Yang Su**[6,7] · **Wei Chen**[6] · **Säm Krucker**[1,8]

✉ D.F. Ryan
daniel.ryan@fhnw.ch; daniel.ryan@ucl.ac.uk

P. Massa
paolo.massa@fhnw.ch

A.F. Battaglia
andrea.battaglia@fhnw.ch

E.C.M. Dickson
ewan.dickson@uni-graz.at

Y. Su
yang.su@pmo.ac.cn

W. Chen
w.chen@pmo.ac.cn

S. Krucker
krucker@berkeley.edu

1 University of Applied Sciences and Arts Northwestern Switzerland, Bahnhofstrasse 6, Windisch 5210, Switzerland

2 Present address: Mullard Space Science Laboratory, University College London, Dorking RH5 6NT, UK

3 Institute for Particle Physics and Astrophysics (IPA), Swiss Federal Institute of Technology in Zurich (ETHZ), Wolfgang-Pauli-Strasse 27, 8039 Zurich, Switzerland

4 Present address: Istituto ricerche solari Aldo e Cele Daccò (IRSOL), Faculty of informatics, Università della Svizzera italiana, Locarno, Switzerland

5 University of Graz, Institute of Physics & Kanzelhöhe Observatory, Universitätsplatz 5, 8010 Graz, Austria

6 Key Laboratory of Dark Matter and Space Astronomy, Purple Mountain Observatory, Chinese Academy of Sciences, Nanjing 210023, China

7 School of Astronomy and Space Science, University of Science and Technology of China, Hefei 230026, China

8 Space Sciences Lab, UC Berkeley, 7 Gauss Way, Berkeley, CA 94708, USA

Part 4: LST Related Research

Solar Physics (2024) 299:72
https://doi.org/10.1007/s11207-024-02310-1

RESEARCH

Automatic Solar Flare Detection Using the Solar Disk Imager Onboard the ASO-S Mission

Lei Lu[1] · Zhengyuan Tian[1,2] · Li Feng[1,2] · Jiahui Shan[1,2] · Hui Li[1,2] · Yang Su[1,2] · Ying Li[1,2] · Yu Huang[1,2] · Youping Li[1,2] · Jingwei Li[1] · Jie Zhao[1] · Beili Ying[1] · Jianchao Xue[1] · Ping Zhang[1] · Dechao Song[1] · Shuting Li[1,2] · Guanglu Shi[1,2] · Yingna Su[1,2] · Qingmin Zhang[1] · Yunyi Ge[1] · Bo Chen[3] · Qiao Li[1,2] · Gen Li[1,2] · Yue Zhou[1,2] · Jun Tian[1,2] · Xiaofeng Liu[1,2] · Zhichen Jing[1,2] · Weiqun Gan[1,4] · Kefei Song[3] · Lingping He[3] · Shijun Lei[1]

Received: 23 February 2024 / Accepted: 27 April 2024 / Published online: 29 May 2024

Abstract
We present an automated solar flare detection software tool to automatically process solar observed images, detect and track solar flares, and finally compile an event catalog. It can identify and track flares that happen simultaneously or temporally close together. The method to identify a flare is based on the local intensity changes in macropixels. The basic characteristics, such as the time and location information of a flare, are determined with a triple-threshold scheme, with the first threshold (global threshold) to determine the occurrence (location) of the flare and the second and third thresholds (local thresholds) to determine the real start and end times of the flare. We have applied this tool to one month of continuous solar ultraviolet (UV) images obtained by the *Solar Disk Imager* (SDI) onboard the *Advanced Space-based Solar Observatory* (ASO-S), which show active phenomena such as flares, filaments or prominences, and solar jets. Our automated tool efficiently detected a total number of 226 solar events. After a visual inspection, we found that only one event was misidentified (unrelated to an active event). We compared the detected events with the GOES X-ray flare list and found that our tool can detect 81% of GOES M-class and above flares (29 out of 36), from which we conclude that the intensity increase in SDI UV images can be considered as a good indicator of a solar flare.

✉ L. Lu
leilu@pmo.ac.cn

✉ L. Feng
lfeng@pmo.ac.cn

1 Key Laboratory of Dark Matter and Space Astronomy, Purple Mountain Observatory, CAS, Nanjing 210033, P.R. China

2 School of Astronomy and Space Science, University of Science and Technology of China, Hefei 230026, P.R. China

3 Changchun Institute of Optics, Fine Mechanics and Physics, Chinese Academy of Sciences, Changchun 130033, China

4 University of Chinese Academy of Sciences, Nanjing 211135, China

Keywords Solar flare · Solar ultraviolet emission · Automatic detection

1. Introduction

Solar flares, characterized by sudden releases of energy and radiation from the Sun's surface on timescales of a few minutes to hours (Shibata and Yokoyama, 2002; Benz, 2008), are a key process to understand the Sun's dynamic behavior and its influence on space weather. They are widely believed to result from a magnetic reconnection process (Shibata and Magara, 2011), which is still not fully understood. Over the last two decades, many space-based solar telescopes and missions such as the *Solar and Heliospheric Observatory* (SOHO: Domingo, Fleck, and Poland, 1995), *Transition Region and Coronal Explorer* (TRACE: Handy et al., 1999), *Hinode* (Kosugi et al., 2007), *Solar Dynamics Observatory* (SDO: Pesnell, Thompson, and Chamberlin, 2012), *Interface Region Imaging Spectrograph* (IRIS: De Pontieu et al., 2014), and *Solar Orbiter* (SolO: Müller et al., 2020) have been designed and launched to observe the Sun in various wavelengths with high spatial and temporal resolutions. The data collected by instruments onboard these satellites enhance our knowledge of solar processes and their impact on the solar system.

Recently, China successfully sent its first comprehensive and dedicated solar observatory – the *Advanced Space-based Solar Observatory* (ASO-S) into space (Gan et al., 2019; Gan, Feng, and Su, 2022; Gan et al., 2023). The *Solar Disk Imager* (SDI) is one of the scientific instruments carried by ASO-S, designed to capture images of the solar disk and the lower solar atmosphere (0 to 1.2 solar radii) in the Lyman-α (Ly-α, 121.6 nm) passband (Li et al., 2019; Chen et al., 2019; Feng et al., 2019). The regular temporal cadence of SDI images is 60 s but could be up to 6 s in burst mode (Lu et al., 2020). As the volume of data generated by SDI continues to expand ($\approx$ 80 GB of data per day), developing efficient and accurate algorithms for automatically detecting solar flares becomes increasingly necessary.

Several algorithms for automatically detecting flares from solar images have been proposed and implemented in recent decades. The flare detection tool of the *Hinode/X-ray telescope* (XRT) segments its X-ray images into a small number of macropixels and detects flares by thresholding the relative difference between an image and its reference image (the latter is calculated from a collection of the previous images) (Kano et al., 2008). Later on, analysis of the intensity changes in macropixels was also used for flare detection in EUV images (Grigis et al., 2010; Bonte et al., 2013). To obtain a more precise flare location, the Solar Demon flare detection algorithm utilizes a region-based paradigm, tracking intensity increases in the original image rather than the macropixels (Kraaikamp and Verbeeck, 2015). Moreover, neural networks (Fernandez Borda et al., 2002) and support vector machines (Qu et al., 2003) have also proven to have great prospects for automatically detecting solar flares.

While previous algorithms offer a systematic and rapid approach to identifying and characterizing solar flare events, there are still some problems. For example, the *Hinode* flare detection tool was designed to detect flares onboard, but there were difficulties in identifying flares that happen simultaneously or temporally close together since a new event cannot be detected until the end of the previous one is found. Moreover, many flares are generally not so intense at the beginning, which means that the start time of a flare is usually not well determined with a simple intensity threshold. Finally, because of the use of macropixels, some uncertainties are often involved in determining the flare location.

In this article, we describe our Automated Solar Flare Detection (ASFD), a software tool that automatically processes solar images from ASO-S/SDI, detects and tracks solar flares, and finally compiles an event catalog. In Section 2, we describe the input data used.

In Section 3, we describe the detection algorithm. The detection results are presented in Section 4. Conclusions and discussion are provided in Section 5.

2. Input Data

Solar flare detection here is performed on SDI Ly-α (121.6 Å) images, which correspond to the most intense emission in the ultraviolet (UV) part of the solar electromagnetic spectrum. In the Sun, Ly-α emission occurs primarily in the chromosphere and transition region of the solar atmosphere when an electron of a hydrogen atom transitions from a higher energy level to the $n = 1$ (ground) energy level. Therefore, observations of the Sun in Ly-α are crucial for studying the outer layers of the solar atmosphere and understanding various processes, such as heating mechanisms and dynamics in the chromosphere. Currently, ongoing space-based observatories equipped with instruments capable of detecting solar Ly-α emissions include the *Extreme Ultraviolet Sensor* (EUVS) on the GOES series satellites, the *Large Yield RAdiometer* (LYRA) on the *PRoject for OnBoard Autonomy 2* (PROBA2), the *Extreme ultraviolet Variability Experiment* (EVE) on the *Solar Dynamics Observatory* (SDO), the *Extreme Ultraviolet Imager* (EUI) on *Solar Orbiter* and the *Solar Disk Imager* (SDI) on ASO-S. EUVS, LYRA, and EVE provide only disk-integrated solar irradiance, while EUI is designed to take partial field-of-view (FOV) Ly-α images of the solar disk, making it difficult to detect solar flares. SDI provides Ly-α images of the Sun with a spatial resolution of around $9.5''$ in two observation modes, i.e., routine mode and burst mode. In the routine mode, SDI takes full-FOV images of the Sun with a temporal cadence of 1 minute and can be automatically shifted to the burst mode once a local emission enhancement is detected. As soon as the burst mode is initiated, solar images are taken at a higher temporal resolution (6 seconds), but only data in a cutout window of 1024×1024 pixels around the flaring region are transmitted to the ground. Meanwhile, full-FOV images are still collected at a temporal cadence of around 1 minute. In this work, only the full-FOV images of the Sun are used as input by our detection tool.

3. Detection Tool

Similar to previous algorithms, the algorithm behind ASFD also takes advantage of the intensity changes in local macropixels. It consists of four basic modules: the first is data preprocessing, which optimizes the input images for flare detection; the second calculates a feature map on which the detection is performed; the third makes an initial detection by searching for pixels with significant intensity enhancements and refine the detection result once an event is detected; the fourth tracks the detected event until the end threshold is met.

3.1. Data Preprocessing

Solar telescopes working in outer space are vulnerable to the bombardment of cosmic rays or energetic particle flows, especially when the satellites travel through the South Atlantic Anomaly (SAA). This could lead to isolated and sudden brightenings on the observed image, which are sometimes misidentified as solar flares. The raw images are first passed through a median filter to avoid this, suppressing sharp noise features based on the ranking statistics theory. Secondly, to improve the computing efficiency as well as to further suppress the background noise, the input images are segmented into a small number (32×32) of

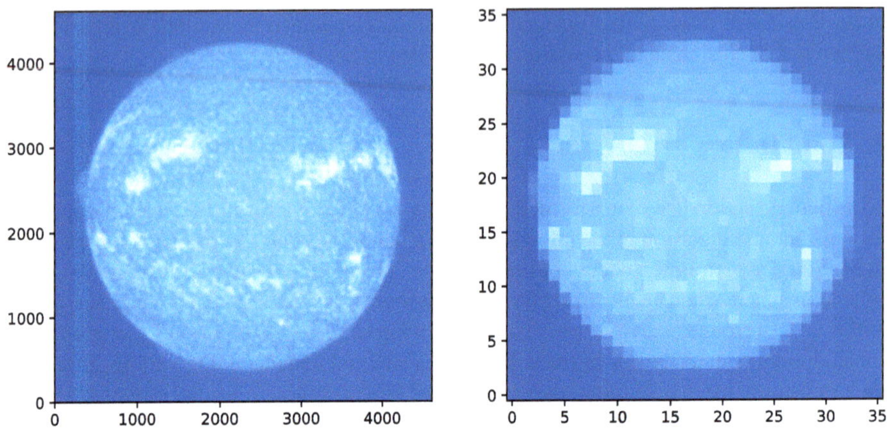

Figure 1 Left: The raw SDI image processed with the median filter. Right: The corresponding macropixel image generated with a rebinning operation.

macropixels by applying a rebinning operation. Figure 1 shows the preprocessing of an input SDI image. In this figure, the left panel shows the SDI image processed with the median filter; the right panel shows the corresponding macropixel image generated through a rebinning operation.

3.2. The Feature Map

The detection of a solar flare is performed on a feature image, which is actually a map (hereafter $\gamma - map$) that represents the local intensity changes between an image and its previous reference image. It can be calculated using the following equation:

$$\gamma_n(i, j) = \frac{[F_n(i, j) - F_{ref}(i, j)]^2}{F_{ref}(i, j) + c} \tag{1}$$

where $F_n(i, j)$ represents an image to be examined, $F_{ref}(i, j)$ represents its reference image generated from the median of a running collection of previous images, c is a supplement constant to avoid a zero denominator, and i and j are macropixel coordinates ranging from 0 to 31 along horizontal and vertical directions, respectively. Note that the intensity changes between an image and its reference are squared to enhance the flaring signals, which helps improve the sensitivity of the algorithm to weak flare events.

3.3. Detection

Flare detection is a three-stage process. The first stage is to find the over-threshold pixels where $\gamma_n(i, j)$ exceeds the pre-defined threshold value γ_1. Note that images with a large number of over-threshold pixels are possibly caused by cosmic rays or particle storms rather than a solar flare. To avoid this, we set an upper limit of the over-threshold pixels (NP_{max}), meaning that only images with the number of over-threshold pixels less than NP_{max} are considered in our detection. Once the over-threshold pixels are identified, the second stage clusters them into different groups using morphological dilation. Each group potentially contains a flare. After clustering, we label each group as an ongoing flare, and the flare

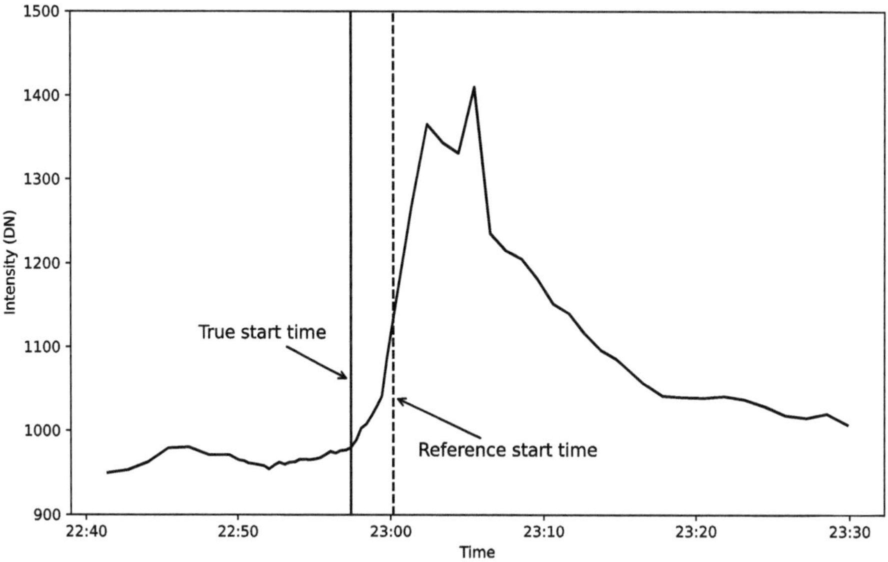

Figure 2 Time profile of the emission from the flare on 20 May 2023, locally integrated from the flaring region in SDI images. The solid vertical line marks the true start time of the flare identified by our software tool, and the dashed line shows the reference start time determined by γ_1.

location is estimated as the intensity-weighted centroid of the group. To avoid a flare from being repeatedly labeled, the third stage is to compare each of the newly labeled flares with those flares found previously but still ongoing. If the location of a flare is found inside 0.1 solar radii of any of the ongoing flares labeled previously, then the flare is considered the same flare, and its label will be dropped. Otherwise, it is considered as a new event. In this way, we are able to detect events that happen temporally close together from different disk locations.

Here γ_1 is a good indicator of the occurrence of a flare. It directly relates to the number and the intensity levels of the flares we will detect (higher γ_1, meaning to filter out events with small intensity changes). However, it usually fails to determine the true start of a flare since many flares are generally not so intense at the beginning. One example of these can be seen in Figure 2, which shows the time profile of the emission from the flare on 20 May 2023. The vertical dashed line indicates the time when the gamma value reaches γ_1, and the vertical solid line indicates the true start time of the flare. It can be seen that the start time of the flare estimated via γ_1 is significantly later than the true start of the flare. This could cause systematical underestimations when calculating flare quantities such as flare duration and energy.

To obtain the true start time of a flare, the start time found according to γ_1 is considered a reference start time. Once a reference start time is found, the true start time of the flare is estimated by tracking the flaring region (macropixels inside 0.1 solar radii of the flare) backward to the earliest time when the gamma value drops below a threshold value γ_2. This time is considered the flare true start time, and γ_2 is defined as the flare true start threshold, which can be estimated according to the maximum gamma value observed during the non-flare period.

Finally, to locate the flare more accurately, once a flare is detected, its initial position (x_0, y_0) is first roughly estimated from the macropixel image, then a circular cutout centered on (x_0, y_0) from the corresponding original image is created. The cutout image subtracted from its previous image is used to calculate the centroid position of the flare by applying a two-dimensional Gaussian (plus a constant) fitting function.

3.4. Tracking

Once a flare or several flares is/are detected and labeled as an ongoing event or ongoing events, our detection tool starts to track each of them separately until the end criteria are met. Meanwhile, the time profile of each flare is recorded. The flare end time is determined by images taken after the flare peak (after-peak images), so the end criteria include:

1. The time of the reference image is after the flare peak time.
2. The gamma value of the flare region drops below a threshold value γ_3.

The flare peak is defined as the maximum flux observed during the flare. γ_3 is defined as the end threshold of a flare. To avoid the same flare event from being repeatedly detected, γ_3 is required to be less than γ_1. Once the end criteria are met, the time of the reference image is defined as the end time of the flare. Similar to γ_2, γ_3 is estimated by observing the non-flare gamma values after a flare. Considering that the background variation after a flare is generally larger than the variation before the flare, γ_3 is usually set to be slightly higher than $\gamma 2$.

3.5. Output

Once a sequence of solar images is given, ASFD will iterate through each image, identifying a set of events that occurred on the solar disk. For each event, summary information, including the occurring date, start time, peak time, end time, duration, position, and significance of the event, is recorded in a human-readable "txt" file, as shown in Table 1. Here, the significance of an event (E_{signif}) is defined as the maximum intensity increase relative to a pre-event background,

$$E_{\text{signif}} = \frac{I_{\max} - I_{\text{bkg}}}{I_{\text{bkg}}} \tag{2}$$

where $I_{\max}$ represents the maximum intensity observed during the event, and I_{bkg} represents the pre-event background intensity. It gives a general idea of the strength and importance of an event.

Along with the "txt" file, for each event, our software tool also outputs a "png" image, showing a snapshot of the event around its peak time, and an "mp4" movie, showing the whole evolution of the event. In Figure 3, we show the snapshot image of the flare on 3 May 2023. The top panel displays the time profile of the Ly-α emission from the flare, showing how the flare evolves (the flare start, peak, and end times are marked with red, green, and blue vertical lines, respectively). The panel includes detailed textual information about the flare in the upper-left corner. The bottom left panel displays the full-FOV image of the Sun, showing the flare location relative to the whole solar disk, and the bottom right panel displays a zoomed-in view of the flare region, providing a closer look at the flare. This combination allows us to understand the event in temporal and spatial dimensions.

4. Results

The ASFD is tested using SDI full-FOV observations and compared with the GOES soft X-ray flare list for the same period. Since the launch of the ASO-S mission (8 October 2022),

Table 1 Event list generated from ASO-S/SDI images with ASFD.

Date	Start	Peak	End	Dur (m)	Pos (pixel)	Signif (%)
2023-02-11	00:35:41	00:39:41	00:48:41	13.0	(1728,1728)	5.4
2023-02-11	07:50:41	07:54:41	07:57:41	7.0	(576,1856)	12.6
2023-02-11	07:58:41	07:59:41	08:05:41	7.0	(832,1216)	4.7
2023-02-11	08:00:41	08:05:41	08:21:41	21.0	(576,1856)	10.9
2023-02-11	09:25:41	09:28:41	09:35:41	10.0	(704,1216)	10.4
2023-02-11	09:13:41	09:19:41	09:37:41	24.0	(3648,2496)	12.7
2023-02-11	09:24:41	09:35:41	09:45:41	21.0	(3904,1600)	12.7
2023-02-11	10:45:41	10:52:41	11:01:41	16.0	(3776,2880)	9.8
2023-02-11	11:30:41	11:32:41	11:37:41	7.0	(704,1216)	14.0
2023-02-11	11:28:41	11:32:41	11:38:41	10.0	(1344,1728)	2.3
2023-02-11	11:53:41	12:08:41	12:26:41	33.0	(3520,2368)	32.1
2023-02-11	12:20:41	12:22:41	12:28:41	8.0	(704,1216)	7.7
2023-02-11	12:25:41	12:35:41	12:45:41	20.0	(448,1856)	10.0
2023-02-11	13:44:41	13:47:41	13:55:41	11.0	(3904,2880)	4.1
2023-02-11	14:16:41	14:19:41	14:24:41	8.0	(1344,1728)	4.5
2023-02-11	14:37:41	14:40:41	14:44:41	7.0	(3520,2624)	3.8
2023-02-11	15:22:41	15:27:41	15:29:41	7.0	(448,1856)	6.9
2023-02-11	15:54:41	15:57:41	16:02:41	8.0	(448,1856)	5.2
2023-02-11	15:42:41	15:48:41	16:28:41	46.0	(3392,2496)	85.9
2023-02-11	16:56:41	16:58:41	17:05:41	9.0	(3520,2496)	2.8
2023-02-11	17:16:41	17:20:41	17:26:41	10.0	(448,1856)	8.3
2023-02-11	18:26:41	18:33:41	18:40:41	14.0	(448,1856)	6.7
2023-02-11	19:10:41	19:14:41	19:16:41	6.0	(1472,1600)	6.1
2023-02-11	21:37:41	21:41:41	21:46:41	9.0	(1344,1600)	3.8
2023-02-11	22:13:41	22:18:41	22:25:41	12.0	(448,1856)	9.6
2023-02-11	22:45:41	22:52:41	23:06:41	21.0	(3648,2624)	10.6
2023-02-11	23:59:41	00:08:41	00:23:41	24.0	(960,1216)	6.8

SDI has undergone an extended in-orbit testing phase due to the need for precise calibration, optimization of algorithms, and unexpected issues about spatial resolution, which makes the observations that suit flare detections intermittent and very limited. For this test, we used a one-month dataset obtained in February 2023, when the Sun was relatively active (51 M-class and above flares were observed), and SDI was conducting continuous scientific observations.

Before the test, it is important to know that the output of ASFD is significantly affected by the choice of the parameter values of the tool. For example, a larger γ_1 would cause more small-scale events to be filtered out, and γ_2/γ_3 directly determine the accuracy of the start/end times of the events. For this test, parameter values are chosen to maximize the number of solar events associated with significant Ly-α emission enhancements. Table 2 lists values of the main parameters used in the detection.

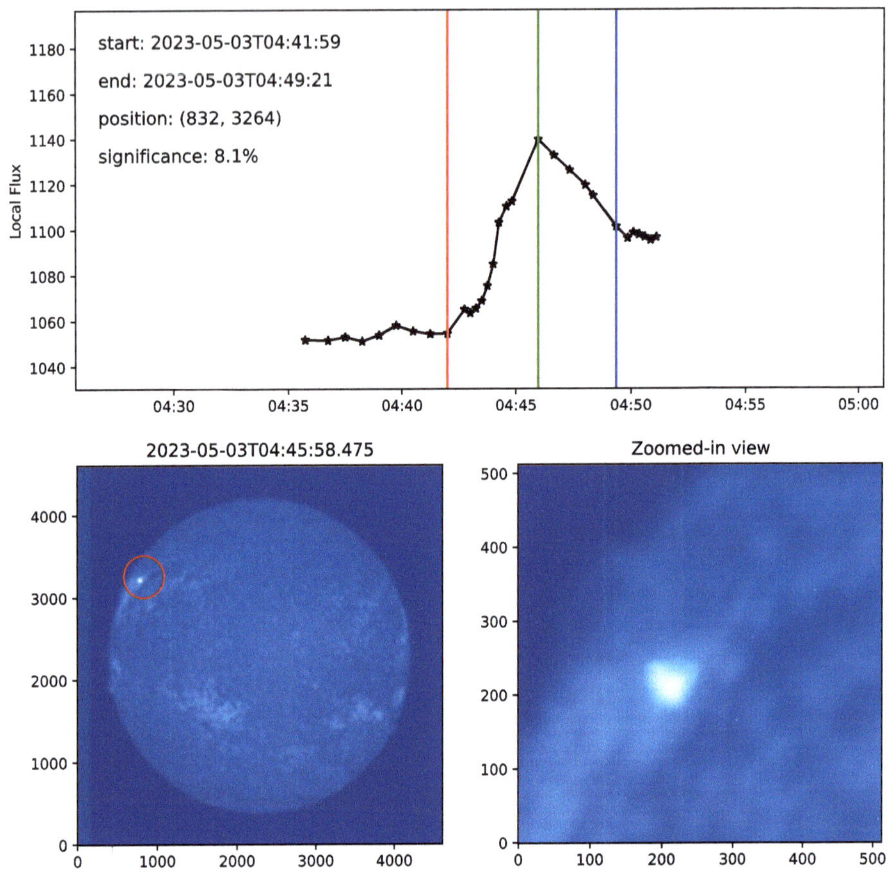

Figure 3 Snapshot image of the flare on 3 May 2023. Top: A time profile of the emission from the flare region, showing how the flare evolves over time. Bottom left: Full-FOV image of the Sun, showing the flare location relative to the whole solar disk. Bottom right: Zoomed-in view of the flare region.

Table 2 The main parameter values used in our software tool.

Parameter	Parameter value	Description
γ_1	100	trigger threshold gamma value
γ_2	3	start threshold gamma value
γ_3	5	end threshold gamma value
NP_{max}	10	allowed maximum number of over-thresholding pixels
c	10	background supplementary constant

4.1. Event List and Visual Inspection

For the observations in February 2023, the ASFD finds a total number of 226 events. After checking the snapshot images and the corresponding movies of each event, we found that only one event was misidentified (due to a data gap in the observations), and all the others (225 events) are related to active phenomena. Among the 225 events, 91% (204/225) are so-

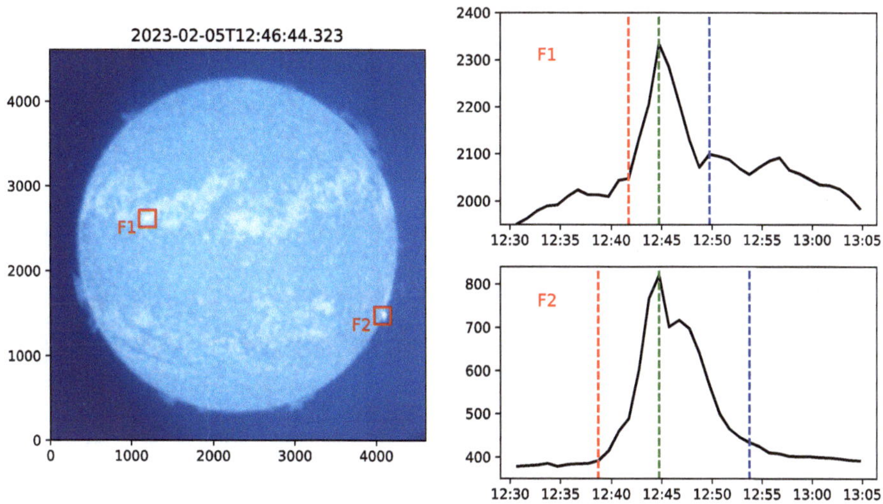

Figure 4 Two Ly-α flares occurred temporally close together on 5 February 2023 but were successfully identified by our software tool. Left: The locations of the two flares on the solar disk (marked with red boxes). Right: Time profiles of the emission from the two flare regions. The red, dark green, and blue lines indicate the start, peak, and end times of the flares, respectively, as identified by ASFD.

lar flare events, 5% (12/225) are associated with prominence or filament eruptions, and 4% (9/225) are related to solar jets or other solar activity. For each detected event, characteristics such as the start/end times and the position are found to be well determined. Figure 4 shows flares that occurred temporally close together from different solar disk locations but were successfully separated by ASFD. Figure 5 shows a double-ribbon flare in L-yα on 11 February 2023, characterized by the appearance of two distinct, parallel ribbons of enhanced brightness on the solar surface. Figure 6 shows an example of Ly-α emission enhancement associated with an eruptive prominence on 9 February 2023. This prominence was observed above the solar limb and evolved into a fast coronal mass ejection (CME) in its later phase.

4.2. Comparison with GOES X-ray Flare List

To further validate the output of ASFD, we compare our event list (as described in Section 4.1) with the X-ray flare list detected by GOES. Figure 7 shows a typical one-day detection of ASFD, along with the time profile of GOES X-ray flux on 11 February 2023. For most of the X-ray flares on this day, especially those M-class and above flares, ASFD shows highly consistent detection results with the GOES X-ray flare list. Due to our spatial resolution, ASFD can identify and distinguish events that happened temporally close together (see Figure 4 for an example).

For the whole month of February 2023, the NOAA solar X-ray flare list from the GOES satellite has reported 255 flares from C1 and above. After excluding events with missing SDI observations, it leaves a total number of 181 flare events for comparison, i.e., 145 C-class flares, 34 M-class flares, and 2 X-class flares. Because of the large number of events, a manual comparison between the event lists is difficult, so an automated method was established and implemented. It involves looking at the starting and ending times of the GOES flares as well as the peak time of the Ly-α flares. If the peak time of a Ly-α flare falls in between the starting and ending times of a GOES flare, then the flare is thought to have a Ly-α correspondence.

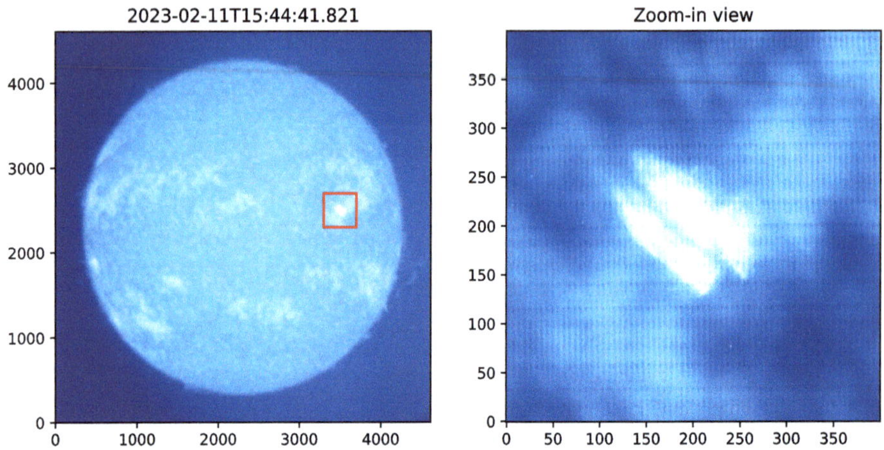

Figure 5 Left: SDI full solar disk image, showing a double-ribbon Ly-α flare on 11 February 2023. Right: Zoomed-in view of this flare.

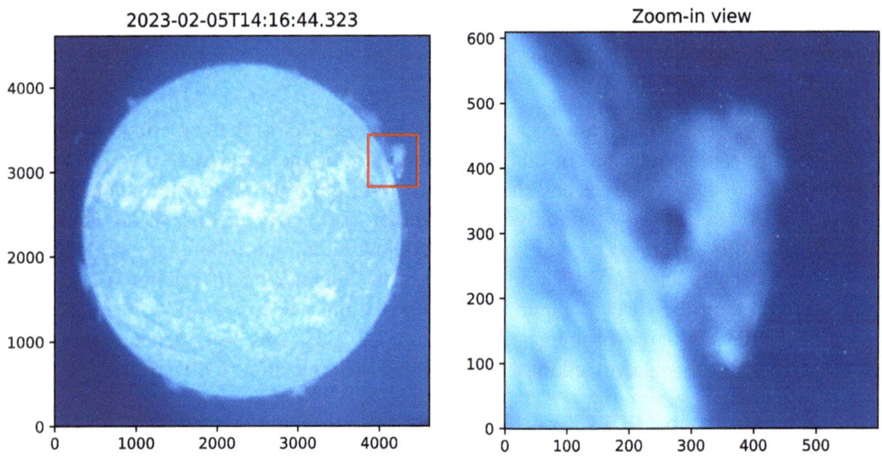

Figure 6 Similar to Figure 5, but for an eruptive solar prominence in Ly-α, captured by SDI on 9 February 2023.

Our analysis yields the following results. For the 181 C-class and above flares detected by GOES, 116 out of them were found to have corresponding Ly-α enhancements. For the 36 M-class and above flares, 29 of them have corresponding Ly-α emission enhancements. As can be seen, some soft X-ray flares are not accompanied by Ly-α emission enhancements. The lack of Ly-α emission is possibly due to the center-to-limb variation (CLV) as suggested by Milligan et al. (2020), according to which the flares closer to the solar limb tend to have weaker Ly-α emission (optically thick). On the other hand, the soft X-ray emission is not affected by this effect due to its optically thin characteristic. To verify this, we examined the location of the seven M-class and above flares not accompanied by Ly-α emission enhancements and found that six are located close to the solar limb (one event close to the disk center). According to a statistical study by Jing et al. (2020), most of the Ly-α emission enhancements tend to have a nonthermal origin (nonthermal electron beam heat-

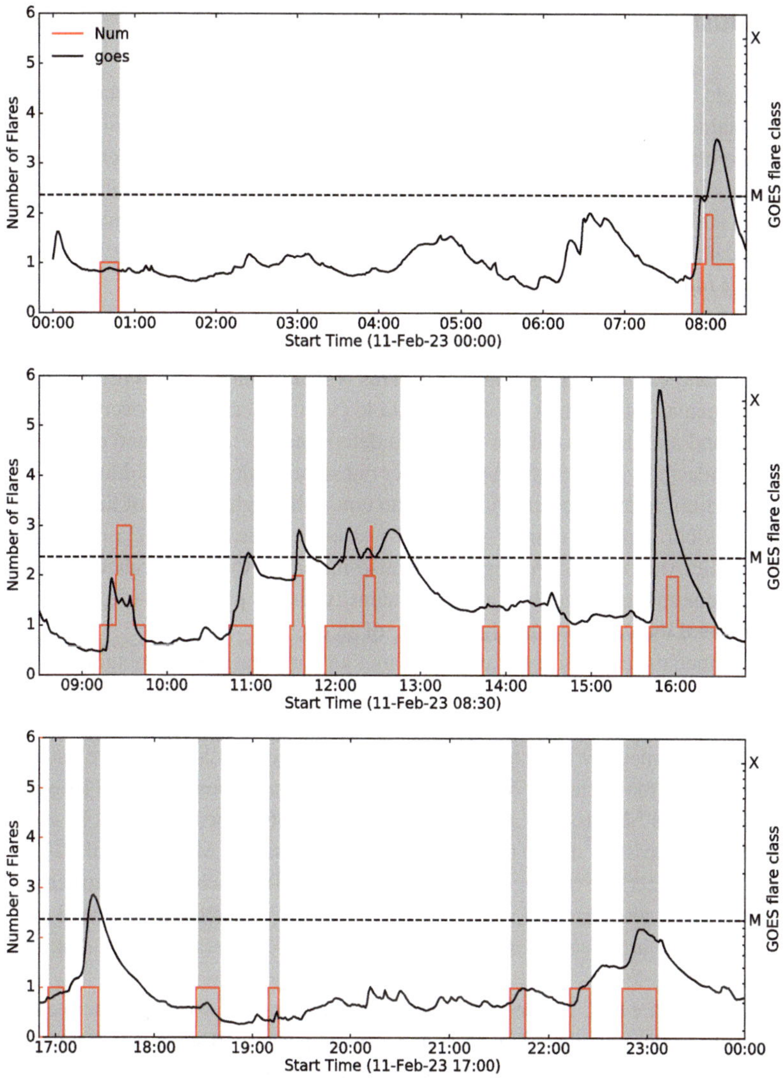

Figure 7 SDI Ly-α flare detections (red histogram: the height of a rectangle represents the number of Ly-α flares detected) with GOES soft X-ray flux (black lightcurve) for observations on 11 February 2023 (from top to bottom). The grey shade areas indicate the time periods of Ly-α emission enhancements.

ing), which means that those thermal-dominated flares, even located close to the disk center, could also produce weak Ly-α emission. Further studies regarding the CLV effect and the thermal/nonthermal nature of Ly-α flares will be presented in a future work.

Finally, we also detect Ly-α flare events that are not registered as GOES soft X-ray flares, here referred to as non-GOES events. These non-GOES events are possibly because our detections are based on the local intensity changes (thus more sensitive to small-scale events) while GOES monitors the disk-integrated solar flux.

5. Conclusions and Discussion

In this article, we described our algorithm for automatically detecting solar flares in Ly-α images obtained by ASO-S/SDI. It is designed to detect and track not only a single event but also events that happened simultaneously or temporally close together. The algorithm steps mainly involve data preprocessing, feature calculation, detection, tracking, parameter extracting, and finally, an event catalog compiling. For each detected event, statistics, such as start, peak, and end times, location, and others, are generated, along with a snapshot image and a movie. This kind of information greatly helps in scientific analysis and practical applications in space weather forecasting.

Compared with previous algorithms, our algorithm has obvious advantages in determining the start and end times of a flare event. This is mainly attributed to the use of a triple-threshold scheme, with the first (global) threshold (γ_1) to determine the occurrence of a flare and the second and third (local) thresholds to determine the real start and end times of the flare. By increasing γ_1, our algorithm can filter out small-scale events. Additionally, we also set an upper limit to exclude detections with an unusually high number of hits in one image, which makes our algorithm rarely affected by satellite jitter and particle noise during the South Atlantic Anomaly (SAA) region. We also introduce the significance parameter, defined as the contrast between the flare peak intensity and its previous background intensity, giving a general indication of the significance of an event.

We examined the capabilities of our software tool with a test dataset obtained in February 2023 and found that only one event was misidentified. Among the detected events, 91% are solar flare events, and 9% are related to solar prominences or filaments, solar jets, or other solar active phenomena. We compared our results with the GOES X-ray flare list and found that our software tool detected 64% of GOES C-class and above flares (116 out of 181) and 81% of GOES M-class and above flares (29 out of 36). The missing of Ly-α emission enhancements is possibly due to the CLV effect as proposed by Milligan et al. (2020) or the nonthermal characteristic of Ly-α flares (Jing et al., 2020), which will be further studied in a future work. We also detected some events that are not registered as GOES flares. These non-GOES events are possibly due to the low signal-to-noise ratio in the disk-integrated X-ray flux.

With our software tool, we are able to produce large catalogs with reproducible and quantifiable results without requiring much computational power and show how interesting solar dynamic regions are detected. In the future, we are actively progressing towards the development of our tool to perform near-real-time detections. This means that as soon as the data from the ASO-S/SDI instrument becomes available, our software tool will automatically retrieve them, apply the detection algorithm, and publish the results daily via our online catalog.

Finally, the use of Ly-α images, while informative, presents challenges in solar flare detection due to the inclusion of various solar phenomena such as eruptive solar filaments or prominences and solar jets. In the future, we plan to apply our software tool to SDO/AIA 9.4 nm images, which show only very hot flaring regions on the Sun, resulting in high contrast and thus easier detection of flares.

Acknowledgments The authors thank the reviewer very much for the valuable comments and suggestions that helped improve the manuscript. We acknowledge the use of data from the *Advanced Space-based Solar Observatory* (ASO-S) and GOES. The ASO-S mission is supported by the Strategic Priority Research Program on Space Science, Chinese Academy of Sciences.

Author contributions L.L. led the project and wrote the main manuscript text. L.F. supervised the project, led the discussions, and improved the manuscript. Z.Y.T. performed the comparison between the SDI and GOES flare list and prepared Figure 7. All authors reviewed the manuscript, discussed the results, and contributed to manuscript preparation.

Funding This work is Supported by the National Key R&D Program of China under grants 2022YFF0711402 (2022YFF0711400), the Strategic Priority Research Program of the Chinese Academy of Sciences (Grant No. XDB0560000), the National Natural Science Foundation of China (Grant Nos. 12103090, 11973012, 11921003, 12233012, 11820101002, 12203102), National Key R&D Program of China under grants 2022YFF0503003(2022YFF0503000), the mobility program (M-0068) of the Sino-German Science Center.

Data Availability No datasets were generated or analysed during the current study.

Declarations

Competing interests The authors declare no competing interests.

References

Benz, A.O.: 2008, Flare observations. *Living Rev. Solar Phys.* **5**, 1. DOI. ADS.

Bonte, K., Berghmans, D., De Groof, A., Steed, K., Poedts, S.: 2013, SoFAST: automated flare detection with the PROBA2/SWAP EUV imager. *Solar Phys.* **286**, 185. DOI. ADS.

Chen, B., Li, H., Song, K.-F., Guo, Q.-F., Zhang, P.-J., He, L.-P., Dai, S., Wang, X.-D., Wang, H.-F., Liu, C.-L., Zhang, H.-J., Zhang, G., Wang, Y., Liu, S.-J., Zhang, H.-X., Liu, L., Mao, S.-L., Liu, Y., Peng, J.-H., Wang, P., Sun, L., Liu, Y., Han, Z.-W., Wang, Y.-L., Wu, K., Ding, G.-X., Zhou, P., Zheng, X., Xia, M.-Y., Wu, Q.-W., Xie, J.-J., Chen, Y., Song, S.-M., Wang, H., Zhu, B., Chu, C.-B., Yang, W.-G., Feng, L., Huang, Y., Gan, W.-Q., Li, Y., Li, J.-W., Lu, L., Xue, J.-C., Ying, B.-L., Sun, M.-Z., Zhu, C., Bao, W.-M., Deng, L., Yin, Z.-S.: 2019, The Lyman-Alpha Solar Telescope (LST) for the ASO-S mission - II. Design of LST. *Res. Astron. Astrophys.* **19**, 159. DOI. ADS.

De Pontieu, B., Title, A.M., Lemen, J.R., Kushner, G.D., Akin, D.J., Allard, B., Berger, T., Boerner, P., Cheung, M., Chou, C., Drake, J.F., Duncan, D.W., Freeland, S., Heyman, G.F., Hoffman, C., Hurlburt, N.E., Lindgren, R.W., Mathur, D., Rehse, R., Sabolish, D., Seguin, R., Schrijver, C.J., Tarbell, T.D., Wülser, J.-P., Wolfson, C.J., Yanari, C., Mudge, J., Nguyen-Phuc, N., Timmons, R., van Bezooijen, R., Weingrod, I., Brookner, R., Butcher, G., Dougherty, B., Eder, J., Knagenhjelm, V., Larsen, S., Mansir, D., Phan, L., Boyle, P., Cheimets, P.N., DeLuca, E.E., Golub, L., Gates, R., Hertz, E., McKillop, S., Park, S., Perry, T., Podgorski, W.A., Reeves, K., Saar, S., Testa, P., Tian, H., Weber, M., Dunn, C., Eccles, S., Jaeggli, S.A., Kankelborg, C.C., Mashburn, K., Pust, N., Springer, L., Carvalho, R., Kleint, L., Marmie, J., Mazmanian, E., Pereira, T.M.D., Sawyer, S., Strong, J., Worden, S.P., Carlsson, M., Hansteen, V.H., Leenaarts, J., Wiesmann, M., Aloise, J., Chu, K.-C., Bush, R.I., Scherrer, P.H., Brekke, P., Martinez-Sykora, J., Lites, B.W., McIntosh, S.W., Uitenbroek, H., Okamoto, T.J., Gummin, M.A., Auker, G., Jerram, P., Pool, P., Waltham, N.: 2014, The Interface Region Imaging Spectrograph (IRIS). *Solar Phys.* **289**, 2733. DOI. ADS.

Domingo, V., Fleck, B., Poland, A.I.: 1995, The SOHO mission: an overview. *Solar Phys.* **162**, 1. DOI. ADS.

Feng, L., Li, H., Chen, B., Li, Y., Susino, R., Huang, Y., Lu, L., Ying, B.-L., Li, J.-W., Xue, J.-C., Yang, Y.-T., Hong, J., Li, J.-P., Zhao, J., Gan, W.-Q., Zhang, Y.: 2019, The Lyman-Alpha Solar Telescope (LST) for the ASO-S mission - III. Data and potential diagnostics. *Res. Astron. Astrophys.* **19**, 162. DOI. ADS.

Fernandez Borda, R.A., Mininni, P.D., Mandrini, C.H., Gómez, D.O., Bauer, O.H., Rovira, M.G.: 2002, Automatic solar flare detection using neural network techniques. *Solar Phys.* **206**, 347. DOI. ADS.

Gan, W.Q., Feng, L., Su, Y.: 2022, A Chinese solar observatory in space. *Nat. Astron.* **6**, 165. DOI. ADS.

Gan, W.-Q., Zhu, C., Deng, Y.-Y., Li, H., Su, Y., Zhang, H.-Y., Chen, B., Zhang, Z., Wu, J., Deng, L., Huang, Y., Yang, J.-F., Cui, J.-J., Chang, J., Wang, C., Wu, J., Yin, Z.-S., Chen, W., Fang, C., Yan, Y.-H., Lin, J., Xiong, W.-M., Chen, B., Bao, H.-C., Cao, C.-X., Bai, Y.-P., Wang, T., Chen, B.-L., Li, X.-Y., Zhang, Y., Feng, L., Su, J.-T., Li, Y., Chen, W., Li, Y.-P., Su, Y.-N., Wu, H.-Y., Gu, M., Huang, L., Tang, X.-J.: 2019, Advanced Space-Based Solar Observatory (ASO-S): an overview. *Res. Astron. Astrophys.* **19**, 156. DOI. ADS.

Gan, W., Zhu, C., Deng, Y., Zhang, Z., Chen, B., Huang, Y., Deng, L., Wu, H., Zhang, H., Li, H., Su, Y., Su, J., Feng, L., Wu, J., Cui, J., Wang, C., Chang, J., Yin, Z., Xiong, W., Chen, B., Yang, J., Li, F., Lin, J., Hou, J., Bai, X., Chen, D., Zhang, Y., Hu, Y., Liang, Y., Wang, J., Song, K., Guo, Q., He, L., Zhang, G., Wang, P., Bao, H., Cao, C., Bai, Y., Chen, B., He, T., Li, X., Zhang, Y., Liao, X., Jiang, H., Li, Y., Su, Y., Lei, S., Chen, W., Li, Y., Zhao, J., Li, J., Ge, Y., Zou, Z., Hu, T., Su, M., Ji, H., Gu, M., Zheng, Y., Xu, D., Wang, X.: 2023, The Advanced Space-Based Solar Observatory (ASO-S). *Solar Phys.* **298**, 68. DOI. ADS.

Grigis, P., Davey, A., Martens, P., Testa, P., Timmons, R., Su, Y., SDO Feature Finding Team: 2010, The SDO flare detective. In: *American Astronomical Society Meeting Abstracts #216*, 402.08. ADS.

Handy, B.N., Acton, L.W., Kankelborg, C.C., Wolfson, C.J., Akin, D.J., Bruner, M.E., Caravalho, R., Catura, R.C., Chevalier, R., Duncan, D.W., Edwards, C.G., Feinstein, C.N., Freeland, S.L., Friedlaender, F.M., Hoffmann, C.H., Hurlburt, N.E., Jurcevich, B.K., Katz, N.L., Kelly, G.A., Lemen, J.R., Levay, M., Lindgren, R.W., Mathur, D.P., Meyer, S.B., Morrison, S.J., Morrison, M.D., Nightingale, R.W., Pope, T.P., Rehse, R.A., Schrijver, C.J., Shine, R.A., Shing, L., Strong, K.T., Tarbell, T.D., Title, A.M., Torgerson, D.D., Golub, L., Bookbinder, J.A., Caldwell, D., Cheimets, P.N., Davis, W.N., Deluca, E.E., McMullen, R.A., Warren, H.P., Amato, D., Fisher, R., Maldonado, H., Parkinson, C.: 1999, The transition region and coronal explorer. *Solar Phys.* **187**, 229. DOI. ADS.

Jing, Z., Pan, W., Yang, Y., Song, D., Tian, J., Li, Y., Cheng, X., Hong, J., Ding, M.D.: 2020, The Lyα emission in solar flares. I. A statistical study on its relationship with the $1-8$ Å Soft X-ray emission. *Astrophys. J.* **904**, 41. DOI. ADS.

Kano, R., Sakao, T., Hara, H., Tsuneta, S., Matsuzaki, K., Kumagai, K., Shimojo, M., Minesugi, K., Shibasaki, K., DeLuca, E.E., Golub, L., Bookbinder, J., Caldwell, D., Cheimets, P., Cirtain, J., Dennis, E., Kent, T., Weber, M.: 2008, The Hinode X-Ray Telescope (XRT): camera design, performance and operations. *Solar Phys.* **249**, 263. DOI. ADS.

Kosugi, T., Matsuzaki, K., Sakao, T., Shimizu, T., Sone, Y., Tachikawa, S., Hashimoto, T., Minesugi, K., Ohnishi, A., Yamada, T., Tsuneta, S., Hara, H., Ichimoto, K., Suematsu, Y., Shimojo, M., Watanabe, T., Shimada, S., Davis, J.M., Hill, L.D., Owens, J.K., Title, A.M., Culhane, J.L., Harra, L.K., Doschek, G.A., Golub, L.: 2007, The Hinode (solar-B) mission: an overview. *Solar Phys.* **243**, 3. DOI. ADS.

Kraaikamp, E., Verbeeck, C.: 2015, Solar demon - an approach to detecting flares, dimmings, and EUV waves on SDO/AIA images. *J. Space Weather Space Clim.* **5**, A18. DOI. ADS.

Li, H., Chen, B., Feng, L., Li, Y., Huang, Y., Li, J.-W., Lu, L., Xue, J.-C., Ying, B.-L., Zhao, J., Yang, Y.-T., Gan, W.-Q., Fang, C., Song, K.-F., Wang, H., Guo, Q.-F., He, L.-P., Zhu, B., Zhu, C., Deng, L., Bao, H.-C., Cao, C.-X., Yang, Z.-G.: 2019, The Lyman-Alpha Solar Telescope (LST) for the ASO-S mission — I. Scientific objectives and overview. *Res. Astron. Astrophys.* **19**, 158. DOI. ADS.

Lu, L., Li, H., Huang, Y., Feng, L., Zhu, B., Wang, P., Song, D.C., Gan, W.Q.: 2020, The trigger and termination scheme for the event mode of the Lyman-Alpha Solar Telescope (LST) onboard the ASO-S mission. *Acta Astron. Sin.* **61**, 38. ADS.

Milligan, R.O., Hudson, H.S., Chamberlin, P.C., Hannah, I.G., Hayes, L.A.: 2020, Lyman-alpha variability during solar flares over solar cycle 24 using GOES-15/EUVS-E. *Space Weather* **18**, e02331. DOI. ADS.

Müller, D., St. Cyr, O.C., Zouganelis, I., Gilbert, H.R., Marsden, R., Nieves-Chinchilla, T., Antonucci, E., Auchère, F., Berghmans, D., Horbury, T.S., Howard, R.A., Krucker, S., Maksimovic, M., Owen, C.J., Rochus, P., Rodriguez-Pacheco, J., Romoli, M., Solanki, S.K., Bruno, R., Carlsson, M., Fludra, A., Harra, L., Hassler, D.M., Livi, S., Louarn, P., Peter, H., Schühle, U., Teriaca, L., del Toro Iniesta, J.C., Wimmer-Schweingruber, R.F., Marsch, E., Velli, M., De Groof, A., Walsh, A., Williams, D.: 2020, The solar orbiter mission. Science overview. *Astron. Astrophys.* **642**, A1. DOI. ADS.

Pesnell, W.D., Thompson, B.J., Chamberlin, P.C.: 2012, The Solar Dynamics Observatory (SDO). *Solar Phys.* **275**, 3. DOI. ADS.

Qu, M., Shih, F.Y., Jing, J., Wang, H.: 2003, Automatic solar flare detection using MLP, RBF, and SVM. *Solar Phys.* **217**, 157. DOI. ADS.

Shibata, K., Magara, T.: 2011, Solar flares: magnetohydrodynamic processes. *Living Rev. Solar Phys.* **8**, 6. DOI. ADS.

Shibata, K., Yokoyama, T.: 2002, A Hertzsprung-Russell-like diagram for solar/stellar flares and corona: emission measure versus temperature diagram. *Astrophys. J.* **577**, 422. DOI. ADS.

Solar Physics (2024) 299:11
https://doi.org/10.1007/s11207-024-02251-9

RESEARCH

A Statistical Study of Solar White-Light Flares Observed by the *White-Light Solar Telescope* of the *Lyman-Alpha Solar Telescope* on the *Advanced Space-Based Solar Observatory* (ASO-S/LST/WST) at 360 nm

Zhichen Jing[1,2] · **Ying Li**[1,2] · **Li Feng**[1,2] · **Hui Li**[1,2] · **Yu Huang**[1,2] · **Youping Li**[1,2] · **Yang Su**[1,2] · **Wei Chen**[1,2] · **Jun Tian**[1,2] · **Dechao Song**[1,2] · **Jingwei Li**[1] · **Jianchao Xue**[1] · **Jie Zhao**[1] · **Lei Lu**[1] · **Beili Ying**[1] · **Ping Zhang**[1] · **Yingna Su**[1,2] · **Qingmin Zhang**[1,2] · **Dong Li**[1,2] · **Yunyi Ge**[1] · **Shuting Li**[1,2] · **Qiao Li**[1,2] · **Gen Li**[1,2] · **Xiaofeng Liu**[1,2] · **Guanglu Shi**[1,2] · **Jiahui Shan**[1,2] · **Zhengyuan Tian**[1,2] · **Yue Zhou**[1,2] · **Weiqun Gan**[1,3]

Received: 22 September 2023 / Accepted: 6 January 2024 / Published online: 23 January 2024

Abstract

Solar white-light flares (WLFs) are those accompanied by brightenings in the optical continuum or integrated light. The *White-light Solar Telescope* (WST), as an instrument of the *Lyman-alpha Solar Telescope* (LST) on the *Advanced Space-based Solar Observatory* (ASO-S), provides continuous solar full-disk images at 360 nm, which can be used to study WLFs. We analyze 205 major flares above M1.0 from October 2022 to May 2023 and identify 49 WLFs at 360 nm from WST observations, i.e. with an occurrence rate of 23.9%. The percentages of WLFs for M1 – M4 (31 out of 180), M5 – M9 (11 out of 18), and above X1 (7 for all) flares are 17.2%, 61.1%, and 100%, respectively, namely the larger the flares, the more likely they are WLFs at 360 nm. We further analyze 39 WLFs among the identified WLFs and investigate their properties such as white-light enhancement, duration, and brightening area. It is found that the relative enhancement of the white-light emission at 360 nm is mostly (>90%) less than 30% and the mean enhancement is 19.4%. The WLFs' duration at 360 nm is mostly (>80%) less than 20 minutes and its mean is 10.3 minutes. The brightening area at 360 nm is mostly (>75%) less than 500 arcsecond^2 and the median value is 225. We find that there exist good correlations between the white-light enhancement/duration/area and the peak soft X-ray (SXR) flux of the flare, with correlation coefficients of 0.68, 0.58, and 0.80, respectively. In addition, the white-light emission in most WLFs peaks around the same time as the temporal derivative of SXR flux as well as the hard X-ray emission at 20 – 50 keV, indicative of the Neupert effect. It is also found that the limb WLFs are more likely to have a greater enhancement, which is consistent with numerical simulations.

Keywords Flares · White-light · Center-limb observations · X-ray bursts · Association with flares

Extended author information available on the last page of the article

1. Introduction

Solar flares are a kind of energetic activity with sudden brightening in the solar atmosphere (e.g. Fletcher et al., 2011; Shibata and Magara, 2011). White-light flares (WLFs) are the flares with an enhancement in the optical continuum or integrated light (Švestka, 1970; Neidig, 1989; Hudson, 2011). Although the first flare observed in 1859 was a WLF (Carrington, 1859; Hodgson, 1859), the number of WLFs is very small. Only about 100 WLFs were reported before 2010 (Cheng, Ding, and Carlsson, 2010). From 2011 to 2016, about 50 WLFs were observed by the *Solar Optical Telescope* (SOT) on *Hinode* (Watanabe, Kitagawa, and Masuda, 2017) and no more than 150 WLFs were reported using the data of *Helioseismic and Magnetic Imager* (HMI) on the *Solar Dynamics Observatory* (SDO) (e.g. Kuhar et al., 2016; Song and Tian, 2018; Castellanos Durán and Kleint, 2020).

Many WLFs are found to have a very strong correlation with the hard X-ray (HXR) and radio emissions in time and space (e.g. Hudson, Wolfson, and Metcalf, 2006; Krucker et al., 2011). This implies that nonthermal electrons are the source of WLFs' energy. However, only the most energetic electrons with more than 900 keV can reach the photosphere (Neidig, 1989), where the white-light (WL) emission is usually emitted (e.g. Watanabe, Masuda, and Segawa, 2012), so it is hard for the nonthermal electrons to heat the photosphere directly. To resolve this problem, much effort has been made to explore the energy transfer and radiative-heating process in the low atmosphere (e.g. Aboudarham and Henoux, 1986a; Tian et al., 2022) and many heating sources or mechanisms for WLFs have been proposed, including electron beams (Hudson, 1972; Aboudarham and Henoux, 1986b; Watanabe and Imada, 2020), proton beams (Machado, Emslie, and Brown, 1978; Procházka, 2019), chromospheric backwarming (Machado, Emslie, and Avrett, 1989; Ding et al., 2003), soft X-ray (SXR) irradiation (Machado, 1978), EUV irradiation (Somov and Syrovatskiĭ, 1976), chromospheric condensations (Gan and Mauas, 1994; Gan, Hénoux, and Fang, 2000), and Alfvén waves (Fletcher and Hudson, 2008). In fact, the enhancement of the WL continuum during a flare is likely caused by the joint action of several heating mechanisms (Xu et al., 2010; Song et al., 2023).

WLFs are observed not only with ground-based telescopes, but also by spacecraft, which can provide detailed information. Compared with ground-based telescopes, space telescopes have better and more stable observation conditions. The *Soft X-ray Telescope* (SXT) on *Yohkoh* can provide solar images in the waveband at 430.8 nm with a bandwidth of $\approx$3 nm (Tsuneta et al., 1991). With the SXT data, WLFs were observed from space for the first time (Hudson et al., 1992). The *Transition Region and Coronal Explorer* (TRACE) has a very broad WL channel from 170 nm to $\approx$1000 nm (Metcalf et al., 2003). *Hinode*/SOT can provide observations at 430.5 nm with a bandwidth of 0.8 nm (Kosugi et al., 2007). SDO/HMI gives solar full-disk images at Fe I 617.3 nm (Scherrer et al., 2012). It should be mentioned that there have been no prior systematic imaging observations in the Balmer continuum (i.e. below the Balmer limit at 364.6 nm) for WLFs, at least not for the wavelength range above 300 nm.

The *Advanced Space-based Solar Observatory* (ASO-S: Gan et al., 2023) is the first comprehensive solar mission in China, which was launched on 9 October 2022, having a Sun-synchronous orbit with an altitude of $\approx$720 km. The primary aim of ASO-S is to study the magnetic field, solar flares, coronal mass ejections, and their relationships (Gan et al., 2019). ASO-S has three payloads: the *Full-disk vector MagnetoGraph* (FMG: Su et al., 2019), the *Hard X-ray Imager* (HXI: Zhang et al., 2019), and the *Lyman-alpha Solar Telescope* (LST: Li et al., 2019). LST consists of three scientific instruments (Chen et al., 2019): a *Solar Disk Imager* (SDI), a *Solar Corona Imager* (SCI), and a *White-light Solar*

Telescope (WST). WST can provide continuous full-disk images in the 360±2 nm waveband (Feng et al., 2019), which helps study WLFs at 360 nm, i.e. in the Balmer continuum. Here we still call the flares with 360 nm brightenings WLFs, the same as in Neidig and Cliver (1983).

To investigate the physical properties of the WL emission at 360 nm for large flares, we study all the M- and X-class flares (i.e. major flares) from 12 October 2022 (when the first M-class flare was observed by WST) to 31 May 2023. 205 M- and X-class flares were collected and 49 of them (23.9%) were identified as WLFs at 360 nm. We further analyze 39 WLFs among them in detail to investigate the relative enhancement, duration, and brightening area of the WL emission at 360 nm. In the following, we first introduce the observational data and flare data set in Section 2. Then we introduce the methods in Section 3. Our observational results are presented in Section 4. The summary and discussions are given in Section 5.

2. Observational Data and Flare Data Set

ASO-S/LST/WST provides the main data in this study. WST works in the 360±2 nm waveband and has a field of view of 1.2 $R_\odot$. The images taken by WST have a size of 4608 × 4608 pixels and the pixel size is ≈0.5″. Note that the spatial resolution of the images (defined as full width at half maximum) is estimated to be about 4″ and the point spread function is still under investigation. The cadence of thc images is two minutes in routine mode, while it can be as high as one or two seconds in a burst mode. Since the burst mode of WST was under test during the period of our flare data set, the cadence of most flare events is two minutes. The level 2.5 data of WST are used here, which have been corrected for dark current and flat field, as well as made North up and radiometric calibration applied. We further use drot_map.pro in the SolarSoftWare (SSW) to remove the influence of solar rotation. It should be pointed out that the WST observations were partially affected by Earth eclipses starting around May 2023 and some flares lack available observations unfortunately.

Considering a strong relationship between the WL and HXR emissions in WLFs, we also use the HXR data from ASO-S/HXI to obtain the peak time of the HXR emission at 20 – 50 keV for the WLFs under study. HXI has an energy range of ≈10 – 300 keV. Its spatial resolution is 3.1″ and the temporal resolution can be as high as 0.125 seconds. Note that due to the influence of the South Atlantic Anomaly (SAA) or the radiation belt and so on, not all the WLFs at 360 nm have HXR data from HXI.

The SXR 1 – 8 Å data from the *Geostationary Operational Environmental Satellite* (GOES) are used in this study as well. The GOES series spacecraft (from GOES-1 to GOES-18) have provided continuous data on solar flares since 1975. The solar X-ray irradiances in 0.5 – 4 Å and 1 – 8 Å are observed by the *X-ray Sensor* (XRS: Hanser and Sellers, 1996), and the latter one is widely used to define the flare magnitude from A- to X-class.

In this work, we collect 205 major flares occurring from 12 October 2022 to 31 May 2023 that have WST observations at 360 nm. There contain 198 M-class and 7 X-class flares with the class from M1.0 to X2.3. Note that there are additional 39 M-class flares observed by GOES but not well observed by WST, mainly due to Earth eclipses, which are not analyzed here. From these 205 flares, we identify 49 (23.9%) WLFs at 360 nm. In our analysis, all the 205 flares collected are used to study the occurrence rate of WLFs at 360 nm as well as compare the SXR duration and spatial distribution of the WLFs flares without WL emission (non-WLFs) (in Sections 4.1 and 4.2, respectively). Furthermore, for 39 out of the identified 49 WLFs, the physical properties including relative enhancement, duration, brightening area, and total flux of the WL emission at 360 nm are studied (in Sections 4.3 – 4.5). In the

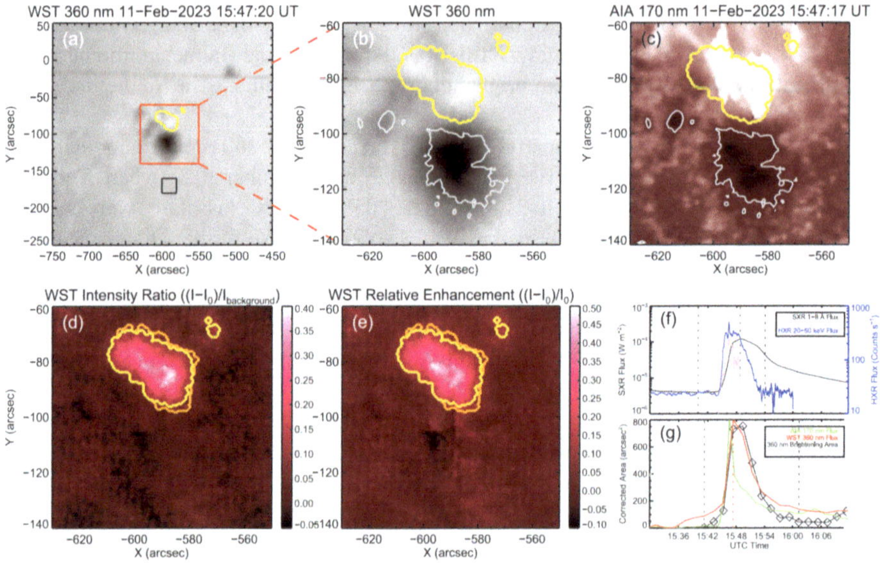

Figure 1 An example of a WLF with a GOES class of X1.1 on 11 February 2023. (**a**) WST 360 nm image around the flare peak time. The *red box* indicates the flaring region in Panel **b**. The *black box* denotes a quiet-Sun region that is used to calculate $I_{\mathrm{background}}$. (**b**) WST image for the flaring region. (**c**) AIA 170 nm image for the flaring region. The *gray contour* marks the sunspot feature. (**d**) and (**e**) Maps of the intensity ratio and relative enhancement of the WL emission at 360 nm. The *yellow curve* (same in Panels **a** – **e**) denotes the brightening area of WL emission at this time, while the *orange curve* (same in Panel **d**) shows the total WL brightening area of this flare. (**f**) and (**g**) Temporal profiles of the SXR 1 – 8 Å flux, its temporal derivative, the HXR 20 – 50 keV flux from HXI, AIA 170 nm emission (normalized), WST 360 nm emission (normalized), and the brightening area at 360 nm. The *gray curve* in Panel **g** shows the intensity fluctuation from the quiet-Sun region marked by the *black box* in Panel **a**. The *three vertical dotted lines* in Panel **f** denote the GOES start, peak, and end times of the flare. While the *three vertical lines* in Panel **g** indicate the start, peak, and end times of the WL emission at 360 nm.

latter, 10 WLFs near the limb are excluded because their brightenings extend to or overlap the solar limb, which leads to an underestimation of the WL area. It should be mentioned that 9 of these excluded 10 WLFs have a longitude of larger than 85°.

3. Methods

3.1. Identification of WLFs at 360 nm

We use the intensity ratio defined as $(I - I_0)/I_{\mathrm{background}}$ to identify a WLF (see Figure 1d for an example WLF), where I, I_0 and $I_{\mathrm{background}}$ are the intensity at 360 nm for a certain pixel in the flare time (i.e. the period between the GOES start and end times of the flare, as indicated in Figure 1f), the mean intensity from the same pixel before the flare onset (30 minutes before the GOES start time, i.e. an interval comparable to the typical duration of a flare), and the mean intensity from a quiet-Sun region before the flare onset (same as the former one), respectively. Note that this ratio is unaffected by the WL brightening region in or outside the sunspot with distinct intensities. It is only affected by a selection of the quiet-Sun region. When selecting the quiet-Sun region, we give preference to those near the flaring active region (see the black box in Figure 1a) to avoid or reduce the limb-darkening effect at 360 nm. Note that the selected quiet-Sun region has a size of $20'' \times 20''$, namely $\approx$25 times the spatial resolution of WST. The intensity fluctuation of the quiet-Sun region is checked to be within $\pm 2\%$ (i.e. the gray line in Figure 1g) for each WLF. For all the analyzed 39 WLFs, the average level of background fluctuations is $\approx$1.5%, which is actually similar to the background level of HMI (1.3%) for the 20 WLFs studied by Song et al. (2018).

Different thresholds of the intensity ratio are used in previous studies for WLFs. For example, a threshold of 5% was adopted for HMI WLFs at 617.3 nm (Song et al., 2018). In the present study, we firstly consider the background fluctuations from a quiet-Sun region, i.e. within 2%. We also consider the fluctuations in the facular region around the flaring active region. In order to make sure that the 360 nm enhancement is caused by the flare itself, we finally adopt 8% (larger than three times the background fluctuation) as the threshold after some tests to determine a WLF at 360 nm. Note that this threshold is somewhat strict and may have some influence on the WL emission properties, which will be discussed in Section 5.

3.2. Calculation of the Brightening Area of WLFs at 360 nm

When the intensity ratio defined above at a flaring pixel is greater than the threshold of 8%, the pixel is taken into account when calculating the brightening area for the WLF (see the region marked by a yellow curve in Figures 1a–e). We further use the morphological-opening operation for the WL brightening region to remove the noise points (Maurya and Ambastha, 2010). Note that the brightening area is a function of time and can be written as S_1, S_2,..., S_n, where n represents the time frame of observations. The final brightening area for a WLF at 360 nm is defined as $S = S_1 \bigcup S_2 \bigcup \cdots \bigcup S_n$ (see the region marked by an orange curve in Figures 1d and e). Note that for a better understanding the WL area and its relationship with some other WL parameters, here S [in units of arcsec^2] is corrected for projection effect by using the heliocentric angle [θ], i.e. $S = S_0 / \cos(\theta)$, where S_0 is the originally measured brightening area and θ is derived from the flare location, as listed in Table 2.

3.3. Definition of the Duration of WLFs at 360 nm

Firstly, we use the temporal profile of WL brightening area (i.e. the black curve in Figure 1g) to determine the start and end times of a WLF at 360 nm. The start time is defined as the first time when the WL brightening area is >10 pixels after the flare onset (see the left red vertical line in Figure 1g), while the end time is defined as the time when the brightening area becomes zero or shows no obvious trend of decline (see the right red vertical line in Figure 1g). Note that the end time of WL emission can be later than that of the SXR emission. Then, the interval between the start and end times of WL emission is the duration [τ] of WLFs at 360 nm. Considering that the temporal resolution of most WLFs is two minutes, τ has an uncertainty of four minutes. In addition, the peak time of a WLF at 360 nm is defined as the time when the WL emission reaches its maximum (as marked by the middle red vertical line in Figure 1g).

3.4. Definition of the Relative Enhancement of the WL Emission at 360 nm

The relative enhancement r of the WL emission at 360 nm at a certain pixel is defined as $r = (I - I_0)/I_0$ (e.g. Hao et al., 2017; Castellanos Durán and Kleint, 2020), as shown in Figure 1e, where I and I_0 have the same meanings as mentioned above. Note that r usually has

Table 1 Occurrence rate of WLFs at 360 nm.

Flare class	WLF	non-WLF	Total	Occurrence rate
M1.0 – M4.9	31	149	180	17.2%
M5.0 – M9.9	11	7	18	61.1%
Above X1.0	7	0	7	100%
Total	49	156	205	23.9%

a larger value for the penumbra/umbra pixels compared with the pixels outside the sunspot, as the former pixels have a relatively lower intensity. In the following analysis, we adopt two maximum enhancements for each WLF. One is called "maximum pixel enhancement" [r_{p}], representing the maximum enhancement for the brightening pixels during the flare. The other is called "maximum mean enhancement" [r_{m}], representing the maximum of the mean enhancement for the WL brightening region during the flare.

3.5. An Example WLF

Figure 1 shows an example WLF with a GOES class of X1.1 on 11 February 2023. From Figures 1a and b, we can see that the identified brightening region at 360 nm (marked by a yellow curve) is near a sunspot, which well matches the brightening kernel at 170 nm (Figure 1c) observed by the *Atmospheric Imaging Assembly* (AIA) on SDO. The selected quiet-Sun region (denoted by the black box in Figure 1a) has an intensity fluctuation of less than $\pm 1\%$, as seen from Figure 1g. For this flare, the intensity-ratio map used to identify the WL brightening (Figure 1d) looks similar to the enhancement map (Figure 1e), although the value in the latter one is mostly larger than that from the former one. From Figures 1f and g, it is seen that the WL emission at 360 nm (integrated over the brightening area, i.e. S, as denoted by the orange curve in Figures 1d and e) is synchronous with the emission at 170 nm roughly, both of which peak around the same time as the HXR 20 – 50 keV emission and the temporal derivative of SXR 1 – 8 Å, i.e. the Neupert effect (Neupert, 1968). It is also seen that the temporal profile of WL brightening area peaks at a later time than the WL emission.

4. Results

4.1. Occurrence Rate of WLFs at 360 nm

Based on the above method, we identify 49 WLFs at 360 nm from the collected 205 M- and X-class flares, i.e. with an occurrence rate of 23.9%. We further check the dependence of the occurrence rate on flare magnitude. As shown in Table 1, the percentage of WLFs for M1.0 – M4.9 (31 out of 180), M5.0 – M9.9 (11 out of 18), and above X1 (7 for all) flares are 17.2%, 61.1%, and 100%, respectively. In other words, the larger the flares are, the more likely they are to be WLFs at 360 nm. In particular, all the X-class flares are WLFs at 360 nm in our data set.

4.2. SXR Duration and Spatial Distribution of the WLFs at 360 nm Compared with Flares without WL Emission (Non-WLFs)

Figures 2a and b show the histogram of the SXR duration (i.e. the temporal interval between the GOES start and end times) and its scatter plot versus peak SXR flux for all the collected

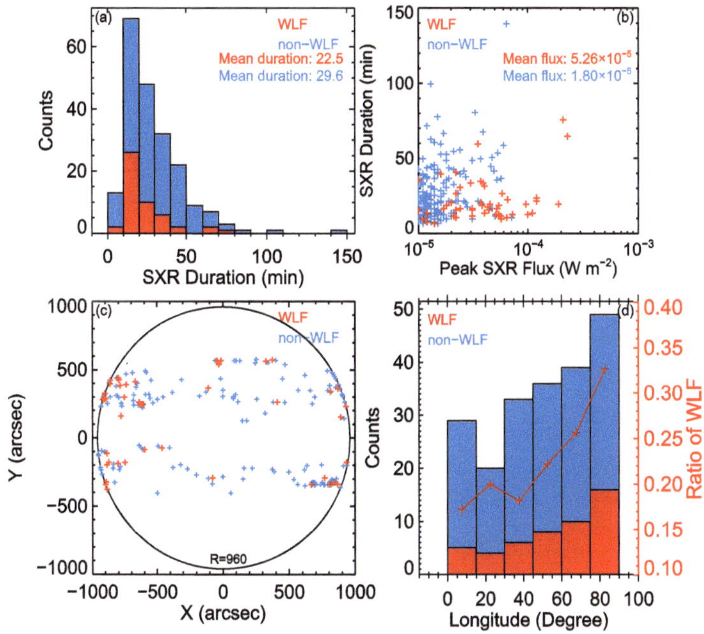

Figure 2 Comparisons of WLFs at 360 nm and non-WLFs. (**a**) Histogram of the SXR duration for all 205 collected major flares. (**b**) Scatter plot of the SXR duration versus peak SXR flux. (**c**) Spatial distribution of the WLFs and non-WLFs on the solar disk. (**d**) Distribution of WLFs and non-WLFs versus solar longitude. The *red curve* plots the ratio of the WLFs. In each panel, the *red and blue colors* represent the WLFs and non-WLFs, respectively.

major flares, in which WLFs and non-WLFs are indicated by red and blue colors, respectively. One can see that the SXR durations for all the flares are a few to tens of minutes. Compared with non-WLFs that have a mean SXR duration of 29.6 minutes, the WLFs tend to have a shorter SXR duration, with a mean value of 22.5 minutes. In addition, the WLFs tend to have a larger peak SXR flux (with a mean value of 5.26×10^{-5} W m^{-2}) compared with non-WLFs (with a mean value of 1.80×10^{-5} W m^{-2}).

Figure 2c shows the spatial distribution on solar disk for all the collected major flares, with the WLFs and non-WLFs plotted in red and blue plus symbols, respectively. There are 120 flares on the northern hemisphere and 32 out of them are WLFs at 360 nm, i.e. with a percentage of 26.7%. For the remaining 85 major flares that are located at the southern hemisphere, there are 17 WLFs at 360 nm among them, i.e. 20%. Therefore, the spatial distribution of WLFs seems to have no dependence on solar northern or southern hemisphere compared with non-WLFs, at least for our flare data set. However, we note that there are many WLFs located near the solar limb. We further investigate the dependence of the WLF's distribution on solar longitude, as shown in Figure 2d. It is seen that the number of WLFs exhibits an increasing tendency, say from several to more than ten, from disk center to limb. Although there are more major flares occurring near the limb in our data set, the percentage of WLFs (i.e. the red curve) still increases from ≈17% to ≈33% over longitude. This indicates that more WLFs take place or can be detected near the solar limb.

Table 2 The 39 WLFs analyzed in detail.

Observation Date [DD.MM.YYYY]	Flare Location[1]	GOES Class	GOES Peak [UT]	HXR Peak[2] [UT]	WL Start [UT]	WL Peak [UT]	WL End [UT]	WL τ^3 [minute]	WL r_p [%]	WL r_m [%]	WL S^4 [arcsec^2]	WL F^5 [erg]
07.11.2022	N12E56	M5.2	00:11	00:05	00:07	00:07	00:13	6.0	16.9	9.94	404±20	$(5.33\pm0.09)\times10^{26}$
14.12.2022	S21W48	M4.5	22:06	–	22:07	22:15	22:39	32	24.6	13.8	194±6.4	$(1.71\pm0.03)\times10^{29}$
16.12.2022	S21W64	M3.5	02:01	–	01:40	01:54	01:48	8.0	25.8	15.2	225±10	$(4.22\pm0.07)\times10^{28}$
16.12.2022	S20W68	M4.0	10:19	10:18	10:16	10:20	10:36	20	28.7	17.0	233±30	$(5.40\pm0.08)\times10^{28}$
30.12.2022	N20E07	M3.7	19:38	19:37	19:36	19:36	19:36	<2.0	10.9	9.47	13.0±0.65	$(8.04\pm0.13)\times10^{26}$
09.01.2023	N24E79	X1.9	18:50	–	18:46	18:50	19:02	16	101	31.9	$(2.62\pm0.33)\times10^{3}$	$(1.91\pm0.03)\times10^{30}$
10.01.2023	S13E71	M5.1	00:16	00:14	00:14	00:14	00:24	10	34.7	18.1	$(1.65\pm0.32)\times10^{3}$	$(2.62\pm0.06)\times10^{29}$
10.01.2023	S12E68	M1.0	02:16	02:13	02:14	02:16	02:18	4.0	23.0	12.8	163±19	$(1.95\pm0.03)\times10^{28}$
10.01.2023	S13E59	M1.2	17:48	17:47	17:48	17:48	17:48	<2.0	30.6	20.3	68.9±6.0	$(5.92\pm0.10)\times10^{27}$
10.01.2023	N22E62	X1.0	22:47	–	22:46	22:48	23:00	14	83.1	31.8	720±92	$(3.28\pm0.06)\times10^{29}$
11.01.2023	N25E56	M2.4	00:59	–	00:52	00:54	01:10	18	16.0	10.4	176±19	$(1.41\pm0.02)\times10^{28}$
11.01.2023	S13E53	M5.6	01:56	01:53	01:54	01:54	01:58	4.0	34.2	17.6	410±21	$(4.97\pm0.08)\times10^{28}$
11.01.2023	N22E56	M3.1	08:33	–	08:32	08:34	08:38	6.0	71.9	24.8	276±6.3	$(5.42\pm0.10)\times10^{28}$
12.01.2023	N23E49	M1.6	06:50	–	06:56	07:00	07:06	10	21.1	12.5	127±7.2	$(1.03\pm0.02)\times10^{28}$
13.01.2023	S11W83	M4.0	10:15	–	10:13	10:13	10:15	2.0	33.1	17.4	$(1.59\pm0.33)\times10^{3}$	$(3.18\pm0.05)\times10^{29}$
07.02.2023	N28E02	M3.8	22:58	–	22:53	22:55	23:03	10	61.0	13.3	65.4±1.9	$(6.99\pm0.11)\times10^{27}$
07.02.2023	N30E03	M6.4	23:07	23:06	23:07	23:07	23:09	2.0	55.5	22.8	230±3.5	$(2.53\pm0.04)\times10^{28}$
08.02.2023	N30E01	M2.1	02:53	02:53	02:53	02:53	02:53	<2.0	31.6	16.0	67.0±1.2	$(5.06\pm0.08)\times10^{27}$
09.02.2023	S09E69	M1.1	07:17	07:17	07:17	07:17	07:19	2.0	23.2	15.0	273±21	$(2.58\pm0.04)\times10^{28}$
10.02.2023	N30W23	M3.7	03:03	–	02:57	02:59	03:07	10	22.5	13.8	89.1±3.8	$(6.26\pm0.10)\times10^{27}$
10.02.2023	N31W27	M1.4	08:05	08:05	08:05	08:05	08:07	2.0	26.0	14.8	82.7±3.7	$(7.95\pm0.12)\times10^{27}$
11.02.2023	N07W69	M2.2	08:08	08:06	08:03	08:05	08:07	4.0	16.9	11.9	18.3±0.41	$(3.07\pm0.05)\times10^{27}$
11.02.2023	S10E39	X1.1	15:48	15:46	15:41	15:47	16:01	20	51.8	23.9	841±15	$(2.71\pm0.05)\times10^{29}$

Table 2 (*Continued*)

Observation Date [DD.MM.YYYY]	Flare Location[1]	GOES Class	GOES Peak [UT]	HXR Peak[2] [UT]	WL Start [UT]	WL Peak [UT]	WL End [UT]	WL τ^3 [minute]	WL r_p [%]	WL r_m [%]	WL S^4 [arcsec^2]	WL F^5 [erg]
12.02.2023	S10E29	M3.1	08:48	08:44	08:45	08:45	08:51	6.0	25.8	16.1	98.8±1.8	(1.36±0.02)×10^{28}
17.02.2023	N25E67	X2.3	20:16	20:02	19:58	20:02	20:42	44	61.6	24.4	(1.77±0.25)×10^3	(1.45±0.02)×10^{30}
03.03.2023	N21W76	X2.1	17:52	–	17:50	17:50	18:06	16	203	62.1	(2.32±0.43)×10^3	(5.83±0.10)×10^{29}
06.03.2023	N18W63	M5.8	02:28	02:21	02:20	02:22	02:32	12	32.3	13.1	222±41	(1.94±0.03)×10^{29}
29.03.2023	S21W59	X1.2	02:33	–	02:30	02:32	02:40	10	45.6	17.3	748±126	(3.36±0.05)×10^{29}
30.03.2023	S22W77	M5.4	07:37	–	07:33	07:35	07:37	4.0	25.9	16.6	(1.08±0.11)×10^3	(1.02±0.05)×10^{29}
27.04.2023	S22E05	M1.8	11:14	11:12	11:13	11:13	11:19	6.0	29.5	17.4	74.4±0.81	(2.19±0.04)×10^{28}
03.05.2023	N13E43	M4.3	09:27	–	09:23	09:25	09:31	8.4	36.0	20.0	141±2.1	(2.03±0.04)×10^{28}
03.05.2023	N12E42	M3.1	10:14	10:12	10:09	10:11	10:11	1.7	16.5	12.6	32.4±0.30	(5.82±0.12)×10^{26}
03.05.2023	N12E42	M7.2	10:45	10:40	10:44	10:44	10:52	7.8	67.9	28.7	234±2.0	(6.25±0.12)×10^{28}
03.05.2023	N12E41	M1.7	12:35	12:34	12:30	12:35	12:44	15	30.2	21.9	76.3±0.64	(1.24±0.02)×10^{28}
03.05.2023	N12E40	M2.2	13:50	13:47	13:47	13:47	13:53	6.6	38.7	20.6	125±1.0	(2.55±0.05)×10^{28}
09.05.2023	N13W26	M6.5	03:54	03:51	03:47	03:53	04:11	24	35.9	20.0	111±6.1	(4.86±0.08)×10^{28}
18.05.2023	N17E77	M3.9	20:23	20:22	20:22	20:22	20:28	6.0	43.3	22.9	417±64	(1.53±0.03)×10^{29}
19.05.2023	N17E79	M5.4	00:48	00:44	00:44	00:44	00:50	6.0	32.7	22.1	298±45	(7.26±0.13)×10^{28}
20.05.2023	N17E55	M5.7	15:00	14:59	14:52	14:58	15:14	22	56.7	25.8	396±7.0	(1.33±0.03)×10^{29}

[1]The flare location is derived from the GOES flare catalog

[2]The HXR peak means the peak time of the HXI 20 – 50 keV emission. Note that some WLFs have no available data from HXI, which are tagged as "-"

[3]There are three flares whose WL duration is less than two minutes. This is because their WL brightenings only appear in one time frame

[4]The uncertainty comes from the maximum difference between the WL brightening locations and the flare location in the process of correcting project effect

[5]The uncertainty of total flux consists of two parts, one from the average photometric fluctuation of 1.5% and the other from the uncertainty of the radiometric calibration factor that is obtained via a comparison between the observed counts of WST and standard irradiance at 360 nm from the American Society for Testing and Materials (ASTM) G173-03 (www.nrel.gov/grid/solar-resource/spectra-am1.5.html)

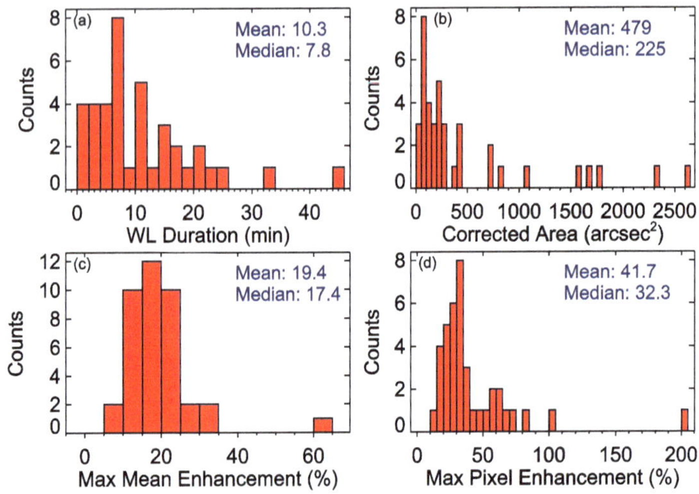

Figure 3 WL parameters at 360 nm, including WL duration (**a**), corrected area (**b**), r_m (**c**), and r_p (**d**) for the selected 39 WLFs.

4.3. WL Duration, Brightening Area, and Relative Enhancement of WLFs at 360 nm

In the following, we focus on 39 WLFs (including 6 X- and 33 M-class flares, as listed in Table 2) among the identified 49 WLFs to further study their physical properties. Note that the other 10 WLFs are excluded here due to their brightenings extending to the solar limb, which could lead to an underestimation of the WL area, as mentioned above.

Figure 3 shows the histogram of WLFs' parameters, including WL duration [τ], brightening area [S], corrected for projection effect), maximum mean enhancement [r_m], and maximum pixel enhancement [r_p]. The values of these parameters are also given in Table 2. From Figure 3a we can see that most WLFs ($\approx$85%) have a duration of less than 20 minutes. The longest duration is 44 minutes coming from the X2.3 WLF on 17 February 2023, i.e. the largest flare in our data set. The mean and median durations for all the 39 WLFs are 10.3 and 7.8 minutes, respectively. From Figure 3b it is seen that most WLFs ($\approx$75%) have a brightening area of smaller than 500 arcsec^2 and the mean/median area of all WLFs is 479/225 arcsec^2. For r_m and r_p shown in Figures 3c and d, one can see that most WLFs have a r_m of less than 40% and a r_p of less than 75%. The mean r_m and r_p of all WLFs are 19.4% and 41.7%, respectively, and the medians are 17.4% and 32.3%, respectively. Note that the X2.1 WLF on 3 March 2023 (the second largest flare in our data set) near the limb has the largest enhancement, with a r_p of 203% and a r_m of 62.1%.

The relationships among the above WL parameters are further studied by calculating their correlation coefficients (cc), as shown in Figure 4. We can see that r_m and r_p of all the 39 WLFs have a strong positive correlation with cc = 0.95 (see the scatter plot in Figure 4a). r_m also has a good relationship with the corrected area (cc = 0.62, Figure 4b). While r_m has no obvious relationship with the WL duration (cc = 0.25, Figure 4c). In addition, the WL duration has a poor correlation with the corrected area (cc = 0.33, Figure 4d).

Figure 5a shows the center-to-limb distribution (in terms of the heliocentric angle [θ] from 0 to 90°) of r_m for all the 39 WLFs. It can be seen that as θ increases, r_m reaches a higher value, although a small r_m can also show up near the limb. The cc between r_m and θ for all the 39 WLFs is 0.24, i.e. showing a weak relevance. However, if we only choose

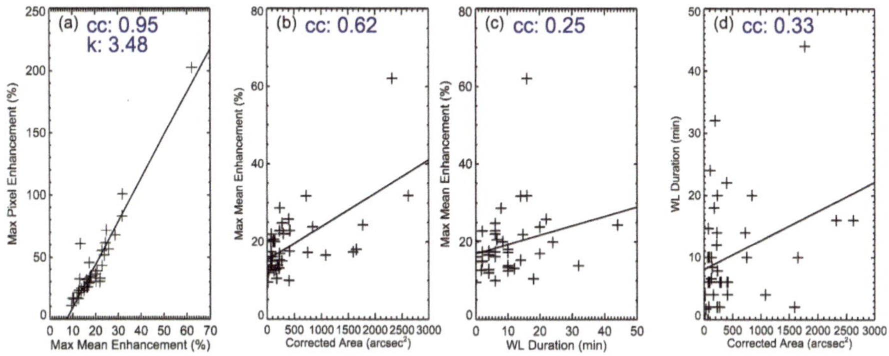

Figure 4 WL parameters and the linear fit for their relationships.

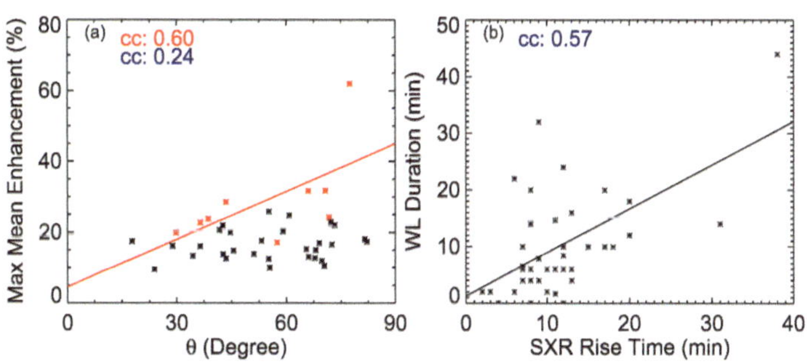

Figure 5 (**a**) r_m versus heliocentric angle. The *red symbols* denote the WLFs above M6 and the *red line* is their linear fit. (**b**) WL duration versus SXR rise time with a linear fit.

WLFs above M6.0 (see the red star symbols), there exists a much stronger correlation (cc = 0.60) between r_m and θ. This suggests that r_m has a center-to-limb variation, especially for large WLFs, which will be discussed in Section 5.

In addition, we check the relationship between WL duration and SXR rise time (i.e. the time period from flare onset to peak) for the 39 WLFs, as shown in Figure 5b. There shows a good positive correlation (with cc = 0.57) between these two parameters. Note that the WL duration has a comparable value with the SXR rise time. This implies that the WL emission is mainly caused by the heating process, as indicated by the SXR rise time of a flare.

The dependencies of WL parameters at 360 nm on flare magnitude or peak SXR flux are shown in Figure 6. We can see that all the parameters, r_p, r_m, τ, and S, have a good relationship with the peak SXR flux, with cc of 0.74, 0.68, 0.58 and 0.80, respectively. This is reasonable, since a larger flare tends to have a more energy, which can lead to a stronger WL brightening.

4.4. Relationships of the WL Peak Time with SXR and HXR Peak Times

Comparing the WL peak time to the SXR or HXR peak time can help us understand the physical origin of WL emission in flares. Figure 7a plots the histogram of the time difference between WL and SXR peak times for all the 39 WLFs. Note that the pink and red colors refer to the WLFs with and without available HXR 20 – 50 keV observations from

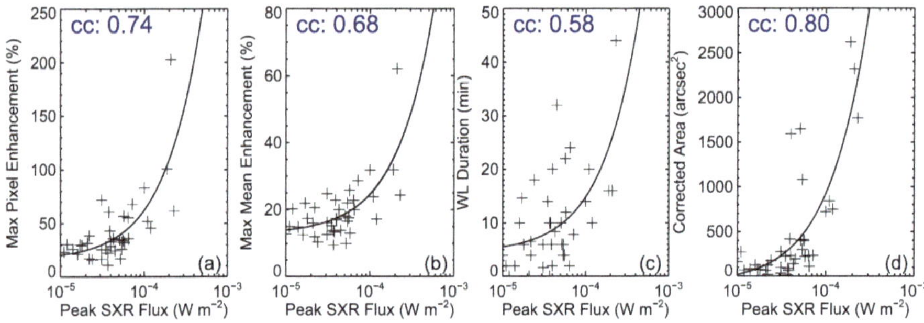

Figure 6 WL parameters versus peak SXR flux with a linear fit.

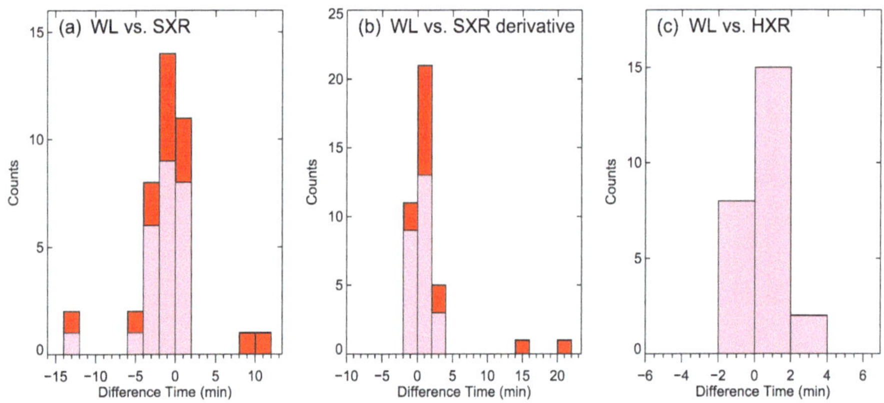

Figure 7 Differences between WL and SXR peak times (**a**), between WL peak time and peak time of SXR temporal derivative (**b**), and between WL and HXR peak times (**c**). In each panel, the *pink and red colors* represent the WLFs with and without HXI data available, respectively.

ASO-S/HXI, respectively. It can be seen that two thirds (26) of the WLFs have an earlier WL peak time compared to the SXR peak time. Considering an uncertainty of ± 2 minutes for the WL peak time, there are still 12 WLFs (about one-third) having a time difference of less than -2 minutes, while only two WLFs have that of more than two minutes. Therefore, the WL emission tends to reach its peak earlier than the SXR emission. This also implies that the WL emission is mainly associated with the heating process of the flare, i.e. similar to the indications of the relationship between the WL duration and SXR rise time, as mentioned above. Note that there are two WLFs (an M4.5 and an M1.6) whose WL emissions peak much later ($\approx$ten minutes) than the SXR emission. These gradual-phase WL emissions might be more related to the cooling process of the flare.

We further check the difference between the WL peak time and the peak time of SXR temporal derivative, as shown in Figure 7b. We can see that most of the WLFs (95%) have a peak time of WL roughly the same as that of the SXR temporal derivative, with the time difference of ± 4 minutes, including all the 25 WLFs with the HXR data available (see the pink color). This indicates that the WL emission at 360 nm exhibits the Neupert effect (Neupert, 1968) in principle. The two exceptional WLFs are those having a later WL peak time, as mentioned before. They also have a later WL peak time than that of the SXR temporal derivative.

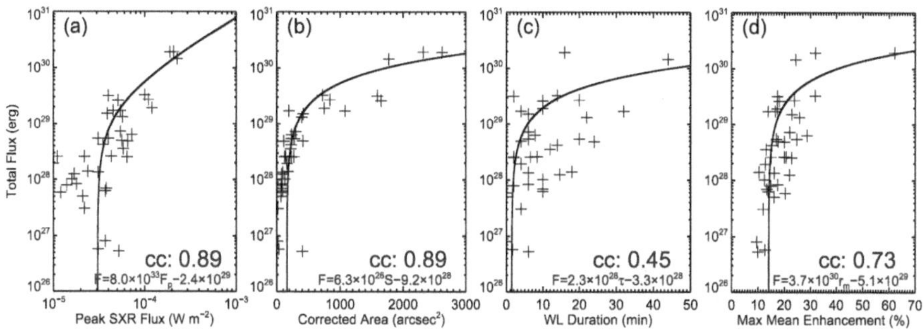

Figure 8 Total flux of the WLFs versus peak SXR flux (**a**), corrected area (**b**), WL duration (**c**), and r_{m} (**d**) with a linear fit. The fit expressions are given in the panels, where F_{p} in Panel **a** represents the peak SXR flux of WLFs.

For the 25 WLFs with HXR data available, we further plot the histogram of the difference between WL and HXR peak times in Figure 7c. One can see that all these WLFs have a similar peak time for the WL and HXR emissions within a range of ± 4 minutes and particularly most (92%) are in the range of ± 2 minutes. This confirms that the WL emission at 360 nm basically follows the Neupert effect and thus implies that the 360 nm emission is closely related to nonthermal electron-beam heating, at least for most WLFs under study.

4.5. Total Flux of WLFs at 360 nm

To describe the absolute enhancement of WLFs at 360 nm, we define the total flux (or total energy) as: $F = \sum_t \sum_{S_t} (I - I_0)$ (Castellanos Durán and Kleint, 2020), i.e. integrating the enhanced WL intensity over corrected area $[S_t]$ and time $[t]$. Note that here the WL intensities of I and I_0 have been calibrated to physical units of erg s^{-1} cm^{-2} sr^{-1}. In addition, the limb-darkening effect at 360 nm has been corrected by using an empirical formula of $\frac{I(\mu)}{I(\mu=1)} = u_\lambda + (1 - u_\lambda)\mu + v_\lambda \mu \ln \frac{2\mu}{\mu+1} + w_\lambda \mu(\mu \ln \frac{\mu+1}{\mu} - \ln 2)$, where u_λ, v_λ, and w_λ are 0.2887, 2.2194, and -5.6933, respectively (Makarova, Roshchina, and Sarychev, 1991). This formula can fit the WST data quite well when $\mu > 0.11$. It should be mentioned that all the analyzed 39 WLFs have $\mu \geq 0.13$. The final F is listed in Table 2. It is seen that the total flux at 360 nm with a bandwidth of ± 2 nm for all the 39 WLFs is about $10^{26} - 10^{30}$ erg. The relationships between the total flux and peak SXR flux (or flare magnitude), brightening area, WL duration, and maximum mean relative enhancement are shown in Figure 8. We can see that the total flux has a good positive correlation with these four parameters in general, with cc of 0.89, 0.89, 0.45, and 0.73, respectively. This result is reasonable and can be predicted. Here we also give the expression of F based on the linear fitting (see each of the panels), which could be used to estimate the radiated energy at 360 ± 2 nm via the peak SXR flux only and further estimate the other WL parameters at 360 nm roughly.

5. Summary and Discussions

In this article, we present a statistical study on the WLFs at 360 nm observed by ASO-S/WST from October 2022 to May 2023. The occurrence rate of WLFs and comparisons between WLFs and non-WLFs are provided. The WLFs' parameters including relative enhancement (r_{m} and r_{p}), WL duration, brightening area, and total flux are given together with their mutual relationships. The WL peak time is also compared to the SXR and HXR peak times, which is helpful to understand the physical origin of WL emission. Our results are summarized below.

i) The occurrence rate of WLFs at 360 nm is 23.9% for major flares above M1.0, which increases with the flare magnitude. It should be mentioned that all seven X-class flares in our data set are WLFs at 360 nm.
ii) Compared with non-WLFs, the WLFs at 360 nm tend to have a shorter SXR duration and a larger peak SXR flux. The spatial distribution of WLFs does not appear to be dependent on solar northern or southern hemisphere, at least for our flare data set. However, the number of WLFs increases from disk center to limb, and the relative enhancement of WL emission can be larger for the WLFs near the limb.
iii) The mean values of WL duration, brightening area, r_{m}, and r_{p} of WLFs at 360 nm are 10.3 minutes, 479 arcsec^2, 19.4%, and 41.7%, respectively. r_{m} is found to have a strong correlation with r_{p} and some positive correlation with the area. However, it has no obvious correlation with the duration. In addition, there is no obvious correlation between the duration and area.
iv) All the WL parameters, including the duration, area, r_{p}, r_{m}, and total flux, show a good positive relationship with the peak SXR flux.
v) The WL emission at 360 nm reaches its peak around the same time as the temporal derivative of SXR flux as well as the HXR 20 – 50 keV emission for most WLFs. This indicates that the WL emission basically follows the Neupert effect. It should be noted that there are two WLFs showing a notable gradual-phase emission at 360 nm.
vi) The total flux at 360 nm for the 39 WLFs under study is about $10^{26} – 10^{30}$ erg. It has a good positive correlation with the peak SXR flux, WL enhancement, and brightening area.

Our study presents a continuous observation of major flares at 360 nm for more than half a year from ASO-S/WST as well as provides the occurrence rate of WLFs at 360 nm for the first time. Although there are some other telescopes observing the Sun in the 360 nm waveband, such as the *Optical and Near-infrared Solar Eruption Tracer* (ONSET: Fang et al., 2013), the advantages of WST are the continuity of its observation in a seeing-free condition from space and its full-disk field of view, which is beneficial for flare studies. Nearly 50 WLFs have been identified in the period of less than one year, which is roughly equivalent to the number of WLFs observed by HMI at 617 nm for about two – three years. Note that our threshold for identifying a WLFs at 360 nm is relatively high and the burst mode of WST for a very high cadence was still under test. Therefore, there should be more than 49 WLFs detected by WST during this period. Here we should also mention that WST and HMI work in different wavebands and the occurrence rate of WLFs in these two wavebands can be different.

Based on some statistical studies on HMI WLFs at 617 nm in the literature, here we can roughly compare the properties of WL emissions between WST 360 nm and HMI 617 nm. For instance, the mean enhancement of the emission at 617 nm is ≈15% (e.g. Song et al., 2018, Castellanos Durán and Kleint, 2020), and r_{m} of the emission at 360 nm is ≈19%. Considering that a higher threshold of intensity ratio for WLFs, i.e. 8% in the present study, would overestimate a r_{m}, we may say that the mean enhancement at 360 nm is comparable to that at 617 nm in WLFs. In addition, the duration and area of WLFs at 617 nm have median values of ≈5 minutes and <50 arcsec^2 (without a correction for the project effect; Song et al., 2018), respectively, both of which are smaller than those of WLFs at 360 nm. Note that a higher threshold for WLFs here would underestimate the brightening area and duration of the emission at 360 nm. A detailed comparison between the emissions at WST 360 nm (in the Balmer continuum) and HMI 617 nm (in the Paschen continuum) deserves a further study for WLFs. In fact, the relationship between the 360 nm emission and the emission at >400 nm, i.e. the true WL, is worthwhile to be explored.

In the present study, the WL emission at 360 nm reaches its peak around the same time as the temporal derivative of the SXR 1 – 8 Å flux as well as the HXR 20 – 50 keV emission in most WLFs, i.e. exhibiting the Neupert effect. This suggests that the 360 nm emission can be closely related to a nonthermal electron beam heating, i.e. it has a nonthermal origin. It should be noted that there are two exceptional WLFs in our data set, whose emissions at 360 nm have a later peak time than the SXR 1 – 8 Å flux. This kind of emission appearing in the decay phase of the flare (or gradual-phase emissions) might be produced by a thermal process (say, thermal conduction or SXR backwarming) rather than a nonthermal electron-beam heating that usually occurs in the rise phase. These are worthy of a detailed study in the future.

In our study, the WL emission at 360 nm exhibits a center-to-limb variation, which is manifested as a larger enhancement plus more WLFs identified near the limb. This is consistent with previous studies on WL emission, both in observations and numerical simulations. A statistical work on 86 WLFs showed that the WLFs prefer to occur at the solar limb (Neidig, Wiborg, and Gilliam, 1993). Some numerical simulations revealed that the electron-beam heating leads to a larger enhancement in the continuum near the solar limb than at the disk center (e.g. Ding, 2007; Cheng, Ding, and Carlsson, 2010). The authors gave an explanation that the formation height of continuum emission is higher in limb flares and it is easier to heat. Our study provides an observational evidence for the center-to-limb effect of the 360 nm emission from a statistical analysis. Furthermore, we find that the center-to-limb variation is more obvious for larger WLFs.

At the time of writing, WST is beginning to provide WL images with a very rapid cadence of one or two seconds in a burst mode. This can bring some new sights on WLFs at 360 nm. Firstly, a high temporal resolution can provide precise start, peak, and end times of WL emission as well as their relationships with the corresponding times of SXR and HXR emissions, which helps understand the physical origin of the 360 nm emissions. A rapid cadence of WL images might also make it possible to detect WL waves at 360 nm accompanied by WLFs (like sunquakes observed from HMI; Wu, Dai, and Ding, 2023). This would be very helpful to explore the conditions in the WL emission source. As more and more WLFs are collected from WST observations, it is worthwhile doing a comprehensive statistical work combining the data from some other instruments such as HMI, ONSET, and the *Chinese Hα Solar Explorer* (CHASE: Li et al., 2022). On the other hand, radiative-hydrodynamic simulations are needed to help explain the observations in terms of formation height and contribution function of the WL emission. All these should improve our understanding of WLFs at 360 nm.

Acknowledgments The authors thank the reviewer very much for the valuable and detailed suggestions and comments that helped to improve the manuscript greatly. The ASO-S mission is supported by the Strategic Priority Research Program on Space Science, Chinese Academy of Sciences. SDO is a mission for NASA's Living With a Star program.

Author contributions Z.C. Jing analyzed the data and wrote the first manuscript. Y. Li proposed the initial idea, helped with the data analysis, and revised the manuscript. J. Tian and D.C. Song provided suggestions on the data analysis. W.Q. Gan is PI of ASO-S. H. Li and L. Feng are PI and Co-PI of LST, respectively. Y. Huang, Y.P. Li, J.W. Li, J.C. Xue, J. Zhao, L. Lu, B.L. Ying, P. Zhang, Y.N. Su, Q.M. Zhang, D. Li, Y.Y. Ge, S.T. Li, Q. Li, G. Li, X.F. Liu, G.L. Shi, J.H. Shan, Z.Y. Tian, and Y. Zhou contributed to the pipeline and release of WST data. Y. Su and W. Chen provided the HXI data. All authors reviewed the manuscript.

Funding The authors are supported by the Strategic Priority Research Program of the Chinese Academy of Sciences (Grant No. XDB0560000), the National Natural Science Foundation of China (Grant Nos. 12273115, 12233012, and 11921003), and the Ministry of Science and Technology of China (Grant No. 2022YFF0503004).

Data Availability Data of ASO-S before 1 April 2023 are under test and not publicly available, but they are available from the corresponding author on reasonable request. Data of ASO-S after 1 April 2023 are publicly available and can be downloaded from the official website of ASO-S at aso-s.pmo.ac.cn/sodc/dataArchive.jsp. Data of SDO used in this work are publicly available at jsoc.stanford.edu.

Declarations

Competing interests The authors declare no competing interests.

References

Aboudarham, J., Henoux, J.C.: 1986a, Energy deposit at temperature minimum level and white light flares. *Astron. Astrophys.* **156**, 73. ADS.

Aboudarham, J., Henoux, J.C.: 1986b, Non-thermal excitation and ionization of hydrogen in solar flares. I. Effects on a flaring chromosphere. *Astron. Astrophys.* **168**, 301. ADS.

Carrington, R.C.: 1859, Description of a singular appearance seen in the Sun on September 1, 1859. *Mon. Not. Roy. Astron. Soc.* **20**, 13. DOI. ADS.

Castellanos Durán, J.S., Kleint, L.: 2020, The statistical relationship between white-light emission and photospheric magnetic field changes in flares. *Astrophys. J.* **904**, 96. DOI. ADS.

Chen, B., Li, H., Song, K.-F., Guo, Q.-F., Zhang, P.-J., He, L.-P., Dai, S., Wang, X.-D., Wang, H.-F., Liu, C.-L., Zhang, H.-J., Zhang, G., Wang, Y., Liu, S.-J., Zhang, H.-X., Liu, L., Mao, S.-L., Liu, Y., Peng, J.-H., Wang, P., Sun, L., Liu, Y., Han, Z.-W., Wang, Y.-L., Wu, K., Ding, G.-X., Zhou, P., Zheng, X., Xia, M.-Y., Wu, Q.-W., Xie, J.-J., Chen, Y., Song, S.-M., Wang, H., Zhu, B., Chu, C.-B., Yang, W.-G., Feng, L., Huang, Y., Gan, W.-Q., Li, Y., Li, J.-W., Lu, L., Xue, J.-C., Ying, B.-L., Sun, M.-Z., Zhu, C., Bao, W.-M., Deng, L., Yin, Z.-S.: 2019, The Lyman-alpha Solar Telescope (LST) for the ASO-S mission – II. Design of LST. *Res. Astron. Astrophys.* **19**, 159. DOI. ADS.

Cheng, J.X., Ding, M.D., Carlsson, M.: 2010, Radiative hydrodynamic simulation of the continuum emission in solar white-light flares. *Astrophys. J.* **711**, 185. DOI. ADS.

Ding, M.D.: 2007, The origin of solar white-light flares. In: Heinzel, P., Dorotovič, I., Rutten, R.J. (eds.) *The Physics of Chromospheric Plasmas* **CS-368**, Astron. Soc. Pacific, San Francisco, 417. ADS.

Ding, M.D., Liu, Y., Yeh, C.-T., Li, J.P.: 2003, Interpretation of the infrared continuum in a solar white-light flare. *Astron. Astrophys.* **403**, 1151. DOI. ADS.

Fang, C., Chen, P.-F., Li, Z., Ding, M.-D., Dai, Y., Zhang, X.-Y., Mao, W.-J., Zhang, J.-P., Li, T., Liang, Y.-J., Lu, H.-T.: 2013, A new multi-wavelength solar telescope: Optical and Near-infrared Solar Eruption Tracer (ONSET). *Res. Astron. Astrophys.* **13**, 1509. DOI. ADS.

Feng, L., Li, H., Chen, B., Li, Y., Susino, R., Huang, Y., Lu, L., Ying, B.-L., Li, J.-W., Xue, J.-C., Yang, Y.-T., Hong, J., Li, J.-P., Zhao, J., Gan, W.-Q., Zhang, Y.: 2019, The Lyman-alpha Solar Telescope (LST) for the ASO-S mission – III. Data and potential diagnostics. *Res. Astron. Astrophys.* **19**, 162. DOI. ADS.

Fletcher, L., Hudson, H.S.: 2008, Impulsive phase flare energy transport by large-scale Alfvén waves and the electron acceleration problem. *Astrophys. J.* **675**, 1645. DOI. ADS.

Fletcher, L., Dennis, B.R., Hudson, H.S., Krucker, S., Phillips, K., Veronig, A., Battaglia, M., Bone, L., Caspi, A., Chen, Q., Gallagher, P., Grigis, P.T., Ji, H., Liu, W., Milligan, R.O., Temmer, M.: 2011, An observational overview of solar flares. *Space Sci. Rev.* **159**, 19. DOI. ADS.

Gan, W.Q., Hénoux, J.C., Fang, C.: 2000, On the origin of solar white-light flares. *Astron. Astrophys.* **354**, 691. ADS.

Gan, W.Q., Mauas, P.J.D.: 1994, Atmospheric heating in solar flares by chromospheric condensation. *Astrophys. J.* **430**, 891. DOI. ADS.

Gan, W.-Q., Ding, M.-D., Huang, Y., Su, Y.-N.: 2019, Preface: Advanced Space-based Solar Observatory (ASO-S). *Res. Astron. Astrophys.* **19**, 155. DOI. ADS.

Gan, W., Zhu, C., Deng, Y., Zhang, Z., Chen, B., Huang, Y., Deng, L., Wu, H., Zhang, H., Li, H., Su, Y., Su, J., Feng, L., Wu, J., Cui, J., Wang, C., Chang, J., Yin, Z., Xiong, W., Chen, B., Yang, J., Li, F., Lin, J., Hou, J., Bai, X., Chen, D., Zhang, Y., Hu, Y., Liang, Y., Wang, J., Song, K., Guo, Q., He, L., Zhang, G., Wang, P., Bao, H., Cao, C., Bai, Y., Chen, B., He, T., Li, X., Zhang, Y., Liao, X., Jiang, H., Li, Y., Su, Y., Lei, S., Chen, W., Li, Y., Zhao, J., Li, J., Ge, Y., Zou, Z., Hu, T., Su, M., Ji, H., Gu, M., Zheng, Y., Xu, D., Wang, X.: 2023, The Advanced Space-Based Solar Observatory (ASO-S). *Solar Phys.* **298**, 68. DOI. ADS.

Hanser, F.A., Sellers, F.B.: 1996, Design and calibration of the GOES-8 solar X-ray sensor: the XRS. In: Washwell, E.R. (ed.) *GOES-8 and Beyond, Soc. Photo-Opt. Instrum. Eng. (SPIE)* **CS–2812**, 344. DOI. ADS.

Hao, Q., Yang, K., Cheng, X., Guo, Y., Fang, C., Ding, M.D., Chen, P.F., Li, Z.: 2017, A circular white-light flare with impulsive and gradual white-light kernels. *Nat. Commun.* **8**, 2202. DOI. ADS.

Hodgson, R.: 1859, On a curious appearance seen in the Sun. *Mon. Not. Roy. Astron. Soc.* **20**, 15. DOI. ADS.

Hudson, H.S.: 1972, Thick-target processes and white-light flares. *Solar Phys.* **24**, 414. DOI. ADS.

Hudson, H.S.: 2011, Global properties of solar flares. *Space Sci. Rev.* **158**, 5. DOI. ADS.

Hudson, H.S., Wolfson, C.J., Metcalf, T.R.: 2006, White-light flares: a TRACE/RHESSI overview. *Solar Phys.* **234**, 79. DOI. ADS.

Hudson, H.S., Acton, L.W., Hirayama, T., Uchida, Y.: 1992, White-light flares observed by YOHKOH. *Publ. Astron. Soc. Japan* **44**, L77. ADS.

Kosugi, T., Matsuzaki, K., Sakao, T., Shimizu, T., Sone, Y., Tachikawa, S., Hashimoto, T., Minesugi, K., Ohnishi, A., Yamada, T., Tsuneta, S., Hara, H., Ichimoto, K., Suematsu, Y., Shimojo, M., Watanabe, T., Shimada, S., Davis, J.M., Hill, L.D., Owens, J.K., Title, A.M., Culhane, J.L., Harra, L.K., Doschek, G.A., Golub, L.: 2007, The Hinode (Solar-B) mission: an overview. *Solar Phys.* **243**, 3. DOI. ADS.

Krucker, S., Hudson, H.S., Jeffrey, N.L.S., Battaglia, M., Kontar, E.P., Benz, A.O., Csillaghy, A., Lin, R.P.: 2011, High-resolution imaging of solar flare ribbons and its implication on the thick-target beam model. *Astrophys. J.* **739**, 96. DOI. ADS.

Kuhar, M., Krucker, S., Martínez Oliveros, J.C., Battaglia, M., Kleint, L., Casadei, D., Hudson, H.S.: 2016, Correlation of hard X-ray and white light emission in solar flares. *Astrophys. J.* **816**, 6. DOI. ADS.

Li, H., Chen, B., Feng, L., Li, Y., Huang, Y., Li, J.-W., Lu, L., Xue, J.-C., Ying, B.-L., Zhao, J., Yang, Y.-T., Gan, W.-Q., Fang, C., Song, K.-F., Wang, H., Guo, Q.-F., He, L.-P., Zhu, B., Zhu, C., Deng, L., Bao, H.-C., Cao, C.-X., Yang, Z.-G.: 2019, The Lyman-alpha Solar Telescope (LST) for the ASO-S mission — I. Scientific objectives and overview. *Res. Astron. Astrophys.* **19**, 158. DOI. ADS.

Li, C., Fang, C., Li, Z., Ding, M., Chen, P., Qiu, Y., You, W., Yuan, Y., An, M., Tao, H., Li, X., Chen, Z., Liu, Q., Mei, G., Yang, L., Zhang, W., Cheng, W., Chen, J., Chen, C., Gu, Q., Huang, Q., Liu, M., Han, C., Xin, H., Chen, C., Ni, Y., Wang, W., Rao, S., Li, H., Lu, X., Wang, W., Lin, J., Jiang, Y., Meng, L., Zhao, J.: 2022, The Chinese Hα Solar Explorer (CHASE) mission: an overview. *Sci. China Phys. Mech. Astron.* **65**, 289602. DOI. ADS.

Machado, M.E.: 1978, Soft X-ray emission and chromospheric flares. *Solar Phys.* **60**, 341. DOI. ADS.

Machado, M.E., Emslie, A.G., Avrett, E.H.: 1989, Radiative backwarming in white-light flares. *Solar Phys.* **124**, 303. DOI. ADS.

Machado, M.E., Emslie, A.G., Brown, J.C.: 1978, The structure of the temperature minimum region in solar flares and its significance for flare heating mechanisms. *Solar Phys.* **58**, 363. DOI. ADS.

Makarova, E.A., Roshchina, E.M., Sarychev, A.P.: 1991, Average data on solar limb darkening in the 300-NM to 2400-NM quasicontinuum. *Soviet Astron.* **35**, 441. ADS.

Maurya, R.A., Ambastha, A.: 2010, A technique for automated determination of flare ribbon separation and energy release. *Solar Phys.* **262**, 337. DOI. ADS.

Metcalf, T.R., Alexander, D., Hudson, H.S., Longcope, D.W.: 2003, TRACE and Yohkoh Observations of a white-light flare. *Astrophys. J.* **595**, 483. DOI. ADS.

Neidig, D.F.: 1989, The importance of solar white-light flares. *Solar Phys.* **121**, 261. DOI. ADS.

Neidig, D.F., Cliver, E.W.: 1983, The occurrence frequency of white light flares. *Solar Phys.* **88**, 275. DOI. ADS.

Neidig, D.F., Wiborg, P.H., Gilliam, L.B.: 1993, Physical properties of white-light flares derived from their center-to-limb distribution. *Solar Phys.* **144**, 169. DOI. ADS.

Neupert, W.M.: 1968, Comparison of solar X-ray line emission with microwave emission during flares. *Astrophys. J. Lett.* **153**, L59. DOI. ADS.

Procházka, O.: 2019, Analysis and Modelling of a Type II White-light Solar Flare. PhD thesis, Queen's University Belfast. DOI. ADS.

Scherrer, P.H., Schou, J., Bush, R.I., Kosovichev, A.G., Bogart, R.S., Hoeksema, J.T., Liu, Y., Duvall, T.L., Zhao, J., Title, A.M., Schrijver, C.J., Tarbell, T.D., Tomczyk, S.: 2012, The Helioseismic and Magnetic Imager (HMI) investigation for the Solar Dynamics Observatory (SDO). *Solar Phys.* **275**, 207. DOI. ADS.

Shibata, K., Magara, T.: 2011, Solar flares: magnetohydrodynamic processes. *Liv. Rev. Solar Phys.* **8**, 6. DOI. ADS.

Somov, B.V., Syrovatskiĭ, S.I.: 1976, Reviews of topical problems: physical processes in the solar atmosphere associated with flares. *Sov. Phys. Usp.* **19**, 813. DOI. ADS.

Song, Y., Tian, H.: 2018, Investigation of white-light emission in circular-ribbon flares. *Astrophys. J.* **867**, 159. DOI. ADS.

Song, Y.L., Tian, H., Zhang, M., Ding, M.D.: 2018, Observations of white-light flares in NOAA active region 11515: high occurrence rate and relationship with magnetic transients. *Astron. Astrophys.* **613**, A69. DOI. ADS.

Song, D.-C., Tian, J., Li, Y., Ding, M.D., Su, Y., Yu, S., Hong, J., Qiu, Y., Rao, S., Liu, X., Li, Q., Chen, X., Li, C., Fang, C.: 2023, Spectral observations and modeling of a solar white-light flare observed by CHASE. *Astrophys. J. Lett.* **952**, L6. DOI. ADS.

Su, J.-T., Bai, X.-Y., Chen, J., Guo, J.-J., Liu, S., Wang, X.-F., Xu, H.-Q., Yang, X., Song, Y.-L., Deng, Y.-Y., Ji, K.-F., Deng, L., Huang, Y., Li, H., Gan, W.-Q.: 2019, Data reduction and calibration of the FMG onboard ASO-S. *Res. Astron. Astrophys.* **19**, 161. DOI. ADS.

Švestka, Z.: 1970, The phase of particle acceleration in the flare development. *Solar Phys.* **13**, 471. DOI. ADS.

Tian, J., Hong, J., Li, Y., Ding, M.D.: 2022, An evaluation of different recipes for chromospheric radiative losses in solar flares. *Astron. Astrophys.* **668**, A96. DOI. ADS.

Tsuneta, S., Acton, L., Bruner, M., Lemen, J., Brown, W., Caravalho, R., Catura, R., Freeland, S., Jurcevich, B., Morrison, M., Ogawara, Y., Hirayama, T., Owens, J.: 1991, The soft X-ray telescope for the SOLAR-A mission. *Solar Phys.* **136**, 37. DOI. ADS.

Watanabe, K., Imada, S.: 2020, White-light emission and chromospheric response by an X1.8-class flare on 2012 October 23. *Astrophys. J.* **891**, 88. DOI. ADS.

Watanabe, K., Kitagawa, J., Masuda, S.: 2017, Characteristics that produce white-light enhancements in solar flares observed by Hinode/SOT. *Astrophys. J.* **850**, 204. DOI. ADS.

Watanabe, K., Masuda, S., Segawa, T.: 2012, Hinode flare catalogue. *Solar Phys.* **279**, 317. DOI. ADS.

Wu, H., Dai, Y., Ding, M.D.: 2023, Highly energetic electrons accelerated in strong solar flares as a preferred driver of sunquakes. *Astrophys. J. Lett.* **943**, L6. DOI. ADS.

Xu, Y., Cao, W., Jing, J., Wang, H.: 2010, High resolution observations of white-light emissions from the opacity minimum during an X-class flare. *Astron. Nachr.* **331**, 596. DOI. ADS.

Zhang, Z., Chen, D.-Y., Wu, J., Chang, J., Hu, Y.-M., Su, Y., Zhang, Y., Wang, J.-P., Liang, Y.-M., Ma, T., Guo, J.-H., Cai, M.-S., Zhang, Y.-Q., Huang, Y.-Y., Peng, X.-Y., Tang, Z.-B., Zhao, X., Zhou, H.-H., Wang, L.-G., Song, J.-X., Ma, M., Xu, G.-Z., Yang, J.-F., Lu, D., He, Y.-H., Tao, J.-Y., Ma, X.-L., Lv, B.-G., Bai, Y.-P., Cao, C.-X., Huang, Y., Gan, W.-Q.: 2019, Hard X-ray Imager (HXI) onboard the ASO-S mission. *Res. Astron. Astrophys.* **19**, 160. DOI. ADS.

Authors and Affiliations

Zhichen Jing[1,2] · Ying Li[1,2] · Li Feng[1,2] · Hui Li[1,2] · Yu Huang[1,2] · Youping Li[1,2] · Yang Su[1,2] · Wei Chen[1,2] · Jun Tian[1,2] · Dechao Song[1,2] · Jingwei Li[1] · Jianchao Xue[1] · Jie Zhao[1] · Lei Lu[1] · Beili Ying[1] · Ping Zhang[1] · Yingna Su[1,2] · Qingmin Zhang[1,2] · Dong Li[1,2] · Yunyi Ge[1] · Shuting Li[1,2] · Qiao Li[1,2] · Gen Li[1,2] · Xiaofeng Liu[1,2] · Guanglu Shi[1,2] · Jiahui Shan[1,2] · Zhengyuan Tian[1,2] · Yue Zhou[1,2] · Weiqun Gan[1,3]

✉ Y. Li
yingli@pmo.ac.cn

[1] Key Laboratory of Dark Matter and Space Astronomy, Purple Mountain Observatory, Chinese Academy of Sciences, Nanjing 210023, China

[2] School of Astronomy and Space Science, University of Science and Technology of China, Hefei 230026, China

[3] University of Chinese Academy of Sciences, Nanjing 211135, China

Solar Physics (2024) 299:73
https://doi.org/10.1007/s11207-024-02313-y

RESEARCH

Spectral and Imaging Observations of a C2.3 White-Light Flare from the *Advanced Space-Based Solar Observatory* (ASO-S) and the *Chinese Hα Solar Explorer* (CHASE)

Qiao Li[1,2] · Ying Li[1,2] · Yang Su[1,2] · Dechao Song[1] · Hui Li[1,2] · Li Feng[1,2] · Yu Huang[1,2] · Youping Li[1,2] · Jingwei Li[1] · Jie Zhao[1] · Lei Lu[1] · Beili Ying[1] · Jianchao Xue[1] · Ping Zhang[1] · Jun Tian[1,2] · Xiaofeng Liu[1,2] · Gen Li[1,2] · Zhichen Jing[1,2] · Shuting Li[1,2] · Guanglu Shi[1,2] · Zhengyuan Tian[1,2] · Wei Chen[1,2] · Yingna Su[1,2] · Qingmin Zhang[1,2] · Dong Li[1,2] · Yunyi Ge[1] · Jiahui Shan[1,2] · Yue Zhou[1,2] · Shijun Lei[1] · Weiqun Gan[1,3]

Received: 4 February 2024 / Accepted: 30 April 2024 / Published online: 29 May 2024

Abstract
Solar white-light flares are characterized by an enhancement in the optical continuum, which are usually large flares (X- and M-class flares). Here, we report a small C2.3 white-light flare (SOL2022-12-20T04:10) observed by the *Advanced Space-based Solar Observatory* and the *Chinese Hα Solar Explorer* (CHASE). This flare exhibits an increase of $\approx 6.4\%$ in the photospheric Fe I line at 6569.2 Å and $\approx 3.2\%$ in the nearby continuum. The continuum at 3600 Å also shows an enhancement of $\approx 4.7\%$. The white-light bright kernels are mainly located at the flare ribbons and co-spatial with nonthermal hard X-ray sources, which implies that the enhanced white-light emissions are related to nonthermal electron-beam heating. At the bright kernels, the Fe I line displays an absorption profile that has a good Gaussian shape, with a redshift up to $\approx 1.7\ \mathrm{km\,s^{-1}}$, while the Hα line shows an emission profile having a central reversal. The Hα line profile also shows a red or blue asymmetry caused by plasma flows with a velocity of several to tens of $\mathrm{km\,s^{-1}}$. It is interesting to find that the Hα asymmetry is opposite at the conjugate footpoints. It is also found that the CHASE continuum increase seems to be related to the change in the photospheric magnetic field. Our study provides comprehensive characteristics of a small white-light flare that help understand the energy release process of white-light flares.

Keywords Flares · White-light · Relation to magnetic field · X-ray bursts · Spectral line · Continuum

✉ Y. Li
yingli@pmo.ac.cn

[1] Key Laboratory of Dark Matter and Space Astronomy, Purple Mountain Observatory, Chinese Academy of Sciences, Nanjing 210023, China

[2] School of Astronomy and Space Science, University of Science and Technology of China, Hefei 230026, China

[3] University of Chinese Academy of Sciences, Nanjing 211135, China

1. Introduction

Solar white-light flares (WLFs) are identified by a sudden increase in the visible continuum (Neidig and Cliver, 1983; Neidig, 1989). The first WLF (also the first solar flare) was reported by Carrington (1859) and Hodgson (1859). Most WLFs occur in the vicinity of sunspots and manifest as bright kernels, but some can also occur in a region almost without a sunspot (e.g. Hudson et al., 1994) and appear as loop-like structures (e.g. Hudson et al., 1992; Hudson, Wolfson, and Metcalf, 2006; Jejčič, Kleint, and Heinzel, 2018). While it is generally accepted that WLFs constitute a minority in the overall flare family (Fang et al., 2013) with X- and M-class flares (i.e., major flares) being the most energetic, there are different views suggesting that all flares may exhibit white-light brightenings (e.g. Matthews et al., 2003; Jess et al., 2008; Song et al., 2018). This hypothesis gains support from discovering white-light continuum enhancements in small flares.

Up to now, there have been only about 20 C-class WLFs reported including several low C-class ones from both statistical studies (e.g. Matthews et al., 2003; Hudson, Wolfson, and Metcalf, 2006; Song et al., 2018; Song and Tian, 2018; Castellanos Durán and Kleint, 2020) and case analyses (e.g. Jess et al., 2008; Yurchyshyn et al., 2017; Song et al., 2020). Statistically, C-class WLFs exhibit an average increase of $\approx 10\%$ in the visible continuum. Some C-class WLFs with a lower flux could display a relatively higher white-light enhancement (e.g. Hudson, Wolfson, and Metcalf, 2006; Song et al., 2018; Song and Tian, 2018), which seems to deviate from the behavior observed in major flares. Note that this discrepancy may be attributed to sample bias and instrumental limitation. For case studies, Jess et al. (2008) observed a C2.0 WLF showing a white-light kernel with a diameter of $\approx 0.4''$ using the *Swedish Solar Telescope* (SST), among the highest resolution telescopes. This small-scale WLF demonstrates an unusually high increase of $\approx 300\%$ in the blue continuum, surpassing the typical enhancement in WLFs. However, degrading the spatial resolution to $1''$ decreases the enhancement to only 1%. In another case study, Song et al. (2020) presented a C2.3 WLF with an enhancement of $\approx 18\%$ observed by the *Helioseismic and Magnetic Imager* (HMI: Scherrer et al., 2012) onboard the *Solar Dynamics Observatory* (SDO: Pesnell, Thompson, and Chamberlin, 2012). This particular flare is associated with weak hard X-ray (HXR) emissions, with the white-light enhancement unlikely caused by accelerated electron beams. Its magnetic field morphology and evolution are more supportive of magnetic reconnection in the lower atmosphere (e.g. Ding, Fang, and Yun, 1999; Chen, Fang, and Ding, 2001) rather than in the corona. This suggests that both the nonthermal particle injection and the evolution of magnetic fields need to be considered when studying WLFs. Given the scarcity of studies on small WLFs, more attention should be paid to the WLFs with a low flux.

Multiple studies have shown a good correlation between the white-light (including the Balmer continuum) and HXR emissions, in terms of time and space (e.g. Matthews et al., 2003; Metcalf et al., 2003; Hudson, Wolfson, and Metcalf, 2006; Wang, 2009; Krucker et al., 2011; Martínez Oliveros et al., 2012; Heinzel and Kleint, 2014; Heinzel et al., 2017). These results support an electron-beam bombardment (e.g. Hudson, 1972; Aboudarham and Henoux, 1987) and subsequent chromospheric condensation (e.g. Gan et al., 1992; Gan and Mauas, 1994; Kowalski et al., 2015) and radiative backwarming (e.g. Machado, Emslie, and Avrett, 1989) processes related to the white-light brightenings. Some studies further reveal that the footpoint or ribbon with a stronger white-light enhancement shows a stronger HXR flux (e.g. Metcalf et al., 2003; Krucker et al., 2011). However, Chen and Ding (2005) reported a stronger white-light kernel accompanied by a weaker nonthermal energy flux in comparison between the two footpoints. Some cases even show no relationship between the white-light and HXR emissions (e.g. Ryan et al., 1983; Ding et al., 1994; Sylwester and

Sylwester, 2000; Song et al., 2020). On the other hand, some other heating mechanisms can also play a role in WLFs, such as magnetic reconnection in the lower atmosphere (e.g. Ding, Fang, and Yun, 1999; Chen, Fang, and Ding, 2001), energy dissipation of Alfvén waves (e.g. Emslie and Sturrock, 1982; Fletcher and Hudson, 2008), and soft X-ray (SXR)/ultraviolet (UV) irradiation (e.g. Poland, Milkey, and Thompson, 1988).

Previous studies have indicated that permanent changes in the photospheric magnetic field commonly appear in flares (e.g. Sudol and Harvey, 2005; Castellanos Durán, Kleint, and Calvo-Mozo, 2018). These changes mean that the magnetic components undergo abrupt transformations and do not return to their original levels for a long time. The coronal implosion process (Hudson, 2000) predicts that photospheric fields become more horizontal during flares, thereby changing the line-of-sight (LoS) magnetic field strength [B_{LoS}] (Hudson, Fisher, and Welsch, 2008). This has been used for an interpretation of the appearance of B_{LoS} changes [ΔB_{LoS}] (e.g. Petrie and Sudol, 2010; Gosain, 2012; Sun et al., 2012). However, differences in ΔB_{LoS} between the chromosphere and photosphere suggested that the chromospheric ΔB_{LoS} may not support a contraction of the field line in flare loops (e.g. Kleint, 2017). Recently, a statistical study (Castellanos Durán and Kleint, 2020) showed that white-light brightenings and ΔB_{LoS} in the photosphere can overlap less than 60%, and the areas of these two parameters appear to have a power-law relation. The relationship between ΔB_{LoS} and white-light brightenings deserves further study, especially for small WLFs.

Spectral analysis is a good way to study the dynamics of plasma in WLFs, especially via some Fe I lines formed in the photosphere (e.g. Babin and Koval, 1992; Lin, Zhang, and Zhang, 1996; Hong et al., 2018). Jurčák et al. (2018) reported a WLF with the response of Fe I lines at 6301.5 Å and 6302.5 Å, which was estimated to originate in the chromosphere instead of the photosphere. In addition, the Hα line at 6562.8 Å formed from the photosphere to the chromosphere can help us understand the physical process of WLFs. This line mainly behaves as an emission profile with a central reversal and a red asymmetry during WLFs (e.g. Babin and Koval, 1992; Zhou, Fang, and Hu, 1997).

In this work, we report a small C2.3 WLF (SOL2022-12-20T04:10) observed by the *Advanced Space-based Solar Observatory* (ASO-S: Gan et al., 2023) and the *Chinese Hα Solar Explorer* (CHASE: Li et al., 2022). ASO-S provides images in the white-light continuum at 3600 Å and also HXR spectra at $\approx 10-300$ keV. CHASE has spectral observations in the Fe I 6569.2 Å and Hα 6562.8 Å lines. All these allow us to study the comprehensive characteristics of this C2.3 WLF when combined with some other multiwavelength observations. In the following Section 2, we describe the data reduction. Our results are presented in Section 3, followed by a summary and discussion in Section 4.

2. Data Analysis

2.1. Data

The data used in this work are obtained by multiple telescopes. The main telescopes include the *Lyman-alpha Solar Telescope* (LST: Li et al., 2019; Chen et al., 2019; Feng et al., 2019) onboard ASO-S, CHASE, and HMI on SDO. LST consists of three instruments, one of which is the *White-light Solar Telescope* (WST). The WST can provide full-disk images in the continuum at 3600 ± 20 Å. The pixel size of the images is $\approx 0.5''$ while the spatial resolution is about $4''$ (Jing et al., 2024). The cadence is two minutes in a routine mode. The level-2.5 (L2.5) data of WST are used here, which have been obtained for scientific calibration and image registration. CHASE provides spectral data of the Fe I line at 6569.2 Å

(6567.8 – 6570.6 Å) and the Hα line at 6562.8 Å (6559.7 – 6565.9 Å) via a raster scanning mode (RSM). The data have a cadence of ≈ 1 minute, a pixel size of ≈ 1″, and a spectral resolution of 0.048 Å pixel^{-1}, with a two-binning mode in both space and spectrum. The calibrated L1.5 data are used here (Qiu et al., 2022). The full-disk images of Fe I 6173 Å and the line of sight (LoS) magnetogram presented in the work are obtained from SDO/HMI, which have a pixel size of 0.5″ and a cadence of 45 seconds.

The *Hard X-ray Imager* (HXI: Zhang et al., 2019; Su et al., 2022) onboard ASO-S is an HXR imaging spectrometer observing the Sun in an energy range of ≈ 10 – 300 keV with energy and angular resolutions of ≈ 16.5% at 32 keV and 3.1″, respectively. The HXI L1 data are used in the present study. The *Spectrometer Telescope for Imaging X-rays* (STIX: Krucker et al., 2020) on *Solar Orbiter* (SolO: Müller et al., 2020) has energy and angular resolutions of ≈ 1 at 6 keV and ≥ 7″, respectively, in a range from 4 keV to 150 keV. Its L1 pixel and spectrogram data are used here. At the observation time of the flare event, STIX had a separation angle of 19.2° with the Earth and a distance of 0.9 AU to the Sun. The *Atmospheric Imaging Assembly* (AIA: Lemen et al., 2012) on SDO provides the images for flare loops in the extreme ultraviolet (EUV) channel at 131 Å with a pixel size of 0.6″ and a cadence of 12 seconds. The SXR 1 – 8 Å light curve is from the *X-Ray Sensor* (XRS: Hanser and Sellers, 1996) onboard the *Geostationary Operational Environmental Satellite* (GOES) with a high cadence of 1 second.

2.2. Data Analysis

We coalign the images from different instruments via the routines in SolarSoftWare (SSW: Freeland and Handy, 1998). Firstly, the AIA and HMI images are registered using aia_prep.pro. The solar rotation has been removed via drot_map.pro. Then, the white-light images from WST and CHASE are coaligned with the HMI continuum images using xy-off.pro based on cross correlation for sunspot features. The uncertainty of the coalignment is estimated to be about 1″.

For HXI light curves at 20 – 50 keV and 50 – 100 keV, we use the combined data of three total flux detectors (D92, D93, and D94) with a cadence of four seconds. The images are reconstructed using the HXI CLEAN algorithm (Su et al., 2019b) with sub-collimators G3 – G10, which correspond to a spatial resolution of ≈ 6.5″. We also use STIX data to produce the light curves at 4 – 10 keV and 10 – 20 keV with a cadence of four seconds and to reconstruct images via the Expectation Maximization (EM) algorithm (Massa et al., 2019). In addition, we make spectral fittings from STIX via a thermal component plus a nonthermal thick target component. Note that the light travel time at the Earth with a correction of 35 s for STIX has been considered.

To remove non-flaring contributions and calculate the relative enhancements caused by the flare, we subtract a background or base image from flaring images for different wavebands. For HMI and WST images, an average of 30 minutes before the flare onset is adopted as the background. For CHASE images, the time at 04:19:48 UT, i.e., around the flare end, is selected as the background, which is relatively quiet during the whole observation of CHASE. Note that this may lead to an underestimation of the CHASE continuum enhancement to some extent. When estimating the uncertainty of relative enhancements for each waveband emission, we choose a quiet-Sun region with a size of $100'' \times 16''$ ($X = [-799'', -699'']$, $Y = [483'', 499'']$) to calculate its relative intensity fluctuations, i.e., the error bars as plotted in Figure 3a and b. Note that this quiet-Sun region is out of the field of view of the images as shown in Figure 1d – i.

Using the CHASE spectral data, we can measure the Doppler velocities of Fe I and Hα lines and also calculate the Hα asymmetry. As mentioned above, the reference line centers

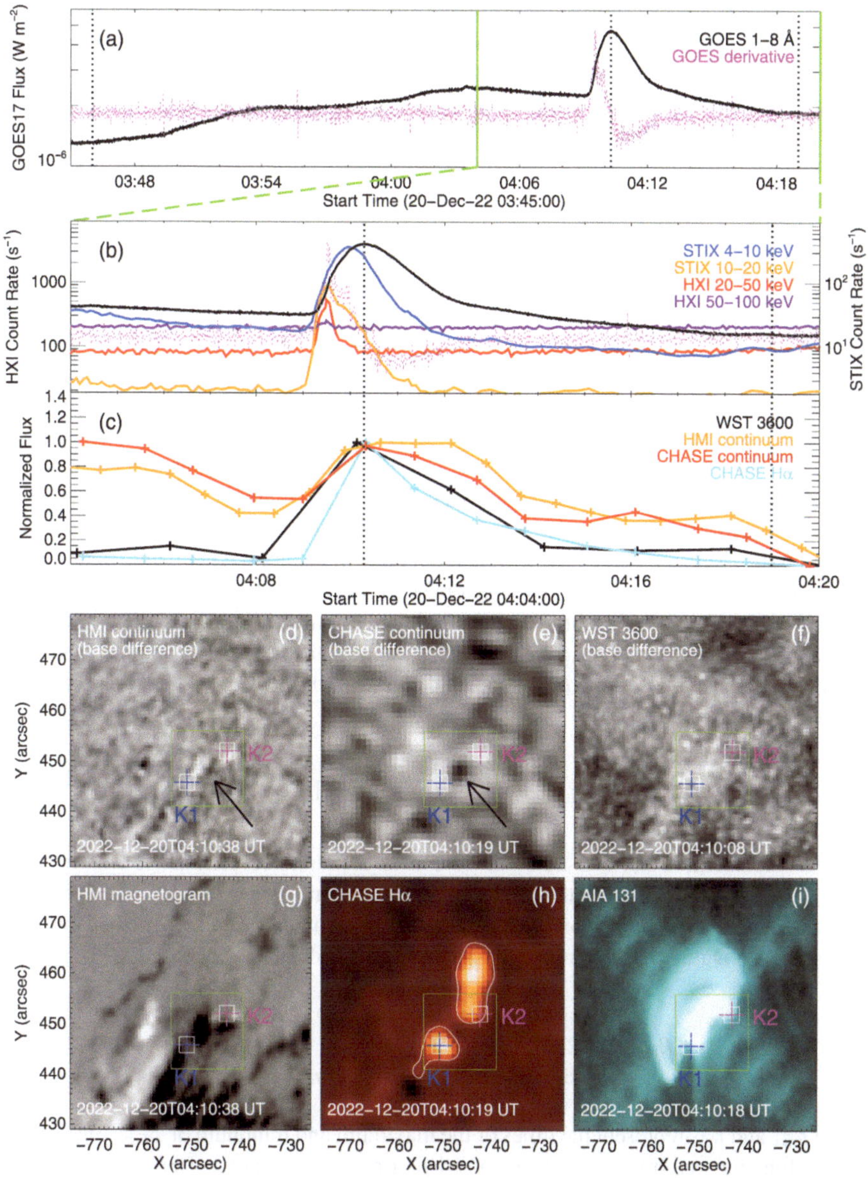

Figure 1 (**a**) Full-disk GOES SXR 1 – 8 Å light curve and its temporal derivative for the C2.3 WLF. The *three vertical dotted lines* mark the GOES onset, peak, and end times. The *two green vertical lines* denote the CHASE observation period as shown in Panels b and c. (**b**) Full-disk light curves of STIX HXR 4 – 10 keV, STIX HXR 10 – 20 keV, HXI HXR 20 – 50 keV, and HXI HXR 50 – 100 keV, together with GOES SXR 1 – 8 Å flux and its temporal derivative. (**c**) Normalized light curves in multiple wavebands from WST, HMI, and CHASE continua and CHASE Hα line (summed over $\pm$ 2.3 Å). The *two vertical dotted lines* in Panels b and c represent the GOES peak and end times. (**d – i**) Base-difference images of HMI, CHASE, and WST continua, HMI magnetogram, CHASE Hα image, and AIA 131 image near the flare peak time. The *two pluses* in each panel denote the two white-light kernels K1 and K2. The *green box* in Panels d – i marks the flaring region used to make the light curves in Panel c. The *black arrow* in Panels d and e denotes a small pore. The *white contours* in Panel h outline the Hα flare ribbons with excess intensities larger than 500 DNs.

of Fe I and Hα are determined by averaging the profiles over the quiet-Sun region. The excess profile, i.e., after subtracting the background profile, is adopted to obtain the Doppler velocity. For Fe I, a single Gaussian fit with a linear background is used, and the background average is further used to define the nearby continuum. It is found that the intensity of this defined continuum is approximately equal to that at the wavelength of 6568.6 Å. For Hα, bisector (e.g. Chae et al., 2013) and moment analyses are applied to derive its Doppler velocity. The velocity uncertainties of Fe I and Hα lines are estimated to be within ± 0.6 and ± 3.9 km s^{-1}, respectively. The Hα asymmetry is calculated based on its original profiles. Referring to Equation 1 in Asai et al. (2012), the asymmetry is expressed as follows:

$$RA = \frac{I_{\text{rp}} - I_{\text{bp}}}{I_{\text{rp}} + I_{\text{bp}}}, \tag{1}$$

where I_{rp} and I_{bp} denote the peaks of Hα red and blue wings, respectively. A positive/negative RA represents a red/blue asymmetry. The uncertainty of the asymmetry is estimated to be about ± 0.014. When showing the profiles of Fe I and Hα (Figure 3c – f and Figure 4a – d), we make a normalization using the nearby quiet-Sun continuum, i.e., in a contrasting way. More specifically, the nearby quiet-Sun continuum intensity of the Hα line is determined by averaging the intensities of the ten wavelength points in its bluest wing, while that of the Fe I line is defined as the average of the linear background of this line.

To investigate the relationships of white-light emission with B_{LoS} and ΔB_{LoS} in the photosphere, the magnetograms from HMI are reshaped into the same pixel size as CHASE images. Then, a stepwise function (Sudol and Harvey, 2005) describes the temporal evolution of B_{LoS} as expressed by:

$$B_{\text{LoS}}(t) = a + bt + c\left\{1 + \frac{2}{\pi} tan^{-1}[n(t - t_0)]\right\}, \tag{2}$$

where t represents time, and a, b, c, n, and t_0 are the free parameters of the fitting. From this stepwise function, the characterized start time $[t_s = t_0 - \pi n^{-1}]$ and end time $[t_e = t_0 + \pi n^{-1}]$ can also be obtained. ΔB_{LoS} is derived mainly using the method described in Castellanos Durán, Kleint, and Calvo-Mozo (2018) and can be expressed as:

$$\Delta B_{\text{LoS}} = \begin{cases} sgn(t_{\max} - t_{\min})[B_{\text{LoS}}(t_{\max}) - B_{\text{LoS}}(t_{\min})] \\ 2c \\ B_{\text{LoS}}(t_e) - B_{\text{LoS}}(t_s) - 2\pi b n^{-1} \end{cases}, \tag{3}$$

where $t_{\max}$ and $t_{\min}$ represent the times of the maximum and minimum of B_{LoS}, respectively, and the function $sgn(x)$ represents the positive or negative sign of x. Here, we also modify some criteria as follows:

1. In this short-duration event, the fitting time range is reduced to 60 minutes centered at the GOES peak time.
2. The duration parameter $[\pi n^{-1}]$ must be smaller than the fitting period, otherwise $\Delta B_{\text{LoS}} = 0$.
3. t_s must be earlier than t_e, otherwise $\Delta B_{\text{LoS}} = 0$.
4. Both t_s and t_e must be within the fitting period, otherwise $\Delta B_{\text{LoS}} = 0$.

Note that the noise level of B_{LoS} and ΔB_{LoS} is defined to be approximately 10 G (Liu et al., 2012).

3. Results

3.1. Overview of the C2.3 Event

The event under study is a GOES C2.3 flare (SOL2022-12-20T04:10) occurring in NOAA active region 13171 (N26E59). This flare began at 03:46 UT, peaked at 04:10 UT, and ended at 04:19 UT on 20 December 2022 (see the GOES SXR 1 – 8 Å light curve, as shown in Figure 1a), showing a relatively long slow rise before 04:04 UT. CHASE observations covered the main period of the flare from 04:04 to 04:20 UT (denoted by two green vertical lines in Figure 1a), which we focus on in the following. Figure 1b plots the HXR light curves at different energy bands from STIX and HXI, together with the SXR 1 – 8 Å flux and its temporal derivative. We can see that the HXR 10 – 20 as well as 20 – 50 keV emissions peak around the same time as the SXR temporal derivative, indicating the Neupert effect (Neupert, 1968). Note that the flare has no response in HXR emission above ≈ 50 keV. Figure 1c shows multiple emission curves integrated over the flaring core region as marked by the green box in Figure 1d – i. It is seen that the emissions at WST 3600 Å and CHASE Hα exhibit a notable rise followed by a decay during the flare. Note that their emission peaks may not represent the true peaks due to a low cadence of more than one minute. Interestingly, the CHASE and HMI continuum emissions show a similar trend, though having multiple peaks during the observation period. It should be mentioned that the main peaks of CHASE and HMI emissions appear in the main phase of the flare.

Figure 1d – i displays multiwavelength images of the flaring region around the flare peak time. We find two main white-light kernels, referred to as K1 and K2, from the base-difference images of HMI, CHASE, and WST continua (Figure 1d – f). These two kernels are located near a small pore (indicated by a black arrow), as seen in the HMI and CHASE continuum images in Figure 1d and e. From the HMI magnetogram in Figure 1g, we can see that K1 is located in negative polarities with a relatively strong magnetic field, while K2 seems to lie in a mix of positive and negative polarities with a weak field, which may be influenced by a projection effect. These two kernels are co-spatial with the Hα ribbons, as shown in Figure 1h. K1 is located in the center of the south ribbon with a bright Hα source, while K2 lies on the edge of the north ribbon with a relatively weak Hα brightening. From the AIA 131 Å image in Figure 1i, one can see that K1 and K2 connect the same set of flare loops, likely to be a pair of conjugate footpoints.

3.2. Spatial Relationship of the White-Light Kernels with Nonthermal HXR Sources

Figure 2 shows the HXR spectral imaging and fitting from STIX and HXI for the C2.3 flare. From the spectral fitting (Figure 2e), we can see a nonthermal component above ≈ 20 keV. Thus, we reconstruct the HXR images at 20 – 35 keV from HXI for the main nonthermal sources. In addition, we reconstruct the HXR images at 16 – 28 keV and 4 – 10 keV from STIX for nonthermal and also thermal sources. These HXR sources corresponding to the flare ribbons and also loops are overplotted on AIA 131 Å, WST 3600 Å, and CHASE continuum and Hα images (Figure 2a – d). At an earlier time of 04:09:22 UT (Figure 2a and b), the nonthermal HXR sources at 20 – 35 keV are mainly co-spatial with the flare ribbons. In particular, the two white-light kernels, K1 and K2, match the nonthermal sources well. It is noted that K1 is close to the nonthermal source with a larger HXR flux while K2 corresponds to the one with a lower flux. At a later time of 04:09:38 UT (Figure 2c and d), the HXR sources evolve a little while K1 is still close to one of the nonthermal HXR sources. These results suggest that the white-light emissions are related to nonthermal electron-beam heating, although the timings of white-light emissions and HXR sources are not conclusive due to a low cadence of the white-light data from WST and CHASE.

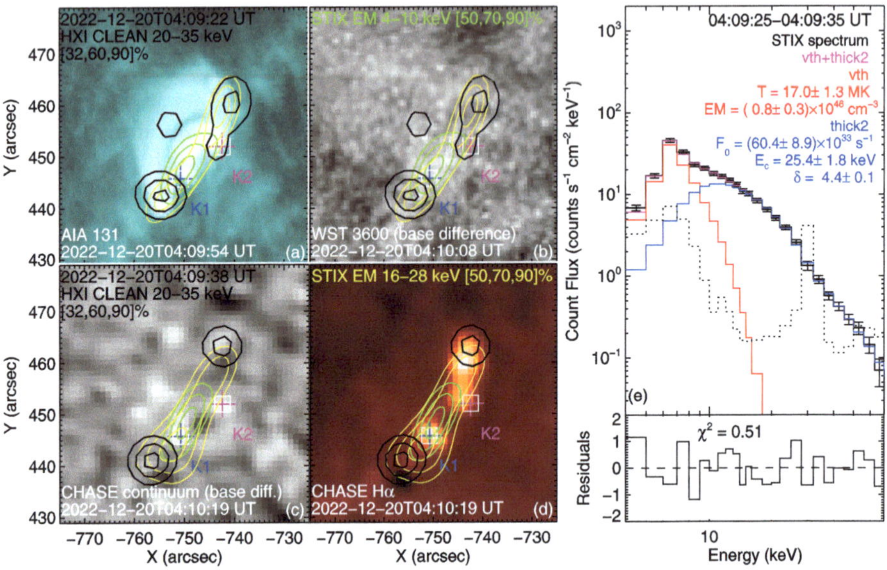

Figure 2 (**a**–**d**) HXR images from HXI and STIX for two times overlaid on AIA 131 Å, WST 3600 Å, CHASE continuum, and CHASE Hα images. The nonthermal and thermal sources are marked by the *black* (HXI 20–35 keV), *yellow* (STIX 16–28 keV), and *green* (STIX 4–10 keV) contours. The *two pluses* in each panel represent the two white-light kernels K1 and K2. (**e**) STIX HXR background-substracted count spectrum (*black solid curve*) fitted by an isothermal (*red curve*) plus a nonthermal thick target (*blue curve*) model for the peak time of the HXR 20–50 keV emission. The *gray dashed curve* is the background emission taken during non-flaring times close to the event.

3.3. Spectral Features of the White-Light Kernels

Figure 3a and b shows the temporal evolution of the relative enhancements of continuum and spectral line emissions from HMI, WST, and CHASE, which are integrated over an area of $3'' \times 3''$ for the two white-light kernels K1 and K2. One can see that at K1 (Figure 3a), all the enhancement curves exhibit an evident rise and then a decay during the flare. The maximum enhancements or increases of HMI, CHASE, and WST continua at $\approx$04:10 UT are 3.9%, 3.2%, and 4.7%, respectively. For comparison, the CHASE Fe I line core and integrated Hα line over $\pm$ 2.3 Å have larger maximum increases of 6.4% and 38.0%, respectively. At K2 (Figure 3b), the maximum increases of HMI, CHASE, and WST continua, as well as CHASE Fe I core and Hα line for their main peaks at $\approx$04:10 UT are 4.3%, 3.0%, 1.9%, 2.2%, and 20.2%, respectively, most of which are smaller than the corresponding values at K1. Note that the other peaks in the HMI and CHASE continua, say those before 04:08 UT and after 04:15 UT, might be unrelated to the flare itself, as no obvious response is found in the HXR emissions.

Figure 3c–f exhibits the temporal evolution of CHASE Fe I and Hα line profiles averaged over an area of $3'' \times 3''$ at K1 and K2. One can see that at both kernels, the Fe I line remains an absorption profile with a symmetric Gaussian shape during the flare (Figure 3c–d). Its line core and the nearby continuum increase, followed by a decrease in intensity over time. The Hα line varies from an absorption to an emission and then back to an absorption profile (Figure 3e–f). Its emission profiles display a central reversal with double peaks in line wings, which are not symmetric. K1 shows a red asymmetry, but K2 has a blue asymmetry; namely, the conjugate footpoints have an opposite asymmetry in Hα.

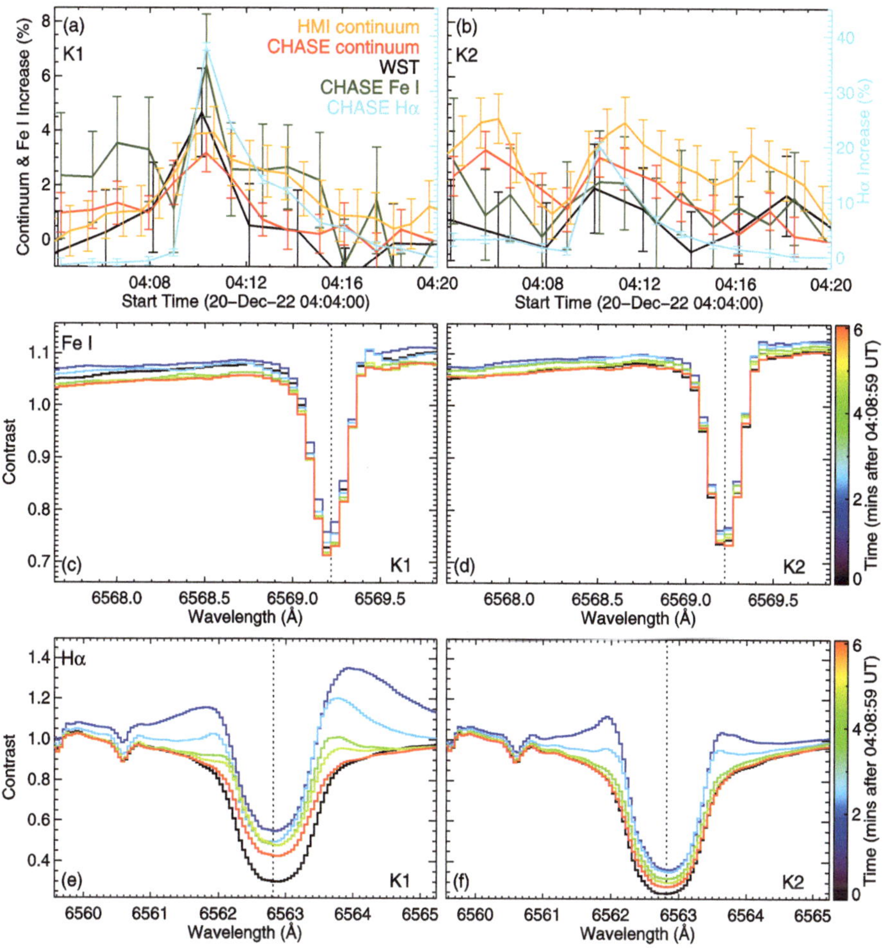

Figure 3 (**a**) and (**b**) Temporal evolution of the relative enhancements or increases of HMI continuum, CHASE continuum, WST continuum, Fe I line core, integrated Hα line (± 2.3 Å) at the white-light kernels K1 and K2 averaged over an area of 3″ × 3″. The *error bars* represent the uncertainties. (**c**–**f**) Temporal evolution of the normalized Fe I and Hα line profiles at K1 and K2 averaged over an area of 3″ × 3″. The *vertical dotted line* in each panel denotes the line center of Fe I or Hα.

We further select the Fe I and Hα line profiles with maximum intensities at 04:10:19 UT to derive the Doppler velocity as shown in Figure 4a–d. Note that the excess profiles are adopted to do a Gaussian fit for Fe I and a moment plus a bisector analysis for Hα. One can see that the Fe I line at K1 exhibits a trivial redshift velocity, which is actually within the uncertainty (Figure 4a). By contrast, the excess Fe I emission at K2 has an evident redshift with a velocity of 1.70 $\mathrm{km\,s^{-1}}$ (Figure 4b), indicative of a plasma downflow at the lower atmospheric layer. The excess Hα profiles with a red or blue asymmetry at K1 and K2 show a much larger redshift or blueshift velocity (Figure 4c–d). At K1, the measured Doppler velocity is 14.67/20.81 $\mathrm{km\,s^{-1}}$ (i.e., redshift) from a moment/bisector method. Meanwhile, at K2, the velocity is -3.35/-3.59 $\mathrm{km\,s^{-1}}$ (blueshift) via moment/bisector. Note that the velocity from the bisector is a median of the ones obtained at different intensity levels.

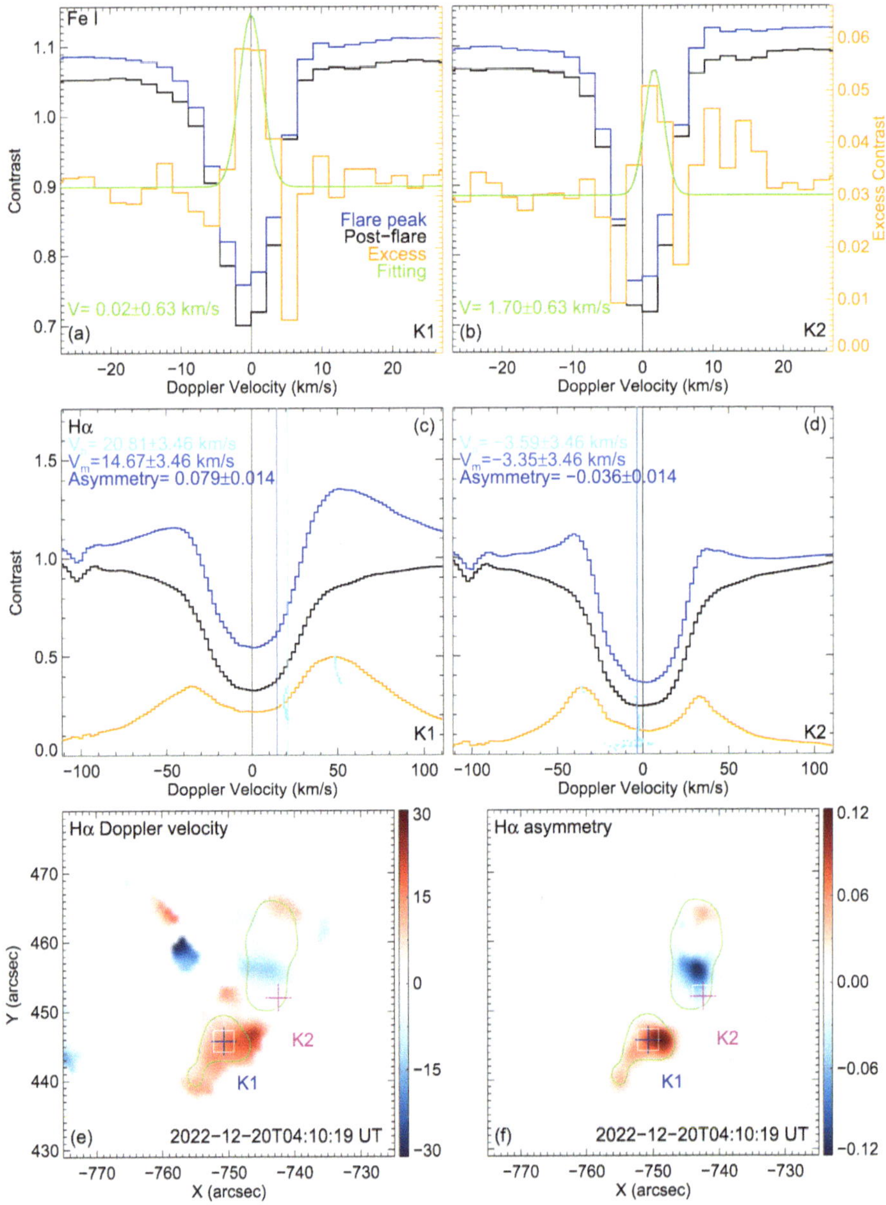

Figure 4 Normalized line profiles of Fe I (**a** and **b**) and Hα (**c** and **d**) from the white-light kernels K1 and K2 averaged over an area of $3'' \times 3''$ at the flare peak time of 04:10:19 UT. The *blue*, *black*, and *orange curves* in each panel indicate the flaring, background, and excess profiles, respectively. The *black vertical line* in each panel denotes the line center of Fe I or Hα. The *green curve* in Panels a and b represents a Gaussian fit to the excess Fe I profile. The *royal blue* and *cyan vertical lines* in Panels c and d mark the Doppler velocities derived from the moment and bisector methods, respectively, the latter of which is a medium of the velocities obtained at different intensity levels. (**e**) and (**f**) Maps of the Hα Doppler velocity and asymmetry at the flare peak time. The Doppler velocity is calculated using the moment method. The *two pluses* in each panel represent the two white-light kernels K1 and K2. The *green contours* denote the Hα flare ribbons whose excess intensities are larger than 500 DNs.

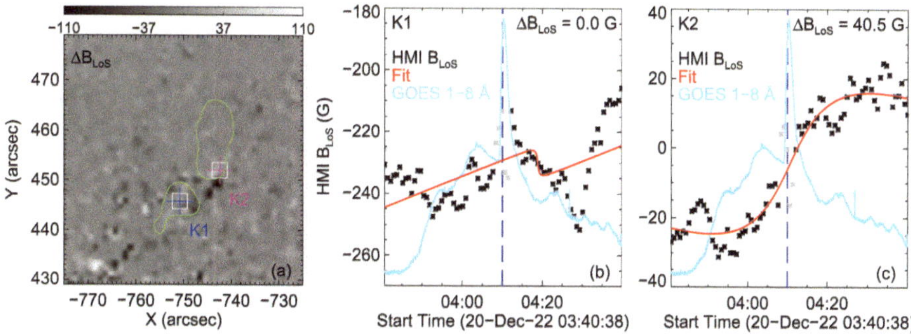

Figure 5 (**a**) Photospheric ΔB_{LoS} map for the flare active region. The *two pluses* indicate the two white-light kernels K1 and K2, and the *green contours* denote the Hα ribbons whose excess intensities are larger than 500 DNs. (**b**) and (**c**) Temporal evolution of HMI B_{LoS} (*black dots*) at K1 and K2 averaged over an area of $3'' \times 3''$, together with the GOES SXR 1 – 8 Å light curve (*sky blue curve*). In each panel, the *red line* denotes a stepwise function fit to B_{LoS} excluding the five data points (*gray dots*) around the flare peak time (marked by the *blue vertical line*).

Moreover, we derive the Hα Doppler velocity map via moment for the time at 04:10:19 UT as shown in Figure 4e. To get a reliable result, we artificially set the velocity to 0 when the peak intensity of the excess profile in a pixel is less than 100 DNs. It is seen that evident Doppler velocities mainly come from the two brightened flare ribbons as marked by green contours (same as the white contours in Figure 1h), whose excess intensities are larger than 500 DNs. In particular, Hα can show redshift and blueshift velocities at the conjugate footpoints, which has been revealed from the profiles at K1 and K2 as described above. We also plot the asymmetry map in Figure 4f. One can see that only at the flare ribbons, the Hα profiles show a notable red or blue asymmetry with an obvious line-wing enhancement. Note that the asymmetry distribution looks quite similar to that of the Doppler velocity at the two ribbons, which can also be seen in the scatter plot in Figure 6i. However, one should caution that the red (blue) asymmetry of Hα profiles, or the measured redshift (blueshift) velocity, is not necessarily caused by a plasma downflow (upflow), as already revealed in previous studies (e.g. Gan, Rieger, and Fang, 1993; Heinzel et al., 1994; Ding and Fang, 1996; Kuridze et al., 2015). It is also interesting to see that the asymmetry is reversed for the conjugate footpoints at K1 and K2. As given in Figure 4c and d, the asymmetries of Hα profiles at K1 and K2 are 0.079 and -0.036, respectively.

3.4. Photospheric Magnetic Field Changes at the White-Light Kernels

Figure 5a shows the map of the photospheric magnetic field change, i.e., ΔB_{LoS}, for the flare region. We can see some notable changes in B_{LoS} (high ΔB_{LoS}) near/at the white-light kernels K1 and K2. Note that some prominent changes of B_{LoS} also appear in some other regions without white-light brightenings, which is actually similar to the study of Castellanos Durán and Kleint (2020). We further plot the temporal evolution of B_{LoS} at K1 and K2 as well as the fitting using a stepwise function in Figure 5b and c. It is seen that there seems to be no obvious change in B_{LoS}, i.e., with a poor fit, at K1, while at K2, there exists an evident change in B_{LoS} with a value of 40.5 G during the flare. In fact, the relationship between the white-light emission and B_{LoS} in this small WLF is somewhat similar to the results in Castellanos Durán and Kleint (2020).

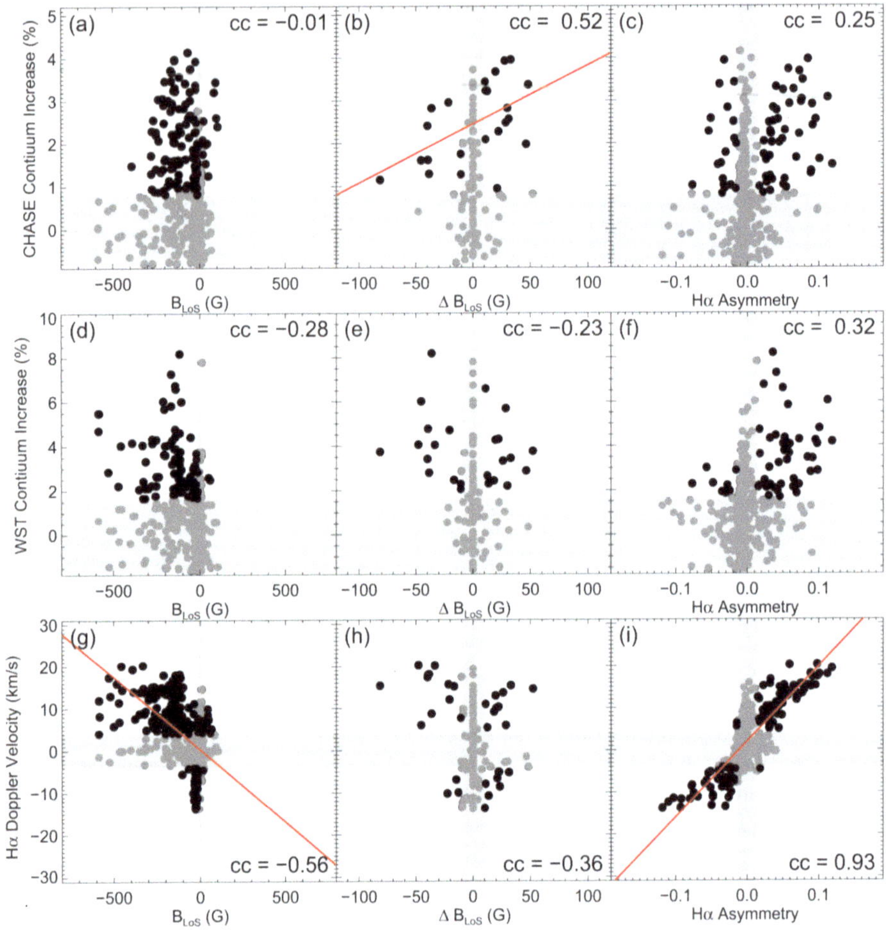

Figure 6 (**a** – **c**) Scatter plots of CHASE continuum increase with HMI B_{LoS}, ΔB_{LoS}, and Hα asymmetry from flare ribbons. (**d** – **f**) Scatter plots of WST 3600 Å increase with the above-mentioned three parameters. (**g** – **i**) Scatter plots of Hα Doppler velocity with the three parameters. The *gray area* in each panel denotes the uncertainty of each parameter. We do a linear fitting (*red line*) if two parameters among them have a good correlation (cc ≥ 0.5).

3.5. Statistical Relationships Among Different Parameters from Flare Ribbons

We check the statistical relationships among different parameters within the two flare ribbons (marked by the contours in Figure 1h, Figure 4e and f, and Figure 5a). To increase the reliability, we collect all the available time frames in observations, namely four frames for the CHASE continuum around its maximum, WST continuum, HMI B_{LoS}, and CHASE Hα asymmetry, but only one frame for HMI ΔB_{LoS}. Figure 6a – c shows the scatter plots of CHASE continuum increase versus photospheric B_{LoS}, ΔB_{LoS}, and Hα asymmetry. Note that the data points in the shaded area, i.e., within uncertainties, are excluded to further calculate the linear Pearson correlation coefficient (cc). It is seen that the CHASE continuum increase has only some correlation with ΔB_{LoS} (cc $= 0.52$). In contrast, it has no obvious relationship with B_{LoS} (cc $= -0.01$) or Hα asymmetry (cc $= 0.25$). Figure 6d – f shows the scatter plots of the WST continuum increase versus $B_{\rm LoS}$, $\Delta B_{\rm LoS}$, and Hα asymmetry. We find that the continuum increase at WST 3600 Å has no linear relationships with $B_{\rm LoS}$ (cc = −0.28), $\Delta B_{\rm LoS}$ (−0.23), or Hα asymmetry (cc = 0.32). Figure 6g – i exhibits the scatter plots of Hα Doppler velocity versus $B_{\rm LoS}$, $\Delta B_{\rm LoS}$, and Hα asymmetry. We can see that the Hα Doppler velocity has good relationships with $B_{\rm LoS}$ (cc = −0.56) as well as Hα asymmetry (cc = 0.93). By contrast, it has no good correlation with $\Delta B_{\rm LoS}$ (cc = −0.36). Note that the negative correlation between the Hα velocity and $B_{\rm LoS}$ is owing to negative magnetic fields at most ribbon locations. The implications or explanations of these relationships among different parameters will be discussed in Section 4.

4. Summary and Discussion

In this article, we perform a comprehensive study of a small C2.3 WLF observed by CHASE and ASO-S, combined with the multiband observations from SDO and SolO. Our main results are summarized as follows:

1. This small flare shows relative enhancements or increases of 4.3%, 3.2%, 4.7%, and 6.4% in the HMI, CHASE, and WST continua and CHASE Fe I line, respectively. The white-light kernels are mainly located near a small pore as well as within the Hα flare ribbons.
2. At the white-light kernels, the CHASE Fe I line remains an absorption profile with a symmetric Gaussian shape during the flare. It can show a redshift velocity up to 1.7 $\rm km\,s^{-1}$. By contrast, the Hα line changes from an absorption to an emission profile, having a central reversal with a line-wing enhancement. In addition, the Hα emission profile displays a red or blue asymmetry corresponding to a velocity of several to tens of $\rm km\,s^{-1}$.
3. Due to a low cadence of HMI, CHASE, and WST observations, the peak times of white-light emissions and their temporal relationship with the HXR emission are inconclusive in this flare. However, the white-light kernels are found to be co-spatial with the nonthermal HXR sources at 20 – 35 keV. This suggests that the white-light emissions are related to nonthermal electron-beam heating, whose radiative backwarming effect could not be excluded.
4. One of the identified white-light kernels is accompanied by a change in photospheric magnetic field [$\Delta B_{\rm LoS}$], while the other is not. Only the CHASE continuum increase at the flare ribbons has a good relationship with $\Delta B_{\rm LoS}$. These confirm that the continuum emission can be linked to $\Delta B_{\rm LoS}$ but in an unknown way (Castellanos Durán and Kleint, 2020).
5. The Hα Doppler velocity at the flare ribbons has a good correlation with $B_{\rm LoS}$ as well as the Hα asymmetry in this small WLF.

From a statistical perspective (Section 3.5), we find a good correlation between the CHASE continuum increase and photospheric $\Delta B_{\rm LoS}$ at flare ribbons, while this is not the case between the WST continuum increase at 3600 Å and $\Delta B_{\rm LoS}$. This may be explained by the different formation heights of these two continua. The CHASE continuum near the Fe I line at 6569.2 Å can be formed lower (say in either the photosphere or chormosphere, see e.g., Kleint et al., 2016; Kerr, 2017) than the WST 3600 Å continuum (i.e., Balmer continuum), the latter of which mainly originates from the chromosphere during a flare (e.g. Avrett, Machado, and Kurucz, 1986). As regards the Hα line, it is formed from the photosphere to the chromosphere from its wing to the core. In particular, this line is sensitive to nonthermal heating (e.g. Henoux, Fang, and Gan, 1993; Rubio da Costa et al., 2016; Capparelli et al.,

2017). Therefore, its Doppler velocity could reflect chromospheric condensation and probably also evaporation (e.g. Song et al., 2023) caused by nonthermal electron beams. The Hα velocity shows some correlation with the photospheric B_{LoS}. This implies that the chromospheric condensation/evaporation can be influenced by the magnetic field in the lower atmosphere, particularly in WLFs, as in a WLF, the position of chromospheric condensation/evaporation could be relatively lower in height due to a lower energy deposition height compared with non-WLFs.

This small C2.3 WLF has some properties similar to those of large or major WLFs. (1) The CHASE Fe I and Hα lines at the white-light kernels in this small WLF display similar profiles, i.e., symmetric Gaussian profiles still in absorption in Fe I and asymmetric emission profiles with a central reversal in Hα, to those in an X1.0 WLF as reported by Song et al. (2023), although their relative increases in the intensity are much lower compared with the X1.0 WLF. (2) The white-light emissions in this small WLF are related to a nonthermal electron-beam heating, which is the same as the case in many large WLFs (e.g. Kuhar et al., 2016; Lee et al., 2017; Song et al., 2023). (3) Some white-light bright points in this small WLF are accompanied by a photospheric ΔB_{LoS} while some are not. This is consistent with the results in X- and M-class flares (Castellanos Durán and Kleint, 2020). All these may hint that the small WLFs with C and even B class may be the same as the large WLFs with M and X class in nature.

More small WLFs are worthy of study in the future, especially with comprehensive observations. Firstly, combining with HMI continuum images at Fe I 6173 Å (in the Paschen continuum) and WST images at 3600 Å (in the Balmer continuum), we can investigate the types of WLFs, i.e., type I or II with or without a Balmer jump (e.g. Fang and Ding, 1995), as well as the origins of these continuum emissions in small WLFs. In particular, WST provides 3600 Å images with a very high cadence of one second in a burst mode, which can capture some rapid variations of white-light brightenings during a WLF. Secondly, using the spectroscopic observations of Fe I and Hα lines from CHASE, we can determine the dynamics from the photosphere to the chromosphere induced by the energy deposition of a WLF. The Doppler velocity and also line width from conjugate footpoints deserve further study, which could provide some insights into the heating or energy deposition in a WLF. Thirdly, HXR spectral and imaging observations from HXI and/or STIX are necessary to determine the heating mechanisms of a WLF. Some quantitative relationships between the HXR flux (or even low-energy cutoff or spectral index of the spectrum) and continuum increase further help to establish a detailed relation between the electron-beam heating and white-light emissions in observations. Finally, magnetic field observations from HMI and also the *Full-disk vector MagnetoGraph* (FMG: Su et al., 2019a) on ASO-S are important to understand how the magnetic field influences the white-light emissions, which have been unknown even in large WLFs. Overall, on the one hand, a statistical study of small WLFs is worthwhile to be performed using high-resolution observations. On the other hand, with the aid of sophisticated radiative hydrodynamic simulations, some case studies are required to understand the energy transport and deposition as well as emission mechanisms for both large and small WLFs.

Acknowledgements We are very grateful to the anonymous reviewer for his/her valuable suggestions on this manuscript. The ASO-S mission is supported by the Strategic Priority Research Program on Space Science, Chinese Academy of Sciences. The CHASE mission is supported by the China National Space Administration. SDO is a mission of NASA's Living With a Star Program. Solar Orbiter is a space mission of international collaboration between ESA and NASA, operated by ESA. The STIX instrument is an international collaboration between Switzerland, Poland, France, Czech Republic, Germany, Austria, Ireland, and Italy. We thank Shihao Rao, Xiaoshuai Zhu, Bing Wang, Zhentong Li, and Ye Qiu for their help in data processing.

Author contributions Q. Li carried out the data analysis and wrote the first manuscript. Y. Li conceived the idea of the work and revised the manuscript. Y. Su provided the HXI data and imaging. D.C. Song provided suggestions on data analysis. W.Q. Gan is PI of ASO-S. H. Li and L. Feng are PI and Co-PI of LST, respectively. Y. Huang, Y.P. Li, J.W. Li, J. Zhao, L. Lu, B.L. Ying, J.C. Xue, P. Zhang, J. Tian, X.F. Liu, G. Li, J.C. Jing, S.T. Li, G.L. Shi, Z.Y. Tian, W. Chen, Y.N. Su, Q.M. Zhang, D. Li, Y.Y. Ge, J.H. Shan, Y. Zhou, and S.J. Lei contributed to the pipeline and release of ASO-S data. All authors reviewed the manuscript.

Funding The authors are supported by the National Natural Science Foundation of China (NSFC) under grants 12233012 and 12273115, the Strategic Priority Research Program of the Chinese Academy of Sciences under grant XDB0560000, the National Key R&D Program of China under grants 2022YFF0503004 and 2022YFF0503002, and NSFC under grants 12333010, 11921003 and 12203102.

Data Availability The ASO-S data used in this work are not publicly available. They were observed during the commissioning phase (from 9 October 2022 to 31 March 2023). However, they are available from the corresponding author on reasonable request. The ASO-S data since 1 April 2023 are publicly available and can be accessed from the official website of ASO-S at aso-s.pmo.ac.cn/sodc/dataArchive.jsp. SDO data are publicly accessible at jsoc.stanford.edu. STIX data are publicly available at datacenter.stix.i4ds.net.

Declarations

Competing interests The authors declare no competing interests.

References

Aboudarham, J., Henoux, J.C.: 1987, Non-thermal excitation and ionization of hydrogen in solar flares. II - effects on the temperature minimum region energy balance and white light flares. *Astron. Astrophys.* **174**, 270. ADS.

Asai, A., Ichimoto, K., Kita, R., Kurokawa, H., Shibata, K.: 2012, A study on red asymmetry of Hα flare ribbons using a narrowband filtergram in the 2001 April 10 solar flare. *Publ. Astron. Soc. Japan* **64**, 20. DOI. ADS.

Avrett, E.H., Machado, M.E., Kurucz, R.L.: 1986, Chromospheric flare models. In: Neidig, D.F., Machado, M.E. (eds.) *The Lower Atmosphere of Solar Flares*, 216. ADS.

Babin, A.N., Koval, A.N.: 1992, Spectral observations of the white-light solar flare of June 15, 1991. *Sov. Astron. Lett.* **18**, 294. ADS.

Capparelli, V., Zuccarello, F., Romano, P., Simões, P.J.A., Fletcher, L., Kuridze, D., Mathioudakis, M., Keys, P.H., Cauzzi, G., Carlsson, M.: 2017, Hα and Hβ emission in a C3.3 solar flare: comparison between observations and simulations. *Astrophys. J.* **850**, 36. DOI. ADS.

Carrington, R.C.: 1859, Description of a singular appearance seen in the Sun on September 1, 1859. *Mon. Not. Roy. Astron. Soc.* **20**, 13. DOI. ADS.

Castellanos Durán, J.S., Kleint, L.: 2020, The statistical relationship between white-light emission and photospheric magnetic field changes in flares. *Astrophys. J.* **904**, 96. DOI. ADS.

Castellanos Durán, J.S., Kleint, L., Calvo-Mozo, B.: 2018, A statistical study of photospheric magnetic field changes during 75 solar flares. *Astrophys. J.* **852**, 25. DOI. ADS.

Chae, J., Park, H.-M., Ahn, K., Yang, H., Park, Y.-D., Cho, K.-S., Cao, W.: 2013, Doppler shifts of the Hα line and the Ca II 854.2 nm line in a quiet region of the Sun observed with the FISS/NST. *Solar Phys.* **288**, 89. DOI. ADS.

Chen, Q.R., Ding, M.D.: 2005, On the relationship between the continuum enhancement and hard X-ray emission in a white-light flare. *Astrophys. J.* **618**, 537. DOI. ADS.

Chen, P.-F., Fang, C., Ding, M.-D.D.: 2001, Ellerman bombs, type II white-light flares and magnetic reconnection in the solar lower atmosphere. *Chin. J. Astron. Astrophys.* **1**, 176. DOI. ADS.

Chen, B., Li, H., Song, K.-F., Guo, Q.-F., Zhang, P.-J., He, L.-P., Dai, S., Wang, X.-D., Wang, H.-F., Liu, C.-L., Zhang, H.-J., Zhang, G., Wang, Y., Liu, S.-J., Zhang, H.-X., Liu, L., Mao, S.-L., Liu, Y., Peng, J.-H., Wang, P., Sun, L., Liu, Y., Han, Z.-W., Wang, Y.-L., Wu, K., Ding, G.-X., Zhou, P., Zheng, X., Xia, M.-Y., Wu, Q.-W., Xie, J.-J., Chen, Y., Song, S.-M., Wang, H., Zhu, B., Chu, C.-B., Yang, W.-G., Feng, L., Huang, Y., Gan, W.-Q., Li, Y., Li, J.-W., Lu, L., Xue, J.-C., Ying, B.-L., Sun, M.-Z., Zhu, C., Bao, W.-M., Deng, L., Yin, Z.-S.: 2019, The Lyman-alpha Solar Telescope (LST) for the ASO-S mission - II. Design of LST. *Res. Astron. Astrophys.* **19**, 159. DOI. ADS.

Ding, M.D., Fang, C.: 1996, On a possible explanation of chromospheric line asymmetries of solar flares. *Solar Phys.* **166**, 437. DOI. ADS.

Ding, M.D., Fang, C., Yun, H.S.: 1999, Heating in the lower atmosphere and the continuum emission of solar white-light flares. *Astrophys. J.* **512**, 454. DOI. ADS.

Ding, M.D., Fang, C., Gan, W.Q., Okamoto, T.: 1994, Optical spectra and semi-empirical model of a white-light flare. *Astrophys. J.* **429**, 890. DOI. ADS.

Emslie, A.G., Sturrock, P.A.: 1982, Temperature minimum heating in solar flares by resistive dissipation of Alfvén waves. *Solar Phys.* **80**, 99. DOI. ADS.

Fang, C., Ding, M.D.: 1995, On the spectral characteristics and atmospheric models of two types of white-light flares. *Astron. Astrophys. Suppl. Ser.* **110**, 99. ADS.

Fang, C., Chen, P.-F., Li, Z., Ding, M.-D., Dai, Y., Zhang, X.-Y., Mao, W.-J., Zhang, J.-P., Li, T., Liang, Y.-J., Lu, H.-T.: 2013, A new multi-wavelength solar telescope: Optical and Near-infrared Solar Eruption Tracer (ONSET). *Res. Astron. Astrophys.* **13**, 1509. DOI. ADS.

Feng, L., Li, H., Chen, B., Li, Y., Susino, R., Huang, Y., Lu, L., Ying, B.-L., Li, J.-W., Xue, J.-C., Yang, Y.-T., Hong, J., Li, J.-P., Zhao, J., Gan, W.-Q., Zhang, Y.: 2019, The Lyman-alpha Solar Telescope (LST) for the ASO-S mission - III. Data and potential diagnostics. *Res. Astron. Astrophys.* **19**, 162. DOI. ADS.

Fletcher, L., Hudson, H.S.: 2008, Impulsive phase flare energy transport by large-scale Alfvén waves and the electron acceleration problem. *Astrophys. J.* **675**, 1645. DOI. ADS.

Freeland, S.L., Handy, B.N.: 1998, Data analysis with the SolarSoft system. *Solar Phys.* **182**, 497. DOI. ADS.

Gan, W.Q., Mauas, P.J.D.: 1994, Atmospheric heating in solar flares by chromospheric condensation. *Astrophys. J.* **430**, 891. DOI. ADS.

Gan, W.Q., Rieger, E., Fang, C.: 1993, Semiempirical flare models with chromospheric condensation. *Astrophys. J.* **416**, 886. DOI. ADS.

Gan, W.Q., Rieger, E., Zhang, H.Q., Fang, C.: 1992, The role of chromospheric condensations in the continuum emission of white-light flares. *Astrophys. J.* **397**, 694. DOI. ADS.

Gan, W., Zhu, C., Deng, Y., Zhang, Z., Chen, B., Huang, Y., Deng, L., Wu, H., Zhang, H., Li, H., Su, Y., Su, J., Feng, L., Wu, J., Cui, J., Wang, C., Chang, J., Yin, Z., Xiong, W., Chen, B., Yang, J., Li, F., Lin, J., Hou, J., Bai, X., Chen, D., Zhang, Y., Hu, Y., Liang, Y., Wang, J., Song, K., Guo, Q., He, L., Zhang, G., Wang, P., Bao, H., Cao, C., Bai, Y., Chen, B., He, T., Li, X., Zhang, Y., Liao, X., Jiang, H., Li, Y., Su, Y., Lei, S., Chen, W., Li, Y., Zhao, J., Li, J., Ge, Y., Zou, Z., Hu, T., Su, M., Ji, H., Gu, M., Zheng, Y., Xu, D., Wang, X.: 2023, The Advanced Space-based Solar Observatory (ASO-S). *Solar Phys.* **298**, 68. DOI. ADS.

Gosain, S.: 2012, Evidence for collapsing fields in the corona and photosphere during the 2011 February 15 X2.2 flare: SDO/AIA and HMI observations. *Astrophys. J.* **749**, 85. DOI. ADS.

Hanser, F.A., Sellers, F.B.: 1996, Design and calibration of the GOES-8 solar X-ray sensor: the XRS. In: Washwell, E.R. (ed.) *GOES-8 and Beyond*, *SPIE Conf. Ser.* **2812**, 344. DOI. ADS.

Heinzel, P., Kleint, L.: 2014, Hydrogen Balmer continuum in solar flares detected by the Interface Region Imaging Spectrograph (IRIS). *Astrophys. J. Lett.* **794**, L23. DOI. ADS.

Heinzel, P., Karlicky, M., Kotrc, P., Svestka, Z.: 1994, On the occurrence of blue asymmetry in chromospheric flare spectra. *Solar Phys.* **152**, 393. DOI. ADS.

Heinzel, P., Kleint, L., Kašparová, J., Krucker, S.: 2017, On the nature of off-limb flare continuum sources detected by SDO/HMI. *Astrophys. J.* **847**, 48. DOI. ADS.

Henoux, J.C., Fang, C., Gan, W.Q.: 1993, Diagnostics of non-thermal processes in chromospheric flares. II. H alpha and CaII K line profiles of an atmosphere bombarded by 100 KeV-1 MeV protons. *Astron. Astrophys.* **274**, 923. ADS.

Hodgson, R.: 1859, On a curious appearance seen in the Sun. *Mon. Not. Roy. Astron. Soc.* **20**, 15. DOI. ADS.

Hong, J., Ding, M.D., Li, Y., Carlsson, M.: 2018, Non-LTE calculations of the Fe I 6173 Å line in a flaring atmosphere. *Astrophys. J. Lett.* **857**, L2. DOI. ADS.

Hudson, H.S.: 1972, Thick-target processes and white-light flares. *Solar Phys.* **24**, 414. DOI. ADS.

Hudson, H.S.: 2000, Implosions in coronal transients. *Astrophys. J. Lett.* **531**, L75. DOI. ADS.

Hudson, H.S., Fisher, G.H., Welsch, B.T.: 2008, Flare energy and magnetic field variations. In: Howe, R., Komm, R.W., Balasubramaniam, K.S., Petrie, G.J.D. (eds.) *Subsurface and Atmospheric Influences on Solar Activity*, *Astron. Soc. Pacific Conf. Ser.* **383**, 221. ADS.

Hudson, H.S., Wolfson, C.J., Metcalf, T.R.: 2006, White-light flares: a TRACE/RHESSI overview. *Solar Phys.* **234**, 79. DOI. ADS.

Hudson, H.S., Acton, L.W., Hirayama, T., Uchida, Y.: 1992, White-light flares observed by Yohkoh. *Publ. Astron. Soc. Japan* **44**, L77. ADS.

Hudson, H.S., Strong, K.T., Dennis, B.R., Zarro, D., Inda, M., Kosugi, T., Sakao, T.: 1994, Impulsive behavior in solar soft X-radiation. *Astrophys. J. Lett.* **422**, L25. DOI. ADS.

Jejčič, S., Kleint, L., Heinzel, P.: 2018, High-density off-limb flare loops observed by SDO. *Astrophys. J.* **867**, 134. DOI. ADS.

Jess, D.B., Mathioudakis, M., Crockett, P.J., Keenan, F.P.: 2008, Do all flares have white-light emission? *Astrophys. J. Lett.* **688**, L119. DOI. ADS.

Jing, Z., Li, Y., Feng, L., Li, H., Huang, Y., Li, Y., Su, Y., Chen, W., Tian, J., Song, D., Li, J., Xue, J., Zhao, J., Lu, L., Ying, B., Zhang, P., Su, Y., Zhang, Q., Li, D., Ge, Y., Li, S., Li, Q., Li, G., Liu, X., Shi, G., Shan, J., Tian, Z., Zhou, Y., Gan, W.: 2024, A statistical study of solar white-light flares observed by the white-light solar telescope of the Lyman-alpha solar telescope on the Advanced Space-based Solar Observatory (ASO-S/LST/WST) at 360 nm. *Solar Phys.* **299**, 11. DOI. ADS.

Jurčák, J., Kašparová, J., Švanda, M., Kleint, L.: 2018, Heating of the solar photosphere during a white-light flare. *Astron. Astrophys.* **620**, A183. DOI. ADS.

Kerr, G.S.: 2017, Observations and modelling of the chromosphere during solar flares. PhD thesis, University of Glasgow, UK. ADS.

Kleint, L.: 2017, First detection of chromospheric magnetic field changes during an X1-flare. *Astrophys. J.* **834**, 26. DOI. ADS.

Kleint, L., Heinzel, P., Judge, P., Krucker, S.: 2016, Continuum enhancements in the ultraviolet, the visible and the infrared during the X1 flare on 2014 March 29. *Astrophys. J.* **816**, 88. DOI. ADS.

Kowalski, A.F., Hawley, S.L., Carlsson, M., Allred, J.C., Uitenbroek, H., Osten, R.A., Holman, G.: 2015, New insights into white-light flare emission from radiative-hydrodynamic modeling of a chromospheric condensation. *Solar Phys.* **290**, 3487. DOI. ADS.

Krucker, S., Hudson, H.S., Jeffrey, N.L.S., Battaglia, M., Kontar, E.P., Benz, A.O., Csillaghy, A., Lin, R.P.: 2011, High-resolution imaging of solar flare ribbons and its implication on the thick-target beam model. *Astrophys. J.* **739**, 96. DOI. ADS.

Krucker, S., Hurford, G.J., Grimm, O., Kögl, S., Gröbelbauer, H.-P., Etesi, L., Casadei, D., Csillaghy, A., Benz, A.O., Arnold, N.G., Molendini, F., Orleanski, P., Schori, D., Xiao, H., Kuhar, M., Hochmuth, N., Felix, S., Schramka, F., Marcin, S., Kobler, S., Iseli, L., Dreier, M., Wiehl, H.J., Kleint, L., Battaglia, M., Lastufka, E., Sathiapal, H., Lapadula, K., Bednarzik, M., Birrer, G., Stutz, S., Wild, C., Marone, F., Skup, K.R., Cichocki, A., Ber, K., Rutkowski, K., Bujwan, W., Juchnikowski, G., Winkler, M., Darmetko, M., Michalska, M., Seweryn, K., Białek, A., Osica, P., Sylwester, J., Kowalinski, M., Ścisłowski, D., Siarkowski, M., Stęślicki, M., Mrozek, T., Podgórski, P., Meuris, A., Limousin, O., Gevin, O., Le Mer, I., Brun, S., Strugarek, A., Vilmer, N., Musset, S., Maksimović, M., Fárník, F., Kozáček, Z., Kašparová, J., Mann, G., Önel, H., Warmuth, A., Rendtel, J., Anderson, J., Bauer, S., Dionies, F., Paschke, J., Plüschke, D., Woche, M., Schuller, F., Veronig, A.M., Dickson, E.C.M., Gallagher, P.T., Maloney, S.A., Bloomfield, D.S., Piana, M., Massone, A.M., Benvenuto, F., Massa, P., Schwartz, R.A., Dennis, B.R., van Beek, H.F., Rodrí-Pacheco, J., Lin, R.P.: 2020, The Spectrometer/Telescope for Imaging X-rays (STIX). *Astron. Astrophys.* **642**, A15. DOI. ADS

Kuhar, M., Krucker, S., Martínez Oliveros, J.C., Battaglia, M., Kleint, L., Casadei, D., Hudson, H.S.: 2016, Correlation of hard X-ray and white light emission in solar flares. *Astrophys. J.* **816**, 6. DOI. ADS.

Kuridze, D., Mathioudakis, M., Simões, P.J.A., Rouppe van der Voort, L., Carlsson, M., Jafarzadeh, S., Allred, J.C., Kowalski, A.F., Kennedy, M., Fletcher, L., Graham, D., Keenan, F.P.: 2015, Hα line profile asymmetries and the chromospheric flare velocity field. *Astrophys. J.* **813**, 125. DOI. ADS.

Lee, K.-S., Imada, S., Watanabe, K., Bamba, Y., Brooks, D.H.: 2017, IRIS, Hinode, SDO, and RHESSI observations of a white light flare produced directly by nonthermal electrons. *Astrophys. J.* **836**, 150. DOI. ADS.

Lemen, J.R., Title, A.M., Akin, D.J., Boerner, P.F., Chou, C., Drake, J.F., Duncan, D.W., Edwards, C.G., Friedlaender, F.M., Heyman, G.F., Hurlburt, N.E., Katz, N.L., Kushner, G.D., Levay, M., Lindgren, R.W., Mathur, D.P., McFeaters, E.L., Mitchell, S., Rehse, R.A., Schrijver, C.J., Springer, L.A., Stern, R.A., Tarbell, T.D., Wuelser, J.-P., Wolfson, C.J., Yanari, C., Bookbinder, J.A., Cheimets, P.N., Caldwell, D., Deluca, E.E., Gates, R., Golub, L., Park, S., Podgorski, W.A., Bush, R.I., Scherrer, P.H., Gummin, M.A., Smith, P., Auker, G., Jerram, P., Pool, P., Soufli, R., Windt, D.L., Beardsley, S., Clapp, M., Lang, J., Waltham, N.: 2012, The Atmospheric Imaging Assembly (AIA) on the Solar Dynamics Observatory (SDO). *Solar Phys.* **275**, 17. DOI. ADS.

Li, H., Chen, B., Feng, L., Li, Y., Huang, Y., Li, J.-W., Lu, L., Xue, J.-C., Ying, B.-L., Zhao, J., Yang, Y.-T., Gan, W.-Q., Fang, C., Song, K.-F., Wang, H., Guo, Q.-F., He, L.-P., Zhu, B., Zhu, C., Deng, L., Bao, H.-C., Cao, C.-X., Yang, Z.-G.: 2019, The Lyman-alpha Solar Telescope (LST) for the ASO-S mission — I. Scientific objectives and overview. *Res. Astron. Astrophys.* **19**, 158. DOI. ADS.

Li, C., Fang, C., Li, Z., Ding, M., Chen, P., Qiu, Y., You, W., Yuan, Y., An, M., Tao, H., Li, X., Chen, Z., Liu, Q., Mei, G., Yang, L., Zhang, W., Cheng, W., Chen, J., Chen, C., Gu, Q., Huang, Q., Liu, M., Han, C., Xin, H., Chen, C., Ni, Y., Wang, W., Rao, S., Li, H., Lu, X., Wang, W., Lin, J., Jiang, Y., Meng, L., Zhao, J.: 2022, The Chinese Hα Solar Explorer (CHASE) mission: an overview. *Sci. China Ser. G, Phys. Mech. Astron.* **65**, 289602. DOI. ADS.

Lin, Y., Zhang, H., Zhang, W.: 1996, A solar flare in the Fe I 5324 line on 24 June, 1993. *Solar Phys.* **168**, 135. DOI. ADS.

Liu, Y., Hoeksema, J.T., Scherrer, P.H., Schou, J., Couvidat, S., Bush, R.I., Duvall, T.L., Hayashi, K., Sun, X., Zhao, X.: 2012, Comparison of line-of-sight magnetograms taken by the Solar Dynamics Observatory/Helioseismic and Magnetic Imager and Solar and Heliospheric Observatory/Michelson Doppler Imager. *Solar Phys.* **279**, 295. DOI. ADS.

Machado, M.E., Emslie, A.G., Avrett, E.H.: 1989, Radiative backwarming in white-light flares. *Solar Phys.* **124**, 303. DOI. ADS.

Martínez Oliveros, J.-C., Hudson, H.S., Hurford, G.J., Krucker, S., Lin, R.P., Lindsey, C., Couvidat, S., Schou, J., Thompson, W.T.: 2012, The height of a white-light flare and its hard X-ray sources. *Astrophys. J. Lett.* **753**, L26. DOI. ADS.

Massa, P., Piana, M., Massone, A.M., Benvenuto, F.: 2019, Count-based imaging model for the Spectrometer/Telescope for Imaging X-rays (STIX) in solar orbiter. *Astron. Astrophys.* **624**, A130. DOI. ADS.

Matthews, S.A., van Driel-Gesztelyi, L., Hudson, H.S., Nitta, N.V.: 2003, A catalogue of white-light flares observed by Yohkoh. *Astron. Astrophys.* **409**, 1107. DOI. ADS.

Metcalf, T.R., Alexander, D., Hudson, H.S., Longcope, D.W.: 2003, TRACE and Yohkoh observations of a white-light flare. *Astrophys. J.* **595**, 483. DOI. ADS.

Müller, D., St. Cyr, O.C., Zouganelis, I., Gilbert, H.R., Marsden, R., Nieves-Chinchilla, T., Antonucci, E., Auchère, F., Berghmans, D., Horbury, T.S., Howard, R.A., Krucker, S., Maksimovic, M., Owen, C.J., Rochus, P., Rodriguez-Pacheco, J., Romoli, M., Solanki, S.K., Bruno, R., Carlsson, M., Fludra, A., Harra, L., Hassler, D.M., Livi, S., Louarn, P., Peter, H., Schühle, U., Teriaca, L., del Toro Iniesta, J.C., Wimmer-Schweingruber, R.F., Marsch, E., Velli, M., De Groof, A., Walsh, A., Williams, D.: 2020, The solar orbiter mission. Science overview. *Astron. Astrophys.* **642**, A1. DOI. ADS.

Neidig, D.F.: 1989, The importance of solar white-light flares. *Solar Phys.* **121**, 261. DOI. ADS.

Neidig, D.F., Cliver, E.W.: 1983, The occurrence frequency of white light flares. *Solar Phys.* **88**, 275. DOI. ADS.

Neupert, W.M.: 1968, Comparison of solar X-ray line emission with microwave emission during flares. *Astrophys. J. Lett.* **153**, L59. DOI. ADS.

Pesnell, W.D., Thompson, B.J., Chamberlin, P.C.: 2012, The Solar Dynamics Observatory (SDO). *Solar Phys.* **275**, 3. DOI. ADS.

Petrie, G.J.D., Sudol, J.J.: 2010, Abrupt longitudinal magnetic field changes in flaring active regions. *Astrophys. J.* **724**, 1218. DOI. ADS.

Poland, A.I., Milkey, R.W., Thompson, W.T.: 1988, Hydrogen and helium excitation by extreme ultraviolet radiation for the production of white-light flares. *Solar Phys.* **115**, 277. DOI. ADS.

Qiu, Y., Rao, S., Li, C., Fang, C., Ding, M., Li, Z., Ni, Y., Wang, W., Hong, J., Hao, Q., Dai, Y., Chen, P., Wan, X., Xu, Z., You, W., Yuan, Y., Tao, H., Li, X., He, Y., Liu, Q.: 2022, Calibration procedures for the CHASE/HIS science data. *Sci. China Ser. G, Phys. Mech. Astron.* **65**, 289603. DOI. ADS.

Rubio da Costa, F., Kleint, L., Petrosian, V., Liu, W., Allred, J.C.: 2016, Data-driven radiative hydrodynamic modeling of the 2014 March 29 X1.0 solar flare. *Astrophys. J.* **827**, 38. DOI. ADS.

Ryan, J.M., Chupp, E.L., Forrest, D.J., Matz, S.M., Rieger, E., Reppin, C., Kanbach, G., Share, G.H.: 1983, Gamma-ray observational constraints on the origin of the optical continuum emission from the white-light flare of 1980 July 1. *Astrophys. J. Lett.* **272**, L61. DOI. ADS.

Scherrer, P.H., Schou, J., Bush, R.I., Kosovichev, A.G., Bogart, R.S., Hoeksema, J.T., Liu, Y., Duvall, T.L., Zhao, J., Title, A.M., Schrijver, C.J., Tarbell, T.D., Tomczyk, S.: 2012, The Helioseismic and Magnetic Imager (HMI) investigation for the Solar Dynamics Observatory (SDO). *Solar Phys.* **275**, 207. DOI. ADS.

Song, Y., Tian, H.: 2018, Investigation of white-light emission in circular-ribbon flares. *Astrophys. J.* **867**, 159. DOI. ADS.

Song, Y.L., Tian, H., Zhang, M., Ding, M.D.: 2018, Observations of white-light flares in NOAA active region 11515: high occurrence rate and relationship with magnetic transients. *Astron. Astrophys.* **613**, A69. DOI. ADS.

Song, Y., Tian, H., Zhu, X., Chen, Y., Zhang, M., Zhang, J.: 2020, A white-light flare powered by magnetic reconnection in the lower solar atmosphere. *Astrophys. J. Lett.* **893**, L13. DOI. ADS.

Song, D.-C., Tian, J., Li, Y., Ding, M.D., Su, Y., Yu, S., Hong, J., Qiu, Y., Rao, S., Liu, X., Li, Q., Chen, X., Li, C., Fang, C.: 2023, Spectral observations and modeling of a solar white-light flare observed by CHASE. *Astrophys. J. Lett.* **952**, L6. DOI. ADS.

Su, J.-T., Bai, X.-Y., Chen, J., Guo, J.-J., Liu, S., Wang, X.-F., Xu, H.-Q., Yang, X., Song, Y.-L., Deng, Y.-Y., Ji, K.-F., Deng, L., Huang, Y., Li, H., Gan, W.-Q.: 2019a, Data reduction and calibration of the FMG onboard ASO-S. *Res. Astron. Astrophys.* **19**, 161. DOI. ADS.

Su, Y., Liu, W., Li, Y.-P., Zhang, Z., Hurford, G.J., Chen, W., Huang, Y., Li, Z.-T., Jiang, X.-K., Wang, H.-X., Xia, F.-X.-Y., Chen, C.-X., Yu, W.-H., Yu, F., Wu, J., Gan, W.-Q.: 2019b, Simulations and software development for the hard X-ray imager onboard ASO-S. *Res. Astron. Astrophys.* **19**, 163. DOI. ADS.

Su, Y., Zhang, Z., Gan, W., Wu, J., Jiang, X.: 2022, The Hard X-ray Imager (HXI) on the Advanced Space-based Solar Observatory (ASO-S). In: Bambi, C., Santangelo, A. (eds.) *Handbook of X-ray and Gamma-ray Astrophysics* **88**. DOI. ADS.

Sudol, J.J., Harvey, J.W.: 2005, Longitudinal magnetic field changes accompanying solar flares. *Astrophys. J.* **635**, 647. DOI. ADS.

Sun, X., Hoeksema, J.T., Liu, Y., Wiegelmann, T., Hayashi, K., Chen, Q., Thalmann, J.: 2012, Evolution of magnetic field and energy in a major eruptive active region based on SDO/HMI observation. *Astrophys. J.* **748**, 77. DOI. ADS.

Sylwester, B., Sylwester, J.: 2000, Evolution of white-light flares observed by Yohkoh. *Solar Phys.* **194**, 305. DOI. ADS.

Wang, H.-M.: 2009, Study of white-light flares observed by Hinode. *Res. Astron. Astrophys.* **9**, 127. DOI. ADS.

Yurchyshyn, V., Kumar, P., Abramenko, V., Xu, Y., Goode, P.R., Cho, K.S., Lim, E.K.: 2017, High-resolution Observations of a White-light Flare with NST. *Astrophys. J.* **838**, 32. DOI. ADS.

Zhang, Z., Chen, D.-Y., Wu, J., Chang, J., Hu, Y.-M., Su, Y., Zhang, Y., Wang, J.-P., Liang, Y.-M., Ma, T., Guo, J.-H., Cai, M.-S., Zhang, Y.-Q., Huang, Y.-Y., Peng, X.-Y., Tang, Z.-B., Zhao, X., Zhou, H.-H., Wang, L.-G., Song, J.-X., Ma, M., Xu, G.-Z., Yang, J.-F., Lu, D., He, Y.-H., Tao, J.-Y., Ma, X.-L., Lv, B.-G., Bai, Y.-P., Cao, C.-X., Huang, Y., Gan, W.-Q.: 2019, Hard X-ray Imager (HXI) onboard the ASO-S mission. *Res. Astron. Astrophys.* **19**, 160. DOI. ADS.

Zhou, X., Fang, C., Hu, J.: 1997, Analysis of high time resolution spectra of the 1991-10-24 white-light flare. *Chin. Astron. Astrophys.* **21**, 99. DOI. ADS.

Solar Physics (2024) 299:104
https://doi.org/10.1007/s11207-024-02349-0

RESEARCH

Asymmetric Hard X-ray Radiation of Two Ribbons in a Thermal-Dominated C-Class Flare

Guanglu Shi[1,2] · Li Feng[1,2] · Jun Chen[1] · Beili Ying[1] · Shuting Li[1,2] · Qiao Li[1,2] · Hui Li[1,2] · Ying Li[1,2] · Kaifan Ji[3] · Yu Huang[1,2] · Youping Li[1,2] · Jingwei Li[1] · Jie Zhao[1] · Lei Lu[1] · Jianchao Xue[1] · Ping Zhang[1] · Dechao Song[1] · Zhengyuan Tian[1,2] · Yingna Su[1,2] · Qingmin Zhang[1] · Yunyi Ge[1] · Jiahui Shan[1,2] · Yue Zhou[1,2] · Jun Tian[1,2] · Gen Li[1,2] · Xiaofeng Liu[1,2] · Zhichen Jing[1,2] · Shijun Lei[1] · Weiqun Gan[1,4]

Received: 14 March 2024 / Accepted: 17 July 2024 / Published online: 30 July 2024

Abstract
The asymmetry in hard X-ray (HXR) emission at the footpoints (FPs) of flare loops is a ubiquitous feature closely associated with nonthermal electron transport. In this study, we analyze the asymmetric HXR radiation at two flare ribbons, which is thermal-dominated during a long-duration C4.4 flare that occurred on March 20, 2023, combining multi-view and multi-waveband observations from the Advanced Space-based Solar Observatory (ASO-S), Solar Orbiter (SolO), and Solar Dynamics Observatory (SDO) spacecraft. We find that the H I Lyman-alpha (Lyα) emission presents similar features to the He II λ304 emission, both in the light curve and spatio-temporal evolution of a pair of conjugate flare ribbons. The spectra and imaging analysis of the HXR emission, detected by the Spectrometer Telescope for Imaging X-rays (STIX) in 4-18 keV, reveal that the two-ribbon flare radiation is thermal dominated by over 95%, and the radiation source mainly concentrates on the northern ribbon, leading to an asymmetric distribution. To understand the underlying reasons for the HXR radiation asymmetry, we extrapolate the magnetic field within the active region using the nonlinear force-free field (NLFFF) model. For 78% of the magnetic field lines starting from the northern flare ribbon, their lengths from the loop-tops (LTs) to the northern FPs are shorter than those to the southern FPs. For 62% of the field lines, their magnetic-field strengths at the southern FPs exceed those at the northern FPs. In addition, considering the larger density, $\approx 1.0 \times 10^{10}$ cm^{-3}, of the low-lying flare loops (< 32 Mm), we find that the shorter path from the LT to the northern FP enables more electrons to reach the northern FP more easily after collisions with the surrounding plasma. Therefore, in this thermal-dominated C-class flare, the asymmetric location of the flare LT relative to its two FPs plays a dominant role in the HXR radiation asymmetry, while such asymmetry is also slightly influenced by the magnetic mirror effect, resulting in larger HXR radiation at the FPs with weaker magnetic strength. Our study enriches the understanding of particle transport processes during solar flares.

Keywords Flares · Energetic particles · Heating · In flares · Active regions · Magnetic fields · Spectrum · X-ray

Extended author information available on the last page of the article

1. Introduction

Solar flares are known to accelerate particles and heat plasma, producing hard X-ray (HXR) bursts and gamma-ray bursts (Hudson and Ryan 1995; Holman et al. 2011). To investigate the underlying physical processes of plasma heating, particle acceleration, and magnetic reconnection, spectra and imaging analysis of the HXR emission during flares serve as crucial methods (Masuda et al. 1994; Liu, Chen, and Petrosian 2013; Su et al. 2013; Krucker and Battaglia 2014; Li et al. 2022b). The enhancement of X-ray fluxes mainly originates from the thermal emission that heats the plasma to a high temperature and nonthermal bremsstrahlung emission associated with the energy release of energetic particles (Brown 1971; Fletcher et al. 2011; Lin 2011; Raymond et al. 2012). HXR sources are often observed at the footpoints (FPs), loop-top (LT), and above the LT of flare loops, which have been extensively studied using instruments such as those on board Yohkoh (Ogawara et al. 1992) and the Reuven Ramaty High-Energy Solar Spectroscopic Imager (RHESSI, Lin et al. 2002).

In observations, the HXR sources located at the FPs of solar flares typically present asymmetry, which can be classified into two main types: S-type (Sakao type, Sakao 1994) and N-type (non-Sakao type). These asymmetric phenomena are generally believed to be associated with the transport processes of nonthermal electrons in the flare loop. The S-type asymmetry of the HXR sources is often explained by the magnetic mirror effect (e.g. Sakao 1994; Kundu et al. 1995; Li et al. 1997; Aschwanden et al. 1999; Alexander and Metcalf 2002; Ding et al. 2003; Siarkowski and Falewicz 2004; Falewicz, Siarkowski, and Berlicki 2006; Silva et al. 2020; Naus et al. 2022; Qiu et al. 2023). In this scenario, the brighter HXR FP is located in a region where the strength of the photospheric magnetic field is weak. The magnetic mirror effect plays a role in guiding and trapping nonthermal electrons, leading them to concentrate and emit more X-rays at one FP than the other. The N-type asymmetry is featured as the brighter FP of the HXR emission corresponds to the region with a stronger magnetic field, which can be interpreted as the reconnection site generated by interactions with other flare loops being closer to that FP (e.g. Asai et al. 2002; Goff et al. 2004; Falewicz and Siarkowski 2007; Joshi et al. 2017). The magnetic reconnection reduces the effect of magnetic-field convergence, allowing numerous electrons to precipitate at the brighter FP.

The Advanced Space-based Solar Observatory (ASO-S, Gan et al. 2023), launched in October 2022, is the first comprehensive solar space observatory in China. The scientific objectives of ASO-S are focused on "1M2B" (Gan et al. 2019), namely the magnetic fields and two types of violent eruptions on the Sun – solar flares and coronal mass ejections (CMEs). There are three coordinated payloads on board the ASO-S, the Full-disk vector MagnetoGraph (FMG, Deng et al. 2019; Su et al. 2019), the Lyman-alpha (Lyα) Solar Telescope (LST, Li et al. 2019; Chen et al. 2019; Feng et al. 2019), and the Hard X-ray Imager (HXI, Zhang et al. 2019; Su et al. 2019). Three payloads can be used to investigate the relationships among solar flares, CMEs, and magnetic fields. The LST payload consists of three instruments: the Solar Disk Imager (SDI), the Solar Corona Imager (SCI), and the White-light Solar Telescope (WST). These instruments enable simultaneous dual-waveband observations from the solar disk to the low corona, focusing on the studies of flares, prominence and CME eruptions.

In this article, we analyze the asymmetric HXR sources in a two-ribbon flare dominated by over 95% thermal radiation during the very long rise phase of about two hours in a long-duration event (LDE). We further explain this phenomenon by combining the results of a nonlinear force-free field (NLFFF) extrapolation. Section 2 provides an overview of the LDE observations. Section 3 presents our analysis and results: (1) STIX HXR spectra and imaging analysis, (2) comparisons between H I Lyα and He II 304 Å emissions, (3) the configuration of the magnetic field extrapolated by the NLFFF model. Our conclusions and discussion are given in Section 4.

2. Overview of Observations

The two-ribbon flare (SOL2023-03-20T15:34) studied in this article was produced by the eruption of a hot channel and filament system in the active region (AR) NOAA 13258. This event was observed by various instruments, including the SDI instrument of the LST payload on board the ASO-S, the Spectrometer Telescope for Imaging X-rays (STIX, Krucker et al. 2020) on board Solar Orbiter (SolO, Müller et al. 2020), the Atmospheric Imaging Assembly (AIA, Lemen et al. 2012) on board the Solar Dynamics Observatory (SDO, Pesnell, Thompson, and Chamberlin 2012), the Geostationary Operational Environmental Satellite (GOES), and the Chinese Hα Solar Explorer (CHASE, Li et al. 2022). SolO was located at a longitude separation of 18.4° from Earth and had a heliocentric distance of 0.51 AU. Multi-view and multi-waveband observations provide detailed insights into the dynamics of the flare and enable a more comprehensive understanding of the associated particle behaviors.

This event is a GOES C4.4 flare that started at 13:30 UT and peaked at 15:34 UT on March 20, 2023, shown in Figure 1a, it is a long-duration event (LDE) that lasted for an impressive time of 11 hours. The curve of the time derivative (purple line) presented in panel a was derived from the GOES data (Thomas, Starr, and Crannell 1985). The light curves of HXR emissions at photon energies of 4–10 keV (green line) and 10–18 keV (magenta line), detected by STIX, are shown in panel b. Panel c displays the light curves of AIA He II 304 Å (black line), AIA 94 Å (cyan line), and SDI H I Lyα (blue line), obtained by integrating over the entire flare region. All of these light curves display a significant enhancement during the flare. This event is classified as a relatively small C-class flare, only low-energy photons (4–18 keV) were detected by STIX at a distance of about 0.51 AU. Additionally, the peak time of the SXR emission, marked by a red dashed line, is close to that of the HXR emission. It implies that the energy released during this event is probably not dominated by nonthermal electrons (further evidence is presented in Section 3.1). Unfortunately, due to the instrument being in the test status, SDI only has available observations starting from 13:57 UT during this period. Despite this limitation, it is still possible to observe a good temporal correlation between the HXR emissions and the light curves of H I Lyα and He II 304 Å, implying that they might have similar radiation origins. In the flare rising phase, Lyα and 304 Å radiation were from flare ribbons. However, clear distinctions are evident between the HXR and 94 Å.

Figure 2 presents the flare-related structures observed by multiple instruments during this LDE. Panels a and b clearly show the associated filament and hot channel observed by CHASE Hα and AIA 94 Å, respectively. As the filament rotates and rapidly rises, it subsequently develops into a CME accompanied by the formation of an elongated current sheet (CS), a series of blobs, and supra-arcade downflows (SADs). Simultaneous observations from AIA He II 304 Å and SDI H I Lyα capture the spatio-temporal evolution of a pair of conjugate flare ribbons, as shown in panels c and d, presenting a good consistency with each other. During the flare rising phase, the two ribbons rapidly separate towards the northern and southern directions, respectively. The flare ribbons can also be observed in the Hα images. There is no significant brightening in the AIA 1600 Å and 1700 Å images, while brightening in these two passbands is generally believed to be associated with the motion of nonthermal electrons in the flare loops. These phenomena indicate that the flare ribbons are

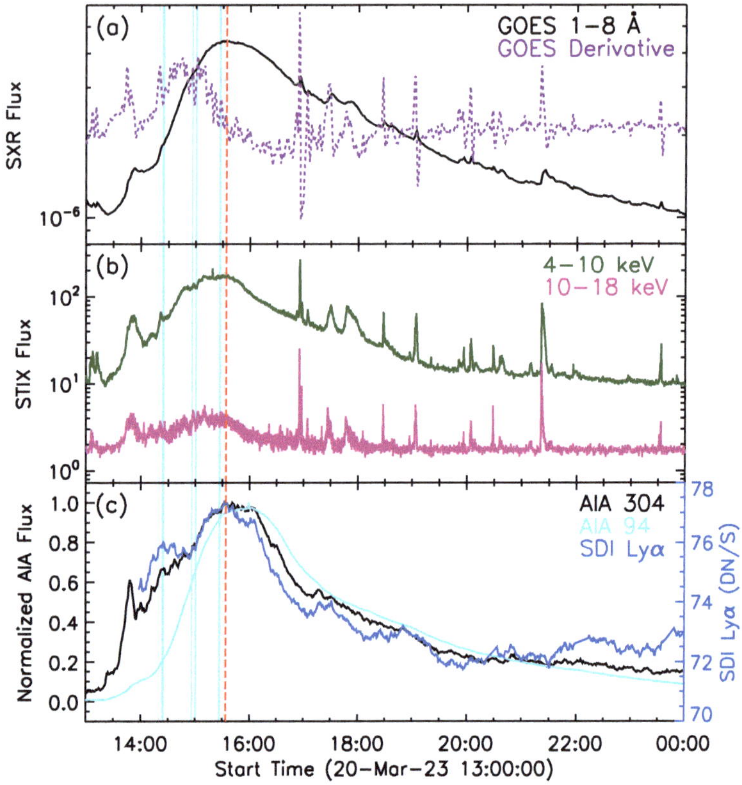

Figure 1 Light curves from multiple wavebands for the long-duration event (LDE) on March 20, 2023. (**a**) GOES SXR light curve ranging from 1.0-8.0 Å along with its time derivative (purple line). (**b**) STIX HXR light curves at 4-10 keV (green line) and 10-18 keV (magenta line). (**c**) Integrated light curves of AIA He II 304 Å (black line), AIA 94 Å (cyan line) and SDI H I Lyα (blue line) in the flare region. Pink and blue shadowed regions represent the integration times for HXR spectra and imaging analysis shown in Figure 3 and Figure 4, respectively. The red vertical line marks the SXR peak time at 15:34 UT.

mainly dominated by thermal emission, and formed in the middle-to-upper chromosphere and transition region.

3. Analysis and Results

3.1. HXR Spectra and Imaging Analysis

To determine the nonthermal/thermal nature of the two-ribbon flare and clarify the origin of the HXR emission, we conduct an analysis of both the spectra and imaging of the HXR detected by STIX during the flare rising phase. Figure 3 and Figure 4 present the results for the selected time intervals marked by the pink and blue shadowed regions in Figure 1, respectively.

First, we use a double thermal model that describes optically thin thermal bremsstrahlung radiation to fit the observed HXR spectra. Figure 3 presents the spatially integrated count spectra (black solid lines) detected by STIX for various time intervals from 2 to 4 minutes,

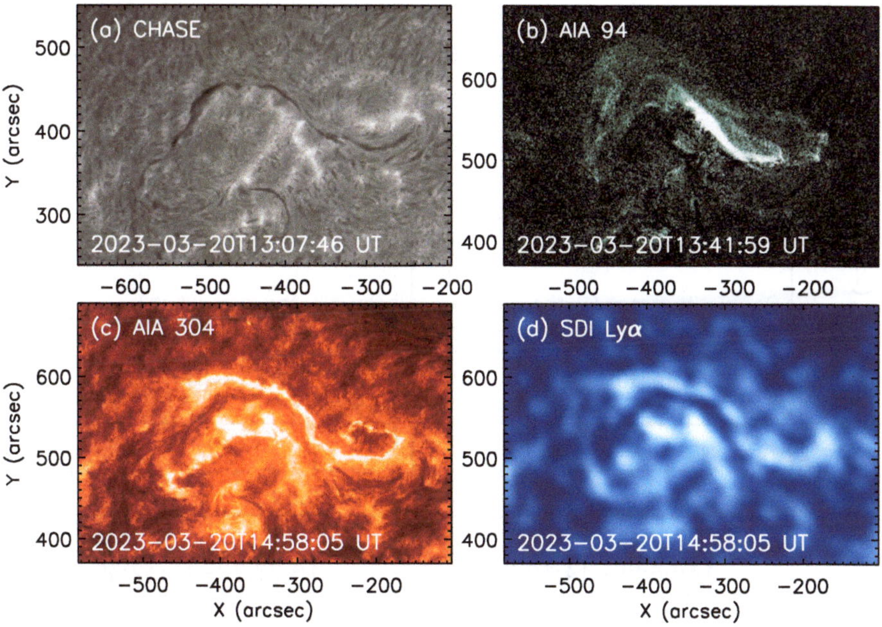

Figure 2 An overview of the observations of the LDE. (**a**) A long filament before the flare onset observed in Hα by CHASE. (**b**) Eruption of the hot channel observed in the AIA base-difference image in the 94 Å passband. (**c**) and (**d**) Separation of the two ribbons captured in AIA He II 304 Å (**c**) and SDI H I Lyα (**d**) towards the northern and southern directions.

after subtracting the pre-flare background (black dashed lines) from 13:20 to 13:22 UT. The chosen time intervals are to ensure sufficient count rates in 4 to 18 keV. Considering the extremely long rising phase of about 2 hours for this flare, the temperature may not have a significant change in 2 to 4 minutes. Blue lines in each panel represent the optimal fitting curves. Free parameters of the double thermal functions, including electron temperature (T) and emission measure (EM), are indicated in each panel. The residual distribution and the coefficient chi-square (χ^2) quantitatively evaluate the level of consistency between the fitting function and the HXR spectra. We also attempted to use a standard combination of thermal and nonthermal power-law functions to fit the HXR spectra. It turns out that their χ^2 values are at least twice as large as those in the double-thermal fittings. The thick-target spectral indices are about 8 to 10 during the flare rising phase, and the ratios of integrated nonthermal to thermal count fluxes are less than 5%. Therefore, based on these two different fitting models, the spectral fitting analysis demonstrates that the two-ribbon flare radiation is thermal dominated by over 95%. Panel a presents the spectral fitting results at 13:51 UT of the sub-peak preceding the main peak of the HXR light curves, which corresponds to the filament eruption. In panels b-d, it can be seen that during the rising phase, the weighted mean temperature range of the flare is from 6.83 MK (14:21 UT) to 7.76 MK (15:30 UT).

Then, we reconstruct the HXR maps detected by STIX and superpose them on AIA He II 304 Å (left column) and SDI H I Lyα (right column) images, as shown in Figure 4. Panel c additionally presents the composite image combining AIA 304 Å (red) and AIA 94 Å (cyan) at 14:58 UT. To generate these images, the visibilities are computed by subcollimators labeled from 3 to 10, with an integration time of 6 minutes centered on the observation time of AIA, and applied to the MEM_GE algorithm (Massa et al. 2020) for the thermal sources in

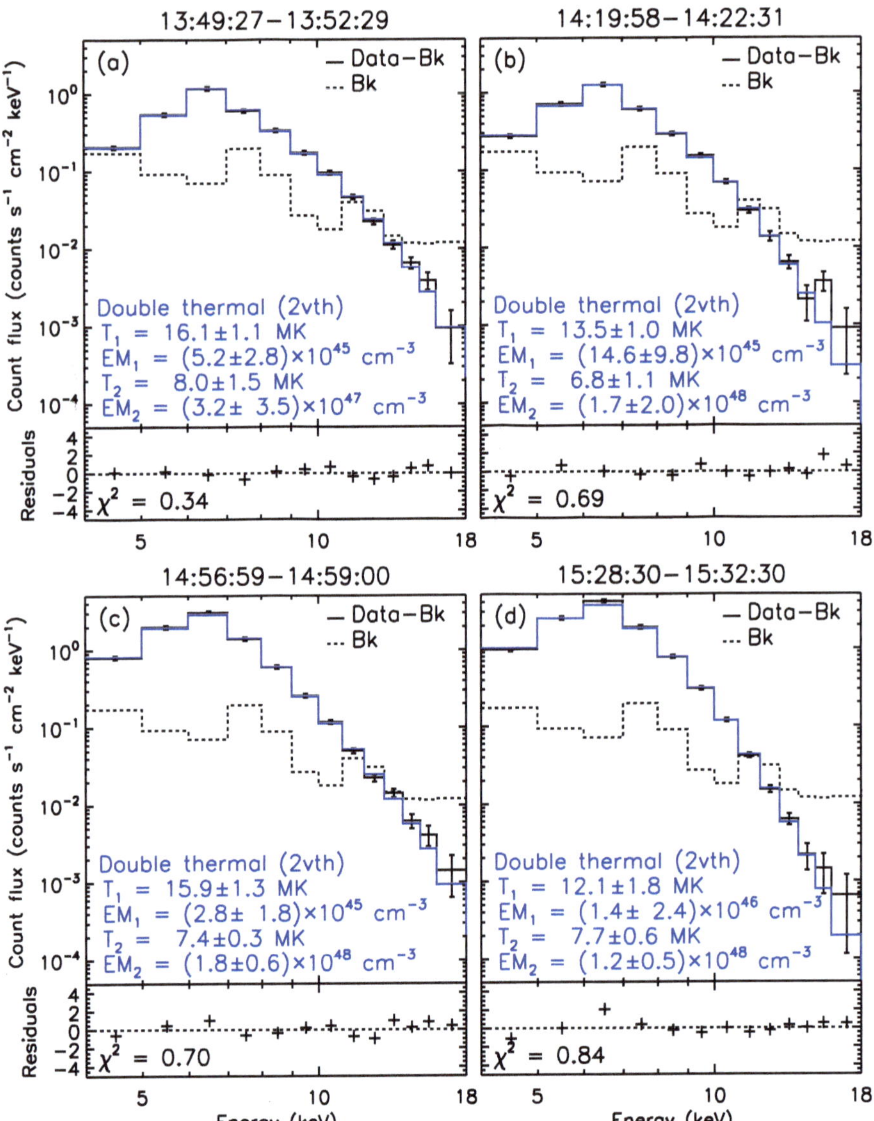

Figure 3 HXR spectral fitting for the times marked by the pink shadowed regions in Figure 1 using a double thermal model. The fitting parameters, including electron temperature (T), emission measure (EM), and chi-square values (χ^2) are displayed in each panel. Black solid lines represent the count rate of the HXR emissions detected by STIX after subtraction of the pre-flare background (dashed lines), while blue lines represent the optimal fitting curves.

the 4-18 keV energy range, resulting in an angular resolution of $\approx 14.6''$. The co-alignment of STIX images was done with the Full Sun Imager (FSI) 304 Å images observed by the Extreme Ultraviolet Imager (EUI, Rochus et al. 2020) on board the SolO. Subsequently, we reprojected the STIX images from the SolO view to Earth view, assuming that all photons originate from almost the same altitude, i.e., the middle-to-upper chromosphere or transition

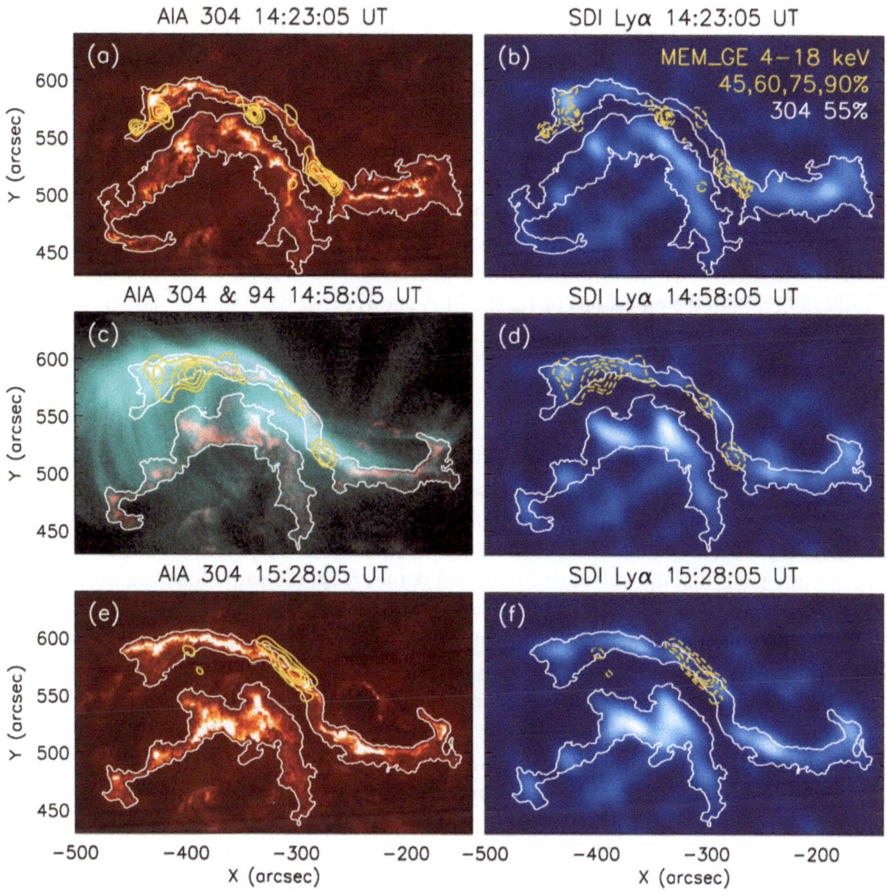

Figure 4 Evolution of two ribbons in the AIA He II 304 Å (left column) and SDI H I Lyα (right column) observations for the times selected by the blue shadowed regions in Figure 1. Panel c presents the composite image of AIA 304 Å (red) and 94 Å (cyan). The overlaid gold contours represent HXR maps of STIX reconstructed by the MEM_GE algorithm at 4-18 keV, using levels at 45%-90% of the maximum intensity with a step of 15% in each panel. The white contour indicates 55% of the maximum intensity of He II 304 Å.

region. The reason for selecting a 6-minute integration time is that it can provide the HXR images in a more consistent way with the flare ribbons observed in AIA 304 Å and SDI Lyα compared to the 4-minute images during the initial stage of the flare, and they have similar imaging results near the peak time of the GOES flux. There were no SDI Lyα observations before 13:57 UT, and the parallel sections of the flare ribbons formed after 14:00 UT. Therefore, STIX imaging results at 13:51 UT are not presented in Figure 4. The gold contours in each panel indicate 45%-90% of the maximum HXR intensity, with a step of 15%. The HXR radiation mainly originates from the flare FPs and presents an asymmetric distribution, primarily concentrated along the parallel section of the northern ribbon. As the flare evolves towards the peak time of the SXR flux, a weak HXR source begins to appear at the southern FP (not shown in Figure 4). The white contours in each panel represent 55% of the maximum intensity of the AIA He II 304 Å. From comparisons, it is evident that the He II 304 Å and H I Lyα emissions present a good spatial consistency in the evolution of the flare FPs, consistent with the study of Li et al. (2022a).

Note that the HXR sources are predominantly from the flare ribbons instead of the projection of the higher coronal source. From the light curves in Figure 1, the STIX light curve has more similarity to that of AIA 304 Å than AIA 94 Å. Moreover, if we compare the morphology details of the STIX HXR source with that of the flare ribbons observed in AIA 304 Å and that of flare loops observed in AIA 94 Å, we find that the HXR source resembles the northern ribbon more, especially at the upper-left end in Figure 4a and c.

The spectra and imaging analysis further support the conclusion that the HXR emission from the flare FPs is dominated by the collision of thermal electrons injected downward from the magnetic reconnection site with the surrounding plasma in the dense flare loop. Furthermore, both the emissions of He II 304 Å and H I Lyα have similar radiation features produced by the thermal process that occurred in the two ribbons of the flare.

3.2. H I Lyα and He II 304 Å Emissions

To quantitatively investigate the relationship between the radiation intensities of He II 304 Å and H I Lyα, we conduct a pixel-by-pixel comparison in the two-ribbon regions and calculate their Pearson correlation coefficients (CCs).

Aligning the SDI images with AIA using the method developed by Cai et al. (2022) is crucial to ensure that the calculated radiation comes from the same flare region, enabling accurate point-to-point comparisons of radiation intensities between He II 304 Å and H I Lyα. We then uniformly sample points within the white contour (representing 55% of the 304 Å intensity) shown in Figure 4 and extract their corresponding radiation intensities on a pixel-by-pixel basis. Figure 5 presents scatter diagrams of the intensity comparisons using logarithmic axes between He II 304 Å and H I Lyα in the southern (left column) and northern (right column) ribbon regions from 14:23 UT to 15:28 UT. One can see that the H I Lyα radiation has a moderate intensity correlation with He II 304 Å, while the CCs of the southern ribbon (0.60–0.71) are greater than those of the northern ribbon (0.44–0.60). The calculated difference in CCs between the two ribbons may be attributed to the distinct environmental conditions in each region. Meanwhile, as the flare ribbons evolve, there is a gradual increase in the correlation between H I Lyα and He II 304 Å emissions, which may be closely associated with the heating process in the flare.

3.3. NLFFF Extrapolation

To understand the asymmetric distribution of the HXR emission and the morphology of the two ribbons, we conduct the NLFFF extrapolation based on the optimization method (Wiegelmann and Neukirch 2003) in the flare region. The structure of the magnetic field is presented in Figure 6.

The Space-Weather HMI Active Region Patches (SHARP, Bobra et al. 2014) map, derived from the Helioseismic and Magnetic Imager (HMI, Hoeksema et al. 2014) on board the SDO, is used as the boundary condition for the NLFFF model. The 180-degree ambiguity is corrected, and the map is projected onto the cylindrical equal-area (CEA) coordinate. Figure 6a presents the B_z component of the SHARP map at 14:58 UT with a pixel scale of 0.36 Mm. The gold contour represents 30% of the maximum HXR intensity projected from the helioprojective-Cartesian (HPC) coordinates in Figure 4c and d to the CEA coordinate, and red dots are uniformly sampled within it as the start points for tracing magnetic field lines (green lines). The dark green lines represent magnetic field lines, where the magnetic-field strengths at the southern FPs are larger than those at the northern FPs. In panel c, the distribution of traced field lines is displayed along with the ribbon contours of AIA He II

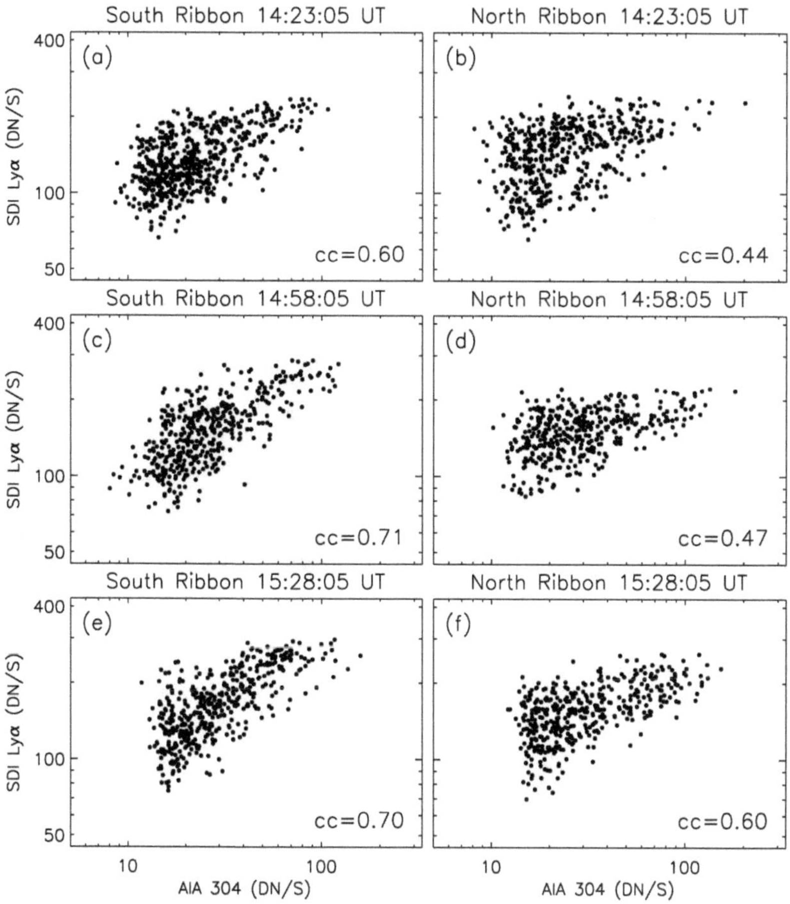

Figure 5 Scatter diagrams of the pixel-by-pixel comparisons between the AIA He II 304 Å (horizontal axis) and SDI H I Lyα (vertical axis) intensities within regions of the southern ribbon (left column) and the northern ribbon (right column) at different times. Correlation coefficients are calculated in each panel.

304 Å (55%, red lines) and HXR (30%, gold lines) on the HMI line-of-sight (LoS) magnetogram. The blue rectangle represents the field-of-view (FoV) of the SHARP map shown in panel a. One can see that the configuration of the magnetic field follows a horn-like shape, featuring a convergence field rooted in the southern region with negative flux and a divergent distribution in the northern region with positive flux.

Based on the extrapolated magnetic field, we calculate the squashing factor Q (Titov, Hornig, and Démoulin 2002) to quantify the magnetic connectivity using the FastQSL code developed by Zhang et al. (2022). Figure 6b presents the distribution of the signed log Q, slog $Q = \mathrm{sgn}(B_z) \cdot \log Q$, at the photosphere. The magnetic field lines originating from the red dots marked in the high-Q region (red) in the northern region are all connected to the high-Q region (blue) in the southern region. To gain insight into the three-dimensional structure of the magnetic field, particularly from the quasi-separatrix layers (QSL), we calculate Q within the 3D box region. Panel d shows its distribution on a cross-section along the dashed line in panel b and perpendicular to the XOY plane. From the QSL map, we can see the positions of the LT and FPs (F1, F2) of the flare loop. The LT is located at a height

Figure 6 Magnetic configuration in the two-ribbon region. (**a**) HMI B_z component of NOAA AR 13258 in the CEA coordinate applied for the NLFFF extrapolation, with a linear saturating at 300 G. Green field lines root at the red dots within the gold contour that represents the 30% level of the maximum intensity of HXR. Dark green lines indicate a stronger magnetic field in the southern FP compared to the northern FP. (**b**) Signed log Q map, slog $Q = \mathrm{sgn}(B_Z) \cdot \log Q$, on the NLFFF photospheric boundary. (**c**) HMI line-of-sight (LOS) magnetogram overlay the red contour (55% of the maximum intensity of He II 304 Å) outlining the flare ribbons; the blue rectangle indicates the field-of-view of the SHARP map in panel a. (**d**) Log Q map, with its axes along the dashed line and Z direction in panel b, respectively. F1 and F2 denote footpoints of the flare loop, while LT represents the flare loop-top.

of ≈ 32 Mm above the photosphere, while F1 and F2 are rooted in the northern and southern ribbons, respectively. One can see that the distance from the flare LT to the two FPs is asymmetric, indicating that the length from the LT to F1 is shorter than that to F2. Such magnetic-field configuration suggests that for the flare loops at the eastern side, the energy release of particles is located closer to F1.

In addition, we conduct a statistical analysis on the relationship between the distance from the LT to the FPs, magnetic field strength B, and HXR intensity I for each traced field line, as shown in Figure 7. In panel a, we calculate the ratio of the de-projected length L_S/L_N from the LT to the northern FP L_N and the southern FP L_S, as well as the ratio of the distance D_S/D_N from the projected points of the LT onto the XOY plane to the two FPs, respectively. The subscript 'S' denotes the southern FP, while 'N' denotes the northern FP. We find that 78% of the traced magnetic field lines have $L_S > L_N$, while 22% of the remaining ones have $L_S < L_N$. Similarly, 79% of magnetic field lines have $D_S > D_N$, whereas 21% have $D_S < D_N$. The statistical relationship quantitatively demonstrates that the position of the flare LT is asymmetric. We then compare the relationship between the de-projected length ratio L_S/L_N from the LT to the FPs and the ratio of the magnetic-field strength B_S/B_N at FPs, which is presented in panel b. Based on the distribution of B_S/B_N, there is a slight difference in the magnetic-field strength between the northern FPs and the southern FPs (62% of magnetic-field lines have $B_S > B_N$ at their FPs). The comparisons

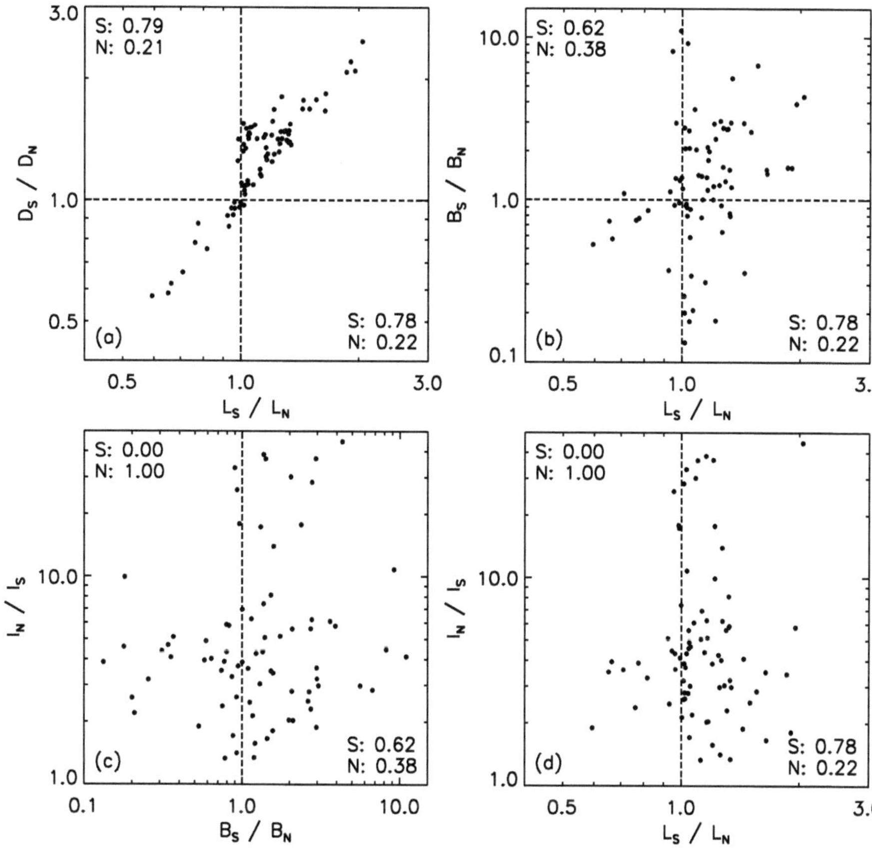

Figure 7 (**a**) Scatter diagram of the ratios between the de-projected length L_S/L_N and the projected length D_S/D_N from the LT to the FPs for each extrapolated field line. L represents the integrated length from the LT to the FPs along each field line, while D represents the distance between the projected point of the LT onto the XOY plane and the FPs. (**b**) Comparisons between the ratios of the integrated length L_S/L_N from the LT to the FPs and the magnetic field strength B_S/B_N at the FPs. Comparisons of the ratios of the HXR intensity I_N/I_S with the magnetic field strength B_S/B_N at FPs (**c**) and length L_S/L_N from the LT to the FPs (**d**). B and I represent the magnetic-field strength and the HXR radiation at the FPs, respectively. The subscript 'S' denotes the southern FP, while 'N' denotes the northern FP. The value after the letter 'S' ('N') in the corners indicates the percentage of the magnetic field lines where the physical quantity at the southern FP is larger (smaller) than that at the northern FP.

reveal that the asymmetry of the LT position is mainly attributed to the horn-shaped magnetic field configuration and is slightly influenced by the magnetic field strength at the FPs.

Furthermore, we compare the ratio of the HXR intensity I_N/I_S at the FPs with the magnetic field strength B_S/B_N and the loop length L_S/L_N, as shown in Figure 7c and d. Due to the uniform sampling of the northern FP for each magnetic field line within the contour of 30% maximum HXR intensity, the HXR intensity ratios between the northern and southern FPs are always greater than 1, i.e., $I_N > I_S$. Similarly, one can see that there is a small difference in the magnetic field strength between the northern FP (38% field lines of $B_S < B_N$) and the southern FP (62% field lines of $B_S > B_N$), while 78% of LTs are closer to the northern FP, $L_S > L_N$. Therefore, we conclude that the asymmetric distribution of the HXR radiation is mainly attributed to the asymmetry of the flare LT, and is also slightly influenced by the magnetic mirror effect.

4. Conclusions and Discussions

In this work, we investigate the asymmetry of the HXR radiation at two ribbons in a C4.4 flare by combining multi-view and multi-waveband observations from the ASO-S, SDO, and SolO spacecraft. The LDE observed on March 20, 2023, was caused by the eruption of a filament system, accompanied by a CME and an elongated CS. The light curves of both the SXR and HXR emissions present a consistent peak time, with no brightening observed in the 1600 Å and 1700 Å passbands, implying that thermal electrons dominate this event. The light curves of HXR, SDI H I Lyα and AIA He II 304 Å during the flare rising phase exhibit a consistent trend, indicating a common radiation origin primarily from flare ribbons formed in the middle-to-upper chromosphere and transition region.

Thanks to the 24-hour uninterrupted full-disk observations from Earth's view provided by the SDI on board the ASO-S, we can study the spatio-temporal features of the two ribbons in H I Lyα waveband. These observations, along with those in the He II 304 Å, reveal similar evolution of the two ribbons, presenting rapid separation towards the northern and southern directions. The pixel-by-pixel comparisons between the SDI and AIA images further quantitatively demonstrate the moderate correlation between H I Lyα and He II 304 Å. Meanwhile, their emissions may be closely associated with the heating process in the flare.

The spectral fitting of the HXR emission detected by STIX further confirms that the flare mainly releases energy through the process of thermal bremsstrahlung emission, implying that the radiations in both He II 304 Å and H I Lyα are produced by a thermal process. Additionally, the MEM_GE algorithm is applied to reconstruct the HXR maps observed by STIX. We find that the HXR radiation presents an asymmetric distribution, with the majority of the emission concentrated at the northern ribbon.

Using the method of differential emission measure (DEM, Cheung et al. 2015; Su et al. 2018), we estimate the electron density n_e inside the flare loops. The total emission measure EM and the average electron number density are deduced via,

$$\mathrm{EM} = \int_{T_{\min}}^{T_{\max}} \mathrm{DEM}(T)dT, \tag{1}$$

$$n_e = \sqrt{\frac{\mathrm{EM}}{l}}\ (\mathrm{cm}^{-3}), \tag{2}$$

where l represents the effective depth along LoS. We further assume that the flare loops have similar extents in depth and width, and we estimate l to be $\approx 2.3 \times 10^8$ cm by measuring the loop width observed by AIA 94 Å. The value of EM can be directly derived from the DEM result. Therefore, we determine the density $n_e \approx 1.0 \times 10^{10}$ cm^{-3}, indicating that the flare loop is relatively dense.

By combining the NLFFF extrapolation with observations, we explore the magnetic configuration and physical parameters within this active region. This approach allows us to understand the magnetic field structure and derive important physical properties within the two ribbons. The characteristic loop length L, derived from the traced field lines, is $\approx 5.9 \times 10^9$ cm. Based on the magnetic topology, we further demonstrate that the asymmetric HXR emission is produced by the asymmetric position of the flare LT and the magnetic mirror effect. The statistical parameters of the extrapolated field lines indicate that the deprojected length from the LT to the northern FP is shorter than that to the southern FP, and the magnetic-field strength at the southern FP is stronger than that at the northern FP, which promotes thermal electrons to collide with the plasma inside the dense flare loop and easily reach the northern ribbon. At the same time, considering the physical conditions such as the lower height, ≈ 32 Mm, of the LT above the photosphere, as well as the denser plasma inside the flare loops, $\approx 1.0 \times 10^{10}$ cm^{-3}, more thermal electrons are more easily able to reach the northern FP during the collision process with the surrounding plasma, resulting in a stronger HXR emission compared to the southern FP.

Our work provides a novel approach for investigating the asymmetric HXR radiation in the two-ribbon flare dominated by over 95% thermal radiation. This approach involves combining the NLFFF extrapolation with observations to understand the magnetic-field structure and physical properties within the flare region. Unfortunately, due to no obvious HXR emissions above ≈ 18 keV in this flare, the HXI on board the ASO-S did not receive sufficient counts for data analysis. Future research endeavors will statistically analyze asymmetric events having HXR emission at the FPs, using the HXR data observed by the HXI combined with the unique H I Lyα waveband full-disk observations from the SDI, to conduct more detailed studies with this kind of asymmetry in HXR radiation.

Acknowledgements We sincerely thank the anonymous referee for providing valuable suggestions that helped us improve the quality of the manuscript. We thank Shangbin Yang and Zhentong Li for their helpful discussions on the NLFFF model and STIX data processing. The authors thank the teams of SolO/STIX, SDO, GOES, CHASE for their open-data use policy. The ASO-S mission is supported by the Strategic Priority Research Program on Space Science, Chinese Academy of Sciences. SolO is a space mission of international collaboration between ESA and NASA, operated by ESA. The STIX instrument is an international collaboration between Switzerland, Poland, France, Czech Republic, Germany, Austria, Ireland, and Italy. SDO is a mission of NASA's Living with a Star (LWS) Program. The CHASE mission is supported by China National Space Administration.

Author contributions G.L. Shi wrote the main manuscript, analyzed the data, and generated figures. L. Feng discovered the asymmetric HXR radiation event, conceived the study, and revised the manuscript. J. Chen, B.L. Ying, S.T. Li, and Q. Li discussed results and provided helpful suggestions on this work. W.Q. Gan is PI of the ASO-S. H. Li and L. Feng are PI and Co-PI of the LST, respectively. K.F. Ji provided an alignment code for the AIA and SDI images. Y. Li, Y. Huang, Y.P. Li, J.W. Li, J. Zhao, L. Lu, J.C. Xue, P. Zhang, D.C. Song, Z.Y. Tian, Y.N. Su, Q.M. Zhang, Y.Y. Ge, J.H. Shan, Y. Zhou, J. Tian, G. Li, X.F. Liu, Z.C. Jing, and S.J. Lei contributed to the in-orbit testing, pipeline, and release of LST data. All authors reviewed the manuscript.

Funding This work is supported by the National Key R&D Program of China 2022YFF0503003 (2022YFF-0503000), the Strategic Priority Research Program of the Chinese Academy of Sciences, Grant No. XDB0560000, NSFC (grant Nos. 11973012, 11921003, 12103090, 12203102, 12233012, 12373115), the mobility program (M-0068) of the Sino-German Science Center.

Data Availability No datasets were generated or analyzed during the current study.

Declarations

Competing interests The authors declare no competing interests.

References

Alexander, D., Metcalf, T.R.: 2002, Energy dependence of electron trapping in a solar flare. *Solar Phys.* **210**, 323. DOI. ADS.

Asai, A., Masuda, S., Yokoyama, T., Shimojo, M., Isobe, H., Kurokawa, H., Shibata, K.: 2002, Difference between spatial distributions of the Hα kernels and hard X-ray sources in a solar flare. *Astrophys. J. Lett.* **578**, L91. DOI. ADS.

Aschwanden, M.J., Fletcher, L., Sakao, T., Kosugi, T., Hudson, H.: 1999, Deconvolution of directly precipitating and trap-precipitating electrons in solar flare hard X-rays. III. Yohkoh hard X-ray telescope data analysis. *Astrophys. J.* **517**, 977. DOI. ADS.

Bobra, M.G., Sun, X., Hoeksema, J.T., Turmon, M., Liu, Y., Hayashi, K., Barnes, G., Leka, K.D.: 2014, The Helioseismic and Magnetic Imager (HMI) vector magnetic field pipeline: SHARPs – space-weather HMI active region patches. *Solar Phys.* **289**, 3549. DOI. ADS.

Brown, J.C.: 1971, The deduction of energy spectra of non-thermal electrons in flares from the observed dynamic spectra of hard X-ray bursts. *Solar Phys.* **18**, 489. DOI. ADS.

Cai, Y.F., Yang, X., Xiang, Y.Y., Yan, X.L., Jin, Z.Y., Liu, H., Ji, K.F.: 2022, The co-alignment of winged Hα data observed by the new vacuum solar telescope. *Res. Astron. Astrophys.* **22**, 065010. DOI. ADS.

Chen, B., Li, H., Song, K.F., Guo, Q.F., Zhang, P.J., He, L.P., Dai, S., Wang, X.D., Wang, H.F., Liu, C.L., Zhang, H.J., Zhang, G., Wang, Y., Liu, S.J., Zhang, H.X., Liu, L., Mao, S.L., Liu, Y., Peng, J.H., Wang, P., Sun, L., Liu, Y., Han, Z.W., Wang, Y.L., Wu, K., Ding, G.X., Zhou, P., Zheng, X., Xia, M.Y., Wu, Q.W., Xie, J.J., Chen, Y., Song, S.M., Wang, H., Zhu, B., Chu, C.B., Yang, W.G., Feng, L., Huang, Y., Gan, W.Q., Li, Y., Li, J.W., Lu, L., Xue, J.C., Ying, B.L., Sun, M.Z., Zhu, C., Bao, W.M., Deng, L., Yin, Z.S.: 2019, The Lyman-alpha Solar Telescope (LST) for the ASO-S mission – II. Design of LST. *Res. Astron. Astrophys.* **19**, 159. DOI. ADS.

Cheung, M.C.M., Boerner, P., Schrijver, C.J., Testa, P., Chen, F., Peter, H., Malanushenko, A.: 2015, Thermal diagnostics with the atmospheric imaging assembly on board the solar dynamics observatory: a validated method for differential emission measure inversions. *Astrophys. J.* **807**, 143. DOI. ADS.

Deng, Y.Y., Zhang, H.Y., Yang, J.F., Li, F., Lin, J.B., Hou, J.F., Wu, Z., Song, Q., Duan, W., Bai, X.Y., Wang, D.G., Lv, J., Ge, W., Wang, J.N., Zheng, Z.Y., Wang, C.J., Wang, N.G., Ni, H.K., Zeng, Y.Z., Zhang, Y., Yang, X., Sun, Y.Z., Zhang, Z.Y., Wang, X.F.: 2019, Design of the Full-disk MagnetoGraph (FMG) onboard the ASO-S. *Res. Astron. Astrophys.* **19**, 157. DOI. ADS.

Ding, M.D., Chen, Q.R., Li, J.P., Chen, P.F.: 2003, Hα and hard X-ray observations of a two-ribbon flare associated with a filament eruption. *Astrophys. J.* **598**, 683. DOI. ADS.

Falewicz, R., Siarkowski, M.: 2007, On the causes of hard X-ray asymmetry in solar flares. *Astron. Astrophys.* **461**, 285. DOI. ADS.

Falewicz, R., Siarkowski, M., Berlicki, A.: 2006, Hard X-ray emission at the footpoints of solar flares. *Adv. Space Res.* **38**, 956. DOI. ADS.

Feng, L., Li, H., Chen, B., Li, Y., Susino, R., Huang, Y., Lu, L., Ying, B.L., Li, J.W., Xue, J.C., Yang, Y.T., Hong, J., Li, J.P., Zhao, J., Gan, W.Q., Zhang, Y.: 2019, The Lyman-alpha Solar Telescope (LST) for the ASO-S mission – III. Data and potential diagnostics. *Res. Astron. Astrophys.* **19**, 162. DOI. ADS.

Fletcher, L., Dennis, B.R., Hudson, H.S., Krucker, S., Phillips, K., Veronig, A., Battaglia, M., Bone, L., Caspi, A., Chen, Q., Gallagher, P., Grigis, P.T., Ji, H., Liu, W., Milligan, R.O., Temmer, M.: 2011, An observational overview of solar flares. *Space Sci. Rev.* **159**, 19. DOI. ADS.

Gan, W.Q., Zhu, C., Deng, Y.Y., Li, H., Su, Y., Zhang, H.Y., Chen, B., Zhang, Z., Wu, J., Deng, L., Huang, Y., Yang, J.F., Cui, J.J., Chang, J., Wang, C., Wu, J., Yin, Z.S., Chen, W., Fang, C., Yan, Y.H., Lin, J., Xiong, W.M., Chen, B., Bao, H.C., Cao, C.X., Bai, Y.P., Wang, T., Chen, B.L., Li, X.Y., Zhang, Y., Feng, L., Su, J.T., Li, Y., Chen, W., Li, Y.P., Su, Y.N., Wu, H.Y., Gu, M., Huang, L., Tang, X.J.: 2019, Advanced Space-based Solar Observatory (ASO-S): an overview. *Res. Astron. Astrophys.* **19**, 156. DOI. ADS.

Gan, W.Q., Zhu, C., Deng, Y., Zhang, Z., Chen, B., Huang, Y., Deng, L., Wu, H., Zhang, H., Li, H., Su, Y., Su, J., Feng, L., Wu, J., Cui, J., Wang, C., Chang, J., Yin, Z., Xiong, W., Chen, B., Yang, J., Li, F., Lin, J., Hou, J., Bai, X., Chen, D., Zhang, Y., Hu, Y., Liang, Y., Wang, J., Song, K., Guo, Q., He, L., Zhang, G., Wang, P., Bao, H., Cao, C., Bai, Y., Chen, B., He, T., Li, X., Zhang, Y., Liao, X., Jiang, H., Li, Y., Su, Y., Lei, S., Chen, W., Li, Y., Zhao, J., Li, J., Ge, Y., Zou, Z., Hu, T., Su, M., Ji, H., Gu, M., Zheng, Y., Xu, D., Wang, X.: 2023, The Advanced Space-based Solar Observatory (ASO-S). *Solar Phys.* **298**, 68. DOI. ADS.

Goff, C.P., Matthews, S.A., van Driel-Gesztelyi, L., Harra, L.K.: 2004, Relating magnetic field strengths to hard X-ray emission in solar flares. *Astron. Astrophys.* **423**, 363. DOI. ADS.

Hoeksema, J.T., Liu, Y., Hayashi, K., Sun, X., Schou, J., Couvidat, S., Norton, A., Bobra, M., Centeno, R., Leka, K.D., Barnes, G., Turmon, M.: 2014, The Helioseismic and Magnetic Imager (HMI) vector magnetic field pipeline: overview and performance. *Solar Phys.* **289**, 3483. DOI. ADS.

Holman, G.D., Aschwanden, M.J., Aurass, H., Battaglia, M., Grigis, P.C., Kontar, E.P., Liu, W., Saint-Hilaire, P., Zharkova, V.V.: 2011, Implications of X-ray observations for electron acceleration and propagation in solar flares. *Space Sci. Rev.* **159**, 107. DOI. ADS.

Hudson, H., Ryan, J.: 1995, High-energy particles in solar flares. *Annu. Rev. Astron. Astrophys.* **33**, 239. DOI. ADS.
Joshi, B., Kushwaha, U., Veronig, A.M., Dhara, S.K., Shanmugaraju, A., Moon, Y.J.: 2017, Formation and eruption of a flux rope from the sigmoid active region NOAA 11719 and associated M6.5 flare: a multi-wavelength study. *Astrophys. J.* **834**, 42. DOI. ADS.
Krucker, S., Battaglia, M.: 2014, Particle densities within the acceleration region of a solar flare. *Astrophys. J.* **780**, 107. DOI. ADS.
Krucker, S., Hurford, G.J., Grimm, O., Kögl, S., Gröbelbauer, H.P., Etesi, L., Casadei, D., Csillaghy, A., Benz, A.O., Arnold, N.G., Molendini, F., Orleanski, P., Schori, D., Xiao, H., Kuhar, M., Hochmuth, N., Felix, S., Schramka, F., Marcin, S., Kobler, S., Iseli, L., Dreier, M., Wiehl, H.J., Kleint, L., Battaglia, M., Lastufka, E., Sathiapal, H., Lapadula, K., Bednarzik, M., Birrer, G., Stutz, S., Wild, C., Marone, F., Skup, K.R., Cichocki, A., Ber, K., Rutkowski, K., Bujwan, W., Juchnikowski, G., Winkler, M., Darmetko, M., Michalska, M., Seweryn, K., Białek, A., Osica, P., Sylwester, J., Kowalinski, M., Ścisłowski, D., Siarkowski, M., Stkeślicki, M., Mrozek, T., Podgórski, P., Meuris, A., Limousin, O., Gevin, O., Le Mer, I., Brun, S., Strugarek, A., Vilmer, N., Musset, S., Maksimović, M., Fárník, F., Kozáček, Z., Kašparová, J., Mann, G., Önel, H., Warmuth, A., Rendtel, J., Anderson, J., Bauer, S., Dionies, F., Paschke, J., Plüschke, D., Woche, M., Schuller, F., Veronig, A.M., Dickson, E.C.M., Gallagher, P.T., Maloney, S.A., Bloomfield, D.S., Piana, M., Massone, A.M., Benvenuto, F., Massa, P., Schwartz, R.A., Dennis, B.R., van Beek, H.F., Rodríguez-Pacheco, J., Lin, R.P. (eds.): 2020, The Spectrometer/Telescope for Imaging X-rays (STIX). *Astron. Astrophys.* **642**, A15. DOI. ADS.
Kundu, M.R., Nitta, N., White, S.M., Shibasaki, K., Enome, S., Sakao, T., Kosugi, T., Sakurai, T.: 1995, Microwave and hard X-ray observations of footpoint emission from solar flares. *Astrophys. J.* **454**, 522. DOI. ADS.
Lemen, J.R., Title, A.M., Akin, D.J., Boerner, P.F., Chou, C., Drake, J.F., Duncan, D.W., Edwards, C.G., Friedlaender, F.M., Heyman, G.F., Hurlburt, N.E., Katz, N.L., Kushner, G.D., Levay, M., Lindgren, R.W., Mathur, D.P., McFeaters, E.L., Mitchell, S., Rehse, R.A., Schrijver, C.J., Springer, L.A., Stern, R.A., Tarbell, T.D., Wuelser, J.P., Wolfson, C.J., Yanari, C., Bookbinder, J.A., Cheimets, P.N., Caldwell, D., Deluca, E.E., Gates, R., Golub, L., Park, S., Podgorski, W.A., Bush, R.I., Scherrer, P.H., Gummin, M.A., Smith, P., Auker, G., Jerram, P., Pool, P., Soufli, R., Windt, D.L., Beardsley, S., Clapp, M., Lang, J., Waltham, N.: 2012, The Atmospheric Imaging Assembly (AIA) on the Solar Dynamics Observatory (SDO). *Solar Phys.* **275**, 17. DOI. ADS.
Li, J., Metcalf, T.R., Canfield, R.C., Wülser, J.P., Kosugi, T.: 1997, What is the spatial relationship between hard X-ray footpoints and vertical electric currents in solar flares? *Astrophys. J.* **482**, 490. DOI. ADS.
Li, H., Chen, B., Feng, L., Li, Y., Huang, Y., Li, J.W., Lu, L., Xue, J.C., Ying, B.L., Zhao, J., Yang, Y.T., Gan, W.Q., Fang, C., Song, K.F., Wang, H., Guo, Q.F., He, L.P., Zhu, B., Zhu, C., Deng, L., Bao, H.C., Cao, C.X., Yang, Z.G.: 2019, The Lyman-alpha Solar Telescope (LST) for the ASO-S mission — I. Scientific objectives and overview. *Res. Astron. Astrophys.* **19**, 158. DOI. ADS.
Li, C., Fang, C., Li, Z., Ding, M., Chen, P., Qiu, Y., You, W., Yuan, Y., An, M., Tao, H., Li, X., Chen, Z., Liu, Q., Mei, G., Yang, L., Zhang, W., Cheng, W., Chen, J., Chen, C., Gu, Q., Huang, Q., Liu, M., Han, C., Xin, H., Chen, C., Ni, Y., Wang, W., Rao, S., Li, H., Lu, X., Wang, W., Lin, J., Jiang, Y., Meng, L., Zhao, J.: 2022, The Chinese Hα Solar Explorer (CHASE) mission: an overview. *Sci. China Phys. Mech. Astron.* **65**, 289602. DOI. ADS.
Li, Y., Li, Q., Song, D.C., Battaglia, A.F., Xiao, H.L., Krucker, S., Schühle, U., Li, H., Gan, W.Q., Ding, M.D.: 2022a, The Lyα emission in a C1.4 solar flare observed by the extreme ultraviolet imager aboard solar orbiter. *Astrophys. J.* **936**, 142. DOI. ADS.
Li, Z., Su, Y., Veronig, A.M., Kong, S., Gan, W., Chen, W.: 2022b, Detailed thermal and nonthermal processes in an A-class microflare. *Astrophys. J.* **930**, 147. DOI. ADS.
Lin, R.P.: 2011, Energy release and particle acceleration in flares: summary and future prospects. *Space Sci. Rev.* **159**, 421. DOI. ADS.
Lin, R.P., Dennis, B.R., Hurford, G.J., Smith, D.M., Zehnder, A., Harvey, P.R., Curtis, D.W., Pankow, D., Turin, P., Bester, M., Csillaghy, A., Lewis, M., Madden, N., van Beek, H.F., Appleby, M., Raudorf, T., McTiernan, J., Ramaty, R., Schmahl, E., Schwartz, R., Krucker, S., Abiad, R., Quinn, T., Berg, P., Hashii, M., Sterling, R., Jackson, R., Pratt, R., Campbell, R.D., Malone, D., Landis, D., Barrington-Leigh, C.P., Slassi-Sennou, S., Cork, C., Clark, D., Amato, D., Orwig, L., Boyle, R., Banks, I.S., Shirey, K., Tolbert, A.K., Zarro, D., Snow, F., Thomsen, K., Henneck, R., Mchedlishvili, A., Ming, P., Fivian, M., Jordan, J., Wanner, R., Crubb, J., Preble, J., Matranga, M., Benz, A., Hudson, H., Canfield, R.C., Holman, G.D., Crannell, C., Kosugi, T., Emslie, A.G., Vilmer, N., Brown, J.C., Johns-Krull, C., Aschwanden, M., Metcalf, T., Conway, A. (eds.): 2002, The Reuven Ramaty High-Energy Solar Spectroscopic Imager (RHESSI). *Solar Phys.* **210**, 3. DOI. ADS.
Liu, W., Chen, Q., Petrosian, V.: 2013, Plasmoid ejections and loop contractions in an eruptive M7.7 solar flare: evidence of particle acceleration and heating in magnetic reconnection outflows. *Astrophys. J.* **767**, 168. DOI. ADS.

Massa, P., Schwartz, R., Tolbert, A.K., Massone, A.M., Dennis, B.R., Piana, M., Benvenuto, F.: 2020, MEM_GE: a new maximum entropy method for image reconstruction from solar X-ray visibilities. *Astrophys. J.* **894**, 46. DOI. ADS.

Masuda, S., Kosugi, T., Hara, H., Tsuneta, S., Ogawara, Y.: 1994, A loop-top hard X-ray source in a compact solar flare as evidence for magnetic reconnection. *Nature* **371**, 495. DOI. ADS.

Müller, D., St. Cyr, O.C., Zouganelis, I., Gilbert, H.R., Marsden, R., Nieves-Chinchilla, T., Antonucci, E., Auchère, F., Berghmans, D., Horbury, T.S., Howard, R.A., Krucker, S., Maksimovic, M., Owen, C.J., Rochus, P., Rodriguez-Pacheco, J., Romoli, M., Solanki, S.K., Bruno, R., Carlsson, M., Fludra, A., Harra, L., Hassler, D.M., Livi, S., Louarn, P., Peter, H., Schühle, U., Teriaca, L., del Toro Iniesta, J.C., Wimmer-Schweingruber, R.F., Marsch, E., Velli, M., De Groof, A., Walsh, A., Williams, D.: 2020, The solar orbiter mission. Science overview. *Astron. Astrophys.* **642**, A1. DOI. ADS.

Naus, S.J., Qiu, J., DeVore, C.R., Antiochos, S.K., Dahlin, J.T., Drake, J.F., Swisdak, M.: 2022, Correlated spatio-temporal evolution of extreme-ultraviolet ribbons and hard X-rays in a solar flare. *Astrophys. J.* **926**, 218. DOI. ADS.

Ogawara, Y., Acton, L.W., Bentley, R.D., Bruner, M.E., Culhane, J.L., Hiei, E., Hirayama, T., Hudson, H.S., Kosugi, T., Lemen, J.R., Strong, K.T., Tsuneta, S., Uchida, Y., Watanabe, T., Yoshimori, M.: 1992, The status of Yohkoh in orbit: an introduction to the initial scientific results. *Publ. Astron. Soc. Japan* **44**, L41. ADS.

Pesnell, W.D., Thompson, B.J., Chamberlin, P.C.: 2012, The Solar Dynamics Observatory (SDO). *Solar Phys.* **275**, 3. DOI. ADS.

Qiu, J., Alaoui, M., Antiochos, S.K., Dahlin, J.T., Swisdak, M., Drake, J.F., Robison, A., DeVore, C.R., Uritsky, V.M.: 2023, The role of magnetic shear in reconnection-driven flare energy release. *Astrophys. J.* **955**, 34. DOI. ADS.

Raymond, J.C., Krucker, S., Lin, R.P., Petrosian, V.: 2012, Observational aspects of particle acceleration in large solar flares. *Space Sci. Rev.* **173**, 197. DOI. ADS.

Rochus, P., Auchère, F., Berghmans, D., Harra, L., Schmutz, W., Schühle, U., Addison, P., Appourchaux, T., Aznar Cuadrado, R., Baker, D., Barbay, J., Bates, D., BenMoussa, A., Bergmann, M., Beurthe, C., Borgo, B., Bonte, K., Bouzit, M., Bradley, L., Büchel, V., Buchlin, E., Büchner, J., Cabé, F., Cadiergues, L., Chaigneau, M., Chares, B., Choque Cortez, C., Coker, P., Condamin, M., Coumar, S., Curdt, W., Cutler, J., Davies, D., Davison, G., Defise, J.M., Del Zanna, G., Delmotte, F., Delouille, V., Dolla, L., Dumesnil, C., Dürig, F., Enge, R., François, S., Fourmond, J.J., Gillis, J.M., Giordanengo, B., Gissot, S., Green, L.M., Guerreiro, N., Guilbaud, A., Gyo, M., Haberreiter, M., Hafiz, A., Hailey, M., Halain, J.P., Hansotte, J., Hecquet, C., Heerlein, K., Hellin, M.L., Hemsley, S., Hermans, A., Hervier, V., Hochedez, J.F., Houbrechts, Y., Ihsan, K., Jacques, L., Jérôme, A., Jones, J., Kahle, M., Kennedy, T., Klaproth, M., Kolleck, M., Koller, S., Kotsialos, E., Kraaikamp, E., Langer, P., Lawrenson, A., Le Clech', J.C., Lenaerts, C., Liebecq, S., Linder, D., Long, D.M., Mampaey, B., Markiewicz-Innes, D., Marquet, B., Marsch, E., Matthews, S., Mazy, E., Mazzoli, A., Meining, S., Meltchakov, E., Mercier, R., Meyer, S., Monecke, M., Monfort, F., Morinaud, G., Moron, F., Mountney, L., Müller, R., Nicula, B., Parenti, S., Peter, H., Pfiffner, D., Philippon, A., Phillips, I., Plesseria, J.Y., Pylyser, E., Rabecki, F., Ravet-Krill, M.F., Rebellato, J., Renotte, E., Rodriguez, L., Roose, S., Rosin, J., Rossi, L., Roth, P., Rouesnel, F., Roulliay, M., Rousseau, A., Ruane, K., Scanlan, J., Schlatter, P., Seaton, D.B., Silliman, K., Smit, S., Smith, P.J., Solanki, S.K., Spescha, M., Spencer, A., Stegen, K., Stockman, Y., Szwec, N., Tamiatto, C., Tandy, J., Teriaca, L., Theobald, C., Tychon, I., van Driel-Gesztelyi, L., Verbeeck, C., Vial, J.C., Werner, S., West, M.J., Westwood, D., Wiegelmann, T., Willis, G., Winter, B., Zerr, A., Zhang, X., Zhukov, A.N. (eds.): 2020, The solar orbiter EUI instrument: the extreme ultraviolet imager. *Astron. Astrophys.* **642**, A8. DOI. ADS.

Sakao, T.: 1994, Characteristics of solar flare hard X-ray sources as revealed with the Hard X-ray Telescope aboard the Yohkoh satellite. PhD thesis, University of Tokyo. ADS.

Siarkowski, M., Falewicz, R.: 2004, Variations of the hard X-ray footpoint asymmetry in a solar flare. *Astron. Astrophys.* **428**, 219. DOI. ADS.

Silva, D.F., Simões, P.J.A., Hidalgo Ramírez, R.F., Válio, A.: 2020, Inferring the magnetic field asymmetry of solar flares from the degree of polarisation at millimetre wavelengths. *Solar Phys.* **295**, 73. DOI. ADS.

Su, Y., Veronig, A.M., Holman, G.D., Dennis, B.R., Wang, T., Temmer, M., Gan, W.: 2013, Imaging coronal magnetic-field reconnection in a solar flare. *Nat. Phys.* **9**, 489. DOI. ADS.

Su, Y., Veronig, A.M., Hannah, I.G., Cheung, M.C.M., Dennis, B.R., Holman, G.D., Gan, W., Li, Y.: 2018, Determination of differential emission measure from solar extreme ultraviolet images. *Astrophys. J. Lett.* **856**, L17. DOI. ADS.

Su, J.T., Bai, X.Y., Chen, J., Guo, J.J., Liu, S., Wang, X.F., Xu, H.Q., Yang, X., Song, Y.L., Deng, Y.Y., Ji, K.F., Deng, L., Huang, Y., Li, H., Gan, W.Q.: 2019, Data reduction and calibration of the FMG onboard ASO-S. *Res. Astron. Astrophys.* **19**, 161. DOI. ADS.

Su, Y., Liu, W., Li, Y.P., Zhang, Z., Hurford, G.J., Chen, W., Huang, Y., Li, Z.T., Jiang, X.K., Wang, H.X., Xia, F.X.Y., Chen, C.X., Yu, W.H., Yu, F., Wu, J., Gan, W.Q.: 2019, Simulations and software development for the hard X-ray imager onboard ASO-S. *Res. Astron. Astrophys.* **19**, 163. DOI. ADS.

Thomas, R.J., Starr, R., Crannell, C.J.: 1985, Expressions to determine temperatures and emission measures for solar X-ray events from goes measurements. *Solar Phys.* **95**, 323. DOI. ADS.

Titov, V.S., Hornig, G., Démoulin, P.: 2002, Theory of magnetic connectivity in the solar corona. *J. Geophys. Res.* **107**, 1164. DOI. ADS.

Wiegelmann, T., Neukirch, T.: 2003, Computing nonlinear force free coronal magnetic fields. *Nonlinear Process. Geophys.* **10**, 313. DOI. ADS.

Zhang, Z., Chen, D.Y., Wu, J., Chang, J., Hu, Y.M., Su, Y., Zhang, Y., Wang, J.P., Liang, Y.M., Ma, T., Guo, J.H., Cai, M.S., Zhang, Y.Q., Huang, Y.Y., Peng, X.Y., Tang, Z.B., Zhao, X., Zhou, H.H., Wang, L.G., Song, J.X., Ma, M., Xu, G.Z., Yang, J.F., Lu, D., He, Y.H., Tao, J.Y., Ma, X.L., Lv, B.G., Bai, Y.P., Cao, C.X., Huang, Y., Gan, W.Q.: 2019, Hard X-ray Imager (HXI) onboard the ASO-S mission. *Res. Astron. Astrophys.* **19**, 160. DOI. ADS.

Zhang, P.J., Chen, J., Liu, R., Wang, C.B.: 2022, FastQSL: a fast computation method for quasi-separatrix layers. *Astrophys. J.* **937**, 26. DOI. ADS.

Authors and Affiliations

Guanglu Shi[1,2] · Li Feng[1,2] · Jun Chen[1] · Beili Ying[1] · Shuting Li[1,2] · Qiao Li[1,2] · Hui Li[1,2] · Ying Li[1,2] · Kaifan Ji[3] · Yu Huang[1,2] · Youping Li[1,2] · Jingwei Li[1] · Jie Zhao[1] · Lei Lu[1] · Jianchao Xue[1] · Ping Zhang[1] · Dechao Song[1] · Zhengyuan Tian[1,2] · Yingna Su[1,2] · Qingmin Zhang[1] · Yunyi Ge[1] · Jiahui Shan[1,2] · Yue Zhou[1,2] · Jun Tian[1,2] · Gen Li[1,2] · Xiaofeng Liu[1,2] · Zhichen Jing[1,2] · Shijun Lei[1] · Weiqun Gan[1,4]

✉ L. Feng
lfeng@pmo.ac.cn

1 Key Laboratory of Dark Matter and Space Astronomy, Purple Mountain Observatory, Chinese Academy of Sciences, Nanjing 210023, China

2 School of Astronomy and Space Science, University of Science and Technology of China, Hefei 230026, China

3 Yunnan Observatories, Chinese Academy of Sciences, Kunming 650216, China

4 University of Chinese Academy of Sciences, Nanjing 211135, China

Solar Physics (2024) 299:64
https://doi.org/10.1007/s11207-024-02306-x

RESEARCH

Rising of Two Crossing Prominences and the Resulting Oscillations

Hengyuan Wei[1] · Zhenghua Huang[1] · Qingmin Zhang[2] · Ying Li[2,3] · Hui Fu[1] · Ming Xiong[4,5] · Lidong Xia[1] · Li Feng[2,3] · Hui Li[2,3] · Weiqun Gan[2,6]

Received: 28 January 2024 / Accepted: 19 April 2024 / Published online: 16 May 2024

Abstract

Prominences are important features in the solar atmosphere. Their activities often develop into solar eruptions, such as flares and/or coronal mass ejections. We report here on observations of activities of two crossing prominences and the resulting oscillations observed with the Advanced Space-based Solar Observatory (ASO-S) and the Solar Dynamics Observatory. We observed the two crossing prominences rising simultaneously with a speed of about 100 km s^{-1}. The lower-lying prominence consists of threads that show increase of writhe during the rising process. We find evidence that the writhe of the lower-lying prominence is transferred into the overlying one. This transfer of writhe leads to a failure of the eruption of the lower-lying prominence and a shearing motion of the legs of the overlying prominence. The failed eruption of the lower-lying prominence also triggers kink oscillations of its threads, which show periods of about 300 s and amplitudes of less than 10 Mm. Such oscillations are considered to be intrinsic mode and can help to probe the magnetic field of the prominence. Our observations support the idea that the transfer and release of writhe play an important role in confining eruptions of a prominence, and interactions among prominences/filaments might be a crucial aspect of a solar eruption.

Keywords Solar prominence · Solar eruption · Solar corona

✉ Z. Huang
z.huang@sdu.edu.cn

1 Shandong Key Laboratory of Optical Astronomy and Solar-Terrestrial Environment, Institute of Space Sciences, Shandong University, Weihai 264209, Shandong, China

2 Key Laboratory of Dark Matter and Space Astronomy, Purple Mountain Observatory, CAS, Nanjing 210023, China

3 School of Astronomy and Space Science, University of Science and Technology of China, Hefei 230026, China

4 Key Laboratory of Solar Activity and Space Weather, National Space Science Center, Chinese Academy of Sciences, Beijing, 100190, China

5 College of Earth and Planetary Sciences, University of Chinese Academy of Sciences, Beijing, 100049, China

6 University of Chinese Academy of Sciences, Nanjing 211135, China

1. Introduction

Prominences, also called filaments when appearing against the solar disk, are common features suspended in the solar corona with cool, dense, and partially ionized plasma. They are aligned with polarity inversion lines (PILs) and anchor to opposite magnetic polarities in the photosphere. Destabilization of the prominences may lead to their eruptions accompanied by solar flares and/or coronal mass ejections (CMEs; Shibata and Magara, 2011; Chen, 2011; Shen, Liu, and Su, 2012; Parenti, 2014; Huang et al., 2014; Benz, 2017; Vashalomidze et al., 2022). The eruptive behavior of filaments depends on their length (Filippov, 2022). Their eruptions can be divided into three types: full, partial, and failed/confined eruptions (Gilbert, Alexander, and Liu, 2007).

The dynamics of prominences, such as oscillations, is also widely studied in the literature. Generally, oscillations are triggered by external disturbances related to coronal waves and CMEs/flares (e.g., Ramsey and Smith, 1966; Jing et al., 2003, 2006; Vršnak et al., 2007; Luna and Karpen, 2012) or internal disturbances related to flux emergence or disappearance (e.g., Feynman and Martin, 1995). Filament eruptions, especially those providing a strong horizontal impulse, can cause Moreton waves and the EUV waves can trigger horizontal and vertical oscillations of nearby quiescent filament (e.g., Dai et al., 2023). These oscillasions usually have an exponentially decaying amplitude (Molowny-Horas et al., 1999) and their periods are highly sensitive to the physical parameters of prominences (e.g., Bashkirtsev and Mashnich, 1984; Mashnich and Bashkirtsev, 1990; Oliver and Ballester, 2002). Some physical parameters of prominences can be derived from oscillations, such as the Alfvén speed and magnetic field strength (e.g., Régnier, Solomon, and Vial, 2001; Lin et al., 2009). Using the parameters of simultaneous longitudinal and transverse oscillations of filament threads after their failed eruption, the magnetic field strength of the filament is estimated to be around 22 G (Mazumder et al., 2020).

A few possible models have been proposed to interpret failed eruptions of prominences. The coronal eruptions are caused by evolution of kinking motions which lead to a catastrophic loss of equilibrium (Liu, Alexander, and Gilbert, 2007). The eruption of prominences triggered by kink instability can be confined by sufficiently strong overlying fields (e.g., Ji et al., 2003; Török and Kliem, 2005; Zhang et al., 2021b; Guo et al., 2021; Xu et al., 2020; Zheng et al., 2019). The magnetic tension of flux ropes can play a role in confining flux ropes in an expanding magnetized volume if no reconnection occurs (Archontis and Török, 2008). So the filament can erupt after the confinement of overlying background field decreases (e.g., Kang et al., 2023; Liu and Su, 2021; Liu et al., 2021). In a filament's system, the newly formed magnetic flux ropes have an equilibrium at a relatively higher height than that of the original filament, like the double-decker flux rope system (Shen et al., 2024). The amount of released energy is also another crucial factor to determine whether the prominence eruption succeeds (Shen, Liu, and Liu, 2011). Reconnection with the overlying field can also confine the eruption of prominences, leading to lateral expansion after the prominence reaches its maximum height and reconnects with overlying fields (e.g., Netzel et al., 2012; Hassanin and Kliem, 2016; Joshi et al., 2022; Hou et al., 2020). The relative angle between emerging magnetic flux and preexisting coronal dipole field can play a key role in determining whether eruptions succeed (Leake, Linton, and Antiochos, 2022). Besides, the mass unloading and loading can also cause eruptions of flux ropes and confine their eruptions (e.g., Zhang et al., 2021a; Jenkins et al., 2019). Gravity might also terminate the rise of filaments and balance Lorentz force (Filippov, 2020).

Many observational and numerical studies have been performed to explore the evolution of kink and torus instability and interaction between erupting prominences and overlying fields in confined prominence eruptions. Coronal flux ropes can rise and erupt as a result of kink instability and they show apparent writhing during their eruption process (Török and Kliem, 2005). Strong observational evidence has been found that kink instability and torus instability are triggers of filament eruptions (Dai et al., 2021; Rees-Crockford et al., 2020). The erupting prominence shows obvious writhing motion after it reaches a certain height (Zhou, Ji, and Zhang, 2023). In prominence eruptions with apparent rotation, the twist of a kink-unstable magnetic flux rope can be transformed into writhe (Zhou et al., 2023). The kink-unstable erupting flux rope will not escape into interplanetary space if it does not reach the critical height for the occurrence of torus instability (Hassanin et al., 2022). The critical height of torus instability is usually at the height where the decay index is 1.5, though that of short eruptive filaments is lower than unity (Filippov, 2021a) and there exist some exceptions such as at 0.2 in a prominence eruption (Filippov, 2021b).

The flux tube can reconnect with overlying arcades after injection of sufficient twist (Amari and Luciani, 1999). The confined prominence eruption comprises two phases of reconnection in some cases: reconnection with overlying flux ropes after a strong writhe and reconnection between the legs of original flux rope (Hassanin and Kliem, 2016). The rise of erupting magnetic flux ropes can be halted by downward magnetic tension of overlying fields even though they have exceeded the critical height of torus instability, and the flux ropes can reconnect with overlying fields after their rotation (Jiang et al., 2023). The magnetic field direction of surrounding fields can also determine whether the filament eruptions are confined, when their direction is the same as the eruptive filament, the filament will be confined (Ding, Zhang, and Hong, 2022).

The magnetic flux rope rotates during its rising process, and this results in decrease of nonlocal writhe and increase of twist, leading to transformation of flux rope from sigmoid to arcade (Zhou et al., 2022). The transformation of flux ropes from sigmoid to near-potential loop arcades is also proved in observations prior to flares (Zhou et al., 2017). The unwrithing phase of an erupting flux rope is accompanied by rotation or slipping motion of its footpoints (Liu et al., 2022). The rotating motion in prominence eruptions also has important impacts on determining whether eruptions succeed or fail (Zhou et al., 2019).

To our knowledge, there is no observational evidence of writhe transfer and release among different prominences. In this paper, we report on a failed eruption of a prominence system consisting of two crossing prominences with multiwavelength imaging observations. In these observations, we find evidence of transfer of writhe between a rising and an overlying prominence, which leads to a failure of the prominence eruption. In what follows, we will give the description of observations in Section 2, results in Section 3, discussion in Section 4, and summary in Section 5.

2. Observations

The data used here were taken by the space-based Lyman-alpha Solar Telescope (LST, Li, 2016; Li et al., 2019) onboard the Advanced Space-based Solar Observatory (ASO-S, Gan et al., 2019, 2023), the Atmospheric Imaging Assembly (AIA, Lemen et al., 2012) onboard Solar Dynamics Observatory (SDO, Pesnell, Thompson, and Chamberlin, 2012) and the Sun Earth Connection Coronal and Heliospheric Investigation (SECCHI, Howard et al., 2008) onboard STEREO-A (Kaiser et al., 2008). All the data are available publicly. The LST data can be obtained in the ASO-S official website,[1] and the SDO/AIA data can be found in the

[1] http://aso-s.pmo.ac.cn/sodc/dataArchive.jsp.

JSOC website.[2] The studied event occurs on the north-eastern limb from 13:00 UT to 15:00 UT on April 2, 2023, and the data we analyzed consist of images taken at Lyman-α passband by LST and images taken at EUV passbands including 94, 131, 171, 193, 211, 304, and 335 Å by AIA. We also use the observation of SECCHI in 195 Å at 13:25:00 UT to estimate the writhe of the lower-lying prominence and images at 13:15:00, 13:27:30, and 14:32:30 UT to show the region of interest at times before, during, and after this event in other perspective. The separation angle between STEREO and the Earth is about 11 degrees.

The spatial pixel scale of the AIA EUV data is 0.6 arcsec per pixel and that of LST is about 0.5 arcsec per pixel. The cadence of the AIA and LST data is 12 and 60 s, respectively. The level 1 data of AIA and LST observations are used for analysis. They are calibrated by the standard procedure of *aia_prep.pro* in *solarsoft* programmed by the instrument team of AIA and the standard procedure of *lst_prep.pro* provided by the instrument team of LST, respectively.

3. Analysis and Results

The two prominences were principally located on the far side of the Sun and were not observed until the eruption had started and they appeared over the north-eastern limb after 13:17 UT. So we cannot follow its triggering process from the very beginning. We show the whole evolution of this event in AIA 304 and 171 Å passbands in Figure 1 and the associated animation. In this figure, we observe that both erupting prominences consist of multiple threads. For the lower-lying prominence, its western part connects to the same footpoint while its eastern part connects to a different footpoint than the overlying prominence (see black dashed lines in Figure 1(d)). The multithreads of the erupting prominences are also clear from the other perspective (see observation from STEREO in Figure 2(b)). The overall shape and appearance of the erupting prominences in different perspectives are similar because the separation angle between SDO and STEREO was small. The two crossing prominences rise during the event. The lower-lying prominence rises to its maximum height at around 13:35 UT, while the overlying one continues rising. We find that the upper and lower threads of the lower-lying prominence show different evolution. During the rising process of the lower-lying prominence, its upper threads experience a writhing process (see the black dashed lines in Figure 1(a)–(b)) while its lower threads do not. The writhe of the upper threads of the lower-lying prominence increases first but it disappears after 13:31:53 UT. The lower-lying prominence expands during the disappearance of the writhe. After that, the upper threads of the lower-lying prominence make contact and merge with its lower ones, forming a more compact prominence (see the associated animation of Figure 1).

We observed an unwinding process of the overlying prominence after the disappearance of writhe mentioned above, while its legs apparently separate. The unwinding process can be observed more clearly in Lyman-α observations of LST, as indicated by white arrows in Figure 3(c)–(h). We make a slice–time plot along the slice "S" (see the white dashed line in Figure 3(h)). We find that the overlying prominence drifts toward the north-west with a speed of 35 km s^{-1}. After 13:35 UT (see black dashed line in Figure 3(i)), the legs of overlying prominence start a motion of separating – joining – separating with a speed of 39, 67, and 10 km s^{-1}, respectively. This indicates that the overlying prominence rotates at least three quarters of a full turn if we assume it is rigid.

[2] http://jsoc.stanford.edu.

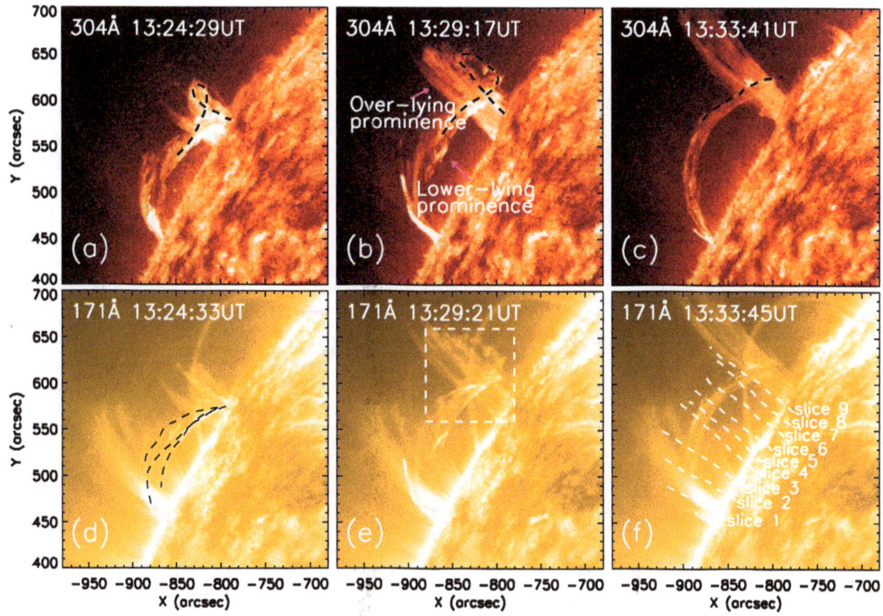

Figure 1 Evolution of the studied region in AIA 304 and 171 Å passbands: **(a)–(c)** Snapshots in AIA 304 Å at the time marked in the panels. The black dashed lines outline the trajectory of threads in the lower-lying prominence; **(d)–(f)** Snapshots in AIA 171 Å at the time marked in the panels. The black dashed lines in (d) outline the trajectories of several threads in the lower-lying prominence. The white dashed rectangle in (e) marks the region shown in Figure 7. The white dashed lines in (f) mark the slices where we obtain slice–time plots in Figure 5. An associated animation can be obtained online.

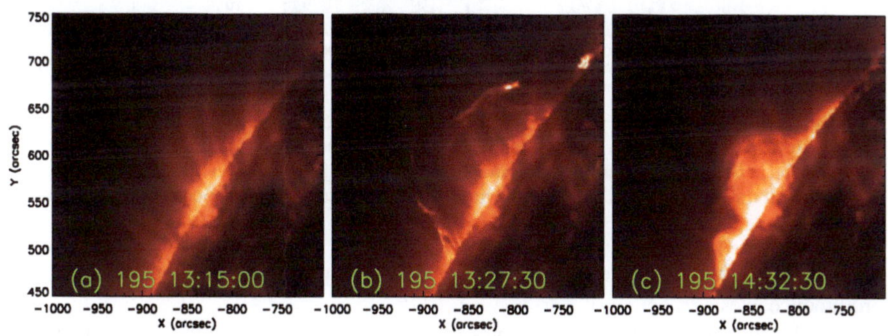

Figure 2 Snapshots of the region of interest as viewed by SECCHI/EUVI 195 Å passband at time before (13:15:00 UT), during (13:27:30 UT), and after (14:32:30 UT) the prominence eruption.

During the eruption, a bright plasma cloud appears underneath the middle of the lower-lying prominence (see Figure 4, indicated by white arrows in panels (b), (e), and (f)), which only appears in high temperature passbands such as AIA 94 Å . The DEM analyses indicate hot plasma in this feature (see text below for more details). Also, several localized EUV brightenings appear at both legs of the lower-lying prominence (see white arrows in Figure 4(c)), which is consistent with the features analyzed by Wei et al. (2023) during the activation of a filament, which might be caused by internal reconnections of braided mag-

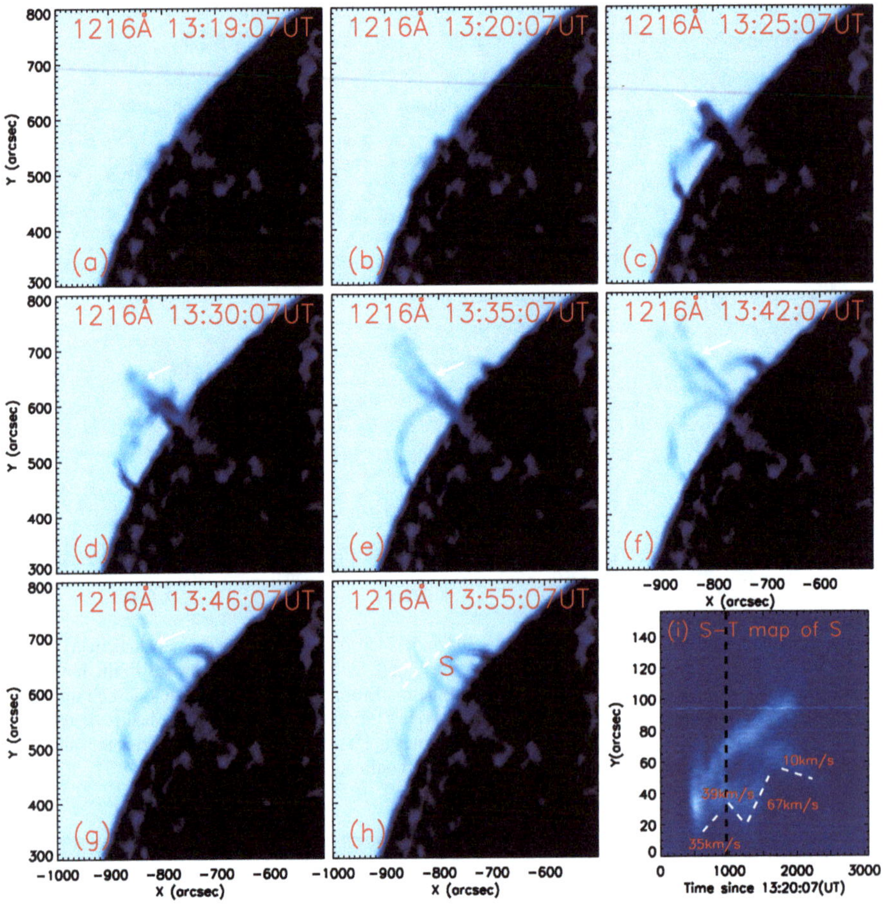

Figure 3 **(a)–(h)** Snapshots in Lyman-α at the time marked at the top of their relative panels. The white arrows in (c)–(h) point at the overlying prominence. The white dashed line in (h) marked by "S" is the place where we obtain slice–time plot in (i); **(i)** The slice–time plot along the slice "S" in (h). The black dashed line marks the start time of rotation of the overlying prominence. An associated animation is provided online.

netic fields. At the end of the eruption, materials in both prominences fall down along their prominence axes.

In order to study the thermal dynamics of this eruption, we apply the differential emission measure technique (DEM, Cheung et al., 2015; Su et al., 2018) with the observations in AIA 94, 131, 171, 193, 211, and 335 Å and obtain the emission measure (EM) of the region at several temperature ranges. We then calculate the average temperature distribution in the region weighted by EM. We notice that the overlying prominence and the middle part of the lower-lying prominence do not show any obvious temperature increase (see Figure 4(d)–(f)). We find evidence that two ends of the lower-lying prominence are heated during its eruption (see Figure 4(e)–(f)). It shows that the southern end is heated to $5.7-9.3$ MK, and its northern end is heated to $1-8$ MK. The upper threads of its northern end are heated to higher temperature than the lower threads (see Figure 4(e)). For the EUV brightenings at both ends of the lower-lying prominence, we find that their temperatures are much higher than those of their surroundings. Besides, the bright feature beneath the lower-lying prominence appears

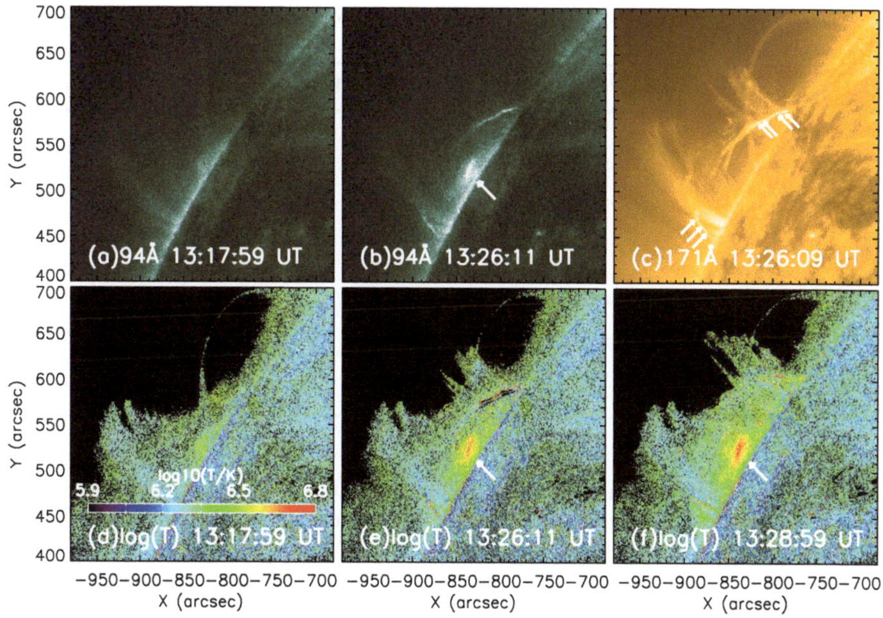

Figure 4 **(a)–(b)** Snapshots in AIA 94 Å passband at 13:17:59 UT and 13:26:11 UT. The white arrow in (b) points at the hot plasma under the lower-lying prominence; **(c)** Snapshot in AIA 171 Å passband at 13:26:09 UT. The white arrows point at the localized brightenings at the legs of the lower-lying prominence; **(d)–(f)** The temperature distribution at 13:17:59 UT, 13:26:11 UT, and 13:28:59 UT, the temperature maps are masked by the AIA 171 Å observations and the pixels with signals lower than 100 DN are filtered out. The white arrows in (e) and (f) point at the hot plasma under the lower-lying prominence.

to be heated to at least 4.5 MK during the eruption process (see white arrows in Figure 4, (b), (e), and (f)).

To study dynamics of the rising threads of the prominences, we make slice–time plots along nine slices perpendicular to the prominence axis (marked as $1-9$ with white dashed lines in Figure 1(f)). We find that the middle part of the lower-lying prominence firstly rises into view while its two ends rise later, as we first observe the erupting lower-lying prominence in slices 4 and 5. The lower-lying prominence rises with a speed ranging from 80 to 160 km s^{-1} (see Table 1).

After the lower-lying prominence reaches its maximum height, we find that most of its threads apparently oscillate, which can be clearly seen in AIA 304 Å and 171 Å images. Its two ends start oscillating prior to its middle part as they first reach their maximum heights. Most threads of the lower-lying prominence show identical oscillations in both 171 and 304 Å passbands (see Figure 5), where the white dashed lines outline the oscillations apparent in AIA 171 Å passband and the cyan dashed lines outline the oscillations apparent in AIA 304 Å passband. Along slice 7, however, the oscillations in the two passbands show different behaviors, which consist of a long period and a short one (see white and cyan dashed lines Figure 5, (g3) and (g4)). We found that the amplitudes of the oscillations are decaying (see Figure 5). To obtain their periods and amplitudes, we then fit the spatial displacements (y) of these oscillations with an exponentially decaying sine function of

$$y = A_0 e^{A_1 t} \sin(A_2 t + A_3) + A_4,$$

Table 1 The rising speeds and the oscillation parameters determined at nine slices perpendicular to the prominence.

slice	rise speed (km s^{-1})	period (s)	amplitude (Mm)	decaying time (s)	phase (rad)
slice1	88	319 ± 32	3.2 ± 2.0	599	–0.25 ± 0.45
slice2	113	289 ± 36	2.5 ± 2.0	1818	–1.13 ± 0.72
slice3	157	–	–	–	–
slice4	108	269 ± 48	1.4 ± 2.2	2671	–1.11 ± 1.54
slice5	118	254 ± 32	1.6 ± 2.2	4426	–1.05 ± 1.17
slice6	101	265 ± 55	1.3 ± 2.1	5152	1.26 ± 1.77
slice7-long	100	554 ± 48	4.0 ± 2.3	1785	–0.54 ± 0.45
slice8	81	350 ± 216	7.6 ± 11.7	100	0.30 ± 0.90
slice9	96	283 ± 45	10.6 ± 5.6	127	0.16 ± 0.30

where t is time, and A_0–A_4 are corresponding to amplitude, decaying index, angular frequency, initial phase, and equilibrium position. The periods of these oscillations are around 250 to 360 s and their amplitudes are about from 1.3 Mm to 10 Mm, with the exception of slice 7, which has a larger period at about 554 s. Such oscillations are considered to be short period oscillations (Oliver and Ballester, 2002). The decay time scales are found to be 100 – 5000 s. These parameters can be found in Table 1. The uncertainty comes from one sigma error of fitting results. Please note that the fittings to slit 3 and the short-period oscillation of slit 7 do not converge, and thus their results are not reliable. At different locations of the lower-lying prominence, we do not find any obvious phase delay of the oscillations. Since we do not observe any external exciters, these oscillations should be intrinsic mode and thus be related to physical properties of the prominence.

4. Discussion

The event observed here shows complex erupting process involving two crossing prominences. The lower-lying prominence shows different features than those failed eruption events reported earlier (e.g., Ji et al., 2003; Zhang et al., 2022; Monga et al., 2021), in which threads of the erupting prominences show similar evolution. In this event, the two prominences first rise with a speed of around 100 km s^{-1}, and we observe the upper and lower threads of the lower-lying prominence show different evolutions. For the lower-lying prominence, during its rise and eruption, its upper threads first go through an increase of writhe (see the black dashed lines in Figure 1, (a) and (b)), which is often observed in prominence eruptions, but its lower threads do not show any writhe. We use the SECCHI observation at 13:25 UT and AIA observation at corresponding time to estimate the writhe with the method described by Zhou et al. (2023) and Prior and Neukirch (2016). We first use the scc_measure (Thompson, 2009) to construct 3D configuration of the lower-lying prominence, and we use the codes integrated into the Writhe Application Toolkit (WAT[3]) to calculate writhe. We find that the writhe of the lower-lying prominence increases to around -0.49 ± 0.27. The uncertainty comes from the uncertainty of locations of feature points (as those marked by blue diamonds in Figure 6). The writhe of the upper threads disappears after 13:31 UT (the structures marked by cyan dashed lines in Figure 7, (a) and (b), disappear). After the

[3] http://sysu-pearl.cn:8080/writhecode.

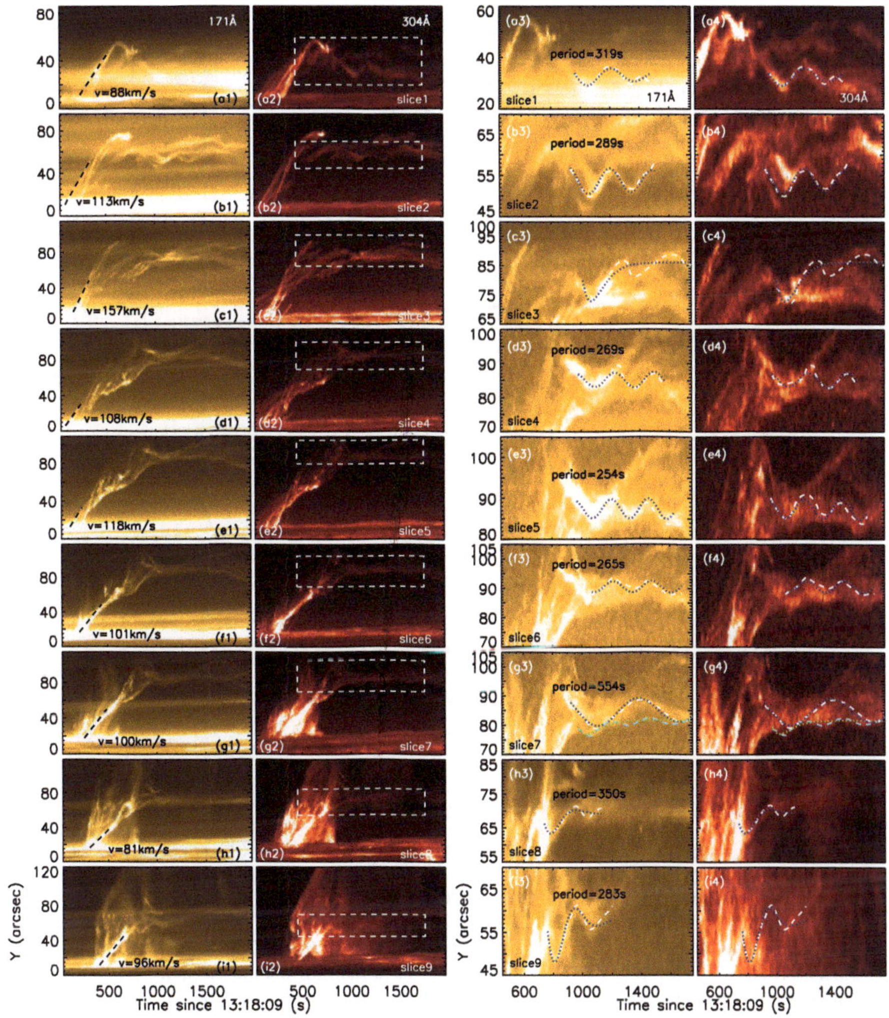

Figure 5 The slice–time plots along slices 1 – 9 marked in Figure 1 (f). **(a1)–(i1)** The slice–time plots in AIA 171 Å passband. The black dashed lines show the slopes that we use to derive the speeds of the rising prominence; **(a2)–(i2)** The slice–time plots in AIA 304 Å passband. The white dashed rectangle marks the region shown in (a3)–(i3) and (a4)–(i4); **(a3)–(i3)** The slice–time plots in AIA 171 Å passband in the region marked in (a2)–(i2). The white and cyan dashed lines outline trajectories of the oscillations and the blue and green dashed lines show their fitting results with an exponentially decaying sine function; **(a4)–(i4)** The slice–time plots in AIA 304 Å passband in the region marked in (a2)–(i2). The white and cyan dashed lines outline trajectories of the oscillations and the blue and green dashed lines show their fitting results with an exponentially decaying sine function.

disappearance of the writhe, we observe that the upper threads connect to the solar surface directly (see red dashed lines in Figure 7, (b) and (c)). During the disappearance of the writhe, the lower-lying prominence expands, which is consistent with lateral expansion of prominence when it reconnects with overlying fields (e.g., Netzel et al., 2012; Hassanin and Kliem, 2016; Joshi et al., 2022). We also observe a brightening (indicated by white arrows in Figure 7(d)–(f)) propagating outward along the overlying prominence. Besides, after the

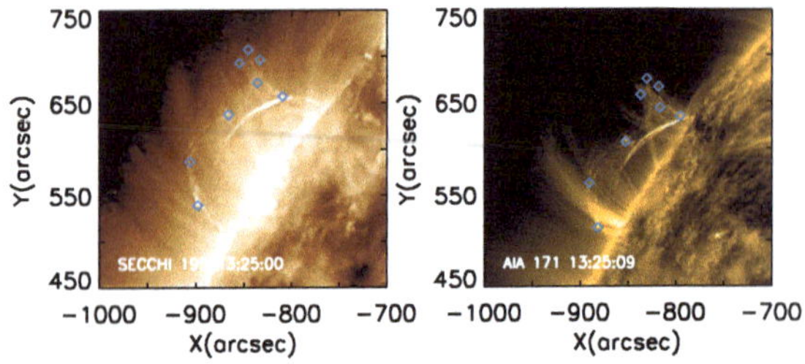

Figure 6 SECCHI 195 Å and AIA 171 Å snapshots of the region of interest viewed at 13:25:00 UT and 13:25:09 UT, respectively. The blue diamonds mark feature points used to calculate the writhe.

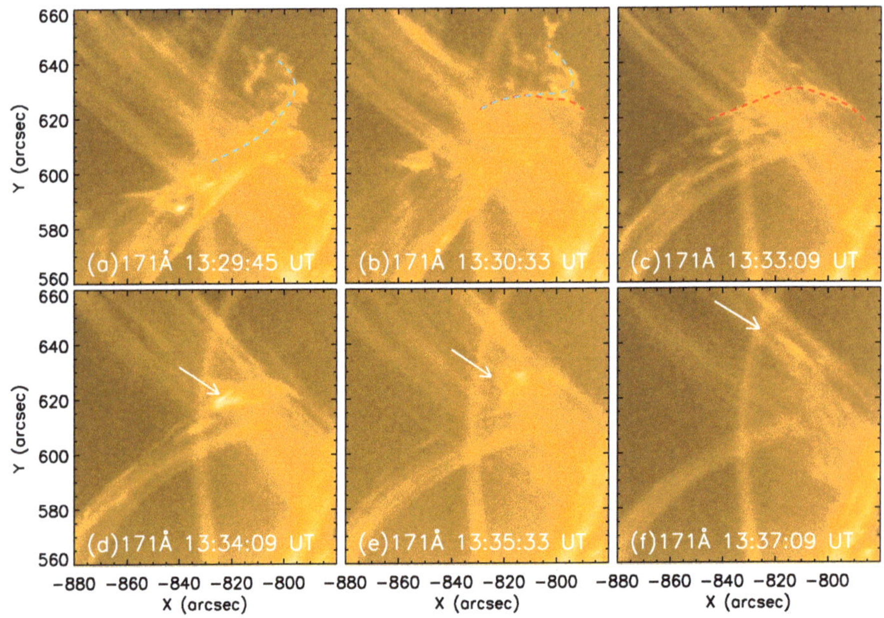

Figure 7 Snapshots in AIA 171 Å passband at 13:29:45 UT, 13:30:33 UT, 13:33:09 UT, 13:34:09 UT, 13:35:33 UT, and 13:37:09 UT. The cyan dashed lines in (a) and (b) outline the trajectory of threads in the lower-lying prominence with writhe. The red dashed lines in (b) and (c) outline trajectory of threads in the lower-lying prominence without writhe. The white arrows in (d)–(f) point at a brightening moving along the overlying prominence. An associated animation can be accessed online.

disappearance of the writhe, the overlying prominence starts to rotate and show shearing motions while its legs separate. These observations suggest that the disappearance of the writhe might be caused by reconnection between the lower-lying and overlying prominences, and the writhe is then transferred to the overlying prominence. The unwinding of the overlying prominence might be the result of writhe release, which is suggested to be an important factor resulting in the failure of a prominence eruption (Zhou et al., 2019). Multiperspective satellites or combining observations from multiple satellites observing from different perspectives should effectively contribute to analyses of this kind of prominence eruptions.

After that the lower-lying prominence reaches its maximum height, and we observe prominent kink oscillations (see Figure 5). These oscillations are similar to that reported by Kumar et al. (2022). In that event, the oscillation starts after the erupting filament reaches overlying loops, and is interpreted as the fundamental standing kink mode of the filament. In our case, as the oscillations form after the threads reach their maximum heights, the magnetic tension and gravity might play an important role in restoring forces. When the threads rise or fall about their position of equilibrium, the threads will be pulled back, which leads to oscillations. The similar amplitudes and periods of oscillations indicate their same origin, and the physical parameters of different threads should be similar because the periods of oscillations are highly sensitive to the physical parameters of the threads (Bashkirtsev and Mashnich, 1984; Mashnich and Bashkirtsev, 1990; Oliver and Ballester, 2002). As a result, the different oscillation periods at slice 7 can be interpreted by different physical parameters in different threads, similar to those reported in loops and filaments with different periods by Li et al. (2023). In addition, we do not observe any apparent phase delay in these oscillations at different locations of the prominence. Therefore, they are not caused by a disturbance at a particular place, which will then propagate along the axis (Liu et al., 2009). The magnetic field strength can be estimated with the help of oscillations, which is related to density, length, width, local sound speed within the prominence and in its surroundings (Joarder and Roberts, 1993; Régnier, Solomon, and Vial, 2001; Lin et al., 2009). In this event, the ratio of length and width of the prominence is around 15.5. The electron density of a prominence is around 10^{10} cm^{-3} and the sound speed is in the order of 100 km s^{-1}. With these parameters, we find a magnetic field strength of several gauss in the prominence while we assume the oscillations are Alfvén mode or sausage mode using the following equation given by Régnier, Solomon, and Vial (2001):

$$B_0^2(\rho_o) = \mu_0 \rho_o c_{so}^2 (\frac{1}{\eta^2} - 1),$$

where B_0, μ_0, ρ_o, c_{so}, η are the magnetic field strength, permeability of vacuum, density, sound speed in the prominence, and the ratio of length and width. This estimation of magnetic field strength here is in agreement with previous studies (e.g., Hirayama, 1985; Joarder and Roberts, 1993; Lin et al., 2009).

During the eruption, we observe a brightening underneath the middle of the lower-lying prominence with a temperature of 4.5 MK. This brightening only appears in passbands representing high temperature such as AIA 94 Å, but we do not find any direct connectivity between this brightening and the prominence. We suspect that this high-temperature brightening might be caused by reconnection below the prominence, similarly to the classic theory of CMEs and flares (e.g., Gosling, 1993; Antiochos, DeVore, and Klimchuk, 1999; Lin and Forbes, 2000; Chen, 2011).

5. Summary

Using observations from ASO-S and SDO, we analyze an active prominence system with two crossing prominences at the north-eastern limb. The two prominences rise with a speed of around 100 km s^{-1}. During the rising process, the upper threads of the lower-lying prominence experience an apparent writhing motion. The writhe of lower-lying prominence first increases and then disappears, which is thought to be possibly transferred to the overlying prominence through reconnection between the two prominences. After the disappearance of writhe, the overlying prominence starts clear rotations and rotates three quarters of a full turn, which leads to separating (39 km s^{-1}) – joining (67 km s^{-1}) – separating (10 km s^{-1}) motion of its legs, most clearly seen in Lyman-α observations projected on the plane of sky. The writhe transferred to the overlying prominence might be released via this rotation. The lower-lying prominence stops rising after the disappearance of its writhe, so the overlying prominence confines the eruption of the lower-lying prominence and the transfer and release of writhe can result in the failed prominence eruption.

In this event, we do not observe effective heating in the overlying prominence and middle part of the lower-lying prominence. The legs of the lower-lying prominence are heated to 1 – 9.3 MK, with some localized brightenings (see white arrows in Figure 4(c)), suggesting internal activities have occurred. In addition, we observe temperature increment under the middle part of the lower-lying prominence (see Figure 4, (b), (e), and (f)), which is heated to 4.5 MK. This heating can be interpreted by reconnection under the erupting prominence, which is proposed in the classic flare and CME theory (Gosling, 1993; Antiochos, DeVore, and Klimchuk, 1999; Lin and Forbes, 2000; Chen, 2011). In the late phase of this eruption, we also observe that materials in the lower-lying and overlying prominences fall back along their axes.

After the lower-lying prominence reaches its maximum height, its threads apparently start to oscillate. Most threads show identical oscillations in both AIA 171 and 304 Å passbands, except along slice 7. This exception can be interpreted by oscillations in threads with different physical parameters. Their amplitudes, periods and decay time scales are found to be around 1 – 10 Mm, 250 – 360 s, and 100 – 5000 s, respectively. Hence these oscillations are classified as short period kink oscillations. We think that the restoring forces of these oscillations are possibly the magnetic tension and gravity. These oscillations do not show any apparent phase delay along the prominence axis, so we suggest that these oscillations do not originate from one particular place and propagate along the prominence.

Supplementary Information The online version contains supplementary material available at https://doi.org/10.1007/s11207-024-02306-x.

Acknowledgements We are grateful to the anonymous referee for very helpful suggestions and comments. We would like to thank Dr. Yuandeng Shen for valuable discussions. This work is supported by National Natural Science Foundation of China (42230203, 42174201, 12373065, 42074208), the National Key R&D Program of China No. 2021YFA0718600 and the Strategic Priority Research Program of the Chinese Academy of Sciences (XDB0560000). ASO-S mission is supported by the Strategic Priority Research Program on Space Science, the Chinese Academy of Sciences, Grant No. XDA15320000. The AIA data are used by courtesy of NASA/SDO, the AIA and HMI teams and JSOC.

Author contributions Z.H. planed and led the study; H.W. carried out the data analyses and wrote the manuscript. All authors contributed on the discussion and reviewed the manuscript.

Data Availability No datasets were generated or analysed during the current study.

Declarations

Competing interests The authors declare no competing interests.

References

Amari, T., Luciani, J.F.: 1999, Confined disruption of a three-dimensional twisted magnetic flux tube. *Astrophys. J. Lett.* **515**, L81. DOI. ADS.

Antiochos, S.K., DeVore, C.R., Klimchuk, J.A.: 1999, A model for solar coronal mass ejections. *Astrophys. J.* **510**, 485. DOI. ADS.

Archontis, V., Török, T.: 2008, Eruption of magnetic flux ropes during flux emergence. *Astron. Astrophys.* **492**, L35. DOI. ADS.

Bashkirtsev, V.S., Mashnich, G.P.: 1984, Oscillatory processes in prominences. *Solar Phys.* **91**, 93. DOI. ADS.

Benz, A.O.: 2017, Flare observations. *Living Rev. Solar Phys.* **14**, 2. DOI. ADS.
Chen, P.F.: 2011, Coronal mass ejections: models and their observational basis. *Living Rev. Solar Phys.* **8**, 1. DOI. ADS.
Cheung, M.C.M., Boerner, P., Schrijver, C.J., Testa, P., Chen, F., Peter, H., Malanushenko, A.: 2015, Thermal diagnostics with the atmospheric imaging assembly on board the solar dynamics observatory: a validated method for differential emission measure inversions. *Astrophys. J.* **807**, 143. DOI. ADS.
Dai, J., Ji, H., Li, L., Zhang, J., Chen, H.: 2021, The formation and eruption of a sigmoidal filament driven by rotating network magnetic fields. *Astrophys. J.* **906**, 66. DOI. ADS.
Dai, J., Zhang, Q., Qiu, Y., Li, C., Li, Z., Li, S., Su, Y., Ji, H.: 2023, Simultaneous horizontal and vertical oscillation of a quiescent filament observed by CHASE and SDO. *Astrophys. J.* **959**, 71. DOI. ADS.
Ding, T., Zhang, J., Hong, J.: 2022, Are the magnetic field directions of surrounding loops a key parameter for confining a solar filament eruption? *Astrophys. J. Lett.* **933**, L38. DOI. ADS.
Feynman, J., Martin, S.F.: 1995, The initiation of coronal mass ejections by newly emerging magnetic flux. *J. Geophys. Res.* **100**, 3355. DOI. ADS.
Filippov, B.: 2020, Failed prominence eruptions near 24 cycle maximum. *Mon. Not. Roy. Astron. Soc.* **494**, 2166. DOI. ADS.
Filippov, B.: 2021a, Critical decay index for eruptions of 'short' filaments. *Mon. Not. Roy. Astron. Soc.* **503**, 3926. DOI. ADS.
Filippov, B.P.: 2021b, Low values of the critical decay index of the magnetic field in prominence eruptions. *Geomagn. Aeron.* **61**, S16. DOI. ADS.
Filippov, B.: 2022, Dependence of the eruptive filaments dynamics on their length. *Mon. Not. Roy. Astron. Soc.* **509**, 5713. DOI. ADS.
Gan, W.-Q., Zhu, C., Deng, Y.-Y., Li, H., Su, Y., Zhang, H.-Y., Chen, B., Zhang, Z., Wu, J., Deng, L., Huang, Y., Yang, J.-F., Cui, J.-J., Chang, J., Wang, C., Wu, J., Yin, Z.-S., Chen, W., Fang, C., Yan, Y.-H., Lin, J., Xiong, W.-M., Chen, B., Bao, H.-C., Cao, C.-X., Bai, Y.-P., Wang, T., Chen, B.-L., Li, X.-Y., Zhang, Y., Feng, L., Su, J.-T., Li, Y., Chen, W., Li, Y.-P., Su, Y.-N., Wu, H.-Y., Gu, M., Huang, L., Tang, X.-J.: 2019, Advanced Space-based Solar Observatory (ASO-S): an overview. *Res. Astron. Astrophys.* **19**, 156. DOI. ADS.
Gan, W., Zhu, C., Deng, Y., Zhang, Z., Chen, B., Huang, Y., Deng, L., Wu, H., Zhang, H., Li, H., Su, Y., Su, J., Feng, L., Wu, J., Cui, J., Wang, C., Chang, J., Yin, Z., Xiong, W., Chen, B., Yang, J., Li, F., Lin, J., Hou, J., Bai, X., Chen, D., Zhang, Y., Hu, Y., Liang, Y., Wang, J., Song, K., Guo, Q., He, L., Zhang, G., Wang, P., Bao, H., Cao, C., Bai, Y., Chen, B., He, T., Li, X., Zhang, Y., Liao, X., Jiang, H., Li, Y., Su, Y., Lei, S., Chen, W., Li, Y., Zhao, J., Li, J., Ge, Y., Zou, Z., Hu, T., Su, M., Ji, H., Gu, M., Zheng, Y., Xu, D., Wang, X.: 2023, The Advanced Space-Based Solar Observatory (ASO-S). *Solar Phys.* **298**, 68. DOI. ADS.
Gilbert, H.R., Alexander, D., Liu, R.: 2007, Filament kinking and its implications for eruption and re-formation. *Solar Phys.* **245**, 287. DOI. ADS.
Gosling, J.T.: 1993, The solar flare myth. *J. Geophys. Res.* **98**, 18937. DOI. ADS.
Guo, Y., Zhong, Z., Ding, M.D., Chen, P.F., Xia, C., Keppens, R.: 2021, Data-constrained magnetohydrodynamic simulation of a long-duration eruptive flare. *Astrophys. J.* **919**, 39. DOI. ADS.
Hassanin, A., Kliem, B.: 2016, Helical kink instability in a confined solar eruption. *Astrophys. J.* **832**, 106. DOI. ADS.
Hassanin, A., Kliem, B., Seehafer, N., Török, T.: 2022, A model of homologous confined and ejective eruptions involving kink instability and flux cancellation. *Astrophys. J. Lett.* **929**, L23. DOI. ADS.
Hirayama, T.: 1985, Modern observations of solar prominences. *Solar Phys.* **100**, 415. DOI. ADS.
Hou, Y.J., Li, T., Song, Z.P., Zhang, J.: 2020, External reconnection and resultant reconfiguration of overlying magnetic fields during sympathetic eruptions of two filaments. *Astron. Astrophys.* **640**, A101. DOI. ADS.
Howard, R.A., Moses, J.D., Vourlidas, A., Newmark, J.S., Socker, D.G., Plunkett, S.P., Korendyke, C.M., Cook, J.W., Hurley, A., Davila, J.M., Thompson, W.T., St Cyr, O.C., Mentzell, E., Mehalick, K., Lemen, J.R., Wuelser, J.P., Duncan, D.W., Tarbell, T.D., Wolfson, C.J., Moore, A., Harrison, R.A., Waltham, N.R., Lang, J., Davis, C.J., Eyles, C.J., Mapson-Menard, H., Simnett, G.M., Halain, J.P., Defise, J.M., Mazy, E., Rochus, P., Mercier, R., Ravet, M.F., Delmotte, F., Auchere, F., Delaboudiniere, J.P., Bothmer, V., Deutsch, W., Wang, D., Rich, N., Cooper, S., Stephens, V., Maahs, G., Baugh, R., McMullin, D., Carter, T. (eds.): 2008, Sun Earth Connection Coronal and Heliospheric Investigation (SECCHI). *Space Sci. Rev.* **136**, 67. DOI. ADS.
Huang, Z., Madjarska, M.S., Koleva, K., Doyle, J.G., Duchlev, P., Dechev, M., Reardon, K.: 2014, Hα spectroscopy and multiwavelength imaging of a solar flare caused by filament eruption. *Astron. Astrophys.* **566**, A148. DOI. ADS.
Jenkins, J.M., Hopwood, M., Démoulin, P., Valori, G., Aulanier, G., Long, D.M., van Driel-Gesztelyi, L.: 2019, Modeling the effect of mass-draining on prominence eruptions. *Astrophys. J.* **873**, 49. DOI. ADS.

Ji, H., Wang, H., Schmahl, E.J., Moon, Y.-J., Jiang, Y.: 2003, Observations of the failed eruption of a filament. *Astrophys. J. Lett.* **595**, L135. DOI. ADS.
Jiang, C., Duan, A., Zou, P., Zhou, Z., Bian, X., Feng, X., Zuo, P., Wang, Y.: 2023, A model of failed solar eruption initiated and destructed by magnetic reconnection. *Mon. Not. Roy. Astron. Soc.* **525**, 5857. DOI. ADS.
Jing, J., Lee, J., Spirock, T.J., Xu, Y., Wang, H., Choe, G.S.: 2003, Periodic motion along a solar filament initiated by a subflare. *Astrophys. J. Lett.* **584**, L103. DOI. ADS.
Jing, J., Lee, J., Spirock, T.J., Wang, H.: 2006, Periodic motion along solar filaments. *Solar Phys.* **236**, 97. DOI. ADS.
Joarder, P.S., Roberts, B.: 1993, The modes of oscillation of a prominence. III. The slab in a skewed magnetic field. *Astron. Astrophys.* **277**, 225. ADS.
Joshi, R., Mandrini, C.H., Chandra, R., Schmieder, B., Cristiani, G.D., Mac Cormack, C., Démoulin, P., Cremades, H.: 2022, Analysis of the evolution of a multi-ribbon flare and failed filament eruption. *Solar Phys.* **297**, 81. DOI. ADS.
Kaiser, M.L., Kucera, T.A., Davila, J.M., St. Cyr, O.C., Guhathakurta, M., Christian, E.: 2008, The STEREO mission: an introduction. *Space Sci. Rev.* **136**, 5. DOI. ADS.
Kang, K., Guo, Y., Li, Y., Wang, J., Lin, J.: 2023, Breaking of large-scale filament due to magnetic reconnection and consequent partial eruption. *Res. Astron. Astrophys.* **23**, 095018. DOI. ADS.
Kumar, P., Nakariakov, V.M., Karpen, J.T., Richard DeVore, C., Cho, K.-S.: 2022, Kink oscillation of a flux rope during a failed solar eruption. *Astrophys. J. Lett.* **932**, L9. DOI. ADS.
Leake, J.E., Linton, M.G., Antiochos, S.K.: 2022, The role of reconnection in the onset of solar eruptions. *Astrophys. J.* **934**, 10. DOI. ADS.
Lemen, J.R., Title, A.M., Akin, D.J., Boerner, P.F., Chou, C., Drake, J.F., Duncan, D.W., Edwards, C.G., Friedlaender, F.M., Heyman, G.F., Hurlburt, N.E., Katz, N.L., Kushner, G.D., Levay, M., Lindgren, R.W., Mathur, D.P., McFeaters, E.L., Mitchell, S., Rehse, R.A., Schrijver, C.J., Springer, L.A., Stern, R.A., Tarbell, T.D., Wuelser, J.-P., Wolfson, C.J., Yanari, C., Bookbinder, J.A., Cheimets, P.N., Caldwell, D., Deluca, E.E., Gates, R., Golub, L., Park, S., Podgorski, W.A., Bush, R.I., Scherrer, P.H., Gummin, M.A., Smith, P., Auker, G., Jerram, P., Pool, P., Soufli, R., Windt, D.L., Beardsley, S., Clapp, M., Lang, J., Waltham, N.: 2012, The Atmospheric Imaging Assembly (AIA) on the Solar Dynamics Observatory (SDO). *Solar Phys.* **275**, 17. DOI. ADS.
Li, H.: 2016, The Lyman-α Solar Telescope (LST) for the ASO-S mission. In: Kosovichev, A.G., Hawley, S.L., Heinzel, P. (eds.) *Solar and Stellar Flares and Their Effects on Planets 320*, 436. DOI. ADS.
Li, H., Chen, B., Feng, L., Li, Y., Huang, Y., Li, J.-W., Lu, L., Xue, J.-C., Ying, B.-L., Zhao, J., Yang, Y.-T., Gan, W.-Q., Fang, C., Song, K.-F., Wang, H., Guo, Q.-F., He, L.-P., Zhu, B., Zhu, C., Deng, L., Bao, H.-C., Cao, C.-X., Yang, Z.-G.: 2019, The Lyman-alpha Solar Telescope (LST) for the ASO-S mission — I. Scientific objectives and overview. *Res. Astron. Astrophys.* **19**, 158. DOI. ADS.
Li, D., Hou, Z., Bai, X., Li, C., Fang, M., Zhao, H., Wang, J., Ning, Z.: 2023, Simultaneous detection of flare-associated kink oscillations and extreme-ultraviolet waves. ArXiv e-prints. DOI. ADS. arXiv.
Lin, J., Forbes, T.G.: 2000, Effects of reconnection on the coronal mass ejection process. *J. Geophys. Res.* **105**, 2375. DOI. ADS.
Lin, Y., Soler, R., Engvold, O., Ballester, J.L., Langangen, Ø., Oliver, R., Rouppe van der Voort, L.H.M.: 2009, Swaying threads of a solar filament. *Astrophys. J.* **704**, 870. DOI. ADS.
Liu, R., Alexander, D., Gilbert, H.R.: 2007, Kink-induced catastrophe in a coronal eruption. *Astrophys. J.* **661**, 1260. DOI. ADS.
Liu, T., Su, Y.: 2021, Tether-cutting and overlying magnetic reconnections in an MHD simulation of prominence-cavity system. *Astrophys. J.* **915**, 55. DOI. ADS.
Liu, W., Berger, T.E., Title, A.M., Tarbell, T.D.: 2009, An intriguing chromospheric jet observed by hinode: fine structure kinematics and evidence of unwinding twists. *Astrophys. J. Lett.* **707**, L37. DOI. ADS.
Liu, L., Liu, J., Chen, J., Wang, Y., Wang, G., Zhou, Z., Cui, J.: 2021, The configuration and failed eruption of a complex magnetic flux rope above a δ sunspot region. *Astron. Astrophys.* **648**, A106. DOI. ADS.
Liu, Y., Su, Y., Liu, R., Chen, J., Liu, T., Ji, H.: 2022, Apparent footpoint rotation and writhe of double hot channels in a solar flare. *Astrophys. J.* **930**, 130. DOI. ADS.
Luna, M., Karpen, J.: 2012, Large-amplitude longitudinal oscillations in a solar filament. *Astrophys. J. Lett.* **750**, L1. DOI. ADS.
Mashnich, G.P., Bashkirtsev, V.S.: 1990, Observations of Doppler velocity oscillations of mass motion in a quiescent prominence during three consecutive days. *Astron. Astrophys.* **235**, 428. ADS.
Mazumder, R., Pant, V., Luna, M., Banerjee, D.: 2020, Simultaneous longitudinal and transverse oscillations in filament threads after a failed eruption. *Astron. Astrophys.* **633**, A12. DOI. ADS.
Molowny-Horas, R., Wiehr, E., Balthasar, H., Oliver, R., Ballester, J.L.: 1999, Prominence Doppler oscillations. In: *JOSO Annu. Rep., 1998*, 126. ADS.

Monga, A., Sharma, R., Liu, J., Cid, C., Uddin, W., Chandra, R., Erdélyi, R.: 2021, On the partial eruption of a bifurcated solar filament structure. *Mon. Not. Roy. Astron. Soc.* **500**, 684. DOI. ADS.

Netzel, A., Mrozek, T., Kołomański, S., Gburek, S.: 2012, Extreme-ultraviolet and hard X-ray signatures of electron acceleration during the failed eruption of a filament. *Astron. Astrophys.* **548**, A89. DOI. ADS.

Oliver, R., Ballester, J.L.: 2002, Oscillations in quiescent solar prominences observations and theory (invited review). *Solar Phys.* **206**, 45. DOI. ADS.

Parenti, S.: 2014, Solar prominences: observations. *Living Rev. Solar Phys.* **11**, 1. DOI. ADS.

Pesnell, W.D., Thompson, B.J., Chamberlin, P.C.: 2012, The Solar Dynamics Observatory (SDO). *Solar Phys.* **275**, 3. DOI. ADS.

Prior, C.B., Neukirch, S.: 2016, The extended polar writhe: a tool for open curves mechanics. *J. Phys. A, Math. Gen.* **49**, 215201. DOI. ADS.

Ramsey, H.E., Smith, S.F.: 1966, Flare-initiated filament oscillations. *Astron. J.* **71**, 197. DOI. ADS.

Rees-Crockford, T., Bloomfield, D.S., Scullion, E., Park, S.-H.: 2020, 2D and 3D analysis of a torus-unstable quiet-sun prominence eruption. *Astrophys. J.* **897**, 35. DOI. ADS.

Régnier, S., Solomon, J., Vial, J.C.: 2001, Oscillations in an active region filament: observations and comparison with MHD waves. *Astron. Astrophys.* **376**, 292. DOI. ADS.

Shen, Y.-D., Liu, Y., Liu, R.: 2011, A time series of filament eruptions observed by three eyes from space: from failed to successful eruptions. *Res. Astron. Astrophys.* **11**, 594. DOI. ADS.

Shen, Y., Liu, Y., Su, J.: 2012, Sympathetic partial and full filament eruptions observed in one solar breakout event. *Astrophys. J.* **750**, 12. DOI. ADS.

Shen, Y., Liu, D., Yao, S., Zhou, C., Tang, Z., Qu, Z., Zhou, X., Duan, Y., Tan, S., Ibrahim, A.A.: 2024, Double-decker pair of flux ropes formed by two successive tether-cutting Eruptions. ArXiv e-prints. DOI. ADS. arXiv.

Shibata, K., Magara, T.: 2011, Solar flares: magnetohydrodynamic processes. *Living Rev. Solar Phys.* **8**, 6. DOI. ADS.

Su, Y., Veronig, A.M., Hannah, I.G., Cheung, M.C.M., Dennis, B.R., Holman, G.D., Gan, W., Li, Y.: 2018, Determination of differential emission measure from solar extreme ultraviolet images. *Astrophys. J. Lett.* **856**, L17. DOI. ADS.

Thompson, W.T.: 2009, 3D triangulation of a Sun-grazing comet. *Icarus* **200**, 351. DOI. ADS.

Török, T., Kliem, B.: 2005, Confined and ejective eruptions of kink-unstable flux ropes. *Astrophys. J. Lett.* **630**, L97. DOI. ADS.

Vashalomidze, Z., Zaqarashvili, T.V., Kukhianidze, V., Ramishvili, G., Hanslmeier, A., Gömöry, P.: 2022, Prominence instability and CMEs triggered by massive coronal rain in the solar atmosphere. *Astron. Astrophys.* **658**, A18. DOI. ADS.

Vršnak, B., Veronig, A.M., Thalmann, J.K., Žic, T.: 2007, Large amplitude oscillatory motion along a solar filament. *Astron. Astrophys.* **471**, 295. DOI. ADS.

Wei, H., Huang, Z., Li, C., Hou, Z., Qiu, Y., Fu, H., Bai, X., Xia, L.: 2023, Internal activities in a solar filament and heating in its threads. *Astrophys. J.* **958**, 116. DOI. ADS.

Xu, H., Su, J., Chen, J., Ruan, G., Awasthi, A.K., Zhang, H., Zhang, M., Ji, K., Zhang, Y., Liu, J.: 2020, Multiwavelength observation of a failed eruption from a helical kink-unstable prominence. *Astrophys. J.* **901**, 121. DOI. ADS.

Zhang, Q., Liu, R., Wang, Y., Li, X., Lyu, S.: 2021a, Confined and eruptive catastrophes of solar magnetic flux ropes caused by mass loading and unloading. *Astrophys. J.* **921**, 172. DOI. ADS.

Zhang, Y., Tan, B., Tan, C., Huang, J., Yan, Y.: 2021b, Multi-wavelength observations of a failed filament eruption and associated hovered coronal mass ejection. *Universe* **7**, 405. DOI. ADS.

Zhang, Y., Zhang, Q., Dai, J., Li, D., Ji, H.: 2022, Multiwavelength observations of a partial filament eruption on 13 June 2011. *Solar Phys.* **297**, 138. DOI. ADS.

Zheng, R., Yang, S., Rao, C., Liu, Y., Zhong, L., Wang, B., Song, H., Li, Z., Chen, Y.: 2019, A confined partial eruption of double-decker filaments. *Astrophys. J.* **875**, 71. DOI. ADS.

Zhou, Y., Ji, H., Zhang, Q.: 2023, Rapid rotation of an erupting prominence and the associated coronal mass ejection on 13 may 2013. *Solar Phys.* **298**, 35. DOI. ADS.

Zhou, Z., Zhang, J., Wang, Y., Liu, R., Chintzoglou, G.: 2017, Toward understanding the 3D structure and evolution of magnetic flux ropes in an extremely long duration eruptive flare. *Astrophys. J.* **851**, 133. DOI. ADS.

Zhou, Z., Cheng, X., Zhang, J., Wang, Y., Wang, D., Liu, L., Zhuang, B., Cui, J.: 2019, Why do torus-unstable solar filaments experience failed eruptions? *Astrophys. J. Lett.* **877**, L28. DOI. ADS.

Zhou, Z., Jiang, C., Liu, R., Wang, Y., Liu, L., Cui, J.: 2022, The rotation of magnetic flux ropes formed during solar eruption. *Astrophys. J. Lett.* **927**, L14. DOI. ADS.

Zhou, Z., Jiang, C., Song, H., Wang, Y., Hao, Y., Cui, J.: 2023, Quantification of the writhe number of the evolution of solar filament axes. *Astrophys. J.* **944**, 175. DOI. ADS.

Solar Physics (2024) 299:89
https://doi.org/10.1007/s11207-024-02336-5

RESEARCH

Association Between a Failed Prominence Eruption and the Drainage of Mass from Another Prominence

Jian-chao Xue[1] · Li Feng[1,2] · Hui Li[1,2] · Ping Zhang[1] · Jun Chen[1] · Guang-lu Shi[1,2] · Kai-fan Ji[3] · Ye Qiu[4] · Chuan Li[4,5] · Lei Lu[1] · Bei-li Ying[1] · Ying Li[1,2] · Yu Huang[1,2] · You-ping Li[1] · Jing-wei Li[1] · Jie Zhao[1] · De-chao Song[1] · Shu-ting Li[1,2] · Zheng-yuan Tian[1,2] · Ying-na Su[1,2] · Qing-min Zhang[1] · Yun-yi Ge[1] · Jia-hui Shan[1,2] · Qiao Li[1,2] · Gen Li[1,2] · Yue Zhou[1,2] · Jun Tian[1,2] · Xiao-feng Liu[1,2] · Zhi-chen Jing[1,2] · Bo Chen[6] · Ke-fei Song[6] · Ling-ping He[6] · Shi-jun Lei[1] · Wei-qun Gan[1,7]

Received: 26 February 2024 / Accepted: 6 June 2024 / Published online: 20 June 2024

Abstract
Sympathetic eruptions of solar prominences have been studied for decades, however, it is usually difficult to identify their causal links. Here, we present two failed prominence eruptions on 26 October 2022 and explore their connections. Using stereoscopic observations, the South prominence (PRO-S) erupts with untwisting motions, flare ribbons occur underneath, and new connections are formed during the eruption. The North prominence (PRO-N) rises up along with PRO-S, and its upper part disappears due to catastrophic mass draining along an elongated structure after PRO-S failed eruption. We suggest that the eruption of PRO-S initiates due to a kink instability, and fails to erupt due to reconnection with surrounding fields. The elongated structure connecting PRO-N overlies PRO-S, which causes the rising up of PRO-N along with PRO-S and mass drainage after PRO-S eruption. This study suggests that a prominence may end its life through mass drainage forced by an eruption underneath.

Keywords Magnetic reconnection · Observational signatures · Prominences · Active · Flares · Models

1. Introduction

Solar prominences, or filaments when they are seen on the disk, are dense structures extending into the solar corona (Vial and Engvold, 2015). Cold and dense prominence mass is generally thought to be suspended by the magnetic-tension force in magnetic dips (Kippenhahn and Schlüter, 1957; Kuperus and Raadu, 1974). Based on the statistics of Ouyang et al. (2017), 11% of filaments have a magnetic flux-rope configuration, while others appear as sheared arcades. A magnetic flux rope refers to a set of magnetic-field lines winding around a common axis more than once (Cheng, Guo, and Ding, 2017). They may suffer from ideal magnetohydrodynamic (MHD) instabilities, mainly a kink instability (Hood and Priest, 1981) and a torus instability (Kliem and Török, 2006). These instabilities cause eruptions accompanied by large solar flares and coronal mass ejections (CMEs). For a uniformly

Extended author information available on the last page of the article

twisted flux rope, the kink instability sets in if the twist number exceeds 1.25 (Hood and Priest, 1981). The torus instability sets in if the external poloidal field decreases fast enough, which is evaluated by a decay index. In theory, the critical value of this index for a circular and a straight flux rope is 1.5 and 1, respectively (Démoulin and Aulanier, 2010).

Solar filament eruptions are generally classified into full eruptions, partial eruptions, and failed or confined eruptions (Gilbert, Alexander, and Liu, 2007). The former two types develop into CMEs with significant numbers of high-energy particles and amounts of magnetic plasma ejected into the interplanetary space. However, for a failed eruption (Ji et al., 2003), neither the prominence mass nor the supporting magnetic structure escapes after an initial acceleration. Causes for the failed eruption of a flux rope can be classified into four types (Wang et al., 2023; Chen et al., 2023b): (1) not reaching the torus instability, where the eruption initiates due to a kink instability (failed-kink regime, Török and Kliem, 2005), or the decay index has a saddle-like profile (Guo et al., 2010); (2) failed-torus regime, where a torus-unstable flux rope fails to erupt due to a downward tension force induced by the external toroidal field (Myers et al., 2015; Wang et al., 2023) or a downward Lorentz force by a nonaxisymmetry of the flux rope (Zhong, Guo, and Ding, 2021; Zhang et al., 2023b); (3) magnetic reconnection-caused destruction of the flux rope in a breakout-like configuration, where a quadrupolar field is necessary (DeVore and Antiochos, 2008; Netzel et al., 2012; Chen et al., 2023a,b); (4) magnetic reconnection-caused destruction due to the flux rope writhing, the process may happen between the flux rope and the external field, or between the flux-rope legs, where the magnetic-field configuration could be dipolar (Hassanin and Kliem, 2016; Yang et al., 2019; Zhou et al., 2019; Jiang et al., 2023). In the regimes mentioned above, an eruption initiates in general, though not always, due to MHD instabilities and further accelerates with flare reconnection below the flux rope (Karpen, Antiochos, and DeVore, 2012; Wang et al., 2023). In the breakout configuration, the breakout reconnection initially promotes the eruption through removing the strapping field. However, when the magnetic reconnection between the external field and the flux rope starts, the flux rope is destroyed and the upward hoop force decreases, which results in the failure of the eruption (Chen et al., 2023a).

Some eruptions may trigger another eruption, or have a common origin. Solar sympathetic (homologous) eruptions refer to the eruptions that occur at different (the same) sites during a relatively short interval with a certain physical connection (Moon et al., 2003). However, for sympathetic eruptions, it is generally difficult to determine whether there is just a chance coincidence. Since solar eruptions in the corona are often dominated by the magnetic field, the causal links between the sympathetic eruptions are generally thought to be of a magnetic nature (Tang et al., 2021). Magnetic reconnection at separatrices, separators, and quasiseparatrix layers is the most common cause for sympathetic eruptions in several works (Liu et al., 2009; Török et al., 2011; Wang et al., 2018; Tang et al., 2021). Magnetic reconnection, along with the expansion of an erupting flux rope (Shen, Liu, and Su, 2012; Yang et al., 2023), mainly plays a role in eroding the overlying field that provides a confining force (Liu et al., 2009; Török et al., 2011; Cheng et al., 2013; Li et al., 2017), and exchanging magnetic flux between flux ropes (Wang et al., 2018). Other authors suggest that perturbations caused by surges, CMEs, and waves can also result in sympathetic eruptions (Wang et al., 2001; Jiang et al., 2011; Dai et al., 2021).

In this work, we report two prominence eruptions on 26 October 2022, where a failed eruption is followed by the catastrophic drainage of mass of another prominence. We analyze the causes of the two eruptions and explore their associations. We introduce the observations in Section 2. Since this is the first light from the Lyman-alpha Solar Telescope (LST, Li et al., 2019; Chen et al., 2019; Feng et al., 2019) on board the Advanced Space-based Solar

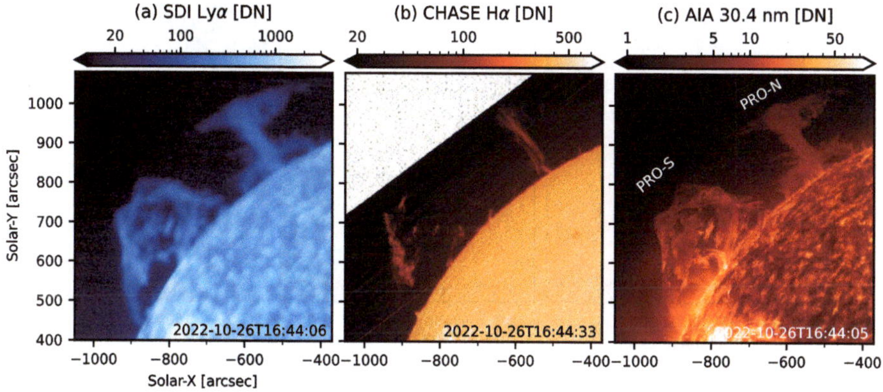

Figure 1 Prominence images of SDI Lyα, CHASE Hα, and AIA 30.4 nm. An animation of SDI, CHASE Hα line center, and Hα Doppler images from 16:41 to 19:00 UT is attached as supplementary material, see movie_a.mp4. For the animation of AIA 30.4, 17.1, and 19.3 nm observations were taken from 16:26 to 20:30 UT, see movie_b.mp4.

Observatory (ASO-S, Gan et al., 2023), emissions in H I Lyα, He II 30.4 nm, and H I Hα lines are compared briefly. In Section 3, the erupting processes of the two prominences are shown. Their eruption causes and association are discussed in Section 4.

2. Observations and Analysis

On 26 October 2022, two solar prominences on the northeastern limb erupted sequentially. Neither flare (by checking GOES soft X-ray light curve) nor associated CME (in the field-of-view of the SOHO/LASCO coronagraph) were found. They were observed by the ASO-S/LST during the commissioning phase. The LST consists of three instruments: a solar-disk imager (SDI) in H I Lyα (Lyα for short), a white-light solar telescope (WST) in the 360 ± 2 nm waveband, and a solar-corona imager (SCI) in both Lyα and white light (700 ± 32 nm). SCI started working in December 2022, hence SCI observations are not available. SDI has a pixel scale of 0.5 arcsec, however, the spatial resolution is worse than that of the Atmospheric Imaging Assembly (AIA, Lemen et al., 2012), i.e., 1.5 arcsec. The SDI cadence is 10 s during first light, and data between 15:44 UT and 16:41 UT are lost. SDI level-1 data are dark-field and flat-field corrected. They are aligned to AIA 30.4 nm using crosscorrelation and optical flow methods (Cai et al., 2022).

This event was also observed by the Chinese Hα Solar Explorer (CHASE, Li et al., 2022; Qiu et al., 2022) Hα Imaging Spectrograph (HIS). In raster scanning mode, HIS scans the solar disk in 1.2 min with a pixel size of 1.04 arcsec and spectral resolution of 0.024 Å (0.048 Å in binning mode for this observation). CHASE data are dark-field, slit-image-curvature, and flat-field corrected; wavelength calibration is conducted using two photospheric absorption lines in a quiet region around the solar-disk center. CHASE maps are coaligned with the Helioseismic and Magnetic Imager (HMI, Scherrer et al., 2012) continuum images using the scale-invariant feature transform (SIFT, Lowe, 2004; Ji et al., 2017). Figure 1 shows snapshots of the prominence eruptions in SDI Lyα, CHASE Hα, and AIA 30.4 nm. Although SDI has a lower spatial resolution than AIA 30.4 nm images, SDI has larger counts, hence a higher signal-to-noise ratio.

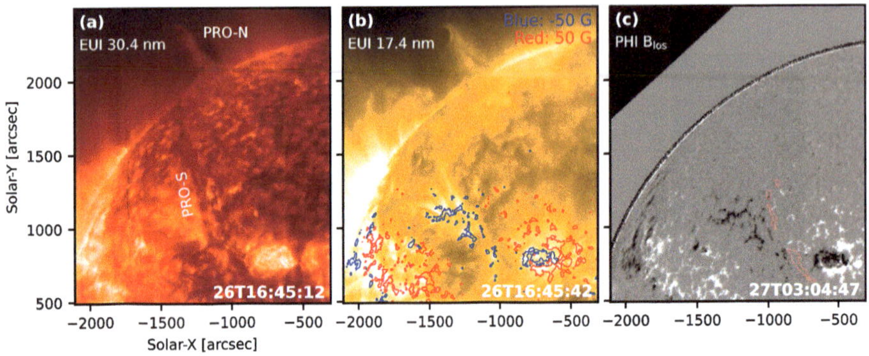

Figure 2 Maps in Solar Orbiter view. (a) and (b) EUI-FSI 30.4 and 17.4 nm maps at 16:45 UT on 26 October 2022. (c) PHI B_{los} at 3:04 UT on 27 October 2022. The blue and red contours in (b) represent B_{los} of -50 and 50 G, respectively. The contours in (c) mark the PRO-S location, which are extracted from EUI 30.4 nm at 2:00 UT on 26 October 2022. An animation of EUI 30.4 and 17.4 nm observations from 10:05 UT on 26 October to 4:25 UT on 27 October is available as supplementary material, see movie_c.mp4.

The Lyα lines of both H I at 121.6 nm and He II at 30.4 nm are among the brightest lines of the solar spectrum. On the basis of irradiance and imaging observations, it is found that their intensities have a close relationship (Auchère, 2005; Gordino et al., 2022). For the emission of 121.6 nm and 30.4 nm lines from prominences, resonant scattering of chromospheric radiation is an important contributor (Zhao et al., 2022). Compared with them, the H I Hα line at 656.28 nm is relatively optically thin and mainly contributed from the prominence core (Gouttebroze, Heinzel, and Vial, 1993). This is the main reason why the prominences in Figure 1 look thinner in Hα than those seen in Lyα and 30.4 nm images. Using a non-LTE radiative-transfer simulation, Gouttebroze, Heinzel, and Vial (1993) found that the Hα brightness is mainly determined by the prominence emission measure (EM $= \int n_e^2 \mathrm{d}z$, where n_e is the electron density and z is distance along the line-of-sight). Hence, the left (south) leg of the South prominence, PRO-S in abbreviation, should have a larger EM than its spine and right leg during its eruption. This brightness distribution suggests that the PRO-S eruption is asymmetric.

With respect to Earth view, the Solar Orbiter (García Marirrodriga et al., 2021) is at the eastern side with an angle of around 46°. The prominences observed by the Full Sun Imager (FSI) of the Extreme Ultraviolet Imager (EUI, Rochus et al., 2020; SolO/EUI Data Release 6.0 January, 2023) are shown in Figure 2, where the erupting PRO-S is seen as a solar filament, and the North prominence, PRO-N, is still above the solar limb. EUI-FSI 30.4 and 17.4 nm images have a pixel scale of 4.44 arcsec, corresponding to 1.32 Mm pixel^{-1} as the distance of Solar Orbiter from the Sun is ≈ 0.41 AU. EUI-FSI has a lower cadence of 10 or 20 min than AIA EUV images of 12 s. EUI 17.4 nm images are mainly contributed by the emission from Fe X and Fe IX, similar to AIA 17.1 nm channel dominated by Fe IX lines with characteristic emission temperature of 0.63 MK. There is no observation from the EUI High Resolution Imager (HRI). Observations of the Polarimetric and Helioseismic Imager (PHI, Solanki et al., 2020) related to this event started from 27 October 2022 at 3:04 UT. A snapshot of the PHI line-of-sight magnetic field (B_{los}) is shown in Figure 2c, where white (black) represents positive (negative) polarities, and the overlaid contours mark the PRO-S location at 2:00 UT on 26 October 2022 extracted from an EUI 30.4 nm image. PHI has a pixel scale of 3.57 arcsec, and the vector magnetogram from PHI is not available for this event. The difference in light travel time between Solar Orbiter and Earth-orbit telescopes

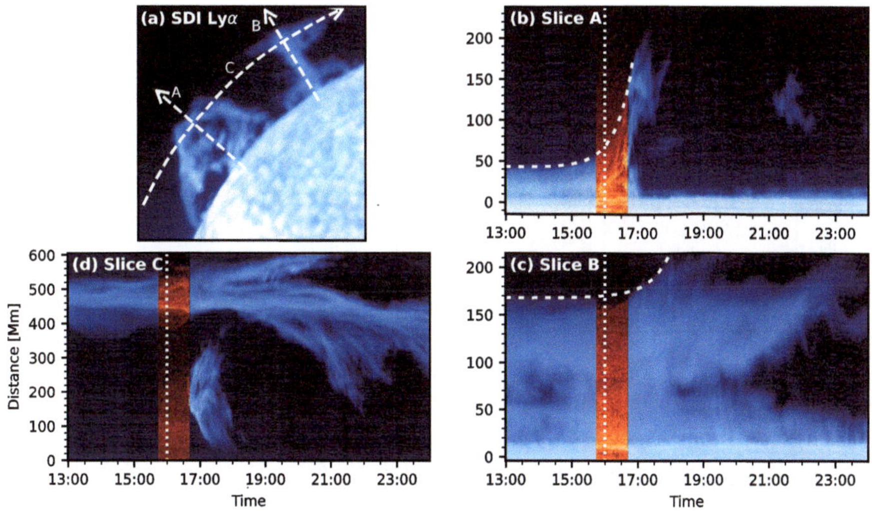

Figure 3 Time–distance diagrams of prominence eruptions. (a) SDI map with the location of artificial slices indicated, and arrows representing directions of the slices. (b)–(d) Time–distance diagrams along slices A–C, the blue parts are from SDI images and the orange parts are AIA 30.4 nm observations.

is corrected and the time at Earth is used in this article. Figure 2 shows that PRO-S is an intermediate filament and PRO-N is a polar-crown one.

3. Results

3.1. Time–Distance Diagrams of the Erupting Prominences

To trace the dynamical evolution of the prominences, time–distance diagrams are synthesized along three artificial slices (Figure 3). Among them, slices A and B are perpendicular to the solar limb and almost along the rising trajectories of PRO-S and PRO-N, respectively. Slice C is parallel to the solar limb. Considering that SDI images have a larger signal-to-noise ratio, we mainly used SDI observations and the missing parts are replaced by AIA 30.4 nm images (shown in blue and orange, respectively, in Figures 3b–d). The height variations of PRO-S and PRO-N, with respect to the solar limb, are fitted using a function composed of a linear part and an exponential part (Cheng et al., 2013, dashed lines in Figure 3b and c). The rising speeds of PRO-S and PRO-N are calculated to be around 70 and $45\,\mathrm{km\,s^{-1}}$, respectively, at the end of the fitting. The time 16:00 UT is marked with vertical dotted lines to compare the eruptions of the two prominences.

From Figures 3b and c, it is seen that PRO-S starts erupting before 16:00 UT and PRO-N rises up following PRO-S. Although no CME is found following their eruptions, PRO-S is fully destroyed but the bottom part of PRO-N survives. In Figure 3d, the lower (South) structure is PRO-S and the upper (North) one is PRO-N. During the PRO-S eruption, part of PRO-N mass flows northwards in the plane-of-sky (PoS); after PRO-S eruption, PRO-N mass flows oppositely. The motion of PRO-N is likely to be associated with the PRO-S eruption.

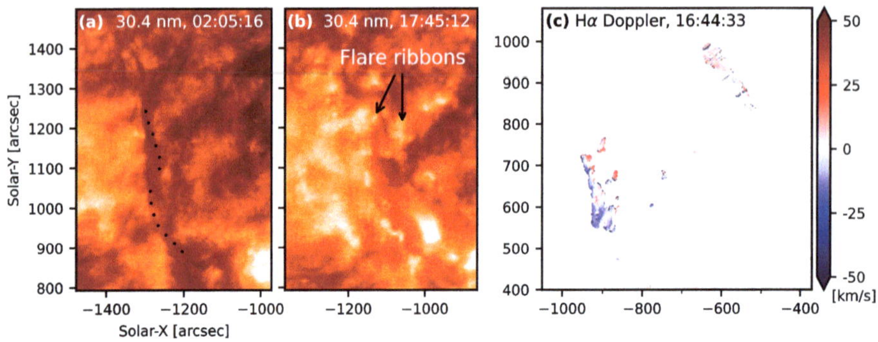

Figure 4 Clues of causes of PRO-S eruption. (a) and (b) EUI 30.4 nm snapshots on 26 October 2022 at 2:05 and 17:45 UT, respectively. (c) Hα Doppler map on 26 October 2022 at 16:44 UT.

3.2. Phenomena Related to PRO-S Eruption

As mentioned in Section 1, the initiation and failing of an eruption can be related to a kink instability, a torus instability, and reconnection below or/and around the flux rope, in addition to external triggers. We explore the eruption of PRO-S in relation to these mechanisms.

Figures 4a and b show snapshots of EUI 30.4 nm maps of PRO-S. Around 14 h before the PRO-S eruption (Figure 4a), a flux rope with a twist number of nearly 2 is seen (delineated with dotted curves), and the helicity is negative. Its southern leg roots in the periphery of an active region (AR), and the northern leg finishes in bifurcated ends. During the PRO-S eruption (Figure 4b), flare ribbons are seen underneath. Figure 4c shows a Doppler map from CHASE Hα observations during the PRO-S eruption. A single Gaussian fitting is used to derive the Doppler speeds for simplification (Xue et al., 2021), hence the derived speeds should be treated as an averaged result, or represent the motions of the main component if there are multipeaks in one Hα profile. Relative to the PRO-S leg axis, the left part is mainly blueshifted and the right part is redshifted with velocities within around $\pm 17\,\mathrm{km\,s^{-1}}$. Compared with the negative helicity shown in Figure 4a and assuming that the footpoint is tied, the Doppler map in Figure 4c indicates an untwisting motion.

The threshold of the torus instability is usually given by a critical decay index. The decay index n is defined by

$$n = -\frac{\mathrm{d}\ln B_{\mathrm{ep}}(R)}{\mathrm{d}\ln R}, \tag{1}$$

where R denotes the distance to the center of an approximately toroidal flux rope, and B_{ep} denotes the external poloidal field (Chen et al., 2023a). For simplification, we use height to replace R and B_{ep} is assumed to be nearly perpendicular to the polarity-inversion line (PIL) based on a potential-field approach. Through comparing B_{los} of PHI and HMI, the magnetic field around PRO-S does not vary significantly in the following days. Hence, the synoptic Carrington magnetogram of the normal component (B_{r}) from HMI is used to compute the potential field with the *pfsspy* package (Stansby, Yeates, and Badman, 2020). The extrapolated potential-field configuration is shown in Figure 5a over an AIA 30.4 nm map, where only closed field lines are plotted. The blue and cyan contours mark HMI B_{los} of -100 and $+100\,\mathrm{G}$, respectively. Although PRO-S erupts on 26 October, the filament channel is clearly seen (indicated by a black arrow, no corresponding filament is seen in the Hα image), which is enveloped by a series of loops. The South footpoint of PRO-S lies on a positive polarity,

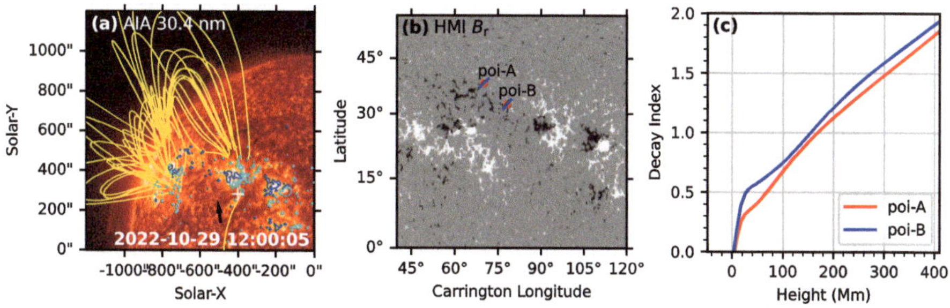

Figure 5 Potential-field configuration and decay index. (a) AIA 30.4 nm snapshots on 29 October 2022 at 12:00 UT and extrapolated magnetic-field configuration, where only the closed field is plotted as golden lines. Blue and cyan contours represent $B_{\rm los}$ of -100 and 100 G, respectively. (b) $B_{\rm r}$ Carrington map from HMI observations. (c) Decay index along height.

it can be seen from the overlaid $B_{\rm los}$ contours, and could also be derived from the fact that a negative-helicity structure locates along a dextral channel (Martin, Bilimoria, and Tracadas, 1994). The decay index is calculated at two points around the middle point of the projected PRO-S axis along the PIL, marked as poi-A and poi-B in Figure 5b, and its changes with height are shown in Figure 5c. From Figure 3b, the axis of PRO-S reaches a height of around 185 Mm before its disappearance, which should be the lower limit considering the projection effect. The decay index at 185 Mm is ≈ 1, reaching the critical value of the torus instability for a straight current channel, but smaller than 1.5 for a circular shape (Démoulin and Aulanier, 2010). Considering the errors of measurements in this work and the limitations in models, we cannot say whether PRO-S keeps torus stable during the rising process.

Signatures of magnetic reconnection between PRO-S and the ambient coronal loops are explored in Figure 6, which mainly shows sequential images of PRO-S failed eruption in AIA 30.4 nm. The first signature is the appearance of the new legs formed during the eruption. The two new legs are marked with magenta arrows in Figures 6b and d. They are filled with PRO-S dropping mass. The second signature is that the rising PRO-S is not regular and a "peak" is marked (green arrow) in Figure 6c, which is also bright in AIA 17.1 and 19.3 nm images, suggesting that the plasma is heated to coronal temperature. However, because the emission of the PRO-S eruption is relatively weak and no X-ray flare is detected by GOES, there is no significant brightening around PRO-S. A brightening in X-rays and EUV is usually used as an indication of magnetic reconnection (Ji et al., 2003; Netzel et al., 2012; Hassanin and Kliem, 2016; Chen et al., 2023b). In Figure 6d, the optical flows (marked with white arrows, calculated using two 30.4 nm images with an interval of 2 min) suggest that the PRO-S mass is mainly dropping along the legs and the increase in height almost stops.

3.3. Phenomena Related to PRO-N Eruption

The end of PRO-N is in the form of mass drainage after PRO-S failed eruption. Figure 7 shows the draining process in EUI 30.4 and 17.4 nm. Among the panels, the white arrows in panels a, b, e, and f point to the erupting PRO-S; the green arrows point to a cluster of mass that rises faster than the PRO-N main body, and stops around the horn-like structure of PRO-N; and the magenta arrows point to the mass draining along the horn. Horn-like structures refer to curved extensions that protrude from the top of quiescent prominences into the cavities (Su et al., 2015). In our case, the horn is clearly seen in 17.4 nm images, and the curvature reduces during the PRO-S rise. After PRO-S failed eruption, the flow along

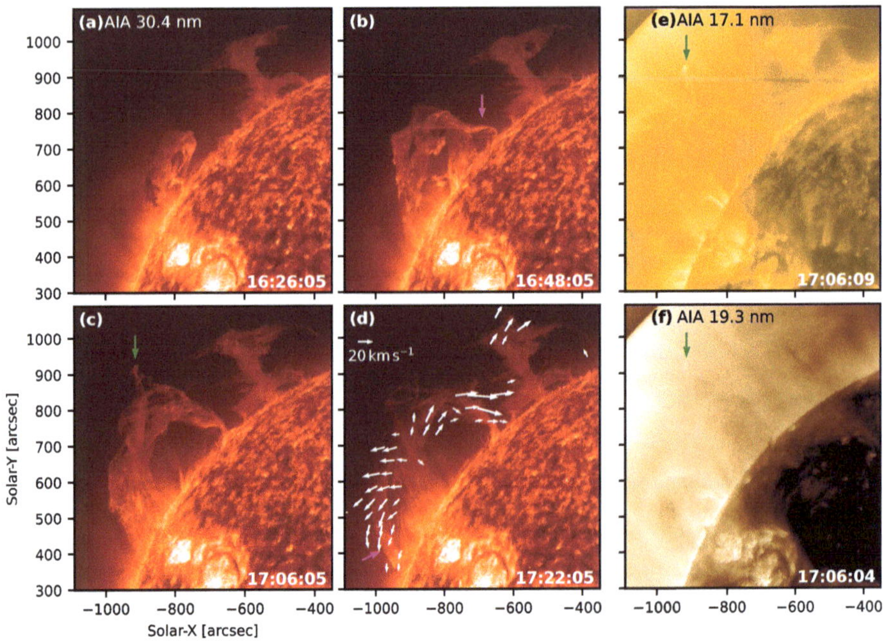

Figure 6 Process of the PRO-S failed eruption. (a)–(d) AIA 30.4 nm images. White arrows in (d) represent flow in the plane-of-sky. (e) AIA 17.1 nm image. (f) AIA 19.3 nm image. Observation times are marked at the lower-right corner in each panel.

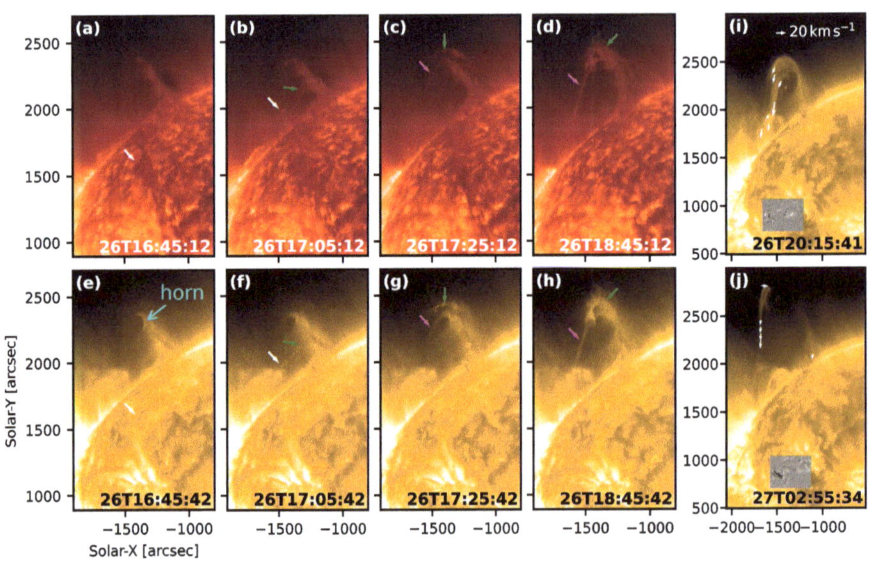

Figure 7 Time-sequence images of EUI 30.4 nm (a–d) and 17.4 nm (e–j) showing PRO-N mass drainage. In (i)–(j), the white arrows represent flows, and the gray insets are difference images.

the horn becomes significant. Figures 7i–j show violent drainage of PRO-N mass along the horn (white arrows mark the optical flow) and the brightening on the disk (difference image of the inset). Due to the projection effect, we cannot identify the other end of the flux tube that connects the PRO-N horn. However, the brightening on the disk has a good temporal correlation with the violent mass drainage (see movie_c.mp4). Hence, we suspect that the PRO-N horn extends southwards and overlies the filament channel where PRO-S is located before the eruption.

4. Discussion and Conclusion

Through the time–distance diagrams, we have seen that the motion of PRO-N is closely related to PRO-S eruption: PRO-N rises up along with PRO-S, and PRO-N mass moves northwards during PRO-S expansion and moves back after PRO-S failed eruption.

Then, we studied the causes of PRO-S failed eruption, and mainly explored the evidence of whether kink instability, torus instability, flare reconnection, and reconnection with ambient magnetic field play roles. PRO-S has a twist number of nearly 2 and shows an untwisting motion during the eruption (Figure 4). Therefore, we suggest that a kink instability happens. A trigger is generally necessary to excite the kink instability, and the trigger resulting in loss of equilibrium could be an oscillation and mass drainage (Bi et al., 2014; Fan, 2020), magnetic reconnection in the rope system (Zhang et al., 2023a) including the one happening in the breakout model (Kumar et al., 2023). The decay index is calculated using an HMI Carrington magnetogram and is found to be around 1 at the height of PRO-S, just reaching the lower limit for the torus instability (1 to 1.5). Hence, it is possible that the external field, independent of the poloidal or toroidal field once PRO-S rotates, plays a role in confining the eruption. There are some shortcomings if a potential-field model is used as an approximation of the real external field. In addition to the problems mentioned in Zhang et al. (2023b), the ambient magnetic field may be disturbed by frequent activity. Before the event we study, a faint but large filament eruption occurs around 1:00 UT on 26 October 2022 (see movie_c.mp4). Flare reconnection under a flux rope is thought to be the main cause of explosive CME acceleration (Karpen, Antiochos, and DeVore, 2012). We observe flare ribbons but no posteruption loops, which suggests that the flare reconnection happens but is very weak during PRO-S eruption. The reconnection between PRO-S and the ambient field is also explored, and new legs filled with dropping prominence mass probably support the occurrence of magnetic reconnection. However, due to the low emission of the eruption (no GOES X-ray flare), the reconnection location cannot be identified from X-ray or EUV radiation. A difference of PRO-S from the generally reported confined eruptions (Chen et al., 2023b) is that no clear deceleration is detected before the disappearance of PRO-S along the rising trajectory (Figure 3b). We suggest that PRO-S is destroyed by the magnetic reconnection between PRO-S and the overlying field before an obvious deceleration. Considering the observed untwisting motion and expected writhing motion (to keep the conservation of magnetic helicity), it is speculated that the magnetic reconnection happens (see the fourth regime for failed eruptions in Section 1).

A horn-like structure is at the top of PRO-N, along which PRO-N mass drains southwards. Brightening occurs on some regions on the solar disk, and these regions may connect with PRO-N horn through flux tubes. Prominent horn structures are generally thought to be flux ropes or hyperbolic flux tubes holding prominences (Fan, 2012; Su et al., 2015). From the EUI 30.4 nm image in Figure 2a, PRO-S has an orientation southwest to northeast. We suspect the flux tube connecting PRO-N horn is over PRO-S, which could explain both the rising up of PRO-N along with PRO-S, and mass drainage after PRO-S failed eruption due to evacuation under the flux tube. Once the mass flow starts, it leads to a catastrophic drainage of PRO-N mass because of a siphon effect. Plasma gravity plays an important role in stabilizing a prominence and continuous mass drainage may initiate an eruption (Fan, 2020; Wang et al., 2023). From Figures 7i–j, PRO-N rises during the violent mass drainage but does not erupt.

The relation between PRO-S and PRO-N is different from the flux-rope systems of the double-decker above the same PIL (Liu et al., 2012) or side-by-side over different PILs. Due to the limitation of observations, we cannot identify exactly the location of the flux tube that connects with PRO-N horn. Horn structures, not rare for quiescent prominences (Wang et al., 2016), are a part of long flux tubes, including both flux ropes and sheared arcades. Prominence and flux tubes that connect with the horns compose the large-scale prominence-cavity systems. They suffer frequent disturbance because of activity underneath including jet and filament eruptions (Chen et al., 2021). From EUI observations, PRO-S is likely to locate below the flux tube that holds PRO-N.

In summary, the sympathetic eruptions in this work could be divided into three stages:

1. PRO-S rises due to a helical kink instability (PRO-S is twisted in the EUI 30.4 nm image and an untwisting motion is observed in the Hα Doppler map). Magnetic reconnection occurs under PRO-S (flare ribbons) and provides a positive feedback to the eruption. As the flux tube that holds PRO-N is over PRO-S, PRO-N also rises up and is pushed away from PRO-S due to the expansion of the latter (time–distance diagrams in Figure 3).
2. PRO-S reconnects with the overlying magnetic field (new legs appear in Figure 6), and the upward hoop force decreases along with the decay of the poloidal field of PRO-S. It is possible that the external magnetic field plays a role in confining the eruption.
3. PRO-N mass flow along the horn starts due to the evacuation after PRO-S failed eruption, which leads to the catastrophic mass drainage due to a siphon effect (Figure 7). In the end, only the bottom part remains.

Mass draining is a usual form in which prominences end their lives, and this work suggests that catastrophic mass draining may result from a filament eruption underneath.

Supplementary Information The online version contains supplementary material available at https://doi.org/10.1007/s11207-024-02336-5.

Acknowledgments The authors thank the reviewer very much for the valuable comments and suggestions that helped improve the manuscript. The ASO-S is supported by the Strategic Priority Research Program on Space Science, Chinese Academy of Sciences. The CHASE mission is supported by the China National Space Administration. SDO is a mission for NASA's Living with a Star Program. Solar Orbiter is a space mission of international collaboration between ESA and NASA, operated by ESA. The EUI instrument was built by CSL, IAS, MPS, MSSL/UCL, PMOD/WRC, ROB, LCF/IO with funding from the Belgian Federal Science Policy Office, the Centre National d'Etudes Spatiales (CNES), the UK Space Agency (UKSA), the Bundesministerium für Wirtschaft und Energie (BMWi), and the Swiss Space Office (SSO).

Author contributions J.X. led the project and wrote the main manuscript text. L.F. and H.L. selected the topic, supervised the project, and led the discussions. P.Z. joined the data reduction. J.C. and G.S. joined the magnetic extrapolation. K.J. prepared the SDI data. Y.Q. and C.L. prepared the CHASE data. All authors reviewed the manuscript, discussed the results, and contributed to manuscript preparation.

Funding This work is supported by the National Natural Science Foundation of China (NSFC) 12233012, the Strategic Priority Research Program of the Chinese Academy of Sciences XDB0560000, National Key R&D Program of China 2022YFF0503003 (2022YFF0503000), the mobility program (M-0068) of the Sino-German Science Center, NSFC (grant Nos. 11973012, 11921003, 12203102, 12103090, 12333009), the Jiangsu Funding Program for Excellent Postdoctoral Talent, the Youth Talent Cultivation Program of the PMO E2ZC161111.

Data Availability No datasets were generated, and only open datasets are analyzed.

Declarations

Competing interests The authors declare no competing interests.

References

Auchère, F.: 2005, Effect of the H I Lyα chromospheric flux anisotropy on the total intensity of the resonantly scattered coronal radiation. *Astrophys. J.* **622**(1), 737. DOI. ADS.

Bi, Y., Jiang, Y., Yang, J., Hong, J., Li, H., Yang, D., Yang, B.: 2014, Solar filament material oscillations and drainage before eruption. *Astrophys. J.* **790**(2), 100. DOI. ADS.

Cai, Y.-F., Yang, X., Xiang, Y.-Y., Yan, X.-L., Jin, Z.-Y., Liu, H., Ji, K.-F.: 2022, The co-alignment of winged hα data observed by the new vacuum solar telescope. *Res. Astron. Astrophys.* **22**(6), 065010. DOI. ADS.

Chen, B., Li, H., Song, K.-F., Guo, Q.-F., Zhang, P.-J., He, L.-P., et al.: 2019, The Lyman-alpha Solar Telescope (LST) for the ASO-S mission – II. Design of LST. *Res. Astron. Astrophys.* **19**(11), 159. DOI. ADS.

Chen, C., Su, Y., Xue, J., Gan, W., Huang, Y.: 2021, Solar prominence bubble and plumes caused by an eruptive magnetic flux rope. *Astrophys. J. Lett.* **923**(1), L10. DOI. ADS.

Chen, Y., Cheng, X., Chen, J., Dai, Y., Ding, M.: 2023b, Observations of a failed solar filament eruption involving external reconnection. *Astrophys. J.* **959**(2), 67. DOI. ADS.

Chen, J., Cheng, X., Kliem, B., Ding, M.: 2023a, A model for confined solar eruptions including external reconnection. *Astrophys. J. Lett.* **951**(2), L35. DOI. ADS.

Cheng, X., Guo, Y., Ding, M.: 2017, Origin and structures of solar eruptions I: magnetic flux rope. *Sci. China Earth Sci.* **60**, 1383. DOI. ADS.

Cheng, X., Zhang, J., Ding, M.D., Olmedo, O., Sun, X.D., Guo, Y., Liu, Y.: 2013, Investigating two successive flux rope eruptions in a solar active region. *Astrophys. J. Lett.* **769**(2), L25. DOI. ADS.

Dai, J., Zhang, Q., Zhang, Y., Xu, Z., Su, Y., Ji, H.: 2021, Oscillations and mass draining that lead to a sympathetic eruption of a quiescent filament. *Astrophys. J.* **923**(1), 74. DOI. ADS.

Démoulin, P., Aulanier, G.: 2010, Criteria for flux rope eruption: non-equilibrium versus torus instability. *Astrophys. J.* **718**(2), 1388. DOI. ADS.

DeVore, C.R., Antiochos, S.K.: 2008, Homologous confined filament eruptions via magnetic breakout. *Astrophys. J.* **680**(1), 740. DOI. ADS.

Fan, Y.: 2012, Thermal signatures of tether-cutting reconnections in pre-eruption coronal flux ropes: hot central voids in coronal cavities. *Astrophys. J.* **758**(1), 60. DOI. ADS.

Fan, Y.: 2020, Simulations of prominence eruption preceded by large-amplitude longitudinal oscillations and draining. *Astrophys. J.* **898**(1), 34. DOI. ADS.

Feng, L., Li, H., Chen, B., Li, Y., Susino, R., Huang, Y., et al.: 2019, The Lyman-alpha Solar Telescope (LST) for the ASO-S mission – III. Data and potential diagnostics. *Res. Astron. Astrophys.* **19**(11), 162. DOI. ADS.

Gan, W., Zhu, C., Deng, Y., Zhang, Z., Chen, B., Huang, Y., et al.: 2023, The Advanced Space-Based Solar Observatory (ASO-S). *Solar Phys.* **298**(5), 68. DOI. ADS.

García Marirrodriga, C., Pacros, A., Strandmoe, S., Arcioni, M., Arts, A., Ashcroft, C., et al.: 2021, Solar Orbiter: mission and spacecraft design. *Astron. Astrophys.* **646**, A121. DOI. ADS.

Gilbert, H.R., Alexander, D., Liu, R.: 2007, Filament kinking and its implications for eruption and reformation. *Solar Phys.* **245**(2), 287. DOI. ADS.

Gordino, M., Auchère, F., Vial, J.-C., Bocchialini, K., Hassler, D.M., Bando, T., et al.: 2022, Empirical relations between the intensities of Lyman lines of H and He^{+}. *Astron. Astrophys.* **657**, A86. DOI. ADS.

Gouttebroze, P., Heinzel, P., Vial, J.C.: 1993, The hydrogen spectrum of model prominences. *Astron. Astrophys. Suppl. Ser.* **99**, 513. ADS.

Guo, Y., Ding, M.D., Schmieder, B., Li, H., Török, T., Wiegelmann, T.: 2010, Driving mechanism and onset condition of a confined eruption. *Astrophys. J. Lett.* **725**(1), L38. DOI. ADS.

Hassanin, A., Kliem, B.: 2016, Helical kink instability in a confined solar eruption. *Astrophys. J.* **832**(2), 106. DOI. ADS.

Hood, A.W., Priest, E.R.: 1981, Critical conditions for magnetic instabilities in force-free coronal loops. *Geophys. Astrophys. Fluid Dyn.* **17**(1), 297. DOI. ADS.

Ji, H., Wang, H., Schmahl, E.J., Moon, Y.-J., Jiang, Y.: 2003, Observations of the failed eruption of a filament. *Astrophys. J. Lett.* **595**(2), L135. DOI. ADS.

Ji, K., Liu, H., Jin, Z., Shang, Z., Qiang, Z.: 2017, An automatic approach of mapping the solar high-resolution image to Helioprojective-Cartesian coordinates system. *Chin. Sci. Bull.* **64**, 1738. DOI.
Jiang, Y., Yang, J., Hong, J., Bi, Y., Zheng, R.: 2011, Sympathetic filament eruptions connected by coronal dimmings. *Astrophys. J.* **738**(2), 179. DOI. ADS.
Jiang, C., Duan, A., Zou, P., Zhou, Z., Bian, X., Feng, X., et al.: 2023, A model of failed solar eruption initiated and destructed by magnetic reconnection. *Mon. Not. Roy. Astron. Soc.* **525**(4), 5857. DOI. ADS.
Karpen, J.T., Antiochos, S.K., DeVore, C.R.: 2012, The mechanisms for the onset and explosive eruption of coronal mass ejections and eruptive flares. *Astrophys. J.* **760**(1), 81. DOI. ADS.
Kippenhahn, R., Schlüter, A.: 1957, Eine Theorie der solaren Filamente. Mit 7 Textabbildungen. *Z. Astrophys.* **43**, 36. ADS.
Kliem, B., Török, T.: 2006, Torus instability. *Phys. Rev. Lett.* **96**(25), 255002. DOI. ADS.
Kumar, P., Karpen, J.T., Antiochos, S.K., DeVore, C.R., Wyper, P.F., Cho, K.-S.: 2023, Plasmoids, flows, and jets during magnetic reconnection in a failed solar eruption. *Astrophys. J.* **943**(2), 156. DOI. ADS.
Kuperus, M., Raadu, M.A.: 1974, The support of prominences formed in neutral sheets. *Astron. Astrophys.* **31**, 189. ADS.
Lemen, J.R., Title, A.M., Akin, D.J., Boerner, P.F., Chou, C., Drake, J.F., et al.: 2012, The Atmospheric Imaging Assembly (AIA) on the Solar Dynamics Observatory (SDO). *Solar Phys.* **275**(1–2), 17. DOI. ADS.
Li, S., Su, Y., Zhou, T., van Ballegooijen, A., Sun, X., Ji, H.: 2017, High-resolution observations of sympathetic filament eruptions by NVST. *Astrophys. J.* **844**(1), 70. DOI. ADS.
Li, H., Chen, B., Feng, L., Li, Y., Huang, Y., et al.: 2019, The Lyman-alpha Solar Telescope (LST) for the ASO-S mission — I. Scientific objectives and overview. *Res. Astron. Astrophys.* **19**(11), 158. DOI. ADS.
Li, C., Fang, C., Li, Z., Ding, M., Chen, P., Qiu, Y., et al.: 2022, The Chinese Hα Solar Explorer (CHASE) mission: an overview. *Sci. China Phys. Mech. Astron.* **65**(8), 289602. DOI. ADS.
Liu, C., Lee, J., Karlický, M., Prasad Choudhary, D., Deng, N., Wang, H.: 2009, Successive solar flares and coronal mass ejections on 2005 September 13 from NOAA AR 10808. *Astrophys. J.* **703**(1), 757. DOI. ADS.
Liu, R., Kliem, B., Török, T., Liu, C., Titov, V.S., Lionello, R., Linker, J.A., Wang, H.: 2012, Slow rise and partial eruption of a double-decker filament. I. Observations and interpretation. *Astrophys. J.* **756**(1), 59. DOI. ADS.
Lowe, D.G.: 2004, Distinctive Image Features from Scale-Invariant Keypoints. *Int. J. Comput. Vis.* 60(91). DOI.
Martin, S.F., Bilimoria, R., Tracadas, P.W.: 1994, Magnetic field configurations basic to filament channels and filaments. In: Rutten, R.J., Schrijver, C.J. (eds.) *Solar Surface Magnetism, NATO Advanced Study Institute (ASI) Series C* **433**, 303. ADS.
Moon, Y.-J., Choe, G.S., Wang, H., Park, Y.D.: 2003, Sympathetic coronal mass ejections. *Astrophys. J.* **588**(2), 1176. DOI. ADS.
Myers, C.E., Yamada, M., Ji, H., Yoo, J., Fox, W., Jara-Almonte, J., Savcheva, A., Deluca, E.E.: 2015, A dynamic magnetic tension force as the cause of failed solar eruptions. *Nature* **528**(7583), 526. DOI. ADS.
Netzel, A., Mrozek, T., Kołomański, S., Gburek, S.: 2012, Extreme-ultraviolet and hard X-ray signatures of electron acceleration during the failed eruption of a filament. *Astron. Astrophys.* **548**, A89. DOI. ADS.
Ouyang, Y., Zhou, Y.H., Chen, P.F., Fang, C.: 2017, Chirality and magnetic configurations of solar filaments. *Astrophys. J.* **835**(1), 94. DOI. ADS.
Qiu, Y., Rao, S., Li, C., Fang, C., Ding, M., Li, Z., et al.: 2022, Calibration procedures for the CHASE/HIS science data. *Sci. China Phys. Mech. Astron.* **65**(8), 289603. DOI. ADS.
Rochus, P., Auchère, F., Berghmans, D., Harra, L., Schmutz, W., Schühle, U., et al.: 2020, The Solar Orbiter EUI instrument: the extreme ultraviolet imager. *Astron. Astrophys.* **642**, A8. DOI. ADS.
Scherrer, P.H., Schou, J., Bush, R.I., Kosovichev, A.G., Bogart, R.S., Hoeksema, J.T., Liu, Y., Duvall, T.L., Zhao, J., Title, A.M., Schrijver, C.J., Tarbell, T.D., Tomczyk, S.: 2012, The Helioseismic and Magnetic Imager (HMI) Investigation for the Solar Dynamics Observatory (SDO). *Solar Phys.* **275**(1–2), 207. DOI. ADS.
Shen, Y., Liu, Y., Su, J.: 2012, Sympathetic partial and full filament eruptions observed in one solar breakout event. *Astrophys. J.* **750**(1), 12. DOI. ADS.
Solanki, S.K., del Toro Iniesta, J.C., Woch, J., Gandorfer, A., Hirzberger, J., Alvarez-Herrero, A., et al.: 2020, The polarimetric and helioseismic imager on Solar Orbiter. *Astron. Astrophys.* **642**, A11. DOI. ADS.
SolO/EUI Data Release 6.0 January 2023. DOI.
Stansby, D., Yeates, A., Badman, S.: 2020, pfsspy: a python package for potential field source surface modelling. *J. Open Sour. Soft.* **5**(54), 2732. DOI. ADS.

Su, Y., van Ballegooijen, A., McCauley, P., Ji, H., Reeves, K.K., DeLuca, E.E.: 2015, Magnetic structure and dynamics of the erupting solar polar crown prominence on 2012 March 12. *Astrophys. J.* **807**(2), 144. DOI. ADS.

Tang, Z., Shen, Y., Zhou, X., Duan, Y., Zhou, C., Tan, S., Elmhamdi, A.: 2021, Sympathetic standard and blowout coronal jets observed in a polar coronal hole. *Astrophys. J. Lett.* **912**(1), L15. DOI. ADS.

Török, T., Kliem, B.: 2005, Confined and ejective eruptions of kink-unstable flux ropes. *Astrophys. J. Lett.* **630**(1), L97. DOI. ADS.

Török, T., Panasenco, O., Titov, V.S., Mikić, Z., Reeves, K.K., Velli, M., Linker, J.A., De Toma, G.: 2011, A model for magnetically coupled sympathetic eruptions. *Astrophys. J. Lett.* **739**(2), L63. DOI. ADS.

Vial, J.-C., Engvold, O.: 2015, Solar Prominences. *Astrophys. Space Sci. Libr.* 415. DOI. ADS.

Wang, H., Chae, J., Yurchyshyn, V., Yang, G., Steinegger, M., Goode, P.: 2001, Inter-active region connection of sympathetic flaring on 2000 February 17. *Astrophys. J.* **559**(2), 1171. DOI. ADS.

Wang, B., Chen, Y., Fu, J., Li, B., Li, X., Liu, W.: 2016, Dynamics of a prominence-horn structure during its evaporation in the solar corona. *Astrophys. J. Lett.* **827**(2), L33. DOI. ADS.

Wang, D., Liu, R., Wang, Y., Gou, T., Zhang, Q., Zhou, Z., Zhang, M.: 2018, Unraveling the links among sympathetic eruptions. *Astrophys. J.* **869**(2), 177. DOI. ADS.

Wang, C., Chen, F., Ding, M., Lu, Z.: 2023, Radiative magnetohydrodynamic simulation of the confined eruption of a magnetic flux rope: unveiling the driving and constraining forces. *Astrophys. J.* **956**(2), 106. DOI. ADS.

Xue, J.-C., Vial, J.-C., Su, Y., Li, H., Xu, Z., Su, Y.-N., Zhou, T.-H., Li, Z.-T.: 2021, High-resolution observations of prominence plume formation with the new vacuum solar telescope. *Res. Astron. Astrophys.* **21**(9), 222. DOI. ADS.

Yang, S., Zhang, J., Song, Q., Bi, Y., Li, T.: 2019, Two-step evolution of a rising flux rope resulting in a confined solar flare. *Astrophys. J.* **878**(1), 38. DOI. ADS.

Yang, L., Yan, X., Xue, Z., Wang, J., Yang, L., Li, Q., et al.: 2023, Sympathetic partial filament eruptions caused by the interaction between two nearby filaments. *Astrophys. J.* **943**(1), 62. DOI. ADS.

Zhang, Q., Cheng, X., Liu, R., Song, A., Li, X., Wang, Y.: 2023a, Influence of magnetic reconnection on the eruptive catastrophes of coronal magnetic flux ropes. *Front. Astron. Space Sci.* **9**, 434. DOI. ADS.

Zhang, X., Guo, J., Guo, Y., Ding, M., Keppens, R.: 2023b, Rotation and Confined Eruption of a Double Flux-Rope System. ArXiv e-prints. DOI. ADS. arXiv.

Zhao, J., Zhang, P., Gibson, S.E., Fan, Y., Feng, L., Yu, F., Li, H., Gan, W.Q.: 2022, Synthetic Lyman-α emissions for the coronagraph aboard the ASO-S mission. I. An eruptive prominence-cavity system. *Astron. Astrophys.* **665**, A39. DOI. ADS.

Zhong, Z., Guo, Y., Ding, M.D.: 2021, The role of non-axisymmetry of magnetic flux rope in constraining solar eruptions. *Nat. Commun.* **12**, 2734. DOI. ADS.

Zhou, Z., Cheng, X., Zhang, J., Wang, Y., Wang, D., Liu, L., Zhuang, B., Cui, J.: 2019, Why do torus-unstable solar filaments experience failed eruptions? *Astrophys. J. Lett.* **877**(2), L28. DOI. ADS.

Authors and Affiliations

Jian-chao Xue[1] · Li Feng[1,2] · Hui Li[1,2] · Ping Zhang[1] · Jun Chen[1] · Guang-lu Shi[1,2] · Kai-fan Ji[3] · Ye Qiu[4] · Chuan Li[4,5] · Lei Lu[1] · Bei-li Ying[1] · Ying Li[1,2] · Yu Huang[1,2] · You-ping Li[1] · Jing-wei Li[1] · Jie Zhao[1] · De-chao Song[1] · Shu-ting Li[1,2] · Zheng-yuan Tian[1,2] · Ying-na Su[1,2] · Qing-min Zhang[1] · Yun-yi Ge[1] · Jia-hui Shan[1,2] · Qiao Li[1,2] · Gen Li[1,2] · Yue Zhou[1,2] · Jun Tian[1,2] · Xiao-feng Liu[1,2] · Zhi-chen Jing[1,2] · Bo Chen[6] · Ke-fei Song[6] · Ling-ping He[6] · Shi-jun Lei[1] · Wei-qun Gan[1,7]

✉ H. Li
nj.lihui@pmo.ac.cn

J. Xue
xuejc@pmo.ac.cn

L. Feng
lfeng@pmo.ac.cn

1 Key Laboratory of Dark Matter and Space Astronomy, Purple Mountain Observatory, Chinese Academy of Sciences, 10 Yuanhua Road, Nanjing 210023, People's Republic of China

2 School of Astronomy and Space Science, University of Science and Technology of China, 96 Jinzhai Road, Hefei 230026, People's Republic of China

3 Yunnan Observatories, Chinese Academy of Sciences, Kunming 650216, People's Republic of China

4 School of Astronomy and Space Science, Nanjing University, Nanjing 210023, People's Republic of China

5 Key Laboratory for Modern Astronomy and Astrophysics (Nanjing University), Ministry of Education, Nanjing 210023, People's Republic of China

6 Changchun Institute of Optics, Fine Mechanics and Physics, Chinese Academy of Sciences, 3888 Dong Nanhu Road, Changchun 130033, People's Republic of China

7 University of Chinese Academy of Sciences, Nanjing 211135, People's Republic of China

Solar Physics (2024) 299:92
https://doi.org/10.1007/s11207-024-02330-x

RESEARCH

Parameter Effects on the Total Intensity of H I Lyα Line for a Modeled Coronal Mass Ejection and Its Driven Shock

Beili Ying[1] · Guanglu Shi[1,2] · Li Feng[1,2] · Lei Lu[1] · Jianchao Xue[1] · Shuting Li[1,2] · Weiqun Gan[1,3] · Hui Li[1,2]

Received: 25 December 2023 / Accepted: 23 May 2024 / Published online: 25 June 2024

Abstract

The combination of the H I Lyα (121.6 nm) line formation mechanism with ultraviolet (UV) Lyα and white-light (WL) observations provides an effective method for determining the electron temperature of coronal mass ejections (CMEs). A key to ensuring the accuracy of this diagnostic technique is the precise calculation of theoretical Lyα intensities. This study performs a modeled CME and its driven shock via the three-dimensional numerical magneto-hydrodynamic simulation. Then, we generate synthetic UV and WL images of the CME and shock within a few solar radii to quantify the impact of different assumptions on the theoretical Lyα intensities, such as the incident intensity of the solar chromospheric Lyα line (I_{disk}), the geometric scattering function ($p(\theta)$), and the kinetic temperature (T_n) assumed to be equal to either the proton (T_p) or electron (T_e) temperature. By comparing differences of the Lyα intensities of the CME and shock under these assumptions, we find that: (1) Using the uniform or Carrington maps of the disk Lyα emission underestimates the corona Lyα intensity (with relative uncertainties below 10%) compared to the synchronic map, except for a slight overestimate (<4%) observed in the partial CME core. The Carrington map yields lower uncertainties than the uniform disk. (2) Neglecting the geometric scattering process has a relatively minor impact on the Lyα intensity, with a maximum relative uncertainty of no more than 5%. The Lyα intensity is underestimated for the most part but overestimated in the CME core. (3) Compared to the assumption $T_n = T_p$, using $T_n = T_e$ leads to more complex relative uncertainties in CME Lyα intensity. The CME core and void are both overestimated, with the maximum relative uncertainty in the core exceeding 50% and in the void remaining below 35%. An appropriate increasing proton-to-electron temperature ratio can reduce the uncertainty in the CME core and void. In the CME front, both overestimates and underestimates exist with relative uncertainties of less than 35%. The electron temperature assumption has a smaller impact on the shock, with an underestimated relative uncertainty of less than 20%.

Keywords Coronal mass ejections · Corona · Waves · Shock · Magnetohydrodynamics · Ultraviolet radiation

1. Introduction

Coronal mass ejections (CMEs) are highly energetic eruptions that occur in the solar atmosphere. Various instruments have been used to observe CMEs, revealing their distinct

Extended author information available on the last page of the article

characteristics in different wavelengths. When the velocity of a CME surpasses the local fast magnetosonic velocity, it can drive a shock wave (e.g., Gopalswamy et al., 2012; Ying et al., 2018).

Several studies have demonstrated the valuable insights gained by combining white-light (WL) coronagraphic images with ultraviolet (UV) observations from the UV Coronagraph Spectrometer (UVCS) (Kohl et al., 1995). This combination offers a unique capability to investigate various aspects of CMEs, including plasma temperatures, kinematic properties, and elemental distributions (Ciaravella et al., 2003; Ciaravella, Raymond, and Kahler, 2006; Raymond et al., 2003; Bemporad et al., 2007; Ying et al., 2020; Bemporad, 2022). Furthermore, it enables the study of numerous CME-related phenomena, such as CME-driven shocks (Ciaravella et al., 2005; Ciaravella, Raymond, and Kahler, 2006; Bemporad and Mancuso, 2010), post-CME current sheets (Bemporad, 2008; Shi et al., 2020; Bemporad et al., 2024), and CME-driven reconnections (Bemporad and Mancuso, 2010). In UV Lyα images, fast CMEs can exhibit a dark front caused by the Doppler dimming (DD) effect (Bemporad, Pagano, and Giordano, 2018; Ying et al., 2020). This effect arises from the Doppler shift of the coronal absorption profile relative to the solar disk incident profile, resulting in a reduced efficiency of atomic excitation according to the formation mechanism of the H I Lyα line. The DD technique has been developed to determine solar wind speed (Noci, Kohl, and Withbroe, 1987; Dolei et al., 2018, 2019), and it has also been applied to estimate the electron temperature of CME plasma, even without slit-spectroscopic data, by combining WL and UV H I Lyα line intensities (Susino and Bemporad, 2016; Bemporad, Pagano, and Giordano, 2018; Ying et al., 2020; Bemporad, 2022). The crucial aspect of this method lies in acquiring an accurate calculation of the theoretical Lyα intensity. By comparing the observed Lyα intensity with the theoretical value, the electron temperature of the CME can be determined. However, many of these studies simplify the calculation of the theoretical Lyα intensity in the DD method by (1) assuming a uniform incident Lyα intensity spectrum from the solar disk, (2) neglecting the geometric scattering process and treating the Sun as a point source, and (3) considering the neutral hydrogen kinetic temperature to be equal to the electron temperature in the absence of spectroscopic observations. These simplifications can potentially lead to inaccuracies in the theoretical Lyα intensity, consequently impacting the accuracy of the estimated electron temperatures.

Several new instruments have been developed to enable simultaneous imaging of the solar corona using WL and UV H I Lyα lines, eliminating the need for slit-spectroscopic observations. Notable examples include the Lyα Solar Telescope (LST, Li et al., 2019; Feng et al., 2019) on board the Chinese Advanced Space-based Solar Observatory mission (ASO-S; Gan et al., 2019, 2023), and the Metis coronagraph on board the ESA Solar Orbiter mission (Antonucci et al., 2017, 2020). The LST comprises three instruments: a Solar Disk Imager (SDI) capable of imaging the solar disk up to 1.2 $R_\odot$ in the Lyα line, a White-light Solar Telescope (WST), and a Solar Corona Imager (SCI) with a field-of-view (FoV) ranging from 1.1 to 2.5 $R_\odot$ in both WL and Lyα passbands. The Metis coronagraph has a FoV of $1.6° - 2.9°$ (corresponding to projected altitudes from 1.7 to 3.1 $R_\odot$when the spacecraft reaches its closest approach at 0.28 AU). With the deployment of these new instruments, the DD method becomes a valuable tool for diagnosing the physical parameters of CMEs and shocks. However, when the instrument's FoV approaches the Sun, particularly in the case of LST, it becomes necessary to investigate the effects of geometric scattering angles and incident chromospheric H I Lyα intensity on the total Lyα radiation from CMEs and shocks when applying the DD technique. We can consider that the most Lyα intensity originates from the chromosphere, due to the narrow width of the reversal center of the Lyα line formed around the transition region, and the fact that the line wings are formed in the mid-chromosphere (Vernazza, Avrett, and Loeser, 1981; Hong et al., 2019). In a study by Auchère (2005), they investigated the impact of chromospheric Lyα intensity anisotropy on the total Lyα intensity of the corona due to resonant scattering. They discovered that during the solar minimum, assuming a uniform disk distribution leads to a systematic overestimation of the total intensity in the polar regions by an average of 15%. Additionally, Dolei et al. (2019) examined the influence of nonuniform chromospheric Lyα radiation on determining the H I outflow velocity in the corona. They found that assuming a uniform chromospheric intensity underestimates the irradiance in the equatorial regions, with a maximum underestimation of 15% at 1.5 $R_\odot$, decreasing to 5% at 4 $R_\odot$. Conversely, in the solar polar regions, the total irradiance is overestimated. However, the impact of these assumptions on the H I Lyα intensities of CMEs and their driven shocks still remains a question. Therefore, this study aims to quantify the effects of different assumptions on the H I Lyα intensities of a CME and its driven shocks. This will be accomplished by comparing the results obtained under the following conditions: (1) whether the incident intensity of the solar chromospheric H I Lyα line is uniform or not, (2) whether the influence of the geometric scattering angle is taken into account, and (3) whether the kinetic temperature of neutral hydrogen is assumed to be equal to the electron temperature.

This work uses a two-temperature (2T, proton and electron temperatures) magneto-hydrodynamic (MHD) model to simulate a fast CME and its driven shock. A brief overview of the 2T MHD model is provided in Section 2. Subsequently, using the parameters obtained from the MHD simulation, we synthesize the CME and shock images in H I Lyα line under different assumptions in Section 3. The synthetic results are then analyzed and compared in Section 4. Finally, Section 5 shows discussions and conclusions.

2. Two-Temperature MHD Simulation

To obtain reasonable physical parameters of a CME for synthesizing the UV H I Lyα intensity map, we make a CME simulation based on an event that occurred on March 7, 2011, originating from active region 11164. The simulation methodology is similar to the one conducted by Jin et al. (2017). In this simulation, the solar corona component is generated using the Alfvén wave damping to heat the corona and considers separate electron and proton temperatures with electron heat conduction (van der Holst et al., 2010; Jin et al., 2012; Sokolov et al., 2013; Oran et al., 2013; Jin et al., 2013, 2017). The initiation of the CME is achieved by utilizing the Gibson–Low (GL) analytical flux rope model (Gibson and Low, 1998). This model allows for the initiation of a CME from a state of force imbalance, improving computational efficiency. Further details regarding the simulation can be found in Jin et al. (2017). In our study, we decrease the propagated speed of the CME compared to the simulation conducted by Jin et al. (2017). This adjustment is made to mitigate the impact of the DD effect when synthesizing the Lyα image while still ensuring the generation of the CME-driven shock. It is important to emphasize that the aim of this work is to investigate the effect of different assumptions on the H I Lyα intensities of the CME and its driven shock. Therefore, the agreement between the simulated CME events and observational data does not affect the results of this study.

The simulation provides various physical parameters, including the electron number density (n_e), the three velocity components along the x, y, and z axes in Cartesian coordinates, as well as the electron (T_e) and proton (T_p) temperatures, among others. The data cube dimensions used in this work are interpolated into a uniform grid of $700 \times 300 \times 300$ along the x, y, and z axes, respectively. The direction of the line of sight (LoS) is aligned with the

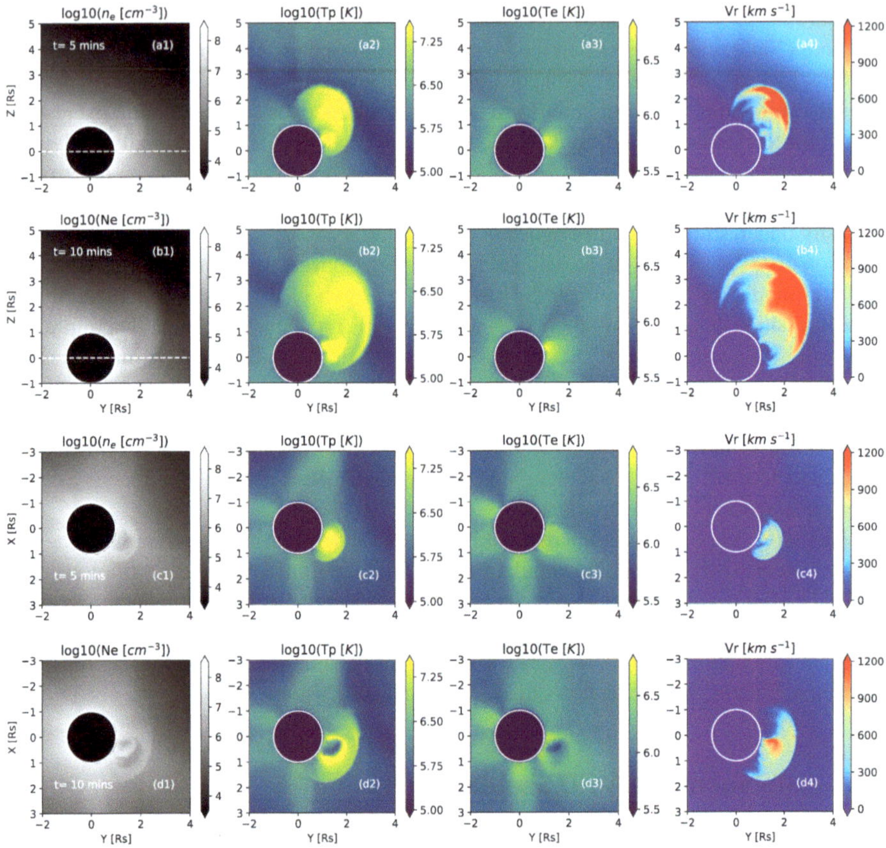

Figure 1 (a)–(b) Parameters of the CME and shock showing the plasma density, proton temperature, electron temperature, and radial speed on the PoS (plane YOZ) from left to right panels at $t = 5$ minutes (first row) and $t = 10$ minutes (second row) after Gibson–Low flux rope implementations, respectively. (c)–(d) Parameter cuts of the CME and shock in the plane XOY as seen from the north pole, corresponding to the dashed lines marked in the plane YOZ, as shown in panels (a1) and (b1).

x-axis. The y-axis ranges from -2 to 4 $R_\odot$, the z-axis ranges from -1 to 5 $R_\odot$, and the x-axis spans from -7 to 7 $R_\odot$. Figure 1 displays the parameters of the simulation on the plane of the sky (PoS, panels (a) and (b)) and in the plane XOY seen from the north pole (panels (c) and (d)) at $t = 5$ and $t = 10$ minutes, respectively. Panels (c) and (d) of Figure 1 show the parameters in the range of -3 to 3 $R_\odot$ along the x-axis. It is obvious that the parameters of the CME are asymmetrical in 3D space. The CME front's radial speed (V_r) on the PoS is approximately 1200 km s^{-1}. In Figure 2, the first column shows the electron column density ($N_e = \int n_e dz$). The second to fourth columns represent the integrated proton temperature, electron temperature, and radial velocity along the LoS weighted by the electron number density, which can be expressed as $p_{LOS} = \int p\, n_e dz / \int n_e dz$, where the parameter p can be T_p, T_e, and V_r. The CME has a three-part structure, as shown in Figure 2, and the core is made of a flux rope. CME fronts and cores are marked by white dots, CME voids are marked by a white dashed line, and shock fronts are denoted by red dots. These structures are identified from the WL and UV H I Lyα synthetic images (as shown in Figure 4) in the next section, based on the physical parameters obtained from the simulation and assum-

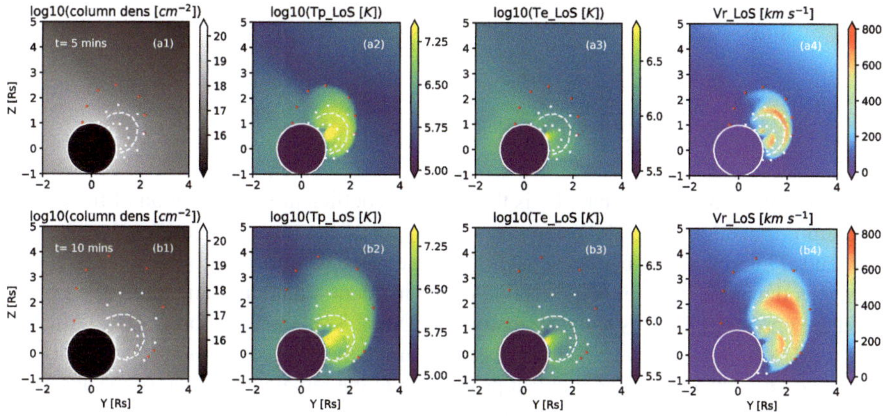

Figure 2 (a)–(b) Parameters of the CME and shock showing the electron column density, integrated proton temperature, electron temperature, and radial velocity along the LoS weighted by the electron number density from left to right panels. CME fronts and cores are marked by white dots, the periphery of CME voids are marked by a white dashed line, and shock fronts are denoted by red dots, which are identified from Figure 4.

ing optical thinness in the corona. Regarding prominences (the core of another category of CME), Zhao et al. (2022) synthesized the Lyα emission of an eruptive prominence through the PRODOP code, which takes into account nonlocal thermodynamic equilibrium radiative transfer processes. However, the discussion of the prominence, considering the optical thickness assumption, is beyond the scope of this study.

3. Synthetic Images Under Different Assumptions

3.1. Formation Mechanism of H I Lyα Lines

The formation mechanism of H I Lyα lines is a combination of radiative and collisional excitations followed by spontaneous emissions (Gabriel, 1971), which can be written as

$$I_{obs} = \int_{LoS} (j_r + j_c)\, dz \ \ [\text{photon s}^{-1}\ \text{cm}^{-2}\ \text{sr}^{-1}], \tag{1}$$

where j_r and j_c are the radiative and collisional emissivities, respectively, from each volume element. The total intensity, I_{obs}, is computed by the integration of all elements along the LoS.

The radiative emissivity,j_r, due to the resonant scattering of radiation from the lower atmosphere by coronal atoms (Noci, Kohl, and Withbroe, 1987; Bemporad, Pagano, and Giordano, 2018) is

$$j_r = \frac{h\, B_{12}\, \lambda_0}{4\,\pi}\, n_i \int_{\Omega} p(\theta)\, d\Omega \int_{-\infty}^{+\infty} I_{ex}(\lambda' - \delta\lambda, \boldsymbol{n})\, \Phi(\lambda' - \lambda_0, \boldsymbol{n})\, d\lambda', \tag{2a}$$

$$p(\theta) = \frac{1}{4\pi}\frac{11 + 3\cos^2\theta}{12}, \tag{2b}$$

$$\Phi(\lambda' - \lambda_0, \boldsymbol{n}) = \frac{1}{\sigma_\lambda(\boldsymbol{n})\sqrt{2\pi}} \exp[-\frac{(\lambda' - \lambda_0)^2}{2\sigma_\lambda^2(\boldsymbol{n})}] \;[\mathrm{cm}^{-1}], \tag{2c}$$

$$\sigma_\lambda(\boldsymbol{n}) = \frac{\lambda_0}{c}\sqrt{\frac{k_B\, T_{\boldsymbol{n}}}{m_p}} \;[\mathrm{cm}], \tag{2d}$$

where h is the Planck constant, B_{12} is the Einstein coefficient for absorption of the considered atom transition, λ_0 is the reference wavelength of the transition in Equation 2a, n_i is the neutral hydrogen number density, $p(\theta)$ is a geometrical function for the scattering process, θ is the angle between the direction of the incident radiation $\boldsymbol{n}$ and along the LoS toward the observer (Beckers and Chipman, 1974; Noci, Kohl, and Withbroe, 1987; Auchère, 2005). Also $I_{ex}(\lambda' - \delta\lambda, \boldsymbol{n})$ is the intensity spectrum of the incident chromospheric radiation, $\delta\lambda = \frac{\lambda_0}{c}\boldsymbol{u}\cdot\boldsymbol{n}$ is the shift of the incident profile due to the radial velocity, $\boldsymbol{u}$, of the absorbing atoms, c is the light speed, $\Phi(\lambda' - \lambda_0, \boldsymbol{n})$ is the normalized coronal absorption profile along the direction of the incident radiation. It can be assumed as a Gaussian profile with a width, $\sigma_\lambda(\boldsymbol{n})$, as shown in Equation 2d. Further, k_B is the Boltzmann constant, and $T_{\boldsymbol{n}}$ is the kinetic temperature along the direction of the incident radiation. In this work, we will assume $T_{\boldsymbol{n}} = T_p$, T_p is the proton temperature provided by the MHD simulation. It is noted that the kinetic temperature of neutral hydrogen can be considered equal to the proton temperature. This is because when a neutral hydrogen atom forms through the recombination of a proton and an electron, its velocity will be close to that of the proton since the proton carries the majority of the momentum. The assumption can be held when the plasma obeys an isotropic thermal velocity distribution.

The collisional component, j_c, of the Lyα emission line in the solar corona, is due to the excitation of coronal atoms by collision with free electrons. The formula can be expressed as

$$j_c = \frac{1}{4\pi}\, n_e\, n_i\, q_{coll}, \tag{3a}$$

$$q_{coll} = 2.73\times\; 10^{-15}\, T_e^{\frac{1}{2}}(E_{12})^{-1}\, f_{12}\,\bar{g}\,\exp^{-\frac{E_{12}}{k_B T_e}} \;[\mathrm{cm}^{-3}\mathrm{s}^{-1}], \tag{3b}$$

$$n_i = 0.833\, n_e\, R_{HI}(T_e), \tag{3c}$$

where n_e is the electron number density, q_{coll} is the collisional coefficient, T_e is the electron temperature, E_{12} is the transition energy, f_{12} is the transition oscillator strength, $\bar{g}$ is the Gaunt factor. Coefficient 0.833 is the proton-to-electron density ratio for a fully ionized gas with a He abundance of 10%; $R_{HI}(T_e)$ is the elemental ionization fraction as a function of the electron temperature under the assumption of ionization equilibrium.

3.2. Lyα Solar Disk Intensity

To generate synthetic UV H I Lyα images of the modeled CME and shock, the intensity spectrum, $I_{ex}(\lambda' - \delta\lambda, \boldsymbol{n})$, of the Ly$\alpha$ radiative component is directly influenced by the chromospheric source radiation; I_{ex} can be separated into two terms, including the normalized chromospheric line profile $f(\lambda' - \delta\lambda)$ and chromospheric disk intensity $I_{disk}(\boldsymbol{n})$. In this work, we assume that $f(\lambda' - \delta\lambda)$ is a unique multi-Gaussian profile proposed by Auchère (2005), which also has been used to estimate the effect of the nonuniform solar chromospheric Lyα radiation on determining the coronal H I outflow velocity (Dolei et al., 2019). Concerning the Lyα solar disk intensity, we build a synchronic map of the Lyα in-

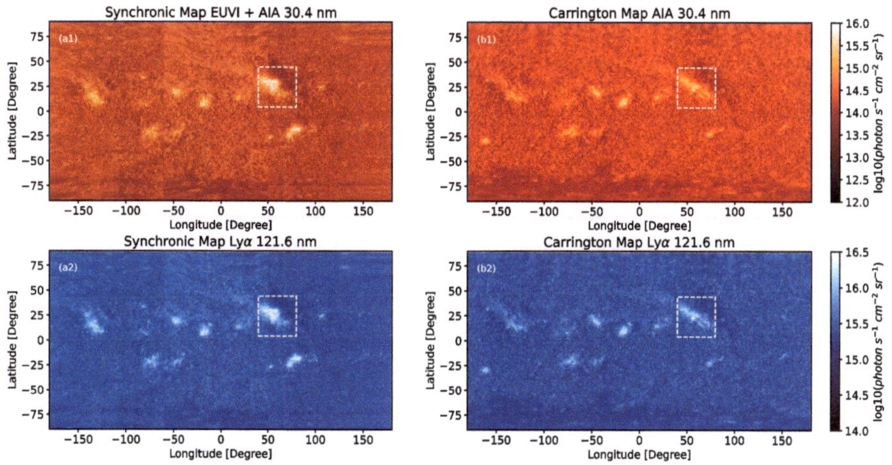

Figure 3 The He II 30.4 nm and H I Lyα 121.6 nm solar disk intensities in synchronic and Carrington maps: (a1) the synchronic map in 30.4 nm was constructed by the SDO/AIA and STEREO/EUVI images at 20:06 UT on March 7, 2011; (a2) the H I Lyα synchronic map is rebuilt based on the correlation between the He II 30.4 nm and H I 121.6 nm in Lyα lines; (b1)–(b2) Carrington maps in He II 30.4 nm and H I Lyα 121.6 nm lines reconstructed by the SDO AIA 30.4 nm images. White boxes denote the source region of the eruptive CME. The missing data above $+78^\circ$ and below -78° latitude has been replaced by data in the latitude $\pm[66^\circ, 78^\circ]$.

tensity based on the correlation between the He II 30.4 nm and H I Lyα 121.6 nm intensities by combining the images of the Atmospheric Imaging Assembly (AIA) onboard the Solar Dynamics Observatory (SDO) and the extreme ultraviolet imager (EUVI) onboard the Heliospheric Imager onboard the Solar TErrestrial RElations Observatory (STEREO) at the same time, as shown in Figure 3(a). The correlation function (Auchère, 2005) is

$$I_{121.6} = 436\,[1 - 0.955\exp(-0.0203\; I_{30.4})], \tag{4}$$

where $I_{121.6}$ and $I_{30.4}$ are in units of [W m^{-2} sr^{-1}]. Gordino et al. (2022) found that the empirical relation between the intensities of the H I 121.6 nm and He II 30.4 nm Lyα lines remains stable over time. However, the accuracy of this relation depends on the spatial resolution of the observations. The contamination of Si XI 30.32 nm line near the He II 30.4 nm and other lines have been subtracted by the differential emission measure technique (Cheung et al., 2015; Su et al., 2018) in the AIA 30.4 nm image, similar to the method used in Gordino et al. (2022). Additionally, we conduct a cross-radiometric calibration by analyzing the overlapping region between the AIA and EUVI 30.4 nm images on the solar disk. We assume that the contamination of the EUVI 30.4 nm image in the active region and quiet Sun is equal to the mean value of that in the AIA 30.4 nm image in the active region and quiet Sun. We also build a Carrington map (a diachronic map created by stitching observations made over one Carrington cycle) in the He II 30.4 nm from the AIA data with the subtraction of the Si XI contamination and then generate the H I Lyα Carrington map (Figure 3(b2)). The mean value of the disk intensity in the H I Lyα line for the synchronic map is 3.20×10^{15} photon s^{-1} cm^{-2} sr^{-1}, and for the Carrington map is 3.15×10^{15} photon s^{-1} cm^{-2} sr^{-1}. The mean value of the synchronic map is used as the value of the uniform map. In the following section, these maps are then used to determine the difference in Lyα intensity between the inputs of the Carrington, uniform, and synchronic maps. The source region of the eruption

Table 1 Different assumptions for the 2D synthetic Lyα intensity: (Column 1) group number; (Column 2) $I_{disk}(\boldsymbol{n})$ is the disk intensity of the incident chromospheric radiation. The value of the uniform map is equal to the mean value of the synchronic map; (Column 3) the geometric function of the scattering process. We consider the geometric function according to Equation 2b and ignore this process when assuming $p(\theta) = 1/4\pi$; (Column 4) $T_{\boldsymbol{n}}$ is the kinetic temperature in Equation 2d.

1 No.	2 $I_{disk}(n)$	3 $p(\theta)$	4 T_n
Group 1	synchronic map	Equation 2b	$T_{\boldsymbol{n}} = T_p$
Group 2	Carrington map	Equation 2b	$T_{\boldsymbol{n}} = T_p$
Group 3	uniform map	Equation 2b	$T_{\boldsymbol{n}} = T_p$
Group 4	synchronic map	$p(\theta) = \frac{1}{4\pi}$	$T_{\boldsymbol{n}} = T_p$
Group 5	synchronic map	Equation 2b	$T_{\boldsymbol{n}} = T_e$
Group 6	uniform map	$p(\theta) = \frac{1}{4\pi}$	$T_{\boldsymbol{n}} = T_e$

is marked by a white box at 20:06 UT on March 7, 2011. The missing data above $+78^\circ$ and below -78° latitude has been replaced by data in the latitude $\pm[66^\circ, 78^\circ]$ in Figure 3.

3.3. Synthetic Images

Based on Equations 2a and 3a and the physical parameters produced by the MHD simulation, we can calculate the H I Lyα emissivity of each plasma in the data cube and obtain the 2D synthetic Lyα image with the intensity integration along the LoS. The composited Lyα images are divided into six groups depending on the different assumptions of three physical parameters in the Lyα radiative component, including the disk intensity $I_{disk}(\boldsymbol{n})$, the geometric scattering function $p(\theta)$, and the kinetic temperature $T_{\boldsymbol{n}}$, as shown in Table 1. Group 1 will be a reference group in what follows. The assumptions of Group 6 have been applied frequently to estimate the CME electron temperatures (Susino and Bemporad, 2016; Bemporad, Pagano, and Giordano, 2018; Ying et al., 2020). Then, we will compare the Lyα intensity difference of these groups to the reference group in the next section.

Examples of the synthetic UV Lyα images under assumptions of Group 1 are shown in Figure 4. From left to right in Figure 4, these panels are UV original, UV base-ratio (normalized by the coronal background image), WL total brightness (tB), and WL base-ratio images, respectively. WL synthetic image is generated by integrating of the electron number density (n_e) along the LoS and multiplying it by geometrical functions that depend solely on the heliocentric distance (Billings, 1966). CME fronts and cores are marked by white dots, CME voids are marked by a white dashed line, and shock fronts are denoted by red dots. Similar to the UV images synthesized by Bemporad, Pagano, and Giordano (2018), CME fronts are darker than the coronal background. The shock regions are indistinguishable in UV Lyα original images and can only be recognizable in base-ratio images, as shown in Figure 4. Previous studies have observed that the Lyα intensity of the shock decreases by $\sim 30\%$ (Ciaravella et al., 2005) and manifests as a dimming region with the shock transition (Bemporad and Mancuso, 2010) in the UVCS slit-spectroscopic observations. In this simulation, the Lyα intensity of the synthetic shock wave is approximately $20-30\%$ lower than that of the coronal background. However, as the shock evolves and the number density decreases (Figure 4(b4)), the Lyα intensity difference between the shock and the background gradually diminishes (Figure 4(b2)).

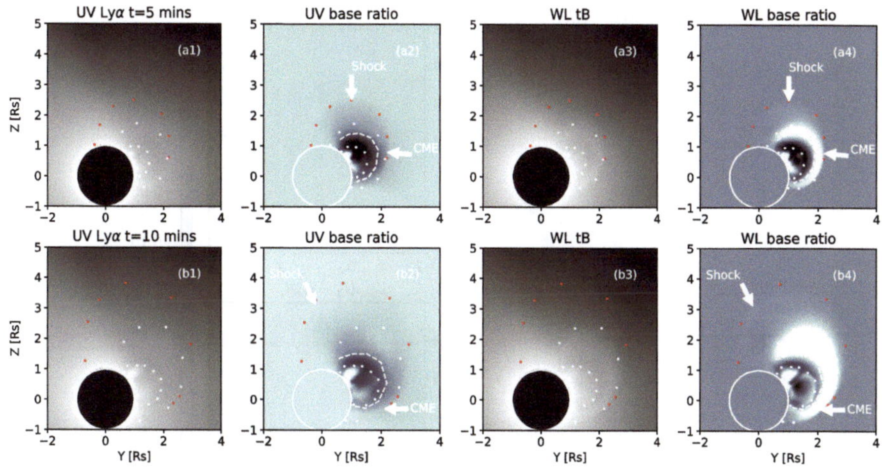

Figure 4 Composited UV H I Lyα (Group 1) and WL images at different simulated times. Panels in columns 1 – 2 denote the UV Lyα original and base-ratio images. Panels in columns 3 – 4 represent the WL total brightness (tB) and base-ratio images. CME fronts and cores are marked by white dots, the periphery of CME voids are marked by a white dashed line, and shock fronts are denoted by red dots.

4. Lyα Intensity Uncertainties Under Different Assumptions

In this section, we derive the relative uncertainties of the H I Lyα intensity by comparing the 2D synthetic intensities of different groups (2 – 6) with respect to Group 1 to quantify the effect of these assumptions on the Lyα total radiation of the CME and shock.

4.1. Uncertainties from Different Solar Disk Maps

Figure 5 illustrates the comparison of the coronal Lyα relative intensities derived with the Carrington map (Group 2) and uniform map (Group 3) with respect to the synchronic map (Group 1). The only differing parameter among these three groups, when calculating the Lyα intensity, is $I_{disk}(\boldsymbol{n})$, as depicted in Table 1. In Figures 5 (a1) and (b1), it can be observed that the calculated Lyα intensities using Carrington map and synchronic map as inputs for $I_{disk}(\boldsymbol{n})$ exhibit small differences in most regions, with relative underestimate uncertainties remaining within 6%. A partial core is overestimated (<4%), whereas the other parts are primarily underestimated. The largest discrepancies between Group 2 and Group 1 are observed in the CME core, with a maximum underestimated uncertainty of approximately 10% (in panel (b1)). This region has relatively lower electron and proton temperatures and higher electron number densities compared to other regions. It is possible that the observed increase in the Lyα intensity of Group 1 is attributed to the influence of a flare present in the synchronic map within the source region (indicated by a white box in Figure 3). However, it should be noted that the impact of the flare is relatively minor or insignificant in comparison. Comparing Group 3 to Group 2, it is evident that Group 3 exhibits an overall increase in relative underestimated uncertainties, with a maximum of 13%. Both the Carrington and uniform maps, when used as inputs for $I_{disk}(\boldsymbol{n})$, underestimate the Lyα intensities of the CME and shock, except for part of the CME core. However, the results from the Carrington map align more closely with those from the synchronic map. Therefore, when estimating the electron temperature of the CME using the DD method in the absence of synchronic map observations, we recommend utilizing the Carrington map as the input for Lyα $I_{disk}(\boldsymbol{n})$, which

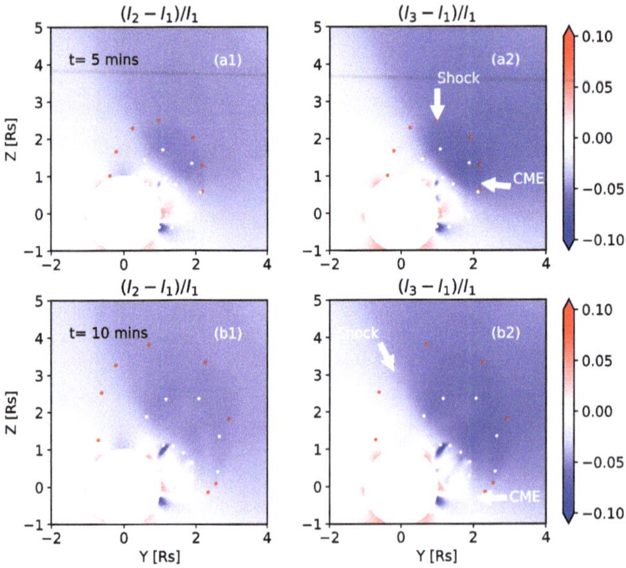

Figure 5 Comparison of the Lyα relative intensities calculated by the Carrington (Group 2, left panels) and uniform (Group 3, right panels) maps with respect to the synchronic (Group 1) map at $t = 5$ minutes (top panels) and $t = 10$ minutes (bottom panels), respectively. Arrows and dots with different colors have the same locations as those in Figure 4.

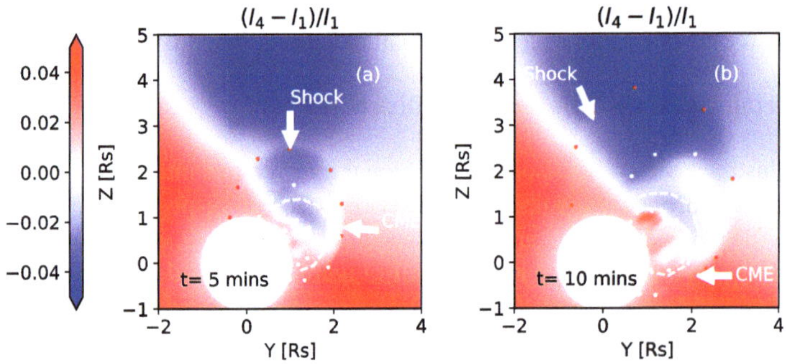

Figure 6 Relative variations of Lyα intensities calculated by assumptions of Group 4 with respect to Group 1 at $t = 5$ minutes and $t = 10$ minutes, respectively. White arrows and dots (red and white) have the same locations as those in Figure 4.

conveniently coincides with the data provided by LST/SDI observations, to obtain the theoretical Lyα intensity. In comparison to the CME, the impact of a uniform distribution of incident radiation on the Lyα intensity of the shock is relatively insignificant.

4.2. Uncertainties from the Geometric Scattering Process

In this subsection, we will assess the impact of neglecting the geometric scattering angle on the theoretical calculation of the Lyα intensity. Figure 6 illustrates the relative variations in H I Lyα intensity between the hypothetical calculations of Group 4 (ignoring geometric scattering function) and Group 1 at $t = 5$ and $t = 10$ minutes, respectively. In Figure 6, we can

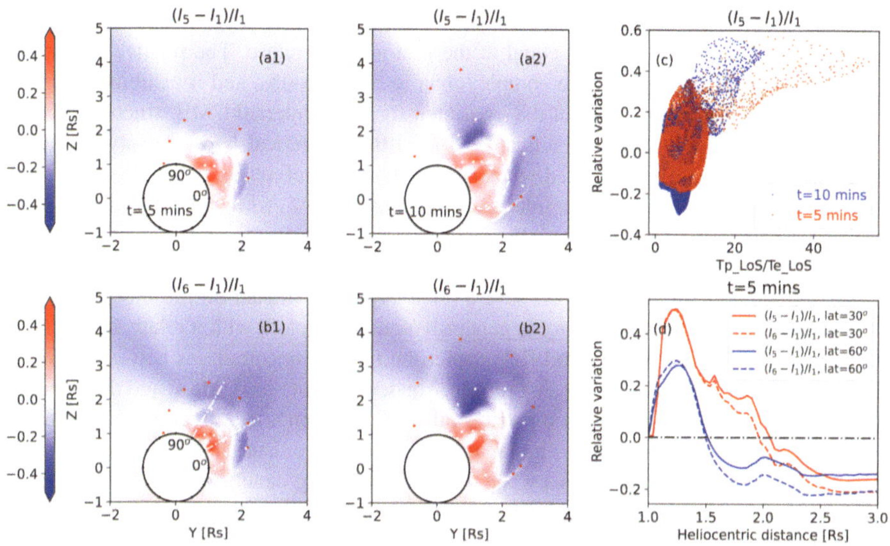

Figure 7 Relative variations of Lyα intensities calculated by assumptions of Groups 5 and 6 with respect to Group 1 at $t = 5$ minutes and $t = 10$ minutes, respectively: (a1)–(a2) the relative variations of Group 5 to Group 1 for the whole CME and shock; (b1)–(b2) the relative variation distributions of Group 6 to Group 1. Dash–dotted lines in (b1) are along radial directions at latitudes of 30° and 60°. Red, white dots, and white dashed curves in panels (a) and (b) have the same locations as those in Figure 4; (c) scatter plot of the relative variation (Group 5) vs. the LoS proton-to-electron temperature ratio. Symbols include all points inside the CME and shock both at $t = 5$ (red) and $t = 10$ (blue) minutes; (d) relative variations of Groups 5 (solid line) and 6 (dashed line) at $t = 5$ minutes along the radial directions at latitude of 30° and 60°. A horizontal line denotes a constant value of zero.

find that the CME core is overestimated in the evolution, and the maximum relative uncertainty is less than 5%. For the CME void and front, both overestimates and underestimates are present, but the underestimation is dominant. Regarding the shock, most parts are underestimated. Then, with the evolution of the shock and the decrease in number density, the underestimation uncertainty increases and becomes comparable to that of the background, as depicted in Figure 6(b). It could be attributed to the lower number density of shocks, where the Lyα intensity is more influenced by the coronal background after the LoS integration. On the other hand, the absolute value of the maximum uncertainty between the two groups does not exceed 5%.

4.3. Uncertainties from Temperature Assumptions

Due to the absence of spectroscopic or in-situ proton temperature observations in the corona, it is common to assume that the proton temperature is equal to the electron temperature when calculating the H I Lyα theoretical intensity (Bemporad, Pagano, and Giordano, 2018). This simplified assumption enables us to diagnose the CME electron temperature using the DD approach. However, it raises the question of how much uncertainty is introduced in calculating the Lyα theoretical intensity under this assumption. In this subsection, we will quantify the impact of the assumption of equal proton and electron temperatures on the uncertainty associated with the computation of the Lyα theoretical intensity.

Figures 7(a1) and 7(a2) illustrate the comparisons between Group 5 and Group 1 at the 5th and 10th minute of the simulation, respectively. In Group 5, we set the kinetic temperature equal to the electron temperature, while Group 1, serving as the reference group, assumes the kinetic temperature equal to the proton temperature. The relative uncertainty distribution in this group is more complex than in other groups, and it exhibits both overestimation and underestimation of the Lyα intensity of the internal CME structure. We find that temperature assumptions significantly impact the theoretical Lyα intensity. The CME and void are always overestimated when compared to the reference group. The CME core demonstrates a maximum relative uncertainty exceeding 50%, while the void exhibits a relative uncertainty of no more than 35%. Regarding the CME front, both overestimation and underestimation are observed, with a relative uncertainty less than 35%. Compared to the CME, the temperature assumptions have a relatively small impact on the shock, and most part of the computed Lyα intensity is generally underestimated, with a relative uncertainty of less than 20%. Panels (b1) and (b2) depict the relative uncertainties between Group 6 (common assumptions for diagnosing the electron temperatures of CMEs) and the reference group. The uncertainty distribution is determined by the combined effect of these three parameters, and the temperature has more significant influence than the other two parameters (I_{disk} and $p(\theta)$) which can be seen when directly comparing panels (a) and (b) in Figure 7. We make two slits along the radial directions at latitudes of 30° and 60° (dash–dotted lines, as shown in Figure 7(b1)), and the relative uncertainties are plotted in the panel (d) of Figure 7; solid lines denote the Lyα intensity uncertainty from Group 5, while dashed lines are from Group 6. Compared to Group 5, Group 6 exhibits decreasing relative uncertainties for most part, indicating a slight reduction in overestimation and an increase in underestimation.

Figure 7(c) presents a scatter plot of the relative variation (from Group 5 at $t = 5$ and $t = 10$ minutes) of the Lyα intensity versus the ratio of proton-to-electron temperature for all data points within the CME and shock. The proton temperature and electron temperature are obtained as the line-of-sight-averaged temperatures weighted by the electron number density (Figure 2, (a2)–(a3) and (b2)–(b3)). The temperature ratio does not show a good correlation with the relative uncertainty and is less than 15 for most plasmas in this simulation. Therefore, it is exceedingly challenging to evaluate directly the uncertainty caused by the equal-temperature assumption through the ratio of proton-to-electron temperatures. The reason is that the variation of the kinetic temperature (T_n) first influences the coronal normalized absorption profile (Equations 2c and 2d), which is then combined with the normalized incident spectral line profile (Section 4.1) incident from the solar disk, collectively affecting the Lyα intensity. We can define it as a DD factor $F_D = \int f(\lambda' - \delta\lambda)\Phi(\lambda' - \lambda_0, \boldsymbol{n})d\lambda'$, when we separate $I_{ex}(\lambda' - \delta\lambda, \boldsymbol{n})$ (in Equation 2a) and extract the chromospheric incident intensity $I_{disk}(\boldsymbol{n})$ from the integral equation. Thus, the change of the DD factor (further the Lyα intensity) will be influenced by not only the kinetic temperature but also the radial velocity of the plasma. Figure 8(c) illustrates the distribution of the DD factor with respect to the radial velocity and kinetic temperature. A dashed line in the figure shows the evolution of the corresponding temperature with velocity when the DD factor is maximized.

Considering that the proton temperature is consistently higher than the electron temperature, we further explore the possibility of adjusting the multiples of the electron temperature, instead of assuming equal proton and electron temperatures, to mitigate the uncertainty in the computed Lyα theoretical intensity. The results are shown in Figure 8. Panels (a1) and (a2) display relative uncertainties of the Lyα intensity calculated by assuming $T_p = 2T_e$ and $T_p = 3T_e$, respectively. Panel (b) indicates the relative uncertainties with different proton-to-electron temperature ratios at $t = 5$ minutes along the radial direction at the latitude of 30°. Compared to Group 5 (Figure 7(a1)), we observe that an appropriate increase in the proton-to-electron temperature ratio leads to a reduction in the relative uncertainty of the CME core and void. Thus, we suggest that when applying the DD method to diagnose the

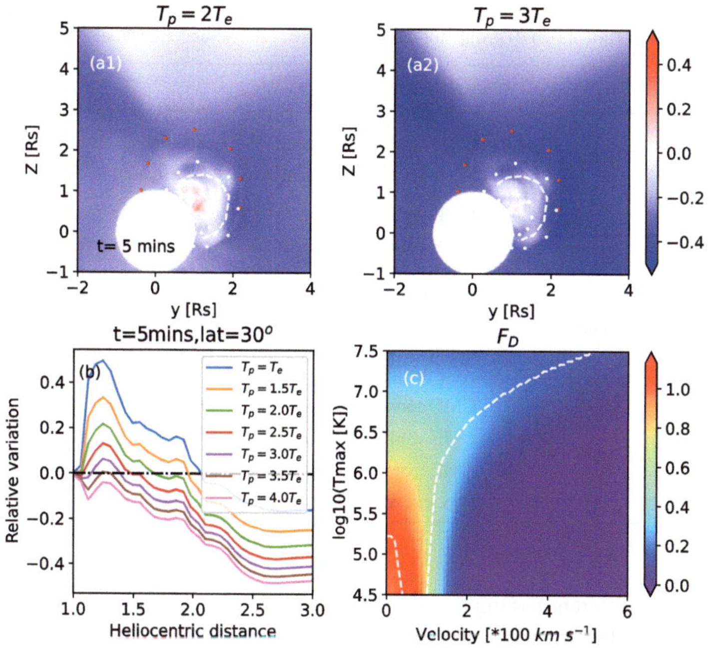

Figure 8 (a1)–(a2) Relative variations of Lyα intensities calculated by assuming $T_p = 2T_e$ and $T_p = 3T_e$, respectively. (b) Relative variations at $t = 5$ minutes along the radial direction at the latitude of 30° (dash-dotted line in Figure 7(b1)) assuming the proton temperature to be equal to various multiples of the electron temperature. (c) Doppler dimming (DD) factor distribution as a function of the radial velocity and temperature. A white dashed line marks the critical temperature corresponding to the maximum DD factor at different speeds.

electron temperature in the CME core (made of the flux rope), the theoretical intensity can be computed by assuming the proton temperature to be a few times greater than the electron temperature, such as $2-3$ times the electron temperature. But for the CME front and the shock with higher speeds, the Lyα theoretical intensity calculated under the assumption of the equal proton and electron temperature is better than that computed by the higher proton-to-electron temperature ratio. The increasing underestimation in the CME front and the shock (Figure 8(b)) can be attributed to the significant influence of the coronal background in those regions, where the DD effect is pronounced due to the high velocity of the shock and CME front. This underestimation effect becomes evident when comparing the backgrounds in Figure 8(a1)–(a2) with Figure 7(a1).

5. Discussion and Conclusions

In this work, we model a fast CME with a three-part structure and its driven shock through a 2T numerical simulation. Based on the simulated physical parameters, we synthesize 2D UV H I Lyα images. We conduct analyses for six groups of different assumptions, with Group 1 serving as the reference group. The assumptions in Group 1 include using a synchronic map as the input for the solar disk intensity ($I_{disk}(\boldsymbol{n})$), considering the geometric scattering process ($p(\theta)$), and assuming the kinetic temperature of hydrogen atoms to be equal to the proton temperature ($T_n = T_p$). By comparing the intensities with Group 1, we assess the

Table 2 Relative uncertainties for different groups shown in Table 1: (Column 1) hypothetical parameters different with Group 1; (Column 2) group number; (Column 3) relative uncertainties in different structures.

Parameter	No.	Relative uncertainties
$I_{disk}(\boldsymbol{n})$	Group 2	CME core: overestimate ($<4\%$) and underestimate ($<10\%$) CME void, front, and shock: underestimate ($<6\%$)
	Group 3	CME core: overestimate ($<4\%$) and underestimate ($<13\%$) CME void, front, and shock: underestimate ($<8\%$)
$p(\theta)$	Group 4	CME core: overestimate ($<5\%$) CME front and void: overestimate ($<3\%$) and underestimate ($<2\%$) Shock: overestimate ($<3\%$) and underestimate ($<5\%$)
T_n	Group 5	CME core: overestimate ($<57\%$) CME void: overestimate ($<35\%$) CME front: overestimate ($<20\%$) and underestimate ($<35\%$) Shock: overestimate ($<10\%$) and underestimate ($<20\%$)

impact of these different assumptions on the theoretical Lyα intensities of the CME and its driven shock. The relative uncertainties in various structures for different groups are summarized in Table 2. The main conclusions are as follows:

1. $I_{disk}(\boldsymbol{n})$. The comparison of the Lyα relative intensities between the Carrington map (Group 2) and the synchronic map (Group 1) exhibit small differences in most regions, with underestimate of relative uncertainties remaining within 6%. The maximum underestimated uncertainty is around 10% in the CME core, with low temperature and high density. A part of the CME core is overestimated but the relative uncertainty is less than 4%. The underestimation of the Lyα intensity, when using the uniform map as the input for $I_{\rm disk}(\boldsymbol{n})$ with a maximum relative uncertainty of approximately 13%, is greater compared to the results obtained from the Carrington map. Therefore, it is preferable to utilize the Lyα Carrington map as the input for $I_{disk}(\boldsymbol{n})$ provided by the LST/SDI (for more details, refer to Li et al., 2024) when calculating the theoretical Lyα intensity of the CME and shock in the absence of synchronic map observations.
2. $p(\theta)$. Ignoring the geometric scattering process has a lesser impact on the results of the H I Lyα intensity for the CME and shock. The absolute value of the maximum relative uncertainty does not exceed 5%. For the most part, the Lyα intensity is underestimated, and the CME core is overestimated. With the evolution of the shock, the underestimation uncertainty increases and becomes comparable to that of the background. This could be attributed to the fact that the lower number density of the shock is more susceptible to the influence of the coronal background. It should be noted that the geometric scattering function is valid primarily in weak magnetic fields ($B < 70$ G) (Beckers and Chipman, 1974), which is generally applicable to most of the quiet corona. However, when plasma is located closer to an active region, particularly near a flare site where the magnetic field strength increases, the geometric scattering function will be changed. Such as the magnetic strength of a small C2.0 flare measured by Landi et al. (2021) can reach up to 500 G. On the other hand, Zhao et al. (2021) conducted a study about the impact of the magnetic field on the linear polarization of the H I Lyα radiation and found that the variation of the magnetic field will not affect the Lyα intensity, but the linear polarization.

Therefore, the estimated relative uncertainty from the geometric scattering assumption in this paper can be valid.

3. T_n. The reference Lyα intensity (Group 1) is calculated by assuming that the hydrogen kinetic temperature is equal to the proton temperature, as the simulation only provides the proton and electron temperatures. We find that the temperature assumptions have a significant impact on the theoretical Lyα intensity. The distribution of the relative uncertainty is more complex than other groups. The CME core and void are overestimated. In the CME core, the maximum value of the relative uncertainty exceeds 50% between the electron (Group 5) and proton (Group 1) temperature assumptions, while the relative uncertainty for the void remains below 35%. We also find that assuming the proton temperature to be a multiple of the electron temperature, such as $2-3$ times the electron temperature, can reduce the uncertainty in the CME core and void. In the CME front, we observe both overestimation and underestimation with uncertainties of less than 35%. In contrast, the temperature assumption has a smaller impact on the shock than the CME, resulting in a generally underestimated uncertainty of approximately 20%. For the CME front and the shock region, assuming equal proton and electron temperatures yields better results compared to those calculated using a higher proton-to-electron temperature ratio.

In this work, the simulated CME has a very high proton temperature, according to the LoS integrated proton temperature weighted by the electron number density, as shown in Figure 2, (a2) and (b2). The ratio between the LoS proton and electron temperatures is very large. The significantly higher proton temperature compared to the electron temperature in the simulated CME, by more than one order of magnitude, may differ from the actual CME's proton temperature. On the other hand, Romeo et al. (2023) reported a CME captured by the Parker Solar Probe (PSP) on September 5, 2022. PSP traversed through the CME and measured many physical parameters, including proton and electron temperatures, at a heliocentric distance of $\sim 14\ R_\odot$. In this CME, the proton temperature is consistently higher than the electron temperature, with the maximum proton temperature reaching over 11 MK while the electron temperature remains below 1 MK. Besides, Reeves and Golub (2011) constrained the temperature of magnetic flux ropes within the range of 5 to 20 MK, indicating the possibility of MFRs attaining flare-like temperatures. These findings support that a CME can possess a high temperature in the core and a significant proton-to-electron temperature ratio in the body. Thus, the uncertainty estimation of this work could be reasonable. In the absence of proton temperature measurements, we recommend assuming that the proton temperature is a multiple of the electron temperature, as we mentioned above. This assumption could help reduce the uncertainty when calculating the Lyα theoretical intensity using the DD method to diagnose the electron temperature.

We ignore the effect of the incident chromospheric Lyα spectral profile and assume it is a unique multi-Gaussian line profile. Capuano et al. (2021) had discussed the impact of the chromospheric Lyα line profile on determining the coronal H I atom speed through the DD method. They considered four parameters of the Lyα line profile, including line width, reversal depth, asymmetry, and distance of the peaks. The result demonstrates that the variation of the Lyα line profile impacts the coronal H I velocity estimation by about 9%–12%. They suggested that it is acceptable to use a unique shape of the Lyα profile over the solar disk that is constant in time.

Furthermore, it is important to note that the calculation of the theoretical Lyα intensity in this study is based on the assumption of ionization equilibrium. However, it is known that the corona is generally not in equilibrium. A freeze-in distance exists for atoms and ions, where lighter elements stop evolving earlier than heavier elements (Landi et al., 2012). An MHD simulation conducted by Pagano, Bemporad, and Mackay (2020) revealed significant nonequilibrium ionization effects in the front of a CME, while the equilibrium ionization assumption remains valid in the core of the CME. Therefore, it becomes necessary to consider the impact of nonequilibrium ionization when calculating the theoretical Lyα intensity.

Appendix: Calculation of the Resonant Component of the H I Lyα Intensity

In this section, we describe calculation details of the resonantly scattered H I Lyα intensity. It mainly refers to Noci, Kohl, and Withbroe (1987), Auchère (2005), Bemporad, Pagano, and Giordano (2018), and Dolei et al. (2019).

For a given plasma at point P in the corona, its emissivity is contributed by the incident Lyα radiation from its corresponding disk region (Figure 9(a)). We can determine its corresponding disk region of the incident Lyα radiation, such as the region marked by the dashed line in the unfolding solar disk in the Stonyhurst heliographic coordinate, as shown in Figure 9(c). For a grid point S in the disk region (panel (a)), the direction of $\boldsymbol{SP}$ indicates the incident direction $\boldsymbol{n}$, and we can calculate $dS = \mathrm{R}_{\odot}{}^2 \cos\phi \; d\phi d\varphi$. The solid angle $d\Omega = dS(\cos\gamma)/d^2$, where d is the modulus of $\boldsymbol{SP}$, and γ is the angle between $\boldsymbol{OS}$ and $\boldsymbol{SP}$. The angle β is determined from $\boldsymbol{u} \cdot \boldsymbol{n}$, and $\boldsymbol{u}$ is the speed of the plasma at point P. Also θ is the geometric scattering angle and can be computed by $\boldsymbol{l} \cdot \boldsymbol{n}$, and $\boldsymbol{l}$ is the direction of the light of sight. In our calculation, $\boldsymbol{l}$ is parallel to the x-axis direction.

As we mentioned in Section 3.2, I_{ex} can be separated into two terms, and $I_{ex}(\lambda' - \delta\lambda, \boldsymbol{n}) = I_{disk}(\boldsymbol{n}) \; f(\lambda' - \delta\lambda)$. We assume that the normalized chromospheric line profile

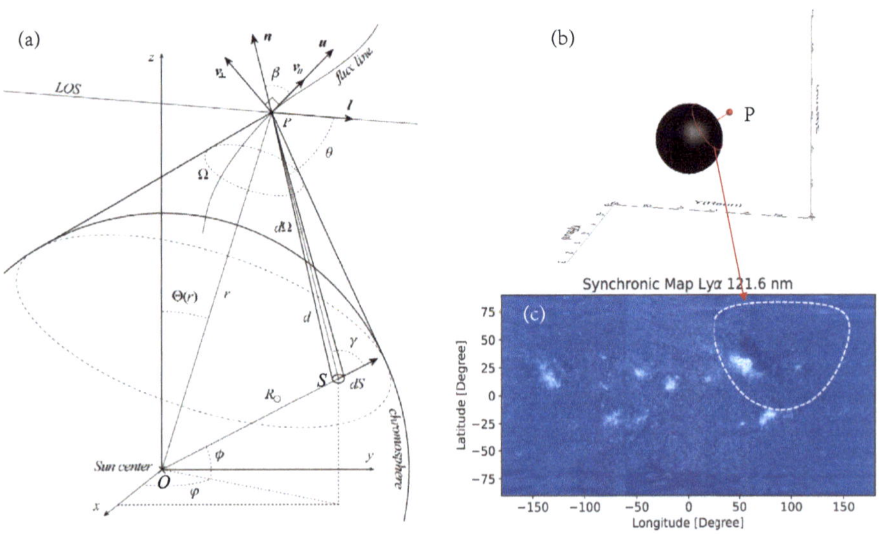

Figure 9 (a) Geometry of the modeling of the resonantly scattered coronal H I Lyα radiation from Auchère (2005). (b) The region of incident intensity (red line) on the solar surface (black ball) for a given point P (red ball, $\varphi = 90°$, $\phi = 35°$, $r = 1.5$ R$_{\odot}$). (c) The synchronic map in the Lyα line of the solar disk, the same as that in Figure 3(a2). A white dashed line indicates the region (red circle, panel (b)) of solar disk incident intensity corresponding to point P after the solar surface unfolding in the Stonyhurst heliographic coordinate.

is a unique multi-Gaussian profile proposed by Auchère (2005), which is

$$f(\lambda' - \delta\lambda) = \sum_{i=1}^{3} f_i(\lambda' - \delta\lambda), \tag{5a}$$

$$f_i(\lambda' - \delta\lambda) = \frac{a_i}{\sigma_i\sqrt{\pi}} \exp\left[-\left(\frac{\lambda' - \delta\lambda - \lambda_0 - \delta\lambda_i'}{\sigma_i}\right)^2\right]. \tag{5b}$$

The values of the parameters a_i, σ_i, and $\delta\lambda_i'$ can be found in Auchère (2005). Then, the Doppler dimming term can be replaced by

$$F_D = \int \sum_{i=1}^{3} f_i(\lambda' - \delta\lambda)\ \Phi(\lambda' - \lambda_0, \boldsymbol{n})\ d\lambda'. \tag{6}$$

According to Bromiley (2014), the product between two normalized Gaussian functions is a normalized Gaussian function, which can be expressed as

$$f(\lambda', \mu_1, \sigma_1)\ \Phi(\lambda', \mu_2, \sigma_2) = \frac{S_{12}}{\sqrt{\pi}\sigma_{12}} \exp\left[-\left(\frac{\lambda' - \mu_{12}}{\sigma_{12}}\right)^2\right], \tag{7}$$

where $\sigma_{12}^2 = \frac{\sigma_1^2\sigma_2^2}{\sigma_1^2+\sigma_2^2}$, $\mu_{12} = \frac{\mu_1\sigma_2^2+\mu_2\sigma_1^2}{\sigma_1^2+\sigma_2^2}$, and $S_{12} = \frac{1}{\sqrt{\pi(\sigma_1^2+\sigma_2^2)}} \exp\left[-\frac{(\mu_1-\mu_2)^2}{\sigma_1^2+\sigma_2^2}\right]$. For Gaussian profiles $f_i(\lambda' - \delta\lambda)$ and $\Phi(\lambda' - \lambda_0, \boldsymbol{n})$, $\mu_1 = \lambda_0 + \delta\lambda + \delta\lambda_i'$, $\mu_2 = \lambda_0$, $\sigma_1 = \sigma_i$, $\sigma_2 = \sqrt{2}\sigma_\lambda$. Then, the integration of Equation 6 gives

$$F_D = \sum_{i=1}^{3} \frac{a_i}{\sqrt{\pi}\sqrt{2\sigma_\lambda^2 + \sigma_i^2}} \exp\left[-\frac{(\delta\lambda_i' + \frac{\lambda_0}{c}\boldsymbol{u}\cdot\boldsymbol{n})^2}{2\sigma_\lambda^2 + \sigma_i^2}\right]. \tag{8}$$

Finally, Equation 2a – 2d can be expressed by

$$\begin{aligned} j_r = {} & \frac{0.833h\ B_{12}\ \lambda_0}{4\,\pi}\ n_e\ R_{HI}(T_e) \\ & \times \int_\phi\int_\varphi \frac{1}{4\pi}\frac{11+3\cos^2\theta}{12}\ I_{disk}(\phi,\varphi)\ R_\odot^2\ (\cos\gamma/d^2)\ \cos\phi \\ & \times \sum_{i=1}^{3}\left\{\frac{a_i}{\sqrt{\pi}\sqrt{2\sigma_\lambda^2+\sigma_i^2}} \exp\left[-\frac{(\delta\lambda_i' + \lambda_0\ u\cos\beta/c)^2}{2\sigma_\lambda^2+\sigma_i^2}\right]\right\}\ d\phi\ d\varphi. \end{aligned} \tag{9}$$

From the data cube of the MHD simulation, we know all parameters, including plasma positions, angles, speeds, temperatures, and electron number densities. Thus, we can compute the emissivity of each plasma (blob), and then obtain the 2D synthetic H I Lyα image by integrating along the x-axis, according to Equation 1.

Acknowledgments We are grateful to Meng Jin for his help in the 2T MHD simulation. We also thank the anonymous referee for suggestions and comments that helped us significantly improve and clarify this paper. This work is supported by the Strategic Priority Research Program of the Chinese Academy of Sciences, Grant No. XDB0560000, National Key R&D Program of China 2022YFF0503003 (2022YFF0503000),

NSFC (grant Nos. 11973012, 11921003, 12203102, 12233012, 12103090), the mobility program (M-0068) of the Sino-German Science Center. This work benefits from the discussions of the ISSI-BJ Team "Solar eruptions: Preparing for the next generation multi-waveband coronagraphs".

Author contributions B.L.Y. was responsible for synthesizing and analyzing the UV and WL data and writing the main manuscript. G.L.S. and L.L. were responsible for the MHD numerical simulation. L.F. contributed ideas to the article and participated in revising its content. J.C.X and S.T.L. were involved in the discussion and revision of the article. W.Q.G. and H.L. participated in revising the article's content. All authors reviewed the manuscript.

Data Availability No datasets were generated or analysed during the current study.

Declarations

Competing interests The authors declare no competing interests.

References

Antonucci, E., Andretta, V., Cesare, S., Ciaravella, A., Doschek, G., Fineschi, S., Giordano, S., Lamy, P., Moses, D., Naletto, G., Newmark, J., Poletto, L., Romoli, M., Solanki, S., Spadaro, D., Teriaca, L., Zangrilli, L.: 2017, METIS, the Multi Element Telescope for Imaging and Spectroscopy: an instrument proposed for the solar orbiter mission. Society of Photo-Optical Instrumentation Engineers (SPIE) Conference Series 10566 105660L. DOI. ADS.

Antonucci, E., Romoli, M., Andretta, V., Fineschi, S., Heinzel, P., Moses, J.D., Naletto, G., Nicolini, G., Spadaro, D., Teriaca, L., Berlicki, A., Capobianco, G., Crescenzio, G., Da Deppo, V., Focardi, M., Frassetto, F., Heerlein, K., Landini, F., Magli, E., Marco Malvezzi, A., Massone, G., Melich, R., Nicolosi, P., Noci, G., Pancrazzi, M., Pelizzo, M.G., Poletto, L., Sasso, C., Schühle, U., Solanki, S.K., Strachan, L., Susino, R., Tondello, G., Uslenghi, M., Woch, J., Abbo, L., Bemporad, A., Casti, M., Dolei, S., Grimani, C., Messerotti, M., Ricci, M., Straus, T., Telloni, D., Zuppella, P., Auchère, F., Bruno, R., Ciaravella, A., Corso, A.J., Alvarez Copano, M., Aznar Cuadrado, R., D'Amicis, R., Enge, R., Gravina, A., Jejčič, S., Lamy, P., Lanzafame, A., Meierdierks, T., Papagiannaki, I., Peter, H., Fernandez Rico, G., Giday Sertsu, M., Staub, J., Tsinganos, K., Velli, M., Ventura, R., Verroi, E., Vial, J.-C., Vives, S., Volpicelli, A., Werner, S., Zerr, A., Negri, B., Castronuovo, M., Gabrielli, A., Bertacin, R., Carpentiero, R., Natalucci, S., Marliani, F., Cesa, M., Laget, P., Morea, D., Pieraccini, S., Radaelli, P., Sandri, P., Sarra, P., Cesare, S., Del Forno, F., Massa, E., Montabone, M., Mottini, S., Quattropani, D., Schillaci, T., Boccardo, R., Brando, R., Pandi, A., Baietto, C., Bertone, R., Alvarez-Herrero, A., García Parejo, P., Cebollero, M., Amoruso, M., Centonze, V.: 2020, Metis: the Solar Orbiter visible light and ultraviolet coronal imager. *Astron. Astrophys.* **642**, A10. DOI. ADS.

Auchère, F.: 2005, Effect of the H I Lyα chromospheric flux anisotropy on the total intensity of the resonantly scattered coronal radiation. *Astrophys. J.* **622**, 737. DOI. ADS.

Beckers, J.M., Chipman, E.: 1974, The profile and polarization of the coronal Lα line. *Solar Phys.* **34**, 151. DOI.

Bemporad, A.: 2008, Spectroscopic detection of turbulence in post-CME current sheets. *Astrophys. J.* **689**, 572. DOI. ADS.

Bemporad, A.: 2022, Temperature and thermal energy of a coronal mass ejection. *Symmetry* **14**, 468. DOI.

Bemporad, A., Mancuso, S.: 2010, First complete determination of plasma physical parameters across a coronal mass ejection-driven shock. *Astrophys. J.* **720**, 130. DOI.

Bemporad, A., Pagano, P., Giordano, S.: 2018, Measuring the electron temperatures of coronal mass ejections with future space-based multi-channel coronagraphs: a numerical test. *Astron. Astrophys.* **619**, A25. DOI. ADS.

Bemporad, A., Raymond, J., Poletto, G., Romoli, M.: 2007, A comprehensive study of the initiation and early evolution of a coronal mass ejection from ultraviolet and white-light data. *Astrophys. J.* **655**, 576. DOI. ADS.

Bemporad, A., Shi, G., Li, S., Ying, B., Feng, L., Lin, J., Abbo, L., Andretta, V., Burtovoi, A., Da Deppo, V., De Leo, Y., Fineschi, S., Frassati, F., Giordano, S., Grimani, C., Jerse, G., Landini, F., Mancuso, S., Naletto, G., Nicolini, G., Pancrazzi, M., Romoli, M., Russano, G., Sasso, C., Spadaro, D., Stangalini, M., Susino, R., Teriaca, L., Uslenghi, M.: 2024, First determination in the extended corona of the 2D thermal evolution of a current sheet after a solar eruption. *Astrophys. J.* **964**, 92. DOI. ADS.

Billings, D.E.: 1966, *A Guide to the Solar Corona*, Academic Press, New York. ADS.

Bromiley, P.A.: 2014, *Products and Convolutions of Gaussian Probability Density Functions*.

Capuano, G.E., Dolei, S., Spadaro, D., Guglielmino, S.L., Romano, P., Ventura, R., Andretta, V., Bemporad, A., Sasso, C., Susino, R., Da Deppo, V., Frassetto, F., Giordano, S.M., Landini, F., Nicolini, G., Pancrazzi, M., Romoli, M., Zangrilli, L.: 2021, Effects of the chromospheric Lyα line profile shape on the determination of the solar wind H I outflow velocity using the Doppler dimming technique. *Astron. Astrophys.* **652**, A85. DOI. ADS.

Cheung, M.C.M., Boerner, P., Schrijver, C.J., Testa, P., Chen, F., Peter, H., Malanushenko, A.: 2015, Thermal diagnostics with the atmospheric imaging assembly on board the solar dynamics observatory: a validated method for differential emission measure inversions. *Astrophys. J.* **807**, 143. DOI. ADS.

Ciaravella, A., Raymond, J.C., Kahler, S.W.: 2006, Ultraviolet properties of halo coronal mass ejections: Doppler shifts, angles, shocks, and bulk morphology. *Astrophys. J.* **652**, 774. DOI.

Ciaravella, A., Raymond, J.C., van Ballegooijen, A., Strachan, L., Vourlidas, A., Li, J., Chen, J., Panasyuk, A.: 2003, Physical parameters of the 2000 February 11 coronal mass ejection: ultraviolet spectra versus white-light images. *Astrophys. J.* **597**, 1118. DOI.

Ciaravella, A., Raymond, J.C., Kahler, S.W., Vourlidas, A., Li, J.: 2005, Detection and diagnostics of a coronal shock wave driven by a partial-halo coronal mass ejection on 2000 June 28. *Astrophys. J.* **621**, 1121. DOI.

Dolei, S., Susino, R., Sasso, C., Bemporad, A., Andretta, V., Spadaro, D., Ventura, R., Antonucci, E., Abbo, L., Da Deppo, V., Fineschi, S., Focardi, M., Frassetto, F., Giordano, S., Landini, F., Naletto, G., Nicolini, G., Nicolosi, P., Pancrazzi, M., Romoli, M., Telloni, D.: 2018, Mapping the solar wind HI outflow velocity in the inner heliosphere by coronagraphic ultraviolet and visible-light observations. *Astron. Astrophys.* **612**, A84. DOI. ADS.

Dolei, S., Spadaro, D., Ventura, R., Bemporad, A., Andretta, V., Sasso, C., Susino, R., Antonucci, E., Da Deppo, V., Fineschi, S., Frassetto, F., Landini, F., Naletto, G., Nicolini, G., Pancrazzi, M., Romoli, M.: 2019, Effect of the non-uniform solar chromospheric Lyα radiation on determining the coronal H I outflow velocity. *Astron. Astrophys.* **627**, A18. DOI. ADS.

Feng, L., Li, H., Chen, B., Li, Y., Susino, R., Huang, Y., Lu, L., Ying, B.-L., Li, J.-W., Xue, J.-C., Yang, Y.-T., Hong, J., Li, J.-P., Zhao, J., Gan, W.-Q., Zhang, Y.: 2019, The Lyman-alpha Solar Telescope (LST) for the ASO-S mission III. Data and potential diagnostics. *Res. Astron. Astrophys.* **19**, 162. DOI. ADS.

Gabriel, A.H.: 1971, Measurements on the Lyman Alpha Corona (Papers presented at the Proceedings of the International Symposium on the 1970 Solar Eclipse, held in Seattle, U. S. A., 18-21 June, 1971.). *Solar Phys.* **21** 392. DOI. ADS.

Gan, W.-Q., Zhu, C., Deng, Y.-Y., Li, H., Su, Y., Zhang, H.-Y., Chen, B., Zhang, Z., Wu, J., Deng, L., Huang, Y., Yang, J.-F., Cui, J.-J., Chang, J., Wang, C., Wu, J., Yin, Z.-S., Chen, W., Fang, C., Yan, Y.-H., Lin, J., Xiong, W.-M., Chen, B., Bao, H.-C., Cao, C.-X., Bai, Y.-P., Wang, T., Chen, B.-L., Li, X.-Y., Zhang, Y., Feng, L., Su, J.-T., Li, Y., Chen, W., Li, Y.-P., Su, Y.-N., Wu, H.-Y., Gu, M., Huang, L., Tang, X.-J.: 2019, Advanced Space-based Solar Observatory (ASO-S): an overview. *Res. Astron. Astrophys.* **19**, 156. DOI. ADS.

Gan, W., Zhu, C., Deng, Y., Zhang, Z., Chen, B., Huang, Y., Deng, L., Wu, H., Zhang, H., Li, H., Su, Y., Su, J., Feng, L., Wu, J., Cui, J., Wang, C., Chang, J., Yin, Z., Xiong, W., Chen, B., Yang, J., Li, F., Lin, J., Hou, J., Bai, X., Chen, D., Zhang, Y., Hu, Y., Liang, Y., Wang, J., Song, K., Guo, Q., He, L., Zhang, G., Wang, P., Bao, H., Cao, C., Bai, Y., Chen, B., He, T., Li, X., Zhang, Y., Liao, X., Jiang, H., Li, Y., Su, Y., Lei, S., Chen, W., Li, Y., Zhao, J., Li, J., Ge, Y., Zou, Z., Hu, T., Su, M., Ji, H., Gu, M., Zheng, Y., Xu, D., Wang, X.: 2023, The Advanced Space-based Solar Observatory (ASO-S). *Solar Phys.* **298**, 68. DOI. ADS.

Gibson, S.E., Low, B.C.: 1998, A time-dependent three-dimensional magnetohydrodynamic model of the coronal mass ejection. *Astrophys. J.* **493**, 460. DOI. ADS.

Gopalswamy, N., Nitta, N., Akiyama, S., Mäkelä, P., Yashiro, S.: 2012, Coronal magnetic field measurement from EUV images made by the solar dynamics observatory. *Astrophys. J.* **744**, 72. DOI. ADS.

Gordino, M., Auchère, F., Vial, J.-C., Bocchialini, K., Hassler, D.M., Bando, T., Ishikawa, R., Kano, R., Kobayashi, K., Narukage, N., Trujillo Bueno, J., Winebarger, A.: 2022, Empirical relations between the intensities of Lyman lines of H and He^+. *Astron. Astrophys.* **657**, A86. DOI. ADS.

Hong, J., Li, Y., Ding, M.D., Carlsson, M.: 2019, The response of the Lyα line in different flare heating models. *Astrophys. J.* **879**, 128. DOI. ADS.

Jin, M., Manchester, W.B., van der Holst, B., Gruesbeck, J.R., Frazin, R.A., Landi, E., Vasquez, A.M., Lamy, P.L., Llebaria, A., Fedorov, A., Toth, G., Gombosi, T.I.: 2012, A global two-temperature corona and inner heliosphere model: a comprehensive validation study. *Astrophys. J.* **745**, 6. DOI. ADS.

Jin, M., Manchester, W.B., van der Holst, B., Oran, R., Sokolov, I., Toth, G., Liu, Y., Sun, X.D., Gombosi, T.I.: 2013, Numerical simulations of coronal mass ejection on 2011 March 7: one-temperature and two-temperature model comparison. *Astrophys. J.* **773**, 50. DOI. ADS.

Jin, M., Manchester, W.B., van der Holst, B., Sokolov, I., Tóth, G., Vourlidas, A., de Koning, C.A., Gombosi, T.I.: 2017, Chromosphere to 1 AU simulation of the 2011 March 7th event: a comprehensive study of coronal mass ejection propagation. *Astrophys. J.* **834**, 172. DOI. ADS.

Kohl, J.L., Esser, R., Gardner, L.D., Habbal, S., Daigneau, P.S., Dennis, E.F., Nystrom, G.U., Panasyuk, A., Raymond, J.C., Smith, P.L., Strachan, L., Van Ballegooijen, A.A., Noci, G., Fineschi, S., Romoli, M., Ciaravella, A., Modigliani, A., Huber, M.C.E., Antonucci, E., Benna, C., Giordano, S., Tondello, G., Nicolosi, P., Naletto, G., Pernechele, C., Spadaro, D., Poletto, G., Livi, S., Von Der Lühe, O., Geiss, J., Timothy, J.G., Gloeckler, G., Allegra, A., Basile, G., Brusa, R., Wood, B., Siegmund, O.H.W., Fowler, W., Fisher, R., Jhabvala, M.: 1995, The Ultraviolet Coronagraph Spectrometer for the solar and heliospheric observatory. *Solar Phys.* **162**, 313. DOI.

Landi, E., Gruesbeck, J.R., Lepri, S.T., Zurbuchen, T.H., Fisk, L.A.: 2012, Charge state evolution in the solar wind. II. Plasma charge state composition in the inner corona and accelerating fast solar wind. *Astrophys. J.* **761**, 48. DOI. ADS.

Landi, E., Li, W., Brage, T., Hutton, R.: 2021, Hinode/EIS coronal magnetic field measurements at the onset of a C2 flare. *Astrophys. J.* **913**, 1. DOI. ADS.

Li, H., Chen, B., Feng, L., Li, Y., Huang, Y., Li, J.-W., Lu, L., Xue, J.-C., Ying, B.-L., Zhao, J., Yang, Y.-T., Gan, W.-Q., Fang, C., Song, K.-F., Wang, H., Guo, Q.-F., He, L.-P., Zhu, B., Zhu, C., Deng, L., Bao, H.-C., Cao, C.-X., Yang, Z.-G.: 2019, The Lyman-alpha Solar Telescope (LST) for the ASO-S mission — I. Scientific objectives and overview. *Res. Astron. Astrophys.* **19**, 158. DOI. ADS.

Li, S., Feng, L., Ying, B., Lu, L., Ji, K., Li, Y., Shi, G., Li, H., Huang, Y., Gan, W.: 2024, Carrington maps in H I Lyα and their relationships with extreme ultraviolet and magnetic field maps. *Astrophys. J.*, in press.

Noci, G., Kohl, J.L., Withbroe, G.L.: 1987, Solar wind diagnostics from Doppler-enhanced scattering. *Astrophys. J.* **315**, 706. DOI. ADS.

Oran, R., van der Holst, B., Landi, E., Jin, M., Sokolov, I.V., Gombosi, T.I.: 2013, A global wave-driven magnetohydrodynamic solar model with a unified treatment of open and closed magnetic field topologies. *Astrophys. J.* **778**, 176. DOI. ADS.

Pagano, P., Bemporad, A., Mackay, D.H.: 2020, Hydrogen non-equilibrium ionisation effects in coronal mass ejections. *Astron. Astrophys.* **637**, A49. DOI. ADS.

Raymond, J.C., Ciaravella, A., Dobrzycka, D., Strachan, L., Ko, Y.-K., Uzzo, M., Raouafi, N.-E.: 2003, Far-ultraviolet spectra of fast coronal mass ejections associated with X-class flares. *Astrophys. J.* **597**, 1106. DOI. ADS.

Reeves, K.K., Golub, L.: 2011, Atmospheric imaging assembly observations of hot flare plasma. *Astrophys. J. Lett.* **727**, L52. DOI. ADS.

Romeo, O.M., Braga, C.R., Badman, S.T., Larson, D.E., Stevens, M.L., Huang, J., Phan, T., Rahmati, A., Livi, R., Alnussirat, S.T., Whittlesey, P.L., Szabo, A., Klein, K.G., Niembro-Hernandez, T., Paulson, K., Verniero, J.L., Lario, D., Raouafi, N.E., Ervin, T., Kasper, J., Pulupa, M., Bale, S.D., Linton, M.G.: 2023, Near-sun in situ and remote-sensing observations of a coronal mass ejection and its effect on the heliospheric current sheet. *Astrophys. J.* **954**, 168. DOI. ADS.

Shi, G.-L., Ying, B.-L., Feng, L., Li, H., Yang, C.-H.: 2020, Study of the streamer current sheet with white-light and UV observations. *Chin. J. Astron. Astrophys.* **44**, 474. DOI. ADS.

Sokolov, I.V., van der Holst, B., Oran, R., Downs, C., Roussev, I.I., Jin, M., Manchester, I., Ward, B., Evans, R.M., Gombosi, T.I.: 2013, Magnetohydrodynamic waves and coronal heating: unifying empirical and MHD turbulence models. *Astrophys. J.* **764**, 23. DOI. ADS.

Su, Y., Veronig, A.M., Hannah, I.G., Cheung, M.C.M., Dennis, B.R., Holman, G.D., Gan, W., Li, Y.: 2018, Determination of differential emission measure from solar extreme ultraviolet images. *Astrophys. J. Lett.* **856**, L17. DOI. ADS.

Susino, R., Bemporad, A.: 2016, Determination of coronal mass ejection physical parameters from a combination of polarized visible light and UV Lyα observations. *Astrophys. J.* **830**, 58. DOI. ADS.

van der Holst, B., Manchester, I.W.B., Frazin, R.A., Vásquez, A.M., Tóth, G., Gombosi, T.I.: 2010, A data-driven, two-temperature solar wind model with Alfvén waves. *Astrophys. J.* **725**, 1373. DOI. ADS.

Vernazza, J.E., Avrett, E.H., Loeser, R.: 1981, Structure of the solar chromosphere. III. Models of the EUV brightness components of the quiet sun. *Astrophys. J. Suppl.* **45**, 635. DOI. ADS.

Ying, B., Feng, L., Lu, L., Zhang, J., Magdalenic, J., Su, Y., Su, Y., Gan, W.: 2018, Properties of a small-scale short-duration solar eruption with a driven shock. *Astrophys. J.* **856**, 24. DOI. ADS.

Ying, B., Bemporad, A., Feng, L., Lu, L., Gan, W., Li, H.: 2020, Extensive study of a coronal mass ejection with UV and white-light coronagraphs: the need for multiwavelength observations. *Astrophys. J.* **899**, 12. DOI. ADS.

Zhao, J., Gibson, S.E., Fineschi, S., Susino, R., Casini, R., Cranmer, S.R., Ofman, L., Li, H.: 2021, Simulating the solar minimum corona in UV wavelengths with forward modeling II. Doppler dimming and microscopic anisotropy effect. *Astrophys. J.* **912**, 141. DOI. ADS.

Zhao, J., Zhang, P., Gibson, S.E., Fan, Y., Feng, L., Yu, F., Li, H., Gan, W.Q.: 2022, Synthetic Lyman-α emissions for the coronagraph aboard the ASO-S mission. I. An eruptive prominence-cavity system. *Astron. Astrophys.* **665**, A39. DOI. ADS.

Authors and Affiliations

Beili Ying[1] · Guanglu Shi[1,2] · Li Feng[1,2] · Lei Lu[1] · Jianchao Xue[1] · Shuting Li[1,2] · Weiqun Gan[1,3] · Hui Li[1,2]

✉ B. Ying
yingbl@pmo.ac.cn

✉ L. Feng
lfeng@pmo.ac.cn

G. Shi
shigl@pmo.ac.cn

L. Lu
leilu@pmo.ac.cn

J. Xue
xuejc@pmo.ac.cn

S. Li
lisht@pmo.ac.cn

W. Gan
wqgan@pmo.ac.cn

H. Li
nj.lihui@pmo.ac.cn

1 Key Laboratory of Dark Matter and Space Astronomy, Purple Mountain Observatory, Chinese Academy of Sciences, Nanjing 210023, People's Republic of China

2 School of Astronomy and Space Science, University of Science and Technology of China, Hefei, Anhui 230026, People's Republic of China

3 University of Chinese Academy of Sciences, Nanjing 211135, People's Republic of China

MIX
Papier aus verantwortungsvollen Quellen
Paper from responsible sources
FSC® C105338

If you have any concerns about our products,
you can contact us on
ProductSafety@springernature.com

In case Publisher is established outside the EU,
the EU authorized representative is:
Springer Nature Customer Service Center GmbH
Europaplatz 3, 69115 Heidelberg, Germany

Printed by Libri Plureos GmbH
in Hamburg, Germany